Minhua Shao (Ed.)

Electrocatalysis in Fuel Cells

MDPI

This book is a reprint of the Special Issue that appeared in the online, open access journal, *Catalysts* (ISSN 2073-4344) in 2015 (available at: http://www.mdpi.com/journal/catalysts/special_issues/electrocatal-fuel-cells).

Guest Editor
Minhua Shao
The Hong Kong University of Science and Technology
Hong Kong

Editorial Office
MDPI AG
Klybeckstrasse 64
Basel, Switzerland

Publisher
Shu-Kun Lin

Managing Editor
Zu Qiu

1. Edition 2016

MDPI • Basel • Beijing • Wuhan • Barcelona

ISBN 978-3-03842-234-1 (Hbk)
ISBN 978-3-03842-219-8 (PDF)

Table of Contents

V

List of Contributors

Radoslav R. Adzic Chemistry Department, Brookhaven National Laboratory, Upton, NY 11973, USA.

Antonino S. Aricò Consiglio Nazionale delle Ricerche, Istituto di Tecnologie Avanzate per l'Energia "Nicola Giordano", Salita S. Lucia sopra Contesse 5, 98126 Messina, Italy.

Vanessa Armel Institut Charles Gerhardt Montpellier UMR 5253, CNRS, Université de Montpellier, 34095 Montpellier Cedex 5, France.

Vincenzo Baglio Consiglio Nazionale delle Ricerche, Istituto di Tecnologie Avanzate per l'Energia "Nicola Giordano", Salita S. Lucia sopra Contesse 5, 98126 Messina, Italy.

Qingguo Bai Key Laboratory for Liquid-Solid Structural Evolution and Processing of Materials (Ministry of Education), School of Materials Science and Engineering, Shandong University, Jingshi Road 17923, Jinan 250061, China.

Zhengyu Bai School of Chemistry and Chemical Engineering, Key Laboratory of Green Chemical Media and Reactions, Ministry of Education, Collaborative Innovation Center of Henan Province for Fine Chemicals Green Manufacturing, Henan Normal University, Xinxiang 453007, China.

Sarah C. Ball Johnson Matthey Technology Centre, Blounts Court, Sonning Common, Reading RG4 9NH, UK.

Dustin Banham Department of Chemistry, University of Calgary, 2500 University Drive NW Calgary, AB T2N1N4, Canada; Ballard Power Systems, 9000 Glenlyon Parkway, Burnaby, BC V5J 5J8, Canada.

Heather M. Barkholtz Chemical Science and Engineering Division, Argonne National Laboratory, Argonne, IL 60439, USA; Department of Chemistry and Biochemistry, Northern Illinois University, DeKalb, IL 60115, USA.

Viola Birss Department of Chemistry, University of Calgary, 2500 University Drive NW Calgary, AB T2N1N4, Canada.

Enric Brillas Laboratori d'Electroquímica dels Materials i del Medi Ambient, Departament de Química Física, Universitat de Barcelona, Martí i Franquès 1-11, 08028 Barcelona, Spain.

Griselda Caballero-Manrique Laboratori d'Electroquímica dels Materials i del Medi Ambient, Departament de Química Física, Universitat de Barcelona, Martí i Franquès 1-11, 08028 Barcelona, Spain.

Pere-Lluís Cabot Laboratori d'Electroquímica dels Materials i del Medi Ambient, Departament de Química Física, Universitat de Barcelona, Martí i Franquès 1-11, 08028 Barcelona, Spain.

Wen-Bin Cai Shanghai Key Laboratory for Molecular Catalysis and Innovative Materials, Collaborative Innovation Center of Chemistry for Energy Materials and Department of Chemistry, Fudan University, Shanghai 200433, China.

Chongjiang Cao Laboratory of Chemical Engineering—Nanomaterials, Catalysis, Electrochemistry, University of Liège (B6a), B-4000 Liège, Belgium; College of Food Science and Engineering, Nanjing University of Finance and Economics, Nanjing 210046, China.

Xiao-Lu Cao Shanghai Key Laboratory of Materials Protection and Advanced Materials in Electric Power, Shanghai University of Electric Power, NO. 2588 Changyang Road, Yangpu District, Shanghai 200090, China.

Francesc Centellas Laboratori d'Electroquímica dels Materials i del Medi Ambient, Departament de Química Física, Universitat de Barcelona, Martí i Franquès 1-11, 08028 Barcelona, Spain.

Jeng-Kuei Chang Institute of Materials Science and Engineering, National Central University, Taoyuan 32001, Taiwan.

Guangyu Chen Chemistry Department, Brookhaven National Laboratory, Upton, NY 11973, USA.

Rong Chen The Key Laboratory of Fuel Cell Technology of Guangdong Province, School of Chemistry and Chemical Engineering, South China University of Technology, Guangzhou 510641, China.

Shengli Chen Hubei Key Laboratory of Electrochemical Power Sources, Department of Chemistry, Wuhan University, Wuhan 430072, China.

Xiaoting Chen Key Laboratory for Liquid-Solid Structural Evolution and Processing of Materials (Ministry of Education), School of Materials Science and Engineering, Shandong University, Jingshi Road 17923, Jinan 250061, China.

Yu-Lun Chen Department of Chemical and Materials Engineering, National Kaohsiung University of Applied Sciences, Kaohsiung 80778, Taiwan.

Xuan Cheng Department of Materials Science and Engineering, College of Materials, Xiamen University, Xiamen 361005, China; Fujian Key Laboratory of Advanced Materials, Xiamen University, Xiamen 361005, China.

Masanobu Chiku Department of Applied Chemistry, Graduate School of Engineering, Osaka Prefecture University, Sakai, Osaka 599-8531, Japan.

Mitsuharu Chisaka Department of Electronics and Information Technology, Hirosaki University, 3 Bunkyo-cho, Hirosaki, Aomori 036-8561, Japan.

Morio Chiwata Special Doctoral Program for Green Energy Conversion Science and Technology, Interdisciplinary Graduate School of Medicine and Engineering, University of Yamanashi, 4 Takeda, Kofu 400-8510, Japan.

Lina Chong Chemical Science and Engineering Division, Argonne National Laboratory, Argonne, IL 60439, USA.

David A. Condit United Technologies Research Center, East Hartford, CT 06108, USA; This author deceased on 4 June 2014.

Hannah Cronk Department of Chemistry, State University of New York at Binghamton, Binghamton, NY 13902, USA.

Yuxuan Dai Jiangsu key laboratory of Key Laboratory of New Power Batteries, Jiangsu Collaborative Innovation Center of Biomedical Functional Materials, School of Chemistry and Materials Science, Nanjing Normal University, Nanjing 210023, China.

Shigehito Deki Fuel Cell Nanomaterials Center, University of Yamanashi, 4 Takeda, Kofu 400-8510, Japan.

E. Bradley Easton Faculty of Science, University of Ontario Institute of Technology, 2000 Simcoe Street North, Oshawa, ON L1H 7K4, Canada.

Lior Elbaz Chemistry Department, Faculty of Exact Sciences, Bar-Ilan University, Ramat-Gan, 5290002, Israel.

Reza Alipour Moghadam Esfahani Department of Applied Science and Technology, Politecnico di Torino, Corso Duca degli Abruzzi 24, 10129 Torino, Italy.

Fangxia Feng Department of Chemistry, University of Calgary, 2500 University Drive NW Calgary, AB T2N1N4, Canada.

Yongjun Feng State Key Laboratory of Chemical Resource Engineering, Beijing University of Chemical Technology, 15 Beisanhuan Eastroad, Beijing 100029, China.

Paulo J. Ferreira Materials Science and Engineering Program, University of Texas, Austin, TX 78712, USA.

Carlotta Francia Department of Applied Science and Technology, Politecnico di Torino, Corso Duca degli Abruzzi 24, 10129 Torino, Italy.

Tobias Fürstenhaupt Microscopy and Imaging Facility, University of Calgary, Health Sciences Centre B159, 3330 Hospital Drive NW, Calgary, AB T2N 4N1, Canada.

Yulai Gao Laboratory for Microstructures, Shanghai University, 99 Shangda Road, Shanghai 200436, China.

José Antonio Garrido Laboratori d'Electroquímica dels Materials i del Medi Ambient, Departament de Química Física, Universitat de Barcelona, Martí i Franquès 1-11, 08028 Barcelona, Spain.

Mohammadreza Zamanzad Ghavidel Faculty of Science, University of Ontario Institute of Technology, 2000 Simcoe Street North, Oshawa, ON L1H 7K4, Canada.

Qiaojuan Gong Department of Applied Chemistry, Yuncheng University, 1155 Fudan West Street, Yun Cheng 04400, China.

Mallika Gummalla United Technologies Research Center, East Hartford, CT 06108, USA.

Julien Hannauer Institut Charles Gerhardt Montpellier UMR 5253, CNRS, Université de Montpellier, 34095 Montpellier Cedex 5, France.

Xiaolei He College of Chemistry & Chemical Engineering, Fujian Normal University, Fuzhou 350007, China.

Eiji Higuchi Department of Applied Chemistry, Graduate School of Engineering, Osaka Prefecture University, Sakai, Osaka 599-8531, Japan.

Yaovi Holade Université de Poitiers, UMR CNRS 7285, "Équipe SAMCat"; 4, rue Michel Brunet, B27, TSA 51106, 86073 Poitiers cedex 09, France.

Jue Hu Chemistry Department, Brookhaven National Laboratory, Upton, NY 11973, USA; Hefei Institutes of Physical Science, Chinese Academy of Sciences, P.O. Box 1126, Hefei 230031, China.

Rumeng Huang School of Chemistry and Chemical Engineering, Key Laboratory of Green Chemical Media and Reactions, Ministry of Education, Collaborative Innovation Center of Henan Province for Fine Chemicals Green Manufacturing, Henan Normal University, Xinxiang 453007, China.

Hiroshi Inoue Department of Applied Chemistry, Graduate School of Engineering, Osaka Prefecture University, Sakai, Osaka 599-8531, Japan.

Akimitsu Ishihara Institute of Advanced Sciences, Yokohama National University, 79-5 Tokiwadai, Hodogaya-ku, Yokohama 240-8501, Japan.

Frédéric Jaouen Institut Charles Gerhardt Montpellier UMR 5253, CNRS, Université de Montpellier, 34095 Montpellier Cedex 5, France.

Nathalie Job Laboratory of Chemical Engineering—Nanomaterials, Catalysis, Electrochemistry, University of Liège (B6a), B-4000 Liège, Belgium.

Pharrah Joseph Department of Chemistry, State University of New York at Binghamton, Binghamton, NY 13902, USA.

Zachary B. Kaiser Chemical Science and Engineering Division, Argonne National Laboratory, Argonne, IL 60439, USA.

Katsuyoshi Kakinuma Fuel Cell Nanomaterials Center, University of Yamanashi, 4 Takeda, Kofu 400-8510, Japan.

Ning Kang Department of Chemistry, State University of New York at Binghamton, Binghamton, NY 13902, USA.

Siwon Kim Center for Clean Energy Engineering, University of Connecticut, Storrs, CT 06269, USA; Department of Material Science and Engineering, University of Connecticut, Storrs, CT 06269, USA.

Yuji Kohno Green Hydrogen Research Center, Yokohama National University, 79-5 Tokiwadai, Hodogaya-ku, Yokohama 240-8501, Japan.

Kouakou B. Kokoh Université de Poitiers, UMR CNRS 7285, "Équipe SAMCat"; 4, rue Michel Brunet, B27, TSA 51106, 86073 Poitiers cedex 09, France.

Ger J.M. Koper Department of Chemical Engineering, Delft University of Technology, Julianalaan 136, 2628BL Delft, The Netherlands.

Chung-Wen Kuo Department of Chemical and Materials Engineering, National Kaohsiung University of Applied Sciences, Kaohsiung 80778, Taiwan.

Kurian A. Kuttiyiel Chemistry Department, Brookhaven National Laboratory, Upton, NY 11973, USA.

Taiki Kuwahara Department of Applied Chemistry, Graduate School of Engineering, Osaka Prefecture University, Sakai, Osaka 599-8531, Japan.

Stéphanie D. Lambert Laboratory of Chemical Engineering—Nanomaterials, Catalysis, Electrochemistry, University of Liège (B6a), B-4000 Liège, Belgium.

Lijuan Le College of Chemistry & Chemical Engineering, Fujian Normal University, Fuzhou 350007, China.

Dianqing Li State Key Laboratory of Chemical Resource Engineering, Beijing University of Chemical Technology, 15 Beisanhuan Eastroad, Beijing 100029, China.

Erling Li Jilin Province Key Laboratory of Low Carbon Chemical Power; State Key Laboratory of Electroanalytical Chemistry, Changchun Institute of Applied Chemistry, Chinese Academy of Science, 5625 Renmin Street, Changchun 130022, China; Graduate University of Chinese Academy of Science, Beijing 100049, China.

Hengyi Li Department of Materials Science and Engineering, College of Materials, Xiamen University, Xiamen 361005, China.

Qiaoxia Li Shanghai Key Laboratory of Materials Protection and Advanced Materials in Electric Power, College of Environmental and Chemical Engineering, Shanghai University of Electric Power, Shanghai 200090, China.

Zhen-Xing Liang Key Laboratory on Fuel Cell Technology of Guangdong Province, School of Chemistry and Chemical Engineering, South China University of Technology, Guangzhou 510641, China.

Shijun Liao The Key Laboratory of Fuel Cell Technology of Guangdong Province, School of Chemistry and Chemical Engineering, South China University of Technology, Guangzhou 510641, China.

Shen Lin College of Chemistry & Chemical Engineering, Fujian Normal University, Fuzhou 350007, China.

Di-Jia Liu Chemical Science and Engineering Division, Argonne National Laboratory, Argonne, IL 60439, USA.

Jing Liu State Key Laboratory of Electroanalytical Chemistry, Changchun Institute of Applied Chemistry, Chinese Academy of Science, 5625 Renmin Street, Changchun 130022, China; Jilin Province Key Laboratory of Low Carbon Chemical Power, Changchun Institute of Applied Chemistry, Chinese Academy of Science, 5625 Renmin Street, Changchun 130022, China.

Mingshuang Liu Shanghai Key Laboratory of Materials Protection and Advanced Materials in Electric Power, College of Environmental and Chemical Engineering, Shanghai University of Electric Power, Shanghai 200090, China.

Qi Liu Shanghai Key Laboratory of Materials Protection and Advanced Materials in Electric Power, College of Environmental and Chemical Engineering, Shanghai University of Electric Power, Shanghai 200090, China.

Oran Lori Chemistry Department, Faculty of Exact Sciences, Bar-Ilan University, Ramat-Gan, 5290002, Israel.

Jin Luo Department of Chemistry, State University of New York at Binghamton, Binghamton, NY 13902, USA.

Ai Ma College of Chemistry & Chemical Engineering, Fujian Normal University, Fuzhou 350007, China.

Radenka Maric Department of Chemical & Biomolecular Engineering; Department of Material Science and Engineering; Center for Clean Energy Engineering, University of Connecticut, Storrs, CT 06269, USA.

Koichi Matsuzawa Green Hydrogen Research Center, Yokohama National University, 79-5 Tokiwadai, Hodogaya-ku, Yokohama 240-8501, Japan.

Hui Meng Department of Physics and Siyuan Laboratory, College of Science and Engineering, Jinan University, Guangzhou 510632, China.

Shigenori Mitsushima Green Hydrogen Research Center; Institute of Advanced Sciences, Yokohama National University, 79-5 Tokiwadai, Hodogaya-ku, Yokohama 240-8501, Japan.

William E. Mustain Center for Clean Energy Engineering, University of Connecticut, Storrs, CT 06269, USA; Department of Chemical & Biomolecular Engineering, University of Connecticut, Storrs, CT 06269, USA.

Deborah J. Myers Chemical Sciences and Engineering Division, Argonne National Laboratory, Lemont, IL 60439, USA.

Timothy D. Myles Center for Clean Energy Engineering, University of Connecticut, Storrs, CT 06269, USA.

Teko W. Napporn Université de Poitiers, UMR CNRS 7285, "Équipe SAMCat"; 4, rue Michel Brunet, B27, TSA 51106, 86073 Poitiers cedex 09, France.

Emanuela Negro Department of Chemical Engineering, Delft University of Technology, Julianalaan 136, 2628BL Delft, The Netherlands.

Lu Niu School of Chemistry and Chemical Engineering, Key Laboratory of Green Chemical Media and Reactions, Ministry of Education, Collaborative Innovation Center of Henan Province for Fine Chemicals Green Manufacturing, Henan Normal University, Xinxiang 453007, China.

Yoshiro Ohgi Kumamoto Industrial Research Institute, 3-11-38 Azuma-cho, Azuma-ku, Kumamoto, Kumamoto 862-0901, Japan.

Luigi Osmieri Department of Applied Science and Technology, Politecnico di Torino, Corso Duca degli Abruzzi 24, 10129 Torino, Italy.

Ken-ichiro Ota Green Hydrogen Research Center, Yokohama National University, 79-5 Tokiwadai, Hodogaya-ku, Yokohama 240-8501, Japan.

Gu-Gon Park Fuel Cell Research Center, Korea Institute of Energy Research, Daejeon 305-343, Korea.

Katie Pei Department of Chemistry, University of Calgary, 2500 University Drive NW Calgary, AB T2N1N4, Canada.

Amra Peles Physical Sciences Department, United Technologies Research Center, East Hartford, CT 06108, USA.

Jean-Paul Pirard Laboratory of Chemical Engineering—Nanomaterials, Catalysis, Electrochemistry, University of Liège (B6a), B-4000 Liège, Belgium.

Lesia Protsailo Pratt & Whitney Program Office, United Technologies Research Center, East Hartford, CT 06108, USA.

Jinli Qiao College of Environmental Science and Engineering, Donghua University, 2999 Ren'min North Road, Shanghai 201620, China.

Xiaochang Qiao The Key Laboratory of Fuel Cell Technology of Guangdong Province, School of Chemistry and Chemical Engineering, South China University of Technology, Guangzhou 510641, China.

Xiaoyu Qiu Jiangsu key laboratory of Key Laboratory of New Power Batteries, Jiangsu Collaborative Innovation Center of Biomedical Functional Materials, School of Chemistry and Materials Science, Nanjing Normal University, Nanjing 210023, China.

Somaye Rasouli Materials Science and Engineering Program, University of Texas, Austin, TX 78712, USA.

Rosa María Rodríguez Laboratori d'Electroquímica dels Materials i del Medi Ambient, Departament de Química Física, Universitat de Barcelona, Martí i Franquès 1-11, 08028 Barcelona, Spain.

Mingbo Ruan Jilin Province Key Laboratory of Low Carbon Chemical Power; State Key Laboratory of Electroanalytical Chemistry, Changchun Institute of Applied Chemistry, Chinese Academy of Science, 5625 Renmin Street, Changchun 130022, China.

Nihat Ege Sahin Université de Poitiers, UMR CNRS 7285, "Équipe SAMCat"; 4, rue Michel Brunet, B27, TSA 51106, 86073 Poitiers cedex 09, France.

Ryotaro Sakai Department of Applied Chemistry, Graduate School of Engineering, Osaka Prefecture University, Sakai, Osaka 599-8531, Japan.

Kotaro Sasaki Chemistry Department, Brookhaven National Laboratory, Upton, NY 11973, USA.

Karine Servat Université de Poitiers, UMR CNRS 7285, "Équipe SAMCat"; 4, rue Michel Brunet, B27, TSA 51106, 86073 Poitiers cedex 09, France.

Shiyao Shan Department of Chemistry, State University of New York at Binghamton, Binghamton, NY 13902, USA.

Minhua Shao Department of Chemical and Biomolecular Engineering, Hong Kong University of Science and Technology, Clear Water Bay, Kowloon, Hong Kong.

Zakiya Skeete Department of Chemistry, State University of New York at Binghamton, Binghamton, NY 13902, USA.

Ping Song Jilin Province Key Laboratory of Low Carbon Chemical Power; State Key Laboratory of Electroanalytical Chemistry, Changchun Institute of Applied Chemistry, Chinese Academy of Science, 5625 Renmin Street, Changchun 130022, China.

Stefania Specchia Department of Applied Science and Technology, Politecnico di Torino, Corso Duca degli Abruzzi 24, 10129 Torino, Italy.

Alessandro Stassi Consiglio Nazionale delle Ricerche, Istituto di Tecnologie Avanzate per l'Energia "Nicola Giordano", Salita S. Lucia sopra Contesse 5, 98126 Messina, Italy.

Dong Su Chemistry Department, Brookhaven National Laboratory, Upton, NY 11973, USA.

Shuhui Sun Institut National de la Recherche Scientifique (INRS), Centre Énergie, Matériaux et Télécommunications, 1650 Boulevard Lionel-Boulet, Varennes, QC J3X 1S2, Canada.

Yuko Tamura Green Hydrogen Research Center, Yokohama National University, 79-5 Tokiwadai, Hodogaya-ku, Yokohama 240-8501, Japan.

Yawen Tang Jiangsu key laboratory of Key Laboratory of New Power Batteries, Jiangsu Collaborative Innovation Center of Biomedical Functional Materials, School of Chemistry and Materials Science, Nanjing Normal University, Nanjing 210023, China; State Key Laboratory of Chemical Resource Engineering, Beijing University of Chemical Technology, 15 Beisanhuan Eastroad, Beijing 100029, China.

Xin Tong Institut National de la Recherche Scientifique (INRS), Centre Énergie, Matériaux et Télécommunications, 1650 Boulevard Lionel-Boulet, Varennes, QC J3X 1S2, Canada.

Bryan Trimm Department of Chemistry, State University of New York at Binghamton, Binghamton, NY 13902, USA.

Makoto Uchida Fuel Cell Nanomaterials Center, University of Yamanashi, 4 Takeda, Kofu 400-8510, Japan.

Hiroyuki Uchida Fuel Cell Nanomaterials Center, University of Yamanashi, 4 Takeda, Kofu 400-8510, Japan; Clean Energy Research Center, University of Yamanashi, 4 Takeda, Kofu 400-8510, Japan.

Alessandro H.A. Monteverde Videla Department of Applied Science and Technology, Politecnico di Torino, Corso Duca degli Abruzzi 24, 10129 Torino, Italy.

Mitsuru Wakisaka Fuel Cell Nanomaterials Center, University of Yamanashi, 4 Takeda, Kofu 400-8510, Japan.

Kai Wan Key Laboratory on Fuel Cell Technology of Guangdong Province, School of Chemistry and Chemical Engineering, South China University of Technology, Guangzhou 510641, China.

Hao Wang College of Chemistry & Chemical Engineering, Fujian Normal University, Fuzhou 350007, China.

Long-Long Wang Shanghai Key Laboratory for Molecular Catalysis and Innovative Materials, Collaborative Innovation Center of Chemistry for Energy Materials and Department of Chemistry, Fudan University, Shanghai 200433, China.

Xiaoying Wang Key Laboratory for Liquid-Solid Structural Evolution and Processing of Materials (Ministry of Education), School of Materials Science and Engineering, Shandong University, Jingshi Road 17923, Jinan 250061, China.

Ya-Jun Wang Shanghai Key Laboratory of Materials Protection and Advanced Materials in Electric Power, Shanghai University of Electric Power, NO. 2588 Changyang Road, Yangpu District, Shanghai 200090, China.

Ye Wang Key Laboratory for Liquid-Solid Structural Evolution and Processing of Materials (Ministry of Education), School of Materials Science and Engineering, Shandong University, Jingshi Road 17923, Jinan 250061, China.

Ying Wang Shanghai Key Laboratory of Materials Protection and Advanced Materials in Electric Power, Shanghai University of Electric Power, NO. 2588 Changyang Road, Yangpu District, Shanghai 200090, China.

Masahiro Watanabe Fuel Cell Nanomaterials Center, University of Yamanashi, 4 Takeda, Kofu 400-8510, Japan.

Qiliang Wei Department of Applied Chemistry, Yuncheng University, 1155 Fudan West Street, Yun Cheng 04400, China; Institut National de la Recherche Scientifique (INRS), Centre Énergie, Matériaux et Télécommunications, 1650 Boulevard Lionel-Boulet, Varennes, QC J3X 1S2, Canada.

Jinfang Wu Department of Chemistry, State University of New York at Binghamton, Binghamton, NY 13902, USA.

Peishan Wu Jiangsu key laboratory of Key Laboratory of New Power Batteries, Jiangsu Collaborative Innovation Center of Biomedical Functional Materials, School of Chemistry and Materials Science, Nanjing Normal University, Nanjing 210023, China.

Tzi-Yi Wu Department of Chemical Engineering and Materials Engineering, National Yunlin University of Science and Technology, Yunlin 64002, Taiwan.

Jingmin Xi State Key Laboratory of Chemical Resource Engineering, Beijing University of Chemical Technology, 15 Beisanhuan Eastroad, Beijing 100029, China.

Fangyan Xie Instrumental Analysis & Research Center, Sun Yat-sen University, Guangzhou 510275, China.

Qunjie Xu Shanghai Key Laboratory of Materials Protection and Advanced Materials in Electric Power, College of Environmental and Chemical Engineering, Shanghai University of Electric Power, Shanghai 200090, China.

Tao Xu Department of Chemistry and Biochemistry, Northern Illinois University, DeKalb, IL 60115, USA.

Weilin Xu Jilin Province Key Laboratory of Low Carbon Chemical Power; State Key Laboratory of Electroanalytical Chemistry, Changchun Institute of Applied Chemistry, Chinese Academy of Science, 5625 Renmin Street, Changchun 130022, China.

Lin Yang School of Chemistry and Chemical Engineering, Key Laboratory of Green Chemical Media and Reactions, Ministry of Education, Collaborative Innovation Center of Henan Province for Fine Chemicals Green Manufacturing, Henan Normal University, Xinxiang 453007, China.

Tae-Hyun Yang Fuel Cell Research Center, Korea Institute of Energy Research, Daejeon 305-343, Korea.

Zhiwei Yang United Technologies Research Center, East Hartford, CT 06108, USA.

Siyu Ye Ballard Power Systems, 9000 Glenlyon Parkway, Burnaby, BC V5J 5J8, Canada.

Chenghang You The Key Laboratory of Fuel Cell Technology of Guangdong Province, School of Chemistry and Chemical Engineering, South China University of Technology, Guangzhou 510641, China.

Kang Yu Materials Science and Engineering Program, University of Texas, Austin, TX 78712, USA.

Zhi-Peng Yu Key Laboratory on Fuel Cell Technology of Guangdong Province, School of Chemistry and Chemical Engineering, South China University of Technology, Guangzhou 510641, China.

Dongrong Zeng Department of Physics and Siyuan Laboratory, College of Science and Engineering, Jinan University, Guangzhou 510632, China.

Juqin Zeng Department of Applied Science and Technology, Politecnico di Torino, Corso Duca degli Abruzzi 24, 10129 Torino, Italy.

Chengxu Zhang Hefei Institutes of Physical Science, Chinese Academy of Sciences, P.O. Box 1126, Hefei 230031, China.

Fengqi Zhang Jiangsu key laboratory of Key Laboratory of New Power Batteries, Jiangsu Collaborative Innovation Center of Biomedical Functional Materials, School of Chemistry and Materials Science, Nanjing Normal University, Nanjing 210023, China.

Gaixia Zhang Institut National de la Recherche Scientifique (INRS), Centre Énergie, Matériaux et Télécommunications, 1650 Boulevard Lionel-Boulet, Varennes, QC J3X 1S2, Canada.

Hanyue Zhang Jiangsu key laboratory of Key Laboratory of New Power Batteries, Jiangsu Collaborative Innovation Center of Biomedical Functional Materials, School of Chemistry and Materials Science, Nanjing Normal University, Nanjing 210023, China.

Jiujun Zhang School of Chemistry and Chemical Engineering, Key Laboratory of Green Chemical Media and Reactions, Ministry of Education, Collaborative Innovation Center of Henan Province for Fine Chemicals Green Manufacturing, Henan Normal University, Xinxiang 453007, China; Department of Chemical & Biochemical Engineering, University of British Columbia, Vancouver, B.C. V6T 1W5, Canada.

Qing Zhang School of Chemistry and Chemical Engineering, Key Laboratory of Green Chemical Media and Reactions, Ministry of Education, Collaborative Innovation Center of Henan Province for Fine Chemicals Green Manufacturing, Henan Normal University, Xinxiang 453007, China.

Shiming Zhang Hubei Key Laboratory of Electrochemical Power Sources, Department of Chemistry, Wuhan University, Wuhan 430072, China.

Xiaofeng Zhang College of Chemistry & Chemical Engineering, Fujian Normal University, Fuzhou 350007, China.

Zhonghua Zhang Key Laboratory for Liquid-Solid Structural Evolution and Processing of Materials (Ministry of Education), School of Materials Science and Engineering, Shandong University, Jingshi Road 17923, Jinan 250061, China.

Wei Zhao Department of Chemistry, State University of New York at Binghamton, Binghamton, NY 13902, USA.

Yinguang Zhao Department of Chemistry, State University of New York at Binghamton, Binghamton, NY 13902, USA.

Qiaoling Zheng Department of Materials Science and Engineering, College of Materials, Xiamen University, Xiamen 361005, China.

Chuan-Jian Zhong Department of Chemistry, State University of New York at Binghamton, Binghamton, NY 13902, USA.

Haihong Zhong State Key Laboratory of Chemical Resource Engineering, Beijing University of Chemical Technology, 15 Beisanhuan Eastroad, Beijing 100029, China.

Shouzhong Zou Department of Chemistry, American University, Washington, DC 20016, USA.

Anthony Zubiaur Laboratory of Chemical Engineering—Nanomaterials, Catalysis, Electrochemistry, University of Liège (B6a), B-4000 Liège, Belgium.

About the Guest Editor

Minhua Shao earned his BS (1999) and MS (2002) degrees in chemistry from Xiamen University, and a PhD degree in materials science and engineering from State University of New York at Stony Brook (2006). He joined UTC Power in 2007 to lead the development of advanced catalysts and supports for PEMFC and PAFC. He was promoted to UTC Technical Fellow and Project Manager in 2012. In 2013, he joined Ford Motor Company to conduct research on lithium-ion batteries for electrified vehicles. He then joined the Hong Kong University of Science and Technology in the Department of Chemical and Biomolecular Engineering as an Associate Professor in 2014. He received the Supramaniam Srinivasan Young Investigator Award from the ECS Energy Technology Division (2014) and the Student Achievement Award from the ECS Industrial Electrochemistry and Electrochemical Engineering Division (2007). His research mainly focuses on electrocatalysis and advanced batteries.

Preface to "Electrocatalysis in Fuel Cells"

Low temperature fuel cells are expected to come into widespread commercial use in the areas of transportation and stationary and portable power generation, and will therefore help solve energy shortage and environmental issues. Despite their great promise, commercialization has been hindered by lower-than-predicted efficiencies and the high cost of the electrocatalysts in the electrodes. The sluggish kinetics of the oxygen reduction reaction (ORR) is one of the main reasons for the high overpotential in a hydrogen proton exchange membrane fuel cell (PEMFC). The introduction of Mirai, the first mass-produced fuel cell vehicles (FCVs), by Toyota in Japan in 2014, and in North America in the following year, has accelerated the development of FCVs by other automotive companies. For instance, Honda and Hyundai announced the mass production of their own FCVs in 2016 and 2017, respectively. The current sale price of a new Mirai is about USD 57,000. One of the main reasons for the high sale price is the high Pt loading in the fuel cell stacks, especially at the cathode electrode, where the ORR occurs. The Pt loading at the anode, where the hydrogen oxidation reaction (HOR) occurs, can be reduced to as low as 0.05 mg cm^{-2} due to the extremely high reaction rate on Pt surfaces, while a much higher Pt loading ($\geq$0.2 mg cm^{-2}) at the cathode is required, using Pt or Pt alloys as the ORR electrocatalysts, in order to achieve a desirable cell performance. Pt is a costly and scarce metal. Thus, reducing its loading or even completely replacing it with abundant and cheap materials would be advantageous to lower the cost of FCVs. Recent research efforts have been focused on developing advanced Pt alloys, core–shell structures, shape-controlled nanocrystals and non-precious-metal (NPM) catalysts.

In addition to ORR activity, one must also consider the durability of the electrode during fuel cell operation in the harsh environment. The life of a fuel cell stack in a FCV has to last at least 10 years in order to compete with the conventional combustion engine. It has been confirmed that the fuel cell performance gradually declined during operation. The main reasons for the degradation of the catalyst layer are the dissolution of the Pt and the corrosion of the carbon support. As a consequence, both catalysts and supports that are more stable than Pt nanoparticles and carbon black may be needed to meet the durability requirements. Promising supports include alternative carbon supports, carbides and oxides.

In other types of low temperature fuel cells, for instance direct alcohol fuel cells (DAFCs), the slow fuel oxidation reactions and fast performance decay, caused by poisonous CO species adsorbed on catalyst surfaces, are the other major contributions to their low performances. Thus, the development of more active catalysts with higher tolerance to CO poisoning is required for high-performance and long-life DAFCs. The most promising DAFCs include DMFC, DEFC and DFAFC which use methanol, ethanol and formic acid as fuel, respectively.

This Special Issue aims to cover recent progress and trends in designing, synthesizing, characterizing and evaluating advanced electrocatalysts and supports for both ORR and small organic molecule oxidation reactions, as well as theoretical understanding in fuel cell reactions. I am honored to be the Guest Editor for this Issue, which includes 34 high quality papers. The nine reviews together with 25 original research papers cover a very broad spectrum of electrocatalysis in fuel cells.

It is obvious that the ORR is still the most important topic in fuel cell electrocatalysis. There are 16 papers that cover some important developments in ORR electrocatalysts including NPM, core-shell and Pt alloy catalysts. A significant effort is focused on completely replacing Pt in the ORR catalysts by developing novel NPM materials. Wei *et al.* summarized the recent progress in design and synthesis of metal-free nitrogen-doped carbon materials, including nitrogen-doped carbon nanotubes (NCNTs) and nitrogen-doped graphene (NG) for ORR in both acidic and alkaline media. Liu *et al.* reviewed the progress made in the past five years in the areas of Fe–N–C electrocatalysts for ORR and understanding the possible active sites in this type of catalyst. The Fe–N–Cs prepared from Fe-doped zeolitic imidazolate frameworks (ZIFs) are among the most active ones in catalyzing the ORR. Barkholtz *et al.* optimized the synthesis and post-treatment protocols of ZIF-based Fe–N–C nanocomposites, as well as the membrane electrode assembly (MEA) fabrication process, and achieved an impressive fuel cell performance of 221.9 mA cm^{-2} at 0.8 V. Armel *et al.* emphasized the importance of the morphology control of Fe–N–C on ORR activity by adjusting the crystal size of ZIF-8, milling speed and heating mode. With the smallest ZIF-8 crystal size (100 nm), the best H_2/O_2 fuel cell performance of 900 mA cm^{-2} at 0.5 V was obtained, which was double the value obtained with previous synthesis protocol. Zhang and Chen developed a novel method to prepare Fe–N–C catalysts by using a cationic surfactant cetyltrimethylammonium bromide (CTAB) as the template and the negatively charged persulfate ions as the oxidative agent to stimulate the aniline polymerization, resulting in a unique one-dimensional (1D) semi-tubular structure of PANI. On the other hand, SBA-15 was used as the template by Wan *et al.* in the synthesis of nitrogen-doped ordered mesoporous carbon. Qiao *et al.* found that P, N dual-doped reduced graphene oxide synthesized by pyrolyzing a mixture of graphite oxide and diammonium hydrogen phosphate was very active for ORR. Non-noble metal oxides and chalcogenides are also promising catalysts for ORR. Some interesting works on the synthesis and evaluation of Ti-Nb oxides, CoS and FeSe$_2$ were also included in this Special Issue.

Core–shell structures consisting of a cheaper core and an atomic thin Pt shell have attracted great attention due to their extremely high Pt utilization and activity enhancement from the core materials. Hu *et al.* designed a core-shell catalyst with a nitride (PdNiN) alloy core and a Pt monolayer. Its stability was dramatically enhanced compared with that of the previously reported structure with a pure Pd core. Inoue *et al.* developed a new method without using any surfactant to synthesize clean Pd nanoparticles as the core for Pt monolayer deposition. On the other hand, Caballero-Manrique *et al.* used Cu nanoparticles as the sacrificing template to prepare core-shell catalysts consisting of Pt and Pt–Ru shells. Amra *et al.* tried to explain the strain and ligand effects from the $Pd_{1-x}Cu_x$ alloy core on the Pt monolayer, based on the first principles density functional theory (DFT) calculations.

Since the discovery of Pt alloys as superior ORR catalysts for fuel cells in the 1980s, they have been considered as the second generation fuel cell catalysts after pure Pt. Pt alloys not only significantly reduce the Pt loading, but also enhance the catalytic activity and stability in comparison with Pt. Shen *et al.* gave a nice review on some of the recent approaches in developing Pt alloy electrocatalysts for the ORR. The particle size effect of Pt alloys on the fuel cell performance and decay rate is very important and has not been systematically studied. Gummalla *et al.* compared the initial performances and decay trends of Pt_3Co/C cathodes in PEMFCs with three different particle sizes (4.9 nm, 8.1 nm, and 14.8 nm), but with the identical Pt loading. The initial mass activity of the 4.9 nm Pt_3Co-based electrode was the highest, as well as the performance decay rate. The impact of PEMFC operating conditions, including upper potential, relative humidity, and temperature, on the alloy catalyst decay trends were also carefully studied for the first time.

Some non-carbon-based materials have been explored as alternative supports for ORR catalysts. Lori and Elbaz summarized the latest studies on ceramic supports including carbides, oxides, nitrides, borides, and some composite materials. Alternative carbon supports including carbon nanotubes, ordered mesoporous carbon, and colloid imprinted carbon were reviewed by Banham *et al.* The importance of carbon wall thickness was highlighted. Functionalized graphitic supports with pyrene carboxylic acid also showed superior durability.

The performance of fuel cell catalysts is certainly dependent on the preparation methods. Holade *et al.* and Job *et al.* summarized the recent advances in the preparation of carbon-supported nanocatalysts based on colloidal methods. The correlation between the structure of the catalysts and their activities and the effects from the synthesis methods were discussed. In addition, the fuel cell performance is also strongly influenced by the composition and fabrication protocols of MEA. A semi-empirical model was developed by Myles *et al.* to understand the performance of the cell as a function of the ratio of Nafion ionomer to carbon support (I/C ratio) in high temperature PEMFCs.

There are three papers in this issue focusing on Pt-based nanocomposites for the methanol oxidation reaction (MOR). Pt/C–Mn_xO_{1+x} and $H_3PMo_{12}O_{40}$–Pt/reduced graphene oxide were found to have better performance than Pt/C and PtRu/C due to the synergetic effects between Pt nanoparticles and hybrid supports. Wu *et al.* synthesized a conductive copolymer based on indole-6-carboxylic acid and 3,4-ethylenedioxythiophene (EDOT) as the support for Pt particles. This nanocomposite also showed good activity for MOR. Chen *et al.* prepared a multi-component nanoporous PtRuCuW electrocatalyst by chemical and mechanical dealloying. The unique ligament/channel nanoporous structure showed an enhanced activity for MOR compared to PtRu/C.

The advances in the study of reaction mechanisms and electrocatalytic materials (mainly Pt- and Pd-based catalysts) for the ethanol oxidation reaction (EOR) were reviewed by Wang *et al.* PdW alloys and hollow PdCu nanocubes, as well as PtMn alloys, showed some improvement over pure Pd or Pt toward EOR.

Meng *et al.* presented a comprehensive review on the Pd-based electrocataysts' formic acid oxidation reaction (FAOR), MOR, EOR and ORR. The high activity of Pd-based materials toward FAOR was also supported by a couple of original research papers included in this issue.

Finally, I would like to thank Keith Hohn, Editor-in-Chief, Mary Fan, Managing Editor, and the staff of the *Catalysts* Editorial Office for their great support during the preparation of this Special Issue. I also thank all the authors for their great contributions and referees for their time reviewing the manuscripts. I believe these excellent papers collected in this Special Issue will make significant contributions to the electrocatalysis' community.

Minhua Shao
Guest Editor

Nitrogen-Doped Carbon Nanotube and Graphene Materials for Oxygen Reduction Reactions

Qiliang Wei, Xin Tong, Gaixia Zhang, Qiaojuan Gong and Shuhui Sun

Abstract: Nitrogen-doped carbon materials, including nitrogen-doped carbon nanotubes (NCNTs) and nitrogen-doped graphene (NG), have attracted increasing attention for oxygen reduction reaction (ORR) in metal-air batteries and fuel cell applications, due to their optimal properties including excellent electronic conductivity, 4e$^-$ transfer and superb mechanical properties. Here, the recent progress of NCNTs- and NG-based catalysts for ORR is reviewed. Firstly, the general preparation routes of these two N-doped carbon-allotropes are introduced briefly, and then a special emphasis is placed on the developments of both NCNTs and NG as promising metal-free catalysts and/or catalyst support materials for ORR. All these efficient ORR electrocatalysts feature a low cost, high durability and excellent performance, and are thus the key factors in accelerating the widespread commercialization of metal-air battery and fuel cell technologies.

Reprinted from *Catalysts*. Cite as: Wei, Q.; Tong, X.; Zhang, G.; Qiao, J.; Gong, Q.; Sun, S. Nitrogen-Doped Carbon Nanotube and Graphene Materials for Oxygen Reduction Reactions. *Catalysts* **2015**, *5*, 1574–1602.

1. Introduction

Developing highly efficient electrocatalysts to facilitate sluggish cathodic oxygen reduction reaction (ORR) is a key issue in metal-air batteries and fuel cells [1–5]. The ORR mechanism includes two different pathways: (i) a four-electron (4e$^-$) process to produce water directly though the reaction of oxygen, electrons and protons, and (ii) a two-electron (2e$^-$) process to create the intermediate compound (hydrogen peroxide) [6]. The 4e$^-$ process is more attractive for cathode catalysts in fuel cells. Although the platinum-based materials are the better choices for the desired 4e$^-$ pathway, the use of very expensive and rare platinum is a major impediment to the development and widespread commercialization of fuel cells. Thus, exploring the substitutes for platinum catalysts by employing non-precious metal catalysts is a very promising direction [7]. In this regard, one-dimensional (1D) carbon nanotubes (CNTs) and two-dimensional (2D) graphene (Figure 1) have attracted a great deal of attention for ORR due to their excellent electronic conductivity, huge specific surface area (SSA), as well as excellent thermal and mechanical properties [8]. Interestingly, when the heteroatoms are incorporated in the carbonaceous skeleton, the ORR performance can be greatly enhanced by effectively modulating the chemisorption energy of O_2, catalytic sites, and the reaction mechanism ($2e^-/4e^-$) of catalysts [9]. Among various possible dopants, N-doped carbon materials are attracting much more attention because of their excellent electrocatalytic performance, low cost, excellent stability, and environmental friendliness, thus setting up a new generation of the metal-free catalysts for ORR. Furthermore, when the nitrogen with excessive valence is introduced to the graphitic plane, more π-electrons can be obtained [10]. This feature, together with the significant difference in the electronegativity of N and C, leads to many unique properties to graphitic carbons, including increased n-type carrier concentration, high surface energy, reduced work-function, as well as tunable polarization [11–14]. As schematically illustrated in Figure 2, three common bonding configurations of N atoms in graphene are demonstrated, including pyrrolic, pyridinic, and graphitic (or quaternary) N [15]. Pyridinic N atoms are located at the edges of graphene planes, and each N atom is bonded to two C atoms and donates one π-electron to the π system. In the case of pyrrolic N atoms, they are incorporated into the heterocyclic rings and each N atom is bonded to two C atoms, contributing two π-electrons to the π system. Graphitic (or quaternary) N refers to the N atoms that replace the carbon atoms in the graphene plane. Such doped N atoms can change the local density state around the Fermi level of N-doped graphitic carbons, which may play a vital role in tailoring the electronic properties and improving their ORR performance [14,16].

On the other hand, metal oxides are also good candidates for ORR catalysts, although they normally suffer from low conductivity, as well as dissolution, sintering, and agglomeration during operation. Consequently, the electrocatalysts show poor electrochemical properties, restricting their applications. NCNTs or NG could effectively buffer the catalyst nanoparticle agglomeration and enhance the electronic conductivity by virtue of their intrinsic excellent conductivity and huge SSA. Therefore, NCNTs and NG can be used as both excellent metal-free electrocatalysts and perfect catalyst support for ORR.

The basic principles and mechanisms behind N doping effectively tailoring the electrical and surface properties of graphitic carbons have been reviewed in some excellent papers [14,17,18]. Here in this review, we place emphasis on the synthesis of NCNTs and NG, and their applications for ORR.

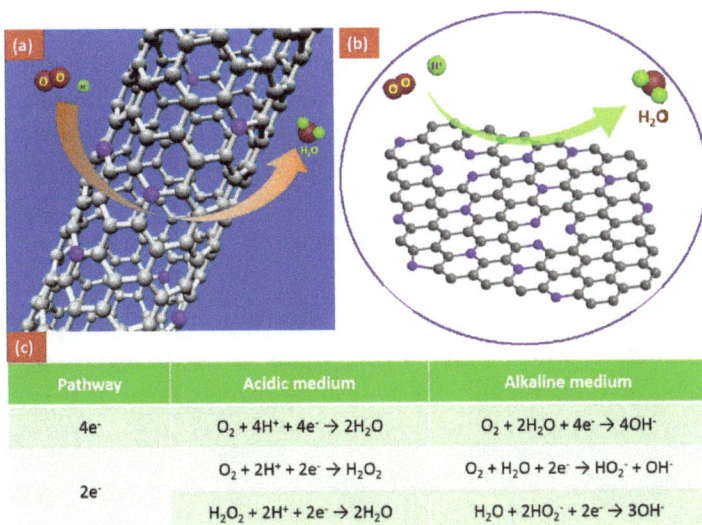

Figure 1. Illustration of ORR on (**a**) NCNTs; (**b**) NG and (**c**) ORR pathway in acid and alkaline medium. Reproduced and adapted in part from [19]. Copyright © 2013, Rights Managed by Nature Publishing Group.

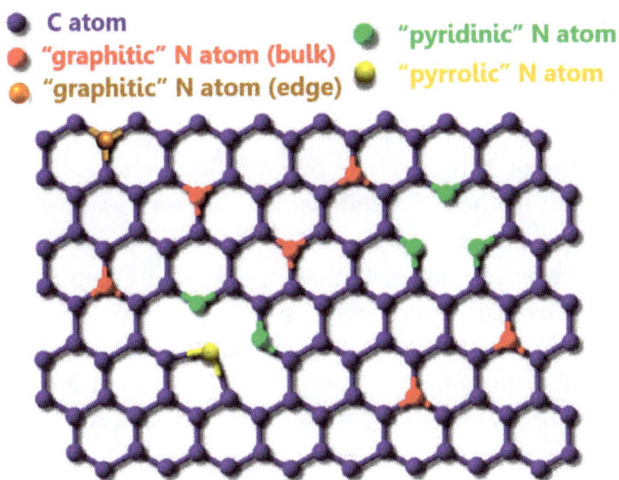

Figure 2. Schematic representation of different types of N atoms (graphitic, pyridinic and pyrrolic N) in NG and NCNTs. Modified with permission from Ref. [20]. Copyright © 2009, American Chemical Society.

2. Synthesis of Nitrogen-Doped Carbon

Nitrogen (N) is a neighboring element of carbon in the periodic table, and its electronegativity (3.04) is larger than that of C (2.55). The incorporation of N atom into a graphene lattice plane could modulate the local electronic properties, as it could form strong bonds with carbon atoms because of its comparable atomic size with carbon. Subsequently, it could generate a delocalized conjugated system between the graphene π-system and the lone pair of electrons from N atom. The introduction of N into carbon nanomaterials could improve both reactivity and electrocatalytic performance. As a result, the N-doped carbon materials have been intensively studied among all the available heteroatoms for doping.

2.1. N-Doped Carbon Nanotubes

NCNTs have become a focus as ORR catalysts due to their high activity and excellent stability. In principle, the N-doping methods can be classified to two categories: *in situ* doping and post-treatment doping [17].

2.1.1. *In Situ* Doping

In situ doping involves the direct incorporation of N heteroatoms into carbon matrix during the preparation process, and it is often used for the preparation of NCNTs. The typical *in situ* doping techniques include high-temperature arc-discharge [21,22], chemical vapor deposition (CVD) [23–27], chemically solvothermal procedures (*ca.* 230–300 °C), [28] and laser ablation methods [29,30]. Thus far, a wide range of N-containing precursors have been used to incorporate N into C matrix with great success. Moreover, the final amount and functionality of N in NCNTs are much more critical for practical applications but could essentially be derived from many different precursors by tuning the synthesis parameters such as temperature of pyrolysis. Among various techniques, CVD is the most promising method to synthesize NCNTs with a different C source (such as methane, acetylene, ethylene, benzene, *etc.*) [31–34] and N source (such as ethylene diamine, dimethylformamide, imidazole, Fe-Phthalocyanine, benzylamine, *etc.*) [34–38]. For instance, recently, by using a co-pyrolysis route of Fe-Phthalocyanine loaded and PEO$_{20}$-PPO$_{70}$-PEO$_{20}$ (P123) retained in mesoporous silica, Wang *et al.* [34] synthesized NCNTs with well-defined morphology and graphitic structure, which exhibited good performance for ORR. Based on CVD, She *et al.* fabricated N-doped 1D macroporous carbonaceous nanotube arrays in anodic alumina oxide (AAO) template, which also showed high performance for ORR [27]. In addition to the precursors and pyrolysis temperatures, for each method, other factors, such as time, gas flow rate, catalysts, also have significant influence on the nitrogen contents and the accurately controlled doping sites [17,28,39,40].

2.1.2. Post-Treatment

NCNTs have also been prepared by various post-treatment methods [41,42]. For instance, Nagaiah *et al.* [41] synthesized NCNTs by post-thermal treating

4

oxidized CNTs with ammonia and used the resultant NCNTs as efficient catalysts for ORR in alkaline medium. However, the post-synthesis treatments [43] normally require high temperature (800–1200 °C) and toxic N precursors (NH_3 or pyridine) which limit their practical application. Moreover, some structural degradation and morphological defects often appear in the materials due to the high temperature treatment [44].

In general, the *in situ* doping tends to form pyrrolic- and/or pyridinic-N atoms, while the graphitic-N in carbon frameworks is normally generated after a high temperature post-treatment [45]. Yet, to obtain an accurate N content and doping sites controllably in these materials is still a challenging problem [17].

2.2. N-Doped Graphene

Compared to doping N into CNTs, the N atom can be more easily introduced into the graphene due to the more open structure in graphene. The N atom could be incorporated into graphene directly during the synthesis of graphene or through post-treatment of graphene oxide (GO) (or graphene). Among numerous methods to produce graphene, CVD, solvothermal fabrication and arc-discharge are normally chosen for *in situ* growth of NG. Compared with the *in situ* synthesis, post-treatment methods which include thermal annealing, plasma or irradiation treatment, or solution treatment are simpler and likely closer to commercialization [46].

2.2.1. *In Situ* Doping

CVD is one of the important methods to prepare NG [20]. In Liu's group, they used Cu/Si as the catalyst, CH_4 as the C source and NH_3 as the N source to produce few-layers NG under 800 °C for the first time (single-layer graphene can be occasionally detected). On the other hand, by using the sole source that contains both C and N (e.g., acetonitrile [47] and pyridine [48]), N atoms can be simultaneously introduced into the graphene lattice during CVD growth of graphene films. The doping amount of N can be adjusted in the range of 1.2–16 at.% by controlling the gas flow rate and the C source to N source ratio [20,49].

A solvothermal process to obtain NG through the reaction between tetrachloromethane and lithium nitride was also developed by Deng *et al.* [50]. It is a one-pot direct synthesis with just placing the reaction reagent in an autoclave and keeping under N_2 and below 350 °C. It allows scalable synthesis and the nitrogen species can be introduced into the graphene structure with 4.5–16.4 at.% of N.

With the presence of pyridine vapor or NH_3, the arc-discharge technique which is commonly used for preparing carbon-based nanomaterials is also employed to fabricate NG. Rao *et al.* [51–53] successfully produced NG with the N content around 0.5–1.5 at.%. However, this process requires complicated purification steps with low yield due to the excessive by-products.

2.2.2. Post-Treatment

Thermal treatment in ammonia atmosphere is an easy and commonly used method to obtain NG by post-modification. Since the N incorporation reactions occur mostly at the defect sites and the edges of graphene, a low N level (e.g., 2.8 at.% in ref.) in graphene is normally obtained in previous reports [54]. In order to get higher N doping, researchers turned their attention to GO which contains a range of reactive oxygen functional groups and more defects to provide more active deposition. In Dai's group [55,56], through thermal annealing of GO under NH_3 atmosphere, the GO nanosheets were reduced and decorated with N simultaneously. At 300 °C, the N-doping process started, while the highest doping level of ~5 at.% N was achieved at 500 °C. The melamine was also used as the N source to synthesize NG and the atomic percentage of N can reach up to 10.1 at.% [57].

Since the chemical defects in graphene play a critical role in the production of NG, some physically based methods such as plasma treatment or ion implantation are used to induce chemical defects [58]. Furthermore, by changing the plasma density or exposure time, the N content can be easily controlled (up to 8.5 at.% N) [59]. For example, Guo *et al.* used N^+-ion irradiation to introduce defects into the plane of the graphene, and then followed by annealing under NH_3 atmosphere to get NG [60]. The level of N doping can also be adjusted by changing the experimental parameters.

In liquid phase environment, the reduction of GO and N doping can be realized simultaneously under the hydrothermal reaction by using N-containing reducing agent such as hydrazine hydrate [61] or urea [62]. At a pH of 10 and temperature of 80 °C, in the presence of hydrazine and ammonia, slightly wrinkled and folded NG sheets (up to 5 at.% N) were obtained. Also, the N-enriched urea could play a key role in the formation of the NG with high N-doping level (10.13 at.%). During the hydrothermal process, NH_3 will release and react with the oxygen-containing groups on GO; meanwhile, the N atoms can dope into a graphene skeleton. Researchers can control the N-doping level through adjusting the experimental parameters, e.g., the mass ratio between GO and the reducing agent, or the reaction temperature.

3. Nitrogen-Doped Carbon Nanotubes (NCNTs) for Oxygen Reduction Reaction (ORR)

3.1. NCNTs as a Metal-Free Catalyst for ORR

The pioneering work of NCNTs as highly efficient electrocatalysts for ORR in alkaline fuel cells was reported by Gong *et al.* in 2009 [6]. A steady-state output potential of −80 mV and a current density of 4.1 mA/cm^2 at −0.22 V were observed in their study, which is superior to that of −85 mV and 1.1 mA/cm^2 at −0.20 V for a Pt/C electrode. Quantum mechanics calculations show that the carbon atoms adjacent to N dopants have very high positive charge density in order to counterbalance the strong electronic affinity of the N atom. Coupled with aligning the NCNTs, the vertically aligned (VA)-NCNTs show an excellent performance of a $4e^-$ pathway for ORR. Following this important study, plenty of research has been conducted to fabricate NCNTs [37,41,62,63] and to investigate their electrocatalytic activity from both mechanistic and experimental perspectives [23,38,64–68]. For example, based on B3LYP (a trustworthy calculation for nanomaterials) [69–71], Hu et al. [69] investigated the adsorption and activation of triplet O_2 on the surface of NCNTs with different diameters and lengths by density functional theory (DFT). The results showed that N doping sufficiently improved the adsorption ability of O_2 on CNTs [69]. Changing the diameter and length of NCNTs has a large effect on the binding energy between O_2 and NCNT and bond length of O_2, and this result further proves that NCNTs are very promising metal-free catalysts for ORR from a theoretical perspective.

From an experimental perspective, in 2009, Y. Tang et al. [72] synthesized NCNTs via the CVD method using acetonitrile or ethanol as precursors and Ar/H_2 as carrier gases. TEM images indicate that the NCNTs are composed of individual nanocups stacked together (Figure 3). Their results indicated that the stacked NCNTs exhibited similar catalytic activity with Pt/CNTs in ORR and they can also be used in the electrochemical detection of H_2O_2 and glucose. Using the CVD method, several other research groups also tried to synthesize NCNTs with different N precursors. Experiments indicate that carbon and N precursors have a significant impact on the morphology and performance of NCNTs. For instance, when ferrocene (catalyst precursor) and imidazole (C and N precursor) were used, the as-synthesized NCNTs had a high N content of 8.54 at.% and a bamboo-like structure [23]; by annealing CNTs and tripyrrolyl[1,3,5]triazine (TPT) mixture in N, the NCNTs annealing at 900 °C exhibited excellent electrochemical performance towards ORR in alkaline medium [73].

In another group, Kundu et al. fabricated NCNTs via the pyrolysis of acetonitrile with cobalt as catalyst at different temperatures in order to control the nitrogen content [63]. The results indicated that NCNTs prepared at lower temperatures had a higher amount of pyridinic groups with more exposed edge planes. Furthermore, they proved that the NCNTs with a higher amount of pyridinic groups possess better catalytic properties for ORR. Later, they synthesized NCNTs using a new approach, i.e., by treating oxidized CNTs with ammonia at 800 °C; the obtained NCNTs exhibited a favorable positive onset potential for ORR, increased reduction current, and excellent stability, demonstrating a very promising cathode catalyst for ORR in alkaline medium [41]. Almost at the same time, Chen and co-workers synthesized NCNTs using various N precursors and/or catalysts [74–77]. It was concluded from their studies that higher N content and more defects in NCNTs lead to higher ORR performance. Similar conclusions were also drawn by Geng et al. [78]. However, others have found that there is no direct correlation between total N content and the ORR performance; for example, a recent study reported, through post-treatment of few-walled carbon nanotubes (FWCNTs) with polyaniline, a much lower N content (~0.5 at.%). Interestingly, the low N-containing FWCNTs exhibited excellent electrocatalytic activity for ORR as well as higher methanol tolerance properties [79]. Therefore, the exact role of N doping in NCNTs for the ORR activity is still under debate. Until recently, Wågberg et al. [45] investigated how a thermal post-treatment on the N-doped MWCNTs can result in the transformation of pyrrolic and pyridinic N sites into quaternary N sites (N-Q$_s$), leading to the improvement of ORR performance. They reached the conclusion that the quaternary N valley sites (N-Q$_{valley}$) are the most active sites in NCNTs for ORR; hence, a $4e^-$ reduction pathway occurs generally on the N edge defects. Based on this fundamental concept, the chemical functionalization becomes an alternative and effective approach to introducing N into complex carbon nanostructures [80]. Accordingly, Tuci et al. reported a systematic study on the synthesis, characterization, and electrocatalytic property of MWCNTs functionalized with a series of well-defined pyridine groups [81]. They also discussed the role of the electronic charge density distribution at the chemically grafted N heterocycles on the ORR performance. This study therein introduced a deep level of complexity to the understanding of the ultimate role of the pyridine groups on ORR in NCNTs.

All these findings introduced above have significant impacts on catalysis and fuel cell domains. However, most of the CNTs used in these reports were synthesized by the pyrolysis of a nitrogen-containing precursor, and the residual catalyst particles of Fe or Co were removed by the electrochemical method. Though great attention has been paid to the purification process, the effects of metal contaminates in NCNTs on the ORR performance are still controversial, unless NCNTs could be obtained by a metal-free synthetic process. In this regard, by employing water-plasma etching SiO$_2$/Si wafers, Dai's group reported a simple but effective approach for the growth of densely packed N-doped single-walled CNTs [82]. Figure 4a shows the schematic illustration of the NCNT fabrication process. Typically, the water-plasma was used to etch the SiO$_2$ coating (30 nm) on the top of the SiO$_2$/Si wafer to produce uniform SiO$_2$ nanoparticles, which will act as the catalysts for NCNT growth during the CVD synthesis. As shown in Figure 4b–e, the produced metal-free NCNTs showed superb electrocatalytic activity and excellent durability toward ORR in acidic medium. For the similar purpose of excluding the possible contribution of metal impurities to ORR catalysis, Wang et al. [64] discovered that, without metal-containing catalysts, N atoms alone show strong promotion for the self-assembly of NCNTs from gaseous carbons. Based on this new discovery, pure metal-free CNTs with a high level of N doping (20 at.%) can be directly synthesized by using melamine as both the carbon and nitrogen precursor, without any post-treatment. More importantly, such intact samples can be used to investigate the intrinsic catalytic activity of NCNTs more clearly; the results indicated that NCNTs indeed performed very well. Furthermore, Li *et al.* reported that the concentration of KOH electrolyte also had a large impact on the ORR performance of the NCNTs [65]. Higher concentration of KOH electrolyte leads to more negative onset potential and lower current densities. For example, when the concentration of KOH increased from 0.1 M to 12 M, the diffusion-limiting current decreased over 100 times. This could be attributed to the very low oxygen solubility in highly concentrated KOH electrolytes. In addition, in 3 M and 6 M KOH electrolytes, NCNTs showed competitive activity with commercial Pt/C catalyst for ORR in alkaline media, and much better activity than the Ag/C catalyst [65].

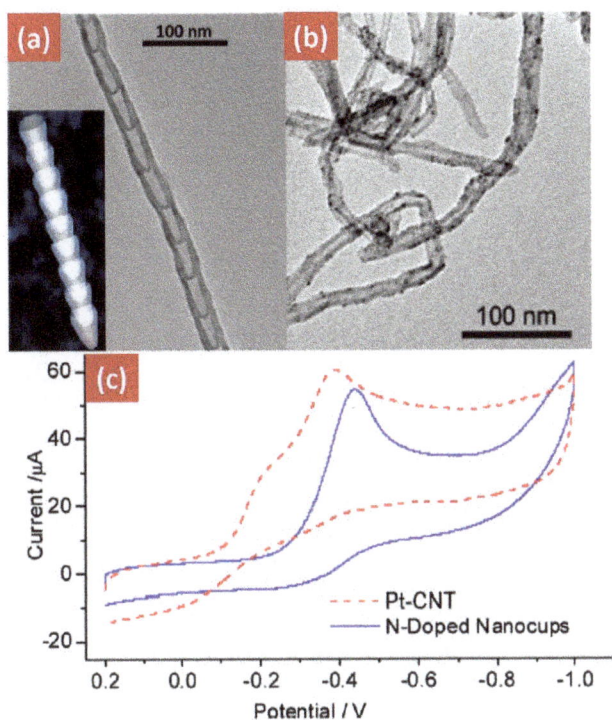

Figure 3. (a,b) TEM image of stacked NCNTs and commercial Pt-CNTs. Inset in (a) is the scheme illustration of the nanocups in stacked NCNTs. (c) CV curves of stacked NCNTs and commercial Pt-CNTs in 0.1 M KOH aqueous solution saturated with O_2. Reprinted with permission from Ref. [72] Copyright © 2009, American Chemical Society.

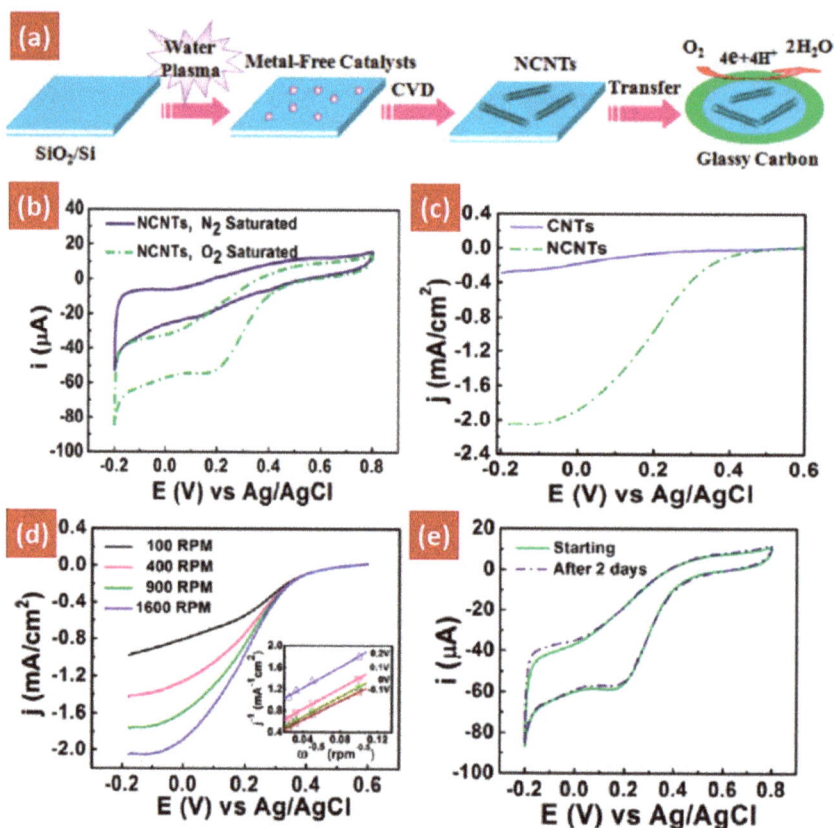

Figure 4. (a) Water-plasma-assisted CVD growth of NCNTs for the ORR; (b) CVs of the NCNTs, 50 mV/s in 0.5 M H_2SO_4 solution saturated with N_2 or O_2; (c) RDE curves of the NCNTs and CNTs in oxygen-saturated 0.5 M H_2SO_4; (d) RDE curves of the NCNT in oxygen-saturated 0.5 M H_2SO_4, inset: Koutecky-Levich plots of the NCNT derived from RDE measurements; (e) The two-day stability measurements of the NCNT by using continuous CV in oxygen-saturated 0.5 M H_2SO_4. Reprinted with permission from [82]. Copyright © 2010, American Chemical Society.

3.2. NCNTs as Catalyst Support Material for ORR

Using CNTs as catalyst supports have attracted significant interest because of their high surface area and excellent electrical conductivity. The N doping creates defects on the surface of pristine CNTs and breaks out its chemical inertness, while preserving its electrical conductivity [83]; moreover, NCNTs contain nitrogenized sites that are electrochemically active. Therefore, NCNTs were also used as excellent supports for catalyst nanoparticles. For instance, Vijayaraghavan *et al.* demonstrated that Pt nanoparticles/NCNTs exhibited enhanced catalytic activity and stability

10

along with N-dopant contents [84]. Later, Sun's group demonstrated that uniform Pt nanoparticles with smaller size and better ORR activity than pure CNTs were obtained from NCNTs [85,86] (Figure 5). The authors also demonstrated that the catalyst stability increased with the increase of N contents in NCNTs [87]. To further take the merits of both carbon and ceramic-based supports for ORR, the Sun group employed the composite nanostructures of NCNTs coated with $TiSi_2O_x$ as Pt catalyst supports, and the results indicated that this composite showed better ORR performance than Pt/NCNT catalysts, thereby illustrating its promise as a catalyst for fuel cells [88]. Chen's group concluded that the NCNTs synthesized from an N-rich precursor solution (ethylenediamine) exhibited superior catalytic activity toward ORR compared with NCNTs grown from a precursor solution with relatively low N content pyridine [89].

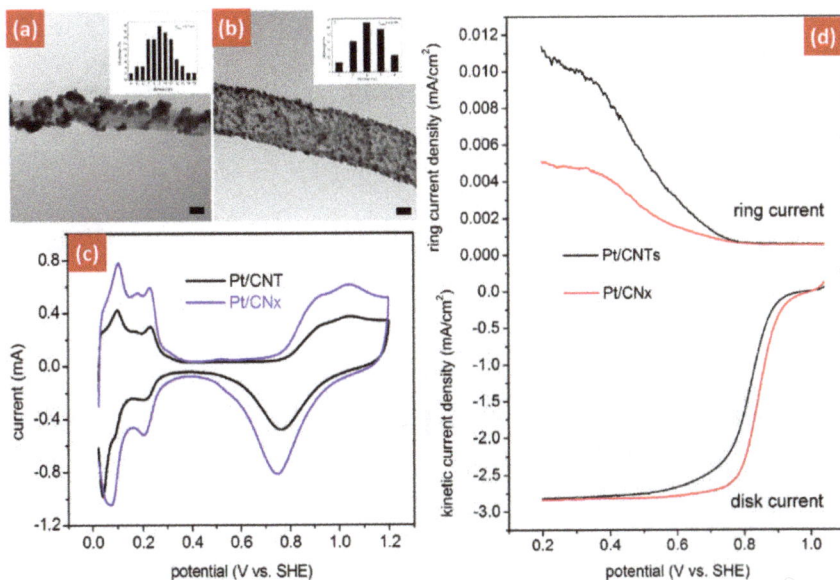

Figure 5. (**a,b**) TEM images and size distribution of Pt/CNTs (**a**) and Pt/CN$_x$ (**b**) (scale bars are 20 nm); (**c**) CVs of Pt/CNTs an Pt/CNx 0.5 M H_2SO_4 with saturated Ar at 50 mV/s; (**d**) RRDE results of Pt/CNTs and Pt/CN$_x$ in 0.5 M H_2SO_4 saturated with O_2 at 5 mV/s at the rotation speed of 1600 rpm at room temperature. Reprinted with permission from [85]. Copyright © 2011, American Chemical Society.

4. Nitrogen-Doped Graphene (NG) for ORR

As discussed above, NCNTs could act as efficient and effective metal-free catalysts for ORR. Carbon atoms adjacent to nitrogen dopants could create a net positive charge density in order to counterbalance the strong electronic affinity of

11

the N atom [6]. Hence the doping of the N atom could readily attract electrons to facilitate the ORR. Similar to NCNTs, coupled with the recent popularity of graphene, NG is also considered an appealing candidate for the applications in ORR where the NCNTs have already been exploited significantly.

4.1. NG as a Metal-Free Catalyst for ORR

Compared with NCNTs, NG has a large surface area and outstanding electrical conductivity; moreover, it also has the unique graphitic basal plane structure that could further facilitate electron transport and supply more active sites.

In 2010, Qu *et al.* first reported the application of NG as catalysts for the ORR [90]. As shown in Figure 6, a free-standing NG film of 4 cm^2 in size consisting of only a few layer sheets was obtained by the CVD method, using gas mixtures of NH_3, CH_4, H_2 and Ar on the Ni catalyst surface. The N content in the as-synthesized NG was *ca.* 4 at.%. The RRDE voltammograms measurements were conducted, in alkaline electrolyte, to investigate the catalytic properties of NG, graphene and Pt/C for ORR. From Figure 6b, it can be seen that the graphene electrode showed a 2 e$^-$ process for ORR with an onset potential of around -0.45 V. After doping with N, the NG electrode exhibited a one-step, 4 e$^-$ pathway for ORR.

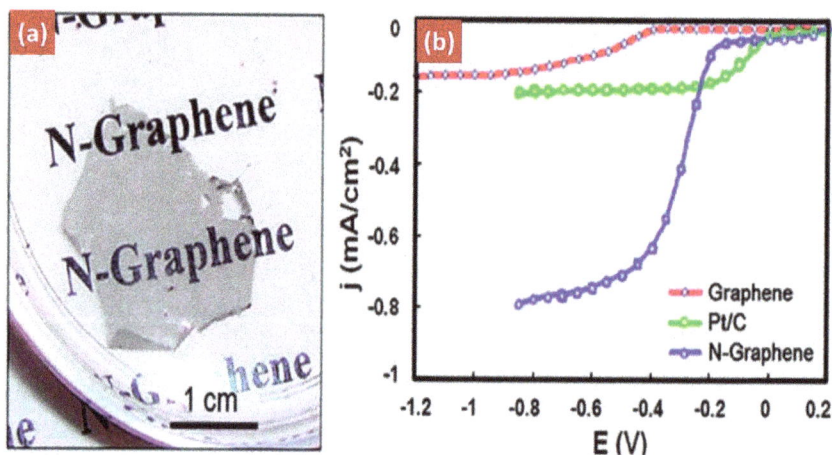

Figure 6. (a) An optical photograph of NG film floating on water; (b) LV curves in 0.1 M KOH saturated with air of different samples. Reprinted with permission from [90]. Copyright © 2010, American Chemical Society.

Calculated by the Koutecky-Levich equation, the transferred electron number per O_2 molecule of the NG was 3.6–4. It was found that the steady-state catalytic current density of the NG electrode was three times higher than the commercial Pt/C electrode. Similar to NCNTs, NG has excellent durability and good selectivity for ORR. The accelerated degradation test (ADT), which was carried out by CV in O_2-saturated electrolyte, is used to estimate the stability of the catalyst. In previous work, the graphene showed obviously more stable catalytic performance than Pt/C. Almost no significant loss in the voltammetric charge was observed after even a 100,000-cycle stability test [91]. Another advantage of NG compared to Pt for ORR is that ORR on NG is not greatly affected by methanol [59,90] and CO [90,92]. For instance, a 40% decrease was observed at the Pt/C electrode on the introduction of 2% (w/w) methanol [90], whereas the NG electrode remained unaffected under the identical condition. The high selectivity of NG toward ORR makes it very attractive for implementation in different kinds of fuel cells.

Based on these results, numerous research studies have been conducted on NG for ORR. Some of the typical works are summarized in Table 1. It is notable that the half-wave potential and onset potential for ORR are important criteria for evaluating the activity of an electrocatalyst, and the number of the electron transfer is determined from RRDE measurements to show that whether the electron transfer mechanism is a $2e^-$ dominated process or $4e^-$ dominated process.

Table 1. Summary of some typical work dedicated to NG as a metal-free catalyst for ORR.

Synthesis Method and Reactants	N-Content (at.%)	Electrocatalytic Performance	Electron Transfer Number	Ref.
Thermal treatment of glucose and urea	33	NG (25 at.%) shows competitive ORR activities with Pt/C and much better crossover resistance and excellent stability	3.2–3.7	[19]
CVD (C source, ethylene; N source, ammonia; Cu)	up to 16	Higher onset potential as compared to Pt/C	close to 2	[49]
Thermal treatment of GO using melamine	10.1	Much higher ORR activity than grapheme	3.4–3.6	[57]
N plasma treatment on graphene	8.5	Higher ORR activity than graphene, and higher durability and selectivity than Pt/C	-	[59]
CVD (C source, methane; N source, ammonia, Cu)	4	Higher activity, better stability and tolerance to crossover than Pt	3.6–4	[90]
Detonation technique with cyanuric chloride and trinitrophenol	12.5	Comparable to that of Pt, more stable and less expensive	3.69	[91]
A resin-based methodology with N-containing resin and metal ions	1.8	The onset potential on the NG electrode is close to that of Pt/C. The current is almost the same for both the Pt/C and NG	2.1–3.9	[92]
Hydrothermal reaction of GO with urea	6.05–7.65	The performance of these NG materials towards ORR is still not as good as that of Pt/C in terms of the half-wave potential and current density	~3	[93]
Covalent functionalize GO using organic molecules and thermal treatment	0.72–4.3	The NG nanosheet exhibited a good electrocatalytic activity through an efficient one-step, 4e$^-$ pathway	3.63	[94]
CVD of N-containing aromatic precursor molecules	2.0–2.7	The N dopants in the graphene reduce the ORR overpotential, thereby enhancing the catalytic activity	3.5–4.0	[95]
GO treatment by ammonia hydroxide, heating under ammonia gas, and reaction with melamine	6.0–6.8	Pyridinic N plays a vital role in ORR	3.2–3.7	[96]
Annealing of GO with ammonia and N-containing polymers	2.91–7.56	The higher limiting current density compared to Pt	2.85–3.65	[97]
Thermal reaction between GO and NH$_3$	2.4–4.6	The onset potential is close to that of Pt/C	~3.8	[98]
Hydrothermal reaction with GO and melamine	26.08	It shows lower ORR activity than Pt/C 40 wt.%	3.2–4.0	[99]
Hydrothermal process using urea and holey GO	8.6	Superb ORR with 4e$^-$ pathway and excellent durability	3.85	[100]
Thermally annealing GO with melamine	8.05	The nG-900 exhibits lower activity and onset potential than Pt/C, albeit higher than graphene; excellent stability	3.3–3.7	[101]
Pyrolyzing GO with urea	7.86	The NG showed a much-higher activity than glassy carbon (GC) and graphene	3.6–4.0	[102]
Redox GO with pyrrole then thermal treatment	6	Shows comparable onset potentials with 40 wt.% Pt/C	3.3	[103]
GO and dicyandiamide under hydrothermal condition	7.78	The onset potentials at rGO-N was lower than that at Pt/C	2.6	[104]
Pyrolysis of graphene oxide and polyaniline	2.4	High activity toward ORR with a superior long-term stability and tolerance to methanol crossover	3.8–3.9	[105]

Table 1. *Cont.*

Synthesis Method and Reactants	N-Content (at.%)	Electrocatalytic Performance	Electron Transfer Number	Ref.
Thermally annealing GO 5-aminotetrazole monohydrate	10.6	Higher current density than Pt/C. Lower onset potential of ORR than that of the commercial Pt/C	3.7	[106]
Pyrolysis of sugar in the presence of urea	3.02–11.2	The NG1000 has comparable ORR half-wave potential to 20 wt.% Pt/C	3.2–3.8	[107]
Hydrothermal reaction of GO with urea	5.8–6.2	NG has higher ORR activity than grapheme, but is not yet comparable to the Pt	3.0–4.0	[108]
Pyrolysis of GO and polydopamine	2.78–3.79	Much more enhanced ORR activities with positive onset potential and larger current density than graphene	3.89	[109]
Pyrolyzing GO with Melamine, urea and dicyandiamide	5	Compared to Pt/C, the half-wave potential of ORR on this NG catalyst was close, wheras the n values are slightly lower	3.5–4	[110]
PANI acting as a N source were deposited on the surface of GNRs via a layer-by-layer approach	4.1–8.3	Very good electrocatalytic activity and stability	3.91	[111]
NG is synthesized by pyrolyzing ion exchange with resin and glycine	0.98–1.65	Doping N in graphene is good to improve the activity for ORR, but still lower than Pt/C catalyst	-	[112]
Microwave heating of graphene under NH_3 flow	4.05–5.47	The doping of graphite N enhanced the activity of the catalysts in the ORR in alkaline solution	3.03–3.3	[113]
Facile hydrothermal method	2.8	Competitive with the commercial Pt/C catalysts in alkaline medium	3.66–3.92	[114]
Gas-phase oxidation strategy using a nitric acid vapor	0.52	The onset potential is (0.755 V *vs.* RHE), comparable to the value of chemically synthesized NG, and the current densities are higher than those demonstrated for NG.	3.2–3.9	[115]
CVD growth of graphene and post-doping with a solid N precursor of graphitic C_3N_4	6.5	Excellent activity, high stability, and very good crossover resistance for ORR in alkaline medium.	3.96-4.05	[116]
A hard templating approach	5.07	Outstanding ORR performance in both acidic and alkaline solutions.	3.9	[117]

In spite of extensive studies, the explanations on the exact catalytic mechanisms of NG (e.g., wherein the N configuration (pyridinic N or graphitic N) is more important for the ORR activity) or even the active sites are still controversial [94,118]. In Sun *et al.*'s research [55], they found that NG containing 0.3892% quaternary N (the highest N content in three samples) showed the best ORR activity and the relationship between ORR activity and graphitic N contents matched very well. It revealed that graphitic type N plays the vital role for ORR activity. Luo *et al.* [49] synthesized the graphene layers doped with nearly 100% pyridinic N through the pyrolysis of methane (CH_4) and NH_3 on Cu substrate, and the as-synthesized pyridinic N-doped graphene mainly exhibited a $2e^-$ transfer process for ORR, indicating that pyridinic N may not, as previously expected, effectively promote the $4e^-$ ORR performance of carbon materials.

On the contrary, in the work of Sheng [57], the NG mainly containing pyridine-like N atoms was obtained by the heat-treatment of GO in the presence of melamine. Since the electrocatalytic activity of the NGs toward ORR is independent of N-doping level, it may indicate that the pyridine-like N in NGs determines its ORR activity. Pyridinic N, which has a lone electron pair in the plane of the carbon matrix, could donate the electron to the π-bond, attract electrons, and therefore be catalytically active. Some results were shown in many previous works [94–96].

In the research of Ruoff's group [97], NG with different N-doping formats was prepared by annealing GO together with different N-containing precursors, such as ammonia and N-containing polymers. It was prone to generate graphitic N and pyridinic N when annealing GO with ammonia, while it tended to form pyridinic and pyrrolic N species when annealing GO with polyaniline or polypyrrole. They found that the total atomic content of N rarely affects the ORR activity under alkaline conditions. Actually, the graphitic N-dominated catalysts exhibit higher catalytic activity and larger limiting current density than that of pyrrolic or pyridinic N-dominated catalysts. However, the pyridinic N could enhance the ORR onset potential and gradually convert the $2e^-$ dominated pass-way to the $4e^-$ dominated process. Also, some researchers [119–122] used the periodic DFT to simulate the ORR at the edge of NG. For example, by taking into account the experimental conditions, *i.e.*, the surface coverage, the water effect, the bias effect and pH, Yu *et al.* [119] presented a systematic theoretical study on the full reaction path of ORR on NG. They concluded that the rate-determining step is the O(ads) removal from the NG surface. From another perspective, by calculating energy variations during each reaction step using DFT, Zhang and Xia [120] demonstrated that the electrocatalytic activity of NG is related to the atomic charge density distribution and electron spin density The reasons for why NG has catalytic capability (while pristine graphene does not) have also been discussed. From Kim *et al.*'s results, [121] doping of N in graphene could promote the oxygen adsorption, the first electron transfer, and the selectivity toward the $4e^-$ reduction pathway. More specifically, they suggested that the outermost graphitic N sites are the main active sites. Meanwhile, they also proposed that the graphitic N site which involves a ring-opening of the cyclic C-N bond at the edge of graphene could result in the pyridinic N, thus, the inter-converts conversion mechanism between pyridinic and graphitic types during the catalytic cycle may reconcile the experimental controversy about what types of N are the ORR active sites for N-doped carbon materials [121].

Besides the doped N species, the morphology of NG also plays a significant role for the ORR properties. During the doping process of graphene, the stacking of graphene sheets is inclined to increase the diffusion resistance of reactants/electrolytes, reduce the specific area, and the exposed active sites. It is thus worth controlling the structure of NG to get more ORR activity. In this regard, there is a great deal of work on the production of N-doped holey graphene [99,100]. For instance, a 3D porous nanostructure which has N-doped holes on individual graphene sheets was synthesized through a hydrothermal process using urea and holey GO by Yu *et al.* [100]. Benefiting from the 3D porous nanostructure, abundant exposed sites, and high-level N doping, the as-prepared material exhibited excellent ORR performance, such as the high limiting current, strong resistance to the methanol crossover, which are competitive with the commercial 20 wt.% Pt/C catalysts.

4.2. NG as Support Material for ORR

The incorporation of N atoms within graphene sheets could contribute more functional groups, higher electron-mobility, and more active sites for catalytic reactions. Also, it is beneficial for facilitating the distribution and uniformity of metal nanoparticles. Moreover, when NG acts as the support, it could enhance the catalytic properties due to the interaction between graphene and metal nanoparticles. Consequently, NG materials have been regarded as one very promising metal catalyst support [123–126].

Typically, NG is proposed to be able to stabilize the noble metal nanoparticles, and improve the durability of the catalysts. Moreover, nitrogen doping could introduce active sites for catalytic reactions and also act as anchoring sites for metal nanoparticle deposition. Yang *et al.* fabricated a composite of Pt-Au alloy nanoparticles on NG sheets by a wet-chemistry method [127]. As shown in Figure 7, the NG was synthesized by thermal treatment of GO powder and melamine. Then the solutions of H_2PtCl_6, $HAuCl_4$, NG in DMF and water underwent the microwave irradiation. The as-prepared Pt_3Au-NPs were found to be well dispersed on the NG sheets (Figure 7b) and the HRTEM image in Figure 7c revealed the lattice fringes of the NPs have an interplanar spacing of 0.232 nm. The fast Fourier transforms (FFTs) shown in Figure 7d indicated the single crystallite nature of the Pt_3Au/NG on (111) plane. Figure 7e,f showed that the corresponding potential for Pt_3Au/NG was much lower than the other two samples at a given oxidation current density. Improved electrocatalytict activity was observed due to the small size, uniform dispersion and a high electrochemical active surface area of the nanocomposites. Recently, more studies on NG- or N-rGO-supported Pt electrocatalysts have also been reported; all these results demonstrate the significant function of N doping in producing highly efficient ORR electrocatalysts [128–130].

Additionally, it was predicted that non-precious-metal-NG hybrid materials would also lead to enhanced catalytic properties. For instance, Chen *et al.* reported a strategy to synthesize ZnSe/NG nanocomposites (NG-ZnSe) [131]. As shown in Figure 8, [ZnSe](DETA)$_{0.5}$ nanobelts were gradually put into the GO solution, and then the sediments were processed by hydrothermal treatment. As shown in Figure 8b, ZnSe nanorods, which were composed of ZnSe nanoparticles, were grown on a graphene surface. It can be seen from Figure 8c that the NG-ZnSe electrode exhibited higher positive onset potential and larger current for ORR. The improved performance can be attributed to the synergetic effects between NG and alloy nanostructures. There are also a number of similar reports using non-precious metal to produce metal/NG composites, showing potential applications [132–136].

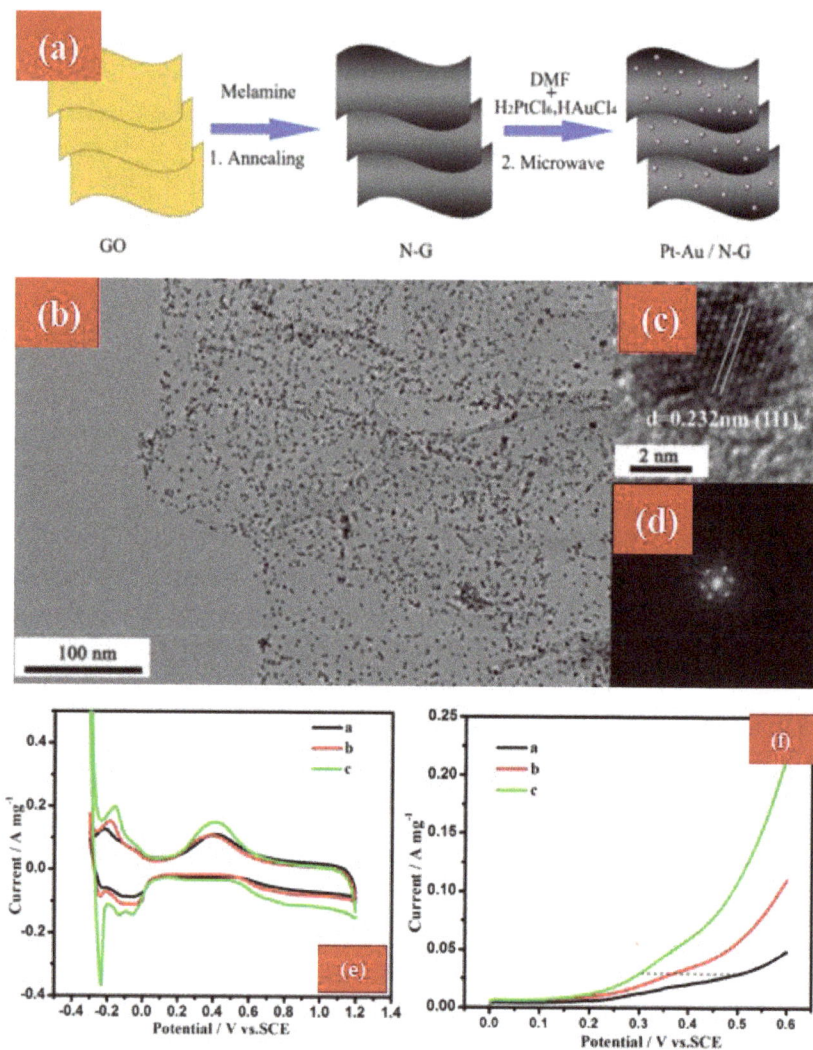

Figure 7. (a) Fabrication of the Pt-Au alloy NPs on the NG sheets; (b) TEM of Pt₃Au/N-G; (c) HRTEM and (d) FFTs of a single Pt₃Au NP on NG; (e) CVs and (f) LSV of Pt/C (a, black), Pt₃Au/G(b, red) and Pt₃Au/N-G catalysts (c, green). Reprinted with permission from [127]. Copyright © 2012, Royal Society of Chemistry.

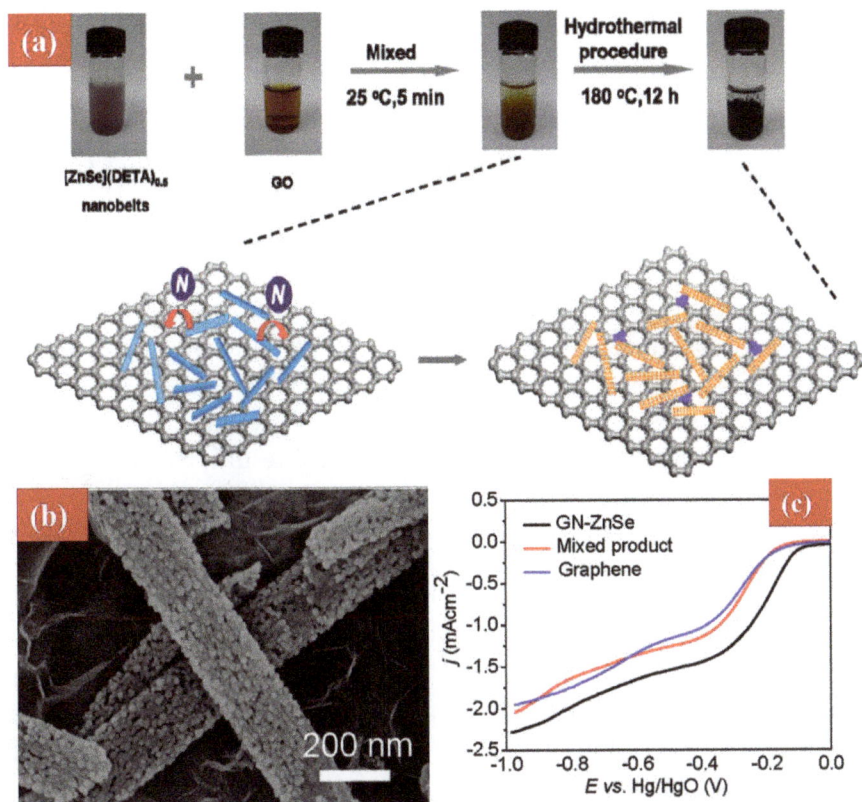

Figure 8. (a) Schematic preparation of NG-ZnSe nanocomposites (blue rods-[ZnSe](DETA)$_{0.5}$ nanobelts; orange rods-ZnSe nanorods; purple balls-N; gray balls-C); (b) SEM photograph of ZnSe/NG; (c) LV curves in 1.0 M KOH solution with saturated O$_2$ of different electrodes. Reprinted with permission from Ref. [131]. Copyright © 2012, American Chemical Society. Note: in the original paper, the authors refer to "nitrogen-doped graphene" as "GN"; here in this review, for consistency, we named it "NG."

5. The Composites of NCNTs and NG for ORR

As a two-dimensional layer structure of sp^2-hybridized carbon, graphene has strong direction-dependent transport properties and is easily agglomerated and restacked to graphite; therefore, when used as a catalyst, it may result in declined activity. A combination of CNT and graphene may be an effective way to solve this problem [137,138]. Dai's group has demonstrated that CNT-graphene complexes can exhibit excellent activity and stability towards ORR in both acidic and basic electrolytes [139]. Furthermore, based on the STEM-HAADF and EELS mapping results, they speculated that the impurities of nitrogen and iron might be the reason for the excellent ORR properties. While, as illustrated in the previous sections, NCNTs and NG have shown excellent electrocatalytic performance for the ORR compared with pure CNTs or graphene. Therefore, , there have recently been efforts to hybridize these two carbon structures (NCNTs and NG) to obtain a synergy effect to further improve their catalytic performance [138,140]. For example, Ma *et al.* fabricated the 3D NCNTs/graphene composite through the pyrolysis of pyridine over the Ni catalyst supported on graphene sheet [140]. The N content in the NCNTs/NG composite was about 6.6 at.%, compared with the undoped CNTs/G; the doped sample showed higher catalytic activity and selectivity for ORR in the alkaline electrolyte. Another example of highly active N-doped G/CNT composite electrocatalyst for ORR is demonstrated by Ratso and coworkers [141]. N-doped few-layer G/CNT composite was fabricated by the pyrolysis of GO/MWCNT with urea and dicyandiamide. Based on the XPS and RDE results, they concluded that the enhanced electrocatalytic activity is due to a higher content of pyridinic N in the samples, and the higher limiting currents of oxygen reduction can be ascribed by the quaternary N. These results are attractive for alkaline fuel cells. However, these methods require high temperature pyrolysis, during which the morphological defects and structural degradation are probably shown up in the final products [17]. In this regard, Chen *et al.* synthesized NG-NCNT nanocomposite through a hydrothermal process at a much lower temperature (*i.e.*, 180 °C) (Figure 9a) [44]. The diameters of the nanotubes are in the range of 9–15 nm, and the atomic percentages of N content are 3.2 at.% and 1.3 at.% for graphene and CNTs, respectively, which confirm the existence of the N element in both graphene and CNTs. This NG-NCNT displayed a $4e^-$ pathway for ORR with more positive onset potential, large peak current, and good durability (Figure 9b–g). Very recently, however, a hybrid of NCNT and graphene prepared by plasma-enhanced CVD showed inferior ORR activity, [142] which is contradictory to the above-mentioned results. The reason for this discrepancy is still not clear, thus extensive and careful research in this area is still needed.

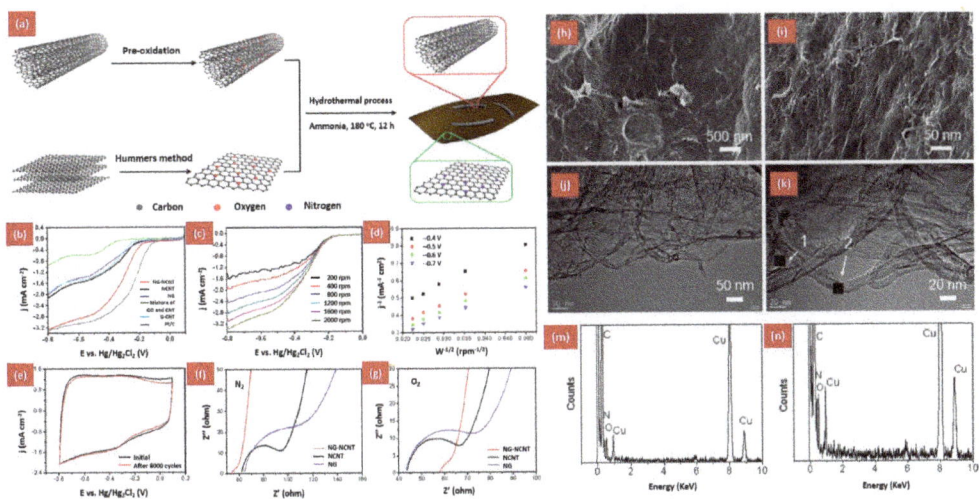

Figure 9. (a) Schematic preparation of the NG-NCNT nanocomposites; (b) LV curves in 0.1 M KOH solution with the rotation speed of 1600 rpm and sweep rate of 20 mV·s^{-1} in oxygen of different samples; (c) LV curves of NG-NCNT with different rotation speeds (sweep rate 20 mV·s^{-1}); (d) K-L plots (i^{-1} *vs.* $\omega^{-1/2}$) at different potentials (*vs.* Hg/Hg$_2$Cl$_2$); (e) CVs of GN-CNT after 8000 cycles with the sweep rate of 150 mV·s^{-1}; (f–g) Impedance data of different samples in 0.1 M KOH solution with saturated N$_2$ and O$_2$, respectively; (h–k) SEM and STEM images of the typical NG-NCNT nanocomposite; (m,n) Elemental analysis image of the NG and NG-NCNT (the area marked with 1 and 2 in Figure (k) respectively. Reprinted with permission from [44]. Copyright © 2013 WILEY-VCH Verlag GmbH & Co. KGaA, Weinheim.

6. Conclusion and Perspectives

ORR plays an essential role in energy-related areas, such as metal-air batteries and fuel cells, and traditionally, the Pt-based catalysts are regarded as the best choice for 4e$^-$ ORR. Due to the prohibitive price and scarcity of Pt, the development of high performance and inexpensive metal-free and non-noble metal catalysts, to replace Pt, are highly desired, and it plays an important role in promoting the large-scale practical applications of these energy devices. Due to their outstanding properties, such as ultrahigh charge carrier mobility, gigantic thermal conductivity, extremely large surface area, exceptional mechanical strength and flexibility, CNTs and graphene have been extensively explored for ORR. The pristine CNTs and graphene mainly exhibit 2e$^-$ pathway for ORR, while N doping has been proved to be a promising way to tailor their properties to promote 4e$^-$ ORR which is much more meaningful for energy applications. For N doping in CNTs or graphene, there are mainly two strategies: the first method is the *in situ* doping where nitrogen can be doped into CNTs or graphene nanosheets during the growing process with the addition of proper carbon and nitrogen sources. The second one is the post-treatment process; in this method, CNTs or GO were firstly synthesized, then annealed at high temperatures together with the nitrogen-containing precursors. Despite much progress, it is still not easy to precisely control the N-doping sites and concentration. All of these characteristics affect the ORR properties of NCNTs and NG in the catalytic applications. Therefore, the development of new and more controllable doping methods is still highly desired. Through N doping, various properties, including the surface energy, work function, carrier concentration, and surface polarization, of CNTs and graphene could be tuned, so that NCNTs and NG have become the most promising metal-free catalysts toward $4e^-$ ORR. In general, three common bonding configurations, including graphitic, pyridinic, and pyrrolic N, are normally achieved when doping nitrogen into CNTs and graphene. Different doping strategies would significantly affect the N-doping levels and N types in NCNTs and NG. For example, the *in situ* doping normally generates pyridinic- and/or pyrrolic-N species, while the post-treatment doping is prone to form graphitic-N in carbon frameworks.

In the applications for ORR, from both theoretical and experimental perspectives, researchers have demonstrated that NCNTs and NG show remarkable electrocatalytic performance. In a theoretical context, through DFT simulations, it was shown that in NCNTs and NG, the carbon atoms with higher spin density usually possess more active sites. Through investigating the reaction mechanisms, it was proved that the removal of $O_{(ads)}$ on the surface of nitrogen-doped carbon determines the reaction rate. In the experimental part, the developments of both NCNTs and NG as metal-free ORR catalysts and as the metal catalyst support for ORR are summarized in detail in this review. All the N-doped carbon materials (NCNTs, NG) exhibit higher catalytic performance compared to their pristine counterparts (CNTs, graphene), indicating a great beneficial effect of N doping on the ORR performance. Moreover, the progresses on NCNTs- and NG-based composites for ORR have also been discussed in this review, demonstrating that it is also a very promising research direction for next-generation non-noble metal or metal-free ORR catalysts. Although much progress has been achieved in the area of NCNTs and NG for ORR catalysts, challenges still exist: (i) New and greener methods are required for the large-scale production of NCNTs and NG; (ii) The control of N doping at specific positions in CNTs and graphene is still lacking; (iii) A careful controlling of nitrogen sites, types and concentration is still highly desired; (iv) The deep understanding of oxygen adsorption and reduction on these NCNTs- and NG-based catalysts is still lacking, and therefore, systematic theoretical simulations are also needed, which may boost the developments of N-doping carbon materials for ORR in the future.

Acknowledgments: We thank the support from Natural Science Foundations of China, the Natural Science Foundation of Shanxi Province, Fonds de Recherche du Québec-Natureet

Technologies (FRQNT), the Natural Sciences and Engineering Research Council of Canada (NSERC), Institut National de la Recherche Scientifique (INRS), and Centre Québécois sur les MateriauxFonctionnels (CQMF) and China Scholarship Council (CSC).

Author Contributions: Qiliang Wei was the leading author from the initial draft writing to the finalization of the manuscript. Qiliang Wei and Xin Tong wrote the first draft of the manuscript. All authors contributed as a team to the manuscript plan, revisions, the literature reading, and the proof reading.

Conflicts of Interest: The authors declare no conflict of interest.

References

1. Dai, L.; Xue, Y.; Qu, L.; Choi, H.-J.; Baek, J.-B. Metal-free catalysts for oxygen reduction reaction. *Chem. Rev.* **2015**, *115*, 4823–4892.

2. Cheng, F.; Chen, J. Metal-air batteries: From oxygen reduction electrochemistry to cathode catalysts. *Chem. Soc. Rev.* **2012**, *41*, 2172–2192.

3. Wang, D.-W.; Su, D. Heterogeneous nanocarbon materials for oxygen reduction reaction. *Energy Environ. Sci.* **2014**, *7*, 576–591.

4. Feng, L.; Yan, Y.; Chen, Y.; Wang, L. Nitrogen-doped carbon nanotubes as efficient and durable metal-free cathodic catalysts for oxygen reduction in microbial fuel cells. *Energy Environ. Sci.* **2011**, *4*, 1892–1899.

5. Suntivich, J.; Gasteiger, H.A.; Yabuuchi, N.; Nakanishi, H.; Goodenough, J.B.; Shao-Horn, Y. Design principles for oxygen-reduction activity on perovskite oxide catalysts for fuel cells and metal–air batteries. *Nat. Chem.* **2011**, *3*, 546–550.

6. Gong, K.; Du, F.; Xia, Z.; Durstock, M.; Dai, L. Nitrogen-doped carbon nanotube arrays with high electrocatalytic activity for oxygen reduction. *Science* **2009**, *323*, 760–764.

7. Wei, W.; Liang, H.; Parvez, K.; Zhuang, X.; Feng, X.; Müllen, K. Nitrogen-doped carbon nanosheets with size-defined mesopores as highly efficient metal-free catalyst for the oxygen reduction reaction. *Angew. Chem.* **2014**, *126*, 1596–1600.

8. Sun, D.M.; Liu, C.; Ren, W.C.; Cheng, H.M. A review of carbon nanotube-and graphene-based flexible thin-film transistors. *Small* **2013**, *9*, 1188–1205.

9. Yu, D.; Nagelli, E.; Du, F.; Dai, L. Metal-free carbon nanomaterials become more active than metal catalysts and last longer. *J. Phys. Chem. Lett.* **2010**, *1*, 2165–2173.

10. Chun, K.-Y.; Lee, H.S.; Lee, C.J. Nitrogen doping effects on the structure behavior and the field emission performance of double-walled carbon nanotubes. *Carbon* **2009**, *47*, 169–177.

11. Bostwick, A.; Speck, F.; Seyller, T.; Horn, K.; Polini, M.; Asgari, R.; MacDonald, A.H.; Rotenberg, E. Observation of plasmarons in quasi-freestanding doped graphene. *Science* **2010**, *328*, 999–1002.

12. Hwang, J.O.; Park, J.S.; Choi, D.S.; Kim, J.Y.; Lee, S.H.; Lee, K.E.; Kim, Y.-H.; Song, M.H.; Yoo, S.; Kim, S.O. Work function-tunable, N-doped reduced graphene transparent electrodes for high-performance polymer light-emitting diodes. *ACS Nano* **2011**, *6*, 159–167.

13. Czerw, R.; Terrones, M.; Charlier, J.-C.; Blase, X.; Foley, B.; Kamalakaran, R.; Grobert, N.; Terrones, H.; Tekleab, D.; Ajayan, P. Identification of electron donor states in N-doped carbon nanotubes. *Nano Lett.* **2001**, *1*, 457–460.

14. Lee, W.J.; Maiti, U.N.; Lee, J.M.; Lim, J.; Han, T.H.; Kim, S.O. Nitrogen-doped carbon nanotubes and graphene composite structures for energy and catalytic applications. *Chem. Commun.* **2014**, *50*, 6818–6830.

15. Liu, H.; Liu, Y.; Zhu, D. Chemical doping of graphene. *J. Mater. Chem.* **2011**, *21*, 3335–3345.

16. Yang, Z.; Nie, H.; Chen, X.A.; Chen, X.; Huang, S. Recent progress in doped carbon nanomaterials as effective cathode catalysts for fuel cell oxygen reduction reaction. *J. Power Sources* **2013**, *236*, 238–249.

17. Zheng, Y.; Jiao, Y.; Jaroniec, M.; Jin, Y.; Qiao, S.Z. Nanostructured metal-free electrochemical catalysts for highly efficient oxygen reduction. *Small* **2012**, *8*, 3550–3566.

18. Vazquez-Arenas, J.; Higgins, D.; Chen, Z.; Fowler, M.; Chen, Z. Mechanistic analysis of highly active nitrogen-doped carbon nanotubes for the oxygen reduction reaction. *J. Power Sources* **2012**, *205*, 215–221.

19. Zhang, Y.W.; Ge, J.; Wang, L.; Wang, D.H.; Ding, F.; Tao, X.M.; Chen, W. Manageable N-doped graphene for high performance oxygen reduction reaction. *Sci. Rep.* **2013**, *3*.

20. Wei, D.; Liu, Y.; Wang, Y.; Zhang, H.; Huang, L.; Yu, G. Synthesis of N-doped graphene by chemical vapor deposition and its electrical properties. *Nano Lett.* **2009**, *9*, 1752–1758.

21. Jung, S.H.; Kim, M.R.; Jeong, S.H.; Kim, S.U.; Lee, O.J.; Lee, K.H.; Suh, J.H.; Park, C.K. High-yield synthesis of multi-walled carbon nanotubes by arc discharge in liquid nitrogen. *Appl. Phys. A* **2003**, *76*, 285–286.

22. Sun, L.; Wang, C.; Zhou, Y.; Zhang, X.; Cai, B.; Qiu, J. Flowing nitrogen assisted-arc discharge synthesis of nitrogen-doped single-walled carbon nanohorns. *Appl. Surf. Sci.* **2013**, *277*, 88–93.

23. Mo, Z.; Liao, S.; Zheng, Y.; Fu, Z. Preparation of nitrogen-doped carbon nanotube arrays and their catalysis towards cathodic oxygen reduction in acidic and alkaline media. *Carbon* **2012**, *50*, 2620–2627.

24. Sharifi, T.; Nitze, F.; Barzegar, H.R.; Tai, C.-W.; Mazurkiewicz, M.; Malolepszy, A.; Stobinski, L.; Wågberg, T. Nitrogen doped multi walled carbon nanotubes produced by CVD-correlating xps and raman spectroscopy for the study of nitrogen inclusion. *Carbon* **2012**, *50*, 3535–3541.

25. Guo, Q.; Zhao, D.; Liu, S.; Chen, S.; Hanif, M.; Hou, H. Free-standing nitrogen-doped carbon nanotubes at electrospun carbon nanofibers composite as an efficient electrocatalyst for oxygen reduction. *Electrochim. Acta* **2014**, *138*, 318–324.

26. Tao, X.Y.; Zhang, X.B.; Sun, F.Y.; Cheng, J.P.; Liu, F.; Luo, Z.Q. Large-scale CVD synthesis of nitrogen-doped multi-walled carbon nanotubes with controllable nitrogen content on a $Co_xMg_{1-x}MoO_4$ catalyst. *Diamond Relat. Mater.* **2007**, *16*, 425–430.

27. She, X.; Yang, D.; Jing, D.; Yuan, F.; Yang, W.; Guo, L.; Che, Y. Nitrogen-doped one-dimensional (1D) macroporous carbonaceous nanotube arrays and their application in electrocatalytic oxygen reduction reactions. *Nanoscale* **2014**, *6*, 11057–11061.

28. Chen, L.; Xia, K.; Huang, L.; Li, L.; Pei, L.; Fei, S. Facile synthesis and hydrogen storage application of nitrogen-doped carbon nanotubes with bamboo-like structure. *Int. J. Hydrogen Energy* **2013**, *38*, 3297–3303.

29. Shi, W.; Venkatachalam, K.; Gavalas, V.; Qian, D.; Andrews, R.; Bachas, L.G.; Chopra, N. The role of plasma treatment on electrochemical capacitance of undoped and nitrogen doped carbon nanotubes. *Nanomater. Energy* **2013**, *2*, 71–81.

30. Du, Z.; Wang, S.; Kong, C.; Deng, Q.; Wang, G.; Liang, C.; Tang, H. Microwave plasma synthesized nitrogen-doped carbon nanotubes for oxygen reduction. *J. Solid State Electrochem.* **2015**, *19*, 1541–1549.

31. Magrez, A.; Seo, J.W.; Smajda, R.; Mionić, M.; Forró, L. Catalytic CVD synthesis of carbon nanotubes: Towards high yield and low temperature growth. *Materials* **2010**, *3*, 4871–4891.

32. Donato, M.G.; Galvagno, S.; Lanza, M.; Messina, G.; Milone, C.; Piperopoulos, E.; Pistone, A.; Santangelo, S. Influence of carbon source and Fe-catalyst support on the growth of multi-walled carbon nanotubes. *J. Nanosci. Nanotechnol.* **2009**, *9*, 3815–3823.

33. Li, J.; Papadopoulos, C.; Xu, J.M.; Moskovits, M. Highly-ordered carbon nanotube arrays for electronics applications. *Appl. Phys. Lett.* **1999**, *75*, 367–369.

34. Wang, Y.; Cui, X.; Li, Y.; Chen, L.; Chen, H.; Zhang, L.; Shi, J. A co-pyrolysis route to synthesize nitrogen doped multiwall carbon nanotubes for oxygen reduction reaction. *Carbon* **2014**, *68*, 232–239.

35. Ayala, P.; Grüneis, A.; Gemming, T.; Grimm, D.; Kramberger, C.; Rümmeli, M.H.; Freire, F.L.; Kuzmany, H.; Pfeiffer, R.; Barreiro, A. Tailoring N-doped single and double wall carbon nanotubes from a nondiluted carbon/nitrogen feedstock. *J. Phys. Chem. C* **2007**, *111*, 2879–2884.

36. Tang, C.; Golberg, D.; Bando, Y.; Xu, F.; Liu, B. Synthesis and field emission of carbon nanotubular fibers doped with high nitrogen content. *Chem. Commun.* **2003**, 3050–3051.

37. Rao, C.V.; Cabrera, C.R.; Ishikawa, Y. In search of the active site in nitrogen-doped carbon nanotube electrodes for the oxygen reduction reaction. *J. Phys. Chem. Lett.* **2010**, *1*, 2622–2627.

38. Rao, C.V.; Ishikawa, Y. Activity, selectivity, and anion-exchange membrane fuel cell performance of virtually metal-free nitrogen-doped carbon nanotube electrodes for oxygen reduction reaction. *J. Phys. Chem. C* **2012**, *116*, 4340–4346.

39. Guo, Q.; Xie, Y.; Wang, X.; Zhang, S.; Hou, T.; Lv, S. Synthesis of carbon nitride nanotubes with the C_3N_4 stoichiometry via a benzene-thermal process at low temperatures. *Chem. Commun.* **2004**, 26–27.

40. Cao, C.; Huang, F.; Cao, C.; Li, J.; Zhu, H. Synthesis of carbon nitride nanotubes via a catalytic-assembly solvothermal route. *Chem. Mater.* **2004**, *16*, 5213–5215.

41. Nagaiah, T.C.; Kundu, S.; Bron, M.; Muhler, M.; Schuhmann, W. Nitrogen-doped carbon nanotubes as a cathode catalyst for the oxygen reduction reaction in alkaline medium. *Electrochem. Commun.* **2010**, *12*, 338–341.

42. Chan, L.H.; Hong, K.H.; Xiao, D.Q.; Lin, T.C.; Lai, S.H.; Hsieh, W.J.; Shih, H.C. Resolution of the binding configuration in nitrogen-doped carbon nanotubes. *Phys. Rev. B* **2004**, *70*, 125408.

43. Vikkisk, M.; Kruusenberg, I.; Ratso, S.; Joost, U.; Shulga, E.; Kink, I.; Rauwel, P.; Tammeveski, K. Enhanced electrocatalytic activity of nitrogen-doped multi-walled carbon nanotubes towards the oxygen reduction reaction in alkaline media. *RSC Adv.* **2015**, *5*, 59495–59505.

44. Chen, P.; Xiao, T.Y.; Qian, Y.H.; Li, S.S.; Yu, S.H. A nitrogen-doped graphene/carbon nanotube nanocomposite with synergistically enhanced electrochemical activity. *Adv. Mater.* **2013**, *25*, 3192–3196.

45. Sharifi, T.; Hu, G.; Jia, X.; Wågberg, T. Formation of active sites for oxygen reduction reactions by transformation of nitrogen functionalities in nitrogen-doped carbon nanotubes. *ACS Nano* **2012**, *6*, 8904–8912.

46. Wood, K.N.; O'Hayre, R.; Pylypenko, S. Recent progress on nitrogen/carbon structures designed for use in energy and sustainability applications. *Energy Environ. Sci.* **2014**, *7*, 1212–1249.

47. Reddy, A.L.M.; Srivastava, A.; Gowda, S.R.; Gullapalli, H.; Dubey, M.; Ajayan, P.M. Synthesis of nitrogen-doped graphene films for lithium battery application. *ACS Nano* **2010**, *4*, 6337–6342.

48. Jin, Z.; Yao, J.; Kittrell, C.; Tour, J.M. Large-scale growth and characterizations of nitrogen-doped monolayer graphene sheets. *ACS Nano* **2011**, *5*, 4112–4117.

49. Luo, Z.; Lim, S.; Tian, Z.; Shang, J.; Lai, L.; MacDonald, B.; Fu, C.; Shen, Z.; Yu, T.; Lin, J. Pyridinic N doped graphene: Synthesis, electronic structure, and electrocatalytic property. *J. Mater. Chem.* **2011**, *21*, 8038–8044.

50. Deng, D.; Pan, X.; Yu, L.; Cui, Y.; Jiang, Y.; Qi, J.; Li, W.-X.; Fu, Q.; Ma, X.; Xue, Q.; *et al.* Toward N-doped graphene via solvothermal synthesis. *Chem. Mater.* **2011**, *23*, 1188–1193.

51. Ghosh, A.; Late, D.J.; Panchakarla, L.S.; Govindaraj, A.; Rao, C.N.R. NO_2 and humidity sensing characteristics of few-layer graphenes. *J. Exp. Nanosci.* **2009**, *4*, 313–322.

52. Panchakarla, L.S.; Subrahmanyam, K.S.; Saha, S.K.; Govindaraj, A.; Krishnamurthy, H.R.; Waghmare, U.V.; Rao, C.N.R. Synthesis, structure, and properties of boron-and nitrogen-doped graphene. *Adv. Mater.* **2009**, *21*, 4726–4730.

53. Subrahmanyam, K.S.; Panchakarla, L.S.; Govindaraj, A.; Rao, C.N.R. Simple method of preparing graphene flakes by an arc-discharge method. *J. Phys. Chem. C* **2009**, *113*, 4257–4259.

54. Geng, D.; Chen, Y.; Chen, Y.; Li, Y.; Li, R.; Sun, X.; Ye, S.; Knights, S. High oxygen-reduction activity and durability of nitrogen-doped graphene. *Energy Environ. Sci.* **2011**, *4*, 760–764.

55. Wang, X.; Li, X.; Zhang, L.; Yoon, Y.; Weber, P.K.; Wang, H.; Guo, J.; Dai, H. N-doping of graphene through electrothermal reactions with ammonia. *Science* **2009**, *324*, 768–771.

56. Li, X.; Wang, H.; Robinson, J.T.; Sanchez, H.; Diankov, G.; Dai, H. Simultaneous nitrogen doping and reduction of graphene oxide. *J. Am. Chem. Soc.* **2009**, *131*, 15939–15944.

57. Sheng, Z.-H.; Shao, L.; Chen, J.-J.; Bao, W.-J.; Wang, F.-B.; Xia, X.-H. Catalyst-free synthesis of nitrogen-doped graphene via thermal annealing graphite oxide with melamine and its excellent electrocatalysis. *ACS Nano* **2011**, *5*, 4350–4358.

58. Wang, H.; Maiyalagan, T.; Wang, X. Review on recent progress in nitrogen-doped graphene: Synthesis, characterization, and its potential applications. *ACS Catal.* **2012**, *2*, 781–794.

59. Shao, Y.; Zhang, S.; Engelhard, M.H.; Li, G.; Shao, G.; Wang, Y.; Liu, J.; Aksay, I.A.; Lin, Y. Nitrogen-doped graphene and its electrochemical applications. *J. Mater. Chem.* **2010**, *20*, 7491–7496.

60. Guo, B.; Liu, Q.; Chen, E.; Zhu, H.; Fang, L.; Gong, J.R. Controllable N-doping of graphene. *Nano Lett.* **2010**, *10*, 4975–4980.

61. Long, D.; Li, W.; Ling, L.; Miyawaki, J.; Mochida, I.; Yoon, S.-H. Preparation of nitrogen-doped graphene sheets by a combined chemical and hydrothermal reduction of graphene oxide. *Langmuir* **2010**, *26*, 16096–16102.

62. Sun, L.; Wang, L.; Tian, C.; Tan, T.; Xie, Y.; Shi, K.; Li, M.; Fu, H. Nitrogen-doped graphene with high nitrogen level via a one-step hydrothermal reaction of graphene oxide with urea for superior capacitive energy storage. *RSC Adv.* **2012**, *2*, 4498–4506.

63. Kundu, S.; Nagaiah, T.C.; Xia, W.; Wang, Y.; Dommele, S.V.; Bitter, J.H.; Santa, M.; Grundmeier, G.; Bron, M.; Schuhmann, W. Electrocatalytic activity and stability of nitrogen-containing carbon nanotubes in the oxygen reduction reaction. *J. Phys. Chem. C* **2009**, *113*, 14302–14310.

64. Wang, Z.; Jia, R.; Zheng, J.; Zhao, J.; Li, L.; Song, J.; Zhu, Z. Nitrogen-promoted self-assembly of N-doped carbon nanotubes and their intrinsic catalysis for oxygen reduction in fuel cells. *ACS Nano* **2011**, *5*, 1677–1684.

65. Li, H.; Liu, H.; Jong, Z.; Qu, W.; Geng, D.; Sun, X.; Wang, H. Nitrogen-doped carbon nanotubes with high activity for oxygen reduction in alkaline media. *Int. J. Hydrogen Energy* **2011**, *36*, 2258–2265.

66. Qiu, Y.; Yin, J.; Hou, H.; Yu, J.; Zuo, X. Preparation of nitrogen-doped carbon submicrotubes by coaxial electrospinning and their electrocatalytic activity for oxygen reduction reaction in acid media. *Electrochim. Acta* **2013**, *96*, 225–229.

67. Wiggins-Camacho, J.D.; Stevenson, K.J. Mechanistic discussion of the oxygen reduction reaction at nitrogen-doped carbon nanotubes. *J. Phys. Chem. C* **2011**, *115*, 20002–20010.

68. Okamoto, Y. First-principles molecular dynamics simulation of O_2 reduction on nitrogen-doped carbon. *Appl. Surf. Sci.* **2009**, *256*, 335–341.

69. Hu, X.; Wu, Y.; Li, H.; Zhang, Z. Adsorption and activation of O_2 on nitrogen-doped carbon nanotubes. *J. Phys. Chem. C* **2010**, *114*, 9603–9607.

70. Barone, V.; Peralta, J.E.; Wert, M.; Heyd, J.; Scuseria, G.E. Density functional theory study of optical transitions in semiconducting single-walled carbon nanotubes. *Nano Lett.* **2005**, *5*, 1621–1624.

71. Nikawa, H.; Yamada, T.; Cao, B.; Mizorogi, N.; Slanina, Z.; Tsuchiya, T.; Akasaka, T.; Yoza, K.; Nagase, S. Missing metallofullerene with C80 cage. *J. Am. Chem. Soc.* **2009**, *131*, 10950–10954.

72. Tang, Y.; Allen, B.L.; Kauffman, D.R.; Star, A. Electrocatalytic activity of nitrogen-doped carbon nanotube cups. *J. Am. Chem. Soc.* **2009**, *131*, 13200–13201.

73. Yang, M.; Yang, D.; Chen, H.; Gao, Y.; Li, H. Nitrogen-doped carbon nanotubes as catalysts for the oxygen reduction reaction in alkaline medium. *J. Power Sources* **2015**, *279*, 28–35.

74. Chen, Z.; Higgins, D.; Tao, H.; Hsu, R.S.; Chen, Z. Highly active nitrogen-doped carbon nanotubes for oxygen reduction reaction in fuel cell applications. *J. Phys. Chem. C* **2009**, *113*, 21008–21013.

75. Chen, Z.; Higgins, D.; Chen, Z. Nitrogen doped carbon nanotubes and their impact on the oxygen reduction reaction in fuel cells. *Carbon* **2010**, *48*, 3057–3065.

76. Higgins, D.; Chen, Z.; Chen, Z. Nitrogen doped carbon nanotubes synthesized from aliphatic diamines for oxygen reduction reaction. *Electrochim. Acta* **2011**, *56*, 1570–1575.

77. Chen, Z.; Higgins, D.; Chen, Z. Electrocatalytic activity of nitrogen doped carbon nanotubes with different morphologies for oxygen reduction reaction. *Electrochim. Acta* **2010**, *55*, 4799–4804.

78. Geng, D.; Liu, H.; Chen, Y.; Li, R.; Sun, X.; Ye, S.; Knights, S. Non-noble metal oxygen reduction electrocatalysts based on carbon nanotubes with controlled nitrogen contents. *J. Power Sources* **2011**, *196*, 1795–1801.

79. Borghei, M.; Kanninen, P.; Lundahl, M.; Susi, T.; Sainio, J.; Anoshkin, I.; Nasibulin, A.; Kallio, T.; Tammeveski, K.; Kauppinen, E. High oxygen reduction activity of few-walled carbon nanotubes with low nitrogen content. *Appl. Catal. B* **2014**, *158*, 233–241.

80. Tuci, G.; Zafferoni, C.; D'Ambrosio, P.; Caporali, S.; Ceppatelli, M.; Rossin, A.; Tsoufis, T.; Innocenti, M.; Giambastiani, G. Tailoring carbon nanotube N-dopants while designing metal-free electrocatalysts for the oxygen reduction reaction in alkaline medium. *ACS Catal.* **2013**, *3*, 2108–2111.

81. Tuci, G.; Zafferoni, C.; Rossin, A.; Milella, A.; Luconi, L.; Innocenti, M.; Truong Phuoc, L.; Duong-Viet, C.; Pham-Huu, C.; Giambastiani, G. Chemically functionalized carbon nanotubes with pyridine groups as easily tunable N-decorated nanomaterials for the oxygen reduction reaction in alkaline medium. *Chem. Mater.* **2014**, *26*, 3460–3470.

82. Yu, D.; Zhang, Q.; Dai, L. Highly efficient metal-free growth of nitrogen-doped single-walled carbon nanotubes on plasma-etched substrates for oxygen reduction. *J. Am. Chem. Soc.* **2010**, *132*, 15127–15129.

83. Sun, S.; Zhang, G.; Zhong, Y.; Liu, H.; Li, R.; Zhou, X.; Sun, X. Ultrathin single crystal Pt nanowires grown on N-doped carbon nanotubes. *Chem. Commun.* **2009**, 7048–7050.

84. Vijayaraghavan, G.; Stevenson, K.J. Synergistic assembly of dendrimer-templated platinum catalysts on nitrogen-doped carbon nanotube electrodes for oxygen reduction. *Langmuir* **2007**, *23*, 5279–5282.

85. Chen, Y.; Wang, J.; Liu, H.; Banis, M.N.; Li, R.; Sun, X.; Sham, T.-K.; Ye, S.; Knights, S. Nitrogen doping effects on carbon nanotubes and the origin of the enhanced electrocatalytic activity of supported Pt for proton-exchange membrane fuel cells. *J. Phys. Chem. C* **2011**, *115*, 3769–3776.

86. Saha, M.S.; Li, R.; Sun, X.; Ye, S. 3-d composite electrodes for high performance pem fuel cells composed of Pt supported on nitrogen-doped carbon nanotubes grown on carbon paper. *Electrochem. Commun.* **2009**, *11*, 438–441.

87. Chen, Y.; Wang, J.; Liu, H.; Li, R.; Sun, X.; Ye, S.; Knights, S. Enhanced stability of Pt electrocatalysts by nitrogen doping in CNTs for PEM fuel cells. *Electrochem. Commun.* **2009**, *11*, 2071–2076.

88. Banis, M.N.; Sun, S.; Meng, X.; Zhang, Y.; Wang, Z.; Li, R.; Cai, M.; Sham, T.-K.; Sun, X. TiSi$_2$O$_x$ coated N-doped carbon nanotubes as Pt catalyst support for the oxygen reduction reaction in PEMFCs. *J. Phys. Chem. C* **2013**, *117*, 15457–15467.

89. Higgins, D.C.; Meza, D.; Chen, Z. Nitrogen-doped carbon nanotubes as platinum catalyst supports for oxygen reduction reaction in proton exchange membrane fuel cells. *J. Phys. Chem. C* **2010**, *114*, 21982–21988.

90. Qu, L.; Liu, Y.; Baek, J.-B.; Dai, L. Nitrogen-doped graphene as efficient metal-free electrocatalyst for oxygen reduction in fuel cells. *ACS Nano* **2010**, *4*, 1321–1326.

91. Feng, L.Y.; Chen, Y.G.; Chen, L. Easy-to-operate and low-temperature synthesis of gram-scale nitrogen-doped graphene and its application as cathode catalyst in microbial fuel cells. *ACS Nano* **2011**, *5*, 9611–9618.

92. He, C.Y.; Li, Z.S.; Cai, M.L.; Cai, M.; Wang, J.Q.; Tian, Z.Q.; Zhang, X.; Shen, P.K. A strategy for mass production of self-assembled nitrogen-doped graphene as catalytic materials. *J. Mater. Chem. A* **2013**, *1*, 1401–1406.

93. Wu, J.J.; Zhang, D.; Wang, Y.; Hou, B.R. Electrocatalytic activity of nitrogen-doped graphene synthesized via a one-pot hydrothermal process towards oxygen reduction reaction. *J. Power Sources* **2013**, *227*, 185–190.

94. Park, M.; Lee, T.; Kim, B.-S. Covalent functionalization based heteroatom doped graphene nanosheet as a metal-free electrocatalyst for oxygen reduction reaction. *Nanoscale* **2013**, *5*, 12255–12260.

95. Yasuda, S.; Yu, L.; Kim, J.; Murakoshi, K. Selective nitrogen doping in graphene for oxygen reduction reactions. *Chem. Commun.* **2013**, *49*, 9627–9629.

96. Xing, T.; Zheng, Y.; Li, L.H.; Cowie, B.C.C.; Gunzelmann, D.; Qiao, S.Z.; Huang, S.M.; Chen, Y. Observation of active sites for oxygen reduction reaction on nitrogen-doped multilayer graphene. *ACS Nano* **2014**, *8*, 6856–6862.

97. Lai, L.; Potts, J.R.; Zhan, D.; Wang, L.; Poh, C.K.; Tang, C.; Gong, H.; Shen, Z.; Lin, J.; Ruoff, R.S. Exploration of the active center structure of nitrogen-doped graphene-based catalysts for oxygen reduction reaction. *Energy Environ. Sci.* **2012**, *5*, 7936–7942.

98. Yang, S.; Zhi, L.; Tang, K.; Feng, X.; Maier, J.; Muellen, K. Efficient synthesis of heteroatom (N or S)-doped graphene based on ultrathin graphene oxide-porous silica sheets for oxygen reduction reactions. *Adv. Funct. Mater.* **2012**, *22*, 3634–3640.

99. Jiang, Z.J.; Jiang, Z.Q.; Chen, W.H. The role of holes in improving the performance of nitrogen-doped holey graphene as an active electrode material for supercapacitor and oxygen reduction reaction. *J. Power Sources* **2014**, *251*, 55–65.

100. Yu, D.S.; Wei, L.; Jiang, W.C.; Wang, H.; Sun, B.; Zhang, Q.; Goh, K.L.; Si, R.M.; Chen, Y. Nitrogen doped holey graphene as an efficient metal-free multifunctional electrochemical catalyst for hydrazine oxidation and oxygen reduction. *Nanoscale* **2013**, *5*, 3457–3464.

101. Lin, Z.Y.; Song, M.K.; Ding, Y.; Liu, Y.; Liu, M.L.; Wong, C.P. Facile preparation of nitrogen-doped graphene as a metal-free catalyst for oxygen reduction reaction. *Phys. Chem. Chem. Phys.* **2012**, *14*, 3381–3387.

102. Lin, Z.Y.; Waller, G.; Liu, Y.; Liu, M.L.; Wong, C.P. Facile synthesis of nitrogen-doped graphene via pyrolysis of graphene oxide and urea, and its electrocatalytic activity toward the oxygen-reduction reaction. *Adv. Energy Mater.* **2012**, *2*, 884–888.

103. Unni, S.M.; Devulapally, S.; Karjule, N.; Kurungot, S. Graphene enriched with pyrrolic coordination of the doped nitrogen as an efficient metal-free electrocatalyst for oxygen reduction. *J. Mater. Chem.* **2012**, *22*, 23506–23513.

104. Zhang, Y.J.; Fugane, K.; Mori, T.; Niu, L.; Ye, J.H. Wet chemical synthesis of nitrogen-doped graphene towards oxygen reduction electrocatalysts without high-temperateure pyrolysis. *J. Mater. Chem.* **2012**, *22*, 6575–6580.

105. Lin, Z.Y.; Waller, G.H.; Liu, Y.; Liu, M.L.; Wong, C.P. Simple preparation of nanoporous few-layer nitrogen-doped graphene for use as an efficient electrocatalyst for oxygen reduction and oxygen evolution reactions. *Carbon* **2013**, *53*, 130–136.

106. Lu, Z.J.; Bao, S.J.; Gou, Y.T.; Cai, C.J.; Ji, C.C.; Xu, M.W.; Song, J.; Wang, R.Y. Nitrogen-doped reduced-graphene oxide as an efficient metal-free electrocatalyst for oxygen reduction in fuel cells. *RSC Adv.* **2013**, *3*, 3990–3995.

107. Pan, F.P.; Jin, J.T.; Fu, X.G.; Liu, Q.; Zhang, J.Y. Advanced oxygen reduction electrocatalyst based on nitrogen-doped graphene derived from edible sugar and urea. *ACS Appl. Mater. Interfaces* **2013**, *5*, 11108–11114.

108. Zheng, B.; Wang, J.; Wang, F.B.; Xia, X.H. Synthesis of nitrogen doped graphene with high electrocatalytic activity toward oxygen reduction reaction. *Electrochem. Commun.* **2013**, *28*, 24–26.

109. Cong, H.P.; Wang, P.; Gong, M.; Yu, S.H. Facile synthesis of mesoporous nitrogen-doped graphene: An efficient methanol-tolerant cathodic catalyst for oxygen reduction reaction. *Nano Energy* **2014**, *3*, 55–63.

110. Vikkisk, M.; Kruusenberg, I.; Joost, U.; Shulga, E.; Kink, I.; Tammeveski, K. Electrocatalytic oxygen reduction on nitrogen-doped graphene in alkaline media. *Appl. Catal. B* **2014**, *147*, 369–376.

111. Liu, M.K.; Song, Y.F.; He, S.X.; Tjiu, W.W.; Pan, J.S.; Xia, Y.Y.; Liu, T.X. Nitrogen-doped graphene nanoribbons as efficient metal-free electrocatalysts for oxygen reduction. *ACS Appl. Mater. Interfaces* **2014**, *6*, 4214–4222.

112. Ouyang, W.; Zeng, D.; Yu, X.; Xie, F.; Zhang, W.; Chen, J.; Yan, J.; Xie, F.; Wang, L.; Meng, H.; *et al.* Exploring the active sites of nitrogen-doped graphene as catalysts for the oxygen reduction reaction. *Int. J. Hydrogen Energy* **2014**, *39*, 15996–16005.

113. Wang, Z.; Li, B.; Xin, Y.; Liu, J.; Yao, Y.; Zou, Z. Rapid synthesis of nitrogen-doped graphene by microwave heating for oxygen reduction reactions in alkaline electrolyte. *Chin. J. Catal.* **2014**, *35*, 509–513.

114. Chen, L.; Du, R.; Zhu, J.; Mao, Y.; Xue, C.; Zhang, N.; Hou, Y.; Zhang, J.; Yi, T. Three-dimensional nitrogen-doped graphene nanoribbons aerogel as a highly efficient catalyst for the oxygen reduction reaction. *Small* **2015**, *11*, 1423–1429.

115. Sun, Z.; Masa, J.; Weide, P.; Fairclough, S.M.; Robertson, A.W.; Ebbinghaus, P.; Warner, J.H.; Tsang, S.C.E.; Muhler, M.; Schuhmann, W. High-quality functionalized few-layer graphene: Facile fabrication and doping with nitrogen as a metal-free catalyst for the oxygen reduction reaction. *J. Mater. Chem. A* **2015**, *3*, 15444–15450.

116. Wu, J.; Ma, L.; Yadav, R.M.; Yang, Y.; Zhang, X.; Vajtai, R.; Lou, J.; Ajayan, P.M. Nitrogen-doped graphene with pyridinic dominance as a highly active and stable electrocatalyst for oxygen reduction. *ACS Appl. Mater. Interfaces* **2015**, *7*, 14763–14769.

117. Zhou, X.; Bai, Z.; Wu, M.; Qiao, J.; Chen, Z. 3-dimensional porous N-doped graphene foam as a non-precious catalyst for the oxygen reduction reaction. *J. Mater. Chem. A* **2015**, *3*, 3343–3350.

118. Kong, X.-K.; Chen, C.-L.; Chen, Q.-W. Doped graphene for metal-free catalysis. *Chem. Soc. Rev.* **2014**, *43*, 2841–2857.

119. Yu, L.; Pan, X.; Cao, X.; Hu, P.; Bao, X. Oxygen reduction reaction mechanism on nitrogen-doped graphene: A density functional theory study. *J. Catal.* **2011**, *282*, 183–190.

120. Zhang, L.; Xia, Z. Mechanisms of oxygen reduction reaction on nitrogen-doped graphene for fuel cells. *J. Phys. Chem. C* **2011**, *115*, 11170–11176.

121. Kim, H.; Lee, K.; Woo, S.I.; Jung, Y. On the mechanism of enhanced oxygen reduction reaction in nitrogen-doped graphene nanoribbons. *Phys. Chem. Chem. Phys.* **2011**, *13*, 17505–17510.

122. Jiao, Y.; Zheng, Y.; Jaroniec, M.; Qiao, S.Z. Origin of the electrocatalytic oxygen reduction activity of graphene-based catalysts: A roadmap to achieve the best performance. *J. Am. Chem. Soc.* **2014**, *136*, 4394–4403.

123. Sun, M.; Liu, H.; Liu, Y.; Qu, J.; Li, J. Graphene-based transition metal oxide nanocomposites for the oxygen reduction reaction. *Nanoscale* **2015**, *7*, 1250–1269.

124. Wang, X.; Sun, G.; Routh, P.; Kim, D.-H.; Huang, W.; Chen, P. Heteroatom-doped graphene materials: Syntheses, properties and applications. *Chem. Soc. Rev.* **2014**, *43*, 7067–7098.

125. Zhou, X.; Qiao, J.; Yang, L.; Zhang, J. A review of graphene-based nanostructural materials for both catalyst supports and metal-free catalysts in pem fuel cell oxygen reduction reactions. *Adv. Energy Mater.* **2014**, *4*.

126. Li, Q.; Pan, H.; Higgins, D.; Cao, R.; Zhang, G.; Lv, H.; Wu, K.; Cho, J.; Wu, G. Metal-organic framework-derived bamboo-like nitrogen-doped graphene tubes as an active matrix for hybrid oxygen-reduction electrocatalysts. *Small* **2015**, *11*, 1443–1452.

127. Yang, G.H.; Li, Y.J.; Rana, R.K.; Zhu, J.J. Pt-Au/nitrogen-doped graphene nanocomposites for enhanced electrochemical activities. *J. Mater. Chem. A* **2013**, *1*, 1754–1762.

128. Imran Jafri, R.; Rajalakshmi, N.; Ramaprabhu, S. Nitrogen doped graphene nanoplatelets as catalyst support for oxygen reduction reaction in proton exchange membrane fuel cell. *J. Mater. Chem.* **2010**, *20*, 7114–7117.

129. Vinayan, B.P.; Nagar, R.; Rajalakshmi, N.; Ramaprabhu, S. Novel platinum–cobalt alloy nanoparticles dispersed on nitrogen-doped graphene as a cathode electrocatalyst for PEMFC applications. *Adv. Funct. Mater.* **2012**, *22*, 3519–3526.

130. Xiong, B.; Zhou, Y.; Zhao, Y.; Wang, J.; Chen, X.; O'Hayre, R.; Shao, Z. The use of nitrogen-doped graphene supporting Pt nanoparticles as a catalyst for methanol electrocatalytic oxidation. *Carbon* **2013**, *52*, 181–192.

131. Chen, P.; Xiao, T.-Y.; Li, H.-H.; Yang, J.-J.; Wang, Z.; Yao, H.-B.; Yu, S.-H. Nitrogen-doped graphene/ZnSe nanocomposites: Hydrothermal synthesis and their enhanced electrochemical and photocatalytic activities. *ACS Nano* **2012**, *6*, 712–719.

132. Parvez, K.; Yang, S.B.; Hernandez, Y.; Winter, A.; Turchanin, A.; Feng, X.L.; Mullen, K. Nitrogen-doped graphene and its iron-based composite as efficient electrocatalysts for oxygen reduction reaction. *ACS Nano* **2012**, *6*, 9541–9550.

133. Bai, J.C.; Zhu, Q.Q.; Lv, Z.X.; Dong, H.Z.; Yu, J.H.; Dong, L.F. Nitrogen-doped graphene as catalysts and catalyst supports for oxygen reduction in both acidic and alkaline solutions. *Int. J. Hydrogen Energy* **2013**, *38*, 1413–1418.

134. Huang, T.; Mao, S.; Pu, H.; Wen, Z.; Huang, X.; Ci, S.; Chen, J. Nitrogen-doped graphene-vanadium carbide hybrids as a high-performance oxygen reduction reaction electrocatalyst support in alkaline media. *J. Mater. Chem. A* **2013**, *1*, 13404–13410.

135. Park, H.W.; Lee, D.U.; Nazar, L.F.; Chen, Z.W. Oxygen reduction reaction using MnO₂ nanotubes/nitrogen-doped exfoliated graphene hybrid catalyst for Li-O₂ battery applications. *J. Electrochem. Soc.* **2013**, *160*, A344–A350.

136. Xiao, J.; Bian, X.; Liao, L.; Zhang, S.; Ji, C.; Liu, B. Nitrogen-doped mesoporous graphene as a synergistic electrocatalyst matrix for high-performance oxygen reduction reaction. *ACS Appl. Mater. Interfaces* **2014**, *6*, 17654–17660.

137. Brownson, D.A.C.; Munro, L.J.; Kampouris, D.K.; Banks, C.E. Electrochemistry of graphene: Not such a beneficial electrode material? *RSC Adv.* **2011**, *1*, 978–988.

138. Choi, C.H.; Chung, M.W.; Kwon, H.C.; Chung, J.H.; Woo, S.I. Nitrogen-doped graphene/carbon nanotube self-assembly for efficient oxygen reduction reaction in acid media. *Appl. Catal. B* **2014**, *144*, 760–766.

139. Li, Y.; Zhou, W.; Wang, H.; Xie, L.; Liang, Y.; Wei, F.; Idrobo, J.-C.; Pennycook, S.J.; Dai, H. An oxygen reduction electrocatalyst based on carbon nanotube-graphene complexes. *Nat. Nanotechnol.* **2012**, *7*, 394–400.

140. Ma, Y.; Sun, L.; Huang, W.; Zhang, L.; Zhao, J.; Fan, Q.; Huang, W. Three-dimensional nitrogen-doped carbon nanotubes/graphene structure used as a metal-free electrocatalyst for the oxygen reduction reaction. *J. Phys. Chem. C* **2011**, *115*, 24592–24597.

141. Ratso, S.; Kruusenberg, I.; Vikkisk, M.; Joost, U.; Shulga, E.; Kink, I.; Kallio, T.; Tammeveski, K. Highly active nitrogen-doped few-layer graphene/carbon nanotube composite electrocatalyst for oxygen reduction reaction in alkaline media. *Carbon* **2014**, *73*, 361–370.

142. Liu, J.Y.; Wang, Z.; Chen, J.Y.; Wang, X. Nitrogen-doped carbon nanotubes and graphene nanohybrid for oxygen reduction reaction in acidic, alkaline and neutral solutions. *J. Nano Res.* **2015**, *30*, 50–58.

Recent Progress on Fe/N/C Electrocatalysts for the Oxygen Reduction Reaction in Fuel Cells

Jing Liu, Erling Li, Mingbo Ruan, Ping Song and Weilin Xu

Abstract: In order to reduce the overall system cost, the development of inexpensive, high-performance and durable oxygen reduction reaction (ORR)N, Fe-codoped carbon-based (Fe/N/C) electrocatalysts to replace currently used Pt-based catalysts has become one of the major topics in research on fuel cells. This review paper lays the emphasis on introducing the progress made over the recent five years with a detailed discussion of recent work in the area of Fe/N/C electrocatalysts for ORR and the possible Fe-based active sites. Fe-based materials prepared by simple pyrolysis of transition metal salt, carbon support, and nitrogen-rich small molecule or polymeric compound are mainly reviewed due to their low cost, high performance, long stability and because they are the most promising for replacing currently used Pt-based catalysts in the progress of fuel cell commercialization. Additionally, Fe-base catalysts with small amount of Fe or new structure of Fe/Fe_3C encased in carbon layers are presented to analyze the effect of loading and existence form of Fe on the ORR catalytic activity in Fe-base catalyst. The proposed catalytically Fe-centered active sites and reaction mechanisms from various authors are also discussed in detail, which may be useful for the rational design of high-performance, inexpensive, and practical Fe-base ORR catalysts in future development of fuel cells.

Reprinted from *Catalysts*. Cite as: Liu, J.; Li, E.; Ruan, M.; Song, P.; Xu, W. Recent Progress on Fe/N/C Electrocatalysts for the Oxygen Reduction Reaction in Fuel Cells. *Catalysts* **2015**, *5*, 1167–1192.

1. Introduction

To meet the increased demand for energy in the world, one of the biggest challenges is the development of technologies that provide inexpensive, readily available, and sustainable energy. Fuel cells are among the most promising candidates for reliable and efficient conversion of alcohols into electric power in automotive and portable electronic applications on a large scale [1,2]. However, the scarcity, high cost, and poor long-term stability of Pt-Based ORR catalysts, the most widely used catalysts for the oxygen reduction reaction (ORR) in fuel cells, are main obstacles for large-scale commercialization of fuel cell technology [3,4]. Since Jasinski reported cobalt phthalocyanine as the ORR electrocatalyst in alkaline electrolytes in 1964 [5], a new era of carbon-supported non-precious metal (Co,

Fe, *etc.*) and metal-free catalyst to replace the expensive Pt-based electrode in fuel cells started [6–12]. Among non-precious metal catalysts, N, Fe-codoped carbon-based (Fe/N/C) electrocatalysts (Fe-based catalysts) are the most promising candidates because some of them exhibit high ORR activity in both acidic and alkaline medium [13–15]. Fe-based catalysts can be obtained through high-temperature pyrolysis of either iron N_4 chelate complexes [16–21], or simple precursors of iron salts, nitrogen-containing components (aromatic [22–24] and aliphatic ligands [25–29] or other nitrogen-rich small molecules [30–36]) on carbon supports. Thus far, the state-of-the art Fe-Based catalysts exhibit much higher ORR activity and durability than those of Pt-Based catalysts in alkaline electrolytes [15,36–40] and comparable ORR activity in acidic media [7,34,41–43].

Along with the achievement of the excellent ORR activity of diverse Fe-based catalysts, the ORR mechanisms on Fe-based catalysts were also widely studied by many groups due to its importance in research and development of high-performance Fe-based ORR catalysts [21,41,44–48]. However, due to different preparation protocols used for Fe-based catalysts, there is still an ongoing debate about the active sites of these materials [45,48–50]. Therefore, there is still a long way to go in order to reach the practical usage and understanding of Fe-Based catalysts in fuel cells applications. This review addresses the current development of Fe-based ORR catalysts with a variety of different structure and properties, along with the proposed catalytically active sites and reaction mechanisms from various authors. By examining the most recent progress and research trends in both theoretical and experimental studies of Fe-based catalysts, this review provides a systematic and comprehensive discussion of the factors influencing catalyst performance as well as the future improvement strategies.

2. Fe-Based Catalysts

Iron, an element of the transition metal group, entered into the world of ORR catalysts in company with nitrogen in 1964 [51]. After that, Fe-based catalysts have gained increasing attention due to their promising catalytic activity for ORR, along with the utilization of abundant, low-cost precursor materials [14]. Research in Fe-based catalysts covers the non-pyrolyzed Fe-based macrocycle compounds [52–55] and pyrolyzed Fe-based macrocycle compounds [18,19,56] or some proper Fe- and N-containing precursor materials [22,57,58]. The former are important in this field of scientific research for fundamental understanding due to their preserved well-defined structure during synthesis procedures, and the latter shows a higher ORR catalytic activity because of the introduction of high temperature heat treatment procedures (~400 to 1000 °C) to the catalyst synthesis process [13]. The structures of active sites on these Fe-based catalysts have been proposed by different groups including the structure of in-plane coordination of an iron atom and four pyridinic or pyrrolic type of nitrogen atoms embedded in a graphene-type matrix (Fe–N_4/C [16,17,56,59,60] or Fe–N_{2+2}/C [61]), the structure of coordination of an iron atom iron and two pyridinic type of nitrogen atoms embedded in a graphene-type matrix (Fe–N_2/C) [60] and N-doped carbon-based structure (N–C) [62,63]. The factors of influence on ORR catalytic activity and stability of Fe-based catalysts have also been studied such as ring substituent group of non-pyrolyzed Fe-based macrocycle compounds [64,65], heat-treatment conditions [64,66,67], Fe content [68] and carbon support properties including surface nitrogen content and microporosity [31,57,69,70]. In order to produce highly active and stable Fe-based catalysts, ample approaches have been used with significant emphasis on introducing the exact effect of synthesis conditions and the nature of the catalytically active sites. Progress in this field of recent research will be divided into three sections and discussed: (1) preparation of Fe-based materials toward ORR; (2) research on structure of Fe-centered ORR active sites and ORR mechanism; and (3) stability of Fe-based ORR catalysts.

2.1. Preparation of Fe-Based Materials toward ORR

In 2011, Chen *et al.* [13] reviewed Fe-based catalysts in detail, so we will lay an emphasis on introducing the development of Fe-based catalysts over the most recent five years. Interestingly, it is worth pointing out that the best performing Fe-based ORR catalyst mentioned by Chen *et al.* [13] was synthesized by Dodelet *et al.* [71], which had a volumetric current density of 99 A cm^{-3} at an *iR*-corrected voltage of 0.8 V, approaching the DOE 2010 target of 130 A cm^{-3}. In fact, soon after that, a more exciting result was reported in August 2011 in a *Nature Communication* by the same group [72]. By using a metal-organic framework consisting of zeolitic Zn (II) imidazolate as the host for Fe and N precursors (iron (II) acetate and 1, 10-phenanthroline (Phen)), they prepared a Fe/Phen/ZIF-8 catalyst with a volumetric activity of 230 A cm^{-3} at 0.8 V (*iR*-free) (Figure 1), a higher catalytic activity compared with that (99 A cm^{-3}) reported in *Science* [71].

In the last five years, Fe-based materials are mainly prepared by the simple pyrolysis of transition metal salt ($FeCl_3$ [34,36,39,41,47,73–80], $Fe(NO_3)_3$ [81–84], FeAc [24,74,85–88], and FeC_2O_4 [42]), carbon support, and nitrogen-rich small molecule [34,36,78,79,85,87–89] or polymeric compound [7,39–41,90]. An important breakthrough was made by Zelenay *et al.* [7] who successfully synthesized Fe/N/C catalysts (PANI–Fe–C) via heat-treatment of polyaniline (PANI), $FeCl_3$ and carbon black (Ketjenblack EC-300J). As displayed in Figure 2a, the PANI–Fe–C catalyst shows a very high ORR onset potential (~0.93 V *vs.* RHE) in 0.5 M H_2SO_4, and very low H_2O_2 yield (<1%) at all potentials. They also carried out a research into effect of heat treatment on catalytic activity of PANI-derived Fe-based catalysts in the range of 400 °C to 1000 °C (Figure 2b). The activity, as measured by the ORR onset and half-wave potential ($E_{1/2}$) in the rotating disk electrode (RDE) polarization plots, increases to the maximum at 900 °C with a very low H_2O_2 yield (<1%) over the potential range from 0.1 to 0.8 V *versus* RHE, signaling virtually complete reduction of O_2 to H_2O in a four-electron process. Although the best-performing catalyst in fuel cell testing is the more active of the two FeCo mixed-metal materials, PANI–FeCo–C, we cannot deny the fact that the ORR onset potential of PANI–Fe–C was the highest at that time [7], marking great progress in Fe-based catalysts. Before long, a new kind of Fe-based catalyst, three-dimensional (3D) N-doped graphene aerogel (N-GA)-supported Fe_3O_4 nanoparticles (Fe_3O_4/N-GAs), is prepared by Wu *et al.* [86]. In studying the effects of carbon support (carbon black, graphene) on the Fe_3O_4 nanoparticles ORR catalysts, they maintained, Fe_3O_4/N-GAs exhibit a more positive onset potential (−0.19 V *vs.* Ag/AgCl), higher cathodic density, lower H_2O_2 yield, and higher electron transfer number for ORR in alkaline media than Fe_3O_4 nanoparticles supported on N-doped carbon black (Fe_3O_4/N-CB) or N-doped graphene sheets (Fe_3O_4/N-GSs), which further verified that choosing a proper carbon support is vital for synthesizing a high-performance ORR catalysts [86]. Recently, Sun *et al.* [34] fabricated a Fe/N/C catalysts with a ORR half-wave potential of 0.75 V (*vs.* RHE) in 0.1 M $HClO_4$ and a low H_2O_2 yield of 2.6% at 0.4 V by pyrolyzing a composite of carbon-supported Fe-doped graphitic carbon nitride (Fe–g–C_3N_4@C) in the optimum conditions of Fe salt/dicyandiamide mass ratio of 1:10 and the pyrolyzed temperature at 750 °C.

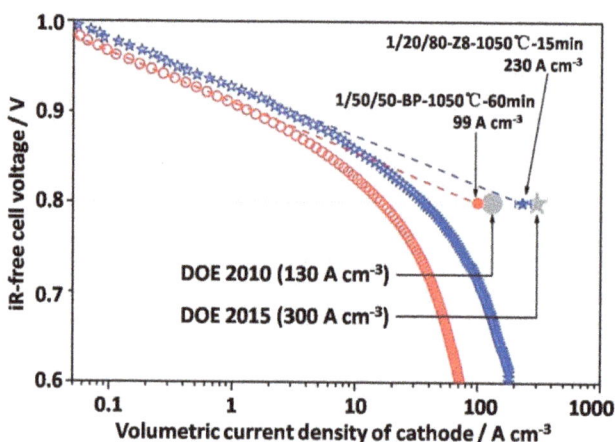

Figure 1. Volumetric current density of the best non-Pt catalysts in H_2/air fuel cell tests at 80 °C and 100% relative humidity for cathodes [71,72] and the U.S. DOE volumetric activity target at 0.8 V (*iR*-free). Red circles: most active iron-based catalyst from previous studies, dashed red line: extrapolation of the linear range to 0.8 V, blue stars: most active iron-based catalyst from the present study, dashed blue line: extrapolation of the linear range to 0.8 V. (Reproduce with permission from Ref. [72]. Copyright © Nature Publishing Group, London, UK, 2011).

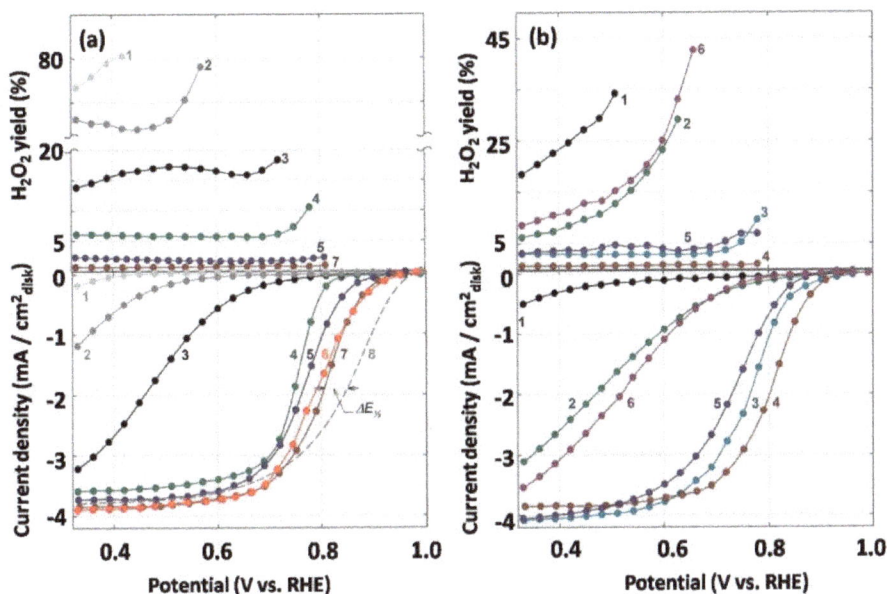

Figure 2. (a) Steady-state ORR polarization plots (**bottom**) and H_2O_2 yield plots (**top**) measured with different PANI-derived catalysts and reference materials: 1, as-received carbon black (Ketjenblack EC-300J); 2, heat-treated carbon black; 3, heat-treated PANI-C; 4, PANI-Co-C; 5, PANI-FeCo-C(1); 6, PANI-FeCo-C(2); 7, PANIFe-C; and 8, E-TEK Pt/C (20 μg_{Pt} cm^{-2}). Electrolyte: O_2-saturated 0.5 M H_2SO_4 (0.1 M $HClO_4$ in experiment involving Pt catalysts (dashed line)); temperature, 25 °C. RRDE experiments were carried out at a constant ring potential of 1.2 V *versus* RHE; RDE/RRDE rotating speed, 900 rpm; and non-precious metal catalyst loading, 0.6 mg cm^{-2}. (b) Steady-state ORR polarization plots (**bottom**) and H_2O_2 yield plots (**top**) measured with a PANI–Fe–C catalyst in 0.5 M H_2SO_4 electrolyte as a function of the heat treatment temperature: 1, 400 °C; 2, 600 °C; 3, 850 °C; 4, 900 °C; 5, 950 °C; and 6, 1000 °C. (Reproduce with permission from Ref. [7]. Copyright © American Association for the Advancement of Science, Washington, DC, USA, 2011).

Sun *et al.* [41] continued their work in synthesizing a Fe/N/C catalyst through high-temperature pyrolysis of the precursor containing poly-m-phenylenediamine (PmPDA) coated carbon black and $FeCl_3$ in which the Fe/N/C catalyst was denoted as PmPDA–Fe–N_x/C. As depicted in Figure 3a,b, the PmPDA–Fe–N_x/C catalysts pyrolyzed at 950 °C possess the highest ORR activity (11.5 A g^{-1} at 0.80 V *vs.* RHE) and the lowest H_2O_2 yield in 0.5 M H_2SO_4. They also carried out preliminary fuel cell test by employing the PmPDA–Fe–N_x/C (950 °C) as cathode catalyst. The maximal power density reached 350 mW cm^{-2} at cell voltage of 0.44 V, current density of

800 mA cm^{-2} and the current density at 0.8 V is about 90 mA cm^{-2} (Figure 3c,d) without back pressure applied during the fuel cell test.

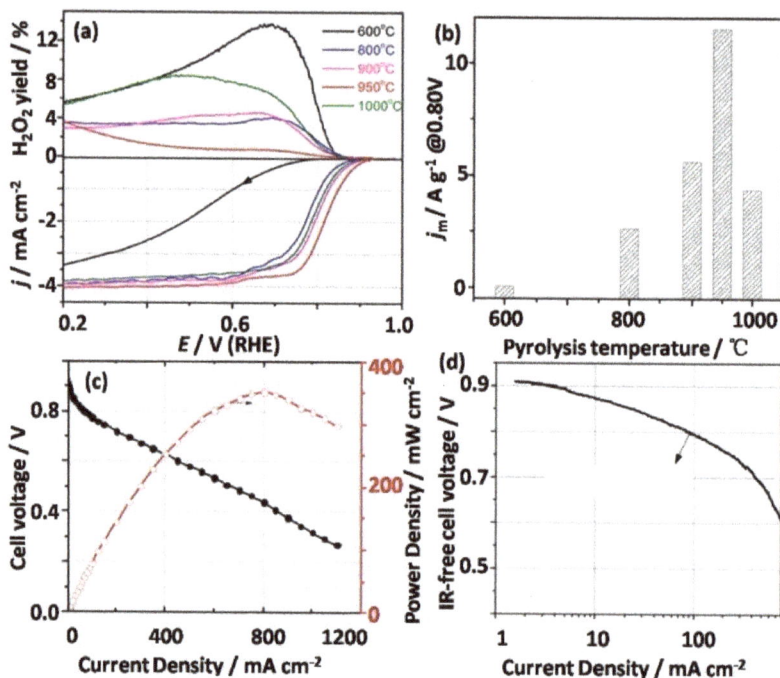

Figure 3. (**a**) ORR polarization curves and H$_2$O$_2$ yield plots of PmPDA–Fe–N$_x$/C catalyst prepared at different pyrolysis temperature, measured in O$_2$-saturated 0.1 M H$_2$SO$_4$. Catalyst loading: 0.6 mg cm^{-2}; Scan rate: 10 mV s^{-1}; and Rotating speed: 900 rpm. (**b**) Variety of ORR mass activity at 0.80 V with pyrolysis temperature. (**c**) Polarization and power density plots for H$_2$O$_2$ single fuel cell with PmPDA–Fe–N$_x$/C as cathode catalyst at 80 °C. MEA active area: 2.0 cm^2; Nafion 211 membrane; cathode catalyst loading: 4 mg cm^{-2}; Anode catalyst: Pt/C (60 wt. %, JM) with Pt loading of 0.5 mg cm^{-2}. No back pressure was applied. (**d**) Plot of *iR*-free cell voltage *versus* the logarithm of current density. (Reproduce with permission from Ref. [41]. Copyright © American Chemical Society, Washington, DC, USA, 2014).

Compared to pyrolysis of simple Fe salt, the Fe-based catalysts prepared via heat-treatment iron phthalocyanines (Pc)/porphyrins and their derivatives supported on carbon materials or some synthesized Fe-based macrocycle compounds have also attracted widely public attention in recent years [15,38,47,55,77,91,92]. Among all the FePc-based catalysts (FePc/SWCNT, FePc/DWCNT, and FePc/MWCNT), synthesized by Morozan *et al.* [55], dispersing iron(II) phthalocyanine on different types of carbon nanotubes (SWCNTs, DWCNTs, MWCNTs), FePc/MWCNT catalysts exhibit the best ORR performance in alkaline electrolyte close to the Pt/C reference. In 2012, by reacting the pyridine-functionalized graphene with iron-porphyrin, a graphene-metalloporphyrin metal organic framework (MOF) with enhanced catalytic activity for ORR was synthesized by Jahan *et al.* [77]. The authors claimed that the addition of pyridine-functionalized graphene changes the crystallization process of iron-porphyrin in the MOF, increase its porosity, and enhances the electrochemical charge transfer rate of iron-porphyrin, and therefore, enhance the ORR catalytic activity of these Fe-based catalysts [77]. After that, an exciting result was reported in *Nature Communication* by the Cho group [91]. A composite of FePc and SWCNTs (FePc–Py–CNTs) from covalent functionalization of SWCNTs, taking advantage of the diazonium reaction, was synthesized by anchoring pyridyl (Py) groups on the walls of CNTs, prior to FePc coordinated to Py–CNTs through the bond formed between nitrogen atom in pyridine and iron center in FePc (Figure 4a) [91]. The as-synthesized composites show a higher ORR catalytic activity with a half-wave potential ($E_{1/2}$) at 0.915 V (*vs.* RHE) than that of the state-of-the-art Pt/C with $E_{1/2}$ value at 0.88 V (Figure 4b). Theoretical calculations made by the authors suggest that the rehybridization of Fe 3d orbitals with the ligand orbitals coordinated from the axial direction results in a significant change in electronic and geometric structure, which greatly increases the ORR catalytic activity of catalysts [91]. Differ from the CNTs used as carbon support of FePc by Cho *et al.*, using chemically reduced graphene as the carbon support of FePc, Chen *et al.* [38] successfully synthesized a g-FePc catalyst through forceful $\pi-\pi$ interaction. The results of electrochemical measurements suggest that g-FePc catalyst possesses prominent ORR catalytic activity, which is comparable with commercial Pt/C in both onset potential and current density in 0.1 M KOH [38]. Furtermore, Liu's group [92,93] and Dai's group [15] also devote themselves to synthesize highly active Fe-based catalysts started from preparation of N-containing Fe-porphyrin complex or the solid-state synthesis of zeoliticimidazolate frameworks. Although the Fe-based catalysts prepared via heat-treatment iron phthalocyanines (Pc)/porphyrins and their derivatives supported on carbon materials or some synthesized Fe-based macrocycle compounds seems to be a little complicated or high-cost relative to pyrolysis of transition metal salt carbon support, and nitrogen-rich small molecule, still plays an important role in the preparation of ORR catalysts and research of ORR active sites.

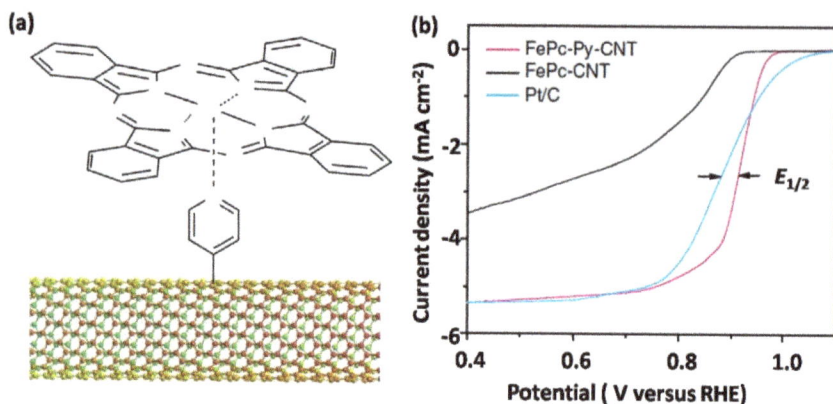

Figure 4. (a) Schematic diagram of the structure of FePc–Py–CNTs composite; and (b) linear scanning voltammograms of FePc–CNTs, FePc–Py–CNTs and commercial Pt/C catalyst. (Reproduce with permission from Ref. [91]. Copyright © Nature Publishing Group, London, UK, 2013).

It is not a unique instance, a new kind of highly active material, N-doped Fe or Fe_3C encapsulated in carbon support (CNTs or Graphitic layers), has been reported by many groups [39,67,87,94–97]. In 2012, Chen *et al.* [94] reported a synthetic strategy that enables synthesis of nitrogen-enriched core-shell structured catalysts with iron-based composite (Fe/Fe_3C) nanorods as the core and graphite carbon as the shell ($N–Fe/Fe_3C@C$) (Figure 5a). The $N–Fe/Fe_3C@C$ shows significantly improved activities and advanced kinetics for ORR in neutral phosphate buffer solution (PBS) compared with the commercial Pt/C catalysts (Pt 10%). The authors proposed that the doped N and core-Fe_3C in the $N–Fe/Fe_3C@C$ play key roles in improving the catalytic performance for ORR [94]. Soon after that, Lee *et al.* [94] found that the Fe/Fe_3C-functionalized melamine foam exhibited good ORR activities in alkaline media. Referring to the possibly important role of the Fe_3C phase in the ORR, Hu *et al.* [94] synthesized a Fe-based catalyst in the form of hollow spheres comprising uniform Fe_3C nanoparticles encased by a graphitic layer (Fe_3C/C) (Figure 5b) via high-pressure pyrolysis. The results of rotating disk electrode and rotating ring disk electrode measurement suggested that Fe_3C/C catalyst exhibited a high ORR activity and stability in both acidic and alkaline media partly due to the activation of the surrounding graphitic layers by the encased carbide nanoparticles, and making the outer surface of carbon layer active towards the ORR [97]. Recently, Xing *et al.* [39] synthesized a Fe-based catalyst with iron carbide encapsulated in N-doped graphitic layers (Fe_3C/NG) (Figure 5c), which also possesses high ORR activity and stability and further affirmed the importance of the structure of Fe/Fe_3C encased in carbon layers. In fact, Fe encapsulated within carbon nanotubes

(Figure 5d,e) as ORR catalysts has also been reported by both the Zelenay group [87] and the Bao group [95]. So we can excitedly find a fact that Fe element will play an important role in the ORR wherever it locates on the surface of N-doped carbon materials bonded with N or is encased by carbon layers.

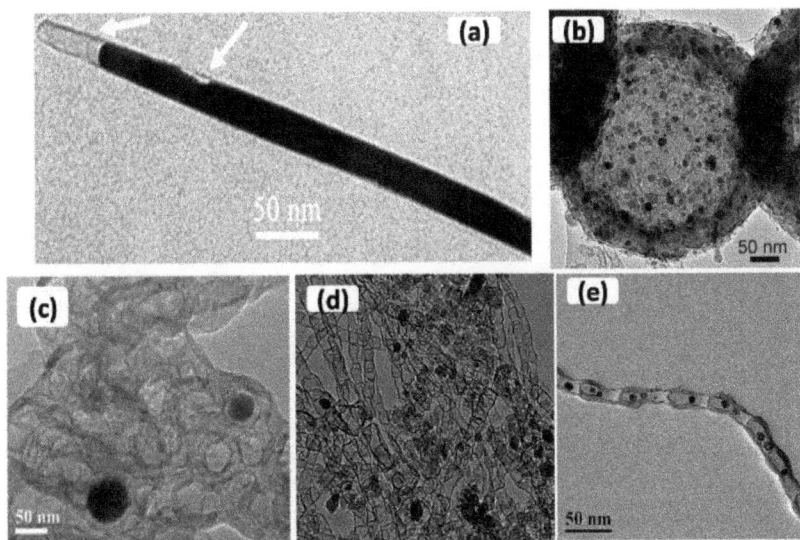

Figure 5. (a) TEM image of simple N-Fe/Fe3C/C nanrod with close-end graphite shell (Reproduce with permission from Ref. [94]. Copyright © WileyY-VCH Verlag GmbH & Co. KGaA, Weinheim, Germany, 2012). (b) TEM image of one typical hollow catalyst sphere of Fe_3C/C (Reproduce with permission from Ref. [97]. Copyright © WileyY-VCH Verlag GmbH & Co. KGaA, Weinheim, Germany, 2014). (c) TEM image of Fe_3C/NG; (Reproduce with permission from Ref. [39]. Copyright © WileyY-VCH Verlag GmbH & Co. KGaA, Weinheim, Germany, 2015). (d) TEM of the N-Fe-CNT/CNP composite (Reproduce with permission from Ref. [87]. Copyright © Nature Publishing Group, 2013). (e) TEM of image of Pod-Fe (Reproduce with permission from Ref. [95]. Copyright © WileyY-VCH Verlag GmbH & Co. KGaA, Weinheim, Germany, 2013).

It is recognized widely that Fe-doping would enhance the performance of N-doped catalyst. Among the transition metals (Mn, Fe, Co, Ni, and Cu), Fe, N-codoped catalysts exhibits the highest ORR activity [98], which fully displays the importance of Fe-base catalysts for ORR. Interestingly, Dai *et al.* [43] synthesized nanotubes-graphene (NT-G) complexed, few-walled carbon nanotubes with the outer wall partially unzipped by harsh oxidation in $KMnO_4/H_2SO_4$, which exhibit a high activity, excellent tolerance to methanol and superior stability in both acidic and alkaline solutions. The authors claimed that the NT-G contains small amount of irons

(1.10 wt. %) originated from nanotube growth seeds, and nitrogen impurities, which facilitate the formation of catalytic sites and boost the activity of the catalyst, and the role of iron in forming active ORR catalytic sites in the NT-G complex is proved by CN^- Poisoning experiments [43]. The role of extremely small amount of iron in ORR was further verified by Xu's group [36]. The authors synthesized a series of Fe-based catalysts by tuning the Fe content and codoping with nitrogen on cheap carbon black (CB) over a wide range from 0.02 to 20 wt. % (Figure 6a) and found that the optimal catalyst with a trace Fe content (0.05 wt. %) showed a superior high performance compared with commercial Pt/C in 0.1 M KOH (Figure 6b). Then after Xu, Pumera's group [99] demonstrated that residual manganese-based metallic impurities in graphene also play an extremely active role in the electrocatalysts of ORR on supposedly metal-free graphene electrode, which indirectly affirmed the role of a small amount of iron in other carbon-based ORR catalysts. Recently, Chen et al. [40] fabricated a series of self-supported N-doped mesoporous carbons with a trace amount of Fe (Fe–N/C). Electrochemical measurements revealed that Fe–N/C with an iron content of 0.24 at. % prepared at 800 °C was the best catalysts (Fe–N/C-800), with a more positive onset potential (0.98 V vs. RHE), higher diffusion-limited current, higher selectivity, higher stability, and stronger tolerance against methanol crossover than commercial Pt/C catalysts in 0.1 M KOH [38]. Interestingly, the results of cyanide poisoning and hot H_2SO_4 leaching for Fe–N/C-800 suggested that ORR was primarily due to iron-free active sites that arose most likely from nitrogen doping and the contributions of Fe-base active sites was small [40]. From all of the above, we can suggest that whether the small amount of iron will form active ORR catalytic sites or not is greatly dependent on the conditions for the preparation of Fe-based catalyst.

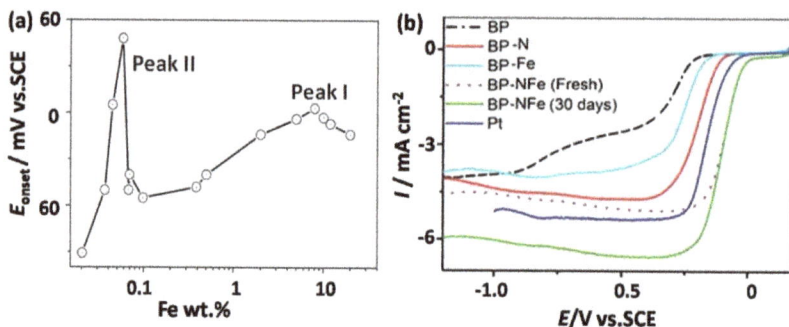

Figure 6. (a) Fe-content dependence of Fe-based catalysts. (b) RDE polarization curves of pure BP, BP–N, BP–Fe, BP–NFe, and Pt/C in O_2-saturated 0.1 M KOH with a scan rate of 5 mV s^{-1} and rotation speed of 1600 rpm. (Reproduce with permission from Ref. [36]. Copyright © WileyY-VCH Verlag GmbH & Co. KGaA, Weinheim, Germany, 2013).

Table 1. Electrocatalytic performance (onset potential (E_O/V *vs.* RHE) and half-wave potential ($E_{1/2}$/V *vs.* RHE)) of recently reported Fe-based catalysts for ORR and the corresponding test results (T/°C, test temperature; OCV/V, open-circuit voltage; MPD/mW cm^{-2}, and maximum power density) of fuel cell (H$_2$-O$_2$ fuel cell, acid/alkaline direct methanol fuel cell (DMFC) or Zn-air fuel cell).

Catalysts	Acid or Alkaline	E_O (V vs. RHE)	$E_{1/2}$ (V vs. RHE)	Cell tests, T(°C), OCV (V) and MPD (mW cm^{-2})	Reference (Year)
PANI–Fe–C	0.1 M HClO$_4$	~0.93	/	H$_2$-O$_2$, 80, ~0.9, <550	Ref. [7] (2011)
C–COP–P–Fe	0.1 M KOH 0.1 M HClO$_4$	ca. 0.98 ca. 0.89	/ /	/ /	Ref. [15] (2014)
BP–NFe	0.1 M KOH	0.045 vs. SCE	−0.089 vs. SCE	DMFC, 60, 0.8, 16.6	Ref. [36] (2013)
	0.5 M H$_2$SO$_4$	0.6 vs. SCE	/	/	
g-FePc	0.1 M KOH	0.98	0.88	/	Ref. [38] (2013)
F$_3$C/NG-800	0.1 M KOH	1.03	0.86	DMFC, 60, 0.75, 31	Ref. [39] (2015)
	0.1 M HClO$_4$	0.92	0.77	DMFC, 60, 0.87, 19	
Fe–N/C-800	0.1 M KOH 0.1 M HClO$_4$	0.98 0.77	/ /	/ /	Ref. [40] (2015)
PmPDA–Fe–N$_x$/C	0.1 M H$_2$SO$_4$	~0.93	/	H$_2$-O$_2$, 80, ~0.9, 350	Ref. [41] (2014)
NT-G	0.1 M KOH 0.1 M HClO$_4$	>1.05 ~0.89	0.87 0.76	/ /	Ref. [43] (2012)
Fe$_3$O$_4$/N-GAs	0.1 M KOH	−0.19 vs. Ag/AgCl	/	/	Ref. [86] (2012)
N–Fe–CNT/CNP	0.1 M NaOH	>1.05	0.93	/	Ref. [87] (2013)
FePc-Py-CNTs	0.1 M KOH	>1.05	0.915	/	Ref. [91] (2013)
Zn(mlm)$_2$TPIP	0.1 M HClO$_4$	0.902	0.76	H$_2$-O$_2$, 80, ~0.95, 620	Ref. [92] (2014)
PFeTTPP-1000	0.1 M HClO$_4$	0.93	0.76	H$_2$-O$_2$, 80, 0.9, 730	Ref. [93] (2013)
N–Fe/Fe$_3$C@C	0.1 M PBS	0.21 vs. Ag/AgCl	/	/	Ref. [94] (2012)
Pod-Fe	0.1 M H$_2$SO$_4$	0.5 vs. Ag/AgCl	0.3 vs. Ag/AgCl	H$_2$-O$_2$, 70, 0.7, /	Ref. [95] (2013)
Ar-800	0.1 M KOH	~0.05 vs. Hg/HgO	/	Zn-air, 1.2, 200	Ref. [96] (2013)
Fe$_3$C/C-800	0.1 M KOH	1.05	0.83	/	Ref. [97] (2014)
	0.1 M HClO$_4$	ca. 0.90	ca. 0.73	/	

Table 1 shows the representational results of electrocatalytic ORR performance and fuel cell tests of recently reported Fe-based catalysts in both acid medium and alkaline medium. Although the ORR performance of Fe-based catalysts in alkaline medium has outperformed that of commercial Pt/C, its ORR performance in acid media is still inferior to that commercial Pt/C. The ORR onset potential of Fe-based catalysts in alkaline has reached the value of 1.05 V (*vs.* RHE) [43,87,91], while the highest value of ORR onset potential to date in acid medium is just close to 0.93 V (*vs.* RHE) [7,39,41,93]. The reason why the Fe-based catalysts own the lower ORR catalytic activity in acid than that of in alkaline medium will be discussed in Section 2.2. In addition, considering the practicality of Fe-based catalysts for ORR in fuel cell, the best performing Fe-based ORR catalysts in acid condition reported to date is synthesized by Liu's group [92,93], who made H_2-O_2 fuel cell tests by using the as-prepared Fe-based materials as cathode catalysts and commercial Pt/C as anode catalysts and get a highly maximum power density of fuel cell tests of 730 mW cm^{-2} [93] and 620 mW cm^{-2} [92] in 2013 and 2014, respectively. However, regarding the ORR activity of Fe-based catalysts in acid, the actual volumetric activity of even the most active Fe-based catalysts needs to be improved. Regarding the ORR stability of Fe-based catalysts in both alkaline media and acid media, the stability tests are generally run at low current density or low power level, which are not real conditions for fuel cell operation. Hence, there is still a long way to go in order to reach the practical usage and understanding of Fe-based catalysts in fuel cells for commercial applications.

In summary, Fe-based catalysts represent a promising family of non-precious metal ORR catalyst candidates. It is obvious that the preparation conditions of Fe-based catalyst have a direct influence on the resulting Fe-based ORR electrocatalyst materials. A proper N-doped carbon support or N-enriched precursor selected for Fe-based catalyst is also vital for the final ORR catalytic activity. Of course, Fe is thus an already profoundly studied dopant for N-doped ORR electrocatalysts and will play a significant role in the further ongoing process within this field.

2.2. Research on Structure of Fe-Centered ORR Active Sites and ORR Mechanism

Considering the most recent progress of Fe-based catalysts, insight into formation mechanisms and structures of ORR active sites is also an ongoing task in the research and development of Fe-based catalysts for fuel cell applications. The current proposed active sites, containing edge plane FeN_2/C and FeN_4/C [60] species as well as basal plane macrocyclic FeN_4/C [19,20] species, are mainly speculated by data obtained from X-ray photoelectron spectroscopy (XPS) [19,100], time-of-flight secondary ion mass spectroscopy (TOF-SIMS) [45,60], X-ray absorption fine structure [18,60], and Mossbauer spectroscopy [17,19]. In 2008, Dodelet's group [61] claimed that the majority of active sites consist of a Fe-N_4/C (labeled by the authors as FeN_{2+2}/C) configuration bridging two adjacent graphene crystallites. Recently, Dodelet *et al.* [73] continue their work to clarify the origin of the enhanced PEM fuel cell performance of catalysts prepared by the procedures described in *Science* [71] and *Nature Communication* [72]. Among all the Fe–N_4-like species they reported, ORR activity is only attributed to Fe–N_4/C and N–Fe–N_{2+2}/C, which are the structure of coordination of nitrogen atoms and the iron atoms of Fe–N_4/C [73]. The former is a well-known site typically found in heat-treated carbon supported or unsupported porphyrins, and the latter is a very new kind of active composite N–Fe–N_{2+2}–NH^+ site, which is the high activity state of N–Fe–N_{2+2}/C. More importantly, Fe–N_4/C and N–Fe–N_{2+2}/C are more available in catalysts pyrolyzed in Ar + NH_3 atmosphere than in only Ar or NH_3 [73]. After the Dodelet's report in 2008 [61], Kramm *et al.* [59] firstly attributed improved ORR kinetics of these Fe–N_4 centers to Fe ion centers with higher electron densities. The authors made a further study on the structure of catalytic site in Fe-based catalysts for ORR via analyzing the Fe-species existed in the Fe-based catalysts prepared by impregnation of iron acetate on carbon black followed by heat-treatment in NH_3 at 950 °C [44]. Five different Fe-species were detected in the Fe-based catalysts containing 0.03 to 1.55 wt. % Fe: three doublets assigned to molecular FeN_4-like sites with their ferrous ions in a low (D1), intermediate (D2) or high (D3) spin state (Figure 7), and two other doublets assigned to a single Fe-species (D4 and D5) consisting of surface oxidized nitride nanoparticles [44]. Among the five Fe-species identified by ^{57}Fe Mossbauer spectroscopy in these catalysts, the authors maintained, only D1 and D3 display catalytic activity for the ORR in acid medium, with D3 featuring a composite structure with a protonated neighbor basic nitrogen and being far from the most active species [44]. These findings reveal that when focusing on the development of Fe-based catalysts with improved active site densities, it is possible to tune the electronic and structure properties of these active site structures, or develop Fe-based catalysts with higher ORR-activity by developing ways to make a larger fraction of the available Fe-atoms form more of the most ORR-active composite N–FeN_{2+2} ⋯ N_{prot}/C (D3) sites.

Recently, Chen *et al.* [45] proposed two possible formation mechanisms for the catalytically active sites occurring during high-temperature pyrolysis treatments through CN^- ions poisoning experiments, dependent on the specific type of precursor and synthesis methods utilized. The proposed structures of high-temperature-treated Fe-based catalysts are depicted in Scheme 1. These active sites include 1,10-phenanthroline (phen)-like iron complexes (A and C) [35,60], single pyridine-like iron complexes (B and E), and macrocyclic-like iron complexes (D and F) [20,45]. The authors claimed that utilizing aromatic iron complex ligands in inert atmospheres, catalytically active sites (C and D) will be formed in the layers of material deposition and will build up on the surface of the carbon support, which will decrease the porosity of surface layer and results in the majority of actives sites being inaccessible, entrapped in the subsurface layers, and such that leading to inhibited reactant and product mass transfer to and from the catalytically active sites. On the contrary, Fe-based catalysts prepared by pyrolyzing nonarmatic ligands, such as NH_3, and aliphatic diamines can result in the simultaneous production of the second active sites (A, B, and E) and well-connected channels [45]. The research provides valuable insight toward the development of Fe-based catalysts with improved ORR activity and stability. In 2013, Kattel*et al.* [50] performed first-principles density functional theory (DFT) calculations to investigate the reaction pathway of ORR on Fe-N_4 catalytic clusters formed between pores in graphene supports. The DFT results indicate that formation of Fe-N_4 clusters at the edges of graphitic pores is energetically feasible and ORR would be proceed on the assuming a pathway that follows the chemical reactions: (1) $O_2 \rightarrow$ *O_2 (adsorption); (2) *$O_2 + (H^+ + e^-) \rightarrow$ *OOH; (3) *OOH + $(H^+ + e^-) \rightarrow 2$*OH; and (4) 2*OH + $2(H^+ + e^-) \rightarrow 2H_2O$. The authors predicted that Fe–N_4 clusters near graphitic pores could promote the $4e^-$ ORR with a single active site contain central Fe atom and four surrounding N atoms due to the split of O–O bond in the reactant O_2 during the interaction of intermediate HOOH with the Fe–N_4 clusters in the above ORR pathway [50]. The theoretical study provides an explanation to the experimentally observed $4e^-$ ORR on heat treated Fe/N/C electrocatalysts and certified the Fe-centered active sites of these Fe/N/C electrocatalysts. More recently, the highly Fe-centered active sites was also verified by Ozkan's group [46] via H_2S poisoning experiments, which suggested that Fe plays a critical role in catalyzing ORR for Fe/N/C catalysts. Interestingly, except the above experimental work mentioned in Section 2.1, in combination with the XRD and XPS results of the pyrolyzed Fe/N/C catalysts, Sun *et al.* [34] propose that the ORR active sites are closely related to Fe_3N and both pyridinic N (which may bond to Fe^{III} to form Fe_3N) and quaternary N in the pyrolyzed Fe/N/C composites are conductive to catalyze the ORR and can serve as catalytically active sites for oxygen reduction in acid media. Through systematic of the effects of a series of inorganic molecules and ions (Cl^-, F^-, Br^-, SCN^-, SO_2 and H_2S) on the ORR activity, they further maintained, the active site of the Fe/N/C in acidic solutions contain Fe element and its valence state is mainly Fe^{III} since this catalyst is not sensitive to CO and NO_x but distinctly sensitive to F^- ion. The new insight into the active site nature of the Fe/N/C through molecule/ion probe is of very useful in rational design of high performance Fe-based catalysts for ORR in acid media [41].

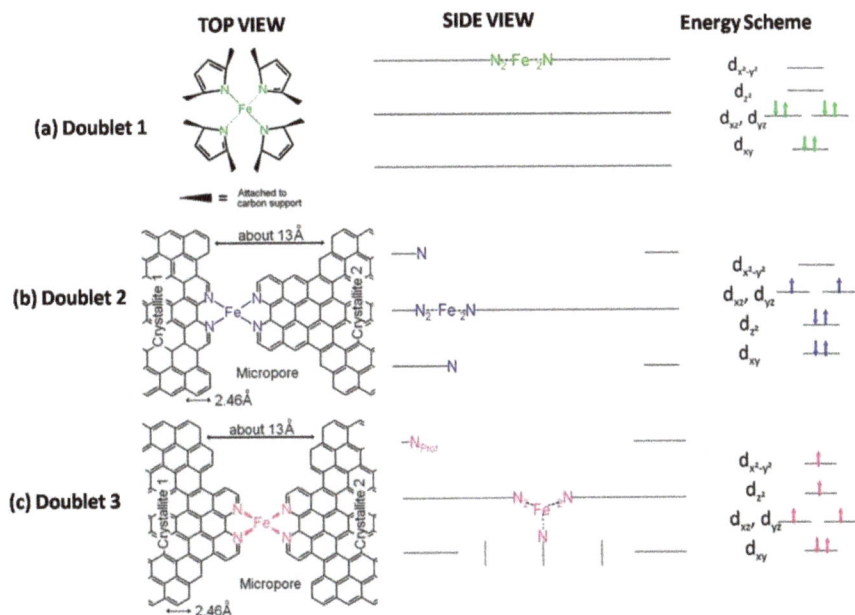

Figure 7. Side views and top views of the proposed structures of: (**a**) the FeN$_4$/C catalytic site in heat-treated, macrocycle-based catalysts assigned to Mossbauer doublet D1; (**b**) the FeN$_{2+2}$-like micropore-hosted site found in the catalyst prepared with iron acetate and heat-treated in ammonia assigned to doublet D2; and (**c**) the N–FeN$_{2+2}$-like composite site, where N–FeN$_{2+2}$ is assigned to doublet D3. In all side views, graphene planes are drawn as lines. (Reproduce with permission from Ref. [44]. Copyright © Royal Society of Chemistry, London, UK, 2012).

Additionally, the ORR catalytic activity of Fe-based catalysts prepared by Mukerjee *et al.* [101] in 2011 through a pyrolysis in NH_3 is mostly imparted by acid-resistant Fe–N$_4$ sites whose turnover frequency for O_2 reduction can be regulated by fine chemical changes of the catalyst surface. The authors claimed that surface N-groups could be protonated at pH 1 and subsequently bind anions, resulting in decreased activity for the O_2 reduction, and the anions can be removed chemically or thermally to restore the activity of acid-resistant Fe–N$_4$ sites [101]. The implications of the findings reported in this work suggested that optimizing the catalyst/electrolyte interface to prevent anion binding is required to combine high activity and durability of Fe-based catalysts. In fact, Mayer *et al.* [21,102,103] has also investigated the selectivity for four-electron reduction to H_2O or two-electron reduction to H_2O_2 of Fe-based catalysts in iron-porphyrin complexes. Using Iron$^{(III)}$ meso-tera(2-carboxyphenyl)-porphine chloride and its isomer as ORR electrocatalysts, the authors found that the Fe-based catalysts containing proton relays closed to the redox center in the second coordination sphere of iron-porphyrin

47

complexes have a high selectivity for four-electron reduction to H_2O, which suggested the importance of catalyst design for selectivity in Fe-based catalysts [102]. Recently, however, the authors verified that the nature of the catalyst film on a carbon electrode has an effect as large as changing the structure of the molecular catalyst itself [21].

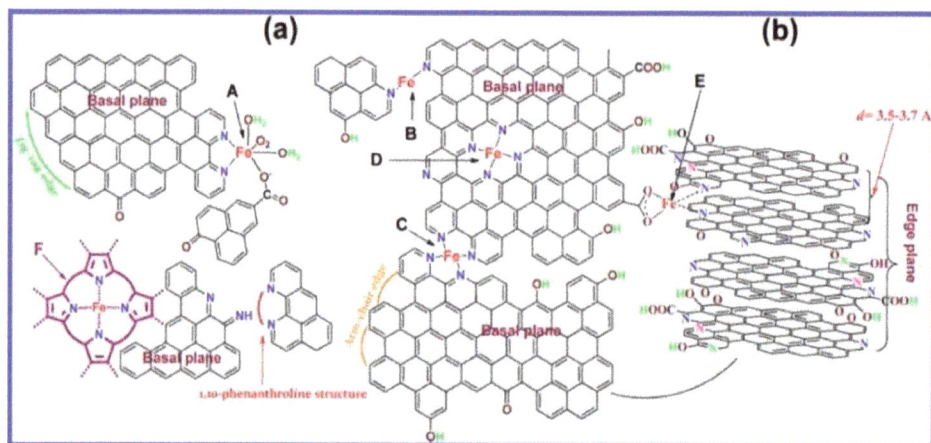

Scheme 1. Possible Iron Active Site Structures on Nanocrystal Graphite: (a) top and (b) side view (Reproduce with permission from Ref. [45]. Copyright © American Chemical Society, Washington, DC, USA, 2012).

Mukerjee *et al.* [104] continued their research on ORR mechanism of pyrolyzed Fe-based catalysts in alkaline medium and identified an activity descriptor based on principles of surface science and coordination chemistry. Using iron(III) meso-tetraphenylporphine chloride (FeTPPCl) as a model system, the authors elucidate inner- *vs.* outer-sphere ORR mechanisms and active-site structure evolution on pyrolyzed Fe-based catalysts. As depicted in Figure 8a, in alkaline media, taking platinum surface as a starting point of illustration, the well-known electrocatalytic inner-sphere electron transfer (ISET) mechanism (Figure 8a, inset i) involves chemisorptions of desolvated O_2 on an oxide-free Pt-site (Figure 8a, when M is represented as Pt) leading to a direct/series $4e^-$ ORR pathway without desorption of reaction intermediates and the coexistence of an outer-sphere electron transfer (OSET) mechanism (Figure 8a, inset ii), wherein the noncovalent hydrogen bonding forces between specifically adsorbed hydroxyl species (OH_{ads} acting as an outer-sphere bridge) and solvated O_2 (localized in outer-Helmholtz plane) promote a $2e^-$ reduction pathway forming HO_2^- anion [104]. Therefore, the goal of promotion of an electrocatalytic inner-sphere reaction mechanism for a complete $4e^-$ ORR process in alkaline electrolytes can be achieved via facilitation of direct adsorption of desolvated O_2 on OH_{ads}-free active sites and avoiding the precipitous outer-sphere reaction of solvated O_2 with OH_{ads} covered active sites [104]. In the system of Fe–N_4/C active sites, the $4e^-$ electrocatalytic inner-sphere electron transfer mechanism in dilute alkaline media is shown in Figure 8b, wherein O_2 displaces the OH^- species and chemisorbs directly on the Fe^{2+} active site [104]. The lability of the axial OH^- anion is due to the redox mechanism of ORR that ensures the reduction of pentacoordinated (H)O–Fe^{3+}–N_4 to the square-planarFe^{2+}–N_4 active site where axial ligation is available for direct O_2 chemisorption. This ensures that the precipitous OSET mechanism is avoided on Fe–N_4/C active sites leading to direct chemisorption of O_2 on the metal center via aninner-sphere mechanism. Once molecular O_2 adsorbs on the Fe^{2+} active site, via the superoxo and the ferric-hydroperoxyl states, the reaction proceeds to the ferrous-hydroperoxyl adduct, which is very critical since its stability determines the product distribution and ORR electrocatalytic activity. For pH > 12, the Lewis basic nature of the anionic hydrogen peroxide intermediate (HO_2^-, $pK_a \approx 11.6$) leads to its apparent stabilization on Lewis acidic Fe^{2+} active sites via the formation of stabilized Lewis acid-base adduct, which ensures that the catalytic cycle in alkaline media undergoes complete $4e^-$ transfer (Figure 8b) to regenerate the active site via the formation of ferric-hydroxyl species. However, in acidic media the analogous ferrous-hydroperoxyl adduct is Fe^{II}–(OHOH), wherein the protonated nature of the hydrogen peroxide intermediate (H_2O_2) negates its Lewis basic character and leads to its apparent destabilization on Fe^{2+}–N_4/C active site, which hence leads to higher overpotential for ORR in acidic media necessitating secondary sites to further reduce or disproportionate H_2O_2. Therefore, the author claimed that Fe–N_4/C active sites are more active for ORR in alkaline media than that of in acid media [104]. In additional, Fe–N_4/C active sites, the authors maintained, which was covalently integrated into the π-conjugated carbon basal plane during the pyrolysis step, could cause a dramatic anodic shift of ~600–900 mV in the metal ion's redox potential. Since the carbon basal plane constitutes an integral part of the active site due to the electron-donating/withdrawing capability of carbon support, the authors further claimed that tuning electron donating/withdrawing capability of the carbon basal plane, conferred upon it by the delocalized π-electrons, (i) causes a downshift of e_g-orbitals ($d_z{}^2$), thereby anodically shifting the metal ion's redox potential, and (ii) optimizes the bond strength between the metal ion and adsorbed reaction intermediates thereby maximizing oxygen-reduction activity [104]. The report makes it being possible to tune the catalytic activity of the class of pyrolyzed Fe-based catalysts by experimentally controlling the degree of π-electron delocalization of the carbonaceous surface and open the door to the development of more active and stable electrocatalysts based on Fe-centered active sites on novel π surfaces. Recently, Mukerjee et al. [48] made a further study on the various structural and functional forms of the active centers in pyrolyzed Fe-based catalysts in both ranges of pH and confirmed the single site $2e^- \times 2e^-$ mechanism in alkaline media on the primary $Fe^{2+}-N_4$ centers and the dual-site $2e^- \times 2e^-$ mechanism in acid media with the significant role of the surface bound coexisting Fe/Fe_xO_y nanoparticles (NPs) as the secondary active sites by employing a combination of *in situ* X-ray spectroscopy and electrochemical methods. From what has been discussed above, we can draw the conclusion that Fe^{3+} is mainly the active sites for ORR in acid media [34,41], while Fe^{3+} and Fe^{2+} are both play vital role for ORR in alkaline [48,104]. On the contrary, surface N-groups protonation is not beneficial for ORR activity [101,102].

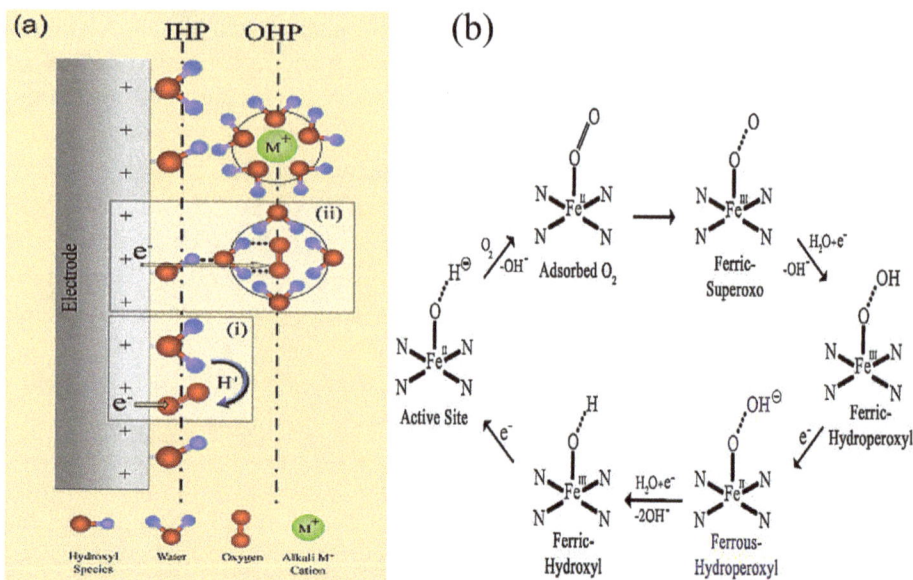

Figure 8. Proposed ORR mechanism. **(a)** Schematic illustration of inner-sphere (inset i) and outer-sphere (inset ii) electron transfer mechanisms during ORR in alkaline media. **(b)** Catalyst cycle showing the redox mechanism involved in ORR on pyrolyzed $Fe-N_4/C$ active sites in dilute alkaline medium; (IHP, inner Helmholtz plane; OHP, outer Helmholtz plane) (Reproduce with permission from Ref. [104]. Copyright © American Chemical Society, 2013).

2.3. Stability of Fe-Based ORR Catalysts

Although Fe-based catalysts with $Fe-N_4$ sites initially exhibit a highly catalytic activity in acidic medium, their durability is still insufficient [105]. Therefore, bridging the gap between the attributes responsible for high activity and high durability has become the main challenge facing Fe-based catalysts. In recent years, the stability of Fe-based ORR catalysts in alkaline medium has shown to be better than that of in acid medium [105]. Xu's group [36] has found the ORR performance in alkaline medium of their Fe-based catalysts containing extremely small amount of iron tend to be improved with larger diffusion-limiting current when the catalysts ink was re-tested after 30 days. The authors attributed the increase of diffusion-limiting current to the increase of oxygen diffusion coefficient in the microenvironment of the catalyst layer or the exposure of more active sites [36]. Compared to the higher stability in alkaline media, the reason for the degradation of Fe-based catalysts in the acidic environment during the ORR process has been attributed to the corrosion/oxidation of the active center and carbon support, attack by hydrogen peroxide of both the Fe and N sites, and the oxidation of the pyridinic active sites [106]. In fact, before the results reported by Xu's group [36], Zelenay's group [7,87] had already reported a phenomenon about the Fe-based catalysts durability, in which the authors make a cycling durability test in O_2-staturated solution in 0.1 M NaOH for their Fe-based ORR catalysts and found that the ORR performance of these catalysts not only did not become poor but shows a positive shift in the $E_{1/2}$ value [87], which is similar to their previous report about the potential shift with cycling observed with non-precious metal ORR catalysts at high current densities in acid medium in proton conducting fuel cells [7]. At that time, the authors attributed this to improved mass transport properties of the catalyst layer due to the loss of inactive species [7]. In order to deeply understand this phenomenon, which is different from the phenomenon of the instability of Fe-based catalysts in acid media reported in previous work [106], they performed further research on the stability of iron species in heat-treated Fe-based catalysts by combining the X-ray absorption near-edge structure (XANES) spectra edge-step analysis and inductively-coupled mass spectrometry measurement with the results of electrochemical measurement [105]. The results obtained by the authors show that Fe was lost from the Fe-based catalyst into the electrochemical environment during the ORR process in acid medium and the kinetic losses of ORR catalytic activity may be attribute to the oxidation of active sites and/or loss of pyridinic-like and pyrrolic-like Fe coordination (Fe–N_2 and Fe–N_4), as well as the mass transport improvement due to the removal of inactive Fe species, predominantly sulfides (FeS and FeS_2), while the durability of this Fe-based catalysts is depend on the stability of the porphyrazin-like Fe coordination [105]. This report elucidates a clear relationship between the electrocatalytic ORR activity and stability of Fe-based catalysts and the Fe species, which has a major significance for designing and preparing the highly stable Fe-base ORR catalysts.

In this section, we may safely draw the conclusion that great progress has been made in exploration of ORR active sites formation mechanisms, structure and stability. Fe-centered active sites possess a unique structure and exhibit very high catalytic activity for ORR, and will show an important role in the development of high-performance Fe-based catalysts. It is worth notice that the Fe-based catalysts synthesized by ACTA S. P. A (An Italy-based company engaged in the development, production and marketing of clean technology products for fuel cells and other hydrogen applications) show outstanding electrochemical ORR performance in alkaline media with a maximum power density of 120 mW cm^{-2} during the direct methanol fuel cell test, which show an important progress in the commercialization of fuel cell and attracting significant interest from several groups with alkaline fuel cell chemistry [107,108]. A long way, however, is still needed in order to reach the practical usage of Fe-based catalysts in the acid system of fuel cells applications. The exact role of the iron-ion center regarding the ORR active site as well as the structures of the active site should be investigated in detail in order to provide further insight into this topic.

3. Conclusions and Perspectives

The current Pt and Pt-based alloy electron catalysts, although they exhibit good ORR activities, suffer from many application challenges, such as high cost and weak durability. Meanwhile, a great advantage of recently developed Fe/N/C electrocatalysts (Fe-based catalysts) is their competitive ORR performance compared with Pt-based materials. Based on the previous report by Zhang et al., this review paper focuses on the progress in this research field over the recent five years. Compared to high-temperature pyrolysis of iron N_4 chelate complexes, Fe-based materials prepared by the simple pyrolysis of transition metal salt carbon support, and nitrogen-rich small molecule polymeric compound are mainly reviewed due to their low cost, high performance, long stability and most promising for replace currently used Pt-based catalysts in the progress of fuel cell commercialization. Additionally, Fe-base catalysts are presented to analyze the effect of Fe loading and existence form on the ORR catalytic activity in Fe-base catalyst. The proposed Fe-centered active sites and reaction mechanisms from various authors are also discussed in detail, which may be of importance for rational designing of high-performance, inexpensive, and practical Fe-base ORR catalysts in future development of fuel cells.

Numerous types of Fe-based ORR catalysts have been developed with ORR catalytic activity comparable with or better than Pt; however, almost all of them only show high catalytic activity in alkaline medium rather than in acidic condition. Due to the limitations of alkaline fuel cells, the acidic fuel cells are more popular. So, in the future research directions, developing of Fe-based catalysts with catalytic activity as highly as that of Pt in acidic condition is more urgent. In order to solve this problem, further study on the catalytic mechanism and kinetics is still needed in order to design and develop rationally carbon-based, Fe-based catalysts with a desirable activity and durability, especially in acidic conditions.

For the final industrial or commercial application, it is also essential to develop simple and cost-efficient methods for the large-scale production of Fe-based catalysts with excellent ORR electrocatalytic activity and long-term operation stability.

Acknowledgments: This work was funded by National Basic Research Program of China (973 Program, 2014CB932700, 2012CB932800 and 2012CB215500), National Natural Science Foundation of China (21273220, 21303180 and 21422307), and "the Recruitment Program of Global youth Experts" of China.

Author Contributions: Jing Liu wrote the most parts of the article, Erling Li contributed to the revise of the review in later stage, Mingbo Ruan and Ping Song were responsible for the mechanism in the article, Weilin Xu checked the article in the process of writing and modification.

Conflicts of Interest: The authors declare no conflict of interest.

References

1. Steele, B.C.H.; Heinzel, A. Materials for fuel-cell technologies. *Nature* **2001**, *414*, 345–352.
2. Higgins, D.; Chen, Z. Recent Development of Non-precious Metal Catalysts. In *Lecture Notes in Energy—Electrocatalysis in Fuel Cells*; Shao, M., Ed.; Springer: New York, NY, USA, 2013; Volume 9, pp. 247–270.
3. Gasteiger, H.A.; Kocha, S.S.; Sompalli, B.; Wagner, F.T. Activity benchmarks and requirements for Pt, Pt-alloy, and non-Pt oxygen reduction catalysts for PEMFCs. *Appl. Catal. B* **2005**, *56*, 9–35.
4. Borup, R.; Meyers, J.; Pivovar, B.; Kim, Y.S.; Mukundan, R.; Garland, N.; Myers, D.; Wilson, M.; Garzon, F.; Wood, D.; *et al.* Scientific Aspects of Polymer Electrolyte Fuel Cell Durability and Degradation. *Chem. Rev.* **2007**, *107*, 3904–3951.
5. Jasinski, R. A New Fuel Cell Cathode Catalyst. *Nature* **1964**, *201*, 1212–1213.
6. Jeon, I.-Y.; Choi, H.-J.; Choi, M.; Seo, J.-M.; Jung, S.-M.; Kim, M.-J.; Zhang, S.; Zhang, L.; Xia, Z.; Dai, L.; *et al.* Facile, scalable synthesis of edge-halogenated graphenenanoplatelets as efficient metal-free eletrocatalysts for oxygen reduction reaction. *Sci. Rep.* **2013**, *3*.
7. Wu, G.; More, K.L.; Johnston, C.M.; Zelenay, P. High-Performance Electrocatalysts for Oxygen Reduction Derived from Polyaniline, Iron, and Cobalt. *Science* **2011**, *332*, 443–447.
8. Gong, K.; Du, F.; Xia, Z.; Durstock, M.; Dai, L. Nitrogen-Doped Carbon Nanotube Arrays with High Electrocatalytic Activity for Oxygen Reduction. *Science* **2009**, *323*, 760–764.
9. Sun, X.; Zhang, Y.; Song, P.; Pan, J.; Zhuang, L.; Xu, W.; Xing, W. Fluorine-Doped Carbon Blacks: Highly Efficient Metal-Free Electrocatalysts for Oxygen Reduction Reaction. *ACS Catal.* **2013**, *3*, 1726–1729.
10. Sun, X.; Song, P.; Zhang, Y.; Liu, C.; Xu, W.; Xing, W. A Class of High Performance Metal-Free Oxygen Reduction Electrocatalysts based on Cheap Carbon Blacks. *Sci. Rep.* **2013**, *3*.
11. Oh, H.-S.; Kim, H. The role of transition metals in non-precious nitrogen-modified carbon-based electrocatalysts for oxygen reduction reaction. *J. Power Sources* **2012**, *212*, 220–225.

12. Jaouen, F. Heat-Treated Transition Metal-N_xC_y Electrocatalysts for the O_2 Reduction Reaction in Acid PEM Fuel Cells. In *Non-Noble Metal Fuel Cell Catalysts*; Wiley-VCH Verlag GmbH & Co. KGaA: Weinheim, Germany, 2014; pp. 29–118.

13. Chen, Z.; Higgins, D.; Yu, A.; Zhang, L.; Zhang, J. A review on non-precious metal electrocatalysts for PEM fuel cells. *Energy Environ. Sci.* **2011**, *4*, 3167–3192.

14. Jaouen, F.; Proietti, E.; Lefèvre, M.; Chenitz, R.; Dodelet, J.-P.; Wu, G.; Chung, H.T.; Johnston, C.M.; Zelenay, P. Recent advances in non-precious metal catalysis for oxygen-reduction reaction in polymer electrolyte fuel cells. *Energy Environ. Sci.* **2011**, *4*, 114–130.

15. Xiang, Z.; Xue, Y.; Cao, D.; Huang, L.; Chen, J.-F.; Dai, L. Highly Efficient Electrocatalysts for Oxygen Reduction Based on 2D Covalent Organic Polymers Complexed with Non-precious Metals. *Angew. Chem. Int. Ed.* **2014**, *53*, 2433–2437.

16. Koslowski, U.I.; Abs-Wurmbach, I.; Fiechter, S.; Bogdanoff, P. Nature of the Catalytic Centers of Porphyrin-Based Electrocatalysts for the ORR: A Correlation of Kinetic Current Density with the Site Density of Fe–N_4 Centers. *J. Phys. Chem. C* **2008**, *112*, 15356–15366.

17. Bouwkamp-Wijnoltz, A.L.; Visscher, W.; van Veen, J.A.R.; Boellaard, E.; van der Kraan, A.M.; Tang, S.C. On Active-Site Heterogeneity in Pyrolyzed Carbon-Supported Iron Porphyrin Catalysts for the Electrochemical Reduction of Oxygen: An *In Situ* Mössbauer Study. *J. Phys. Chem. B* **2002**, *106*, 12993–13001.

18. Bae, I.T.; Tryk, D.A.; Scherson, D.A. Effect of Heat Treatment on the Redox Properties of Iron Porphyrins Adsorbed on High Area Carbon in Acid Electrolytes: An *In Situ* Fe K-Edge X-ray Absorption Near-Edge Structure Study. *J. Phys. Chem. B* **1998**, *102*, 4114–4117.

19. Schulenburg, H.; Stankov, S.; Schünemann, V.; Radnik, J.; Dorbandt, I.; Fiechter, S.; Bogdanoff, P.; Tributsch, H. Catalysts for the Oxygen Reduction from Heat-Treated Iron(III) Tetramethoxyphenylporphyrin Chloride: Structure and Stability of Active Sites. *J. Phys. Chem.* **2003**, *107*, 9034–9041.

20. Arechederra, R.L.; Artyushkova, K.; Atanassov, P.; Minteer, S.D. Growth of Phthalocyanine Doped and Undoped Nanotubes Using Mild Synthesis Conditions for Development of Novel Oxygen Reduction Catalysts. *ACS Appl. Mater. Interfaces* **2010**, *2*, 3295–3302.

21. Rigsby, M.L.; Wasylenko, D.J.; Pegis, M.L.; Mayer, J.M. Medium Effects Are as Important as Catalyst Design for Selectivity in Electrocatalytic Oxygen Reduction by Iron-Porphyrin Complexes. *J. Am. Chem. Soc.* **2015**, *137*, 4296–4299.

22. Bezerra, C.W.B.; Zhang, L.; Lee, K.; Liu, H.; Zhang, J.; Shi, Z.; Marques, A.L.B.; Marques, E.P.; Wu, S.; Zhang, J. Novel carbon-supported Fe–N electrocatalysts synthesized through heat treatment of iron tripyridyltriazine complexes for the PEM fuel cell oxygen reduction reaction. *Electrochim. Acta* **2008**, *53*, 7703–7710.

23. Liu, H.; Shi, Z.; Zhang, J.; Zhang, L.; Zhang, J. Ultrasonic spray pyrolyzed iron-polypyrrole mesoporous spheres for fuel celloxygen reduction electrocatalysts. *J. Mater. Chem.* **2009**, *19*, 468–470.

24. Choi, J.-Y.; Higgins, D.; Chen, Z. Highly Durable GrapheneNanosheet Supported Iron Catalyst for Oxygen Reduction Reaction in PEM Fuel Cells. *J. Electrochem. Soc.* **2011**, *159*, B86–B89.

25. Ye, S.; Vijh, A.K. Non-noble metal-carbonized aerogel composites as electrocatalysts for the oxygen reduction reaction. *Electrochem. Commun.* **2003**, *5*, 272–275.

26. Lalande, G.; Côté, R.; Guay, D.; Dodelet, J.P.; Weng, L.T.; Bertrand, P. Is nitrogen important in the formulation of Fe-based catalysts for oxygen reduction in solid polymer fuel cells? *Electrochim. Acta* **1997**, *42*, 1379–1388.

27. Hsu, R.S.; Chen, Z. Improved Synthesis Method for a Cyanamide Derived Non-Precious ORR Catalyst for PEFCs. *ECS Trans.* **2010**, *28*, 39–46.

28. Choi, J.-Y.; Hsu, R.S.; Chen, Z. Highly Active Porous Carbon-Supported Nonprecious Metal-N Electrocatalyst for Oxygen Reduction Reaction in PEM Fuel Cells. *J. Phys. Chem. C* **2010**, *114*, 8048–8053.

29. Choi, J.-Y.; Hsu, R.S.; Chen, Z. Nanoporous Carbon-Supported Fe/Co–N Electrocatalyst for Oxygen Reduction Reaction in PEM Fuel Cells. *ECS Trans.* **2010**, *28*, 101–112.

30. Chung, H.T.; Johnston, C.M.; Artyushkova, K.; Ferrandon, M.; Myers, D.J.; Zelenay, P. Cyanamide-derived non-precious metal catalyst for oxygen reduction. *Electrochem. Commun.* **2010**, *12*, 1792–1795.

31. Jaouen, F.; Lefèvre, M.; Dodelet, J.-P.; Cai, M. Heat-Treated Fe/N/C Catalysts for O_2 Electroreduction: Are Active Sites Hosted in Micropores? *J. Phys. Chem. B* **2006**, *110*, 5553–5558.

32. Choi, C.H.; Park, S.H.; Woo, S.I. N-doped carbon prepared by pyrolysis of dicyandiamide with various $MeCl_2 \cdot xH_2O$ (Me = Co, Fe, and Ni) composites: Effect of type and amount of metal seed on oxygen reduction reactions. *Appl. Catal. B* **2012**, *119–120*, 123–131.

33. Tian, J.; Morozan, A.; Sougrati, M.T.; Lefèvre, M.; Chenitz, R.; Dodelet, J.-P.; Jones, D.; Jaouen, F. Optimized Synthesis of Fe/N/C Cathode Catalysts for PEM Fuel Cells: A Matter of Iron-Ligand Coordination Strength. *Angew. Chem. Int. Ed.* **2013**, *52*, 6867–6870.

34. Wang, M.-Q.; Yang, W.-H.; Wang, H.-H.; Chen, C.; Zhou, Z.-Y.; Sun, S.-G. Pyrolyzed Fe–N–C Composite as an Efficient Non-precious Metal Catalyst for Oxygen Reduction Reaction in Acidic Medium. *ACS Catal.* **2014**, *4*, 3928–3936.

35. Lefèvre, M.; Dodelet, J.P.; Bertrand, P. O_2 Reduction in PEM Fuel Cells: Activity and Active Site Structural Information for Catalysts Obtained by the Pyrolysis at High Temperature of Fe Precursors. *J. Phys. Chem. B* **2000**, *104*, 11238–11247.

36. Liu, J.; Sun, X.; Song, P.; Zhang, Y.; Xing, W.; Xu, W. High-Performance Oxygen Reduction Electrocatalysts based on Cheap Carbon Black, Nitrogen, and Trace Iron. *Adv. Mater.* **2013**, *25*, 6879–6883.

37. Li, X.; Liu, G.; Popov, B.N. Activity and stability of non-precious metal catalysts for oxygen reduction in acid and alkaline electrolytes. *J. Power Sources* **2010**, *195*, 6373–6378.

38. Jiang, Y.; Lu, Y.; Lv, X.; Han, D.; Zhang, Q.; Niu, L.; Chen, W. Enhanced Catalytic Performance of Pt-Free Iron Phthalocyanine by Graphene Support for Efficient Oxygen Reduction Reaction. *ACS Catal.* **2013**, *3*, 1263–1271.

39. Xiao, M.; Zhu, J.; Feng, L.; Liu, C.; Xing, W. Meso/Macroporous Nitrogen-Doped Carbon Architectures with Iron Carbide Encapsulated in Graphitic Layers as an Efficient and Robust Catalyst for the Oxygen Reduction Reaction in Both Acidic and Alkaline Solutions. *Adv. Mater.* **2015**, *27*, 2521–2527.

40. Niu, W.; Li, L.; Liu, X.; Wang, N.; Liu, J.; Zhou, W.; Tang, Z.; Chen, S. Mesoporous N-Doped Carbons Prepared with Thermally Removable Nanoparticle Templates: An Efficient Electrocatalyst for Oxygen Reduction Reaction. *J. Am. Chem. Soc.* **2015**, *137*, 5555–5562.

41. Wang, Q.; Zhou, Z.-Y.; Lai, Y.-J.; You, Y.; Liu, J.-G.; Wu, X.-L.; Terefe, E.; Chen, C.; Song, L.; Rauf, M.; *et al.* Phenylenediamine-Based FeN$_x$/C Catalyst with High Activity for Oxygen Reduction in Acid Medium and Its Active-Site Probing. *J. Am. Chem. Soc.* **2014**, *136*, 10882–10885.

42. Yin, J.; Qiu, Y.; Yu, J. Onion-like graphitic nanoshell structured Fe–N/C nanofibers derived from electrospinning for oxygen reduction reaction in acid media. *Electrochem. Commun.* **2013**, *30*, 1–4.

43. Li, Y.; Zhou, W.; Wang, H.; Xie, L.; Liang, Y.; Wei, F.; Idrobo, J.-C.; Pennycook, S.J.; Dai, H. An oxygen reduction electrocatalyst based on carbon nanotube-graphene complexes. *Nat. Nanotechnol.* **2012**, *7*, 394–400.

44. Kramm, U.I.; Herranz, J.; Larouche, N.; Arruda, T.M.; Lefevre, M.; Jaouen, F.; Bogdanoff, P.; Fiechter, S.; Abs-Wurmbach, I.; Mukerjee, S.; *et al.* Structure of the catalytic sites in Fe/N/C-catalysts for O$_2$-reduction in PEM fuel cells. *Phys. Chem. Chem. Phys.* **2012**, *14*, 11673–11688.

45. Li, W.; Wu, J.; Higgins, D.C.; Choi, J.-Y.; Chen, Z. Determination of Iron Active Sites in Pyrolyzed Iron-Based Catalysts for the Oxygen Reduction Reaction. *ACS Catal.* **2012**, *2*, 2761–2768.

46. Singh, D.; Mamtani, K.; Bruening, C.R.; Miller, J.T.; Ozkan, U.S. Use of H$_2$S to Probe the Active Sites in FeNC Catalysts for the Oxygen Reduction Reaction (ORR) in Acidic Media. *ACS Catal.* **2014**, *4*, 3454–3462.

47. Zhu, Y.; Zhang, B.; Liu, X.; Wang, D.-W.; Su, D.S. Unravelling the Structure of Electrocatalytically Active Fe–N Complexes in Carbon for the Oxygen Reduction Reaction. *Angew. Chem. Int. Ed.* **2014**, *53*, 10673–10677.

48. Tylus, U.; Jia, Q.; Strickland, K.; Ramaswamy, N.; Serov, A.; Atanassov, P.; Mukerjee, S. Elucidating Oxygen Reduction Active Sites in Pyrolyzed Metal-Nitrogen Coordinated Non-Precious-Metal Electrocatalyst Systems. *J. Phys. Chem. C* **2014**, *118*, 8999–9008.

49. Robson, M.H.; Serov, A.; Artyushkova, K.; Atanassov, P. A mechanistic study of 4-aminoantipyrine and iron derived non-platinum group metal catalyst on the oxygen reduction reaction. *Electrochim. Acta* **2013**, *90*, 656–665.

50. Kattel, S.; Wang, G. A density functional theory study of oxygen reduction reaction on Me–N$_4$ (Me = Fe, Co, or Ni) clusters between graphitic pores. *J. Mater. Chem. A* **2013**, *1*, 10790–10797.

51. Badger, G.; Jones, R.; Laslett, R. Porphyrins. VII. The synthesis of porphyrins by the Rothemund reaction. *Aust. J. Chem.* **1964**, *17*, 1028–1035.

52. Baranton, S.; Coutanceau, C.; Roux, C.; Hahn, F.; Léger, J.M. Oxygen reduction reaction in acid medium at iron phthalocyanine dispersed on high surface area carbon substrate: Tolerance to methanol, stability and kinetics. *J. Electroanal. Chem.* **2005**, *577*, 223–234.

53. Vasudevan, P.; Satya, S.; Mann, N.; Tyagi, S. Transition metal complexes of porphyrins and phthalocyanines as electrocatalysts for dioxygen reduction. *Transit. Met. Chem.* **1990**, *15*, 81–90.

54. Zagal, J.H. Metallophthalocyanines as catalysts in electrochemical reactions. *Coord. Chem. Rev.* **1992**, *119*, 89–136.

55. Morozan, A.; Campidelli, S.; Filoramo, A.; Jousselme, B.; Palacin, S. Catalytic activity of cobalt and iron phthalocyanines or porphyrins supported on different carbon nanotubes towards oxygen reduction reaction. *Carbon* **2011**, *49*, 4839–4847.

56. Faubert, G.; Lalande, G.; Côté, R.; Guay, D.; Dodelet, J.P.; Weng, L.T.; Bertrand, P.; Dénès, G. Heat-treated iron and cobalt tetraphenylporphyrins adsorbed on carbon black: Physical characterization and catalytic properties of these materials for the reduction of oxygen in polymer electrolyte fuel cells. *Electrochim. Acta* **1996**, *41*, 1689–1701.

57. Gupta, S.; Tryk, D.; Bae, I.; Aldred, W.; Yeager, E. Heat-treated polyacrylonitrile-based catalysts for oxygen electroreduction. *J. Appl. Electrochem.* **1989**, *19*, 19–27.

58. Wu, G.; Chen, Z.; Artyushkova, K.; Garzon, F.H.; Zelenay, P. Polyaniline-derived Non-Precious Catalyst for the Polymer Electrolyte Fuel Cell Cathode. *ECS Trans.* **2008**, *16*, 159–170.

59. Kramm, U.I.; Abs-Wurmbach, I.; Herrmann-Geppert, I.; Radnik, J.; Fiechter, S.; Bogdanoff, P. Influence of the Electron-Density of FeN_4-Centers Towards the Catalytic Activity of Pyrolyzed FeTMPPCl-Based ORR-Electrocatalysts. *J. Electrochem. Soc.* **2011**, *158*, B69–B78.

60. Lefèvre, M.; Dodelet, J.P.; Bertrand, P. Molecular Oxygen Reduction in PEM Fuel Cells: Evidence for the Simultaneous Presence of Two Active Sites in Fe-Based Catalysts. *J. Phys. Chem. B* **2002**, *106*, 8705–8713.

61. Charreteur, F.; Jaouen, F.; Ruggeri, S.; Dodelet, J.-P. Fe/N/C non-precious catalysts for PEM fuel cells: Influence of the structural parameters of pristine commercial carbon blacks on their activity for oxygen reduction. *Electrochim. Acta* **2008**, *53*, 2925–2938.

62. Liu, G.; Li, X.; Ganesan, P.; Popov, B.N. Studies of oxygen reduction reaction active sites and stability of nitrogen-modified carbon composite catalysts for PEM fuel cells. *Electrochim. Acta* **2010**, *55*, 2853–2858.

63. Nabae, Y.; Moriya, S.; Matsubayashi, K.; Lyth, S.M.; Malon, M.; Wu, L.; Islam, N.M.; Koshigoe, Y.; Kuroki, S.; Kakimoto, M.-A.; *et al.* RETRACTED: The role of Fe species in the pyrolysis of Fe phthalocyanine and phenolic resin for preparation of carbon-based cathode catalysts. *Carbon* **2010**, *48*, 2613–2624.

64. Baker, R.; Wilkinson, D.P.; Zhang, J. Electrocatalytic activity and stability of substituted iron phthalocyanines towards oxygen reduction evaluated at different temperatures. *Electrochim. Acta* **2008**, *53*, 6906–6919.

65. Li, W.; Yu, A.; Higgins, D.C.; Llanos, B.G.; Chen, Z. Biologically Inspired Highly Durable Iron Phthalocyanine Catalysts for Oxygen Reduction Reaction in Polymer Electrolyte Membrane Fuel Cells. *J. Am. Chem. Soc.* **2010**, *132*, 17056–17058.

66. Liu, G.; Li, X.; Ganesan, P.; Popov, B.N. Development of non-precious metal oxygen-reduction catalysts for PEM fuel cells based on N-doped ordered porous carbon. *Appl. Catal. B* **2009**, *93*, 156–165.

67. Velázquez-Palenzuela, A.; Zhang, L.; Wang, L.; Cabot, P.L.; Brillas, E.; Tsay, K.; Zhang, J. Carbon-Supported Fe–N$_x$ Catalysts Synthesized by Pyrolysis of the Fe(II)–2,3,5,6-Tetra(2-pyridyl)pyrazine Complex: Structure, Electrochemical Properties, and Oxygen Reduction Reaction Activity. *J. Phys. Chem. C* **2011**, *115*, 12929–12940.

68. Zhang, L.; Lee, K.; Bezerra, C.W.B.; Zhang, J.; Zhang, J. Fe loading of a carbon-supported Fe–N electrocatalyst and its effect on the oxygen reduction reaction. *Electrochim. Acta* **2009**, *54*, 6631–6636.

69. Jaouen, F.; Marcotte, S.; Dodelet, J.-P.; Lindbergh, G. Oxygen Reduction Catalysts for Polymer Electrolyte Fuel Cells from the Pyrolysis of Iron Acetate Adsorbed on Various Carbon Supports. *J. Phys. Chem. B* **2003**, *107*, 1376–1386.

70. Nallathambi, V.; Lee, J.-W.; Kumaraguru, S.P.; Wu, G.; Popov, B.N. Development of high performance carbon composite catalyst for oxygen reduction reaction in PEM Proton Exchange Membrane fuel cells. *J. Power Sources* **2008**, *183*, 34–42.

71. Lefevre, M.; Proietti, E.; Jaouen, F.; Dodelet, J.P. Iron-Based Catalysts with Improved Oxygen Reduction Activity in Polymer Electrolyte Fuel Cells. *Science* **2009**, *324*, 71–74.

72. Proietti, E.; Jaouen, F.; Lefèvre, M.; Larouche, N.; Tian, J.; Herranz, J.; Dodelet, J.-P. Iron-based cathode catalyst with enhanced power density in polymer electrolyte membrane fuel cells. *Nat. Commun.* **2011**, *2*, 416.

73. Kramm, U.I.; Lefèvre, M.; Larouche, N.; Schmeisser, D.; Dodelet, J.-P. Correlations between Mass Activity and Physicochemical Properties of Fe/N/C Catalysts for the ORR in PEM Fuel Cell via ^{57}Fe Mössbauer Spectroscopy and Other Techniques. *J. Am. Chem. Soc.* **2014**, *136*, 978–985.

74. Byon, H.R.; Suntivich, J.; Shao-Horn, Y. Graphene-Based Non-Noble-Metal Catalysts for Oxygen Reduction Reaction in Acid. *Chem. Mater.* **2011**, *23*, 3421–3428.

75. Wu, G.; Johnston, C.M.; Mack, N.H.; Artyushkova, K.; Ferrandon, M.; Nelson, M.; Lezama-Pacheco, J.S.; Conradson, S.D.; More, K.L.; Myers, D.J.; et al. Synthesis-structure-performance correlation for polyaniline-Me-C non-precious metal cathode catalysts for oxygen reduction in fuel cells. *J. Mater. Chem.* **2011**, *21*, 11392–11405.

76. Wu, G.; Nelson, M.; Ma, S.; Meng, H.; Cui, G.; Shen, P.K. Synthesis of nitrogen-doped onion-like carbon and its use in carbon-based CoFe binary non-precious-metal catalysts for oxygen-reduction. *Carbon* **2011**, *49*, 3972–3982.

77. Jahan, M.; Bao, Q.; Loh, K.P. Electrocatalytically Active Graphene-Porphyrin MOF Composite for Oxygen Reduction Reaction. *J. Am. Chem. Soc.* **2012**, *134*, 6707–6713.

78. Yang, T.; Han, G. Synthesis of a Novel Catalyst via Pyrolyzing Melamine with Fe Precursor and its Excellent Electrocatalysis for Oxygen Reduction. *Int. J. Electrochem. Sci.* **2012**, *7*, 10884–10893.

79. Xiao, H.; Shao, Z.-G.; Zhang, G.; Gao, Y.; Lu, W.; Yi, B. Fe–N–carbon black for the oxygen reduction reaction in sulfuric acid. *Carbon* **2013**, *57*, 443–451.

80. Lopes, T.; Olivi, P. Non-precious Metal Oxygen Reduction Reaction Catalysts Synthesized via Cyanuric Chloride and *N*-Ethylamine. *Electrocatalysis* **2014**, *5*, 396–401.

81. Qiu, Y.; Yu, J.; Wu, W.; Yin, J.; Bai, X. Fe–N/C nanofiber electrocatalysts with improved activity and stability for oxygen reduction in alkaline and acid solutions. *J. Solid State Electrochem.* **2012**, *17*, 565–573.

82. Serov, A.; Robson, M.H.; Artyushkova, K.; Atanassov, P. Templated non-PGM cathode catalysts derived from iron and poly(ethyleneimine) precursors. *Appl. Catal. B* **2012**, *127*, 300–306.

83. Serov, A.; Robson, M.H.; Halevi, B.; Artyushkova, K.; Atanassov, P. Highly active and durable templated non-PGM cathode catalysts derived from iron and aminoantipyrine. *Electrochem. Commun.* **2012**, *22*, 53–56.

84. Serov, A.; Robson, M.H.; Smolnik, M.; Atanassov, P. Tri-metallic transition metal-nitrogen-carbon catalysts derived by sacrificial support method synthesis. *Electrochim. Acta* **2013**, *109*, 433–439.

85. Nallathambi, V.; Leonard, N.; Kothandaraman, R.; Barton, S.C. Nitrogen Precursor Effects in Iron-Nitrogen-Carbon Oxygen Reduction Catalysts. *Electrochem. Solid-State Lett.* **2011**, *14*, B55–B58.

86. Wu, Z.-S.; Yang, S.; Sun, Y.; Parvez, K.; Feng, X.; Müllen, K. 3D Nitrogen-Doped Graphene Aerogel-Supported Fe_3O_4 Nanoparticles as Efficient Electrocatalysts for the Oxygen Reduction Reaction. *J. Am. Chem. Soc.* **2012**, *134*, 9082–9085.

87. Chung, H.T.; Won, J.H.; Zelenay, P. Active and stable carbon nanotube/nanoparticle composite electrocatalyst for oxygen reduction. *Nat. Commun.* **2013**, *4*, 1922.

88. Kim, B.J.; Lee, D.U.; Wu, J.; Higgins, D.; Yu, A.; Chen, Z. Iron- and Nitrogen-Functionalized Graphene Nanosheet and Nanoshell Composites as a Highly Active Electrocatalyst for Oxygen Reduction Reaction. *J. Phys. Chem. C* **2013**, *117*, 26501–26508.

89. Zhang, S.; Liu, B.; Chen, S. Synergistic increase of oxygen reduction favourable Fe–N coordination structures in a ternary hybrid of carbon nanospheres/carbon nanotubes/graphene sheets. *Phys. Chem. Chem. Phys.* **2013**, *15*, 18482–18490.

90. Mo, Z.; Peng, H.; Liang, H.; Liao, S. Vesicular nitrogen doped carbon material derived from Fe_2O_3 templated polyaniline as improved non-platinum fuel cell cathode catalyst. *Electrochim. Acta* **2013**, *99*, 30–37.

91. Cao, R.; Thapa, R.; Kim, H.; Xu, X.; Gyu Kim, M.; Li, Q.; Park, N.; Liu, M.; Cho, J. Promotion of oxygen reduction by a bio-inspired tethered iron phthalocyanine carbon nanotube-based catalyst. *Nat. Commun.* **2013**, *4*, 2076.

92. Zhao, D.; Shui, J.-L.; Grabstanowicz, L.R.; Chen, C.; Commet, S.M.; Xu, T.; Lu, J.; Liu, D.-J. Electrocatalysts: Highly Efficient Non-Precious Metal Electrocatalysts Prepared from One-Pot Synthesized Zeolitic Imidazolate Frameworks. *Adv. Mater.* **2014**, *26*, 1093–1097.

93. Yuan, S.; Shui, J.-L.; Grabstanowicz, L.; Chen, C.; Commet, S.; Reprogle, B.; Xu, T.; Yu, L.; Liu, D.-J. A Highly Active and Support-Free Oxygen Reduction Catalyst Prepared from Ultrahigh-Surface-Area Porous Polyporphyrin. *Angew. Chem. Int. Ed.* **2013**, *52*, 8349–8353.

94. Wen, Z.; Ci, S.; Zhang, F.; Feng, X.; Cui, S.; Mao, S.; Luo, S.; He, Z.; Chen, J. Nitrogen-Enriched Core-Shell Structured Fe/Fe$_3$C–C Nanorods as Advanced Electrocatalysts for Oxygen Reduction Reaction. *Adv. Mater.* **2012**, *24*, 1399–1404.

95. Deng, D.; Yu, L.; Chen, X.; Wang, G.; Jin, L.; Pan, X.; Deng, J.; Sun, G.; Bao, X. Iron Encapsulated within Pod-like Carbon Nanotubes for Oxygen Reduction Reaction. *Angew. Chem. Int. Ed.* **2013**, *52*, 371–375.

96. Lee, J.-S.; Park, G.S.; Kim, S.T.; Liu, M.; Cho, J. A Highly Efficient Electrocatalyst for the Oxygen Reduction Reaction: N-Doped Ketjenblack Incorporated into Fe/Fe$_3$C-Functionalized Melamine Foam. *Angew. Chem. Int. Ed.* **2013**, *52*, 1026–1030.

97. Hu, Y.; Jensen, J.O.; Zhang, W.; Cleemann, L.N.; Xing, W.; Bjerrum, N.J.; Li, Q. Hollow Spheres of Iron Carbide Nanoparticles Encased in Graphitic Layers as Oxygen Reduction Catalysts. *Angew. Chem. Int. Ed.* **2014**, *53*, 3675–3679.

98. Peng, H.; Liu, F.; Liu, X.; Liao, S.; You, C.; Tian, X.; Nan, H.; Luo, F.; Song, H.; Fu, Z.; et al. Effect of Transition Metals on the Structure and Performance of the Doped Carbon Catalysts Derived From Polyaniline and Melamine for ORR Application. *ACS Catal.* **2014**, *4*, 3797–3805.

99. Wang, L.; Ambrosi, A.; Pumera, M. "Metal-Free" Catalytic Oxygen Reduction Reaction on Heteroatom-Doped Graphene is Caused by Trace Metal Impurities. *Angew. Chem. Int. Ed.* **2013**, *52*, 13818–13821.

100. Jaouen, F.; Herranz, J.; Lefèvre, M.; Dodelet, J.-P.; Kramm, U.I.; Herrmann, I.; Bogdanoff, P.; Maruyama, J.; Nagaoka, T.; Garsuch, A.; et al. Cross-Laboratory Experimental Study of Non-Noble-Metal Electrocatalysts for the Oxygen Reduction Reaction. *ACS Appl. Mater. Interfaces* **2009**, *1*, 1623–1639.

101. Herranz, J.; Jaouen, F.; Lefèvre, M.; Kramm, U.I.; Proietti, E.; Dodelet, J.-P.; Bogdanoff, P.; Fiechter, S.; Abs-Wurmbach, I.; Bertrand, P.; et al. Unveiling N-Protonation and Anion-Binding Effects on Fe/N/C Catalysts for O$_2$ Reduction in Proton-Exchange-Membrane Fuel Cells. *J. Phys. Chem. C* **2011**, *115*, 16087–16097.

102. Carver, C.T.; Matson, B.D.; Mayer, J.M. Electrocatalytic Oxygen Reduction by Iron Tetra-arylporphyrins Bearing Pendant Proton Relays. *J. Am. Chem. Soc.* **2012**, *134*, 5444–5447.

103. Matson, B.D.; Carver, C.T.; von Ruden, A.; Yang, J.Y.; Raugei, S.; Mayer, J.M. Distant protonated pyridine groups in water-soluble iron porphyrin electrocatalysts promote selective oxygen reduction to water. *Chem. Commun.* **2012**, *48*, 11100–11102.

104. Ramaswamy, N.; Tylus, U.; Jia, Q.; Mukerjee, S. Activity Descriptor Identification for Oxygen Reduction on Nonprecious Electrocatalysts: Linking Surface Science to Coordination Chemistry. *J. Am. Chem. Soc.* **2013**, *135*, 15443–15449.

105. Ferrandon, M.; Wang, X.; Kropf, A.J.; Myers, D.J.; Wu, G.; Johnston, C.M.; Zelenay, P. Stability of iron species in heat-treated polyaniline-iron-carbon polymer electrolyte fuel cell cathode catalysts. *Electrochim. Acta* **2013**, *110*, 282–291.

106. Wu, G.; Artyushkova, K.; Ferrandon, M.; Kropf, A.J.; Myers, D.; Zelenay, P. Performance Durability of Polyaniline-derived Non-precious Cathode Catalysts. *ECS Trans.* **2009**, *25*, 1299–1311.

107. Varcoe, J.R.; Atanassov, P.; Dekel, D.R.; Herring, A.M.; Hickner, M.A.; Kohl, P.A.; Kucernak, A.R.; Mustain, W.E.; Nijmeijer, K.; Scott, K.; *et al.* Anion-exchange membranes in electrochemical energy systems. *Energy Environ. Sci.* **2014**, *7*, 3135–3191.

108. Katzfuß, A.; Poynton, S.; Varcoe, J.; Gogel, V.; Storr, U.; Kerres, J. Methylated polybenzimidazole and its application as a blend component in covalently cross-linked anion-exchange membranes for DMFC. *J. Membr. Sci.* **2014**, *465*, 129–137.

Nanoscale Alloying in Electrocatalysts

Shiyao Shan, Jinfang Wu, Ning Kang, Hannah Cronk, Yinguang Zhao, Wei Zhao, Zakiya Skeete, Pharrah Joseph, Bryan Trimm, Jin Luo and Chuan-Jian Zhong

Abstract: In electrochemical energy conversion and storage, existing catalysts often contain a high percentage of noble metals such as Pt and Pd. In order to develop low-cost electrocatalysts, one of the effective strategies involves alloying noble metals with other transition metals. This strategy promises not only significant reduction of noble metals but also the tunability for enhanced catalytic activity and stability in comparison with conventional catalysts. In this report, some of the recent approaches to developing alloy catalysts for electrocatalytic oxygen reduction reaction in fuel cells will be highlighted. Selected examples will be also discussed to highlight insights into the structural and electrocatalytic properties of nanoalloy catalysts, which have implications for the design of low-cost, active, and durable catalysts for electrochemical energy production and conversion reactions.

Reprinted from *Catalysts*. Cite as: Shan, S.; Wu, J.; Kang, N.; Cronk, H.; Zhao, Y.; Zhao, W.; Skeete, Z.; Joseph, P.; Trimm, B.; Luo, J.; Zhong, C.-J. Nanoscale Alloying in Electrocatalysts. *Catalysts* **2015**, *5*, 1465–1478.

1. Introduction

The design of active, stable and low-cost catalysts is essential for many reactions in electrochemical energy production, conversion and storage. Metal nanoparticles, especially alloy nanoparticles, have attracted a great deal of interest in both experimental and theoretical studies [1–3]. It is the nanoscale size range over which metal nanoparticles undergo a transition from metallic to atomic properties which leads to unique electronic and catalytic properties different from their bulk counterparts. Significant advances have been made in harnessing the nanoscale catalytic properties in energy and environmental fronts [3]. However, challenges remain, especially in preparation and characterization of active, robust, low-cost nanocatalysts with controllable sizes, shapes, compositions and structures.

An alloy is a mixture of two or more metallic species, which can exist either in a complete solid solution state exhibiting a single phase characteristics or in a partial or phase-segregated solid solution state with multiple phases. Nanoalloy (NA) differs from bulk alloys in several significant aspects in terms of mixing patterns and geometric shapes. There are different types of mixing patterns, two of which include completely phase segregated NAs where the different phases share either an extended mixed interface or a very limited number of hetero-nuclear metal bonds, and mixed NAs with chemically ordered/disordered structures. The degree of segregation, mixing and atomic ordering depends on a number of factors, including relative strengths of differences in atomic sizes, surface energies of the component element, homoatomic *vs.* heteroatomic bonds, charge transfer between the different atomic species, and strength of binding to surface ligands or support materials. The observed atomic arrangement for a particular alloy depends critically on the balance of the preparation and usage conditions. By alloying, changes including atomic structure, physical and optical properties, and chemical properties could lead to two major effects: (i) ensemble effect where hetero-nuclear metals geometrically arranged in the favor of certain properties (e.g., catalytic process); (ii) ligand effect where electronic charge transfer between hetero-nuclear metals induces the change of functionality of metals (e.g., molecule adsorption properties) [4]. It is noted that the change of geometric arrangement by varying alloy composition or structural ordering/disordering cannot be done without changing electronic atmosphere. In catalysis, alloying hetero-nuclear metals plays a role in several different aspects, including (i) activation of the main components for enhanced activity; (ii) activation of successive reaction for the enhancement of overall reactions; (iii) removal of poisonous intermediates to facilitate the reactions; (iv) inhibition of certain intermediates and byproducts; and (v) reactants storage. [5] In addition, the surface structures of NAs could be very complex due to the enrichment of certain element in the core or shell. In the case of metal dissolution ("dealloying") in the presence of acidic electrolytes, often referred to as Pt-skin structure formation [6,7], the details of the noble metal skin could be influenced strongly by the structural types of the NAs, the understanding of which in terms of structural evolution, noble metal skin or d-band center shift has attracted increasing interests in electrocatalysis.

Supported metal nanoparticles from traditional preparative methods have been well demonstrated for various catalytic reactions [2,3,8]. In the last decade, new approaches to the synthesis of molecularly capped metal nanoparticles for the preparation of catalysts have been rapidly emerging (see Figure 1). While some of the catalysts exploit the functional groups from the capping shell of the nanoparticles, most others explore the surface active sites over the metal nanoparticles either after removing the capping layers [9] or through open channels of the capping layers [10]. In addition, the nanoscale facet is an important factor in catalysis. For noble metal (e.g., Pt, Pd) alloyed with other transitions metals (e.g., Ni, Cu, Co, *etc.*), a volcano curve has been observed for certain metal ratios (e.g., 1:1, 1:3, or 3:1) in binary nanocatalysts [2]. Besides various nanostructural design parameters [11], the understanding of how these factors play an important role on catalytic properties is increasingly important.

In this article, some of the recent findings in the exploration of nanoscale alloying degree for the preparation of the supported nanoalloy catalysts for catalytic and electrocatalytic reactions [12–17] will be highlighted. One important focus is the understanding of the structural correlation of nanoscale alloying properties with the electrocatalytic properties.

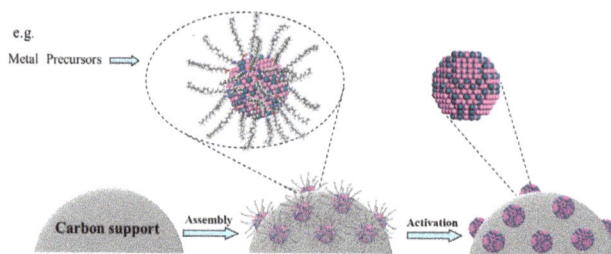

Figure 1. Illustration of the synthesis, assembly, and activation of nanoalloy particles for the preparation of the supported electrocatalysts.

2. Synthesis and Preparation

The synthesis of molecularly-capped metal nanoparticles as building blocks for engineering the nanoscale catalytic materials takes advantage of diverse attributes, including monodispersity, processability, and stability in terms of size, shape, composition, phase and surface properties in comparison with traditional approaches [18]. One important approach involves core-shell type synthesis [19]. The core and shell are from different matters in close interaction, including inorganic/organic and inorganic/inorganic combinations [20,21]. The synthesis of metal nanoparticles in the presence of organic capping agents to form encapsulated metal nanoparticles has demonstrated promises for preparing nanocatalysts [20,21]. The coupling of molecularly-mediated synthesis of nanoparticles and post-synthesis thermochemical processing under controlled temperatures and atmospheres have demonstrated effectiveness in the preparation of nanocatalysts. In comparison with other methods such as plasmatic cleaning or chemical cleaning [22], thermochemical processing strategy is not only effective in removing the encapsulation, but also in refining the nanostructural parameters. The combination of the molecular encapsulation based synthesis and thermochemical processing strategies typically involves a sequence of steps for the preparation of nanoalloy catalysts: (1) chemical synthesis of the metal nanocrystal cores capped with ligands, (2) assembly of the encapsulated nanoparticles on supporting materials (e.g., carbon powders, TiO_2 or SiO_2), and (3) thermal treatment of the supported nanoparticles [12–17]. The size and composition of the nanoparticles produced by thermochemical processing are controllable. As shown for a series of binary and ternary alloy nanoparticle systems in Table 1 [12–17,23–35], the catalysts prepared by the molecularly-mediated synthesis and thermochemical processing methods have demonstrated enhanced catalytic and electrocatalytic properties for oxygen reduction reaction (ORR), methanol oxidation reaction (MOR), and ethanol oxidation reaction (EOR), *etc.*

Table 1. Examples of alloy nanoparticles and catalysts prepared by molecularly-mediated synthesis and thermochemical processing methods.

Catalysts	Synthesis method	Catalytic reactions	Refs
	Bimetallic nanoalloys		
PtNi NPs (2–8 nm)/C, TiO$_2$, or SiO$_2$	*Precursor*: Pt(acac)$_2$, Ni(acac)$_2$; *Capping agent*: oleic acid (OA), oleylamine (OAM); *Reducing agent*: 1,2-hexadecanediol (HDD); *Solvent*: octyl ether (OE)	ORR, CO oxidation	[12,23]
PtCo NPs (2–8 nm)/C, TiO$_2$ or SiO$_2$	*Precursor*: Pt(acac)$_2$, Co$_2$(CO)$_8$; *Capping agent*: oleic acid and oleylamine; *Reducing agent*: 1,2-hexadecanediol; *Solvent*: octyl ether	ORR, CO oxidation	[12,23]
PtRu (~5 nm)/C	*Precursor*: Pt(acac)$_2$, Ru(acac)$_2$; *Capping agent*: oleic acid and oleylamine; *Reducing agent*: 1,2-hexadecanediol; *Solvent*: octyl ether	EOR, CO oxidation	[24]
AuCu (~5 nm)/C, SiO$_2$	*Precursor*: HAuCl$_4$, CuCl$_2$; *Capping agent*: 1-decanethiol (DT); Tetraoctylammonium bromide (TOABr); *Reducing agent*: NaBH$_4$; *Solvent*: H$_2$O and Toluene	CO oxidation	[14]
AuCu (4–8 nm)/C	*Precursor*: Au NPs and Cu NPs; *Capping agent*: 1-decanethiol (DT); Tetraoctylammonium bromide (TOABr); *Method*: Thermally aggregated growth	CO oxidation	[14]
PdNi (7–10 nm)/C	*Precursor*: Pd(acac)$_2$, Ni(acac)$_2$; *Capping agent*: oleic acid and oleylamine; *Reducing agent*: 1,2-hexadecanediol; *Solvent*: octyl ether or benzyl ether	ORR, CO oxidation	[25]
PdCu (7–10 nm)/C	*Precursor*: Pd(acac)$_2$, Cu(acac)$_2$; *Capping agent*: oleic acid and oleylamine; *Reducing agent*: 1,2-hexadecanediol; *Solvent*: octyl ether or benzyl ether	EOR, CO oxidation	[26]
AuPt (~4–5 nm)/C	*Precursor*: HAuCl$_4$, HPtCl$_4$; *Capping agent*: DT, OAM/OA; *Reducing agent*: NaBH$_4$; *Solvent*: H$_2$O and Toluene;	ORR, MOR	[13,27,28]
	Trimetallic Nanoalloys		
PtNiCo (3–5 nm)/C, TiO$_2$, and SiO$_2$	*Precursor*: Pt(acac)$_2$, Ni(acac)$_2$, Co(acac)$_3$; *Capping agent*: oleic acid and oleylamine; *Reducing agent*: 1,2-hexadecanediol; *Solvent*: octyl ether	ORR, CO oxidation	[12,29,30]
PtVCo (3–5 nm)/C	*Precursor*: Pt(acac)$_2$, VO(acac)$_2$, Co(acac)$_3$; *Capping agent*: oleic acid and oleylamine; *Reducing agent*: 1,2-hexadecanediol; *Solvent*: octyl ether	ORR, CO oxidation	[17,31]
PtNiFe (3–5 nm)/C	*Precursor*: Pt(acac)$_2$, Ni(acac)$_2$, Fe(CO)$_5$; *Capping agent*: oleic acid and oleylamine; *Reducing agent*: 1,2-hexadecanediol; *Solvent*: octyl ether	ORR, CO oxidation	[32,33]
PtIrCo (3–5 nm)/C	*Precursor*: Pt(acac)$_2$, Ir$_4$(CO)$_{12}$, Co(acac)$_2$; *Capping agent*: oleic acid and oleylamine; *Reducing agent*: 1,2-hexadecanediol; *Solvent*: octyl ether	ORR, CO oxidation	[15,17,34]
PtVFe (3–5 nm)/C	*Precursor*: Pt(acac)$_2$, VO(acac)$_2$, Fe(CO)$_5$; *Capping agent*: oleic acid and oleylamine; *Reducing agent*: 1,2-hexadecanediol; *Solvent*: octyl ether	ORR,	[35,36]

3. Examples of Nanoalloy Electrocatalysts

Pt- or Pd-based nanoalloys have been extensively explored as electrocatalysts for electrocatalytic ORR, which is an important reaction in proton exchange membrane fuel cell (PEMFC), an electrochemical energy conversion device that converts hydrogen at the anode and oxygen at the cathode through a membrane electrode assembly (MEA) into water and produce electricity. The desired reaction pathway in the cathode of a PEMFC is $4e^-$ reduction reaction of oxygen. During PEMFC reaction process, the voltage is the summation of the thermodynamic potential E_{Nernst}, the activation overpotential η_{act} (from both anode and cathode overpotentials, *i.e.*, $\eta_{act(cathode)} - \eta_{act(anode)}$), and the ohmic overpotential η_{ohmic}. The thermodynamic potential is governed by Nernst equation in terms of the E_0 (1.23 V) and the operating concentrations ($P(H_2)$ and $P(O_2)$), the activation overpotential is dependent on the electrode kinetics in terms of current flow, and the overpotential associated with catalyst activity ($\eta_{act(catalyst)}$). The overpotential $\eta_{act(catalyst)}$ is large mainly attributed to the sluggish activity of ORR. The adsorption of O_2 over Pt surface could produce Pt–O or Pt–OH in a dissociative adsorption, which constitutes a four-electron reduction pathway forming water, or Pt–O_2^- or Pt–O_2H in an associative adsorption which often proceeds in a two-electron reduction pathway forming hydrogen peroxide. Although the understandings based on Pt skin on an alloy or dealloyed surface can explain partially some of the experimental facts, the exploration of how Pt–O or –OH intermediate species would influence the overall ORR by varying their binding strength and the formation and removal of Pt–O/Pt–OH species are known to play a key role in the overall electrocatalytic ORR over Pt-alloy catalysts [37]. The rational design of Pt-alloys involving transition metals (M/M′) could create a bifunctional (or multifunctional) synergy for the formation and removal of Pt–O or Pt–OH species. For ternary catalysts, the introduction of a second M′ into Pt-M alloy may further lead to a manipulation of the surface oxophilicity by maneuvering –O/–OH species over M and M′ sites through structural or compositional manipulation [12,15,17,28–34]. The understanding of how Pt–OH and Pt–O binding energies can be tuned by the M/M′ oxophilicity would aid the design of the alloying metals for synergistic formation and removal of Pt–OH species in correlation with the structural and chemical complexity of the nanoalloys.

3.1. Bimetallic Nanoalloy Catalysts

Bimetallic nanoalloy catalysts derived from combinations of two heterometals often exhibit unique bifunctional or other physical and chemical properties. A strong correlation between the size, structure and catalytic activities was revealed over several interesting systems, e.g., PtNi, PtCo and AuPt [12,13,23,27,28]. Gold-platinum (AuPt) nanoalloys serve as an intriguing system in terms of the unique synergistic properties [12,27,28]. In contrast to the bulk counterpart which displays a miscibility gap at 20%–90% Au, nanoscale AuPt particles synthesized by wet chemical methods has shown alloy properties. The morphology and alloy structures are controllable, as shown by the example of $Au_{22}Pt_{78}$ nanoparticles on carbon support (Figure 2A). The observation of the indicated lattice fringes, 0.235 nm, corresponding to 111 planes, indicates that the carbon-supported nanoparticles are highly crystalline. Carbon-supported AuPt nanoparticles have been shown to exhibit alloying characteristics and possess a uniform distribution of the two metals across the entire nanoparticles. The subtle increase of the particle sizes for the thermochemcally-treated carbon-supported $Au_{22}Pt_{78}$ nanoparticles was due to the thermal sintering of the nanoparticles.

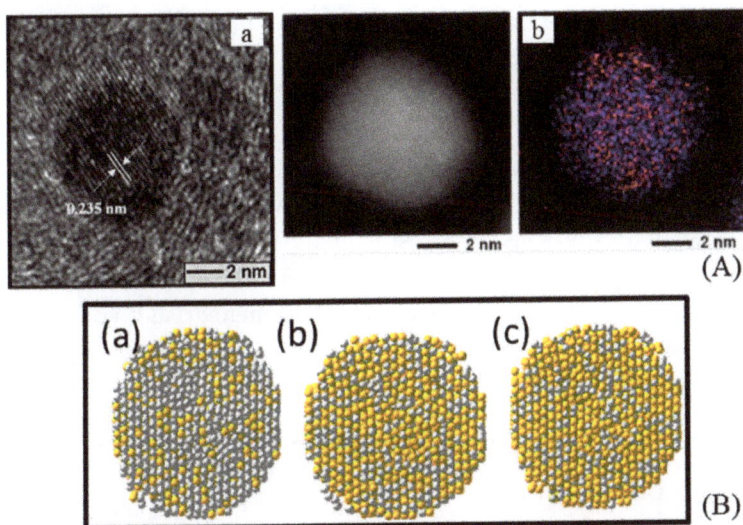

Figure 2. (A) HRTEM (a) and EDS (b) composition mapping for $Au_{22}Pt_{78}$/C (red: Pt, Blue: Au) nanoparticles; (**B**) Cross sections of 5.1 nm Pt-Au particles (about 5000 atoms) with random alloy structure: (**a**) $Pt_{77}Au_{33}$; (**b**) $Pt_{51}Au_{49}$ and (**c**) $Pt_{40}Au_{60}$.Pt atoms are in gray, Au in yellow. ((**B**) reproduced from reference [13] with permission. Copyright 2012, American Chemical Society; (**A**) reproduced from reference [27] with permission. Copyright 2010, American Chemical Society).

The detailed nanoscale alloying characteristics is recently evidenced by studies using element-specific resonant high energy X-ray diffraction coupled to pair distribution function analysis (HE-XRD/PDF) [12]. This technique, aided by Reverse Monte Carlo simulation (RMC) modeling, has provided an atomic-scale insight into the alloy structures of AuPt nanoparticles (Figure 2B). Pure Au and Pt nanoparticles are used to produce model configurations for Au_nPt_{100-n} nanoparticles ($n = 40, 51, 77$) where Au and Pt atoms show various patterns of chemical order-disorder effects.

From the modeling, it was found that the alloying of Pt and Au occurs not only within a wide range of Pt-Au concentrations but is also stable in nanoparticles of different sizes.

The electrocatalytic ORR activities of AuPt nanoalloys have been assessed by rotating disk electrode (RDE) measurements [27]. As shown by RDE curves in Figure 3 for $Au_{22}Pt_{78}$/C and $Au_{49}Pt_{51}$/C catalysts, there are clear differences of the reduction currents in the kinetic region (0.8–0.9 V *vs.* NHE). These differences demonstrated that both the bimetallic composition and the phase structures had profound effects on the electrocatalytic activity. It is evident that the mass activity depends on both thermal treatment temperature and condition (Figure 3 insert). The data for $Au_{22}Pt_{78}$/C showed an increase of mass activity to a maximum at 400 °C and further decrease with increasing temperature. The decrease of the activity with temperature is consistent with the findings of the increased phase segregation and the Pt core-Au shell formation by experimental HRXRD/PDF data. The temperature for the maximum activity was also found to depend on the bimetallic composition, as supported by the observations of a maximum activity at 400 °C for $Au_{22}Pt_{78}$/C and a maximum activity at 600–700 °C for $Au_{49}Pt_{51}$/C. A combination of lattice parameter and surface structural effects as a result of the differences in composition and treatment conditions is believed to be operative. The observed differences between $Au_{22}Pt_{78}$/C and $Au_{49}Pt_{51}$/C catalysts indicate that there exists an optimized surface structure with an appropriate Pt–O bonding strength for achieving the enhanced electrocatalytic activity.

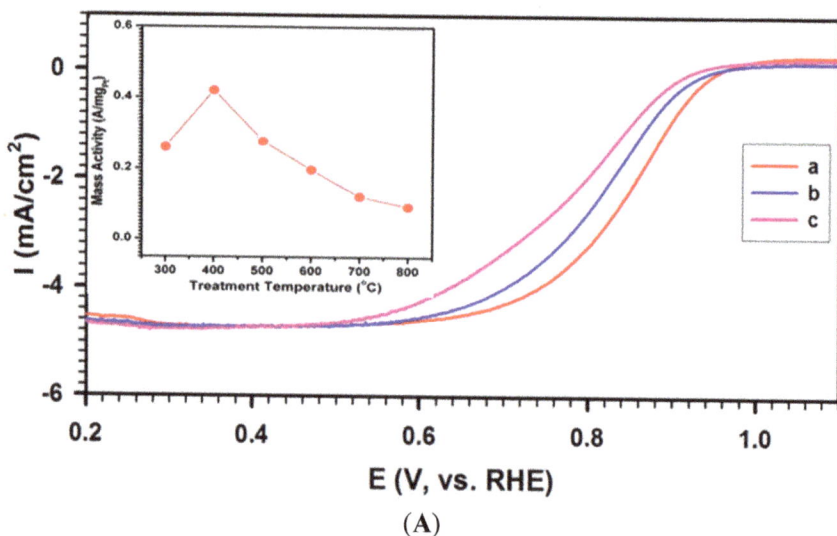

(A)

Figure 3. *Cont.*

68

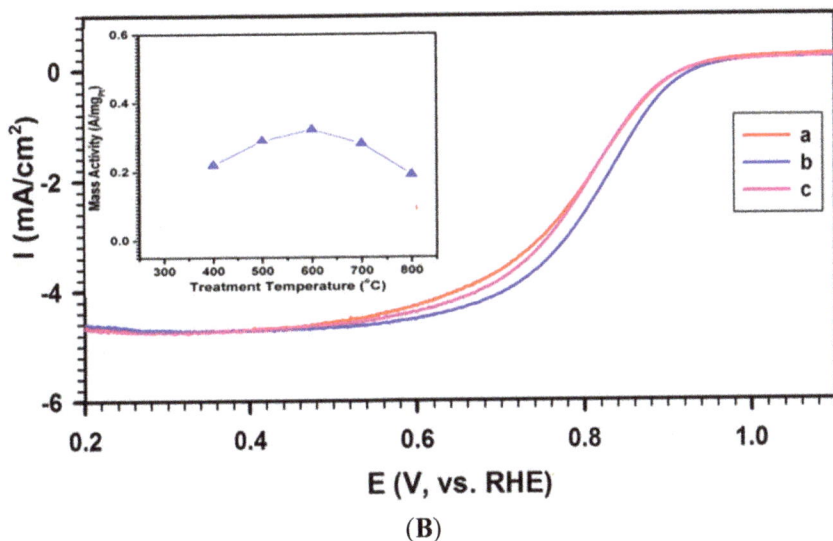

Figure 3. RDE curves for ORR for $Au_{22}Pt_{78}/C$ (**A**) and $Au_{49}Pt_{51}/C$ (**B**) catalysts treated under H_2 for 30 min (normalized for comparison) at 400 (**a**), 600 (**b**), and 800 °C (**c**). (Glassy carbon electrode (0.196 cm^2); 0.5 M H_2SO_4 saturated with O_2; catalyst loading: 10 µg; scan rate: 10 mV/s; speed: 1600 rpm). (reproduced from reference [27] with permission. Copyright 2010, American Chemical Society).

3.2. Trimetallic Nanoalloy Catalysts

In comparison with bimetallic systems, ternary nanoalloy catalysts provide an increased degree of structural tunability. In addition to the obvious tunability in nanoscale alloying, the manipulation of surface oxophilicity is demonstrated by the introduction of a second metal M' into Pt-M alloys. In many cases, ternary PtMM' catalysts, where M, M' = Ni, Co, Fe, V, Ir, *etc.*, have demonstrated enhanced electrocatalytic activities and stabilities in comparison with commercial Pt/C and their binary counterparts [12,15,17,29–36]. These aspects can be illustrated by studies of the ternary nanoalloy of PtIrCo in comparison with its binary counterparts [15,34]. As an example, the $Pt_{25}Ir_{20}Co_{55}$ nanoalloys prepared by the molecularly-mediated synthesis display a size of 2.5 ± 0.2 nm (Figure 4A). Based on HE-XRD/PDF characterization of PtIrCo/C and its binary counterparts (PtIr/C and PtCo/C) treated under H_2 at 400 and 800 °C, the detailed structural ordering and atomic configuration in the nanoparticles can modelled by RMC simulation (Figure 4B). Each of the configurations have the real stoichiometry and size of the nanoalloy and atomic PDFs computed from the configurations match the experimental PDF data very well.

$Pt_{45}Ir_{55}$ catalyst treated at 400 °C is a random alloy of Pt and Ir whereas that at 800 °C tends to segregate into a Ir-core and Pt-surface-enriched structure. This finding is qualitatively in agreement with the XPS based analysis of the relative surface composition, which showed a 16% increase in Pt upon treatment at 800 °C. $Pt_{73}Co_{27}$ catalyst treated at 400 °C features an alloy where Co atoms show some preference to the center of the nanoparticles whereas that at 800 °C, features an alloy with Co atoms being somewhat closer to the surface of the particle. In comparison, the $Pt_{25}Ir_{20}Co_{55}$ catalyst treated at 400 °C features an alloy where Co and Ir species tend to occupy the inner part of the nanoparticles while Pt atoms show some preference to the surface of the nanoparticles. The ternary catalyst treated at 800 °C features a rather random type of alloy where Co, Pt and Ir atoms are almost uniformly distributed across the nanoparticles, a finding that was qualitatively in agreement with the small changes derived from the XPS analysis of the relative surface composition. These results reveal that the atomic distribution across the nanoparticles depends strongly on the binary/ternary composition and the thermochemical treatment temperature.

The electrocatalytic activities of PtIrCo catalysts with different compositions for ORR were measured using the RDE method (e.g., $Pt_{65}Ir_{11}Co_{24}/C(a)$, $Pt_{40}Ir_{28}Co_{32}/C(b)$, and $Pt_{25}Ir_{20}Co_{55}/C(c)$). In comparison with the mass activity for Pt/C catalysts, the mass activity was increased by a factor of 2–4 for these catalysts. There is a clear trend showing the increase of the mass activity with the increase of Co% and the decrease of Pt% in the nanoparticles. In comparison with the specific activity for Pt/C catalysts, the specific activity was increased by a factor of ~3 for the ternary catalysts. In comparison with its binary counterparts (PtCo and PtIr), the mass activity for $Pt_{25}Ir_{20}Co_{55}/C$ nanocatalyst showed an increase by factor of ~2 (Figure 5A) [15]. Based on the detailed atomic structural data, the substantially shorter metal-metal distances in the ternary nanocatalysts are believed to be one of the key factors responsible for the improved catalytic properties. The increase in SA from lower to higher temperature (e.g., from 400 °C to 800 °C) for the ternary nanoalloys is also likely due to the further decrease in the metal-metal distances and the changes in coordination numbers. In addition to a favorable change in Co-Pt first coordination number, there are also changes in Co-Ir, Pt-Ir and Ir-Ir coordination numbers indicating an increased degree of alloying. Moreover, the introduction of Ir in PtCo to form a ternary system was indeed shown to increase stability of the electrocatalytic activity indicating the important role of the addition of Ir. The mass activity is the highest for the ternary catalyst among the three catalysts (see Figure 5A insert). The 2× increase of specific activity for the ternary catalyst in comparison with the relatively small increase for PtIr indicates the importance of adding a third metal with greater oxophilicity to the alloy. The marked enhancement of the activity ternary nanoparticles is believed to be linked to the decrease in the 1st Metal-Metal distances and the formation of alloy featuring either an Co-Ir core with Pt rich surface or a uniform distribution of Co, Pt and Ir species across the entire nanoparticle.

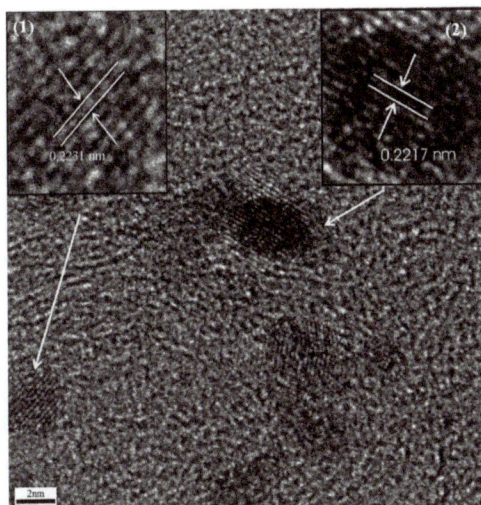

(A)

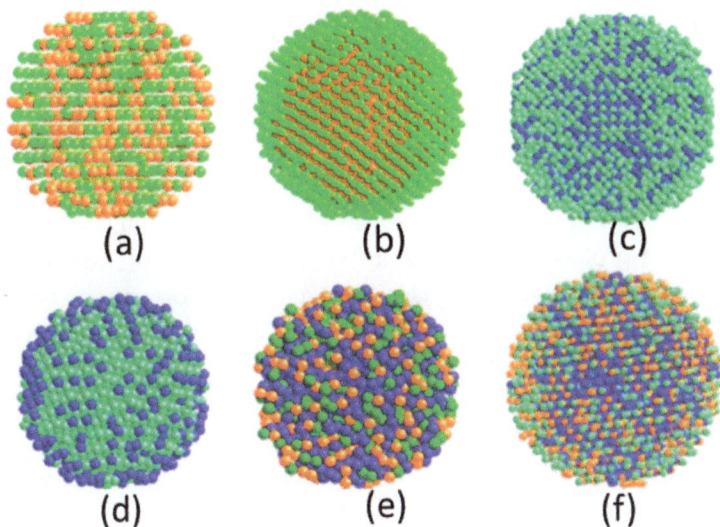

<table>
<tr><td>(a)</td><td>(b)</td><td>(c)</td></tr>
<tr><td>(d)</td><td>(e)</td><td>(f)</td></tr>
</table>

(B)

Figure 4. (A) HR-TEM for $Pt_{25}Ir_{20}Co_{55}$ and (B) RMC constructed models for $Pt_{45}Ir_{55}$, $Pt_{73}Co_{27}$, and $Pt_{25}Ir_{20}Co_{55}$ processed at 400 °C (**a, c, e**, respectively) and 800 °C (**b, d, f**, respectively). (Pt atoms: green, Ir atoms: orange, and Co atoms: blue). Note that the sizes of atoms are drawn not to scale to fit in the picture frame. ((**A**) reproduced from reference [15] with permission. Copyright 2013, American Chemical Society, (**B**) reproduced from reference [34] with permission. Copyright 2012, American Chemical Society).

71

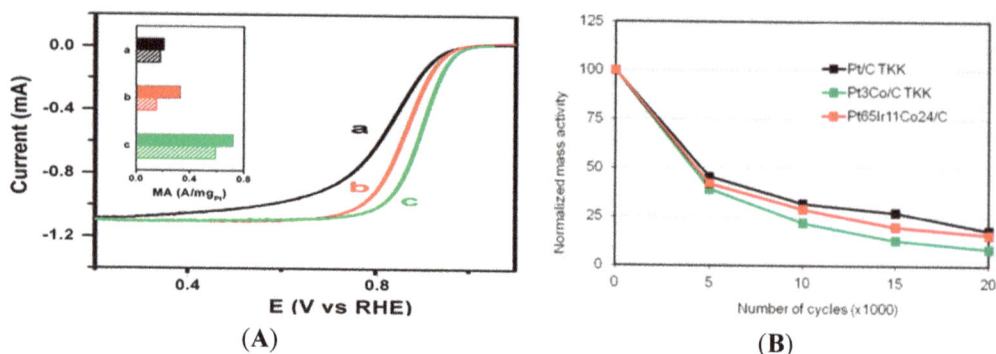

Figure 5. Electrocatalytic activities for ORR in O_2-saturated 0.1 M $HClO_4$: (**A**) RDE curves for 400 °C (solid bar) and 800 °C (dash bar) treated catalysts: $Pt_{45}Ir_{55}/C$ (**a**), $Pt_{73}Co_{27}/C$ (**b**), and $Pt_{25}Ir_{20}Co_{55}/C$ (**c**). Insert: Mass activities extracted from RDE curves for the same catalysts. (**B**) Durability plots of normalized mass activities as a function of the number of potential cycles (ranging from 0 to 20,000 cycles) for different catalysts upon potential cycling ((**A**) reproduced from reference [15] with permission. Copyright 2013, American Chemical Society, (**B**) reproduced from reference [34] with permission. Copyright 2012, American Chemical Society).

The durability of the PtIrCo catalysts was also found to show an improvement in comparison with its binary counterparts. This is substantiated by the durability data for the catalysts in the O_2-saturated 0.1 M $HClO_4$ as a function of square-wave potential cycling protocol [34]. Most of the mass activity loss for all the samples occurred during the initial 5000 cycles (Figure 5B). In comparison with those for the commercial catalysts, the rate of the mass activity loss of $Pt_{65}Ir_{11}Co_{24}/C$ was comparable to that of Pt/C and slightly lower than that of Pt_3Co/C. After 20,000 cycles, the mass activity of $Pt_{65}Ir_{11}Co_{24}/C$ is found to be higher than that of Pt_3Co/C. The ternary nanoalloy catalyst synthesized by the molecularly-mediated synthesis and thermochemical processing method has durability comparable to that of commercial catalysts upon the severe potential cycling.

4. Summary and Perspectives

In summary, the ability to control the nanoscale alloying structures is essential for understanding the enhanced electrocatalytic activities of Pt or Pd based nanoalloys. It is the unique nanoscale phenomena in terms of atomic-scale alloying, interatomic distances, metal coordination structures, structural/chemical ordering, and phase states that operate synergistically in activating oxygen and maneuvering surface oxygenated species. Understanding this synergy is important for the design of catalysts with high activity with a significantly-reduced use of noble metals [38,39]. In addition to studies aimed at further lowering the noble metal content in the nanoalloy catalysts, future work is needed in the area of theoretical computation and modelling to understand how the structural-catalytic synergy are influenced by the binary or ternary metal composition. This understanding will also guide the development of the ability to enhance the stability of metal components in the nanoalloy catalysts under the electrocatalytic operation conditions. In addition, *in situ* experiments will be very useful to probe the structural evolution processes such as the de-alloying process in the electrolyte and atomic-scale rearrangements leading to changes in size, shape, or surface energy. With recent advents in using synchrotron X-ray based techniques for the study of various catalyst systems, new insights are expected for elucidating the detailed factors controlling the activity and stability of nanoalloy catalysts, which will further advance the endeavor of electrochemical energy conversion and storage.

Acknowledgments: The authors express their gratitude to collaborators who have made contributions to the work described in this article. The research work was supported by the DOE-BES (DE-SC0006877) and NSF (CMMI-1100736).

Author Contributions: Shiyao Shan was the leading author from the initial draft writing to the finalization of the manuscript, during which all authors contributed as a team to the manuscript revisions, the literature reading, and the proof reading. All authors have the expertise in the subject area of the materials described in the manuscript.

Conflicts of Interest: The authors declare no conflict of interest.

References

1. Wu, J.B.; Yang, H. Platinum-Based Oxygen Reduction Electrocatalysts. *Acc. Chem. Res.* **2013**, *46*, 1848–1857.

2. Guo, S.J.; Zhang, S.; Sun, S.H. Tuning Nanoparticle Catalysis for the Oxygen Reduction Reaction. *Angew. Chem. Int. Edit.* **2013**, *52*, 8526–8544.

3. Yu, W.T.; Porosoff, M.D.; Chen, J.G.G. Review of Pt-Based Bimetallic Catalysis: From Model Surfaces to Supported Catalysts. *Chem. Rev.* **2012**, *112*, 5780–5817.

4. Tao, F.; Zhang, S.R.; Nguyen, L.; Zhang, X.Q. Action of bimetallic nanocatalysts under reaction conditions and during catalysis: Evolution of chemistry from high vacuum conditions to reaction conditions. *Chem. Soc. Rev.* **2012**, *41*, 7980–7993.

5. Shi, J.L. On the Synergetic Catalytic Effect in Heterogeneous Nanocomposite Catalysts. *Chem. Rev.* **2013**, *113*, 2139–2181.

6. Wang, C.; Markovic, N.M.; Stamenkovic, V.R. Advanced Platinum Alloy Electrocatalysts for the Oxygen Reduction Reaction. *ACS Catal.* **2012**, *2*, 891–898.

7. Watanabe, M.; Tryk, D.A.; Wakisaka, M.; Yano, H.; Uchida, H. Overview of recent developments in oxygen reduction electrocatalysis. *Electrochim. Acta* **2012**, *84*, 187–201.

8. Cui, C.H.; Yu, S.H. Engineering Interface and Surface of Noble Metal Nanoparticle Nanotubes toward Enhanced Catalytic Activity for Fuel Cell Applications. *Acc. Chem. Res.* **2013**, *46*, 1427–1437.

9. Aiken, J.D.; Finke, R.G. A review of modern transition-metal nanoclusters: Their synthesis, characterization, and applications in catalysis. *J. Mol. Catal. A* **1999**, *145*, 1–44.

10. Qian, H.F.; Zhu, M.Z.; Wu, Z.K.; Jin, R.C. Quantum Sized Gold Nanoclusters with Atomic Precision. *Acc. Chem. Res.* **2012**, *45*, 1470–1479.

11. Hostetler, M.J.; Zhong, C.J.; Yen, B.K.H.; Anderegg, J.; Gross, S.M.; Evans, N.D.; Porter, M.; Murray, R.W. Stable, monolayer-protected metal alloy clusters. *J. Am. Chem. Soc.* **1998**, *120*, 9396–9397.

12. Yang, L.F.; Shan, S.Y.; Loukrakpam, R.; Petkov, V.; Ren, Y.; Wanjala, B.N.; Engelhard, M.H.; Luo, J.; Yin, J.; Chen, Y.S.; *et al.* Role of Support-Nanoalloy Interactions in the Atomic-Scale Structural and Chemical Ordering for Tuning Catalytic Sites. *J. Am. Chem. Soc.* **2012**, *134*, 15048–15060.

13. Petkov, V.; Wanjala, B.N.; Loukrakpam, R.; Luo, J.; Yang, L.F.; Zhong, C.J.; Shastri, S. Pt-Au Alloying at the Nanoscale. *Nano Lett.* **2012**, *12*, 4289–4299.

14. Yin, J.; Shan, S.Y.; Yang, L.F.; Mott, D.; Malis, O.; Petkov, V.; Cai, F.; Ng, M.S.; Luo, J.; Chen, B.H.; *et al.* Gold-Copper Nanoparticles: Nanostructural Evolution and Bifunctional Catalytic Sites. *Chem. Mater.* **2012**, *24*, 4662–4674.

15. Loukrakpam, R.; Shan, S.Y.; Petkov, V.; Yang, L.F.; Luo, J.; Zhong, C.J. Atomic Ordering Enhanced Electrocatalytic Activity of Nanoalloys for Oxygen Reduction Reaction. *J. Phys. Chem. C* **2013**, *117*, 20715–20721.

16. Petkov, V.; Ren, Y.; Shan, S.Y.; Luo, J.; Zhong, C.J. A distinct atomic structure-catalytic activity relationship in 3–10 nm supported Au particles. *Nanoscale* **2014**, *6*, 532–538.

17. Shan, S.Y.; Petkov, V.; Yang, L.F.; Mott, D.; Wanjala, B.N.; Cai, F.; Chen, B.H.; Luo, J.; Zhong, C.J. Oxophilicity and Structural Integrity in Maneuvering Surface Oxygenated Species on Nanoalloys for CO Oxidation. *ACS Catal.* **2013**, *3*, 3075–3085.

18. Galow, T.H.; Drechsler, U.; Hanson, J.A.; Rotello, V.M. Highly reactive heterogeneous Heck and hydrogenation catalysts constructed through "bottom-up" nanoparticle self-assembly. *Chem. Comm.* **2002**, 1076–1077.

19. Brust, M.; Walker, M.; Bethell, D.; Schiffrin, D.J.; Whyman, R. Synthesis of Thiol-Derivatized Gold Nanoparticles in a 2-Phase Liquid-Liquid System. *J. Chem. Soc. Chem. Comm.* **1994**, *7*, 801–802.

20. Shao, M.H.; Shoemaker, K.; Peles, A.; Kaneko, K.; Protsailo, L. Pt Mono layer on Porous Pd-Cu Alloys as Oxygen Reduction Electrocatalysts. *J. Am. Chem. Soc.* **2010**, *132*, 9253–9255.

21. Schadt, M.J.; Cheung, W.; Luo, J.; Zhong, C.J. Molecularly tuned size selectivity in thermal processing of gold nanoparticles. *Chem. Mater.* **2006**, *18*, 5147–5149.

22. Zhou, Y.; Wang, Z.Y.; Liu, C.J. Perspective on CO oxidation over Pd-based catalysts. *Catal. Sci. Technol.* **2015**, *5*, 69–81.

23. Loukrakpam, R.; Luo, J.; He, T.; Chen, Y.S.; Xu, Z.C.; Njoki, P.N.; Wanjala, B.N.; Fang, B.; Mott, D.; Yin, J.; *et al.* Nanoengineered PtCo and PtNi Catalysts for Oxygen Reduction Reaction: An Assessment of the Structural and Electrocatalytic Properties. *J. Phys. Chem. C* **2011**, *115*, 1682–1694.

24. Prasai, B.; Ren, B.; Shan, S.; Zhao, Y.; Cronk, H.; Luo., J.; Zhong, C.J.; Petkov, V. Synthesis-atomic structure-properties relationships in metallic nanoparticles by total scattering experiments and 3D computer simulations: case of Pt-Ru nanoalloy catalysts. *Nanoscale* **2015**, *7*, 8122–8134.

25. Shan, S.Y.; Petkov, V.; Yang, L.F.; Luo, J.; Joseph, P.; Mayzel, D.; Prasai, B.; Wang, L.Y.; Engelhard, M.; Zhong, C.J. Atomic-Structural Synergy for Catalytic CO Oxidation over Palladium-Nickel Nanoalloys. *J. Am. Chem. Soc.* **2014**, *136*, 7140–7151.

26. Yin, J.; Shan, S.Y.; Ng, M.S.; Yang, L.F.; Mott, D.; Fang, W.Q.; Kang, N.; Luo, J.; Zhong, C.J. Catalytic and Electrocatalytic Oxidation of Ethanol over Palladium-Based Nanoalloy Catalysts. *Langmuir* **2013**, *29*, 9249–9258.

27. Wanjala, B.N.; Luo, J.; Loukrakpam, R.; Fang, B.; Mott, D.; Njoki, P.N.; Engelhard, M.; Naslund, H.R.; Wu, J.K.; Wang, L.C.; *et al.* Nanoscale Alloying, Phase-Segregation, and Core-Shell Evolution of Gold-Platinum Nanoparticles and Their Electrocatalytic Effect on Oxygen Reduction Reaction. *Chem. Mater.* **2010**, *22*, 4282–4294.

28. Luo, J.; Wang, L.; Mott, D.; Njoki, P.N.; Lin, Y.; He, T.; Xu, Z.; Wanjana, B.N.; Lim, I.I.S.; Zhong, C.J. Core/Shell Nanoparticles as Electrocatalysts for Fuel Cell Reactions. *Adv. Mater.* **2008**, *20*, 4342–4347.

29. Wanjala, B.N.; Loukrakpam, R.; Luo, J.; Njoki, P.N.; Mott, D.; Zhong, C.J.; Shao, M.H.; Protsailo, L.; Kawamura, T. Thermal Treatment of PtNiCo Electrocatalysts: Effects of Nanoscale Strain and Structure on the Activity and Stability for the Oxygen Reduction Reaction. *J. Phys. Chem. C* **2010**, *114*, 17580–17590.

30. Wanjala, B.N.; Fang, B.; Loukrakpam, R.; Chen, Y.S.; Engelhard, M.; Luo, J.; Yin, J.; Yang, L.F.; Shan, S.Y.; Zhong, C.J. Role of Metal Coordination Structures in Enhancement of Electrocatalytic Activity of Ternary Nanoalloys for Oxygen Reduction Reaction. *ACS Catal.* **2012**, *2*, 795–806.

31. Wanjala, B.N.; Fang, B.; Shan, S.Y.; Petkov, V.; Zhu, P.Y.; Loukrakpam, R.; Chen, Y.S.; Luo, J.; Yin, J.; Yang, L.F.; *et al.* Design of Ternary Nanoalloy Catalysts: Effect of Nanoscale Alloying and Structural Perfection on Electrocatalytic Enhancement. *Chem. Mater.* **2012**, *24*, 4283–4293.

32. Wanjala, B.N.; Fang, B.; Luo, J.; Chen, Y.S.; Yin, J.; Engehard, M.H.; Loukrakpam, R.; Zhong, C.J. Correlation between Atomic Coordination Structure and Enhanced Electrocatalytic Activity for Trimetallic Alloy Catalysts. *J. Am. Chem. Soc.* **2011**, *133*, 12714–12727.

33. Chen, G.X.; Zhao, Y.; Fu, G.; Duchesne, P.N.; Gu, L.; Zheng, Y.P.; Weng, X.F.; Chen, M.S.; Zhang, P.; Pao, C.W.; *et al.* Interfacial Effects in Iron-Nickel Hydroxide-Platinum Nanoparticles Enhance Catalytic Oxidation. *Science* **2014**, *344*, 495–499.

34. Loukrakpam, R.; Wanjala, B.N.; Yin, J.; Fang, B.; Luo, J.; Shao, M.H.; Protsailo, L.; Kawamura, T.; Chen, Y.S.; Petkov, V.; *et al.* Structural and Electrocatalytic Properties of Nanoengineered PtIrCo Catalysts for Oxygen Reduction Reaction. *ACS Catal.* **2011**, *1*, 562.

35. Luo, J.; Han, L.; Kariuki, N.N.; Wang, L.Y.; Mott, D.; Zhong, C.J.; He, T. Synthesis and characterization of monolayer-capped PtVFe nanoparticles with controllable sizes and composition. *Chem Mater.* **2005**, *17*, 5282–5290.

36. Luo, J.; Kariuki, N.; Han, L.; Wang, L.Y.; Zhong, C.J.; He, T. Preparation and characterization of carbon-supported PtVFe electrocatalysts. *Electrochim. Acta* **2006**, *51*, 4821–4827.

37. Zhang, J.; Sasaki, K.; Sutter, E.; Adzic, R.R. Stabilization of platinum oxygen-reduction electrocatalysts using gold clusters. *Science* **2007**, *315*, 220–222.

38. Anderson, R.M.; Yancey, D.F.; Zhang, L.; Chill, S.T.; Henkelman, G.; Crooks, R.M. A Theoretical and Experimental Approach for Correlating Nanoparticle Structure and Electrocatalytic Activity. *Acc. Chem. Res.* **2015**, *48*, 1351–1357.

39. Nie, Y.; Li, L.; Wei, Z.D. Recent advancements in Pt and Pt-free catalysts for oxygen reduction reaction. *Chem. Soc. Rev.* **2015**, *44*, 2168–2201.

Recent Advances in Carbon Supported Metal Nanoparticles Preparation for Oxygen Reduction Reaction in Low Temperature Fuel Cells

Yaovi Holade, Nihat Ege Sahin, Karine Servat, Teko W. Napporn and Kouakou B. Kokoh

Abstract: The oxygen reduction reaction (ORR) is the oldest studied and most challenging of the electrochemical reactions. Due to its sluggish kinetics, ORR became the major contemporary technological hurdle for electrochemists, as it hampers the commercialization of fuel cell (FC) technologies. Downsizing the metal particles to nanoscale introduces unexpected fundamental modifications compared to the corresponding bulk state. To address these fundamental issues, various synthetic routes have been developed in order to provide more versatile carbon-supported low platinum catalysts. Consequently, the approach of using nanocatalysts may overcome the drawbacks encountered in massive materials for energy conversion. This review paper aims at summarizing the recent important advances in carbon-supported metal nanoparticles preparation from colloidal methods (microemulsion, polyol, impregnation, Bromide Anion Exchange ...) as cathode material in low temperature FCs. Special attention is devoted to the correlation of the structure of the nanoparticles and their catalytic properties. The influence of the synthesis method on the electrochemical properties of the resulting catalysts is also discussed. Emphasis on analyzing data from theoretical models to address the intrinsic and specific electrocatalytic properties, depending on the synthetic method, is incorporated throughout. The synthesis process-nanomaterials structure-catalytic activity relationships highlighted herein, provide ample new rational, convenient and straightforward strategies and guidelines toward more effective nanomaterials design for energy conversion.

Reprinted from *Catalysts*. Cite as: Holade, Y.; Sahin, N.E.; Servat, K.; Napporn, T.W.; Kokoh, K.B. Recent Advances in Carbon Supported Metal Nanoparticles Preparation for Oxygen Reduction Reaction in Low Temperature Fuel Cells. *Catalysts* **2015**, *5*, 310–348.

1. Introduction

Formerly, electrocatalysis was practiced with metals in the bulk state. Electrochemists became aware very early of a number of obstacles and/or limitations of fundamental and economic order. On a fundamental level, it is difficult to tune the activity and selectivity of the electrodes enabling the design of advanced energy converters. Since catalysis is a surface science [1–3], the use of a bare electrode has a huge economical impact. The recent advances in nanomaterial science enable the control of the nanoparticle growth steps for rational and effective preparation of nanostructures with different shapes and sizes. Seminal papers from different research groups have shown that the electrocatalytic properties of the metals depend quasi-exclusively on their crystallographic structures [2,4]. As electrochemical reactions involve the surface of the electrode, the bulk material is only of fundamental interest and thus serves as a model electrode for reaction testing. Concerning the oxygen reduction reaction, ORR, recently fundamental investigations have effectively and definitively demonstrated that its kinetics strongly depends on the particle size correlated with the corresponding aftermath [5–8]. Especially, Shao and coworkers unexpectedly found that the optimum particle size window for Pt is roughly 2.2 nm during ORR tests in perchloric acid, while the activity decreases drastically for ultra small particles [6].

Nanomaterials have different and unique physicochemical properties compared to the bulk metals from which they result. There are undoubtedly corollaries of their sizes and shapes. They exhibit special optical, magnetic, electronic, and catalytic properties. Consequently, they can be used in various applications either ranging from physics to chemistry or from bionanotechnology to medicine. Plainly speaking about ORR considered as the major contemporary technological hurdle, electrochemists are wondering how to reduce electrode cost without losing performance or durability and, to reduce cathode loadings under 0.1 mg_{PGM} cm^{-2} (United States Department of Energy (DOE) 2015 & 2017 targets [9]), while keeping the same activities. The approaches mostly emerging, focus on the catalytic nanoparticles' surface structure and composition to achieve such gates (Figure 1). For this objective, new nanocatalyst preparation protocols have been introduced. Amongst these synthetic routes, featuring colloidal methods, high-surface area and conducting, carbon-based materials are employed to ultra-disperse platinum group metal (PGM) nanoparticles [10–14]. Figure 1 shows the kinetic activities of Pt-based electrocatalysts prepared from the major chemical synthetic approaches. The catalytic ability of an electrode material toward ORR is performed either by rotating disc electrode (RDE) or in membrane electrode assembly (MEA); it is currently expressed using the real electrochemical active surface (ECSA), the kinetic current density, j_k, (mA cm^{-2}_{PGM}) or the metal content (mA μg^{-1}_{PGM}) (Figure 1a). When ECSA increases, j_k decreases for the same activity expressed in mA μg^{-1}_{PGM}, as can be seen in Figure 1b. The DOE's 2015 target is 0.7 mA cm^{-2}_{PGM}. Currently, the state-of-the-art commercial pure Pt/C catalysts (2–4 nm) have specific activities of 0.15–0.20 mA cm^{-2}_{PGM} and 0.10–0.12 mA μg^{-1}_{PGM} measured in MEAs [9].

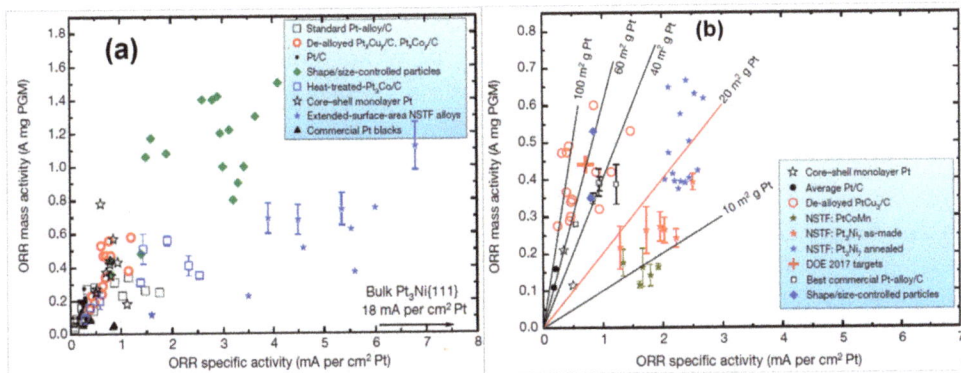

Figure 1. Kinetic activities of the main Pt-based electrocatalyst systems at 0.9 V *vs.* Reversible Hydrogen Electrode (RHE): **(a)** Activities are measured by rotating disc electrode (RDE) and **(b)** Activities are measured in membrane electrode assemblies (MEAs) at 80 °C and 150 kPa saturated O_2. Reprinted with permission from Ref. [9]. Copyright 2012, Nature Publishing Group.

In gas-phase heterogeneous catalysis, the metal loading on the support typically ranges from 0.1–1 wt%. Conversely to these kind of catalysts, the metal content in the electrocatalysts must be at least 10 wt%, due to the reduced mass-transport rates of the reactant molecules in the liquid phase *versus* the gas phase [15]. The most used support material for electrocatalyst preparation is carbon black. It should be emphasized that other carbon-based materials are used as supports e.g., carbon nanotubes [16,17], single/multi-walled carbon nanotubes [18,19], buckypaper [16], carbon nanofibers [20,21]; depending on the synthesis protocols. The exceptional electrical, physical, and thermal properties of these advanced carbon-based nanocomposites make them a preferential choice in electronics [17,22], bionanotechnology [16,17,20], energy conversion and storage [17,23,24]. However, even now, carbon black powder (Vulcan XC 72 or XC 72R) is the preferred support for low temperature FCs applications. In addition to the high metal loading, its relatively inert character implies that the widespread used method in gas-phase heterogeneous catalysts preparation such as ion exchange [25] is not effective and suitable in the electrocatalyst's case. To overcome this, the electrocatalyst synthesis is based on chemical/colloidal methods. In this context, various processes have been developed. Basically, these routes can be classified into two categories: physical and chemical techniques [2,26]. The main advantage of the chemical routes is the facility of controlling and handling the primary structures of metal nanoparticles, such as size, shape and composition (in the case of multimetallic nanomaterials) as well as to achieve large-scale production. For the physical methods, all these crucial operations are more difficult.

This review aims to discuss the recent advances in the preparation of highly dispersed nanoparticles onto carbon-based substrates for low temperature FCs applications. These kinds of electrode materials enable increasing of the surface-to-volume ratio and thus the electrochemical reaction rates [5,27,28]. We focus on the recent developments regarding nanocatalysts preparation from water-in-oil microemulsion [29–34], polyol-based [35], impregnation-reduction and Bromide Anion Exchange synthetic routes [36–38].

2. Strategies to Synthesize Carbon Supported Nanocatalysts

2.1. Heterogeneous Catalysis: The Major Emergency for Supported Nanomaterials Design

Various synthetic routes have been successfully developed over the last twenty years for carbon supported nanoparticle preparation to be used in electrocatalysis, and particularly in ORR science. It is interesting to know why a support is needed during catalyst preparation. In catalysis, it is reported that the direct immobilization of metal nanoparticles onto carbon-based substrates induces a high improvement in the nanoparticles' catalytic performances [2,21,39]. This enhancement has been attributed to the strong interaction between nanoparticles and the support. Free nanoparticles in solution are used in electrocatalysis to find out the intrinsic activity of the electrocatalysts, especially the structure sensitivity [2]. While it is difficult to control, distinguish, and separate the intrinsic activity of the different crystallographic facets, the single crystal (bulk) as well as the shape-controlled nanoparticles constituted cornerstones for the fundamental understanding of electrocatalytic activity. Even if many examples can be found in the literature about the use of these kinds of electrocatalysts toward ORR and related reactions, they are not still yet proven as of any significant interest in FC sciences. Indeed, a support is needed to boost the current density when the catalyst is immobilized for the tests of the FCs. Their preparation and long-term storage processes are less competitive than supported nanoparticles. Besides, it is more difficult to produce these types of materials for large-scale applications such as for FCs.

2.2. Water-in-Oil (w/o) Microemulsion Method

The preparation of metal nanoparticles from the "water-in-oil" microemulsion (w/o) method was initiated by Boutonnet et al. [30] in 1982 when they reported the successful preparation of Pt, Pd, Rh and Ir nanoparticles with a size ranging from 3–5 nm. After this stage of initiation, it is worthy of note that the preparation of nanoparticles from this method was successful, particularly in the field of catalysis. It should be noted that the term "microemulsion" was introduced by the English chemist, J. H. Schuman [29,31,40,41]. According to Clausse and co-workers, it can be assumed to be "a macroscopically monophasic fluid transparent compound made up by mixing water and hydrocarbon in the presence of suitable surface active agents (surfactants)" [42]. Similar definitions can be found in the literature [30,31,43]. It should be mentioned that this solution is optically isotropic and thermodynamically stable [44]. From a macroscopic point of view, the internal structure of the microemulsion appears to be homogeneous, but at the nanoscale, it is heterogeneous, consisting either of nano-spherical monosized droplets or a bi-continuous phase (10–40 nm) [41,44]. Obtaining this droplet is very crucial in order to control the size of the nanoparticles during their preparation.

The microscopic structure of the microemulsion is precisely determined at a given temperature by the ratio of its different constituents, as illustrated in Figure 2. Two systems can be clearly identified in this figure: the water-rich phase and the oil-rich one, determined by the water and oil contents. The water-rich phase is obtained for a high concentration of water where the internal structure of the microemulsion consists of small oil droplets in a continuous water phase (micelles or direct micelles) and is known as "oil-in-water, o/w" microemulsion (ratio o/w $\ll$ 1). Indeed, to stabilize the system, the hydrophilic portion of the surfactant molecules must be oriented toward the aqueous phase (water), while the hydrophobic tail is directed toward the organic phase (oil) to form oil droplets. In nanoparticles' preparation from colloidal methods, it is important to keep in mind that the droplet size is the key parameter for size, shape and control of other factors. Consequently, the system o/w is unusable for nanoparticles preparation because the metal ion (always in the aqueous phase) will be outside the droplet. Indeed, most metal precursors are inorganic salts and are soluble in water, not in oil [32]. At a high oil concentration, the system consists of small water droplets in a continuous oil phase (reversed micelles), also known as a "water-in-oil, w/o" microemulsion (ratio w/o $\ll$ 1). For the solution stabilization, the hydrophilic portion of the surfactant molecules oriented toward the aqueous phase (water) and the hydrophobic tail directed toward the organic phase (oil) form together a water droplet: reversed micelles which are water-in-oil droplets stabilized by a surfactant are obtained. Therefore, the system w/o (emulsions with low water concentration and high oil concentration) is used almost exclusively for nanoparticles' preparation because the metal ion will be inside the droplet. It is a necessary condition for nanoparticles' preparation from metal precursors. Furthermore, between these extreme situations, there is a bicontinuous phase without any clearly defined shape.

Obviously, at fixed water content, the volume of the surfactant plays crucial role for the size of the formed micelles. Basically, their size is inversely proportional to the amount of surfactant present in the microemulsion [32,44–46]. The surfactant can be ionic or non-ionic; its presence is very important for the stability of the microemulsion [31,44,45]. Moreover, the structure of micelles must be flexible to allow the penetration of the reducing agent and the interactions/exchanges between micelles during the collision step. In addition, when the ionic surfactants are used, they can form strong bonds with the surface of the nanoparticles. Thereby, it becomes particularly difficult to remove them during the cleaning step. This is the case of sodium bis(2-ethylhexyl)sulphosuccinate, AOT, which can form thiol bonds with the metal surface, making it more complicated to remove. In order to minimize this kind of phenomenon, the most used non-ionic surfactant is a polyethylene glycoldodecylether known as Brij® 30 [47–50].

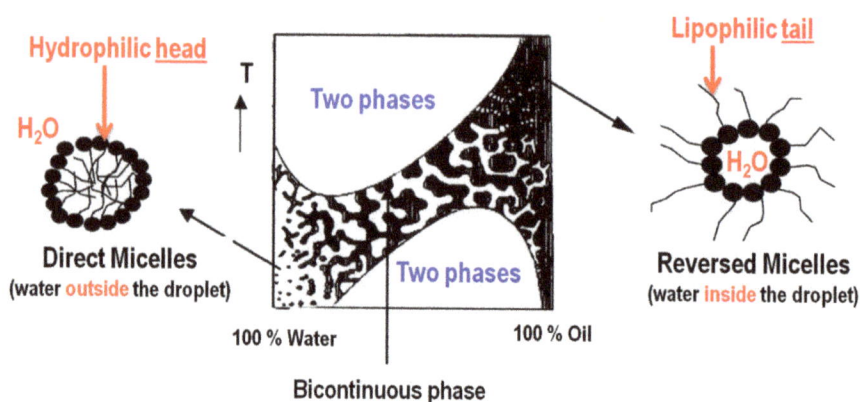

Figure 2. Scheme of the microscopic structure of a microemulsion at a given concentration of surfactant as function of temperature and water concentration, showing the different systems. Reprinted, adapted with permission from Ref. [44]: Copyright 2004, Elsevier and from Ref. [31]: Copyright 1995, American Chemical Society.

Concerning the nanoparticles' preparation, there are currently two main ways: either by preparing a second microemulsion (having the same composition as that containing the metal ions), which contains a reducing agent, or by direct addition of the reducing agent to the microemulsion that already contains metal salt precursors [44,51]. Figure 3 illustrates the formation of nanoparticles following the first procedure. Here, two microemulsions are mixed together, one containing the precursor and the other, the reducing agent. Due to the physical and chemical properties of its different constituents, the colloidal solution is very sensitive to temperature and the synthesis is currently performed at room temperature (20–25 °C) [34,49,50,52–56]. Current reducing agents are sodium borohydride ($NaBH_4$), hydrazine (N_2H_4), gas hydrogen (H_2) *etc.* The organic solvent (oil) can be hexane, n-heptane, cyclohexane or isooctane [44,47,55,56]. The formation of particles occurs in two steps known as the *nucleation process* inside the droplet and the *aggregation process* to form the final particle [44]. The major role of the surfactant is to control the growth via steric effect in order to have well-dispersed and homogeneous particles with a good size distribution.

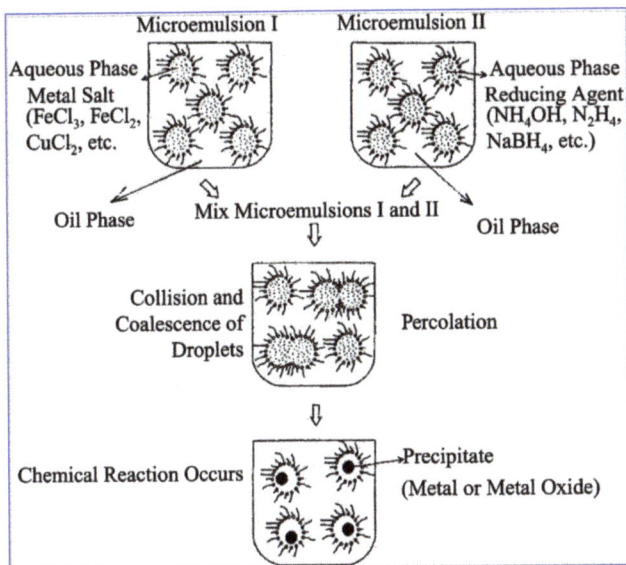

Figure 3. Proposed mechanism for the formation of metal particles from the microemulsion method using the two microemulsions protocol. Reprinted with permission from Ref. [45]. Copyright 2004, Elsevier.

Summarizing, the optimization of this method over more than twenty years has led to the following experimental parameters:

- molar ratio $n_{(water)}/n_{(surfactant)}$, ω: 3.8 [47,48,52,56],
- volume of organic solvent (*n*-heptane) per synthesis reactor: 27.35 mL [50],
- total metal salt concentration in the aqueous solution: 0.1–0.2 mol L^{-1} [47,48,50],
- volume percentage of Brij® 30 in the microemulsion: 16.5% [47,48,52,56],
- molar ratio between the reducing agent and the metal salt: 15 [49,50,52,56],
- synthesis conducted at room temperature [34,49,50,52–56].

In order to have a good dispersion of nanoparticles, conducting and high surface area (BET surface) carbon substrates are currently used. The main roles of the support are to improve nanoparticles dissemination, to reduce the metal content (generally noble metals are used) and to provide good nanoparticles-support interactions [53,54,57]. Vulcan XC 72 and Vulcan XC 72R carbons are the most used substrates for electrocatalytic applications and the metal content rises from 20–40 wt.%. For this purpose, Vulcan carbon is thermally pre-treated (see Section 2.4) to remove the potential undesired contaminants coming from its industrial manufacture such as sulfur [53,54,57,58]. Very recently, it has been shown that this thermal activation highly improved the physicochemical and electrocatalytic properties of Vulcan [57]. We will briefly describe herein a typical synthesis as currently carried out in our research group, from two microemulsions: one containing the dissolved metal(s) precursor(s) and the other one, the reducing agent, $NaBH_4$. Typically, microemulsions I and II (as shown in Figure 3) are mixed in a synthesis reactor thermostatted at 25 °C for about 30 min under magnetic stirring: the mixture gradually darkens, reflecting the formation of metal nanoparticles. Then, an appropriate amount of the Vulcan XC 72 or Vulcan XC 72R carbon is added after the solution has been transferred into an ultrasonic bath for homogenization (15–20 min), followed by additional vigorous stirring for 2 h. Finally the carbon supported metal nanoparticles are thoroughly filtered and washed to remove organic solvent and surfactant using three different solvents, acetone, ethanol and water. The filtration occurs under a vacuum system (Buchner) using a Millipore filter type GV 0.22 μm (also known as GVWP 0.22 μm). The washing steps are carried out strictly in the following order: addition of acetone, then ethanol and finally a mixture of 50 vol% acetone-water. It is strongly recommended to repeat this procedure at least three times. The material is rinsed thoroughly with Milli-Q® Millipore water (18.2 MΩ cm at 20 °C), and finally the filter containing the catalyst is removed and dried in an oven for at least 12 h at 75 °C. Figure 4 shows the low and high resolution TEM images of the bimetallic 40 wt% AuPt/C. From these physicochemical and electrochemical characterizations, Habrioux and co-workers found that the obtained bimetallic nanomaterials exhibit two different behaviors. For high Au-content, the Au-Pt particles exhibited alloy properties, and at low Au-content, atom rearrangement leads to an enrichment of the electrode surface with those of Pt. It should be mentioned that the AuPd/C bimetallic nanomaterials synthesized from w/o by Simões et al. [34] are unalloyed for Au ⩾ 50 at%, leading to two phases: Au islets + AuPd alloy. Various studies have shown that ORR is a structure-sensitive reaction. This means that the crystallographic orientations (hkl) play a major role, more especially the low-index ones [2,4]. The seminal papers on the single crystal revealed that (111) and (110) are the most active facets toward ORR in aqueous media. Particularly, for Pt(hkl), the activity toward ORR increases in the order (100) < (110) ≈ (111) in $HClO_4$; (100) < (110) < (111) in KOH and (111) ≪ (100) < (110) in H_2SO_4 [2]. As highlighted in Figure 4b, the presence of (111) face on the nanoparticles prepared from the w/o approach allows better ORR efficiency to be expected on these catalysts.

Notwithstanding the diversity of applications, this method remains questionable due to the surfactant Brij® 30 that is not completely removed from the surface of the metal nanoparticles [59,60]. The remaining molecules undoubtedly block some catalytic sites, therefore affecting the catalytic performances of the obtained catalysts. In conclusion, the nature of the surfactant (Brij® 30 or others) and its strong adsorption onto the metal nanoparticles' surface constitute the main drawback of the water-in-oil microemulsion method.

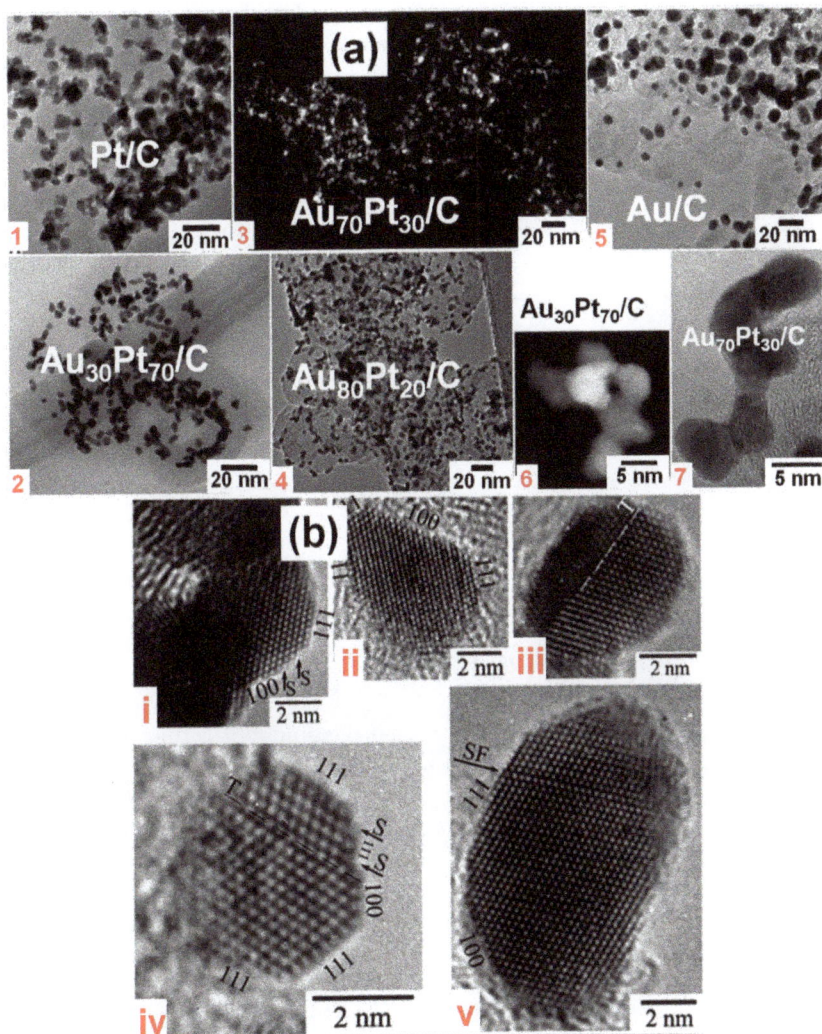

Figure 4. (a) TEM images of: (1) Pt/C, (2) $Au_{30}Pt_{70}$/C, (3) $Au_{70}Pt_{30}$/C, (4) $Au_{80}Pt_{20}$/C, (5) Au. Particle agglomerates are shown in (6) and (7) for $Au_{30}Pt_{70}$/C and $Au_{70}Pt_{30}$/C, respectively. (b) HRTEM images of (i and ii) Pt, (iii) $Au_{30}Pt_{70}$/C, (iv) $Au_{70}Pt_{30}$/C and (v) Au/C highlighting the facets, steps (S), twins (T) and stacking faults (SF). Note that the metal loading was 40 wt%. Reproduced and adapted in part from Ref. [54] with permission of the PCCP Owner Societies. Copyright 2009, Royal Society of Chemistry.

2.3. Polyol Method

The polyol process refers to a polyalcohol that acts both as a solvent and a reducing agent. It has spread for its self-seeding mechanism and lack of required hard or soft templating materials, making it an ideal process for potential industrial scale-up due to the low cost and simplicity of processing. The main superiority of the polyol synthesis is the use of high-boiling alcohols such as ethylene glycol (197 °C), propylene glycol (188 °C), butylene glycol (207 °C), diethylene glycol (244 °C), glycerol (290 °C), tetraethylene glycol (327 °C), benzyl alcohol (205 °C) *etc.*, as a solvent and stabilizing agent for controlling particle nucleation and growth. Each polyol solvent has different oxidation potentials that along with the metal reagent, define the temperature at which particle formation takes place. Especially, ethylene glycol is one of the most widely used solvents for the polyol process owing to its strong reducing capability, relatively high boiling point and high dielectric constant, which increases the solubility of inorganic salts. Therefore, ethylene glycol is more convenient to act as reducing agent for the metal nanoparticles. In the polyol process, a metallic precursor in the form of chlorides, acetates, nitrates, hydroxides, oxides is dissolved in polyol solvent, and then the experimental conditions are optimized to complete the reduction of metallic precursor. To control the shape, size, and distribution of metallic particles, each metal precursor requires modified-experimental conditions.

To identify the mechanism and comprehend the influence of polyol solvents on nucleation and growth kinetics, several studies were reported. Fievet *et al.* [61] focused on the general mechanism of reduction of $Ni(OH)_2$ and $Co(OH)_2$ in ethylene glycol, shown in Scheme 1. They proposed that acetaldehyde is the possible reductant for the synthesis strategies of preparing metal nanoparticles. It is necessary that the precursors require high solubility in polyol solvents to actualize the reaction described in Scheme 1 by steps (1) and (2).

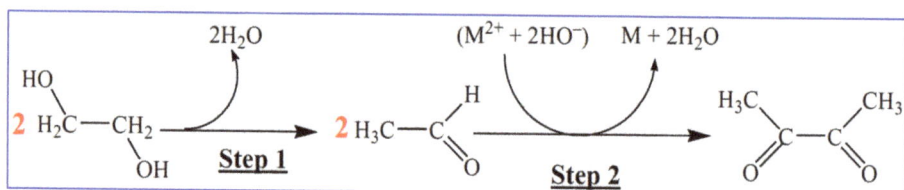

Scheme 1. General mechanism of reduction of $M(OH)_2$ in ethylene glycol.

According to the mechanism proposed by Fievet *et al.* [61], diacetyl appears to be the main oxidation product, which may be explained by a duplicative oxidation of acetaldehyde previously produced by dehydration of ethylene glycol. The metal is generated in the liquid phase and when the super-saturation is high enough, nucleation and the growth steps occur. For better understanding of the metallic reduction mechanism in polyol solvent, some significant researches were undertaken. One of them was based on redox phenomena related to the reduction potential of metal precursor and the oxidation potential of ethylene glycol [62]. The researchers proposed that chemical reduction of noble metal species by ethylene glycol is thermodynamically unfavorable. For this reason, to get a completely reduced metal species, an energy barrier must be overcome by heating the polyol solvent.

Another study on reaction mechanism was proposed by Bock *et al.* [63] for reduction of the metals by the oxidation of ethylene glycol to aldehydes, carboxylic acids, and CO_2, as shown by Scheme 2.

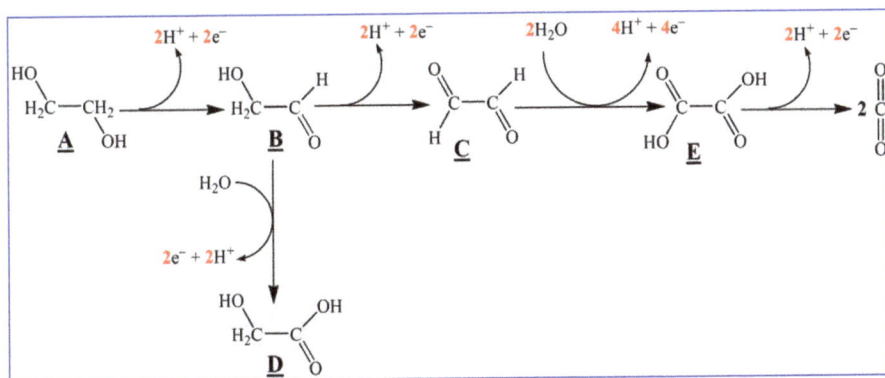

Scheme 2. Ethylene glycol (**A**) oxidation pathways to aldehydes (**B**, **C**), glycolic acid (**D**), oxalic acid (**E**) and further CO_2 and carbonate in alkaline medium owing to interaction of –OH groups of ethylene glycol with metal-ion sites.

The oxidation of the ethylene glycol gives glycolate or glycolic acid (depending on the pH) which acts as a stabilizer for the metal species; the size of the noble metal colloids is thereby controlled through the pH value of the synthesis solution. Oxidation products resulting from the ethylene glycol oxidation reaction interact with the noble metal colloids and hence act as their stabilizers. Additionally, Skrabalak *et al.* [64] reported how metal ions are reduced by the ethylene glycol oxidation to glycolaldehyde, using a spectrophotometric method. They proposed an alternative pathway related to the reduction of metal precursor depending on the reaction atmosphere. For example, heating ethylene glycol between 140 and 160 °C in air, Equation (1), may generate glycolaldehyde as a reductant for the many metal precursors

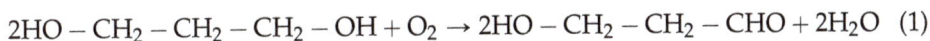

$$2HO - CH_2 - CH_2 - CH_2 - OH + O_2 \rightarrow 2HO - CH_2 - CH_2 - CHO + 2H_2O \quad (1)$$

It was found out from spectroscopic data that glycolaldehyde is the main reductant depending on reaction temperature, atmosphere, and experimental setup [64]. Numerous studies have been reported on synthesizing metal nanoparticles with controlled-size, shape and morphology, and satisfactory stability, as well as high product yield and low environmental contamination [65–67]. Physical properties of nanoparticles can influence catalytic activity [68,69], selectivity [70,71] and durability [72]. Biacchi et al. [73] demonstrated that proper selection of the polyol solvent can be used to manipulate the metal nanoparticles morphology. They reported that the kinetics and thermodynamics of nanoparticles synthesis is critically important for controlling the shape and size when using different polyol solvents such as ethylene glycol, diethylene glycol, triethylene glycol, and tetraethylene glycol. In recent years, methodological development of the synthesis of the highly active electrocatalysts has been one of the major topics in energy converting systems [74–79]. Joseyphus and co-workers [80] investigated the reaction rate of the polyol method using cobalt and its alloys. The reduction limit of polyol solution depends mainly on the concentration of hydroxyl ions (OH^-) for the reduction of metal ions. Although the presence of OH^- ions in the metal ion-polyol system acts as a catalyst in accelerating the formation of precursor complexes, it may decrease the reaction rate by forming metal hydroxide compounds which are not easily reduced. Therefore, it is important to control the degree of complex or hydroxide forms in the presence of OH^- ions by using UV-Visible spectroscopy. Susut and Tong demonstrated that the particle shape could be controlled in the presence of $AgNO_3$ with different concentrations in the reaction mixture before the addition of PVP (poly-vinylpyrrolidone) and Pt precursor solutions [81]. Gonzalez-Quinjano et al. [82] also reported the synthesis of Pt-Sn/C electrocatalysts by using ethylene glycol containing ethanol and water in different ratios. They noticed that the chemical composition, lattice parameter, and degree of alloying depended on the solution ratio between ethylene glycol, ethanol, and water. The average particle size was observed as between 1.8 and 4.7 nm, the smaller particle sizes were reported in the absence of water. Furthermore, Jiang et al. [83] prepared Pt/C and Pt-Sn/C catalysts by slightly modifying the polyol method. In brief, they prepared a tin complex in ethylene glycol at 190 °C for 30 min and then added the required chloroplatinic acid and finally, the mixture was maintained at 130 °C for 2 h under argon gas to remove the oxygen and organic by-products. Lee et al. [84] prepared in ethylene glycol, acid pre-treated carbon supported Pt and Pt-Ni electrocatalysts, which exhibited significantly improved electrocatalytic activity.

In the polyol process, polyalcohols are currently used as both solvent and reducing agent. Alternatively, poly-vinylpyrrolidone (PVP) can be added as surfactant or capping agent [85,86] in association with a variety of reducing agents such as sodium borohydride ($NaBH_4$) [87,88], and formaldehyde (H-CHO) [88] for controlling surface reactivity. These properties significantly influence the electrochemical performance. During their synthesis, metal particles tend to aggregate, particularly in the liquid phase, because of their very high surface energy in the nano level. Therefore, nanoparticles have a natural tendency to be attracted to each other through Van der Waals forces leading to agglomeration. In order to neutralize the Van der Waals interaction, the required repulsive forces have to be improved. As a result, a stabilization procedure is required to ensure the quality of the nanomaterial products. The stability relation of the nanoparticles can be mainly controlled by two kinds of effects in the dispersing medium. One is an electrostatic stabilization (Figure 5a) for developing surface charges to repulse (high positive E_T) aggregated particles, which can be evaluated by zeta potential measurements and the other is a steric stabilization (Figure 5b) for controlling the protective layer, using organic ligands and polymer [89].

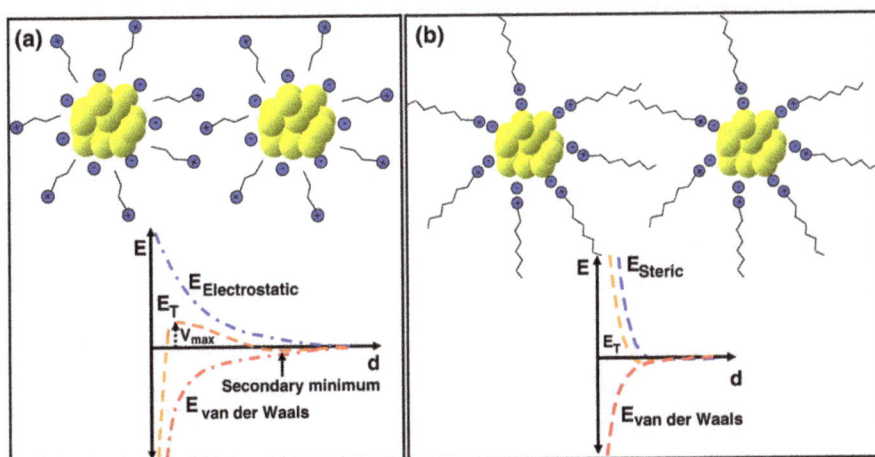

Figure 5. Stabilization of nanoparticles in a dispersing medium; (a) electrostatic stabilization and (b) steric stabilization of metal. Reprinted with permission from Ref. [89]. Copyright 2004, Elsevier.

2.4. Impregnation-Reduction Process

The impregnation-reduction method is one of the simple and straightforward techniques in material science for supported nanocatalysts preparation. It is still widely used in the field of gas-phase heterogeneous catalysis to prepare nanostructured catalysts. Compared to the w/o method, the impregnation technique is one of the green class methods to prepare nanocatalysts due to its environmental friendly solvents *versus* organics ones. The synthesis takes place either at low or room temperatures, thus minimizing energy consumption [32,51]. For a long time, this chemical approach was devoted to Pt-Ru catalyst preparation [90–92]. Basically, it involves two steps: impregnation and reduction. During the impregnation, the support (mainly carbon) is immersed in the aqueous solution containing the desired metal precursors. Then, metal ions are reduced to their metallic state by the addition of an aqueous solution of reducing agent such as $Na_2S_2O_3$, $NaBH_4$, $Na_4S_2O_5$, N_2H_4, H-CHO, or H_2 [51,91,93,94]. It should be noted that the impregnation duration step strongly depends on the nature of the precursors, the targeted metal loading, as well as the nature of the support. Recently, Knani et al. [93,95] optimized an experimental approach for synthesizing methanol tolerant ORR nanocatalysts (10 wt% Pt-Co-Sn/C) preparation using $NaBH_4$ and Vulcan XC 72R carbon as reducing agent and support, respectively. Typically, the mixture made of metal precursors and the support is ultrasonically homogenized for 30 min followed by additional stirring for 2 h before the reduction step at 80 °C. Then, the filtered solid composite is washed several times with ultra pure water and dried in an oven at 110 °C for about 4 h [95]. Miller et al. [96], reported the preparation of noble metal free electrocatalysts based on iron(II) and silver(I) phthalocyanines using an impregnation-reduction method. During this procedure the Ketjenblack EC-600JD used as support was impregnated with the metal precursors by stirring (30 min) and sonicating (30 min) ethanol suspensions (200 mL) of the metal phthalocyanine complexes with the carbon material at room temperature. The obtained mixture was stirred afterwards for 24 h (to improve the impregnation process with the support surface) at room temperature and then sonicated for 30 min again. After these steps, the solvent was removed under reduced pressure to yield a solid residue which was dried under high vacuum. Then, the resulting powder annealed at high temperature (250–800 °C) for 2 h was cooled to room temperature under continued argon flow prior to use. These nanostructures have shown good electrochemical performances toward ORR in alkaline medium. Others impregnation-reduction processes have been reported for the successful preparation of PtRhM/C (M = W, Pd, or Mo) [97]; PtRh/C [98]; PtNi/C [99].

Different metal precursors can be used; chloride, sulfite, nitrate, carbonyl complexes. It should be emphasized that the metal carbonyl complexes are especially interesting since the second step is not required in some cases [32,100]. As any process, many experimental parameters affect the electrochemical activity of the obtained catalysts by controlling their composition, morphology, and dispersion onto the support. From the experimental view, it has been suggested that, the control of the nanoparticles size and as yet the particle distribution are more difficult by the impregnation method, which thus constitutes its major drawback [32,51].

2.5. Bromide Anion Exchange (BAE) Method

Obviously, electrocatalysis is one of the disciplines that requires maximum cleanness. Indeed, from catalyst preparation to initiating chemical reactions, the cleanliness is mandatory throughout the chain. Any impurity can drastically alter the catalytic activity/selectivity through the control of active sites. Thus, taking into account the fact that the reactions involved in electrocatalysis are surface reactions, the electrocatalyst surface state (cleanness) is the key parameter for the best electroactivity. Thereby, the surface of the nanomaterial must be free from impurities such as surfactant molecules and other capping ones. In this way, the development of an advanced synthetic method must limit the use of strong organic molecules, which have an affinity with the nanoparticles surface. To this end, the research group of Prof. B. Kokoh has recently initiated a new clean, easy, and accessible synthetic route called "Bromide Anion Exchange, BAE" [36,38,101]. Most of the metal salts (precursors) used in nanomaterials synthesis are chlorinated. The bromide ion ($r(Br^-) = 195$ nm) is larger than its counterpart chloride ($r(Cl^-) = 181$ nm). Thus, the partial or total substitution of Cl^- by bromide anions must accentuate steric hindrance around the metal cation, which will further play a crucial role during the seed's growth. This advanced method has been successfully used to synthesize Au-based nanocatalysts for glucose electro-oxidation [38,57], Pd-based electrocatalysts for glycerol electro-oxidation [36,37,101] and hybrid/abiotic electrodes for biofuel cell application [102,103].

This convenient and straightforward synthesis approach is an environmentally friendly method and is based on the use of bromide ion as a capping agent, the major gate in the BAE process. It has been reported that halide ions (chloride, bromide, and iodide) could serve as coordination ligands and thus, play the role of capping agent for shape and size control of nanocrystals [67,104–107]. Figure 6 summarizes the different steps for preparing nanocatalysts with the BAE route. The main feature of BAE lies in the simplicity undertaken. By using no organic compounds as surfactants or capping agents, clean, small, and well-dispersed nanoparticles with highly improved catalytic properties are currently obtained. The effects of the different parameters such as the metal salt concentration, the amount of bromide anion and the temperature of the synthesis reactor were scrutinized recently. From these reports, the molar ratio between KBr and total metal(s): $\varphi = n(KBr)/n(\text{metal(s)})$ is 1.46; the total molar concentration of metal salts is 1 mM and the reactor temperature is 25 and 40 °C (meaning 25 °C before the addition of the reducing agent and 40 °C after) [108].

Typically, the metal precursor salts are dissolved in a reactor containing ultrapure water thermostatted at 25 °C under magnetic stirring. Then, an appropriate amount of KBr via the parameter φ is added under vigorous stirring. A suitable amount of carbon support (Vulcan or Ketjenblack) is then added under constant ultrasonic homogenization for 45 min, followed by the drop wise addition of the reducing agent, under vigorous stirring. Afterwards, the reactor temperature is elevated tot 40 °C (to improve the reaction kinetics) for a 2 h period. Finally, the carbon supported nanomaterials are filtered, washed several times with ultra pure water and dried in an oven at 40 °C for 12 h. In the whole BAE procedure, the

91

Vulcan or Ketjenblack supports were thermally pretreated at 400 °C under nitrogen atmosphere for 4 h, in order to improve their physical properties and remove any contaminant coming from their industrial manufacture [57]. Holade *et al.* [57] found surprisingly that the BET surface area of the support was enhanced, being 322 and 1631 m^2 g^{-1} instead of 262 and 1102 m^2 g^{-1} for the as-received Vulcan XC 72R (C) and Ketjenblack EC-600JD (KB) materials, respectively. They also found that Pt/C and Pt/KB exhibited a highly improved specific electrochemical surface area (SECSA). It is worthy of note that all catalysts were prepared with a high chemical synthesis yield ($\varpi > 90$ %), defined as the percentage of the ratio between the experimental mass and the theoretical one based on the initial reactor mixture [102,103]. It should be emphasized that it is the first time that such a synthesis yield has been reported. Indeed, for electrocatalysts preparation, neither the microemulsion method [30,109] nor any of the others [14,35,110,111] yielded this important result, indicating that the BAE method is suitable for nanomaterials preparation.

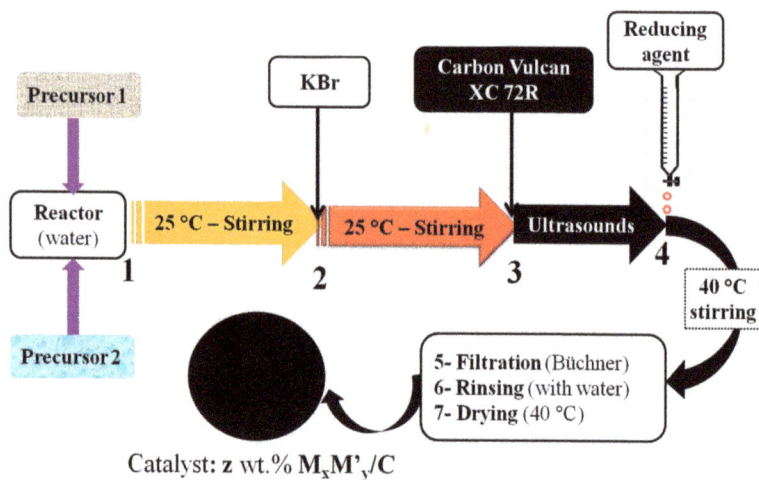

Figure 6. The experimental setup of the Bromide Anion Exchange (BAE) route for nanoscale materials synthesis. Reproduced with permission from Ref. [37] Copyright 2014, The Electrochemical Society.

During the synthesis (before the reduction step), change of the initial solution color was observed after the addition of KBr (see Figure 6: before and after step 2). Color changes can be seen in Figure 7a,c,e. To gain further insights on the origin of this phenomenon, UV-Vis measurements were performed. As can be seen in Figure 7b,d,f, there is a change in the UV-Vis spectra on the addition of KBr. In the case of the Pt salt, the addition of KBr shifts the band at 287 nm toward 298 nm with an intense shoulder around 411 nm. The absorption band due to a ligand-to-metal charge transfer transition of the $[PtCl_6]^{2-}$ ion complex is found to be at 263 nm [112] or 260 nm [46,113,114]. Thus, the shift and intensity of the present bands indicate partial substitution of Cl^- in the $[PtCl_6]^{2-}$ ion complex by Br^-. For the Pd salt, the aqueous solution goes from a clear yellow to a deep yellow, depending on the metal salt concentration. Furthermore, the addition of KBr to this solution drastically changes its appearance. According to the literature, the absorption wavenumber associated with the complex $[PdCl_4]^{2-}$ in aqueous solution may be 425 nm [115] or 415 nm [116]. This value depends undoubtedly on the complex ion concentration in the solution and could be affected by the presence of other species. The metal salt solution without KBr presents a band at *ca.* 300 nm and a shoulder at *ca.* 400 nm. With KBr, in addition to the band at 325 nm, two shoulders at 400 nm and 510 nm can be observed. The slight shift of the peak position and the appearance of the band around 510 nm indicate clearly the insertion of the Br^- ions in the complex $[PdCl_4]^{2-}$. Herein, the ratio $n(Br^-)/n(Pd^{2+})$ is 1.5 *versus* 4 for the complex ion $[PdBr_4]^{2-}$. Thus, there is no complete substitution of chloride by bromide. The change of the color, substantiated with the UV-Vis observations is attributed to the complex ion $[PdCl_{4-x}Br_x]^{2-}$, $0 \leqslant x \leqslant 4$. Klotz *et al.* [117] reported that, in aqueous media, $[PdI_4]^{2-}$ is $10^{8.1}$ times more stable than $[PdBr_4]^{2-}$ which is $10^{4.1}$ times more stable than $[PdCl_4]^{2-}$. Consequently, the complex ions that control the particles size/shape growth after reduction is $[PdCl_{4-x}Br_x]^{2-}$. The latter species provides a more steric environment than $[PdCl_4]^{2-}$. This hypothesis was confirmed when nanoparticles were synthesized without and with different amounts of KBr [108].

Figure 8a,b show the TEM micrographs (with their HRTEM images in the inset) of 20 wt% AuPt nanomaterials and their corresponding particle size distribution when using Vulcan XC 72R or Ketjenblack EC-600JD as supports, respectively. Particles are well dispersed onto the support with a mean particle size between 3–6 nm. The HRTEM images highlight an octahedron-like shape. It has been observed that for Au-based bimetallics, the particle size increases with increasing Au content, which is a well-known phenomenon, coming from the difference in the reduction kinetics of the metal salts. Trimetallics AuPtPd supported on both Vulcan and Ketjenblack were also successfully prepared from BAE [102,103].

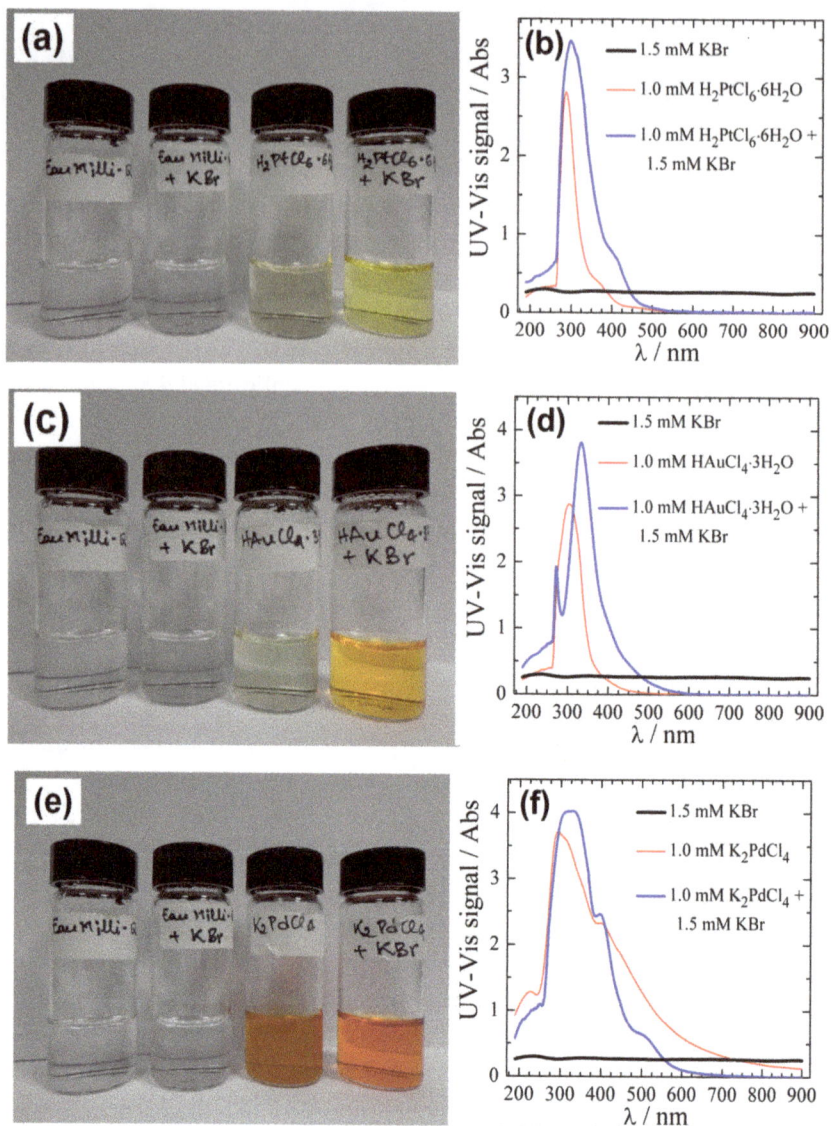

Figure 7. (a), (c) and (e) From left to right in each photograph, images of water containing: no substance, 1.5 mM KBr, 1.0 mM metal salt and 1.5 mM KBr + 1.0 mM metal salt. (b), (d) and (f) Their corresponding UV-Vis absorption spectra of water containing 1.5 mM KBr (black), 1.0 mM H_2PtCl_6 $6H_2O$ (red) and 1.5 mM KBr + 1.0 mM H_2PtCl_6 $6H_2O$ (blue). Note: (a–b) for platinum; (c–d) for gold and (e–f) for palladium.

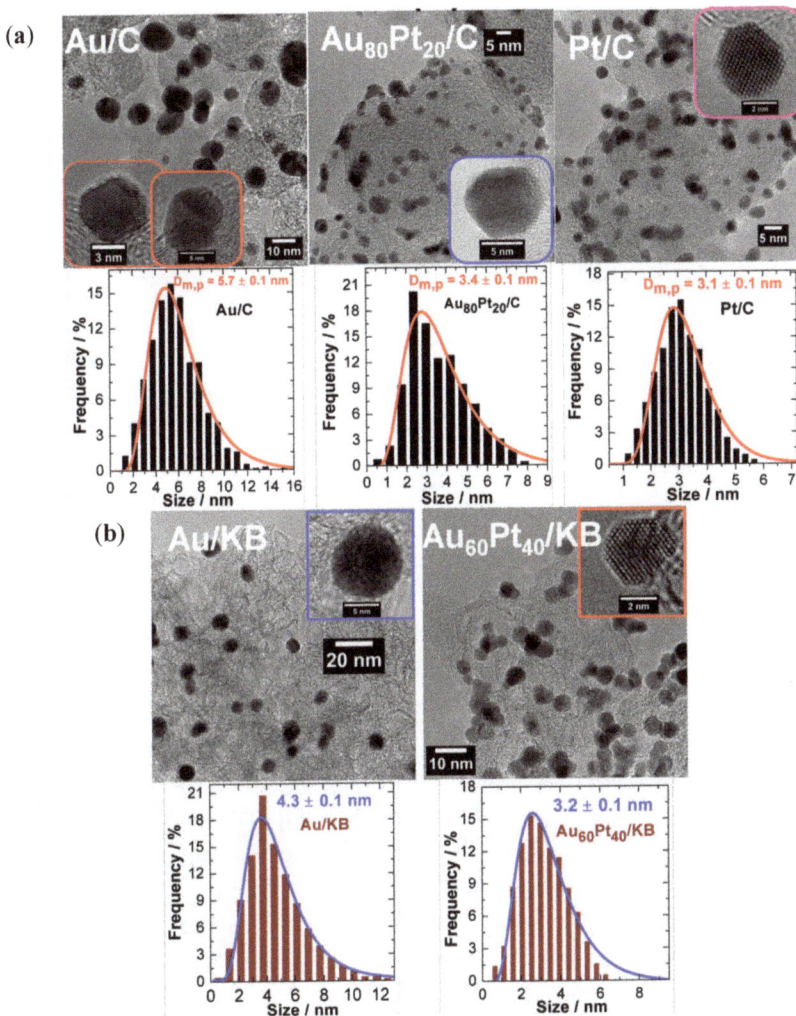

Figure 8. TEM-HRTEM micrographs and their corresponding particle size distribution (histograms were fitted using the log-normal function) of the nanostructured AuPt (20 wt%) supported on (**a**) Vulcan XC 72R and (**b**) Ketjenblack EC-600JD. (a) Reprinted and adapted with permission from Ref. [103]. Copyright 2014, John Wiley & Sons, Inc. (**b**) Reprinted and adapted with permission from Ref. [102]. Copyright 2014, John Wiley & Sons, Inc.

2.6. Other Synthetic Routes

From chemical to physical approaches, a huge number of metal nanoparticles preparation methods have been initiated over the last twenty years. Formerly reserved for the application in physics and related fields [26,39], nanomaterials prepared from physical routes are becoming unavoidable targets for electrocatalysis. Free nanoparticles prepared in solution from radiolysis [118] or laser ablation processes have been successfully tested in electrocatalysis [119]. Up till now, no test has been performed using carbon as support for application in electrocatalysis. The other chemical methods are the historical methods developed by Bönnemann and co-workers [94,120,121]. They have been adapted for electrocatalysts preparation [58,94]. In such a method, the reducing agent is a tetra-alkylammonium triethylborohydride, which acts also as a surfactant after reducing the metal salt in a tetrahydrofuran medium. After addition of the support, e.g., Vulcan XC 72 carbon, the filtered powder is calcined under air at 300 °C. Small nanoparticles about 2–5 nm are currently obtained. However, because of the use of some organic molecules as reducing agents, their removal from the nanoparticles surface is not always effective. Thus, their catalytic performance can be affected.

3. Application of Carbon-Supported Nanocatalysts toward the ORR

3.1. ORR Activity on Various Carbon Supported Nanoparticles Prepared from w/o Method

Because of its natural abundance (20.95 vol%; 23.20 wt% of the earth's atmosphere and roughly 21% in air), dioxygen is the first choice of oxidant used at the cathode in FCs. Already known as an oxidant in the propulsion system, H_2O_2 can supply O_2 as in the case of submarines. From this perspective, FCs were already developed [122,123]. This section will focus on the electrocatalytic performances of carbon-based substrates supporting metal nanoparticles toward the ORR. An emphasis on analyzing data from theoretical models to address the intrinsic and specific electrocatalytic properties depending on the synthetic method is incorporated throughout. The issue of the ORR is as old as that of FCs. Obviously; it is certainly one of the most widely studied processes due to its important applications in the field of clean energy conversion and storage systems. The ORR involves several basic steps. To date, two plausible mechanisms have been proposed in the literature (Figure 9) [124–126]. According to Acre *et al.* [125], the direct O_2 reduction to H_2O (path A) is the result of O_2 adsorption parallel to the catalytic surface plane. This requires the presence of active sites side by side. The other pathway (path B) proceeds by an initial adsorption of O_2 perpendicularly to the electrode surface (by a single atom). However, it should be noted that the second step (reduction of H_2O_2) has a high activation energy, which increases the overpotential.

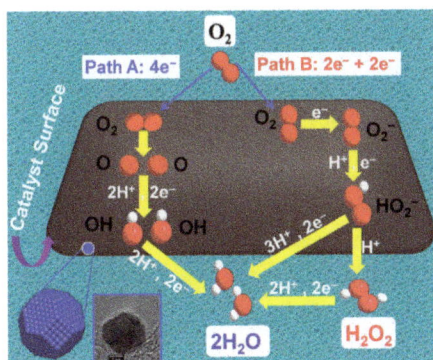

Figure 9. Schematic representation of the oxygen reduction reaction (ORR) mechanism by direct pathway (A: adsorption parallel to the surface) and indirect pathway (B: adsorption perpendicular to the surface). Reprinted and adapted with permission from Ref. [125]. Copyright 1997, Elsevier.

In the early 2000s, most of the catalytic applications of nanomaterials prepared from the water-in-oil method (developed more than 10 years ago) were oxidation of organic molecules in heterogeneous catalysis ranging from batch reactor to electrocatalysis. Tuning the experimental factors that affect the w/o method will be crucial for designing more active low temperature FCs cathodes. In 2008, Demarconnay and co-workers reported the use of the w/o route to prepare various Pt-Bi bimetallic nanomaterials dispersed onto Vulcan XC 72 carbon (metal loading: 50 wt%) [33] for the oxygen reduction reaction in alkaline medium. It is worth mentioning that this metal loading is too high to achieve the condition for commercialization of FCs which is limited to 8 g of platinum group metals (PGM) per vehicle, meaning less than 100 μg_{PGM} cm^{-2} at the cathode [9]. However, it can be helpful to understand the ORR electrocatalysis before thinking about the required three major criteria for FCs MEAs: cost, performance, and durability. In this preliminary ORR investigation using electrocatalyst from the w/o method, the RRDE technique to find out the fundamental data was used. It is worthy of note that, after the preparation of a carbon black catalyst, a black powder is obtained. Before using this powder for electrochemical tests, a catalytic ink is prepared. To this end, different approaches are currently used and are based on the initial method proposed at the beginning of the 1990s using Nafion® suspension [101,127–129]. Figure 10 shows the polarization curves at 5 mV s^{-1} (2500 rpm) at the disc for Pt/C, Vulcan XC 72R and different bimetallic catalysts in 0.2 M NaOH. The reaction starts at *ca.* 1.05 V *vs.* RHE on PtBi/C and 0.87 V *vs.* RHE on Vulcan XC 72R. An onset circuit potential (OCP) of 1.05 V *vs.* RHE reflects the ORR sluggishness because it must be roughly 1.19 V *vs.* RHE. As can be seen in the activation–diffusion mixed region

97

(from 1.0–0.7 V *vs.* RHE) the bimetallics $Pt_{90}Bi_{10}/C$ and $Pt_{80}Bi_{20}/C$ are more active than Pt/C and $Pt_{70}Bi_{30}/C$. This means that the optimum window is obtained when at %Bi < 70 in the PtBi alloy system. The Pt/C catalyst synthesized by this method did not produce any peroxide, yielding roughly 4 as the number of exchanged moles of electron per mole of oxygen as on bulk Pt [130]. The evaluated kinetic current density at 0.95 V *vs.* RHE was 0.60, 1.05, 1.14, and 0.49 mA cm^{-2} on Pt/C, $Pt_{90}Bi_{10}/C$, $Pt_{80}Bi_{20}/C$, and $Pt_{70}Bi_{30}/C$, respectively. Unfortunately, it was not always mentioned whether this kinetic current density, which is free from mass transport, was evaluated using geometrical or active surface area. Therefore, it is difficult to compare it with other values found in the literature. Based on these values, $Pt_{80}Bi_{20}/C$ shows the best kinetic activity. The value of exchange current density (j_0) at the high overpotential region on this electrode material is 23.2 10^{-3} mA cm^{-2}, which is higher than 16.8 and 5.6 10^{-3} mA cm^{-2} on Pt/C and $Pt_{70}Bi_{30}/C$, respectively. Typically the exchange current density is 5.4 10^{-5} mA cm^{-2} on the bulk Pt [130]. The low value of the Tafel slope on $Pt_{70}Bi_{30}/C$ at high overpotential (99 mV dec^{-1}) contrary to theoretical value of 120 mV dec^{-1} has been explained by the presence of bismuth oxides. Table 1 shows the influence of the synthesis method and catalyst composition on the kinetic parameters. More importantly, the authors found that $Pt_{80}Bi_{20}/C$ exhibits a high tolerance for ORR in the presence of 0.1 M ethylene glycol from 1.0–0.9 V *vs.* RHE. Indeed, they investigated the tolerance properties of the catalysts towards the ORR in the presence of ethylene glycol as fuel. The platinum substitution by bismuth up to 20 at% improves the catalyst tolerance by shifting the reduction wave towards higher potentials. These kinds of fuel tolerant cathode materials are promising electrodes for the development of advanced electrocatalysts for direct alcohol fuel cells in which the fuel can crossover the membrane to be mixed with O_2 in the cathodic compartment.

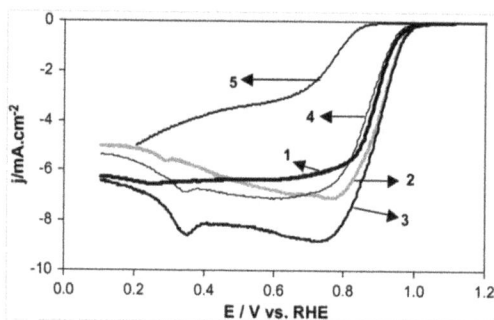

Figure 10. ORR polarization curves recorded in an O_2-saturated 0.2 M NaOH solution at 50 wt.% PtBi/C catalysts prepared from w/o method: (1) Pt/C, (2) $Pt_{90}Bi_{10}/C$, (3) $Pt_{80}Bi_{20}/C$, (4) $Pt_{70}Bi_{30}/C$ and (5) Vulcan XC 72 carbon (5 mV s^{-1}, 2500 rpm, 20 °C). Metal loading on the electrode: 177 µg cm^{-2}. Reprinted with permission from Ref. [33]. Copyright 2004, Elsevier.

Until the early 2000s, it was difficult to decrease the metal content in the electrocatalysts without losing significant catalytic activity. This poses a considerable challenge to the material science community, particularly when considering that for most catalyst systems, durability and high current density work together with the PGM content. Habrioux and co-workers used the same w/o method to design platinum-gold nanoalloys with improved electrocatalytic properties [52]. They managed to reduce the metal charge up to 40 wt% without a significant loss in catalytic activities. Au is known to improve the Pt electrode durability by modifying the Pt–OH bond strength [52]. O_2 reduction starts at *ca.* 0.7 V *vs.* RHE on Vulcan, 0.9 V *vs.* RHE on Au/C and 0.95 V *vs.* RHE on $Au_{70}Pt_{30}$/C, $Au_{20}Pt_{80}$/C, and Pt/C. The peroxide production increases when the Pt content decreases in the electrode materials. From the kinetic parameter, $Au_{70}Pt_{30}$/C has the best exchange current density: j_0 = 300, 300, 700, 100 and 400 $\mu A\ cm^{-2}$ for Pt/C, $Au_{20}Pt_{80}$/C, $Au_{70}Pt_{30}$/C Au/C, and Vulcan carbon, respectively. In order to reduce the cost of the electrocatalysts, while keeping the same reaction kinetics, Pd-based electrodes have been recognized to be excellent candidates. Precisely, the addition of Ni or Ag boots the electroactivity of Pd either for oxidation of organic molecules [36,101,108] or ORR [56,131] in both acid and alkaline media. The synthesis of 2–5 nm of PdAg/C and PdNi/C (20 wt%) from the w/o method using the reversed micelles approach has been reported [56]. The polarization curves for ORR on Pd/C, $Pd_{70}Ni_{30}$/C, and $Pd_{70}Ag_{30}$/C electrode materials are represented in Figure 11. In the activation region (0.95–0.85 V *vs.* RHE), the addition of the second metal to Pd does not induce any benefit in terms of activity. The electrodes' efficiency in terms of oxygen reduction current follows the order Pd/C > $Pd_{70}Ni_{30}$/C > $Pd_{70}Ag_{30}$/C. This is supported by the value of the exchange current density (j_0), which is 11.1, 7.4 and 1.6 $\mu A\ cm^{-2}$ for Pd/C, $Pd_{70}Ni_{30}$/C, and $Pd_{70}Ag_{30}$/C, respectively [56]. In the mixed activation-diffusion limiting control domain (0.85–0.65 V *vs.* RHE, Figure 11), the presence of Ag or Ni increases slightly the limiting current, which is roughly 6.9 mA cm^{-2} (Pd/C), 7.1 mA cm^{-2} ($Pd_{70}Ag_{30}$/C), and 8.0 mA cm^{-2} ($Pd_{70}Ni_{30}$/C). The determined number of exchanged electrons, from Koutecky-Levich plots, is close to 4 for all the catalysts. This shows that O_2 reduction is a four-electron transfer process for the electrode potential centered at *ca.* 0.85 V *vs.* RHE. But, the careful analysis of the two behaviors in the polarization curves (on the disc) in 0.9–0.8 V *vs.* RHE and 0.75–0.6 V *vs.* RHE indicates a 2 + 2 electrons process.

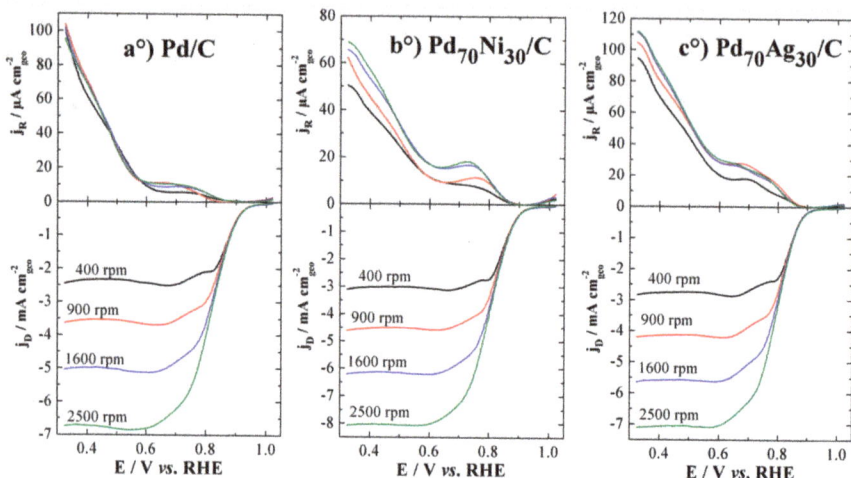

Figure 11. Disc (**bottom**) and ring (**top**) current-potential curves for ORR on (**a**) Pd/C, (**b**) $Pd_{70}Ni_{30}$/C and (**c**) $Pd_{70}Ag_{30}$/C electrocatalysts prepared from w/o method (20 wt%), in O_2-saturated 1 M NaOH at 5 mV s^{-1}. Metal loading on the electrode *ca.* 100 μg cm^{-2}.

3.2. ORR Activity on Various Carbon Supported Nanoparticles Prepared from the Polyol Method

Examination of the oxygen reduction reaction on various metals, including Pt, Pd, Rh, Ir, and Au began in the 1960s. Over the past decade the polyol process has been used to synthesize metallic nanoparticles such as Pt [132,133], Pd [88,134], Au [135], Fe [136], Ni [137,138], Co [139], Ag [140,141], Cu [142,143], Sn [144], and Rh [73]. Platinum is one of the most active metal catalysts toward many electrochemical reactions, such as oxidation of small molecules and reduction of molecular oxygen in PEMFC. Compared to other transition metals, Pt adsorbs oxygen with an intermediate bond strength. That is, Pt adsorbs oxygen strongly enough to be reduced, but not so strongly that the surface oxidizes. Additionally, the transition metals, for instance Ni, Co, Cr and Fe, adsorb oxygen so strongly that the surface may fully oxidize, while Au adsorbs oxygen so weakly that it does not stick to the surface. As platinum exhibits the highest catalytic activity for the oxygen reduction reaction [145], carbon supported platinum-based bimetallic alloys have been investigated as electrocatalysts to reduce the voltage losses associated with the cathode performance. Platinum-based bimetallic catalysts provide high oxygen reduction activity on the basis of d-band modification by the addition of a second metal. Toda *et al.* [146] reported that oxygen adsorption increases in the case of changing of the electronic structure of Pt induced by a transition metal, and then the O-O bond is weakened. For this purpose, various platinum-based bimetallic catalysts, such as Pt-Ni [147,148], Pt-Co [149,150], Pt-Fe [151,152], Pt-Cr [153,154], Pt-Cu [155,156], Pt-V [157,158], Pt-Mn [159], Pt-Bi [35], Pt-Te [160] have been reported. It can be clearly emphasized that modified-platinum catalysts display 1.5–3 times higher catalytic activity than that of pure platinum catalyst. Alvarez et al. [88] reported oxygen reduction activity on carbon supported palladium prepared by using ethylene glycol, sodium borohydride, and formaldehyde. They showed that reduction of the palladium precursor salt in alkaline medium led to small palladium nanoparticles around 5.7 nm at pH 11. It is clearly displayed that the peak position for palladium oxide reduction depends on the nanoparticles size. Indeed, this peak is centered at 0.70 V vs. RHE on Pd/C-ETEK; 0.72 V vs. RHE on Pd/C-CH$_2$O, 0.74 V vs. RHE on Pd/C-NaBH$_4$ and 0.74 V vs. RHE on Pd/C-EG. Rao et al. [152] reported Pt and Pt-M (M: Fe, Co, Cr) alloy catalysts prepared by the polyol method in 1,2-hexadecanediol in the presence of nonanoic acid and nonylamine as protecting agents. The results of linear sweep voltammetry indicated that the Pt alloy catalysts exhibited 1.5–1.7 times higher oxygen reduction activity than that of the as-synthesized and commercial Pt catalyst. Additionally, electrocatalytic activity and stability on graphene supported Pt$_3$-Co and Pt$_3$-Cr alloy catalysts were reported by Rao et al. [150] for the oxygen reduction reaction. The fuel cell performance of the catalysts was evaluated with 0.4 mg$_{Pt}$ cm^{-2} catalyst loading on the cathode and at 353 K and 1 atm. The power densities of 790, 875, 985 mW cm^{-2} were observed for graphene supported Pt, Pt$_3$-Co, and Pt$_3$-Cr catalysts, respectively. The stability of the so-called catalysts were investigated by using continuous potential cyclic voltammetry swept for 500 cycles in O$_2$-saturated 0.5 M H$_2$SO$_4$ and then linear scan voltammetry recorded at 1600 rpm and 5 mV s^{-1}. No obvious decrease in the oxygen reduction activity was observed for graphene supported catalysts after a continuous 500 cycles. Santiago et al. [149] prepared homogeneously dispersed Pt-Co bimetallic catalysts with 1.9 nm particle size, which have a high degree of alloying without thermal treatment. H$_2$/O$_2$ PEM fuel cell polarization curves for oxygen reduction were recorded at 80 °C with a 0.4 mg cm^{-2} total metal loading. The single cell polarization response related to the as-prepared catalysts exhibited superior mass activity compared to commercial Pt/C catalyst. As reported by Chen and co-workers [161], shape controlled Pt-Ni bimetallic nanocrystals exhibit enhanced oxygen reduction activity. Pt$_3$-Ni nanoframe catalysts exhibited in mass activity a factor of 22 and in specific activity a factor of 36, for the enhancement for the oxygen reduction reaction. In addition to the high intrinsic and mass activities, Pt$_3$-Ni nanoframe catalysts showed considerable durability for a duration of 10,000 potential cycles at different scan rates from 2–200 mV s^{-1}. Kumar et al. [162] prepared carbon supported palladium catalysts by using the polyol process for the oxygen reduction reaction. They reported that pretreatment of Vulcan XC-72R carbon support influenced Pd nanoparticle morphology and its activity towards the oxygen reduction reaction in acidic solution. The mass activities, measured at 0.7 V *vs.* RHE, for Pd at 0.07 M H_3PO_4, 10% H_2O_2 and 0.2 M KOH treated carbon supports were superior to that of E-TEK 20% Pd/C. They observed less than 4% H_2O_2 formation on different Pd/C catalysts in the kinetic potential regions. The Tafel slopes of the oxygen reduction reaction on different Pd/C catalysts showed two different regions with two different slopes. These Tafel slope values are about 60 and 120 mV dec^{-1}, at the low current and high current density regions, suggesting different adsorption isotherms of oxygenated species such as Temkin and Langmuir isotherms [163,164], respectively. Carbon supported Pt-Cu bimetallic catalysts were prepared by the polyol process for the oxygen reduction reaction by Tseng *et al.* [155]. The prepared catalysts were exposed under 300, 600, 900 °C for 1 h in a flowing mixture of 90% Ar–10% H_2. They reported that the Pt-Cu/C catalyst treated at 300 °C exhibited superior catalytic activity in terms of mass activity and specific activity than that of Pt/C in 0.1 M $HClO_4$. From the experimental data recorded at 1600 rpm and 5 mV s^{-1}, it was reported that Pt-Cu/C-300 showed the highest mass activity of 651 mA mg^{-1}, and the highest specific activity of 1.33 mA cm^{-2}. In the case of Pt-Cu/C-600 and Pt-Cu/C-900, lower mass and specific activities were observed than that of the unheated Pt-Cu/C catalyst.

As a conclusion, the polyol reduction process permits the preparation of size and shape controlled metal nanoparticles by improving the synthesis parameters for electrochemical applications. Parameters such as reaction duration, reaction temperature, and the pH value of the electrolyte assist to a remarkable extent the reduction kinetics [63,80].

3.3. ORR Activity on Pt/C Electrocatalyst Synthesized from BAE Method

Recently introduced in nanoscale material science, the BAE method enables the preparation of nanoscale electrocatalysts without using organic molecules. Materials from this advanced synthetic route have been primary used as anode materials for organics electro-oxidation [36,38,101] and have been successfully utilized as anode-based electrodes in the glucose hybrid biofuel cell for bionanotechnology applications. More importantly, it has been demonstrated that Au-Pt catalysts prepared with the BAE method exhibit unexpected cathode selectivity-tolerance-durability in mixed reactants and in a poisoning environment, and physiological medium [102]. Indeed, $Au_{60}Pt_{40}$ bimetallic (3.2 nm) supported onto carbon Ketjenblack EC 600-JD was able to selectively reduce oxygen in a membraneless biomedical implantable glucose fuel cell at pH 7.7 in human serum to activate a pacemaker, which constitutes a real application [102]. To some extent, BAE allows the development of advanced low temperature FCs electrocatalysts. In order to check these exceptional behaviors and compare them with the existing methods in acidic and alkaline media using the RRDE technique, we conducted ORR at 20 wt%

Pt/C, as prepared from BAE as a state-of-art catalytic material [9,28,165,166]. In both media, several metal loadings were studied ranging from 6–100 μg_{Pt} cm^{-2}. Durability tests were performed by cycling the electrode potential from 0.05–1.10 V *vs.* RHE for 1000 cyclic voltammograms (CVs). Figure 12a shows the ORR polarization curves recorded in O$_2$-saturated 0.1 M HClO$_4$. Before recording these curves, the RRDE setups were calibrated as illustrated in Figure 9. Then, the electrode was scanned from 0.05–1.1 V *vs.* RHE at 50 mV s^{-1} twenty times followed by 2 CVs at 5 mV s^{-1}. Finally ORR was performed by scanning the disc (glassy carbon, 0.196 cm^2) from 1.1–0.2 V *vs.* RHE at 5 mV s^{-1}, while that of the potential of the ring (platinum, 0.11 cm^2) was fixed at 1.2 V *vs.* RHE to oxidize any peroxide intermediate. The oxygen reduction at the disc starts earlier than 1 V *vs.* RHE, with negligible H$_2$O$_2$ in the whole scanned potential range. As a first qualitative observation, the catalyst displays good kinetics because of the sharp current behavior in the potential range of 1.0–0.8 V *vs.* RHE reaching a half-potential ($E_{1/2}$) of 0.90 and 0.85 *vs.* RHE for 400 and 900 rpm, respectively. Almost the same value of the OCP and $E_{1/2}$ were reached in 0.1 M NaOH. The important diffusion current density obtained herein and compared to that resulting in the w/o method can be assigned to the synthetic method, whereas the magnitude of the metal loading is almost two times lower [33,52]. One of the recurring themes in the FCs performance loss is the decrease of the ORR activity due to the active electrochemical surface area (ECSA) loss over the cycles [9,58,167–170]. The durability test was performed as depicted in Figure 12b. It provides evidence that the BAE method delivers prototype catalyst with impressive durability performances where the ORR curve is superimposed with the initial polarization curve, even if the catalyst loses 12% on its maximum ECSA. Indeed, compared to the current carbon supported nanoparticles, in such a situation, the catalyst is expected to lose more than 50% [58,171,172].

The kinetic parameters of the catalyst were analyzed. Figure 13a shows the Koutecky-Levich plots, highlighting a linear dependence at all potentials. This linearity combined with the parallelism is not surprising and clearly indicates that the oxygen reduction reaction is first-order kinetics with respect to oxygen. The type of plot is crucial for determining the apparent kinetic current density at each potential. Then, all these values are plotted as in Figure 13b to give the limiting current density (j_L). In 0.1 M HClO$_4$, j_L = 150 and 140 mA cm^{-2} for the initial Pt/C and Pt/C after 1000 CVs, respectively. These values are found to be 105 and 97 mA cm^{-2} in 0.1 M NaOH. Considering the likely impact of the electrolyte on the ORR performances at the nanoparticles, Nesselberger *et al.* [5] found in 2011 that the absolute reaction rates decrease in the order HClO$_4$ > KOH > H$_2$SO$_4$. They explained it by the anionic adsorption strength increase (acid solutions), whereas the lower activity of KOH compared to HClO$_4$ might be due to the noncovalent interactions between hydrated K$^+$ and adsorbed OH$^-$ [5,173]. Because of the ORR improved durability and activity in HClO$_4$, this solution is the electrolyte of choice for the electrochemical tests in a three-electrode cell [11,14,173–175].

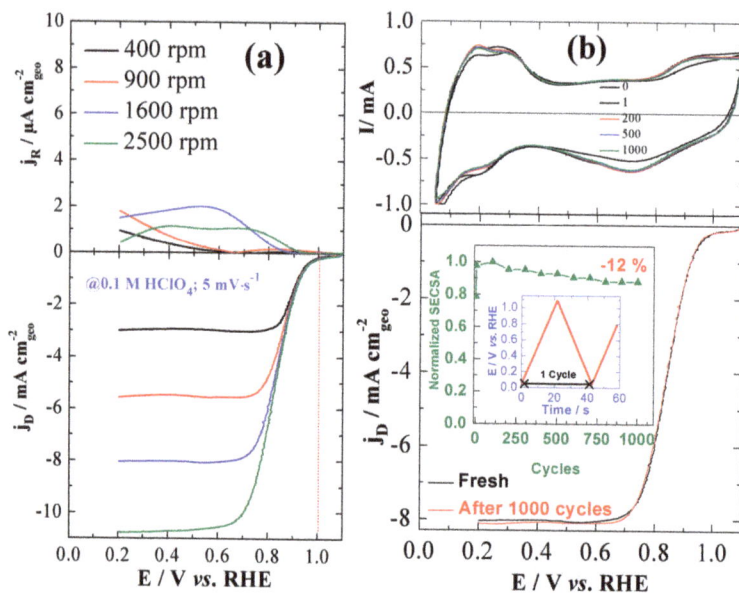

Figure 12. ORR polarization curves recorded in a O$_2$-saturated 0.1 M HClO$_4$ solution at 20 wt% Pt/C catalyst prepared from the BAE method: (**a**) Ring (top) and disc (bottom) current density; (**b**) Electrochemical durability (top) and the ORR polarization after 1000 CVs (bottom). ORR performed at 5 mV s^{-1}, 1600 rpm, and room temperature. Insets show the SECSA decay over the 1000 cycles from 0.05–1.1 V *vs.* RHE at 50 mV s^{-1}. Metal loading on the electrode: 100 μg$_{Pt}$ cm^{-2}.

Two Tafel slopes (inset in Figure 13b) were determined: 125 mV dec^{-1} at low overpotential (PtO$_x$ region) and 63 mV dec^{-1} at high overpotential (Pt free region). On the fresh Pt/C electrode, these values were 130 and 67, respectively. Besides, 126 and 69 mV dec^{-1} were evaluated with the fresh catalyst. Then, 140 and 71 mV dec^{-1} were obtained after the durability test in 0.1 M NaOH. All these determined Tafel slopes are close to the theoretical ones, which are 120 mV dec^{-1} (low η) and 60 mV dec^{-1} (high η) [126]. It should be emphasized that these values are in agreement with those reported both for Temkin adsorption isotherms of oxygenated species (low η), or Langmuir ones for high η (where Pt surface is free of PtO$_x$ species) [176,177].

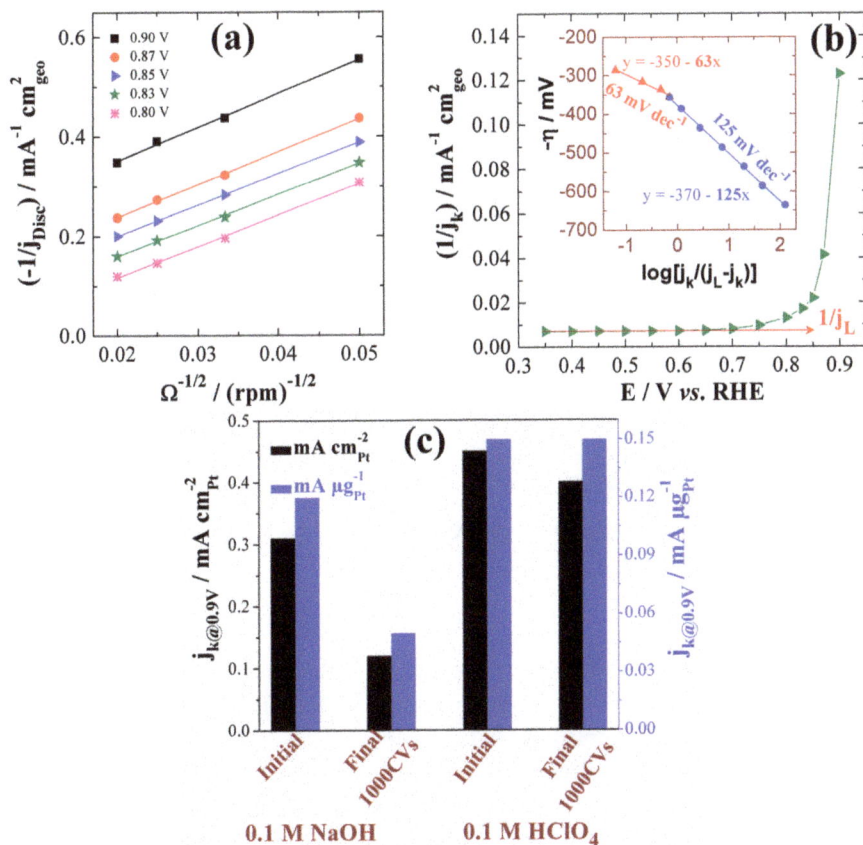

Figure 13. (a) Koutecky–Levich and (b) j_k^{-1} plot for the determination of j_L. Inset in (b) shows the Tafel plots: data are extracted from ORR after the durability test in 0.1 M HClO$_4$. (c) Comparison of j_k at 0.9 V *vs.* RHE in 0.1 M HClO$_4$ and 0.1 M NaOH.

Figure 13c gathers the kinetic current (extracted from Figure 13a) normalized either with ECSA mass (mA cm^{-2}$_{Pt}$) or Pt mass (mA µg^{-1}$_{Pt}$) at 0.9 V *vs.* RHE. Initially, j_k = 0.45 mA cm^{-2}$_{Pt}$ (0.15 mA µg^{-1}$_{Pt}$) in 0.1 M HClO$_4$ and j_k = 0.31 mA cm^{-2}$_{Pt}$ (0.12 mA µg^{-1}$_{Pt}$) in 0.1 M NaOH. After the durability test, they became j_k = 0.40 mA cm^{-2}$_{Pt}$ (0.15 mA µg^{-1}$_{Pt}$) in 0.1 M HClO$_4$ and j_k = 0.14 mA cm^{-2}$_{Pt}$ (0.05 mA µg^{-1}$_{Pt}$) in 0.1 M NaOH, meaning a good stability in acid medium and a performance loss in the alkaline medium. The stability could be improved by the addition of second metals like palladium or gold to platinum. Table 1 summarizes the different parameters. It can be seen that j_0 is 179 10^{-3} mA cm^{-2} (and 153 10^{-3} mA cm^{-2} after stability) in HClO$_4$, and 122 10^{-3} mA cm^{-2} (and 114 10^{-3} mA cm^{-2} after stability) in NaOH, respectively. These values of j_0 are more important than those reported by Demarconnay *et al.* [33] (16.8 mA cm^{-2}) and Habrioux *et al.* [52] (0.3 10^{-3} mA cm^{-2}) on Pt/C synthesized from the w/o method, reflecting an enhanced ORR kinetics at these electrode materials.

Table 1 gathers the experimental data concerning the ORR on carbon-supported nanomaterials prepared from the colloidal method. It would be interesting to discuss each point depending on the used method. Unfortunately, there is missing information in the literature. The ORR occurs with high OCP close to 1 V *vs.* RHE and suitable half-potential (E$_{1/2}$) roughly at 0.85 V *vs.* RHE, as on the most active and advanced Vulcan supported PtCo [3,14], PtNi [3,11] or PtNiCo [13] nanoparticles as well as on the free nanoparticles in solution (unsupported catalysts) [10,65]. The other interesting result from this table concerns the number of exchanged electrons. This value is close to 4, which means that the reaction does not produce any significant peroxide, maximizing the Faradaic yield. Conversely, it is difficult to compare the kinetic activity with j_k due to the fact that some authors did not clearly indicate whether the current was normalized with the geometric or active surface area. For this, we recommend further papers include full information concerning their materials for better comparison.

Table 1. Comparative performances of oxygen reduction reaction (ORR) results from RRDE experiments on various catalysts prepared from colloidal methods. E$_{1/2}$ was graphically determined at 1600 rpm (potential at i = I$_D$/2). Note: OCP, "w/o" and "BAE" refer to open circuit potential, water-in-oil and bromide anion exchange methods, respectively. Empty box (–) means that the original article does not provide such data.

Catalyst	Electrolyte	OCP	E$_{1/2}$	j_k (mA cm^{-2}Pt) At (V *vs.* RHE)		Tafel slope (mV dec^{-1})		j_0 (× 10^{-3} mA cm^{-2})		n_{ex}	Method And Ref.
		V *vs.* RHE		0.90	0.85	Low	High	Low	High		
50 wt% Pt/C	0.2 M NaOH	1.05	0.85	0.60	-	81	126	1.2	16.8	4	w/o [33]
50 wt% Pt$_{80}$Bi$_{20}$/C		1.05	0.87	1.14	-	62	127	0.3	23.2	4	
20 wt% Pd/C	1 M NaOH	0.95	0.83	-	0.07	89	162	-	11.1	3.9	w/o [56]
20 wt% Pd$_{70}$Ni$_{30}$/C		0.95	0.85	-	0.06	76	133	-	1.63	3.8	
20 wt% Pd/C		0.80	0.70	-	-	60	120	-	-	4	Polyol [162]
20 wt% Pt/C	0.5 M H$_2$SO$_4$	0.925	0.84	-	*	-	-	-	-	4	Polyol [152]
20 wt% Pt-Co/C		0.980	0.85	-	-	-	-	-	-	4	Polyol [152]
20 wt% Pt-Cr/C		0.985	0.85	-	-	-	-	-	-	4	Polyol [152]
20 wt% Ag/C	0.1 M NaOH	0.95	0.80	-	-	-	6.3	-	157	3.9	w/o [178]
20 wt% Pt/C	0.1 M HClO$_4$	1.09	0.85	0.45	1.01	67	130	0.45	179	4.0	BAE Here
	0.1 M NaOH	1.08	0.85	0.31	0.77	69	126	0.64	122	4.0	

4. Summary and Perspectives

This review paper focused on the preparation and application of carbon supported nanoparticles in electrocatalysis, and especially on the oxygen reduction reaction (ORR) in low temperature fuel cells (FCs). We examined the recent developments in nanocatalysts preparation science in order to better understand and correlate their catalytic performances toward the ORR. From this survey, we found that the catalytic properties can be precisely and effectively tuned by changing the experimental conditions. Definitively, among the developed methods for carbon-supported nanoparticles, the colloidal ones are the most used in ORR electrochemistry. The state-of-the-art Pt/C electrocatalyst shows poor long-term stability and different co-atoms (Co, Ni, Bi, Ag, Au) have been proposed to improve its performances. The reaction on Pt-based electrodes starts at *ca.* 1.1 V *vs.* RHE, which represents only 100 mV difference compared to the theoretical value (close to 1.2 V *vs.* RHE). More importantly, most of the optimized systems, display sharp behavior between 0.8–0.9 V *vs.* RHE in the potential-current ORR polarization curves, for the range of interest for FCs applications. From the different results, it can be concluded that the oxygen reduction reaction at the metal nanoparticles depends strongly on the electrolyte medium as well as the particle size. Fundamental studies at the laboratory scale reveal that the reaction kinetics decreases in the order $HClO_4$ > NaOH (or KOH) > H_2SO_4. For the particles size effect, the optimum window is 2–3 nm, where the active sites on the corners and edges are more available. Unfortunately, basic but fundamental data are missing in research papers about the kinetic parameters to enable better comparison. In this review paper, we were not able to compare the specific kinetics of the different electrode materials derived from the various preparation methods due to lack of information. Future studies are urged to provide clear and full in depth information on their ORR tests.

These recent advances in low temperature FCs electrocatalysts preparation indicate that the standard and quality of fundamental research in this area needs to continue unabated. Water-in-oil has been the method of choice for heterogeneous catalysis. The inability to clean the nanoparticles surface from the organic molecules used as surfactant affects the catalytic performance of the obtained catalysts. The recent initiated "Bromide Anion Exchange, BAE" method leads to various surfactant-free nanoparticles. Undoubtedly, the performance of such materials in the long term is expected to be of particular importance in fuel cell science. Even catalysts from colloidal methods (water-in-oil, polyol, Bönnemann, BAE) have demonstrated excellent ability toward the ORR; they have not been widely tested in the Membrane-Electrode-Assembly, MEA. Future works in this area should first focus on the performances of carbon supported metal nanoparticles in MEAs for the *in situ* oxygen reduction reaction as well as the FC results. Furthermore, in order to reduce the electrode cost and thus the FC system, incorporating non-noble metals is needed to reduce the amount of precious metals in the electrode materials. The experimental tools provided herein will be useful for early career researchers in FCs and could help in finding suitable ORR methodology.

Acknowledgments: The authors would like to thank financial supports from the French National Research Agency (ANR) through the program "ChemBio-Energy" and "Region Poitou-Charentes".

Author Contributions: Kouakou B. Kokoh conceived the project. Yaovi Holade and Nihat Ege Sahin wrote the first draft of the manuscript which was then improved by Karine Servat, Teko W. Napporn and Kouakou B. Kokoh. All the authors researched the literature.

Conflicts of Interest: The authors declare no conflict of interest.

References

1. Marković, N.M.; Ross, P.N., Jr. Surface science studies of model fuel cell electrocatalysts. *Surf. Sci. Rep.* **2002**, *45*, 117–229.
2. Wieckowski, A.; Savinova, E.R.; Vayenas, C.G. *Catalysis and Electrocatalysis at Nanoparticle Surfaces*; Marcel Dekker, Inc.: New York, NY, USA, 2003; p. 970.
3. Stamenkovic, V.R.; Mun, B.S.; Arenz, M.; Mayrhofer, K.J.J.; Lucas, C.A.; Wang, G.; Ross, P.N.; Markovic, N.M. Trends in electrocatalysis on extended and nanoscale Pt-bimetallic alloy surfaces. *Nat. Mater.* **2007**, *6*, 241–247.
4. Markovic, N.M.; Gasteiger, H.A.; Ross, P.N. Oxygen Reduction on Platinum Low-Index Single-Crystal Surfaces in Sulfuric Acid Solution: Rotating Ring-Pt(hkl) Disk Studies. *J. Phys. Chem.* **1995**, *99*, 3411–3415.
5. Nesselberger, M.; Ashton, S.; Meier, J.C.; Katsounaros, I.; Mayrhofer, K.J.J.; Arenz, M. The Particle Size Effect on the Oxygen Reduction Reaction Activity of Pt Catalysts: Influence of Electrolyte and Relation to Single Crystal Models. *J. Am. Chem. Soc.* **2011**, *133*, 17428–17433.
6. Shao, M.; Peles, A.; Shoemaker, K. Electrocatalysis on Platinum Nanoparticles: Particle Size Effect on Oxygen Reduction Reaction Activity. *Nano Lett.* **2011**, *11*, 3714–3719.
7. Anastasopoulos, A.; Davies, J.C.; Hannah, L.; Hayden, B.E.; Lee, C.E.; Milhano, C.; Mormiche, C.; Offin, L. The Particle Size Dependence of the Oxygen Reduction Reaction for Carbon-Supported Platinum and Palladium. *ChemSusChem* **2013**, *6*, 1973–1982.
8. Hernandez-Fernandez, P.; Masini, F.; McCarthy, D.N.; Strebel, C.E.; Friebel, D.; Deiana, D.; Malacrida, P.; Nierhoff, A.; Bodin, A.; Wise, A.M.; *et al.* Mass-selected nanoparticles of Pt$_x$Y as model catalysts for oxygen electroreduction. *Nat. Chem.* **2014**, *6*, 732–738.
9. Debe, M.K. Electrocatalyst approaches and challenges for automotive fuel cells. *Nature* **2012**, *486*, 43–51.
10. Cui, C.; Gan, L.; Li, H.-H.; Yu, S.-H.; Heggen, M.; Strasser, P. Octahedral PtNi Nanoparticle Catalysts: Exceptional Oxygen Reduction Activity by Tuning the Alloy Particle Surface Composition. *Nano Lett.* **2012**, *12*, 5885–5889.

11. Chen, C.; Kang, Y.; Huo, Z.; Zhu, Z.; Huang, W.; Xin, H.L.; Snyder, J.D.; Li, D.; Herron, J.A.; Mavrikakis, M.; *et al.* Highly Crystalline Multimetallic Nanoframes with Three-Dimensional Electrocatalytic Surfaces. *Science* **2014**, *343*, 1339–1343.

12. Corti, H.; Gonzalez, E.R. *Direct Alcohol Fuel Cells: Materials, Performance, Durability and Applications*; Springer: Dordrecht, The Netherlands, 2014; p. 370.

13. Huang, X.; Zhao, Z.; Chen, Y.; Zhu, E.; Li, M.; Duan, X.; Huang, Y. A Rational Design of Carbon-Supported Dispersive Pt-Based Octahedra as Efficient Oxygen Reduction Reaction Catalysts. *Energy Environ. Sci.* **2014**, *7*, 2957–2962.

14. Wang, D.; Xin, H.L.; Hovden, R.; Wang, H.; Yu, Y.; Muller, D.A.; DiSalvo, F.J.; Abruña, H.D. Structurally ordered intermetallic platinum–cobalt core–shell nanoparticles with enhanced activity and stability as oxygen reduction electrocatalysts. *Nat. Mater.* **2013**, *12*, 81–87.

15. Ross, P.N.; Radmilovic, V.; Markovic, N.M. Physical and Electrochemical Characterization of Bimetallic Nanoparticle Electrocatalysts. In *Catalysis and Electrocatalysis at Nanoparticle Surfaces*; Wieckowski, A., Savinova, E.R., Vayenas, C.G., Eds.; CRC Press: New York, NY, USA, 2003; pp. 311–342.

16. Geim, A.K.; Novoselov, K.S. The rise of graphene. *Nat. Mater.* **2007**, *6*, 183–191.

17. De Volder, M.F. L.; Tawfick, S.H.; Baughman, R.H.; Hart, A.J. Carbon Nanotubes: Present and Future Commercial Applications. *Science* **2013**, *339*, 535–539.

18. Andrews, R.; Jacques, D.; Qian, D.; Rantell, T. Multiwall Carbon Nanotubes: Synthesis and Application. *Acc. Chem. Res.* **2002**, *35*, 1008–1017.

19. Karousis, N.; Tagmatarchis, N.; Tasis, D. Current Progress on the Chemical Modification of Carbon Nanotubes. *Chem. Rev.* **2010**, *110*, 5366–5397.

20. Che, A.-F.; Germain, V.; Cretin, M.; Cornu, D.; Innocent, C.; Tingry, S. Fabrication of free-standing electrospun carbon nanofibers as efficient electrode materials for bioelectrocatalysis. *New J. Chem.* **2011**, *35*, 2848–2853.

21. Andersen, S.M.; Borghei, M.; Lund, P.; Elina, Y.-R.; Pasanen, A.; Kauppinen, E.; Ruiz, V.; Kauranen, P.; Skou, E.M. Durability of carbon nanofiber (CNF) & carbon nanotube (CNT) as catalyst support for Proton Exchange Membrane Fuel Cells. *Solid State Ionics* **2013**, *231*, 94–101.

22. Shulaker, M.M.; Hills, G.; Patil, N.; Wei, H.; Chen, H.-Y.; Wong, H.S.P.; Mitra, S. Carbon nanotube computer. *Nature* **2013**, *501*, 526–530.

23. Gao, F.; Viry, L.; Maugey, M.; Poulin, P.; Mano, N. Engineering hybrid nanotube wires for high-power biofuel cells. *Nat. Commun.* **2010**, *1*, 3.

24. Zebda, A.; Gondran, C.; le Goff, A.; Holzinger, M.; Cinquin, P.; Cosnier, S. Mediatorless high-power glucose biofuel cells based on compressed carbon nanotube-enzyme electrodes. *Nat. Commun.* **2011**, *2*, 370.

25. Sinfelt, J.H.; Lam, Y.L.; Cusumano, J.A.; Barnett, A.E. Nature of ruthenium-copper catalysts. *J. Catal.* **1976**, *42*, 227–237.

26. Ferrando, R.; Jellinek, J.; Johnston, R.L. Nanoalloys: From Theory to Applications of Alloy Clusters and Nanoparticles. *Chem. Rev.* **2008**, *108*, 845–910.

27. Raimondi, F.; Scherer, G.G.; Kötz, R.; Wokaun, A. Nanoparticles in Energy Technology: Examples from Electrochemistry and Catalysis. *Angew. Chem. Int. Ed.* **2005**, *44*, 2190–2209.

28. Katsounaros, I.; Cherevko, S.; Zeradjanin, A.R.; Mayrhofer, K.J.J. Oxygen Electrochemistry as a Cornerstone for Sustainable Energy Conversion. *Angew. Chem. Int. Ed.* **2014**, *53*, 102–121.

29. Hoar, T.; Schulman, J. Transparent water-in-oil dispersions: the oleopathic hydro-micelle. *Nature* **1943**, *152*, 102–103.

30. Boutonnet, M.; Kizling, J.; Stenius, P.; Maire, G. The preparation of monodisperse colloidal metal particles from microemulsions. *Colloids Surf.* **1982**, *5*, 209–225.

31. Schwuger, M.-J.; Stickdorn, K.; Schomaecker, R. Microemulsions in Technical Processes. *Chem. Rev.* **1995**, *95*, 849–864.

32. Bock, C.; Halvorsen, H.; MacDougall, B. Catalyst Synthesis Techniques. In *PEM Fuel Cell Electrocatalysts and Catalyst Layers*; Zhang, J., Ed.; Springer-Verlag: London, UK, 2008; pp. 447–485.

33. Demarconnay, L.; Coutanceau, C.; Léger, J.M. Study of the oxygen electroreduction at nanostructured PtBi catalysts in alkaline medium. *Electrochim. Acta* **2008**, *53*, 3232–3241.

34. Simões, M.; Baranton, S.; Coutanceau, C. Electrooxidation of Sodium Borohydride at Pd, Au, and Pd_xAu_{1-x} Carbon-Supported Nanocatalysts. *J. Phys. Chem. C* **2009**, *113*, 13369–13376.

35. Roychowdhury, C.; Matsumoto, F.; Mutolo, P.F.; Abruña, H.D.; DiSalvo, F.J. Synthesis, Characterization, and Electrocatalytic Activity of PtBi Nanoparticles Prepared by the Polyol Process. *Chem. Mater.* **2005**, *17*, 5871–5876.

36. Holade, Y.; Morais, C.; Arrii-Clacens, S.; Servat, K.; Napporn, T.W.; Kokoh, K.B. New Preparation of PdNi/C and PdAg/C Nanocatalysts for Glycerol Electrooxidation in Alkaline Medium. *Electrocatalysis* **2013**, *4*, 167–178.

37. Holade, Y.; Morais, C.; Napporn, T.W.; Servat, K.; Kokoh, K.B. Electrochemical Behavior of Organics Oxidation on Palladium-Based Nanocatalysts Synthesized from Bromide Anion Exchange. *ECS Trans.* **2014**, *58*, 25–35.

38. Tonda-Mikiela, P.; Napporn, T.W.; Morais, C.; Servat, K.; Chen, A.; Kokoh, K.B. Synthesis of Gold-Platinum Nanomaterials Using Bromide Anion Exchange-Synergistic Electroactivity toward CO and Glucose Oxidation. *J. Electrochem. Soc.* **2012**, *159*, H828–H833.

39. Lahmani, M.; Bréchignac, C.; Houdy, P. *Les Nanosciences: 2. Nanomatériaux et Nanochimie*, 2nd ed.; Belin: Paris, France, 2012; p. 732. (In French)

40. Schulman, J.H.; Stoeckenius, W.; Prince, L.M. Mechanism of Formation and Structure of Micro Emulsions by Electron Microscopy. *J. Phys. Chem.* **1959**, *63*, 1677–1680.

41. Schulman, J.H.; Friend, J.A. Light scattering investigation of the structure of transparent oil-water disperse systems. II. *J. Colloid Sci.* **1949**, *4*, 497–509.

42. Clausse, M.; Peyrelasse, J.; Heil, J.; Boned, C.; Lagourette, B. Bicontinuous structure zones in microemulsions. *Nature* **1981**, *293*, 636–638.

43. Danielsson, I.; Lindman, B. The definition of microemulsion. *Colloids Surf.* **1981**, *3*, 391–392.

44. Eriksson, S.; Nylén, U.; Rojas, S.; Boutonnet, M. Preparation of catalysts from microemulsions and their applications in heterogeneous catalysis. *Appl. Catal. A* **2004**, *265*, 207–219.

45. Capek, I. Preparation of metal nanoparticles in water-in-oil (w/o) microemulsions. *Adv. Colloid Int. Sci.* **2004**, *110*, 49–74.

46. Ingelsten, H.H.; Bagwe, R.; Palmqvist, A.; Skoglundh, M.; Svanberg, C.; Holmberg, K.; Shah, D.O. Kinetics of the Formation of Nano-Sized Platinum Particles in Water-in-Oil Microemulsions. *J. Colloid Int. Sci.* **2001**, *241*, 104–111.

47. Solla-Gullón, J.; Rodes, A.; Montiel, V.; Aldaz, A.; Clavilier, J. Electrochemical characterisation of platinum–palladium nanoparticles prepared in a water-in-oil microemulsion. *J. Electroanal. Chem.* **2003**, *554–555*, 273–284.

48. Solla-Gullón, J.; Vidal-Iglesias, F.J.; Montiel, V.; Aldaz, A. Electrochemical characterization of platinum–ruthenium nanoparticles prepared by water-in-oil microemulsion. *Electrochim. Acta* **2004**, *49*, 5079–5088.

49. Habrioux, A. Préparation et caractérisation de nanoparticules à base d'or et de platine pour l'anode d'une biopile glucose/dioxygène. Ph.D. Thesis, University of Poitiers, France, October 2009.

50. Simões, M. Développement d'électrocatalyseurs anodiques plurimétalliques nanostructurés pour une application en pile à combustible à membrane alcaline solide (SAMFC). Ph.D. Thesis, University of Poitiers, France, March 2011.

51. Liu, H.; Song, C.; Zhang, L.; Zhang, J.; Wang, H.; Wilkinson, D.P. A review of anode catalysis in the direct methanol fuel cell. *J. Power Sources* **2006**, *155*, 95–110.

52. Habrioux, A.; Diabaté, D.; Rousseau, J.; Napporn, T.; Servat, K.; Guétaz, L.; Trokourey, A.; Kokoh, K.B. Electrocatalytic Activity of Supported Au–Pt Nanoparticles for CO Oxidation and O_2 Reduction in Alkaline Medium. *Electrocatalysis* **2010**, *1*, 51–59.

53. Habrioux, A.; Sibert, E.; Servat, K.; Vogel, W.; Kokoh, K.B.; Alonso-Vante, N. Activity of Platinum–Gold Alloys for Glucose Electrooxidation in Biofuel Cells. *J. Phys. Chem. B* **2007**, *111*, 10329–10333.

54. Habrioux, A.; Vogel, W.; Guinel, M.; Guetaz, L.; Servat, K.; Kokoh, B.; Alonso-Vante, N. Structural and electrochemical studies of Au-Pt nanoalloys. *Phys. Chem. Chem. Phys.* **2009**, *11*, 3573–3579.

55. Wu, M.-L.; Chen, D.-H.; Huang, T.-C. Preparation of Au/Pt Bimetallic Nanoparticles in Water-in-Oil Microemulsions. *Chem. Mater.* **2001**, *13*, 599–606.

56. Diabaté, D.; Napporn, T.W.; Servat, K.; Habrioux, A.; Arrii-Clacens, S.; Trokourey, A.; Kokoh, K.B. Kinetic Study of Oxygen Reduction Reaction on Carbon Supported Pd-Based Nanomaterials in Alkaline Medium. *J. Electrochem. Soc.* **2013**, *160*, H302–H308.

57. Holade, Y.; Morais, C.; Servat, K.; Napporn, T.W.; Kokoh, K.B. Enhancing the available specific surface area of carbon supports to boost the electroactivity of nanostructured Pt catalysts. *Phys. Chem. Chem. Phys.* **2014**, *16*, 25609–25620.

58. Grolleau, C.; Coutanceau, C.; Pierre, F.; Léger, J.M. Effect of potential cycling on structure and activity of Pt nanoparticles dispersed on different carbon supports. *Electrochim. Acta* **2008**, *53*, 7157–7165.

59. Solla-Gullón, J.; Montiel, V.; Aldaz, A.; Clavilier, J. Electrochemical characterisation of platinum nanoparticles prepared by microemulsion: How to clean them without loss of crystalline surface structure. *J. Electroanal. Chem.* **2000**, *491*, 69–77.

60. Napporn, T.; Habrioux, A.; Rousseau, J.; Servat, K.; Léger, J.-M.; Kokoh, B. Effect of the Cleaning Step on the Morphology of Gold Nanoparticles. *Electrocatalysis* **2011**, *2*, 24–27.

61. Fievet, F.; Lagier, J.P.; Blin, B.; Beaudoin, B.; Figlarz, M. Homogeneous and heterogeneous nucleations in the polyol process for the preparation of micron and submicron size metal particles. *Solid State Ionics* **1989**, *32–33*(Part 1), 198–205.

62. Bonet, F.; Guéry, C.; Guyomard, D.; Herrera Urbina, R.; Tekaia-Elhsissen, K.; Tarascon, J.M. Electrochemical reduction of noble metal compounds in ethylene glycol. *Int. J. Inorg. Mater.* **1999**, *1*, 47–51.

63. Bock, C.; Paquet, C.; Couillard, M.; Botton, G.A.; MacDougall, B.R. Size-Selected Synthesis of PtRu Nano-Catalysts: Reaction and Size Control Mechanism. *J. Am. Chem. Soc.* **2004**, *126*, 8028–8037.

64. Skrabalak, S.E.; Wiley, B.J.; Kim, M.; Formo, E.V.; Xia, Y. On the Polyol Synthesis of Silver Nanostructures: Glycolaldehyde as a Reducing Agent. *Nano Lett.* **2008**, *8*, 2077–2081.

65. Chen, J.; Lim, B.; Lee, E.P.; Xia, Y. Shape-controlled synthesis of platinum nanocrystals for catalytic and electrocatalytic applications. *Nano Today* **2009**, *4*, 81–95.

66. Xiong, Y.; Xia, Y. Shape-Controlled Synthesis of Metal Nanostructures: The Case of Palladium. *Adv. Mater.* **2007**, *19*, 3385–3391.

67. Grzelczak, M.; Perez-Juste, J.; Mulvaney, P.; Liz-Marzan, L.M. Shape control in gold nanoparticle synthesis. *Chem. Soc. Rev.* **2008**, *37*, 1783–1791.

68. Wang, C.; Daimon, H.; Onodera, T.; Koda, T.; Sun, S. A General Approach to the Size- and Shape-Controlled Synthesis of Platinum Nanoparticles and Their Catalytic Reduction of Oxygen. *Angew. Chem. Int. Ed.* **2008**, *47*, 3588–3591.

69. Alia, S.M.; Zhang, G.; Kisailus, D.; Li, D.; Gu, S.; Jensen, K.; Yan, Y. Porous Platinum Nanotubes for Oxygen Reduction and Methanol Oxidation Reactions. *Adv. Funct. Mater.* **2010**, *20*, 3742–3746.

70. Lee, I.; Delbecq, F.; Morales, R.; Albiter, M.A.; Zaera, F. Tuning selectivity in catalysis by controlling particle shape. *Nat. Mater.* **2009**, *8*, 132–138.

71. Kyriakou, G.; Beaumont, S.K.; Humphrey, S.M.; Antonetti, C.; Lambert, R.M. Sonogashira Coupling Catalyzed by Gold Nanoparticles: Does Homogeneous or Heterogeneous Catalysis Dominate? *ChemCatChem* **2010**, *2*, 1444–1449.

72. Campbell, C.T.; Parker, S.C.; Starr, D.E. The Effect of Size-Dependent Nanoparticle Energetics on Catalyst Sintering. *Science* **2002**, *298*, 811–814.

73. Biacchi, A.J.; Schaak, R.E. The Solvent Matters: Kinetic *versus* Thermodynamic Shape Control in the Polyol Synthesis of Rhodium Nanoparticles. *ACS Nano* **2011**, *5*, 8089–8099.

74. Song, H.; Kim, F.; Connor, S.; Somorjai, G.A.; Yang, P. Pt Nanocrystals: Shape Control and Langmuir–Blodgett Monolayer Formation. *J. Phys. Chem. B* **2004**, *109*, 188–193.

75. Susut, C.; Nguyen, T.D.; Chapman, G.B.; Tong, Y. Particle Size Limit for Concomitant Tuning of Size and Shape of Platinum Nanoparticles. *J. Cluster Sci.* **2007**, *18*, 773–780.

76. Lee, E.; Murthy, A.; Manthiram, A. Comparison of the stabilities and activities of Pt–Ru/C and Pt3–Sn/C electrocatalysts synthesized by the polyol method for methanol electro-oxidation reaction. *J. Electroanal. Chem.* **2011**, *659*, 168–175.

77. Kim, H.J.; Choi, S.M.; Green, S.; Tompsett, G.A.; Lee, S.H.; Huber, G.W.; Kim, W.B. Highly active and stable PtRuSn/C catalyst for electrooxidations of ethylene glycol and glycerol. *Appl. Catal. B: Env.* **2011**, *101*, 366–375.

78. Wang, L.; Zhai, J.-J.; Jiang, K.; Wang, J.-Q.; Cai, W.-B. Pd–Cu/C electrocatalysts synthesized by one-pot polyol reduction toward formic acid oxidation: Structural characterization and electrocatalytic performance. *Int. J. Hydrogen Energ.* **2015**, *40*, 1726–1734.

79. Wang, G.; Takeguchi, T.; Muhamad, E.N.; Yamanaka, T.; Sadakane, M.; Ueda, W. Preparation of Well-Alloyed PtRu/C Catalyst by Sequential Mixing of the Precursors in a Polyol Method. *J. Electrochem. Soc.* **2009**, *156*, B1348.

80. Joseyphus, R.J.; Matsumoto, T.; Takahashi, H.; Kodama, D.; Tohji, K.; Jeyadevan, B. Designed synthesis of cobalt and its alloys by polyol process. *J. Solid State Chem.* **2007**, *180*, 3008–3018.

81. Susut, C.; Tong, Y.J. Size-Dependent Methanol Electro-oxidation Activity of Pt Nanoparticles with Different Shapes. *Electrocatalysis* **2011**, *2*, 75–81.

82. González-Quijano, D.; Pech-Rodríguez, W.J.; Escalante-García, J.I.; Vargas-Gutiérrez, G.; Rodríguez-Varela, F.J. Electrocatalysts for ethanol and ethylene glycol oxidation reactions. Part I: Effects of the polyol synthesis conditions on the characteristics and catalytic activity of Pt–Sn/C anodes. *Int. J. Hydrogen Energ.* **2014**, *39*, 16676–16685.

83. Jiang, L.; Hsu, A.; Chu, D.; Chen, R. Ethanol electro-oxidation on Pt/C and PtSn/C catalysts in alkaline and acid solutions. *Int. J. Hydrogen Energy* **2010**, *35*, 365–372.

84. Lee, S.; Kim, H.J.; Choi, S.M.; Seo, M.H.; Kim, W.B. The promotional effect of Ni on bimetallic PtNi/C catalysts for glycerol electrooxidation. *Appl. Catal. A* **2012**, *429–430*, 39–47.

85. Nghia, N.V.; Truong, N.N.K.; Thong, N.M.; Hung, N.P. Synthesis of Nanowire-Shaped Silver by Polyol Process of Sodium Chloride. *Int. J. Mater. Chem.* **2012**, *2*, 75–78.

86. Chee, S.-S.; Lee, J.-H. Synthesis of tin nanoparticles through modified polyol process and effects of centrifuging and drying on nanoparticles. *Trans. Nonferrous Met. Soc. China* **2012**, *22*, s707–s711.

87. Salgado, J.R.C.; Antolini, E.; Gonzalez, E.R. Structure and Activity of Carbon-Supported Pt–Co Electrocatalysts for Oxygen Reduction. *J. Phys. Chem. B* **2004**, *108*, 17767–17774.

88. Alvarez, G.F.; Mamlouk, M.; Senthil Kumar, S.M.; Scott, K. Preparation and characterisation of carbon-supported palladium nanoparticles for oxygen reduction in low temperature PEM fuel cells. *J. Appl. Electrochem.* **2011**, *41*, 925–937.

89. Pradeep, T.; Anshup. Noble metal nanoparticles for water purification: A critical review. *Thin Solid Films* **2009**, *517*, 6441–6478.

90. Nashner, M.S.; Frenkel, A.I.; Somerville, D.; Hills, C.W.; Shapley, J.R.; Nuzzo, R.G. Core Shell Inversion during Nucleation and Growth of Bimetallic Pt/Ru Nanoparticles. *J. Am. Chem. Soc.* **1998**, *120*, 8093–8101.

91. Takasu, Y.; Fujiwara, T.; Murakami, Y.; Sasaki, K.; Oguri, M.; Asaki, T.; Sugimoto, W. Effect of Structure of Carbon-Supported PtRu Electrocatalysts on the Electrochemical Oxidation of Methanol. *J. Electrochem. Soc.* **2000**, *147*, 4421–4427.

92. Rahsepar, M.; Pakshir, M.; Piao, Y.; Kim, H. Preparation of Highly Active 40 wt.% Pt on Multiwalled Carbon Nanotube by Improved Impregnation Method for Fuel Cell Applications. *Fuel Cells* **2012**, *12*, 827–834.

93. Knani, S.; Chirchi, L.; Baranton, S.; Napporn, T.W.; Léger, J.-M.; Ghorbel, A. A methanol–Tolerant carbon supported Pt–Sn cathode catalysts. *Int. J. Hydrogen Energy* **2014**, *39*, 9070–9079.

94. Coutanceau, C.; Brimaud, S.; Lamy, C.; Léger, J.M.; Dubau, L.; Rousseau, S.; Vigier, F. Review of different methods for developing nanoelectrocatalysts for the oxidation of organic compounds. *Electrochim. Acta* **2008**, *53*, 6865–6880.

95. Knani, S.; Chirchi, L.; Napporn, W.T.; Baranton, S.; Léger, J.M.; Ghorbel, A. Promising ternary Pt–Co–Sn catalyst for the oxygen reduction reaction. *J. Electroanal. Chem.* **2015**, *738*, 145–153.

96. Miller, H.A.; Bevilacqua, M.; Filippi, J.; Lavacchi, A.; Marchionni, A.; Marelli, M.; Moneti, S.; Oberhauser, W.; Vesselli, E.; Innocenti, M.; *et al.* Nanostructured Fe-Ag electrocatalysts for the oxygen reduction reaction in alkaline media. *J. Mater. Chem. A* **2013**, *1*, 13337–13347.

97. Choi, S.M.; Yoon, J.S.; Kim, H.J.; Nam, S.H.; Seo, M.H.; Kim, W.B. Electrochemical benzene hydrogenation using PtRhM/C (M = W, Pd, or Mo) electrocatalysts over a polymer electrolyte fuel cell system. *Appl. Catal. A* **2009**, *359*, 136–143.

98. Kim, H.J.; Choi, S.M.; Nam, S.H.; Seo, M.H.; Kim, W.B. Effect of Rh content on carbon-supported PtRh catalysts for dehydrogenative electrooxidation of cyclohexane to benzene over polymer electrolyte membrane fuel cell. *Appl. Catal. A* **2009**, *352*, 145–151.

99. Kim, H.J.; Choi, S.M.; Nam, S.H.; Seo, M.H.; Kim, W.B. Carbon-supported PtNi catalysts for electrooxidation of cyclohexane to benzene over polymer electrolyte fuel cells. *Catal. Today* **2009**, *146*, 9–14.

100. Longoni, G.; Chini, P. Synthesis and chemical characterization of platinum carbonyl dianions $[Pt_3(CO)_6]_n^{2-}$ (n = apprx.10, 6, 5, 4, 3, 2, 1). A new series of inorganic oligomers. *J. Am. Chem. Soc.* **1976**, *98*, 7225–7231.

101. Holade, Y.; Morais, C.; Servat, K.; Napporn, T.W.; Kokoh, K.B. Toward the Electrochemical Valorization of Glycerol: Fourier Transform Infrared Spectroscopic and Chromatographic Studies. *ACS Catal.* **2013**, *3*, 2403–2411.

102. Holade, Y.; MacVittie, K.; Conlon, T.; Guz, N.; Servat, K.; Napporn, T.W.; Kokoh, K.B.; Katz, E. Pacemaker Activated by an Abiotic Biofuel Cell Operated in Human Serum Solution. *Electroanalysis* **2014**, *26*, 2445–2457.

103. Holade, Y.; Engel, A.B.; Tingry, S.; Servat, K.; Napporn, T.W.; Kokoh, K.B. Insights on Hybrid Glucose Biofuel Cell Based on Bilirubin Oxidase Cathode and Gold-Based Nanomaterials Anode. *ChemElectroChem* **2014**, *1*, 1976–1987.

104. Lim, B.; Kobayashi, H.; Camargo, P.C.; Allard, L.; Liu, J.; Xia, Y. New insights into the growth mechanism and surface structure of palladium nanocrystals. *Nano Res.* **2010**, *3*, 180–188.

105. Zhang, H.; Jin, M.; Xiong, Y.; Lim, B.; Xia, Y. Shape-Controlled Synthesis of Pd Nanocrystals and Their Catalytic Applications. *Acc. Chem. Res.* **2013**, *46*, 1783–1794.

106. Langille, M.R.; Personick, M.L.; Zhang, J.; Mirkin, C.A. Defining Rules for the Shape Evolution of Gold Nanoparticles. *J. Am. Chem. Soc.* **2012**, *134*, 14542–14554.

107. Huang, X.; Li, Y.; Li, Y.; Zhou, H.; Duan, X.; Huang, Y. Synthesis of PtPd Bimetal Nanocrystals with Controllable Shape, Composition, and Their Tunable Catalytic Properties. *Nano Lett.* **2012**, *12*, 4265–4270.

108. Holade, Y.; Servat, K.; Napporn, T.W.; Kokoh, K.B. Electrocatalytic properties of nanomaterials synthesized from "Bromide Anion Exchange" method—Investigations of glucose and glycerol oxidation. *Electrochim. Acta* **2015**.

109. Habrioux, A.; Servat, K.; Tingry, S.; Kokoh, K.B. Enhancement of the performances of a single concentric glucose/O_2 biofuel cell by combination of bilirubin oxidase/Nafion cathode and Au–Pt anode. *Electrochem. Commun.* **2009**, *11*, 111–113.

110. Lee, W.-D.; Lim, D.-H.; Chun, H.-J.; Lee, H.-I. Preparation of Pt nanoparticles on carbon support using modified polyol reduction for low-temperature fuel cells. *Int. J. Hydrogen Energy* **2012**, *37*, 12629–12638.

111. Sakai, G.; Arai, T.; Matsumoto, T.; Ogawa, T.; Yamada, M.; Sekizawa, K.; Taniguchi, T. Electrochemical and ESR Study on Pt-TiO$_x$/C Electrocatalysts with Enhanced Activity for ORR. *ChemElectroChem* **2014**, *1*, 366–370.

112. Teranishi, T.; Kurita, R.; Miyake, M. Shape Control of Pt Nanoparticles. *J. Inorg. Organomet. Polym.* **2000**, *10*, 145–156.

113. Mirdamadi-Esfahani, M.; Mostafavi, M.; Keita, B.; Nadjo, L.; Kooyman, P.; Remita, H. Bimetallic Au-Pt nanoparticles synthesized by radiolysis: Application in electro-catalysis. *Gold Bull.* **2010**, *43*, 49–56.

114. Rivadulla, J.F.; Vergara, M.C.; Blanco, M.C.; López-Quintela, M.A.; Rivas, J. Optical Properties of Platinum Particles Synthesized in Microemulsions. *J. Phys. Chem. B* **1997**, *101*, 8997–9004.

115. Lim, B.; Jiang, M.; Tao, J.; Camargo, P.H.C.; Zhu, Y.; Xia, Y. Shape-Controlled Synthesis of Pd Nanocrystals in Aqueous Solutions. *Adv. Funct. Mater.* **2009**, *19*, 189–200.

116. Wang, Z.-L.; Yan, J.-M.; Wang, H.-L.; Ping, Y.; Jiang, Q. Pd/C Synthesized with Citric Acid: An Efficient Catalyst for Hydrogen Generation from Formic Acid/Sodium Formate. *Sci. Rep.* **2012**, *2*, 598.

117. Klotz, P.; Feldberg, S.; Newman, L. Mixed-ligand complexes of palladium(II) with bromide and chloride in acetonitrile. *Inorg. Chem.* **1973**, *12*, 164–168.

118. Ksar, F.; Ramos, L.; Keita, B.; Nadjo, L.; Beaunier, P.; Remita, H. Bimetallic Palladium–Gold Nanostructures: Application in Ethanol Oxidation. *Chem. Mater.* **2009**, *21*, 3677–3683.

119. Imbeault, R.; Reyter, D.; Garbarino, S.; Roué, L.; Guay, D. Metastable Au_xRh_{100-x} Thin Films Prepared by Pulsed Laser Deposition for the Electrooxidation of Methanol. *J. Phys. Chem. C* **2012**, *116*, 5262–5269.

120. Bonneman, H.; Brijoux, W.; Brinkmann, R.; Fretzen, R.; Joussen, T.; Koppler, R.; Korall, B.; Neiteler, P.; Richter, J. Preparation, characterization, and application of fine metal particles and metal colloids using hydrotriorganoborates. *J. Mol. Catal.* **1994**, *86*, 129–177.

121. Bönnemann, H.; Braun, G.A. Enantioselective Hydrogenations on Platinum Colloids. *Angew. Chem. Int. Ed.* **1996**, *35*, 1992–1995.

122. Choudhury, N.A.; Raman, R.K.; Sampath, S.; Shukla, A.K. An alkaline direct borohydride fuel cell with hydrogen peroxide as oxidant. *J. Power Sources* **2005**, *143*, 1–8.

123. Ma, J.; Choudhury, N.A.; Sahai, Y. A comprehensive review of direct borohydride fuel cells. *Renew. Sustain. Energy Rev.* **2010**, *14*, 183–199.

124. Wroblowa, H.S.; Yen Chi, P.; Razumney, G. Electroreduction of oxygen: A new mechanistic criterion. *J. Electroanal. Chem. Interf. Electrochem.* **1976**, *69*, 195–201.

125. Acres, G.J.K.; Frost, J.C.; Hards, G.A.; Potter, R.J.; Ralph, T.R.; Thompsett, D.; Burstein, G.T.; Hutchings, G.J. Electrocatalysts for fuel cells. *Catal. Today* **1997**, *38*, 393–400.

126. Xing, W.; Yin, G.; Zhang, J. *Rotating Electrode Methods and Oxygen Reduction Electrocatalysts*, 1st ed.; Elsevier: Amsterdam, The Netherlands, 2014; p. 322.

127. Srinivasan, S.; Ticianelli, E.A.; Derouin, C.R.; Redondo, A. Advances in solid polymer electrolyte fuel cell technology with low platinum loading electrodes. *J. Power Sources* **1988**, *22*, 359–375.

128. Wilson, M.S.; Valerio, J.A.; Gottesfeld, S. Low platinum loading electrodes for polymer electrolyte fuel cells fabricated using thermoplastic ionomers. *Electrochim. Acta* **1995**, *40*, 355–363.

129. Gloaguen, F.; Andolfatto, F.; Durand, R.; Ozil, P. Kinetic study of electrochemical reactions at catalyst-recast ionomer interfaces from thin active layer modelling. *J. Appl. Electrochem.* **1994**, *24*, 863–869.

130. Coutanceau, C.; Croissant, M.J.; Napporn, T.; Lamy, C. Electrocatalytic reduction of dioxygen at platinum particles dispersed in a polyaniline film. *Electrochim. Acta* **2000**, *46*, 579–588.

131. Li, B.; Amiruddin, S.; Prakash, J. A Kinetic Study of Oxygen Reduction Reaction on Palladium-Nickel Alloy Surfaces. *ECS Trans.* **2008**, *6*, 139–144.

132. Oh, H.-S.; Oh, J.-G.; Kim, H. Modification of polyol process for synthesis of highly platinum loaded platinum–carbon catalysts for fuel cells. *J. Power Sources* **2008**, *183*, 600–603.

133. Lebègue, E.; Baranton, S.; Coutanceau, C. Polyol synthesis of nanosized Pt/C electrocatalysts assisted by pulse microwave activation. *J. Power Sources* **2011**, *196*, 920–927.

134. Brunel, L.; Denele, J.; Servat, K.; Kokoh, K.B.; Jolivalt, C.; Innocent, C.; Cretin, M.; Rolland, M.; Tingry, S. Oxygen transport through laccase biocathodes for a membrane-less glucose/O$_2$ biofuel cell. *Electrochem. Commun.* **2007**, *9*, 331–336.

135. Silvert, P.Y.; Tekaia-Elhsissen, K. Synthesis of monodisperse submicronic gold particles by the polyol process. *Solid State Ionics* **1995**, *82*, 53–60.

136. Joseyphus, R.J.; Kodama, D.; Matsumoto, T.; Sato, Y.; Jeyadevan, B.; Tohji, K. Role of polyol in the synthesis of Fe particles. *J. Magn. Magn. Mater.* **2007**, *310*, 2393–2395.

137. Couto, G.G.; Klein, J.J.; Schreiner, W.H.; Mosca, D.H.; de Oliveira, A.J.A.; Zarbin, A.J.G. Nickel nanoparticles obtained by a modified polyol process: Synthesis, characterization, and magnetic properties. *J. Colloid Int. Sci.* **2007**, *311*, 461–468.

138. Tzitzios, V.; Basina, G.; Gjoka, M.; Alexandrakis, V.; Georgakilas, V.; Niarchos, D.; Boukos, N.; Petridis, D. Chemical synthesis and characterization of hcp Ni nanoparticles. *Nanotechnology* **2006**, *17*, 3750–3755.

139. Kalyan Kamal, S.S.; Sahoo, P.K.; Premkumar, M.; Rama Rao, N.V.; Jagadeesh Kumar, T.; Sreedhar, B.; Singh, A.K.; Ram, S.; Chandra Sekhar, K. Synthesis of cobalt nanoparticles by a modified polyol process using cobalt hydrazine complex. *J. Alloys Compd.* **2009**, *474*, 214–218.

140. Ducamp-Sanguesa, C.; Herrera-Urbina, R.; Figlarz, M. Synthesis and characterization of fine and monodisperse silver particles of uniform shape. *J. Solid State Chem.* **1992**, *100*, 272–280.

141. Donati, I.; Travan, A.; Pelillo, C.; Scarpa, T.; Coslovi, A.; Bonifacio, A.; Sergo, V.; Paoletti, S. Polyol Synthesis of Silver Nanoparticles: Mechanism of Reduction by Alditol Bearing Polysaccharides. *Biomacromolecules* **2009**, *10*, 210–213.

142. Blosi, M.; Albonetti, S.; Dondi, M.; Martelli, C.; Baldi, G. Microwave-assisted polyol synthesis of Cu nanoparticles. *J. Nanopart. Res.* **2011**, *13*, 127–138.

143. Cuya Huaman, J.L.; Sato, K.; Kurita, S.; Matsumoto, T.; Jeyadevan, B. Copper nanoparticles synthesized by hydroxyl ion assisted alcohol reduction for conducting ink. *J. Mater. Chem.* **2011**, *21*, 7062–7069.

144. Jo, Y.H.; Jung, I.; Choi, C.S.; Kim, I.; Lee, H.M. Synthesis and characterization of low temperature Sn nanoparticles for the fabrication of highly conductive ink. *Nanotechnology* **2011**, *22*, 225701.

145. Nørskov, J.K.; Rossmeisl, J.; Logadottir, A.; Lindqvist, L.; Kitchin, J.R.; Bligaard, T.; Jónsson, H. Origin of the Overpotential for Oxygen Reduction at a Fuel-Cell Cathode. *J. Phys. Chem. B* **2004**, *108*, 17886–17892.

146. Toda, T.; Igarashi, H.; Uchida, H.; Watanabe, M. Enhancement of the Electroreduction of Oxygen on Pt Alloys with Fe, Ni, and Co. *J. Electrochem. Soc.* **1999**, *146*, 3750–3756.

147. Xiong, L.; Manthiram, A. Effect of Atomic Ordering on the Catalytic Activity of Carbon Supported PtM (M = Fe, Co, Ni, and Cu) Alloys for Oxygen Reduction in PEMFCs. *J. Electrochem. Soc.* **2005**, *152*, A697–A703.

148. Yang, H.; Vogel, W.; Lamy, C.; Alonso-Vante, N. Structure and Electrocatalytic Activity of Carbon-Supported Pt−Ni Alloy Nanoparticles Toward the Oxygen Reduction Reaction. *J. Phys. Chem. B* **2004**, *108*, 11024–11034.

149. Santiago, E.I.; Varanda, L.C.; Villullas, H.M. Carbon-Supported Pt–Co Catalysts Prepared by a Modified Polyol Process as Cathodes for PEM Fuel Cells. *J. Phys. Chem. C* **2007**, *111*, 3146–3151.

150. Rao, C.V.; Reddy, A.L.M.; Ishikawa, Y.; Ajayan, P.M. Synthesis and electrocatalytic oxygen reduction activity of graphene-supported Pt₃Co and Pt₃Cr alloy nanoparticles. *Carbon* **2011**, *49*, 931–936.

151. Liu, C.; Wu, X.; Klemmer, T.; Shukla, N.; Yang, X.; Weller, D.; Roy, A.G.; Tanase, M.; Laughlin, D. Polyol Process Synthesis of Monodispersed FePt Nanoparticles. *J. Phys. Chem. B* **2004**, *108*, 6121–6123.

152. Venkateswara Rao, C.; Viswanathan, B. ORR Activity and Direct Ethanol Fuel Cell Performance of Carbon-Supported Pt–M (M = Fe, Co, and Cr) Alloys Prepared by Polyol Reduction Method. *J. Phys. Chem. C* **2009**, *113*, 18907–18913.

153. Min, M.-k.; Cho, J.; Cho, K.; Kim, H. Particle size and alloying effects of Pt-based alloy catalysts for fuel cell applications. *Electrochim. Acta* **2000**, *45*, 4211–4217.

154. Léger, J.M. Preparation and activity of mono- or bi-metallic nanoparticles for electrocatalytic reactions. *Electrochim. Acta* **2005**, *50*, 3123–3129.

155. Tseng, C.-J.; Lo, S.-T.; Lo, S.-C.; Chu, P.P. Characterization of Pt-Cu binary catalysts for oxygen reduction for fuel cell applications. *Mater. Chem. Phys.* **2006**, *100*, 385–390.

156. Xiong, L.; Kannan, A.M.; Manthiram, A. Pt–M (M = Fe, Co, Ni and Cu) electrocatalysts synthesized by an aqueous route for proton exchange membrane fuel cells. *Electrochem. Commun.* **2002**, *4*, 898–903.

157. Cambanis, G.; Chadwick, D. Platinum-vanadium carbon supported catalysts for fuel cell applications. *Appl. Catal.* **1986**, *25*, 191–198.

158. Antolini, E.; Passos, R.R.; Ticianelli, E.A. Electrocatalysis of oxygen reduction on a carbon supported platinum–vanadium alloy in polymer electrolyte fuel cells. *Electrochim. Acta* **2002**, *48*, 263–270.

159. Mukerjee, S.; Srinivasan, S.; Soriaga, M.P.; McBreen, J. Effect of Preparation Conditions of Pt Alloys on Their Electronic, Structural, and Electrocatalytic Activities for Oxygen Reduction—XRD, XAS, and Electrochemical Studies. *J. Phys. Chem.* **1995**, *99*, 4577–4589.

160. Huang, M.; Li, L.; Guo, Y. Microwave heated polyol synthesis of Pt3Te/C catalysts. *Electrochim. Acta* **2009**, *54*, 3303–3308.

161. Chen, C.; Kang, Y.; Huo, Z.; Zhu, Z.; Huang, W.; Xin, H.L.; Snyder, J.D.; Li, D.; Herron, J.A.; Mavrikakis, M.; *et al.* Highly crystalline multimetallic nanoframes with three-dimensional electrocatalytic surfaces. *Science* **2014**, *343*, 1339–1343.

162. Senthil Kumar, S.M.; Soler Herrero, J.; Irusta, S.; Scott, K. The effect of pretreatment of Vulcan XC-72R carbon on morphology and electrochemical oxygen reduction kinetics of supported Pd nano-particle in acidic electrolyte. *J. Electroanal. Chem.* **2010**, *647*, 211–221.

163. Murthi, V.S.; Urian, R.C.; Mukerjee, S. Oxygen Reduction Kinetics in Low and Medium Temperature Acid Environment: Correlation of Water Activation and Surface Properties in Supported Pt and Pt Alloy Electrocatalysts. *J. Phys. Chem. B* **2004**, *108*, 11011–11023.

164. Jiang, L.; Hsu, A.; Chu, D.; Chen, R. Oxygen Reduction Reaction on Carbon Supported Pt and Pd in Alkaline Solutions. *J. Electrochem. Soc.* **2009**, *156*, B370–B376.

165. Gasteiger, H.A.; Kocha, S.S.; Sompalli, B.; Wagner, F.T. Activity benchmarks and requirements for Pt, Pt-alloy, and non-Pt oxygen reduction catalysts for PEMFCs. *Appl. Catal. B* **2005**, *56*, 9–35.

166. Yam, V.W.W. Behind platinum's sparkle. *Nat. Chem.* **2010**, *2*, 790.

167. Borup, R.; Meyers, J.; Pivovar, B.; Kim, Y.S.; Mukundan, R.; Garland, N.; Myers, D.; Wilson, M.; Garzon, F.; Wood, D.; *et al.* Scientific Aspects of Polymer Electrolyte Fuel Cell Durability and Degradation. *Chem. Rev.* **2007**, *107*, 3904–3951.

168. Martins, C.; Fernández, P.; Troiani, H.; Martins, M.; Arenillas, A.; Camara, G. Agglomeration and Cleaning of Carbon Supported Palladium Nanoparticles in Electrochemical Environment. *Electrocatalysis* **2014**, *5*, 204–212.

169. Hiraoka, F.; Matsuzawa, K.; Mitsushima, S. Degradation of Pt/C Under Various Potential Cycling Patterns. *Electrocatalysis* **2013**, *4*, 10–16.

170. Cui, C.-H.; Yu, S.-H. Engineering Interface and Surface of Noble Metal Nanoparticle Nanotubes toward Enhanced Catalytic Activity for Fuel Cell Applications. *Acc. Chem. Res.* **2013**, *46*, 1427–1437.

171. Ma, J.; Habrioux, A.; Alonso-Vante, N. The Effect of Substrates at Cathodes in Low-temperature Fuel Cells. *ChemElectroChem* **2014**, *1*, 37–46.

172. Jiang, Z.-Z.; Wang, Z.-B.; Chu, Y.-Y.; Gu, D.-M.; Yin, G.-P. Ultrahigh stable carbon riveted Pt/TiO$_2$-C catalyst prepared by *in situ* carbonized glucose for proton exchange membrane fuel cell. *Energy Environ. Sci.* **2011**, *4*, 728–735.

173. Strmcnik, D.; Kodama, K.; van der Vliet, D.; Greeley, J.; Stamenkovic, V.R.; Marković, N.M. The role of non-covalent interactions in electrocatalytic fuel-cell reactions on platinum. *Nat. Chem.* **2009**, *1*, 466–472.

174. Strmcnik, D.; Escudero-Escribano, M.; Kodama, K.; Stamenkovic, V.R.; Cuesta, A.; Marković, N.M. Enhanced electrocatalysis of the oxygen reduction reaction based on patterning of platinum surfaces with cyanide. *Nat. Chem.* **2010**, *2*, 880–885.

175. Zhang, J.; Sasaki, K.; Sutter, E.; Adzic, R.R. Stabilization of Platinum Oxygen-Reduction Electrocatalysts Using Gold Clusters. *Science* **2007**, *315*, 220–222.

176. Sepa, D.B.; Vojnovic, M.V.; Vracar, L.M.; Damjanovic, A. Apparent enthalpies of activation of electrodic oxygen reduction at platinum in different current density regions—I. Acid solution. *Electrochim. Acta* **1986**, *31*, 91–96.

177. Damjanovic, A. Electron transfer through thin anodic films in oxygen evolution at Pt electrodes in alkaline solutions. *Electrochim. Acta* **1992**, *37*, 2533–2539.

178. Demarconnay, L.; Coutanceau, C.; Léger, J.M. Electroreduction of dioxygen (ORR) in alkaline medium on Ag/C and Pt/C nanostructured catalysts—effect of the presence of methanol. *Electrochim. Acta* **2004**, *49*, 4513–4521.

Design of Pt/Carbon Xerogel Catalysts for PEM Fuel Cells

Nathalie Job, Stéphanie D. Lambert, Anthony Zubiaur, Chongjiang Cao and
Jean-Paul Pirard

Abstract: The design of efficient catalytic layers of proton exchange membrane
fuel cells (PEMFCs) requires the preparation of highly-loaded and highly-dispersed
Pt/C catalysts. During the last few years, our work focused on the preparation of
Pt/carbon xerogel electrocatalysts, starting from simple impregnation techniques
that were further optimized via the strong electrostatic adsorption (SEA) method
to reach high dispersion and a high metal weight fraction. The SEA method, which
consists of the optimization of the precursor/support electrostatic impregnation
through an adequate choice of the impregnation pH with regard to the support
surface chemistry, leads to very well-dispersed Pt/C samples with a maximum
8 wt.% Pt after drying and reduction under H_2. To increase the metal loading, the
impregnation-drying-reduction cycle of the SEA method can be repeated several
times, either with fresh Pt precursor solution or with the solution recycled from the
previous cycle. In each case, a high dispersion (Pt particle size ~3 nm) is obtained.
Finally, the procedure can be simplified by combination of the SEA technique with
dry impregnation, leading to no Pt loss during the procedure.

Reprinted from *Catalysts*. Cite as: Job, N.; Lambert, S.D.; Zubiaur, A.; Cao, C.;
Pirard, J.-P. Design of Pt/Carbon Xerogel Catalysts for PEM Fuel Cells. *Catalysts*
2015, *5*, 40–57.

1. Introduction

Pt supported on a high surface area carbon support is commonly used in
low-temperature proton exchange membrane fuel cells (PEMFCs) to catalyze the
oxidation of H_2 at the anode and the reduction of O_2 at the cathode [1]. The former
is fast, thus allowing small Pt loading to be used at the anode. However, due to
the sluggish O_2 reduction kinetics, high Pt loading is required at the cathode. In
addition, the thickness of both the cathode and the anode should be as small as
possible to avoid diffusional limitations; this means that electrocatalysts with a high
Pt mass fraction are required. In commercial Pt/carbon black catalysts, the Pt mass
fraction may be increased up to 60 wt.% to cope with these limitations. However,
the electrode structure does not guarantee that each Pt particle is active: indeed,
to be electrochemically active, the Pt particles must be in contact with both the
electrically-conductive carbon support and the membrane, which can be achieved
only by reconstructing an ionomer network (Nafion®) within the porosity of the

catalytic layer. In addition, mass transport of reactants and products within the catalytic layers should be easy: (i) the Pt particles must be accessible to the gas reactant, through the porous structure of the catalytic layer; (ii) protons have to circulate in the ionomer network and reach the membrane; (iii) electrons must be collected by the catalyst support and be driven to the current collector. In most cases, a non-negligible fraction of the Pt particles does not meet all of these requirements, which results in undesirable Pt waste.

To produce efficient electrocatalysts and, thus, decrease the mass of Pt used, significant efforts have been directed towards the synthesis. First, the size of the Pt particles should be appropriate: ~3-nm particles lead to the most active catalyst per mass unit of Pt [2]. Second, mass transport limitations can be decreased by using carbon supports with an appropriate pore texture. This is why research turns towards nanostructured carbons [3]. Finally, the distribution of the electron and the ion components (Pt/C and Nafion®) depends on the processing, which must be optimized. This optimization strongly depends on the support chosen and especially on its pore texture and surface chemistry.

For several years now, our group has been working on the development of new Pt/C electrocatalysts with high specific activity and a support nanostructure that allows for optimal mass transport. The supports studied are carbon xerogels, *i.e.*, texture-controlled synthetic carbon materials prepared by drying and pyrolysis of resorcinol-formaldehyde aqueous gels [4]. Indeed, these supports proved to be excellent materials for heterogeneous catalysis in gas phase reactions [5]: since their pore texture can easily be tuned, from nm- to μm-sized pores, one can design the catalyst support in order to decrease the mass-transport limitations. The same idea was then applied to other catalytic systems, *i.e.*, PEMFCs [6–8]. Since the pore texture and surface chemistry are fully adjustable, one can design the support in order to: (i) improve Pt dispersion; (ii) improve mass transport in the operating conditions; and (iii) improve the Pt-Nafion contact, so as to reach 100% Pt particle utilization, *i.e.*, a configuration in which each Pt particle is electroactive for the oxygen reduction reaction.

One of the objectives of our studies is to rationalize the synthesis procedure, so as to keep it as simple and inexpensive as possible. The metal deposition was first performed by simple wet impregnation, followed by reduction under hydrogen [9]. However, the excellent metal dispersion (particle size ~2 nm) obtained at low Pt loading could not be maintained at a high metal weight percentage [6]. The impregnation was thus studied in depth, with attention paid to the metal precursor-support interactions, to design new procedures allowing for high Pt dispersion and high Pt weight percentage. The synthesis techniques had to remain as simple as possible to: (i) make their industrial scale-up possible; and (ii) avoid metal losses during preparation. The present article consists thus of a review of the different procedures investigated during the last few years in our research group. Its aim is to clearly depict the reasoning leading to optimized procedures and to sum up the results obtained at each step. The properties of the catalysts, as well as the pros and cons of each synthesis technique are described and compared. Finally, the goal of this paper is to open new synthesis routes that will allow for the easy production of new supported metal catalysts with high loading and high dispersion.

2. Synthesis Techniques

2.1. Carbon Support

In all cases, the support was a carbon xerogel prepared following a well-known method [4]. In the present work, the carbon xerogel chosen was a material with a specific surface area of ~600 m^2/g and a total pore volume of ~2.1 cm^3/g (micropore volume ~0.23 cm^3/g), and the average meso-macropore size was ~70 nm (Figure 1). These properties were measured by coupling nitrogen adsorption to mercury porosimetry, following a method fully described elsewhere [4].

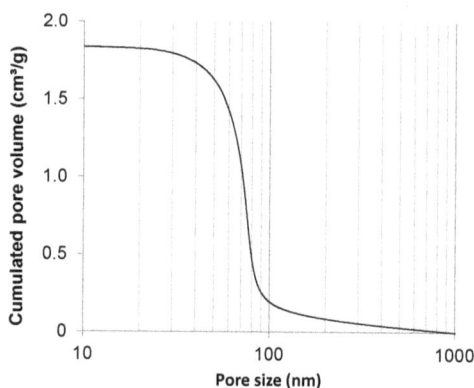

Figure 1. Cumulative pore volume *vs.* pore size of the carbon xerogel support (micropores excluded) calculated from Hg porosimetry data.

Briefly, the gel was obtained by polycondensation of resorcinol with formaldehyde in water. The resorcinol/formaldehyde molar ratio, R/F, was fixed at 0.5; the resorcinol/sodium carbonate molar ratio, R/C, was chosen to be equal to 1000; and the dilution ratio, D, *i.e.*, the solvent/(resorcinol and formaldehyde) molar ratio, was set at 5.7. The resorcinol, formaldehyde, sodium carbonate and water amounts can be found in [4]. The sealed flask was put in an oven at 358 K for gelling and aging for 72 h, then the obtained gel was dried under vacuum, first at 333 K under decreasing pressure (stepwise, from atmospheric pressure down to 10^3 Pa, 8 h), second at 423 K and 10^3 Pa for 12 h. When the sample was dry, it was pyrolyzed at 1073 K under nitrogen flow following a procedure described in another study [4]. After pyrolysis, the xerogel was crushed into a fine powder.

2.2. Wet Impregnation

Pt/carbon xerogel catalysts can be obtained by simple wet impregnation (WI) [9]. One catalyst was prepared by soaking the solid in an H_2PtCl_6 aqueous solution with the appropriate concentration, calculated with regard to the target Pt loading; in this case, one supposes: (i) that all of the metal entering the support porosity remains trapped after drying; and (ii) that no interaction exists between the support and the Pt precursor. Therefore, with the total pore volume equal to 2.1 cm^3/g, one calculates that, to obtain a 1-wt.% Pt/C catalyst, the impregnation suspension should contain, for 1 g of carbon support, 0.64 cm^3 of $H_2PtCl_6 \cdot 6H_2O$ solution (100 g/L) and 4.36 mL of deionized water [9]. The nominal Pt weight percentage, Pt_{th}, i.e., the Pt mass fraction calculated from the two above-mentioned assumptions, is equal to 1 wt.%. The maximum Pt weight percentage, Pt_{max}, i.e., the value reached should all the Pt present in the solution be deposited on the carbon support, equals 2.4 wt.% [9]. The carbon support was simply soaked in the precursor aqueous solution under magnetic stirring for 1 h, at ambient temperature. After impregnation, the excess of solution was removed by filtration. The catalyst was dried under ambient air for 24 h, then under vacuum (10^3 Pa), at 333 K, for another 12 h. The sample was finally reduced under hydrogen flow (0.025 mmol/s) for 3 h at 623 K (heating rate: 350 K/h).

The sample prepared by the wet impregnation technique described in the present section is labelled "WI" (for "wet impregnation").

2.3. Wet Impregnation Coupled to Liquid Phase Reduction

To reach much higher Pt weight percentages, which is undoubtedly necessary for PEMFC applications, one option is to use a precursor solution with high Pt precursor concentration and to reduce the metal directly on the solid by the addition of a reductant in the liquid phase (e.g., $NaBH_4$) [6]. In this case, all of the Pt present in the solution is supposed to be reduced on the support.

The nominal Pt weight percentage, Pt_{th}, was chosen equal to 35 wt.% Pt. The ground carbon xerogel (1 g) was suspended in an H_2PtCl_6 aqueous solution (0.6 g$_{Pt}$/L) for 1 h, at ambient temperature, under magnetic stirring. At the beginning of the impregnation, the pH of the solution was always about 2.2, due to the acidity of H_2PtCl_6. After 24 h of magnetic stirring, $NaBH_4$ was added to reduce the Pt ionic precursor into metallic Pt. A very large excess of $NaBH_4$ (several times the stoichiometric quantity required to reduce the total amount of Pt salt) was used for this; indeed, under these conditions, water is reduced into H_2, and this side-reaction competes with the Pt reduction process. The catalyst sample was washed thoroughly with boiling water. After filtration, the samples was dried in open air at 333 K during

12 h. Finally, the catalyst was reduced in flowing H_2 (0.025 mmol/s) during 3 h at 623 K to ensure the transformation of the last platinum ions into the metallic Pt.

The sample obtained by this technique is labeled "WI-R" (for "wet impregnation-reduction").

2.4. Strong Electrostatic Adsorption

The study of the WI technique shows that interactions exist between the support and the precursor: the amount of Pt deposited on the carbon xerogel is higher than expected [9]. This is due to the electrostatic attraction between the support and the precursor. This effect can be exploited to reach a high Pt weight percentage. Indeed, the electrostatic interactions can be emphasized by choosing an adequate pH of impregnation. This is the principle of the strong electrostatic adsorption technique, inspired from the early work of Brunelle *et al.* [10], who postulated that the adsorption of noble metal complexes onto common oxide supports was essentially Coulombic in nature. Rational synthesis techniques were then developed by Regalbuto *et al.* [11,12], initially to deposit Pt and Pd nanoparticles on inorganic supports. The technique is, however, quite versatile: it was adapted to various supports, like silica [13,14], alumina [15] and carbon [16,17], and can be extended to other metals and to bimetallic nanoparticles [18,19].

The point of zero charge (PZC) of a support corresponds to the pH value at which the electric charge density on the support surface is zero (neutral surface). At a pH lower than its PZC, the support charges positively and adsorbs preferentially anions (e.g., $PtCl_6^{2-}$). On the contrary, at a pH higher than the PZC of the support, the adsorption of cations (e.g., $[Pt(NH_3)_4]^{2+}$) is enhanced. This property can be exploited by the so-called "strong electrostatic adsorption" (SEA) method [11,20], which consists of maximizing the electrostatic interactions, so as to adsorb the maximum amount of Pt at the support surface. The PZC of the support can be measured by the method of Park and Regalbuto (equilibrium pH at high loading, EpHL) [11]. Briefly, the porous solid was soaked in water solutions of various initial pH, and after stabilization, the pH was measured again. The PZC value corresponds to a plateau in a pH_{final} *vs.* initial $pH_{initial}$ plot. For all measurements, the surface loading (SL), *i.e.*, the total carbon surface in solution, was fixed at 10^4 $m^2 \cdot L^{-1}$. Figure 2a shows that the PZC of the carbon xerogel, *i.e.*, the pH_{final} value of the plateau, equals 9.3.

Afterwards, the precursor adsorption curve *vs.* pH was determined. Since the PZC of the carbon xerogels equals 9.3, the adsorption of $PtCl_6^{2-}$ anions is favored for a pH lower than this value. The adsorption curve was measured by contacting 0.042 g of carbon xerogel with 25 mL of H_2PtCl_6 (5.1×10^{-3} mol/L) aqueous solution, the pH of which was adjusted from 1 to 10 with HCl or NaOH. The mass of carbon was chosen so as to fix the surface loading, *i.e.*, the total material surface area in solution, at 10^3 m^2/L. This variable is indeed a key point in both the PZC measurement and the SEA technique, but is kept higher in the former case to enhance the buffering effect of the carbon. Contacted slurries were then placed on a rotary shaker for 1 h, after which the final pHs of these slurries were measured again. Three to 4 mL of the contacted slurries were withdrawn and filtered. The remaining concentration of Pt in the solution was determined by inductively-coupled plasma (ICP) with a Perkin–Elmer (Waltham, MA, USA) Optima 2000 ICP instrument. Platinum uptakes from pH 1.5 to 10 were determined from the difference in Pt concentration between the pre-contacted and post-contacted solutions. The adsorption curve was then reported as the Pt surface density (μmol_{Pt}/m^2) vs. the final pH of the solution (Figure 2b). The adsorption curve shows that the maximum Pt uptake (0.9 μmol_{Pt}/m^2, which corresponds to ~8 wt.%) is obtained for a final pH equal to 2.3 (initial pH = 2.5). Note that the Pt uptake is constant for initial H_2PtCl_6 concentrations higher than ~4 × 10^{-3} mmol/L.

The SEA catalyst was then prepared by adjusting the final impregnation pH to this value. One gram of carbon xerogel was soaked in 0.6 L of H_2PtCl_6 solution (4.1 mmol/L), the pH of which was adjusted to 2.5 with HNO_3 prior to carbon addition. Therefore, the surface loading (SL) was again fixed at 10^3 m^2/L. After 1 h under magnetic stirring at ambient temperature, the slurry was filtrated, and the recovered solid was dried in air at 333 K for 12 h. The catalyst obtained was then reduced under flowing H_2 (0.025 mmol/s) at 473 K for 1 h.

The sample produced by this technique is labeled "SEA" (for "strong electrostatic adsorption").

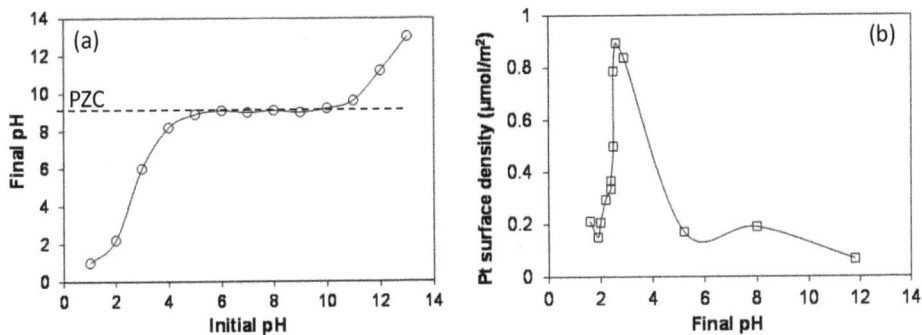

Figure 2. (a) pH equilibrium (point of zero charge (PZC) measurement) for the carbon support at maximum surface loading (SL = 10^4 m^2/L); and (b) final metal precursor uptake vs. pH for the adsorption of $PtCl_6^{2-}$ over carbon xerogel (SL = 10^3 m^2/L, $[H_2PtCl_6]$ = 5.1 × 10^{-3} mol/L). The results are adapted from [20].

2.5. Multiple Strong Electrostatic Adsorption

As mentioned previously, the maximum Pt uptake obtained by maximizing the precursor-support interaction is ~8 wt.% in the case of the H_2PtCl_6/carbon xerogel pair. However, this metal weight fraction is too low for PEMFC applications. To increase the Pt weight fraction, the impregnation-drying-reduction cycle can be repeated several times, using at each step a fresh precursor solution [21]. Therefore, the procedure detailed in Section 2.3 was simply performed several times on the same sample. After reduction under hydrogen flow, the obtained sample was contacted again for 1 h with 0.6 L of fresh H_2PtCl_6 solution (4.1 mmol/L, pH adjusted to 2.5 with HNO_3), filtered and dried in air at 333 K for 12 h before reduction under flowing H_2 (0.025 mmol/s) at 473 K for 1 h. A fraction of the same sample was also reduced at 723 K for 5 h.

The SEA technique requires the use of large amounts of Pt precursor solution (~0.6 L per gram of carbon), and only a small fraction of the Pt is deposited on the support. This obviously induces inacceptable metal losses during the synthesis process, and the SEA technique, although quite elegant in scientific studies, cannot be applied as presented in industrial production. Alternatively, one can, of course, thus imagine re-using the residual solution by re-adjusting its pH and concentration at the required values [22]. Indeed, as long as the concentration of the H_2PtCl_6 solution remains higher than ~4 mmol/L, the Pt uptake remains constant [20]. Therefore, another catalyst was prepared by multiple SEA (M-SEA), but the concentration of the initial impregnation solution was higher, then re-used several times. One gram of carbon xerogel powder was mixed with 567 mL of an H_2PtCl_6 solution at 8.97 mmol/L (*i.e.*, 1.75 g_{Pt}/L) with an initial pH of 2.5. The surface loading (SL) was equal to 10^3 m^2/L. the mixture was mechanically stirred for 1 h, then filtered; the filtrate was stored for re-use in the following impregnation step. The solid was dried in an oven at 333 K during 12 h and reduced at 473 K under H_2 flow (0.04 mmol/s) during 1 h. The "impregnation-drying-reduction" steps were performed two times on the same support. After the second impregnation, the catalyst was reduced under H_2 (0.04 mmol/s), either at 473 K during 1 h or at 723 K during 5 h.

The samples obtained by this techniques are labelled "M-SEA", followed by the temperature of the last reduction treatment (samples M-SEA-473 and M-SEA-723). In the case that the impregnation solution is re-used, an "r" is added at the end of the sample name (M-SEA-r). In the case of this specific sample, the last reduction treatment was performed at 723 K (5 h).

2.6. Charge-Enhanced Dry Impregnation

Another option, which is certainly much more efficient, is to combine the principles of SEA with dry impregnation (*i.e.*, the volume of impregnating solution is equal to the pore volume of the support) [23]. Therefore, after determining the optimal pH conditions for an impregnation slurry, which must be performed for every support-precursor pair, one can use the exact volume of solution (corresponding to the pore volume) in which the maximum amount of Pt ion that can be deposited on the support under those optimal conditions is dissolved.

In a typical synthesis, the mass of metal precursor (H_2PtCl_6) corresponding to the maximum metal uptake of the solid support, first determined by the SEA method, was dissolved in a volume of deionized water corresponding exactly to the amount necessary to wet the solid. Prior to impregnation, the latter volume was measured by dropping deionized water (50 μL at a time) on the carbon until it was just wet. Three mL of H_2PtCl_6 solution with a concentration of 28.7 mol/L were prepared, and the initial pH was adjusted to 2.5 with dilute HNO_3, according to the optimal initial pH of the support/complex pair determined by SEA; this precursor solution was slowly added, 50 μL at a time, to 1 g of carbon xerogel. The sample was then directly dried in air at 298 K for 48 h and reduced in H_2 flow (0.04 mmol/s) at 523 or 723 K for 1 h.

This technique, called the "charge-enhanced dry impregnation" (CEDI) method, was first developed by Zhu *et al.* [24], who recently combined the SEA method with the classical dry impregnation technique to synthesize 2-wt.% Pt catalysts supported on oxidized active carbon or γ-alumina. The samples obtained by this method are labelled "CEDI", followed by the reduction temperature (*i.e.*, CEDI-523 or CEDI-723).

2.7. Characterization

Several physico-chemical techniques were used to characterize the catalysts. The pore texture of the raw support, as well as that of the final catalysts were measured by a combination of nitrogen adsorption and mercury porosimetry [4,25]. This is necessary, because nitrogen adsorption is limited to pores smaller than 50 nm, while mercury porosimetry gives access to pores larger than 7.5 nm [25]. The particle size distribution was determined by image analysis of transmission electron microscopy (TEM) micrographs obtained with a Jeol (Tokyo, Japan) 2010 microscope (200 kV, LaB$_6$ filament) or from scanning transmission electron microscopy (STEM) micrographs (Jeol, Tokyo, Japan, JEM-2010F). The image analysis method used is fully described in [21]. The samples were also analyzed by X-ray diffraction (XRD) with a Siemens (Karlsruhe, Germany) D5000 goniometer using the Cu-K$_\alpha$ line (Ni filter). The average crystallite size, d_{XRD}, was estimated using Scherrer's equation [26]. Note that in the case of well-dispersed Pt particles, the good agreement between d_{XRD} and the particle diameter calculated from TEM images allows us to conclude that the particles are monocrystalline and that d_{XRD} also corresponds to the diameter of the Pt particles. The metal dispersion and surface availability were determined by CO chemisorption using a Fisons (Ipswich, UK) Sorptomatic 1990 equipped with a turbomolecular vacuum pump that allows the reaching of a high vacuum of 10^{-3} Pa. The entire procedure, from the sample preparation to the adsorption measurement, is fully described elsewhere [6]. This technique allowed calculating the accessible Pt surface, $S_{CO-chem}$, and the corresponding Pt particle diameter, d_{CO}, assuming that all of the particles are spheres of equal size.

Samples were also investigated by electrochemical techniques, except the WI catalyst, due to its too low Pt content (see Section 3). The reaction used was the electrooxidation of carbon monoxide adsorbed (CO_{ads}) at the surface of the Pt particles. This reaction, called "CO stripping", allows for the determination of the electrochemically-active Pt surface, $S_{CO-strip}$, which can be compared to the Pt surface detected by CO chemisorption. CO_{ads} stripping was performed in liquid electrolyte (sulfuric acid 1 M, Suprapur-Merck, Overijse, Belgium), at 298 K, using an Autolab-PGSTAT20 potentiostat (Metrohm, Antwerp, Belgium) with a three-electrode cell and a saturated calomel electrode (SCE) as the reference (+0.245 V $vs.$ normal hydrogen electrode, NHE). However, all of the potentials are expressed on the NHE scale hereafter. The procedures, from sample preparation to measurements, are completely described in [21]. Globally, a thin layer of the catalyst was fixed using Nafion® on a rotating disk electrode (EDT 101 Tacussel from Materials Mates, Sarcenas, France). In the case of CO_{ads} stripping measurements, the surface of the Pt nanoparticles was saturated with CO (N47, Alphagaz, Paris, France) by bubbling for 6 min in the solution. Then, the non-adsorbed CO was purged from the cell by Ar bubbling for 39 min. During these two steps, the electrode potential was held at +0.095 V $vs.$ NHE. Voltammetric cycles were recorded between +0.045 and +1.245 V $vs.$ NHE at 0.02 V/s. The active area of platinum, $S_{CO-strip}$, was calculated assuming that the electrooxidation of a full monolayer of adsorbed CO requires 420×10^{-6} C/cm$^2_{Pt}$ [27].

3. Results and Discussion

For all samples, the pore texture analysis was performed and compared to that of the raw support. We do not report detailed results here, but globally, the only effect of metal deposition on the pore texture of the carbon xerogel is a decrease of the specific surface area, S_{BET}, certainly due to a partial blocking of the micropores by nm-sized Pt particles. Depending on the loading, the loss of specific surface area, reported per mass of carbon, ranges from 100 to 200 m^2/g [20,21]. The meso-macropores remain unchanged, both in terms of pore size and pore volume, compared to the pristine carbon xerogel support.

Table 1 regroups the characterization results issued from physico-chemical techniques. The table displays the theoretical and the measured Pt weight percentage of the catalysts, $i.e.$, Pt_{th} and Pt_{ICP}, respectively. From the TEM images, the average particle size, d_{TEM}, and its standard deviation, σ, were calculated. The surface weighted average diameter, d_s, and the volume weighted average diameter, d_v, were also calculated for comparison with Pt particle diameters obtained from surface or volume measurements, respectively. Indeed, since XRD is sensitive to the volume of the particles, the diameter estimated from Scherrer's equation, d_{XRD}, corresponds to a volume weighted average diameter, $d_v = \sum n_i d_i^4 / n_i d_i^3$ [26]; since CO chemisorption and CO stripping are surface phenomena, the diameters calculated by these methods should be compared to a surface weighted average diameter $d_s = \sum n_i d_i^3 / n_i d_i^2$. In both cases, n_i is the number of particles of diameter d_i as observed on TEM or STEM micrographs. Table 1 also shows the particle diameter calculated from XRD patterns using Scherrer's equation, d_{XRD}, and parameters issued from CO chemisorption: $n_{s,m}$ is the amount of CO needed to form a chemisorbed monolayer on surface Pt atoms (mmol/g_{Pt}), D_{Pt} is the Pt dispersion, i.e., the proportion of metal located at the surface of the Pt particles, d_{CO} is the particle diameter leading to a metal surface equivalent to that detected by chemisorption and $S_{CO\text{-chem}}$ is the total surface of the Pt particles. The last three parameters are calculated from $n_{s,m}$ (mmol/g_{Pt}) using the following equations [26]:

$$D_{Pt} = n_{s,m} M_{Pt} X_{Pt\text{-}CO} \times 10^{-3} \qquad (1)$$

$$d_{CO} = \frac{6\,(v_m/a_m)}{D_{Pt}} \qquad (2)$$

$$S_{CO-chem} = 6\frac{V_{Pt}}{d_{CO} m_{Pt}} = 6\frac{1}{d_{CO} \rho_{Pt}} \qquad (3)$$

where M_{Pt} is the atomic weight of Pt (195.09 g/mol), $X_{Pt\text{-}CO}$ represents the chemisorption mean stoichiometry, i.e., the average number of Pt atoms on which one CO molecule is adsorbed, v_m is the mean volume occupied by a metal atom in the bulk of a metal particle (for Pt: $v_m = 0.0151$ nm^3), a_m is the mean surface area occupied by a surface metal atom (for Pt: $a_m = 0.0807$ nm^2) and ρ_{Pt} (21.09 g/cm^3) is the density of Pt. Note that $X_{Pt\text{-}CO}$ was chosen equal to 1.61 for samples containing small Pt particles (<5 nm) and equal to 1.00 for samples containing large particles (>5 nm), according to the conclusions of Rodríguez-Reinoso et al. [28] about the effect of the Pt particle size on the CO adsorption stoichiometry. The values of $X_{Pt\text{-}CO}$ for each sample are mentioned in Table 1 (see the notes below the table). Finally, the electroactive Pt surface detected by CO stripping, $S_{CO\text{-strip}}$, is also mentioned. All of these data are discussed below.

TEM images of several catalysts prepared using the above-mentioned methods are presented in Figure 3. In each case, the support is a raw carbon xerogel (PZC ~9), with a pore size of around 70 nm, and the Pt precursor is H_2PtCl_6, but it is worth noticing that very similar results were found with carbon xerogels oxidized in HNO_3 as the support and $[Pt(NH_3)_4](NO_3)_2$ as the precursor [20,23]; in that case, since the PZC of the oxidized carbon xerogel was equal to 2.4, impregnation was performed under basic conditions (initial pH = 12.5, *i.e.*, optimized conditions for the $[Pt(NH_3)_4](NO_3)_2$/oxidized carbon xerogel pair).

The simple wet impregnation (WI) leads to the obtaining of very well-dispersed catalysts (Figure 3a). In addition, the amount of Pt deposited is higher than expected (Table 1): in the case of sample WI, the target value, Pt_{th}, was 1.0 wt.%, while the measured Pt weight fraction was 1.9 wt.% (to be compared with the maximum possible amount of Pt calculated from the total amount of Pt in the impregnation solution, *i.e.*, 2.4 wt.%). This can be explained by the existence of electrostatic interactions between the support and the chloroplatinic ion ($PtCl_6^{2-}$). Indeed, the PZC of a raw carbon xerogel is around 9.0, which means that it charges positively at pH lower than this value. In the case of a carbon xerogel soaked in an H_2PtCl_6 solution, the pH is acidic, the support charges positively and electrostatic interactions cause the precursor to adsorb on the carbon surface. This property was further used to develop the SEA method. TEM and XRD data are in good agreement, since d_{XRD} compares well to d_v (1.8 and 2.0 nm, respectively). CO chemisorption is in good agreement with TEM, too: d_{CO} and d_s are identical (1.9 nm). The Pt specific surface area obtained from CO chemisorption being very high (153 m^2/g_{Pt}). This type of catalyst shows thus very nice properties, but the Pt loading is obviously far too low for PEMFC catalytic layers.

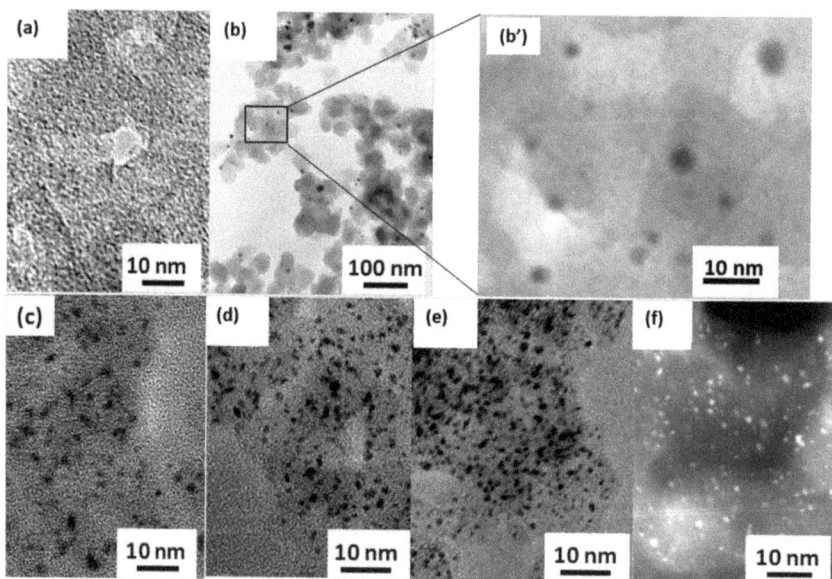

Figure 3. TEM and STEM images of Pt/C catalysts: **(a)** WI (1.9 wt.%); **(b)** WI-R (31.0 wt.%), **(b')** magnified inset of WI-R; **(c)** SEA (7.5 wt.%), **(d)** M-SEA-723 (double SEA, 15.0 wt.%); **(e)** M-SEA-r (double SEA with recycling, 14.7 wt.%); and **(f)** CEDI-473 (10 wt.%), STEM image.

Table 1. Catalyst characterization results. WI, wet impregnation; WI-R, WI-reduction; SEA, strong electrostatic adsorption; M-SEA, multiple SEA; CEDI, charge-enhanced dry impregnation; -r, reused.

Catalyst	Impregnation cycles	$T_{r,final}$	$t_{r,final}$	Pt_{th}	ICP-AES Pt_{ICP}	TEM d_{TEM}	σ	d_s	d_v	XRD d_{XRD}	$n_{s,m}$	CO chemisorption D_{Pt}	d_{CO}	$S_{CO-chem}$	CO stripping $S_{CO-strip}$
	(-)	(K)	(h)	(wt.%)	(wt.%)	(nm)	(nm)	(nm)	(nm)	(nm)	(mmol/g$_{Pt}$)	(-)	(nm)	(m²/g$_{Pt}$)	(m²/g$_{Pt}$)
WI	1	623	3	1.0	1.9	1.6	0.5	1.9	2.0	1.8	1.92	0.60[a]	1.9	153	-[b]
WI-R	1	623	3	35.0	31.0	4.1–17.7[c]	-[d]	-[d]	-[d]	22[e]	0.82	0.16[f]	6.9	41	32
SEA	1	473	1	8.0	7.5	2.0	0.7	2.5	2.7	2.6	1.15	0.36[a]	3.1	92	34
M-SEA-473	2	473	1	16.0	15.0	1.9	0.8	2.5	2.8	2.6	1.10	0.34[a]	3.2	89	37
M-SEA-723	2	723	5	16.0	15.0	2.0	0.7	2.5	2.7	2.7	1.53	0.48[a]	2.3	122	127
M-SEA-r	2	723	5	16.0	14.7	2.2	0.7	2.7	3.0	2.3	-[g]	-[g]	-[g]	-[g]	93
CEDI-523	1	523	1	10.0	10.0	2.0	0.4	1.6	1.8	1.9	-[g]	-[g]	-[g]	-[g]	77
CEDI-723	1	723	1	10.0	10.0	2.0	0.4	1.7	1.8	2.0	-[g]	-[g]	-[g]	-[g]	95

[a] calculated considering $X_{Pt-CO} = 1.61$ (Pt particles < 5 nm); [b] not measured due to too low Pt weight percentage; [c] bi-disperse catalysts: the values represent the average sizes of the two populations; [d] not pertinent (bi-disperse catalyst); [e] size related to the large particles (bidisperse catalyst); [f] calculated considering $X_{Pt-CO} = 1$ (Pt particles > 5 nm); [g] not measured (not accurate due to Cl poisoning). $T_{r,final}$ = final reduction temperature; $t_{r,final}$ = duration of the final reduction treatment; Pt_{ICP} = Pt weight percentage of the catalyst measured by ICP-AES; d_{TEM} = average particle sizes estimated from TEM; σ = standard deviation associated with d_{TEM}; d_s = average surface diameter of Pt particles, $\sum n_i d_i^3 / n_i d_i^2$, estimated from TEM; d_v = mean volume diameter of metal particles, $\sum n_i d_i^4 / n_i d_i^3$, estimated from TEM; d_{XRD} = mean size of Pt particles estimated from X-ray line broadening; $n_{s,m}$ = amount of CO needed to form a chemisorbed monolayer on surface Pt atoms; D_{Pt} = metal dispersion; d_{CO} = equivalent mean Pt particle diameter obtained from CO chemisorption; $S_{CO-chem}$ = accessible Pt surface deduced from CO chemisorption; $S_{CO-strip}$ = accessible Pt surface deduced from CO stripping.

When trying to deposit 35 wt.% in one step via the WI-R method, *i.e.*, by direct reduction of the precursor in the aqueous phase by $NaBH_4$, one obtains a mix of large and small Pt particles (Figure 3b): the precursor is partly adsorbed, which leads to small particles (~2 nm), but a large fraction of $PtCl_6^{2-}$ anions remains in excess. These are directly reduced in the liquid phase, in the pore texture or outside the carbon particles, which leads to the deposition of large Pt particles (~10–30 nm). The Pt particle distribution is clearly bimodal: this is why two values of d_{TEM} (4.1 and 17.7 nm) are mentioned in Table 1. d_{XRD} (22 nm) corresponds to the average size of the large particles. The amount of CO chemisorbed is much lower than in the case of WI, which translates into an equivalent particle diameter, d_{CO}, of 6.9 nm and a lower Pt specific surface area (41 m^2/g_{Pt}). d_{CO} represents an average between the two populations, and the accessible Pt surface decreases due to the presence of large Pt particles. The WI-R technique is efficient to deposit high amounts of Pt in one step, but the Pt particles are badly dispersed, which leads to an inacceptable loss of active surface area. Note, however, that this technique can be improved. Very recently, Alegre *et al.* [29] obtained well-dispersed Pt particles (around 4 nm in diameter) supported on a mesoporous carbon xerogel by impregnation with H_2PtCl_6 followed by reduction with either $NaBH_4$ or formic acid. The loading was 20 wt.%, and the TEM pictures do not show any large particles or aggregates. The main differences between their technique and the WI-R method presented here are: (i) the target loading (20 wt.% instead of 35 wt.%); and (ii) the pH adjustment at a value of 5 with NaOH before $NaBH_4$ addition. Additional investigations could lead to optimal Pt/carbon xerogel catalysts in one or two impregnation steps.

In the SEA technique, the pH is adjusted, so as to maximize the electrostatic interaction and, thus, to adsorb the maximum quantity of Pt precursor at the surface of the support; as a result, the Pt weight percentage increases with regard to the WI technique, but remains limited to max. 8–10 wt.%. Contrary to the WI-R method, the dispersion after drying and reduction remains excellent (Figure 3c). Since the amount of Pt deposited by the SEA method is the maximum quantity that can be adsorbed at the carbon surface, it is clear that trying to obtain 35.0 wt.% in one single impregnation step (WI-R technique) cannot lead to one single Pt particle population. Comparison between the two approaches (WI-R and SEA) clearly confirms that, in WI-R, two phenomena occur during the impregnation-reduction in the liquid phase: (i) adsorption of $PtCl_6^{2-}$ on the carbon support, leading after reduction to very small Pt particles (~2 nm); and (ii) direct reduction in the liquid phase, leading to large Pt particles (~10–30 nm). One may notice, however, some discrepancies in terms of the Pt surface detected by CO chemisorption and CO stripping. Indeed, $S_{CO\text{-}chem}$ is lower for the SEA sample than in the case of WI (92 and 153 m^2/g_{Pt}, respectively). The value obtained by CO_{ads} stripping for the SEA sample is even lower (34 m^2/g_{Pt}). From Equation (3), one finds that a sample containing Pt particles

2.0 nm in diameter should display a Pt surface area close to 140 m^2/g_{Pt}. The low value obtained for the SEA sample, which was reduced at 473 K, was attributed to Cl poisoning of the Pt surface [30]. Indeed, XPS measurements demonstrated that, for too low reduction temperatures, the Pt surface was still partly covered with Cl issued from the decomposition of the Pt precursor ($PtCl_6^{2-}$), which leads to a decrease of the Pt surface detected, both by CO chemisorption and CO_{ads} stripping. CO, which strongly adsorbs onto Pt atoms [31], slowly displaces Cl. This explains why measurements obtained from CO_{ads} stripping (34 m^2/g_{Pt}) and CO chemisorption (92 m^2/g_{Pt}) are not in agreement. In CO chemisorption, the device waits for a pseudo-equilibrium to be reached, the next point being taken when the pressure seems stable; on the contrary, CO_{ads} stripping is always performed according to the same time schedule, without taking into account possible very slow reactions. As a result, the Cl displacement by CO is more complete in the case of CO chemisorption than in the case of CO_{ads} stripping, leading to different values. In any case, this shows that the reduction temperature should be higher than 473 K to clean the Pt surface: 723 K (5 h) leads to almost Cl-free Pt nanoparticles [30]. It also shows that CO chemisorption overestimates the real accessible Pt surface due to the Cl removal. Clearly, for PEMFC applications, the true Pt electroactive surface would be that left free by the Cl species and not the surface calculated from CO chemisorption, after Cl displacement by CO. This is why our further studies rely on CO stripping measurements and not on CO chemisorption to determine the Pt electroactive surface area: M-SEA-r and CEDI samples were not investigated by CO chemisorption.

In order to increase the Pt content of the catalysts, the impregnation-drying-reduction cycle of the SEA method can be performed several times. Figure 3d and Table 1 show that samples M-SEA-473 and M-SEA-723 display very small Pt particles (d_{TEM} = 2.0 and 1.9 nm, respectively). No Pt large particles or agglomerates are visible, which is confirmed by the good agreement between TEM and XRD. Increasing the reduction temperature has thus no effect on the particle size. However, these two samples show significant differences when comparing the Pt specific surface area. $S_{CO\text{-}chem}$ is higher in the case of M-SEA-723 (122 m^2/g_{Pt} vs. 89 m^2/g_{Pt} in the case of M-SEA-473). The difference is even more pronounced for $S_{CO\text{-}strip}$ (127 m^2/g_{Pt} vs. 37 m^2/g_{Pt}, respectively). Good agreement between d_{CO} and d_s is found only for sample M-SEA-723 (2.3 and 2.5 nm, respectively). This result shows again that, at 473 K, the Pt surface is not fully accessible. Again, one can show through XPS characterization that this phenomenon is due to the partial blocking of the Pt surface by Cl [30]; high reduction temperatures only can efficiently clean the Pt surface.

The recycling of the solution (re-use in further impregnation step) does not alter at all the Pt dispersion (Figure 3e). Results obtained for sample M-SEA-r are identical to those of sample M-SEA-723, except for the Pt surface measured by CO_{ads} stripping which is slightly lower for the former sample (M-SEA-r: 93 m^2/g$_{Pt}$; M-SEA-723: 127 m^2/g$_{Pt}$). This could again be due to an incomplete Pt cleaning, even after 5 h at 723 K.

Finally, the CEDI technique fully preserves the optimal metal dispersion (Figure 3f, Table 1) and allows avoiding any Pt losses during the synthesis. Results show again the importance of the reduction temperature on the final electroactive surface area, since the latter increases from 77 to 95 m^2/g$_{Pt}$ when increasing the temperature from 523 (1 h) to 723 K (1 h). It is worth noting that, in principle, the CEDI technique can be developed in multi-steps. For example, double-CEDI impregnation of oxidized carbon xerogels with [Pt(NH$_3$)$_4$](NO$_3$)$_2$ as the precursor was already performed [23]: the effect was to double the Pt weight percentage without affecting the size, surface or electrochemical properties of the Pt nanoparticles. One CEDI step with the [Pt(NH$_3$)$_4$](NO$_3$)$_2$/oxidized carbon xerogel pair leads to ~5 wt.% Pt/C catalysts; so 10 wt.% samples were obtained by double impregnation. Though the number of impregnation-drying-reduction cycles will obviously be higher in that case, the advantage is the absence of Cl, leading to quite clean Pt nanoparticles.

The shape of the CO$_{ads}$ stripping voltammograms are also in good agreement with the above conclusions. Indeed, CO$_{ads}$ stripping voltammetry provides us with information about the electroactive surface area ($S_{CO\text{-}strip}$) and with information about the Pt particle size distribution [32–34] and the presence of poisons at its surface. Figure 4 shows the curves obtained for samples WI-R, M-SEA-473, M-SEA-723 and M-SEA-r. The surface of the electrooxidation peak(s) is obviously directly proportional to the Pt electroactive area.

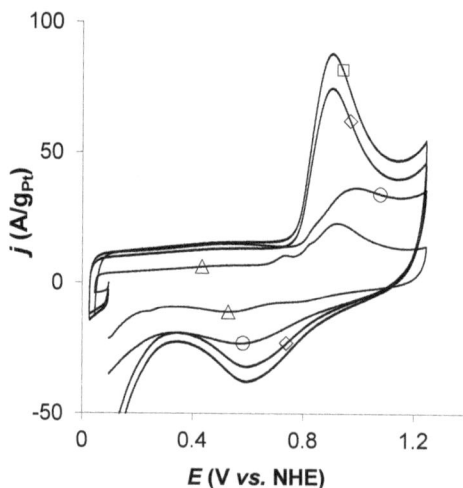

Figure 4. CO$_{ads}$ stripping voltammogram in H$_2$SO$_4$ (1 M) at 298 K; sweep rate of 0.020 V/s. (△) WI-R; (○) M-SEA-473; (□) M-SEA-723; (◇) M-SEA-r.

First, one can note that WI-R displays three CO_{ads} oxidation peaks centered at about +0.73, +0.81 and +0.92 V $vs.$ NHE. The electrooxidation of a CO_{ads} monolayer is a structure-sensitive reaction and provides a wealth of information on the particle size distribution and the presence/absence of particle agglomeration [32–34]. In particular, the position of the CO_{ads} stripping peak strongly depends on the average particle size and is shifted toward positive potential with decreasing of the Pt particle size. Taking into account a sweep rate dependence of 0.080 V dec^{-1} [34], one can consider that the highest oxidation peak (at $ca.$ 0.92 V $vs.$ NHE) corresponds to CO_{ads} electrooxidation at small nanoparticles (d < 1.9 nm) and the lowest oxidation peak at $ca.$ 0.81 V $vs.$ NHE to the CO_{ads} electrooxidation at large particles (d > 3.3 nm) [32]. Finally, the peak located at +0.73 V $vs.$ NHE highlights the presence of Pt particle aggregates [33,34]. These observations remarkably parallel the TEM analysis of sample WI-R.

Second, all of the other samples (M-SEA-473, M-SEA-723 and M-SEA-r) display one single peak corresponding to small Pt particles (~2 nm). In some case, a pre-peak at $ca.$ 0.81 V $vs.$ NHE appears [22], but its intensity always remains low for samples prepared using the SEA method (multiple or not). The position of the peak, however, shifts towards lower potentials when the reduction temperature increases. This reflects the presence of Cl species at the surface of the Pt particles, for instance in the case of sample M-SEA. Indeed, CO_{ads} electrooxidation proceeds via a Langmuir–Hinshelwood mechanism on Pt, which includes water dissociation into oxygen-containing species and recombination of the former species with CO, yielding CO_2 [32,34–36]. The shift towards higher potential values of the peak corresponding to small particles suggests that water and chloride species compete at the Pt catalytic sites.

The CO_{ads} stripping curves obtained by the CEDI method follow exactly the same tendencies: (i) in general, one single peak corresponding to small particles (~2 nm) is visible; (ii) sometimes, a second peak, small in intensity, corresponding to larger Pt particles (>3 nm) appears; (iii) the position of the main electrooxidation peak (small Pt particles) shifts towards lower potentials when the reduction temperature increases. In all cases, the latter effect is not due to any modification in the Pt particle size, but is in our case attributed to the removal of Cl species from the Pt surface as the reduction temperature and duration increase.

Finally, it is worth noticing that samples prepared either by SEA (multiple or not, with fresh or recycled impregnation solution) and CEDI display no difference in terms of electrocatalytic activity towards the oxygen reduction reaction (ORR). The electrocatalytic activity was measured in a three-electrode cell filled with liquid electrolyte (H_2SO_4 aqueous solution), using a rotating disk electrode to eliminate the effect of external diffusion. Since all of the electrocatalytic results obtained in the same conditions were identical and because the present paper is clearly focused on catalyst preparation, the complete data are not fully reproduced here. However, details can be found in [21–23].

4. Conclusions

The synthesis of Pt/C catalysts was rationalized in order to evolve towards catalysts with high metal dispersion and a high Pt weight fraction. The wet impregnation of a carbon xerogel with an H_2PtCl_6 solution, followed by drying and reduction under H_2 flow, leads to small Pt particles (~2 nm) well distributed on the support. However, when trying to deposit directly 35 wt.% Pt, by combining impregnation with reduction in the liquid phase, two populations of Pt particles (centered at ~4 and 20 nm) are obtained, which strongly decrease the reactive Pt surface. The impregnation technique can be optimized by the strong electrostatic adsorption method, which consists of maximizing the electrostatic interactions between the Pt precursor and the carbon support via an adequate choice of the impregnation pH. In this case, the Pt weight fraction obtained is the maximum possible value without affecting the excellent Pt dispersion obtained by impregnation. With the $PtCl_6^{2-}$-carbon xerogel pair, one impregnation-drying-reduction cycle leads to the obtaining of 8 wt.% Pt/C catalysts with a narrow particle size distribution centered at *ca.* 2 nm; this Pt weight percentage is too low for PEMFC applications. This cycle can however be repeated several times in order to increase the metal loading; up to now, it was possible to reach 25 wt.% without decreasing the Pt dispersion [21,22]. In order to lower the Pt losses during the impregnation, the Pt precursor solution can be recycled from one cycle to another without any problem.

Finally, the dry impregnation and SEA techniques can be combined to develop a new method, called the "charge-enhanced dry impregnation" (CEDI); the latter is efficient and avoids any metal losses, since only the amount of Pt precursor that the support is able to fix by electrostatic interactions is present in the impregnation solution. Again, the Pt dispersion is excellent (particles *ca.* 2 nm in size). One must, however, notice that the use of a chlorinated Pt compound as the precursor leads to Cl-covered Pt particles if the reduction temperature is not high enough. Our work now turns towards the use of non-chlorinated Pt complexes to avoid this problem.

To conclude, the work summarized in the present paper shows how the synthesis of a supported metal catalyst can be rationalized in order to fulfil the various criteria from both the economic and performance point of view. The studied case is very specific (Pt nanoparticles supported on carbon), but since the synthesis methods developed are based on very general principles, the same reasoning can be applied to many systems, especially when a relatively high loading of expensive metal is required.

Acknowledgments: Stéphanie D. Lambert is grateful to the Belgian "Fonds de la Recherche Scientifique" (F.R.S.-FNRS) for a Research Associate position. Chongjiang Cao thanks the

F.R.S.-FNRS (Belgium) for a postdoctoral fellowship grant. The authors also thank the Fonds de Recherche Fondamentale Collective (FRFC No. 2.4.542.10.F), the Ministère de la Région Wallonne (projects INNOPEM No. 1117490 and HYLIFE No. 1410135) and the Fonds de Bay for their financial support.

Author Contributions: Nathalie Job and Stéphanie D. Lambert performed the synthesis and characterization of WI, WI-R and SEA catalysts. The M-SEA technique was developed by Anthony Zubiaur, while Chongjiang Cao studied the CEDI method. The whole work was supervised by both Nathalie Job and Jean-Paul Pirard.

Conflicts of Interest: The authors declare no conflict of interest.

References

1. Sopian, K.; Daud, W.R.W. Challenges and future developments in proton exchange membrane fuel cells. *Renew. Energy* **2006**, *31*, 719–727.
2. Kinoshita, K. Particle size effects for oxygen reduction on highly dispersed platinum in acid electrolytes. *J. Electrochem. Soc.* **1990**, *137*, 845–848.
3. Antolini, E. Carbon supports for low-temperature fuel cell catalysts. *Appl. Catal. B* **2009**, *88*, 1–24.
4. Job, N.; Théry, A.; Pirard, R.; Marien, J.; Kocon, L.; Rouzaud, J.-N.; Béguin, F.; Pirard, J.-P. Carbon aerogels, xerogels and cryogels: Influence of the drying method on the textural properties of porous carbon materials. *Carbon* **2005**, *43*, 2481–2494.
5. Job, N.; Heinrichs, B.; Lambert, S.; Pirard, J.-P.; Colomer, J.-F.; Vertruyen, B.; Marien, J. Carbon xerogels as catalyst supports: Study of mass transfer. *AICHE J.* **2006**, *52*, 2663–2676.
6. Job, N.; Marie, J.; Lambert, S.; Berthon-Fabry, S.; Achard, P. Carbon xerogels as catalyst supports for PEM fuel cell cathode. *Energy Convers. Manag.* **2008**, *49*, 2461–2470.
7. Liu, B.; Creager, S. Carbon xerogels as Pt catalyst supports for polymer electrolyte membrane fuel-cell applications. *J. Power Source* **2010**, *195*, 1812–1820.
8. Figueiredo, J.L.; Pereira, M.F.R. Synthesis and functionalization of carbon xerogels to be used as supports for fuel cell catalysts. *J. Energy Chem.* **2013**, *22*, 195–201.
9. Job, N.; Pereira, M.F.R.; Lambert, S.; Cabiac, A.; Delahay, G.; Colomer, J.-F.; Marien, J.; Figueiredo, J.L.; Pirard, J.-P. Highly dispersed platinum catalysts prepared by impregnation of texture-tailored carbon xerogels. *J. Catal.* **2006**, *240*, 160–171.
10. Brunelle, J.P. Preparation of catalysts by metallic complex adsorption on mineral oxides. *Pure Appl. Chem.* **1978**, *50*, 1211–1229.
11. Regalbuto, J.R. Strong Electrostatic adsorption of metals onto catalyst support. In *Catalyst Preparation: Science and Engineering*; Regalbuto, J.R., Ed.; CRC Press, Taylor & Francis Group: Boca Raton, FL, USA, 2007; pp. 297–318.
12. Regalbuto, J.R. Electrostatic adsorption. In *Synthesis of Solid Catalysts*; de Jong, K.P., Ed.; Wiley-VCH: Weinheim, Germany, 2009; pp. 33–58.
13. Schreier, M.; Regalbuto, J.R. A fundamental study of Pt tetraammine impregnation of silica: 1. The electrostatic nature of platinum adsorption. *J. Catal.* **2004**, *225*, 190–202.

14. Miller, J.T.; Schreier, M.; Kropf, A.J.; Regalbuto, J.R. A fundamental study of platinum tetraammine impregnation of silica: 2. The effect of method of preparation, loading, and calcination temperature on (reduced) particle size. *J. Catal.* **2004**, *225*, 203–212.

15. Spieker, W.A.; Liu, J.; Hao, X.; Miller, J.T.; Kropf, A.J.; Regalbuto, J.R. An EXAFS study of the coordination chemistry of hydrogen hexachloroplatinate (IV): 2. Speciation of complexes adsorbed onto alumina. *Appl. Catal. A* **2003**, *243*, 53–66.

16. Hao, X.; Quach, L.; Korah, J.; Spieker, W.A.; Regalbuto, J.R. The control of platinum impregnation by PZC alteration of oxides and carbon. *J. Mol. Catal. A* **2004**, *219*, 97–107.

17. Hao, X.; Barnes, S.; Regalbuto, J.R. A fundamental study of Pt impregnation of carbon: Adsorption equilibrium and particle synthesis. *J. Catal.* **2011**, *279*, 48–65.

18. Feltes, T.E.; Smit, E.D.; D'Souza, L.; Meyer, R.J.; Weckhuysen, B.M.; Regalbuto, J.R. Selective adsorption of manganese onto cobalt for optimized $Mn/Co/TiO_2$ Fischer-Tropsch catalysts. *J. Catal.* **2010**, *270*, 95–102.

19. D'Souza, L.; Regalbuto, J.R. Strong electrostatic adsorption for the preparation of Pt/Co/C and Pd/Co/C bimetallic electrocatalysts. *Stud. Surf. Sci. Catal.* **2010**, *175*, 715–718.

20. Lambert, S.; Job, N.; D'Souza, L.; Pereira, M.F.R.; Pirard, R.; Figueiredo, J.L.; Heinrichs, B.; Pirard, J.-P.; Regalbuto, J.R. Synthesis of very highly dispersed platinum catalysts supported on carbon xerogels by the strong electrostatic adsorption method. *J. Catal.* **2009**, *261*, 23–33.

21. Job, N.; Lambert, S.; Chatenet, M.; Gommes, C.J.; Maillard, F.; Berthon-Fabry, S.; Regalbuto, J.R.; Pirard, J.-P. Preparation of highly loaded Pt/carbon xerogel catalysts for PEM fuel cells by the Strong Electrostatic Adsorption method. *Catal. Today* **2010**, *150*, 119–127.

22. Zubiaur, A.; Chatenet, M.; Maillard, F.; Lambert, S.D.; Pirard, J.-P.; Job, N. Using the Multiple SEA method to synthesize Pt/Carbon xerogel electrocatalysts for PEMFC applications. *Fuel Cells* **2014**, *14*, 343–349.

23. Cao, C.; Yuang, G.; Dubau, L.; Maillard, F.; Lambert, S.D.; Pirard, J.-P.; Job, N. Highly dispersed Pt/C catalysts prepared by the Charge Enhanced Dry Impregnation method. *Appl. Catal. B* **2014**, *150–151*, 101–106.

24. Zhu, X.; Cho, H.-R.; Pasupong, M.; Regalbuto, J.R. Charge-enhanced dry impregnation: A simple way to improve the preparation of supported metal catalysts. *ACS Catal.* **2013**, *3*, 625–630.

25. Job, N.; Pirard, R.; Alié, C.; Pirard, J.-P. Non intrusive mercury porosimetry: Pyrolysis of resorcinol-formaldehyde xerogels. *Part. Part. Syst. Charact.* **2006**, *23*, 72–81.

26. Bergeret, G.; Gallezot, P. Particle size and dispersion measurements. In *Handbook of Heterogeneous Catalysis*; Ertl, G., Knözinger, H., Weitkamp, J., Eds.; Wiley-VCH: Weinheim, Germany, 1997; pp. 439–464.

27. Trasatti, S. Real surface area measurements in electrochemistry. *J. Electroanal. Chem.* **1992**, *327*, 353–376.

28. Rodríguez-Reinoso, F.; Rodríguez-Ramos, I.; Moreno-Castilla, C.; Guerrero-Ruiz, A.; López-González, J.D. Platinum catalysts supported on activated carbons: I. Preparation and characterization. *J. Catal.* **1986**, *99*, 171–183.

29. Alegre, C.; Gálvez, M.E.; Moliner, R.; Baglio, V.; Aricò, A.S.; Lázaro, M.J. Towards an optimal synthesis route for the preparation of highly mesoporous carbon xerogel-supported Pt catalysts for the oxygen reduction reaction. *Appl. Catal. B* **2014**, *147*, 947–957.

30. Job, N.; Chatenet, M.; Berthon-Fabry, S.; Hermans, S.; Maillard, F. Efficient Pt/carbon electrocatalysts for Proton Exchange Membrane fuel cells: Avoid chloride-based Pt salts! *J. Power Sources* **2013**, *240*, 294–305.

31. Holscher, A.A.; Sachtler, W.M.H. Chemisorption and surface corrosion in the tungsten + carbon monoxide system, as studied by field emission and field ion microscopy. *Discuss. Faraday Soc.* **1966**, *41*, 29–42.

32. Maillard, F.; Eikerling, M.; Cherstiouk, O.V.; Schreier, S.; Savinova, E.; Stimming, U. Size effects on reactivity of Pt nanoparticles in CO monolayer oxidation: The role of surface mobility. *Faraday Discuss.* **2004**, *125*, 357–377.

33. Maillard, F.; Schreier, S.; Hanzlik, M.; Savinova, E.R.; Weinkauf, S.; Stimming, U. Influence of particle agglomeration on the catalytic activity of carbon-supported Pt nanoparticles in CO monolayer oxidation. *Phys. Chem. Chem. Phys.* **2005**, *7*, 385–393.

34. Maillard, F.; Savinova, E.R.; Stimming, U. CO monolayer oxidation on Pt nanoparticles: Further insights into the particle size effects. *J. Electroanal. Chem.* **2007**, *599*, 221–232.

35. Maillard, F.; Pronkin, S.; Savinova, E.R. Influence of size on the electrocatalytic activities of supported metal nanoparticles in fuel cells related reactions. In *Handbook of Fuel Cells—Electrocatalysis, Materials, Diagnostics and Durability*; Vielstich, W., Gasteiger, H.A., Yokokawa, H., Eds.; John Wiley & Sons: New York, NY, USA, 2009; Volume 5, pp. 91–111.

36. Andreaus, B.; Maillard, F.; Kocylo, J.; Savinova, E.R.; Eikerling, M. Kinetic modeling of COad monolayer oxidation on carbon-supported platinum nanoparticles. *J. Phys. Chem. B* **2006**, *110*, 21028–21040.

Advances in Ceramic Supports for Polymer Electrolyte Fuel Cells

Oran Lori and Lior Elbaz

Abstract: Durability of catalyst supports is a technical barrier for both stationary and transportation applications of polymer-electrolyte-membrane fuel cells. New classes of non-carbon-based materials were developed in order to overcome the current limitations of the state-of-the-art carbon supports. Some of these materials are designed and tested to exceed the US DOE lifetime goals of 5000 or 40,000 hrs for transportation and stationary applications, respectively. In addition to their increased durability, the interactions between some new support materials and metal catalysts such as Pt result in increased catalyst activity. In this review, we will cover the latest studies conducted with ceramic supports based on carbides, oxides, nitrides, borides, and some composite materials.

Reprinted from *Catalysts*. Cite as: Lori, O.; Elbaz, L. Advances in Ceramic Supports for Polymer Electrolyte Fuel Cells. *Catalysts* **2015**, *5*, 1445–1464.

1. Introduction

The need for advanced alternative energy technologies for transportation, backup-, and main-power applications is undisputable. Of the three available technologies, batteries, solar cells, and fuel cells (FCs), the latter is considered to be the most promising option for such applications due to its low footprint, high energy density, and low maintenance costs. In addition, fuel cells do not require a complex logistical effort and can be easily deployed in any terrain and weather (e.g., fuel cells as backup power for cellular antennas in remote locations). Hence, there is a growing use of fuel cells across industries (e.g., server farms, forklifts, buses, cellular antennas, and cars).

One of the significant hurdles in the mass deployment and commercialization of this technology is the lifetime of the fuel cell, which is mostly limited by the stability and durability of the catalyst support. Further understanding and improvement of this technology is expected to increase fuel cells' lifetime and reliability, and lower their cost.

Carbon is the most common and preferred catalyst support material for polymer electrolyte membrane fuel cells (PEMFCs) and alkaline fuel cells (AFCs). It possesses most of the primary required features: it is abundant, and it has a high surface area and good electrical conductivity. However, the use of carbon is problematic due to its low resistance to corrosion. The electrode integrity and durability is

140

currently a technical barrier [1] in PEMFCs and AFCs, especially for applications that demand high power. This is mainly due to the loss of the fuel cell performance as a consequence of the use of carbon supports. More specifically, the degradation and corrosion of carbon-based electrodes lead to losses in the overall activity of FCs, and this is usually attributed to catalyst dissolution and agglomeration, as illustrated in Figure 1. The complete oxidation of carbon by water, which leads to its corrosion, is a four-electron process with the production of four protons, as follows:

$$C(s) + 2H_2O \rightarrow CO_2 + 4H^+ + 4e^- \qquad (1)$$

$$E^0 = 0.207 \text{ vs. NHE}$$

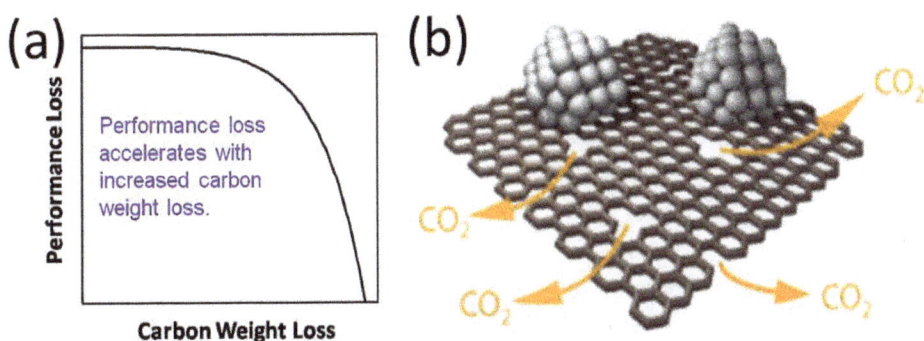

Figure 1. Relationship of carbon corrosion and activity loss in fuel cells (**a**) and illustration of catalyst detachment from corroded carbon (**b**) [2].

This is the potential of most power devices including batteries, solar cells, and fuel cells. When the pH increases, such as in AFCs, the proton activity is lowered. Consequently, the equilibrium shifts to the right in the above reaction and the rate of the carbon corrosion increases. The standard potential for complete oxidation was previously calculated by Pourbaix [3] as follows:

$$E^0 \text{ (V)} = 0.207 - 0.0591 \times pH + 0.0148 \log P_{CO2} \qquad (2)$$

The conditions in PEMFCs and AFCs are oxidizing, especially at the cathode [4–6]. As discussed above, these conditions are detrimental to the carbon electrodes and can significantly shorten the lifetime of the fuel cell [7,8].

Gruver et al. [7] showed that as the carbon support corrodes and turns into CO and CO_2, the catalyst will either wash out or migrate and aggregate. Figure 2 presents images of a pre- and post-mortem sample from a fuel cell, where the loss of carbon (bright material) was observed and the platinum nanoparticles (dark material) have agglomerated (a significant increased in size was observed when compared to the particles in the Figure 2a). The result of both wash-out or aggregation of the catalyst is manifested in a decrease in the catalyst ECSA (Electro-Chemical Surface Area) in the case of precious group metal catalysts (PGMCs), or leaching and dissolution in the case of non-precious group metal catalysts (NPGMCs). One way to improve the stability of electrodes from corrosion in FCs is by the use of graphitic, nano-structured carbon materials, such as graphene nano-sheets, carbon nanotubes, and carbon nanofibers as catalyst supports [9–12]. The high degree of graphitized structures of these compounds provides a higher resistance to chemical and electrochemical oxidation. Another way is to use ceramic materials as catalyst supports [8].

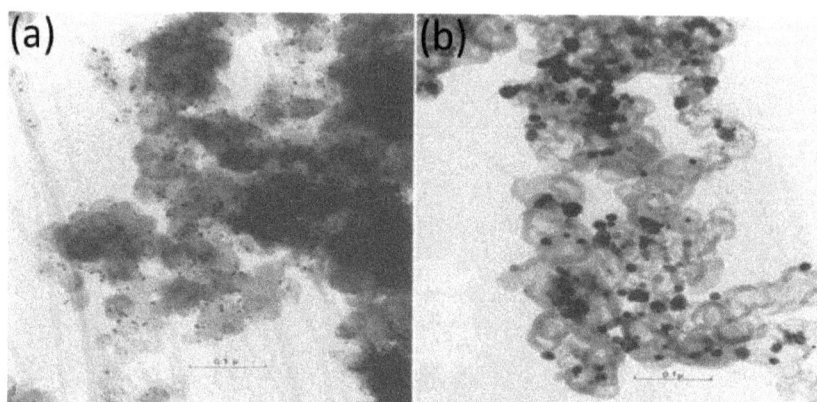

Figure 2. PEM fuel cell catalyst layer with support before (**a**) and after (**b**) use. Darker dots are Pt nanoparticles and the brighter material is carbon. [7] (Adapted from this reference with the permission of the Journal of Electrochemical Society).

Many ceramic supports were developed in order to increase the durability of fuel cells. Supports such as titanium-oxides [13,14], molybdenum-nitride [15], tungsten-oxide [16], and others were synthesized. Most of these possess some of the qualities needed for a good FC electrode (mechanical properties, thermal stability, chemical corrosion resistance), but lack others, such as good electrical conductivity and high surface area.

In this manuscript, we will review the recent advances in ceramic supports for polymer electrolyte fuel cells with a focus on five categories: carbides, oxides, nitrides, borides, and composites.

2. Carbides

The interest in metal carbides in recent years mostly rose from their possible use as catalytic materials. Some similarities between the catalytic behavior of tungsten carbide and platinum was found by Levy and Bourdart [17]. Others have shown studies on the catalytic activity of molybdenum and tungsten carbides as catalysts for methane reforming [18]. The attractiveness of these ceramic materials as catalysts was attributed to their high activity [19–21], lower price when compared to precious metals [22], unique structure [23], and stability [24] in acidic and alkaline mediums. In order for these materials to be good supports for electro-catalysts, they need to have high surface area (equivalent to 300 m^2/g of carbon; with ceramic supports, this surface area may be much lower due to the atomic weight of the metals and comparisons should be made carefully) and good electrical conductivity (circa 4 S/cm) [25]. These requirements narrowed the possible candidates to a handful.

One of the most studied carbide supports is tungsten carbide. Ticianelli *et al.* [26] recently reported a composite of tungsten carbide/carbon (WC/C) synthesized using a chemical vapor deposition (CVD), and used it as a support for Pt in PEMFC anodes. They showed enhanced corrosion resistance when compared with Vulcan XC-72. Roman-Leshkov *et al.* [23] developed a removable ceramic coating method for the synthesis of WC. Using this technique, they were able to produce high-surface-area, electronically conductive WC, which also exhibited superior stability on the anode when compared to Vulcan XC-72. The electronic interaction between Pt and WC and its effect on the catalysis of the oxygen reduction reaction were also studied. It was concluded that the Pt is strongly attached to the support and that the interaction between the two promoted a favorable catalytic activity for the oxygen reduction reaction (ORR) [27]. Although very interesting, in these studies and others that showed similar trends [28], the WC was not exposed to harsh oxidizing conditions and was only studied at the anode where carbon corrosion is less of an issue [29]. Hence, so far, there is not enough information regarding the WC stability under oxidizing conditions.

A different carbide support was recently proposed by Elbaz *et al.* [30], who used the polymer-assisted deposition (PAD) method developed for the synthesis of nanoparticles by Jia *et al.* [31], for the synthesis of molybdenum carbide. They were able to form a Mo_2C/C composite which, similarly to the WC, showed enhanced catalytic activity to ORR. Although their composite material was not completely resistant to corrosion, it did perform better than XC-72. They tied the loss of Pt electrochemical surface area (ECSA) on the Mo_2C/C during their accelerated stress tests (ASTs) to the presence of excess amorphous carbon. In a more recent publication [32], the authors synthesized amorphous carbon-free Pt/Mo_2C, which was found to be very resistant to corrosion, as shown in Figure 3 (less than 10% loss of ECSA *vs.* 90% with Pt/XC-72). In this case the Pt was added to the support during its synthesis and formed non-crystalline small atomic clusters (3–6 atoms of Pt), which the authors called Nano Rafts. This new system exhibited remarkable

ORR activity with a 50% higher mass activity and a higher onset potential, attributed to the electronic interaction between the Pt and the support.

Titanium carbide has previously been tested as a support material of Ir for hydrogen evolution [33] and as a support of Pt for the electro-oxidation of methanol [34], both in acidic media, and showed promising results electrochemically and even in alleviating CO poisoning.

Recently, some studies were published also showing better durability of Pt supported on TiC than Pt/CB under ADTs (0.6–1.2V in 0.1M $HClO_4$ solution) [35] and $Pt_3Pd/TiC@TiO_2$ support [36], which even exceeds that of Pt/TiC (under 0.4–1.2V), attributed to the corrosion resistance of the supports.

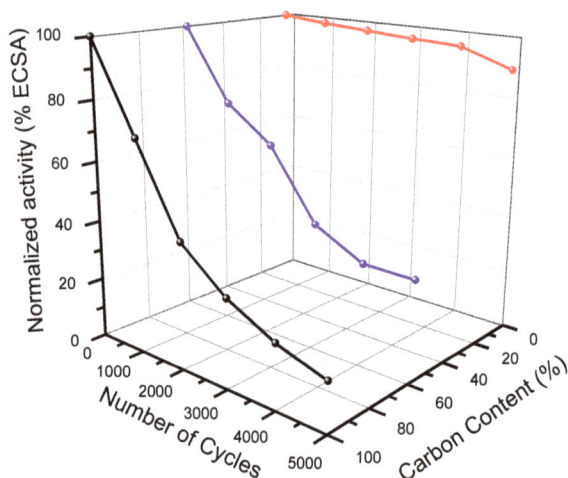

Figure 3. Normalized activity of Pt/C (black), $Pt/Mo_2C/C$ (blue), and Pt/Mo_2C (red) as a function of the number of accelerated test cycles and carbon content.

3. Oxides

Metal Oxides are inorganic compounds that possess several compatible properties for FC applications. Exhibiting properties such as corrosion resistance and mechanical and thermal stability, they show a lot of potential for FCs, although some of them are only semi-conductors or even poor electrical conductors and require certain modifications in order to make sufficient FC materials.

3.1. Titanium Oxide

Titanium oxides received extensive interest in recent years and can be roughly divided into two groups: semi- or non-conducting systems (such as TiO_2) and systems that show high electrical conductivity (Ti_nO_{2n-1} group).

Ti$_n$O$_{2n-1}$ (where n is between 4 and 10) is widely used in photo-catalysis [37–39], water splitting [40], and gas sensing [41] and has attracted much attention in the FC field as a possible catalyst support due to its relatively good conductivity. Recently, Ti$_n$O$_{2n-1}$ has shown promising effects on the durability and catalytic activity of fuel cell catalysts because of its good mechanical resistance and stability in acidic and oxidative environments. Among this series of distinct oxides, Ti$_4$O$_7$ (titanium sub-oxide, TSO) exhibits the highest electrical conductivity, exceeding 1000 S/cm at 25 °C [41].

In a recent study, Wu *et al.* [42] prepared Ti$_n$O$_{2n-1}$ (magneli-phase TSO) and XC-72 to support Pt electro-catalysts for comparison purposes, both loaded with 20%$_{wt}$ Pt. Even though the TSO-based catalyst showed low conductivity ($\sim$10^{-3} S/cm *vs.* 4.78 S/cm for XC-72 at 12 MPa), lower ECSA value (13 m^2/g *vs.* 30 m^2/g), and minimally lower onset potential ($\Delta E = 0.02$ V) compared to Pt/XC-72, their ADT procedure (cycling between -0.5 and 2.0V *vs.* Saturated Calomel Electrode, SCE) was found to cause no change in the onset potential and minimal loss in ECSA (about 12% *vs.* twice as much for Pt/XC-72) after 8000 cycles. TEM images proved that after 8000 cycles the Pt/XC-72 went through considerable Pt aggregation whereas the morphology of Pt/TSO remained intact. Ioroi *et al.* also studied Pt/TSO-based electrodes for PEMFC [43–45]. Although promising, one of the most significant issues that still remains unsolved with the sub-stoichiometric titanium oxide is its relatively low surface area which translates to lower current densities.

In this study, the synthesis, structure, and morphology of Pt/TSO using a laser-irradiated TSO support were investigated, as well as the electrochemical activity for ORR and its stability under high potential conditions for Pt/TSO. They found that the Pt/TSO catalyst had shown a specific activity for the oxygen reduction reaction (ORR) very similar to those of commercial Pt/C catalysts, and much better oxidation resistance under high potential conditions as well. It was also shown that the conductivity of TSO-supported catalysts increased with an increase in Pt loading: 20 wt. % Pt on TSO showed conductivity of *ca.* 8 S/cm at 50 MPa, which was about one-quarter of that of 40 wt. % Pt on XC-72 (30 S/cm) under the same conditions. On the other hand, the calculated ECSA of Pt/TSO was rather small compared to that of Pt/XC-72 (values of 22 and 16 m^2/g for 10 wt. % and 20 wt. % *vs.* 44 m^2/g for Pt/XC-72), indicating a larger diameter of deposited Pt particles, which was consistent with the results of SEM and TEM observations: the Pt particle diameter was between 10 and 20 nm, which is much larger than the common Pt/C catalysts. The performance of the Pt/Ti$_4$O$_7$ cathode was evidently low compared to that of 20% Pt/XC72, mainly due to the smaller ECSA.

Non-stoichiometric mixtures of several titanium oxide phases, mainly Ti$_4$O$_7$ and Ti$_5$O$_9$, known as magneli-phase and by the registered name Ebonex, were investigated as well [46]. Ebonex is considered to be electrochemically stable with a

145

tendency for ORR in acid and base solutions, possesses a high electrical conductivity of 1000 S/cm, and could be a good alternative catalyst support material [41,47]. The Pt/Ebonex catalytic activity was reported to be as much as 10 times higher than that of pure Pt. This was attributed to the increase of active Pt surface area by the reduction of Pt particles sized even below 1 nm and the increase of the number of active sites for oxygen reduction through simple geometric effects, as well as change of oxygen adsorption conditions through the change of the electronic structure of the catalyst caused by the electronic interactions between platinum and the Ebonex. These interactions were rationalized by the Ebonex's hypo-d-electron character which has the ability to interact with Pt that has the hyper-d-electron character. This synergetic effect was explained through the increase of the 5d vacancy of Pt and the decrease of the Pt-Pt bond distances as a result of the interaction with Ebonex, which inhibits the chemisorption of OH^-. It also shifts the PtOH formation to more positive potentials, facilitating the interaction of oxygen with Pt, hence increasing the activity of the catalyst for the oxygen reduction reaction.

Another titanium oxide which was studied as a possible catalyst support for FCs is TiO_2. It is a wide band gap semiconductor and its conductivity is insufficient for a support material without modification such as doping [48–51]. In spite of that, it was widely studied as a ceramic support promoting ORR in PEMFC. In a recent study [13], the synthesis of high surface area TiO_2 and TiO composite materials in a single step was presented. The high surface area conductive titanium oxide was successfully synthesized using the polymer-assisted deposition technique [31]. The TiO_2 and TiO nano-crystalline materials were formed with an average crystallite size on the order of 4 and 8 nm and a BET surface area of 286 and 200 m^2/g for TiO and TiO_2, respectively. Pt was added to the supports, and the calculated ECSA value also showed promise with approximately 60 m^2/g for both phases. This system exhibited better ORR activity when compared to Pt/XC-72. This was again, as in the case of the sub-stoichiometric titanium oxides, attributed to the electronic interaction between the support and the Pt catalyst.

Shanmugam et al. [14] synthesized mesoporous TiO_2 using the sono-chemical method. This synthesis resulted in a spherical globular morphology of the TiO_2 particles, and a size range of 100–200 nm with pores of 4–7 nm which were around the size of the Pt particles deposited. This Pt/TiO_2 was compared to Pt/C. The onset potential of the oxide formation on the Pt/TiO_2 was shifted toward a higher potential, indicating a better resistance nature of Pt/TiO_2 for Pt-OH formation. The Pt/TiO_2 also exhibited superior results towards ORR electro-catalysis than Pt/C. This enhanced activity is attributed to several factors, such as high dispersion, better stabilization, and the modification of the electronic structure of Pt nanoparticles by interaction with the oxide interface, which results in a change in the adsorption characteristics of Pt nanoparticles on TiO_2. They also studied the stability examined by chronoamperometry at an applied voltage of 0.55V and found that the current decay rate was higher for Pt/C than for Pt/TiO$_2$ when only 13% decay was detected after 1 h (80% for conventional Pt/C). Several more studies were conducted with titanium oxide supports and showed very similar results [52–55].

3.2. Tungsten Oxide

Tungsten oxide is an n-type semiconductor with a band gap of a few eVs. Tungsten, which has several oxidation states (usually 2 to 6), appears in many forms, making it compatible with various applications. The conductivity of tungsten oxide comes from its non-stoichiometric composition, causing a donor level formed by oxygen-vacancy defects in the lattice.

Tungsten oxides (predominantly WO$_3$) were studied for quite a long time as a catalyst for DMFC and showed high catalytic activity toward methanol oxidation reaction, possibly due to the formation of tungsten bronzes favoring the dehydrogenation of methanol, and a synergistic effect leading to CO tolerance [56–58]. Park *et al.* [59] showed excellent performance for the use of porous tungsten oxide in thin film fuel cells and also showed good stability in sulfuric acid. In another study [60], Pt was added to commercially available tungsten oxide. The performance of this system was compared to Pt/XC-72 and showed very high stability in acidic conditions.

Nano-sized WO$_3$ was also studied as a possible support material for monolayer Pt ORR electro-catalysts in acid electrolyte [61]. Pt/WO$_3$ exhibited good activity for ORR and superior electron transfer capability compared to conventional Pt/C and Pt. However, a thorough examination of the WO$_3$ support revealed that it can easily turn to water-soluble hydrogen tungsten bronze (H$_x$WO$_3$), facilitating the detachment of Pt nanoparticles as also been discussed elsewhere [62].

Recently, Lu *et al.* [63] studied the electrochemistry, structure, and interaction of nano-sheets of Pd on tungsten oxide (Pd/W$_{18}$O$_{49}$), comparing it to three other systems: Pt/C, Pd/C (both obtained commercially), and support-less Pd NPs. The Pd/W$_{18}$O$_{49}$ was found to have considerably higher electrical conductivity than the support devoid of Pd and had a higher surface area (40 *vs.* 30 m^2/g). Electrochemical studies of ORR catalysis and accelerated life tests in alkaline media showed fairly remarkable results of the tungsten-based catalyst support, outperforming the other examined systems in almost every aspect (ECSA = 48 m^2/g, E$_{1/2}$ = 0.875V *vs.* Reversible Hydrogen Electrode) and mass activity at 0.9 V–0.216 A/mg$_{pt}$) and with very high stability of the W$_{18}$O$_{49}$ nano-sheets system.

Theoretical and experimental studies also revealed that oxygen has a higher affinity for Pd than Pt when on W$_{18}$O$_{49}$ [64], making O$_2$ adsorb better on Pd, enabling the O=O bonds to break more easily. This causes a decrease in the electron density of Pd, which weakens the Pd-O bond and could significantly increase the dissociation of O_2. In fact, introducing $W_{18}O_{49}$, which substantially alters the electronic structure of Pd, and the excess of oxygen vacancies present that might increase the electronic conductivity of $W_{18}O_{49}$ translated into enhanced electro-catalytic activity.

3.3. Tin Oxide

Tin oxide (SnO_2) is a post-transition metal dioxide with a structure resembling rutile TiO_2, and is often referred to as an oxygen-deficient n-type semiconductor that has been studied in the fields of chemical sensors [65] and electronic devices [66].

Pt and Pd supported on SnO_2 were investigated as catalyst systems for various chemical reactions, such as the low temperature oxidation of CO and methane [67], the reduction of NOxs [68], and the electro-oxidation of alcoholic fuels [69,70]. Indeed, pure tin dioxide is a wide band gap semiconductor with electrical conductivity varying from 0.1 to 10^{-6} S/cm. Therefore, if considered as a catalyst support for fuel cells, it must have higher conductivity and, hence, certain modifications such as doping [71–78] or distinct synthesis routes are required.

Various forms of tin oxide were reported, such as SnO_2 nanowires [79] (TONW) synthesized by the thermal evaporation method and meso-porous tin oxide (MPTO) by the neutral-surfactant template-assisted method [80], having high BET surface area (205 m^2/g for the MPTO, very close to the value of XC-72, *ca.* 230 m^2/g) and low particle size (20 nm diameter for TONWs and 6 nm for MPTO). After the addition of Pt to both supports, their ORR activity was compared with that of Pt on carbon and was found to be superior. In terms of durability and stability, the MPTO support outperformed commercial Pt/C during potential cycling, including two steps: the first at a constant potential of 1.2V *vs.* RHE and the second between 0.6 and 1.2 V *vs.* RHE in 0.5M sulfuric acid.

Other metal oxides that are still under investigation are Pt/MnO_2 [81] and Pd/Mn_2O_3 [82], which have shown better catalytic activity than Pt/C even with very low noble metal content, and SiO_2 [83] and NbO_2 [84], which were reported as ORR catalysts in fuel cells and have shown promising and interesting results.

4. Nitrides

Like metal oxides, metal nitrides have also attracted much attention owing to their excellent thermal and chemical stabilities. Recently, some of them were found to have catalytic properties similar to those of noble metals like Pd and Pt [85,86]. However, in terms of fuel cell applications, there have not been many publications dealing with metal nitrides as catalyst supports thus far.

Among these nitrides, TiN is the most studied since it seems to show the most promise. It has a high electrical conductivity of ~4000 S/m [87,88] and is considered to have good corrosion resistance, usually attributed to the oxy-nitride layer, caused by the formation on the surface oxide due to atmospheric oxidation and/or acidic media, which prevents further oxygen diffusion to the bulk [89], in turn making it relatively chemically inert.

However, corrosion still might occur under fuel cell conditions (relatively high temperature, acidic environment, oxygen presence, high water content, and high potential applications), turning TiN into oxide form which leads to catalyst particle growth and significantly lowers the conductivity [89]. Some have tried to use this to their advantage by making oxy-nitride supports [90], employing the high corrosion stability of TiO_2 and the electrical conductivity of TiN. These supports with Pt deposited on them have shown to have an ECSA more than three times higher than the conventional Pt/C catalyst and the activity under prolonged operations (denoted by ECSA) exceeded 50% even after 1000 cycles (close to 0% for Pt/C under the same conditions). Another study on TiON [91] substantiated this increased stability and even showed an improved result of only 20% loss in ECSA after 1000 cycles (80% loss for Pt/C).

More complex morphologies of TiN were also synthesized in order to increase the surface area and lower the TiN content and overall weight. For example, Pan et al. [92] synthesized hollow TiN nanotubes (NTs) by two-step synthesis. The TiN hollow NTs showed better electrochemical activity and stabilty than comercial Pt/C after an accelerated life test.

Since nitrides do not show any significant catalytic activity, a catalyst is usually added to the support. The interaction of the catalyst, in most cases -Pt, with the support is of extreme interest, since the support may change the Pt-Pt bond length and the electron density on the Pt, hence changing its catalytic activity. Zhang et al. [93] investigated the thermodynamics of a single-atom Pt catalyst bonding to the TiN surface and found that Pt atoms prefer to be embedded on the surface of the TiN, at the N vacancy sites, instead of forming Pt clusters. Therefore, under typical PEM fuel cell operation, TiN surface vacancies come into play, anchoring the Pt atom for better catalytic function as illustrated in Figure 4.

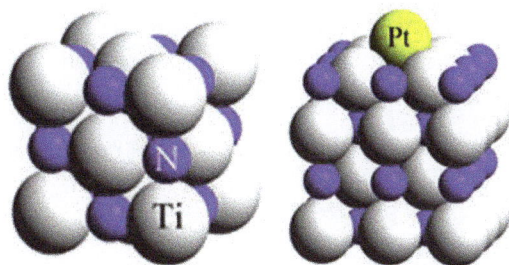

Figure 4. Crystal structure of TiN and the atomic structure model of Pt embedded on an N-vacancy site [93] (adapted from this reference with the permission of RSC publishing).

Although showing good stability when compared to carbon supports, TiNs show degradation under fuel cell operating conditions. This was studied by Avasarala et al. [89,94], who investigated the degradation mechanism of the Pt/TiN catalyst under fuel cell conditions using their accelerated stress test (AST) protocols. They found that out of the three main degradation mechanisms taken into account under these conditions, (1) support oxidation, (2) catalyst agglomeration, and (3) catalyst dissolution, Pt catalyst agglomeration and coalescence were the most dominant. Furthermore, during potential cycling, the oxy-nitride formed on the outer layer tends to dissolve to Ti (IV) hydroxide ions, leading to further passivation of the surface.

Among the other less-studied nitride supports, vanadium nitride (VN) has also been shown to have properties that might be suitable for fuel cells, such as reasonable surface area (55.4 m^2/g) and good electrical conductivity (72 S/cm). It was studied as a support for Pd for formic acid fuel cells and has shown to be efficiently prepared by solid-solid phase separation. Higher catalytic activity promoting formic acid oxidation, dehydrogenation path, and diminished Pd catalyst CO poisoning was also demonstrated [95]. In addition, pure VN showed some catalytic activity for ORR [86], and was also suggested as a promising electrode material for electrochemical super-capacitors [96].

Another interesting nitride support is Mo_2N, which was reported both as a catalyst substitute for Pt on carbon [85,97] and as a catalyst support [15]. In these studies, Mo_2N supports were prepared by polymer-assisted deposition, and subsequently, Pt nanoparticles were grown on it. The ECSA of the Pt was 20.8 m^2/g, lower than commercial Pt/C. Although not much work has been conducted with it, it has shown some initial potential and requires further research and improvement.

Other non-metal ceramics such as carbo-nitrides were also studied as combined catalyst-support arrays with noble and non-noble metals [98,99]. Xu et al. [100] reported the synthesis of graphitic C_3N_4 support by the direct heating of dicyandiamide for the deposition of Ag nanoparticles as the catalyst, forming Ag/g-C_3N_4. They came to the conclusion that this system showed fair ORR catalytic activity that remained almost unchanged after 200 cycles in oxygen atmosphere. C_3N_4 was also examined for durability and showed inconclusive results.

5. Composites/Hybrides

Some ceramic supports possess good resistance to electrochemical oxidation and stability in acidic environments, which leads to the consideration of them as alternative catalyst supports. However, the low electrical conductivity of some of them may prevent their extensive application in fuel cells.

5.1. Carbon-Based Ceramic Composites

Carbon has high electrical conductivity and surface area, therefore it has been introduced into ceramic supports physically and/or synthetically in order to improve supports that are lacking some or both of these critical properties. However, adding carbon may also lower the composite materials' corrosion resistance and enhance the supports' degradation in fuel cells. Therefore, ceramics that contain carbon might possess both carbon-like and ceramic-like properties, improving the support's compatibility for fuel cells but causing it to be more susceptible to corrosion.

Pd catalyst on iron-molybdenum-carbon (Fe_2MoC) composite was synthesized and then compared electrochemically to conventional Pt/C, Pd/C, and Pd/MoC catalysts [101]. The ORR onset potential of the Pd on the composite support was much higher when compared to Pd/C and Pd/MoC, but it did not demonstrate much improvement when compared to Pt/C. However, the calculated mass activity did show detectable improvement for the Pd/Fe_2MoC compared to the Pt/C (146.4 and 124.3 mA/mg_{cat}, respectively), in spite of a smaller catalyst loading ratio (37.6% and 47.6%, respectively). In addition, it showed better stability when compared to Pt/C.

Some other carbon-based composite ceramics that were synthesized and examined for ORR performance are Co_3W_3C [102] and $Co_6Mo_6C_2$ [103]. Loaded with catalysts (Pd for the tungsten-based hybrid and Pt for the molybdenum-based), they were electrochemically characterized, and while the tungsten-based support exhibited only slightly better electrochemical characteristics such as onset potential, mass activity, and stability under prolonged exposure to an oxidizing environment compared with conventional Pt/C, the molybdenum-based support demonstrated substantial improvement in mass activity (more than twice the value of Pt/C) and showed a detectable improved stability with no detectable degradation during the AST (1000 CV cycles between 0.6–1.2V and 0.05–1.1V *vs.* RHE, respectively, in O_2-saturated 0.1 M $HClO_4$ solution).

Tungsten carbide and carbon nanotubes (CNTs) have exhibited ORR catalytic activity with a noble metal catalyst separately [9,10,12,104]. Considering that a hybrid tungsten carbideCNTs was synthesized and tested by Liang *et al.* [105], even with half the Pt loading, the Pt/WC-CNT electro-catalyst had a higher onset potential compared to the Pt/CNT, indicating a synergistic effect between Pt, WC, and the CNTs. Different composites of WC/C were reported. Garcia & Ticianelli [106] mixed tungsten hexacarbonyl with Vulcan XC-72. They tested different samples of the support for ORR activity. The supports were distinguished by the WC to C ratio in the presence of W_2C. They found that all samples had better catalytic activity (denoted by onset potentials and specific activity) compared to Pt/C due to an increase of the Pt 5d-band occupancy, which led to a weaker Pt-OH interaction, resulting in a lower Pt-oxide coverage and thus increasing the kinetics of ORR.

The electrochemical properties and application of tungstate salts such as $SrWO_4$ and $CaWO_4$ mixed with graphite were investigated as well [107,108]. Both catalysts exhibited better electrochemical activity for oxygen reduction reaction in H_2SO_4 solution compared to Pt/graphite, denoted by the onset potential for $SrWO_4$ (0.65V vs. 0.55V for Pt/graphite) and the half-wave potential for $CaWO_4$ (0.51 V vs. 0.45 V for Pt/graphite).

5.2. Other Ceramic Composites

Other composites that do not contain carbon and are usually based on more than one metal were reported as well. An interesting cesium- and tungsten-based composite that has formerly been reported as an electrolyte for fuel cells [109] due to their good proton conductivity and stability was examined in PEMFC [110]. The $Cs_{2.5}H_{0.5}PW_{12}O_{40}$ was shown to have a relatively high surface area (136 m^2/g) diminished to 35–50 m^2/g after the addition of Pt, suggesting Pt saturation of the low dimension pores. The composite also showed better catalytic activity and stability in acidic media than conventional Pt/XC-72, which was attributed to the pores preventing Pt agglomeration.

The widely use dindium tin oxide (ITO) is an n-type semiconductor with a wide band gap, which is produced by replacing In^{3+} by Sn^{4+} in the cubic structure of indium oxide. This replacement produces free electrons enhancing its conductivity and, thus, influences the optical and electrical properties of the ITO film [111]. ITO is a commercially available material often used as a transparent conducting oxide (TCO) for smart windows. Chhina *et al.* [112] fabricated a Pt/ITO catalyst as a potential non-carbon catalyst support and investigated the thermal and electrochemical stability. The Pt on ITO had an average crystallite size of 13 nm. Electrochemical measurements indicated that this catalyst was much more stable than those of both commercially available Hispec 4000 and Pt/XC-72R. In a different study [113], Pt clusters were deposited on the ITO NPs through the galvanic displacement of Cu by Pt. The specific ECSA of Pt/ITO (83.1 m^2/g) was found to be three times that of Pt/C (27.3 m^2/g) and, after 1000 cycles, changes in the Pt ECSA and electrocatalytic activity proved the stability of the Pt/ITO catalyst was far superior to that of Pt/C when Pt/ITO showed no recordable loss of Pt ECSA. However, for Pt/C, only ~65% of the original ECSA remained after potential cycling. However, TEM pictures taken after the stability tests detected several small holes on the ITO surface due to the corrosion and dissolution of the surface Sn.

6. Titanium Diboride

TiB_2 is a relatively novel titanium-based support that has been considered as a base material for a range of different applications. It exhibits good electrical (~10^5 S/cm) and high thermal conductivity (~65 W/mK), excellent thermal stability and corrosion resistance in acidic medium [114], and might be considered a promising candidate for PEMFC catalyst ceramic support.

TiB_2 was first reported in the context of fuel cells by Yin et al. [115,116], who deposited Pt particles on TiB_2 support by colloidal route. Before then, it was reported that smaller particle size leads to the agglomeration of particles and the loss of ECSA due to a higher specific surface energy. Using their synthesis route, the highly dispersed Pt nanoparticles on the ceramic support were dispersed and stabilized by Nafion in order to prevent agglomeration and particle growth. Indeed, the ceramic boride showed similar catalytic activity (similar onset potential) in addition to better durability when compared to conventional Pt/C after 6000 cycles of CVs in 0.5 M sulfuric acid. About 60% loss of ECSA was exhibited while about 80% loss was detected for Pt/C after 5000 cycles.

The electrochemical stability of the Pt/TiB_2 catalyst was approximately four times higher than that of the commercial Pt/C catalyst after cycling between 0 and 1.2 V in 0.5 M H_2SO_4, due to the support characteristics and also possibly from the Nafion stabilization effects, which enhanced both the metal support interaction and the steric hindrance effect of the surface Pt nanoparticles. In several other studies, Pt/TiB_2 was obtained not only by colloidal route [115–117] but also by carbo-thermal reduction [118,119].

Huang et al. [117] investigated the effect of several kinds of pretreatments applied to the support (mostly exposing it to acidic or alkaline media), in an attempt to reduce the influence of TiB_2 particle size on the electrochemical performance of the system. The study showed that among the investigated pretreatments, the hydrogen peroxide pretreatment demonstrated the best results, producing a catalyst with about twice the ECSA compared to other pretreated samples, possibly due to the presence of TiO_2. However, Roth et al. [118] deposited Pt particles on TiB_2 prepared via the carbo-thermal reduction method [116], and showed that although TiB_2 exhibited good stability under normal cycling, its performance, denoted by power density, reduced much more rapidly (it showed less than half the power density after only 100 cycles) under actual fuel cell conditions (oxygen presence and an elevated temperature of 80 °C), claimed to be due to oxide formation which probably led to reduced conductivity.

7. Conclusions

Corrosion is a serious issue in fuel cell technology as it can dramatically reduce the electrode life time, and thus the overall performance of the cell. In order to try to increase the lifetime of the electrodes, various materials have been proposed, the most promising of which are the ceramic materials. In most cases, carbides, oxides, nitrides, borides, and composite materials have shown better stability and durability when compared to the commercially available standard, Pt/XC-72. Unfortunately, due to the lack of standard protocols for the assessment of these parameters, it is impossible to compare between supports and tell which is best under certain conditions.

One very interesting outcome, which most of the researchers in the field seem to agree upon, is that the move from carbon to other supports opens up a wide array of possibilities when it comes to catalyst activity. In fact, in many of the studies surveyed in this review, enhancement of the catalytic properties was shown and was attributed to the favorable interaction of the catalysts, in most cases -Pt, with the support. When choosing ceramic supports for fuel cells, one must consider their conductivity, which in some cases is very low when compared to carbon supports, and the surface area of the support. The latter is an issue that most studies do not tackle yet, but may impact the overall performance significantly.

Acknowledgments: The authors would like to thank MAFAT for funding this work.

Author Contributions: Oran Lori and Lior Elbaz wrote this review together with equal contribution. Oran Lori was responsible for revising it and Lior Elbaz edited its final version.

Conflicts of Interest: The authors declare no conflict of interest.

References

1. Borup, R.; Meyers, J.; Pivovar, B.; Kim, Y.S.; Mukundan, R.; Garland, N.; Myers, D.; Wilson, M.; Garzon, F.; Wood, D. Scientific aspects of polymer electrolyte fuel cell durability and degradation. *Chem. Rev.* **2007**, *107*, 3904–3951.

2. Meier, J.C.; Galeano, C.; Katsounaros, I.; Witte, J.; Bongard, H.J.; Topalov, A.A.; Baldizzone, C.; Mezzavilla, S.; Schueth, F.; Mayrhofer, K.J.J. Design criteria for stable pt/c fuel cell catalysts. *Beilstein J. Nanotechnol.* **2014**, *5*, 44–67.

3. Pourbaix, M. *Atlas of Electrochemical Equilibria in Aqueous Solutions*; National Association of Corrosion: Oxford, UK, 1974.

4. Barth, T.; Lunde, G. The lattice constants of metallic platinum, silver and gold. *Z. Phys. Chem.* **1926**, *121*, 78–102.

5. Clougherty, E.V.; Lothrop, K.H.; Kafalas, J.A. New phase formed by high-pressure treatment. *Nature* **1961**.

6. Rudy, E.; Windisch, S.; Stosick, A.J.; Hoffman, J.R. The constitution of binary molybdenum-carbon alloys. *AIME* **1967**, *239*, 1247–1267.

7. Gruver, G.A. Corrosion of Carbon-black in Phosphoric-acid. *J. Electrochem. Soc.* **1978**, *125*, 1719–1720.

8. Meyers, J.P.; Darling, R.M. Model of carbon corrosion in PEM fuel cells. *J. Electrochem. Soc.* **2006**, *153*, A1432–A1442.

9. Li, W.; Wang, X.; Chen, Z.; Waje, M.; Yan, Y. Pt-Ru Supported on Double-Walled Carbon Nanotubes as High-Performance Anode Catalysts for Direct Methanol Fuel Cells. *J. Phys. Chem. B* **2006**, *110*, 15353–15358.

10. Wu, G.; Xu, B.-Q. Carbon nanotube supported Pt electrodes for methanol oxidation: A comparison between multi- and single-walled carbon nanotubes. *J. Power Sources* **2007**, *174*, 148–158.

11. Zhou, X.; Qiao, J.; Yang, L.; Zhang, J. A Review of Graphene-Based Nanostructural Materials for Both Catalyst Supports and Metal-Free Catalysts in PEM Fuel Cell Oxygen Reduction Reactions. *Adv. Energy Mater.* **2014**, *4*, 1301523:1–1301523:25.

12. Wang, X.; Li, W.; Chen, Z.; Waje, M.; Yan, Y. Durability investigation of carbon nanotube as catalyst support for proton exchange membrane fuel cell. *J. Power Sources* **2006**, *158*, 154–159.

13. Armstrong, K.J.; Elbaz, L.; Bauer, E.; Burrell, A.K.; McCleskey, T.M.; Brosha, E.L. Nanoscale titania ceramic composite supports for PEM fuel cells. *J. Mater. Res.* **2012**, *27*, 2046–2054.

14. Shanmugam, S.; Gedanken, A. Synthesis and electrochemical oxygen reduction of platinum nanoparticles supported on mesoporous TiO_2. *J. Phys. Chem. C* **2009**, *113*, 18707–18712.

15. Blackmore, K.J.; Elbaz, L.; Bauer, E.; Brosha, E.L.; More, K.; McCleskey, T.M.; Burrell, A.K. High Surface Area Molybdenum Nitride Support for Fuel Cell Electrodes. *J. Electrochem. Soc.* **2011**, *158*, B1255–B1259.

16. Cui, X.; Guo, L.; Cui, F.; He, Q.; Shi, J. Electrocatalytic activity and co tolerance properties of mesostructured pt/wo3 composite as an anode catalyst for pemfcs. *J. Phys. Chem. C* **2009**, *113*, 4134–4138.

17. Levy, R.B.; Boudart, M. Platinum-like behavior of tungsten carbide in surface catalysis. *Science* **1973**, *181*, 547–549.

18. York, A.P.E.; Claridge, J.B.; Brungs, A.J.; Brungs, A.J.; Tsang, S.C.; Green, M.L.H. Molybdenum and tungsten carbides as catalysts for the conversion of methane to synthesis gas using stoichiometric feedstocks. *Chem. Commun.* **1997**, 39–40.

19. Burns, S.; Gallagher, J.G.; Hargreaves, J.S.J.; Harris, P.J.F. Direct observation of carbon nanotube formation in Pd/H-ZSM-5 and MoO_3/H-ZSM-5 based methane activation catalysts. *Catal. Lett.* **2007**, *116*, 122–127.

20. Ribeiro, F.H.; Boudart, M.; Dalla Betta, R.A.; Iglesia, E. Catalytic reactions of *n*-Alkanes on β-W_2C and WC: The effect of surface oxygen on reaction pathways. *J. Catal.* **1991**, *130*, 498–513.

21. Ledoux, M.J.; Huu, C.P.; Guille, J.; Dunlop, H. Compared activities of platinum and high specific surface area Mo_2C and WC catalysts for reforming reactions: I. Catalyst activation and stabilization: Reaction of *n*-hexane. *J. Catal.* **1992**, *134*, 383–398.

22. Esposito, D.V.; Chen, J.G. Monolayer platinum supported on tungsten carbides as low-cost electrocatalysts: opportunities and limitations. *Energy Environ. Sci.* **2011**, *4*, 3900–3912.

23. Hunt, S.T.; Nimmanwudipong, T.; Román-Leshkov, Y. Engineering Non-sintered, Metal-Terminated Tungsten Carbide Nanoparticles for Catalysis. *Angew. Chem. Int. Ed.* **2014**, *53*, 5131–5136.

24. Curry, K.E.; Thompson, L.T. Carbon-hydrogen bond activation over tungsten carbide catalysts. *Catal. Today* **1994**, *21*, 171–184.

25. Pantea, D.; Darmstadt, H.; Kaliaguine, S.; Sümmchen, L.; Roy, C. Electrical conductivity of thermal carbon blacks: Influence of surface chemistry. *Carbon* **2001**, *39*, 1147–1158.

26. Hassan, A.; Paganin, V.A.; Ticianelli, E.A. Pt modified tungsten carbide as anode electrocatalyst for hydrogen oxidation in proton exchange membrane fuel cell: CO tolerance and stability. *Appl. Catal. B* **2015**, *165*, 611–619.

27. Poh, C.K.; Lim, S.H.; Lin, J.; Feng, Y.P. Tungsten Carbide Supports for Single-Atom Platinum-Based Fuel-Cell Catalysts: First-Principles Study on the Metal–Support Interactions and O_2 Dissociation on W_xC Low-Index Surfaces. *J. Phys. Chem. C* **2014**, *118*, 13525–13538.

28. Nikolic, V.M.; Perovic, I.M.; Gavrilov, N.M.; Pašti, I.A.; Saponjic, A.B.; Vulic, P.J.; Karic, S.D.; Babic, B.M.; Marceta Kaninski, M.P. On the tungsten carbide synthesis for PEM fuel cell application—Problems, challenges and advantages. *Int. J. Hydrogen Energy* **2014**, *39*, 11175–11185.

29. Tang, H.; Qi, Z.; Ramani, M.; Elter, J.F. PEM fuel cell cathode carbon corrosion due to the formation of air/fuel boundary at the anode. *J. Power Sources* **2006**, *158*, 1306–1312.

30. Elbaz, L.; Kreller, C.R.; Henson, N.J.; Brosha, E.L. Electrocatalysis of oxygen reduction with platinum supported on molybdenum carbide-carbon composite. *J. Electroanal. Chem.* **2014**, *720–721*, 34–40.

31. Jia, Q.X.; McCleskey, T.M.; Burrell, A.K.; Lin, Y.; Collis, G.E.; Wang, H.; Li, A.D.Q.; Foltyn, S.R. Polymer-assisted deposition of metal-oxide films. *Nat. Mater.* **2004**, *3*, 529–532.

32. Elbaz, L.; Phillips, J.; Artyushkova, K.; More, K.; Brosha, E.L. Evidence of High Electrocatalytic Activity of Molybdenum Carbide Supported Platinum Nanorafts. *J. Electrochem. Soc.* **2015**, *162*, H681–H685.

33. Ma, L.; Sui, S.; Zhai, Y. Preparation and characterization of Ir/TiC catalyst for oxygen evolution. *J. Power Sources* **2008**, *177*, 470–477.

34. Ou, Y.; Cui, X.; Zhang, X.; Jiang, Z. Titanium carbide nanoparticles supported Pt catalysts for methanol electrooxidation in acidic media. *J. Power Sources* **2010**, *195*, 1365–1369.

35. Chiwata, M.; Kakinuma, K.; Wakisaka, M.; Uchida, M.; Deki, S.; Watanabe, M.; Uchida, H. Oxygen Reduction Reaction Activity and Durability of Pt Catalysts Supported on Titanium Carbide. *Catalysts* **2015**, *5*, 966–980.

36. Ignaszak, A.; Song, C.; Zhu, W.; Zhang, J.; Bauer, A.; Baker, R.; Neburchilov, V.; Ye, S.; Campbell, S. Titanium carbide and its core-shelled derivative $TiC@TiO_2$ as catalyst supports for proton exchange membrane fuel cells. *Electrochim. Acta* **2012**, *69*, 397–405.

37. Toyoda, M.; Yano, T.; Tryba, B.; Mozia, S.; Tsumura, T.; Inagaki, M. Preparation of carbon-coated Magneli phases Ti_nO_{2n-1} and their photocatalytic activity under visible light. *Appl. Catal. B* **2009**, *88*, 160–164.

38. Toyoda, M.; Yano, T.; Mozia, S.; Tsumura, T.; Itoh, E.; Amao, Y.; Inagaki, M. Development of visible-light sensitive reduced phases of titania, Ti_nO_{2n-1}, through carbon coating. *Tanso* **2005**, *220*, 265–269.

39. Ni, M.; Leung, M.K.; Leung, D.Y.; Sumathy, K. A review and recent developments in photocatalytic water-splitting using TiO_2 for hydrogen production. *Renewable Sustainable Energy Rev.* **2007**, *11*, 401–425.

40. Zhu, Y.; Shi, J.; Zhang, Z.; Zhang, C.; Zhang, X. Development of a gas sensor utilizing chemiluminescence on nanosized titanium dioxide. *Anal. Chem.* **2002**, *74*, 120–124.

41. Bartholomew, R.F.; Frankl, D. Electrical properties of some titanium oxides. *Phys. Rev.* **1969**, *187*.

42. Wu, Q.; Ruan, J.; Zhou, Z.; Sang, S. Magneli phase titanium sub-oxide conductive ceramic Ti_nO_{2n-1} as support for electrocatalyst toward oxygen reduction reaction with high activity and stability. *J. Cent. South Univ.* **2015**, *22*, 1212–1219.

43. Ioroi, T.; Akita, T.; Yamazaki, S.-i.; Siroma, Z.; Fujiwara, N.; Yasuda, K. Corrosion-Resistant PEMFC Cathode Catalysts Based on a Magnéli-Phase Titanium Oxide Support Synthesized by Pulsed UV Laser Irradiation. *J. Electrochem. Soc.* **2011**, *158*, C329–C334.

44. Ioroi, T.; Senoh, H.; Siroma, Z.; Yamazaki, S.-i.; Fujiwara, N.; Yasuda, K. Stability of Corrosion-Resistant Magnéli-Phase Ti_4O_7-Supported PEMFC Catalysts. *ECS Trans.* **2007**, *11*, 1041–1048.

45. Ioroi, T.; Siroma, Z.; Fujiwara, N.; Yamazaki, S.-i.; Yasuda, K. Sub-stoichiometric titanium oxide-supported platinum electrocatalyst for polymer electrolyte fuel cells. *Electrochem. Commun.* **2005**, *7*, 183–188.

46. Vračar, L.M.; Krstajić, N.V.; Radmilović, V.R.; Jakšić, M.M. Electrocatalysis by nanoparticles- oxygen reduction on Ebonex/Pt electrode. *J. Electroanal. Chem.* **2006**, *587*, 99–107.

47. Geng, P.; Su, J.Y.; Miles, C.; Comninellis, C.; Chen, G.H. Highly-Ordered Magneli Ti_4O_7 Nanotube Arrays as Effective Anodic Material for Electro-oxidation. *Electrochim. Acta* **2015**, *153*, 316–324.

48. Du, Q.; Wu, J.; Yang, H. Pt@Nb-TiO_2 catalyst membranes fabricated by electrospinning and atomic layer deposition. *ACS Catal.* **2013**, *4*, 144–151.

49. Chevallier, L.; Bauer, A.; Cavaliere, S.; Hui, R.; Rozière, J.; Jones, D.J. Mesoporous nanostructured Nb-doped titanium dioxide microsphere catalyst supports for PEM fuel cell electrodes. *ACS Appl. Mater. Interfaces* **2012**, *4*, 1752–1759.

50. Bauer, A.; Chevallier, L.; Hui, R.; Cavaliere, S.; Zhang, J.; Jones, D.; Rozière, J. Synthesis and characterization of Nb-TiO_2 mesoporous microsphere and nanofiber supported Pt catalysts for high temperature PEM fuel cells. *Electrochim. Acta* **2012**, *77*, 1–7.

51. Elezović, N.; Babić, B.; Gajić-Krstajić, L.; Radmilović, V.; Krstajić, N.; Vračar, L. Synthesis, characterization and electrocatalytical behavior of Nb-TiO_2/Pt nanocatalyst for oxygen reduction reaction. *J. Power Sources* **2010**, *195*, 3961–3968.

52. Siracusano, S.; Stassi, A.; Modica, E.; Baglio, V.; Aricò, A.S. Preparation and characterisation of Ti oxide based catalyst supports for low temperature fuel cells. *Int. J. Hydrogen Energy* **2013**, *38*, 11600–11608.

53. Huang, S.-Y.; Ganesan, P.; Popov, B.N. Titania supported platinum catalyst with high electrocatalytic activity and stability for polymer electrolyte membrane fuel cell. *Appl. Catal. B* **2011**, *102*, 71–77.

54. Ioroi, T.; Akita, T.; Asahi, M.; Yamazaki, S.; Siroma, Z.; Fujiwara, N.; Yasuda, K. Platinum-titanium alloy catalysts on a Magnéli-phase titanium oxide support for improved durability in Polymer Electrolyte Fuel Cells. *J. Power Sources* **2013**, *223*, 183–189.

55. Huang, D.; Zhang, B.; Bai, J.; Zhang, Y.; Wittstock, G.; Wang, M.; Shen, Y. Pt Catalyst Supported within TiO_2 Mesoporous Films for Oxygen Reduction Reaction. *Electrochim. Acta* **2014**, *130*, 97–103.

56. Micoud, F.; Maillard, F.; Gourgaud, A.; Chatenet, M. Unique CO-tolerance of Pt-WO_x materials. *Electrochem. Commun.* **2009**, *11*, 651–654.

57. Rajeswari, J.; Viswanathan, B.; Varadarajan, T.K. Tungsten trioxide nanorods as supports for platinum in methanol oxidation. *Mater. Chem. Phys.* **2007**, *106*, 168–174.

58. Cui, X.; Shi, J.; Chen, H.; Zhang, L.; Guo, L.; Gao, J.; Li, J. Platinum/mesoporous WO_3 as a carbon-free electrocatalyst with enhanced electrochemical activity for methanol oxidation. *J. Phys. Chem. B* **2008**, *112*, 12024–12031.

59. Park, K.W.; Ahn, K.S.; Choi, J.H.; Nah, Y.C.; Kim, Y.M.; Sung, Y.E. Pt-WO_x electrode structure for thin-film fuel cells. *Appl. Phys. Lett.* **2002**, *81*, 907–909.

60. Chhina, H.; Campbell, S.; Kesler, O. *Ex situ* Evaluation of Tungsten Oxide as a Catalyst Support for PEMFCs. *J. Electrochem. Soc.* **2007**, *154*, B533–B539.

61. Liu, Y.; Shrestha, S.; Mustain, W.E. Synthesis of Nanosize Tungsten Oxide and Its Evaluation as an Electrocatalyst Support for Oxygen Reduction in Acid Media. *ACS Catal.* **2012**, *2*, 456–463.

62. Jaksic, J.M.; Labou, D.; Papakonstantinou, G.D.; Siokou, A.; Jaksic, M.M. Novel spillover interrelating reversible electrocatalysts for oxygen and hydrogen electrode reactions. *J. Phys. Chem. C* **2010**, *114*, 18298–18312.

63. Lu, Y.; Jiang, Y.; Gao, X.; Wang, X.; Chen, W. Strongly Coupled Pd Nanotetrahedron/Tungsten Oxide Nanosheet Hybrids with Enhanced Catalytic Activity and Stability as Oxygen Reduction Electrocatalysts. *J. Am. Chem. Soc.* **2014**, *136*, 11687–11697.

64. Lima, F.H.B.; Zhang, J.; Shao, M.H.; Sasaki, K.; Vukmirovic, M.B.; Ticianelli, E.A.; Adzic, R.R. Catalytic Activity-d-Band Center Correlation for the O_2 Reduction Reaction on Platinum in Alkaline Solutions. *J. Phys. Chem. C* **2007**, *111*, 404–410.

65. Fagan, J.G.; Amarakoon, R. Reliability and reproducibility of ceramic sensors. III: Humidity sensors. *Am. Ceram. Soc. Bull.* **1993**, *72*, 119–130.

66. Pianaro, S.; Bueno, P.; Longo, E.; Varela, J.A. A new SnO_2-based varistor system. *J. Mater. Sci. Lett.* **1995**, *14*, 692–694.

67. Sekizawa, K.; Widjaja, H.; Maeda, S.; Ozawa, Y.; Eguchi, K. Low temperature oxidation of methane over Pd catalyst supported on metal oxides. *Catal. Today* **2000**, *59*, 69–74.

68. Amalric-Popescu, D.; Bozon-Verduraz, F. SnO_2-supported palladium catalysts: activity in deNO_x at low temperature. *Catal. Lett.* **2000**, *64*, 125–128.

69. Du, W.; Yang, G.; Wong, E.; Deskins, N.A.; Frenkel, A.I.; Su, D.; Teng, X. Platinum-Tin Oxide Core-Shell Catalysts for Efficient Electro-Oxidation of Ethanol. *J. Am. Chem. Soc.* **2014**, *136*, 10862–10865.

70. Katayama, A. Electrooxidation of methanol on a platinum-tin oxide catalyst. *J. Phys. Chem.* **1980**, *84*, 376–381.

71. Dou, M.; Hou, M.; Wang, F.; Liang, D.; Zhao, Q.; Shao, Z.; Yi, B. Sb-Doped SnO_2 Supported Platinum Catalyst with High Stability for Proton Exchange Membrane Fuel Cells. *J. Electrochem. Soc.* **2014**, *161*, F1231–F1236.

72. Kakinuma, K.; Chino, Y.; Senoo, Y.; Uchida, M.; Kamino, T.; Uchida, H.; Deki, S.; Watanabe, M. Characterization of Pt catalysts on Nb-doped and Sb-doped $SnO_{2-\delta}$ support materials with aggregated structure by rotating disk electrode and fuel cell measurements. *Electrochim. Acta* **2013**, *110*, 316–324.

73. Senoo, Y.; Kakinuma, K.; Uchida, M.; Uchida, H.; Deki, S.; Watanabe, M. Improvements in electrical and electrochemical properties of Nb-doped $SnO_{2-\delta}$ supports for fuel cell cathodes due to aggregation and Pt loading. *RSC Adv.* **2014**, *4*, 32180–32188.

74. Yin, M.; Xu, J.; Li, Q.; Jensen, J.O.; Huang, Y.; Cleemann, L.N.; Bjerrum, N.J.; Xing, W. Highly active and stable Pt electrocatalysts promoted by antimony-doped SnO_2 supports for oxygen reduction reactions. *Appl. Catal. B* **2014**, *144*, 112–120.

75. Kakinuma, K.; Uchida, M.; Kamino, T.; Uchida, H.; Watanabe, M. Synthesis and electrochemical characterization of Pt catalyst supported on $Sn_{0.96}Sb_{0.04}O_{2-\delta}$ with a network structure. *Electrochim. Acta* **2011**, *56*, 2881–2887.

76. Savych, J.; Subianto, S.; Nabil, Y.; Cavaliere, S.; Jones, D.; Rozière, J. Negligible degradation on *in situ* voltage cycling of a PEMFC with electrospun niobium-doped tin oxide supported Pt cathode. *Phys. Chem. Chem. Phys.* **2015**, *17*, 16970–16976.

77. Chino, Y.; Taniguchi, K.; Senoo, Y.; Kakinuma, K.; Hara, M.; Watanabe, M.; Uchida, M. Effect of Added Graphitized CB on Both Performance and Durability of Pt/Nb-SnO_2 Cathodes for PEFCs. *J. Electrochem. Soc.* **2015**, *162*, F736–F743.

78. Senoo, Y.; Taniguchi, K.; Kakinuma, K.; Uchida, M.; Uchida, H.; Deki, S.; Watanabe, M. Cathodic performance and high potential durability of Ta-$SnO_{2-\delta}$-supported Pt catalysts for PEFC cathodes. *Electrochem. Commun.* **2015**, *51*, 37–40.

79. Saha, M.S.; Li, R.; Cai, M.; Sun, X. High electrocatalytic activity of platinum nanoparticles on SnO_2 nanowire-based electrodes. *Electrochem. Solid State Lett.* **2007**, *10*, B130–B133.

80. Zhang, P.; Huang, S.-Y.; Popov, B.N. Mesoporous Tin Oxide as an Oxidation-Resistant Catalyst Support for Proton Exchange Membrane Fuel Cells. *J. Electrochem. Soc.* **2010**, *157*, B1163–B1172.

81. Wang, X.; Yang, Z.; Zhang, Y.; Jing, L.; Zhao, Y.; Yan, Y.; Sun, K. MnO_2 Supported Pt Nanoparticels with High Electrocatalytic Activity for Oxygen Reduction Reaction. *Fuel Cells* **2014**, *14*, 35–41.

82. Dong, H.-Q.; Chen, Y.-Y.; Han, M.; Li, S.-L.; Zhang, J.; Li, J.-S.; Lan, Y.-Q.; Dai, Z.-H.; Bao, J.-C. Synergistic effect of mesoporous Mn_2O_3-supported Pd nanoparticle catalysts for electrocatalytic oxygen reduction reaction with enhanced performance in alkaline medium. *J. Mater. Chem. A* **2014**, *2*, 1272–1276.

159

83. Seger, B.; Kongkanand, A.; Vinodgopal, K.; Kamat, P.V. Platinum dispersed on silica nanoparticle as electrocatalyst for PEM fuel cell. *J. Electroanal. Chem.* **2008**, *621*, 198–204.

84. Sasaki, K.; Zhang, L.; Adzic, R.R. Niobium oxide-supported platinum ultra-low amount electrocatalysts for oxygen reduction. *Phys. Chem. Chem. Phys.* **2008**, *10*, 159–167.

85. Zhong, H.X.; Zhang, H.M.; Liu, G.; Liang, Y.M.; Hu, J.W.; Yi, B.L. A novel non-noble electrocatalyst for PEM fuel cell based on molybdenum nitride. *Electrochem. Commun.* **2006**, *8*, 707–712.

86. Huang, T.; Mao, S.; Zhou, G.; Wen, Z.; Huang, X.; Ci, S.; Chen, J. Hydrothermal synthesis of vanadium nitride and modulation of its catalytic performance for oxygen reduction reaction. *Nanoscale* **2014**, *6*, 9608–9613.

87. Oyama, S.T. Introduction to the chemistry of transition metal carbides and nitrides. In *The Chemistry of Transition Metal Carbides and Nitrides*; Oyama, S.T., Ed.; Springers: Heidelberg, Germany, 1996; pp. 1–27.

88. Giner, J.; Swette, L. Oxygen Reduction on Titanium Nitride in Alkaline Electrolyte. *Nature* **1966**, *211*, 1291–1292.

89. Avasarala, B.; Haldar, P. On the stability of TiN-based electrocatalysts for fuel cell applications. *Int. J. Hydrogen Energy* **2011**, *36*, 3965–3974.

90. Seifitokaldani, A.; Savadogo, O. Electrochemically Stable Titanium Oxy-Nitride Support for Platinum Electro-Catalyst for PEM Fuel Cell Applications. *Electrochim. Acta* **2015**, *167*, 237–245.

91. Wang, W.; Savadogo, O.; Ma, Z.-F. The oxygen reduction reaction on Pt/TiO$_x$N$_y$-based electrocatalyst for PEM fuel cell applications. *J. Appl. Electrochem.* **2012**, *42*, 857–866.

92. Pan, Z.; Xiao, Y.; Fu, Z.; Zhan, G.; Wu, S.; Xiao, C.; Hu, G.; Wei, Z. Hollow and porous titanium nitride nanotubes as high-performance catalyst supports for oxygen reduction reaction. *J. Mater. Chem. A* **2014**, *2*, 13966–13975.

93. Zhang, R.-Q.; Lee, T.-H.; Yu, B.-D.; Stampfl, C.; Soon, A. The role of titanium nitride supports for single-atom platinum-based catalysts in fuel cell technology. *Phys. Chem. Chem. Phys.* **2012**, *14*, 16552–16557.

94. Avasarala, B.; Haldar, P. Durability and degradation mechanism of titanium nitride based electrocatalysts for PEM (proton exchange membrane) fuel cell applications. *Energy* **2013**, *57*, 545–553.

95. Yang, M.; Cui, Z.; DiSalvo, F.J. Mesoporous vanadium nitride as a high performance catalyst support for formic acid electrooxidation. *Chem. Commun.* **2012**, *48*, 10502–10504.

96. Zhou, X.; Chen, H.; Shu, D.; He, C.; Nan, J. Study on the electrochemical behavior of vanadium nitride as a promising supercapacitor material. *J. Phys. Chem. Solids* **2009**, *70*, 495–500.

97. Cao, B.; Neuefeind, J.C.; Adzic, R.R.; Khalifah, P.G. Molybdenum Nitrides as Oxygen Reduction Reaction Catalysts: Structural and Electrochemical Studies. *Inorg. Chem.* **2015**, *54*, 2128–2136.

98. Di Noto, V.; Negro, E. Development of nano-electrocatalysts based on carbon nitride supports for the ORR processes in PEM fuel cells. *Electrochim. Acta* **2010**, *55*, 7564–7574.

99. Di Noto, V.; Negro, E.; Giffin, G.A. (Keynote Lecture) Multi-Metal Nano-Electrocatalysts Based on Carbon Nitride Supports for the ORR and FOR in PEM Fuel Cells. *ECS Trans.* **2012**, *40*, 3–10.

100. Xu, L.; Li, H.; Xia, J.; Wang, L.; Xu, H.; Ji, H.; Li, H.; Sun, K. Graphitic carbon nitride nanosheet supported high loading silver nanoparticle catalysts for the oxygen reduction reaction. *Mater. Lett.* **2014**, *128*, 349–353.

101. Yan, Z.; Zhang, M.; Xie, J.; Zhu, J.; Shen, P.K. A bimetallic carbide Fe_2MoC promoted Pd electrocatalyst with performance superior to Pt/C towards the oxygen reduction reaction in acidic media. *Appl. Catal. B* **2015**, *165*, 636–641.

102. Li, Z.; Ji, S.; Pollet, B.G.; Shen, P.K. A Co_3W_3C promoted Pd catalyst exhibiting competitive performance over Pt/C catalysts towards the oxygen reduction reaction. *Chem. Commun.* **2014**, *50*, 566–568.

103. Ma, X.; Meng, H.; Cai, M.; Shen, P.K. Bimetallic Carbide Nanocomposite Enhanced Pt Catalyst with High Activity and Stability for the Oxygen Reduction Reaction. *J. Am. Chem. Soc.* **2012**, *134*, 1954–1957.

104. Chhina, H.; Campbell, S.; Kesler, O. Thermal and electrochemical stability of tungsten carbide catalyst supports. *J. Power Sources* **2007**, *164*, 431–440.

105. Liang, C.; Ding, L.; Li, C.; Pang, M.; Su, D.; Li, W.; Wang, Y. Nanostructured WC_x/CNTs as highly efficient support of electrocatalysts with low Pt loading for oxygen reduction reaction. *Energy Environ. Sci.* **2010**, *3*, 1121–1127.

106. Garcia, A.C.; Ticianelli, E.A. Investigation of the oxygen reduction reaction on Pt-WC/C electrocatalysts in alkaline media. *Electrochim. Acta* **2013**, *106*, 453–459.

107. Farsi, H.; Barzgari, Z. Chemical Synthesis of Nanostructured $SrWO_4$ for Electrochemical Energy Storage and Conversion Applications. *Int. J. Nanosci.* **2014**, *13*, 1450013:1–1450013:9.

108. Farsi, H.; Barzgari, Z. Synthesis, characterization and electrochemical studies of nanostructured $CaWO_4$ as platinum support for oxygen reduction reaction. *Mater. Res. Bull.* **2014**, *59*, 261–266.

109. Kukino, T.; Kikuchi, R.; Takeguchi, T.; Matsui, T.; Eguchi, K. Proton conductivity and stability of $Cs_2HPW_{12}O_{40}$ electrolyte at intermediate temperatures. *Solid State Ionics* **2005**, *176*, 1845–1848.

110. Dsoke, S.; Kolary-Zurowska, A.; Zurowski, A.; Mignini, P.; Kulesza, P.J.; Marassi, R. Rotating disk electrode study of $Cs_{2.5}H_{0.5}PW_{12}O_{40}$ as mesoporous support for Pt nanoparticles for PEM fuel cells electrodes. *J. Power Sources* **2011**, *196*, 10591–10600.

111. Yu, H.Y.; Feng, X.D.; Grozea, D.; Lu, Z.H.; Sodhi, R.N.S.; Hor, A.-M.; Aziz, H. Surface electronic structure of plasma-treated indium tin oxides. *Appl. Phys. Lett.* **2001**, *78*, 2595–2597.

112. Chhina, H.; Campbell, S.; Kesler, O. An oxidation-resistant indium tin oxide catalyst support for proton exchange membrane fuel cells. *J. Power Sources* **2006**, *161*, 893–900.

113. Liu, Y.; Mustain, W.E. High Stability, High Activity Pt/ITO Oxygen Reduction Electrocatalysts. *J. Am. Chem. Soc.* **2013**, *135*, 530–533.

114. Munro, R.G. Material properties of titanium diboride. *J. Res. Nat. Inst. Stand. Technol.* **2000**, *105*, 709–720.

115. Yin, S.; Mu, S.; Lv, H.; Cheng, N.; Pan, M.; Fu, Z. A highly stable catalyst for PEM fuel cell based on durable titanium diboride support and polymer stabilization. *Appl. Catal. B* **2010**, *93*, 233–240.

116. Yin, S.; Mu, S.; Pan, M.; Fu, Z. A highly stable TiB$_2$-supported Pt catalyst for polymer electrolyte membrane fuel cells. *J. Power Sources* **2011**, *196*, 7931–7936.

117. Huang, Z.; Lin, R.; Fan, R.; Fan, Q.; Ma, J. Effect of TiB$_2$ Pretreatment on Pt/TiB$_2$ Catalyst Performance. *Electrochim. Acta* **2014**, *139*, 48–53.

118. Roth, C.; Bleith, P.; Schwöbel, C.A.; Kaserer, S.; Eichler, J. Importance of Fuel Cell Tests for Stability Assessment—Suitability of Titanium Diboride as an Alternative Support Material. *Energies* **2014**, *7*, 3642–3652.

119. Bača, L.; Stelzer, N. Adapting of sol-gel process for preparation of TiB$_2$ powder from low-cost precursors. *J. Eur. Ceram. Soc.* **2008**, *28*, 907–911.

Recent Development of Pd-Based Electrocatalysts for Proton Exchange Membrane Fuel Cells

Hui Meng, Dongrong Zeng and Fangyan Xie

Abstract: This review selectively summarizes the latest developments in the Pd-based cataysts for low temperature proton exchange membrane fuel cells, especially in the application of formic acid oxidation, alcohol oxidation and oxygen reduction reaction. The advantages and shortcomings of the Pd-based catalysts for electrocatalysis are analyzed. The influence of the structure and morphology of the Pd materials on the performance of the Pd-based catalysts were described. Finally, the perspectives of future trends on Pd-based catalysts for different applications were considered.

Reprinted from *Catalysts*. Cite as: Meng, H.; Zeng, D.; Xie, F. Recent Development of Pd-Based Electrocatalysts for Proton Exchange Membrane Fuel Cells. *Catalysts* **2015**, 5, 1221–1274.

1. Introduction

Fuel cells convert chemical energy directly into electrical current without combustion. The first article illustrating such a device was published at the end of the 1830s [1], and the interest in this field has been growing since the 1950s [2]. Among various types of the low-temperature fuel cells, proton exchange membrane fuel cells (PEMFCs) are attractive power sources for portable, automotive and stationary applications due to their high energy density, high efficiency and low operating temperature. Comparing with H_2 as fuel, the liquid fuels such as formic acid and ethanol have special advantage in storage and transport, which can find better applications in portable devices and make use of current gasoline system. The direct formic acid fuel cells (DFAFCs) have an open circuit potential of 1.190 V and energy density of 2086 Wh L^{-1}. The formic acid is non-toxic and has small crossover flux. The shortcoming of DFAFCs is their relatively low energy density. Compared with formic acid, ethanol has much higher energy density of 8030 Wh kg^{-1}. However, ethanol suffers from the sluggish reaction kinetics.

The electrochemical oxidation of fuels requires the use of a catalyst to achieve the high current densities for practical applications. Platinum (Pt) is the mostly used catalyst in the PEMFCs. However, the vast commercialization of fuel cell is hindered by the high cost and low reserve of Pt. The kinetics of the oxygen reduction reaction (ORR), which is the cathode reaction of a fuel cell is slow on Pt. Moreover,

Pt is easy to be poisoned without recovery by the intermediates of the reaction of the impurities from the fuel or oxidant. Particularly in some types of PEMFCs, Pt is not the best choice. For example, in direct formic acid fuel cells (DFAFCs), the oxidation of formic acid is quite low due to the poisoning of the Pt by the CO-like intermediates during the reaction. In direct ethanol fuel cells (DEFCs), the poor utilization and the poisoning of Pt catalyst particularly in alkaline solution also limite its applications. The electrocatalytic activities of the ethanol oxidation reaction could be significantly improved in alkaline media on Pd-based catalyst which have comparable or even better electrocatalytic activity than that of Pt-based catalyst. The ORR is one of the key reactions in fuel cells with the higher overpotential compared with the anode reactions. Pt group metal based catalysts are currently used for PEMFCs to reduce the large ORR overpotential. Unfortunately, even on the active Pt surface, the overpotential is over 200 mV at open circuit voltage (OCV). Pd is another active metal for the ORR. Binary Pd-base metal systems have been identified as promising PEMFC cathode catalyst with the enhanced activity for ORR and stability compared with Pd alone [3].

Pd has an electronic configuration identical to Pt and forms a not very strong bond to most absorbates. The key differences are that the d bands of Pd are closer to the cores than that of Pt. There are less d electron densities available for bonding. This leads to weaker interactions with d bonds, which allows unique chemistry to occur. Pd has higher oxidation potential than Pt and the Pd oxides are more stable. Weak inter-atomic bonds between Pd atoms compared with Pt lead to easier formation of the subsurface species. Also, Pd has a very similar lattice constant to that of Pt. The electrocatalysis of formic acid on a Pd single crystal surface could be significantly enhanced as the d-band centre of Pd shifted down with an appropriate value due to the modest lattice compressive strain. All these electronic properties make Pd a promising alternative to Pt or even better than Pt in many situations.

From the recent 5-year price change of the Pd, its price has changed from one forth to two fifth of the Pt [4]. However, Pd still has some advantages considering the reservation and price. As one of the most studied materials, Pd has attracted considerable interest for its applications in many fields. Similar to Pt, most of the Pd is used in the automotive industry for catalytic converters to reduce the toxicity of emissions from a combustion engine. Pd also has vast applications in electronic, dental and jewelry. Only a small percentage of Pd is used in chemistry. In electrochemistry, Pd nanoparticles are very important catalyst, especially for the oxidation of formic acid, ehanol oxidation in alkaline solution, hydrogen oxidation and the ORR. Pd-based catalyst for alcohol oxidation have been reviewed in 2009 [5]. Another review paper in 2009 reviewed the application of Pd in fuel cell anode and cathode [6]. However, the Pd-based fuel cells are still very hot in recent years. We have surveyed the 270 published papers on the Pd-based catalysts in recent four years from 2009 to 2014 and the distribution of the papers was as shown in Figure 1. Seven review papers were published after 2009. Two review papers published in 2010 partly concerned the application of Pd in fuel cells [7,8]. Morozan *et al.*, rediewed the application of Pd in fuel cells [9]. Recently Shao reviewed Pd as catalyst for hydrogen oxidation and oxygen reduction reaction [3]. Zhao *et al.*, reviewed the catalysts for direct methanol fuel cells including the application of Pd [10]. Adams and Chen's review paper focused on the application of Pd in hydrogen storage [11]. The nano-structure of Pd for catalysis and hydrogen storage was also reviewed recently by Zhou in *Chem. Soc. Rev.* [12]. Analyzing recent review papers on Pd-based catalysts, it was found that there was no comprehensive review paper concerning the application of Pd in fuel cells. This review analyzed the latest four years' publications on Pd catalyst for proton exchange membrane fuel cells and provides a comprehensive review on the recent development of the Pd-based catalysts for formic acid oxidation, alcohol oxidation and oxygen reduction reaction.

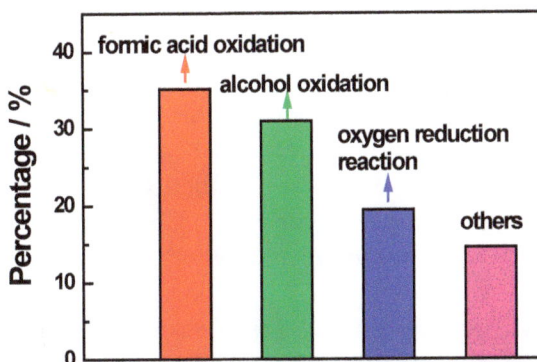

Figure 1. The distribution of Pd in recent four years in fuel cell technologies.

Based on our survey, in all the published papers on Pd-based catalysts, 35.12% is for the the formic acid oxidation, 30.99% for alcohol oxidation and 19.42% for ORR as shown in Figure 1. Therefore, this review article mainly focuses on the recent development of the Pd-based catalysts, particularly, on these three most important reactions. This survey result also shows the possible areas where Pd can compete with or replace Pt in PEMFCs. Besides above aspects, the control of the morphology and crystallography of Pd and the corresponding effect on catalysis are also reviewed.

2. Pd Nanostructures

The development of nanotechnology makes it possible to control the morphology and crystallography of Pd nanosturctures, which has been proven to affect the catalytic activity. Crystalline surfaces with a high density of low-coordinated atoms are generally superior in catalytic activity and stability to flat planes that are composed of closely packed surface atoms. This makes nanoparticle possible to show high activity and stability at very small Pd loadings if the nanoparticles are controllably synthesized as an open-structured surfaces with high density of low-coordinated atoms. The availability of the open-structured surfaces comes from the morphology of the material, so the shape control of the nanocrystals has become one of the crucial challenges.

2.1. The 0-D Pd Strucutres

Zero-dimensional (0-D) Pd structures include nano-particles, quantum dots, atomic cluster and nanoclusters and so on. The nano-particles are the most often used structure for fuel cell applications, Pd can be prepared into nano-particles which are usually called "Pd black" or Pd nano-particles loaded on support materials. Both Pd black and the supported Pd nano-particles are widely used as fuel cell catalysts. Other than the common ball-like particles, the particles with multifacets attracted much attention since their unique crystallographic and morphologic structures, which greatly improved the activity and stability when used as fuel cell catalysts. With a square-wave potential method Tian *et al.* [13] electrodeposited tetrahexahedral Pd nanocrystals with {730} high-index facets. With similar technique, Zhou *et al.* [14] prepared the Pd NCs not limited to the tetrahexahedral crystals. Shen *et al.* [15] used a sonoelectrochemical method to prepare Pd spherical nanoparticles, multitwinned particles, and spherical spongelike particles. Ding *et al.* [16] prepared single crystalline Pd nanocubes with the polyol method. Porous Pd nanoflowers were prepared by a liquid phase approach [17]. The above Pd particles all show improved mass activity or specific activity in fuel cell reactions, for example the tetrahexahedral Pd shows 4-6 times enhancement of specific activity and 1.5–3 times enhancement of mass activity in ethanol oxidation compared with commercial Pd particles on carbon [13].

The morphology of the crystal is determined by the internal features of a crystal, but the relationship between the crystalline structure and crystalline shape is still an unresolved problem. It still needs a correlation between crystallographic structure, morphological evolution and resultant shape of the nanocrystals. Zhou *et al.* [14] proposed the correlation between crystalline planes and nanocrystalline shape. There is a triangle relationship between the unit stereographic of fcc single-crystal and the surface atomic arrangement. There is also an intrinsic triangle that coordinates the crystalline surface index and the shape of the metal, which is shown in Figure 2.

2.2. The 1-D Pd Structures

Compared with nanoparticess, one-dimensional (1-D) materials such as the nanowires, nanothorns and nanotubes offer unique benefits including (1) anisotropic morphology; (2) thin metal catalyst layer which leads to higher mass transport of the reactants; (3) high aspect ratio which is immune to surface energy driven coalescence via crystal migration; (4) less vulnerable to dissolution, Ostwald ripening and aggregation during the electrocatalytic process due to their micrometer-sized length; (5) high electrochemical active areas. Therefore, the 1-D structures are also intensively studied. Meng *et al.* [18,19] synthesized Pd single-crystal nanothorns along the <220> direction with a square wave electrochemical reduction method as shown in Figure 3. The nanothorn was made by a succession of epitaxic dodecahedrons of decreasing sizes aligned in the direction of the (111) plane and the growth of the thorn occurs along the (220) plane. Tian *et al.* [20] obtained five fold twinned Pd nanorods with high-index facets of {hkk} or {hk0}. Patra *et al.* [21] prepared Pd dendrite branches growing along the <110> directions. According to the mechanism proposed by different authors, the Pd thorns obtained from different methods might share similar mechanism in the crystal growth [21]. It has been proposed that the breaking of symmetry leads to the formation of one-dimensional nanostructure, which facilitates the formation of one specific facet as the bounding side facet in the nanostructure. The formation of a tapered structure implies that the specific facet has a high surface step density to accommodate the change in diameter.

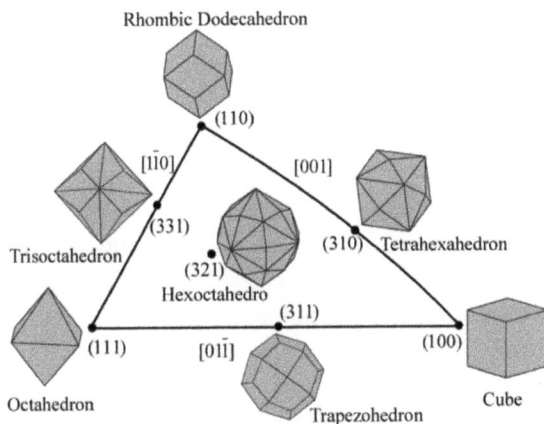

Figure 2. Unit stereographic triangle of polyhedral nanocrystals bounded by different crystal planes. Reproduced with permission from reference [14].

Besides the preparation of nanothorns without templates or sufactants, other 1-D structures such as the nanorods, nanotube and nanowires can be prepared based on the templates such as anodic aluminum oxide (AAO) [22]. Du *et al.* [23] prepared porous Pd-based alloy nanowires with AAO. Lee *et al.* [24] prepared Pd nanotubes with ZnO nanowires as sacrificial templates. The nanotube structures have large surface areas and it is expected to show enhanced catalytic efficiencies. Wen *et al.* [25] used Te NWs as sacrificial template to prepare ultrathin Pd nanowires.

The Pd nanowire or nanotube forms nanoporous metallic structure with high surface area, unique chemical properties and interconnected structures that do not require any support to avoid the corrosion and detachment problems common for carbon supported catalysts.

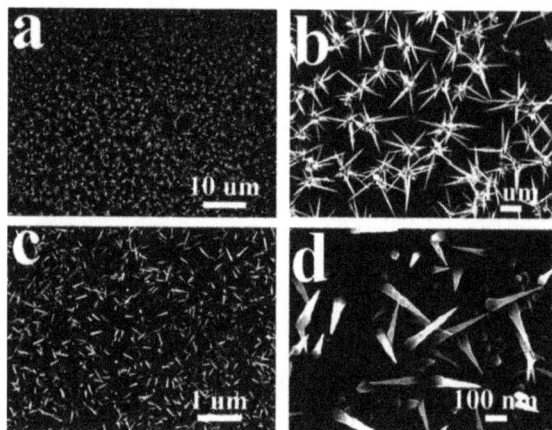

Figure 3. SEM micrographs of pure Pd thorn clusters (**a,b**) and the mixture of Pd thorns and Pd particles (**c,d**). Reproduced with permission from reference [19].

2.3. The 3-D Pd Structures

Compared with two-dimensional (2-D) structure, the three-dimensional (3-D) structure often has high porosity which will lead to higher surface area, especially higher electrochemical active surface area. Usually, the 2-D structure is very few used in the fuel cell application. Jena *et al.* [26] explored the synthesis of 3-D porous Pd nanostructures with a various shapes and morphologies. These structures have high surface roughness and surface steps which can contribute to the increased accessibility of reactant species and are more attractive for enhancing catalytic application. Zhou *et al.* [27] and Yu *et al.* [28] produced dendritic structures by electrochemical deposition. The authors concluded that the morphology of the electrochemically deposited nanostructured Pd could be solely controlled by tuning the depositing potentials. Fang *et al.* [29] reported an electrochemical route to synthesize Pd nanourchins. Li *et al.* [30] synthesized Pd/Au hollow cone-like microstructures by electrodeposition. Ye *et al.* [31,32] fabricated a three-dimensional mesoporous Pd networks by a simple reduction method in solution using a face centered cubic silica super crystal as template. From above analysis, it is concluded that the electrochemical synthesis is a useful tool in the preparation of Pd nanostructures without templates or surfactants. Many structures such as the high index facets enclosed nanoparticles, nanothorns and dendritic structures were prepared by the electrodeposition.

2.4. Hollow or Core-Shell Structures

The hollow nanomaterials have big potentials for further reducing the cost. Moreover, they have distinguished chemical and physical properties resulting from their special shape and composition [33,34]. Liu *et al.* [35] and Bai *et al.* [36] prepared raspberry hollow Pd nanospheres by a galvanic replacement reaction involving Co nanoparticles as sacrificial template. The raspberry surface might be helpful in increasing the surface area for catalysis. The core-shell structure is effective to reduce cost by reducing the amount of Pd used in the catalyst. Fang *et al.* [37] prepared the Au@Pd@Pt structure with a gold core, a Pd shell and Pt clusters on the shell as shown in Figure 4. The optimized structure had only two atomic layers of Pd and a half-monolayer equivalent of Pt. The activity was critically dependent upon the Pd-shell thickness and the Pt-cluster coverage. The high activity originated from the synergistic effect existing between the three different nanostructure components (sphere, shell and islands). Ksar *et al.* [38–40] synthesized bimetallic Pd-Au nanostructures with a core rich in gold and a Pd porous shell. This structure greatly reduced the loading of Pd. The addition of Au to Pd catalysts was not only improved the catalytic activity and selectivity but also enhanced the resistance to poisoning.

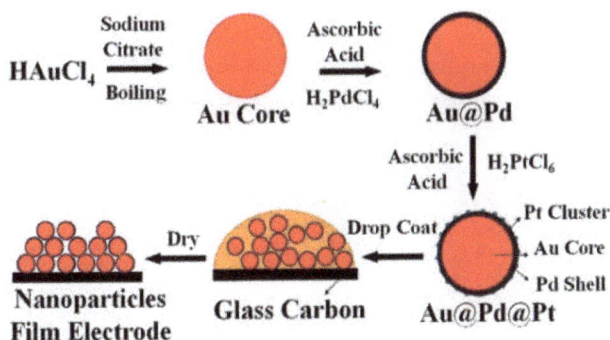

Figure 4. The procedure used to prepare an Au@Pd@Pt NP film on a glass carbon (GC) electrode. Reproduced with permission from reference [37].

2.5. Conclusions and Perspective Discussions

With the development of nanotechnology, researchers are able to control the morphology of Pd in nanoscale. Different morphologies such as particles with multi-facets, cubes, twinned particles, assembled particles, nanorods, nanothorns, nanowires, nanotubes, dendritic structure, networks and hollow/core-shell

169

structures and so on have been successfully prepared. Most of the Pd nanomaterials have high performance in fuel cell half cell characterization, which is caused by the unique crystallography and morphology of the material. Both the activity and stability of the electrochemical reaction could be improved when catalyzed by the Pd nanomaterials to significantly reduce the usage of Pt in fuel cells. However, the research on Pd nanostructures is still in the fundermental stage and great effortsl need to be done to apply these nanomaterials into the real fuel cell applications. With the development of nano-synthesis technology there will be more mophologies with unique advantages prepared, emphasis should be put on the controlled synthesis and application in fuel cell membrane electrode assembly application.

3. Pd-Based Catalysts for the Direct Formic Acid Fuel Cells (DFAFCs)

The direct formic acid fuel cells (DFAFCs) have a high theoretical open circuit potential of 1.450 V compared with 1.229 V for H_2 proton exchange membrane fuel cells (PEMFCs) and 1.190 V for direct methanol fuel cells (DMFCs) at room temperature. The formic acid is a non-toxic liquid fuel and lower crossover fiux than methanol and ethanol. Although the net energy density of formic acid (2086 Wh L^{-1}) is lower than that of methanol (4690 Wh L^{-1}), high concentrated formic acid can be used as fuel, e.g., 20 M (70 wt. %), compared with lower methanol concentration, e.g., 1–2 M. Therefore, formic acid carries more energy per volume than methanol. The formic acid itself is an electrolyte and can facilitate proton transport within anode compartment [41].

It was reported that Pd/C exhibited much better activity compared with Pt/C, however, the activity was not satisfactory and more importantly the durability of Pd/C catalyst was in urgent needed for further improvement. Generally, formic acid oxidation on Pd or Pt surface follows a dual pathway mechanism, namely, a dehydration pathway [42–44]:

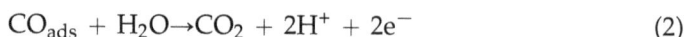

$$HCOOH \rightarrow CO_{ads} + H_2O \tag{1}$$

$$CO_{ads} + H_2O \rightarrow CO_2 + 2H^+ + 2e^- \tag{2}$$

and a dehydrogenation one:

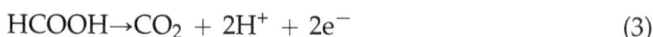

$$HCOOH \rightarrow CO_2 + 2H^+ + 2e^- \tag{3}$$

On Pt catalyst, formic acid oxidation also took place via the third pathway with bridge bonded adsorbed formate intermediate being in equilibrium with the solution formats [36–38].

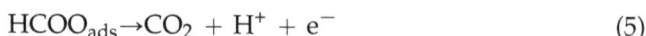

$$HCOOH \rightarrow HCOO_{ads} + H^+ + e^- \tag{4}$$

$$HCOO_{ads} \rightarrow CO_2 + H^+ + e^- \tag{5}$$

Pd has a propensity to break only the O–H bonds of the HCOOH molecule in the entire potential region, while, Pt has a propensity to break the C–O and/or C–H bond (at low overpotential) and the O–H bond (at high overpotential). Consequently, formic acid oxidation on Pd surfaces proceeds exclusively through the dehydrogenation reaction step, whereas, on Pt surfaces the dehydration pathway is predominant at low potentials. It should be noted that one of the major problems in formic acid oxidation is that the intermediary species formed during the oxidation of HCOOH cause catalyst poisoning. These intermediate species could be mostly CO, which strongly interacts with the active sites on the electrode surface and requires a higher overpotential for oxidation to CO_2. On a Pt catalyst, formic acid decomposition proceeds mainly via a dehydration pathway where the strongly adsorbed CO greatly hinders the catalytic activity. On pure Pt surfaces, CO poisoning due to the dehydration of formic acid on at least two or more contiguous Pt atoms hinders the direct dehydrogenation oxidation of formic acid at lower overpotentials. In contrast, formic acid decomposition on a Pd catalyst mainly proceeds in a dehydrogenation pathway. The generation of minor CO on the surfaces of Pd nanoparticles leads to the rapid decay of the catalytic activity. Hence, it is of great importance to design a Pd catalyst with excellent CO tolerance. For HCOOH oxidation, Pd shows superior initial performance compared with Pt. However, the high performance cannot be sustained, as Pd dissolves in acidic solutions and is vulnerable towards intermediate species. Modification of Pd with foreign metal has been considered as an effective method to enhance the activity and durability towards HCOOH oxidation. Studies have shown that the Pd oxidation activity for formic acid is also strongly influenced by the morphology and size of the Pd nanoparticles.

Table 1 listed the available data of the formic acid oxidation on Pd-based catalyst described in the literature. Three factors were compared: the electrochemical active surface area (EASA), the peak potential of formic acid oxidation and the peak current density of formic acid oxidation. According to the results provided by different authors, the EASA varied from the minimum of 15.7 m^2 g^{-1} to the maximum of 208.2 m^2 g^{-1}. The peak current density varied from 3900 to 56.5 mA mg^{-1}_{Pd}, or from 159 to 12.4 mA cm^{-2}. The peak potential analysis showed that the alloying of the Pd with other metals could reduce the overpotential for formic acid oxidation as evidenced by the negative shift of the peak potential on the alloys. Alloying with other metals can modify the electronic structure and induce tensile strain of the Pd clusters, and finally influence their catalytic activities. The highest peak current density observed on $Pd_{0.9}Pt_{0.1}$/C is nearly twice as high as that on Pd/C and six times as that on the commercial Pt/C [45]. The anodic peak current density obtained by using PdCo/MWCNTs as catalyst is 3 and 4.3 times higher than Pd/MWCNTs and Pd/XC-72 [46]. $PdNi_2$ alloy had almost three times the activity of Pd, even if the molar Pd content in $PdNi_2$ alloy was only one third of pure Pd [47].

Table 1. Details of the half cell performance of the Pd-based catalysts.

Catalysts	EASA/m^2 g^{-1}	Peak Potential/V RHE	Peak Current Density/mA mg^{-1}$_{Pd}$	Conditions H$_2$SO$_4$/Formic Acid/Scan Rate	References
Pt@Pd/C	156.5	0.277	3900	0.5 M/2 M/10 mV s^{-1}	[48]
Pt/C	198	0.687	/	0.5 M/2 M/10 mV s^{-1}	[48]
Pd/C	208.2	0.417	1200	0.5 M/2 M/10 mV s^{-1}	[48]
butylphenyl-stabilized Pd	122	0.38	3390	0.1 M/0.1 M/100 mV s^{-1}	[49]
Pd black	33.6	0.46	750	0.1 M/0.1 M/100 mV s^{-1}	[49]
butylphenyl-stabilized Pd	122	0.38	3390	0.1 M/0.1 M/100 mV s^{-1}	[49]
PdPt/C	49	/	2500	0.5 M/0.5 M/50 mV s^{-1}	[50]
Pd/C	/	/	2400	0.1 M/0.5 M/20 mV s^{-1}	[51]
Pd/C	107.2	/	1426.5	0.5 M/0.5 M/50 mV s^{-1}	[52]
PdSn/C	/	/	1420	0.5 M/0.5 M/20 mV s^{-1}	[53]
Pd/C	/	/	610	0.5 M/0.5 M/20 mV s^{-1}	[53]
Pd/CMRT	67.27	/	1140 at 0.44 V *	0.5 M/0.5 M/50 mV s^{-1}	[54]
Pd/RT	41.47	/	670 at 0.44 V *	0.5 M/0.5 M/50 mV s^{-1}	[54]
Pd/C	63.72	/	440 at 0.44 V *	0.5 M/0.5 M/50 mV s^{-1}	[54]
Pd/HPMo-PDDA-MWCNT	/	/	945	0.5 M/0.5 M/20 mV s^{-1}	[55]
Pd/AO-MWCNTs	/	/	554	0.5 M/0.5 M/20 mV s^{-1}	[55]
Pd/C	/	/	373	0.5 M/0.5 M/20 mV s^{-1}	[55]
Pd$_{57}$Ni$_{43}$	/	/	830	0.5 M/0.5 M/50 mV s^{-1}	[23]
Pd/C	/	/	700	0.5 M/0.5 M/50 mV s^{-1}	[23]
Pd-PANI	/	/	822	0.5 M/0.2 M/100 mV s^{-1}	[56]
PdAu	90	/	800	0.5 M/0.5 M/20 mV s^{-1}	[57]
Pd/C	43	/	250	0.5 M/0.5 M/20 mV s^{-1}	[57]
Pd/graphene	72.72	/	446.3	0.5 M/0.5 M/50 mV s^{-1}	[58]
Pd/C	31.85	/	191.9	0.5 M/0.5 M/50 mV s^{-1}	[58]
Pd/graphene	/	0.39	300	0.5 M/0.5 M/50 mV s^{-1}	[59]
Pd/C	/	/	193	0.5 M/0.5 M/50 mV s^{-1}	[59]
Pd networks	/	/	275.4	0.5 M/0.5 M/50 mV s^{-1}	[31]
Nanoporous Pd	23	/	262	0.5 M/0.5 M/10 mV s^{-1}	[60]
Pd/CNT	/	0.44	200 at 0.27 V *	0.5 M/0.5 M/50 mV s^{-1}	[61]
Pt/CNT	/	0.92	30 at 0.27 V *	0.5 M/0.5 M/50 mV s^{-1}	[61]
Pt Pd/CNT	/	0.64	50 at 0.27 V *	0.5 M/0.5 M/50 mV s^{-1}	[61]
Pt$_1$Pd$_3$/CNT	/	0.64	125 at 0.27 V *	0.5 M/0.5 M/50 mV s^{-1}	[61]
PdSn/C	64.7	/	170.2	0.5 M/1 M/10 mV s^{-1}	[62]
Pd/C	39.8	/	102.3	0.5 M/1 M/10 mV s^{-1}	[62]
Pd black	/	/	56.5	0.5 M/0.5 M/50 mV s^{-1}	[31]
Pd/untreated-MWCNT	46.2	0.517	159 mA cm^{-2}	0.5 M/0.5 M/50 mV s^{-1} with glutamate	[63]
Pd/acid-oxidized MWCNT	35.7	/	108 mA cm^{-2}	0.5 M/0.5 M/50 mV s^{-1}	[63]
Pd/untreated-MWCNT	21.4	/	72 mA cm^{-2}	0.5 M/0.5 M/50 mV s^{-1} without glutamate	[63]
PdNi/C	/	0.247	105.1 mA cm^{-2}	0.5 M/1 M/10 mV s^{-1}	[64]
Pd/C	/	0.287	72.9 mA cm^{-2}	0.5 M/1 M/10 mV s^{-1}	[64]
PdCo/MWCNTs	/	0.28	107 mA cm^{-2}	0.5 M/0.1 M/20 mV s^{-1}	[46]
Pd/MWCNTs	/	0.32	35.6 mA cm^{-2}	0.5 M/0.1 M/20 mV s^{-1}	[46]
Pd/XC-72	/	0.38	24.8 mA cm^{-2}	0.5 M/0.1 M/20 mV s^{-1}	[46]
Pd$_{0.9}$Pt$_{0.1}$/C	83	0.35	87.5 mA cm^{-2}	0.5 M/0.5 M/50 mV s^{-1}	[45]
Pt/C	85.6	0.52	15.1 mA cm^{-2}	0.5 M/0.5 M/50 mV s^{-1}	[45]
Pd/C	89.7	0.48	42.5 mA cm^{-2}	0.5 M/0.5 M/50 mV s^{-1}	[45]
Pd-B/C	87.6	/	65.4 mA cm^{-2}	0.5 M/0.5 M/50 mV s^{-1}	[65]
Pd/C	90	/	36.0 mA cm^{-2}	0.5 M/0.5 M/50 mV s^{-1}	[65]
Pd–Au/C	/	0.37	18.6 mA cm^{-2}	0.5 M/0.5 M	[66]
Pd/C	/	0.48	12.4 mA cm^{-2}	0.5 M/0.5 M	[66]
Pd/phen-MWCNTs	37.6	/	/	0.5 M/1 M/50 mV s^{-1}	[67]
Pd/AO-MWCNTs	15.7	/	/	0.5 M/1 M/50 mV s^{-1}	[67]
Pd/graphene	44	0.367	/	1 M/1 M/10 mV s^{-1}	[68]
Pd/Vulcan C	35	0.377	/	1 M/1 M/10 mV s^{-1}	[68]

* The potentials were *vs.* RHE.

3.1. Pd Supported on Carbon Materials

3.1.1. Pd on Carbon Powders

An active catalyst should be dispersed on a convenient support to stabilize the catalytic nanoparticles, to obtain optimum catalyst utilization and to reduce the amount of precious metal used, reducing the catalyst cost. In DFAFCs, highly conductive carbon materials such as Vulcan XC-72 carbon (Cabot, Boston, MA, USA) provide a high dispersion of metal nanoparticles to facilitate electron transfer, resulting in better catalytic activity. The supported electrocatalyst is a practical means to achieve high utilization of expensive noble metals and to maintain good life-time. Other carbon materials such as the carbon fiber are also studied as support material for palladium [69–71].

There are still aspects to improve in the Pd/C as catalyst in fuel cell applications. The works usually focus on the Pd particle size control via novel synthesis techniques to achieve high electrochemical active surface area and improve the utilization efficiency of the Pd catalyst. Cheng *et al.* [72] prepared highly dispersed Pd/C catalyst through an ambient aqueous way instead of the traditional high temperature polyol process in ethylene glycol. The Pd/C catalyst without stabilizer had a higher oxidation activity toward formic acid compared with that of a traditionally prepared Pd/C catalyst. Liang *et al.* [52] synthesized a highly dispersed and ultrafine carbon-supported Pd nanoparticle catalyst which exhibited significantly high electrochemical active surface area and high electrocatalytic performance for formic acid oxidation with four times larger formic acid oxidation current compared with that prepared by general $NaBH_4$ reduction method. The large electrochemical specific surface may be due to the high dispersion and small particle size of Pd/C catalyst. Suo *et al.* [73] used a simple and stabilizer-free ethylene glycol reduction method to prepare Pd/C catalyst. Size-dependent electrochemical property was observed and electrochemical evaluation showed that Pd/C with a particle size of 6.1 nm performed the highest activity for formic acid oxidation. The performance of the Pd/C catalyst for the oxidation of formic acid could be greatly promoted with 3.19 times enhancement in catalytic stability and 1.57 times improvement in the catalytic activity by simply introducing vanadium ions in very low concentration to the electrolyte [51]. The improvement in the catalytic performance may be attributed to the facilitating of formic acid oxidation due to the existence of VO^{2+}/V^{3+} redox pair and the ensemble effect induced by the adsorption of vanadium ions onto the surface of Pd.

3.1.2. Pd alloys on Carbon Powders

Pd displays an initial high activity for the oxidation of HCOOH. However, its long term performance is poor. Deactivation of Pd activity has been assigned to catalyst poisoning by CO, although, other poisoning species such as anions from the electrolyte can also effective. Alloying Pd with the second metal to change the surface electronic state, an ensemble effect could occur, which could possibly reduce the catalyst poisoning and increase the activity and lifetime of the catalyst.

(1) PdPt alloys

It was found $Pd_{0.9}Pt_{0.1}$/C was the optimum catalyst for the desired formic acid oxidation reaction [45]. The highest peak current density observed on $Pd_{0.9}Pt_{0.1}$/C was nearly twice as high as that on Pd/C and six times as that on the commercial Pt/C, which is superior to any reported carbon black-supported Pt/C, Pd/C, and Pt_xPd_{1-x}/C catalysts. There is also a large negative shift (up to 0.21 V) in peak potential for a Pd_xPt_{1-x}/C $versus$ that for Pd/C. The high performance of the $Pd_{0.9}Pt_{0.1}$ nanoalloy can be ascribed to the effectively inhibited CO poisoning at largely separated Pt sites and appropriately lowered d-band centre of Pd sites. The addition of Pt to Pd considerably improved the steady-state activity of Pd [74–76]. Chronoamperometric measurements showed that the most active catalyst was $Pd_{0.5}Pt_{0.5}$ with the particle size of 4 nm. Wu et $al.$ [48] prepared a Pd decorated Pt/C catalyst, Pt@Pd/C, with a small amount of Pt as core. It was found that the catalyst showed excellent activity toward anodic oxidation of formic acid at room temperature and its activity was 60% higher than that of Pd/C. It is speculated that the high performance of Pt@Pd/C may result from the unique core-shell structure and synergistic effect of Pt and Pd at the interface. Wang et $al.$ [50] decorated Pd/C with Pt nanoparticles where the amount of three neighbouring Pt or Pd atoms markedly decreased. As a result, discontinuous Pd and Pt atoms suppressed CO formation and exhibited unprecedented catalytic activity and stability toward formic acid oxidation.

(2) PdSn alloys

Liu et $al.$ [77] prepared Pd and PdSn nanoparticles supported on Vulcan XC-72 carbon by a microwave-assisted polyol process. It was found that the addition of Sn to Pd could increase the lattice parameter of the Pd (fcc) crystal. The PdSn/C catalysts have higher electrocatalytic activity for formic acid oxidation than a comparative Pd/C catalyst. The Pd_2Sn_1/C catalyst exhibited higher current density and enhanced electrocatalytic stability compared with Pd/C. There was also a negative shift of the peak potential on Pd_2Sn_1/C than that of Pd/C. Zhang et $al.$ [53] synthesised PdSn/C catalysts with different atomic ratios of Pd to Sn. The alloy catalysts exhibited significantly higher catalytic activity and stability for formic acid oxidation than that of Pd/C catalyst. Pd was modified by Sn through an electronic effect which could decrease the adsorption strength of the poisonous intermediates on Pd and thus promote the formic acid oxidation. Tu et $al.$ [62] prepared a carbon-supported PdSn

(PdSn/C) catalyst with greatly improved performance for formic acid oxidation compared with that of Pd/C. Adding Sn as a small ratio into the carbon-supported Pd catalyst could largely increase the current density of the formic acid oxidation and shift the onset potential toward the negative compared with that of Pd/C. The reason for the improvement of the catalyst was likely attributed to the high dispersion of the Pd and due to the change in the electronic properties of the Pd.

(3) Other alloys

Zhang *et al.* [66] prepared carbon-supported PdAu catalysts with different alloying degree. The electrocatalytic activity of PdAu/C catalyst for the formic acid oxidation was strongly dependent on the alloying degree of Pd-Au nanoparticles. The PdAu/C catalyst with higher alloying degree showed a higher electrocatalytic activity and stability for the formic acid oxidation compared with the PdAu/C catalyst at lower alloying degree, which can be ascribed to the enhancement of CO tolerance and possible suppression of the dehydration pathway in the course of formic acid oxidation. The catalytic activity of the PdAu/C catalyst was also found to be affected by the nature of the supporting materials [78–81]. Gao *et al.* [64,82,83] synthesized carbon-supported PdNi catalyst by sodium borohydride reduction reaction. The performance of the PdNi/C catalyst for formic acid oxidation was significantly improved compared with that of Pd/C. The potential of the main anodic peak of formic acid at PdNi/C catalyst electrode was about 40 mV more negative than that at Pd/C catalyst electrode. The onset potential of formic acid oxidation at PdNi/C catalyst electrode was 30 mV more negative than that at Pd/C catalyst electrode. The reason for the promotion effect may be due to that Ni can contribute to the adsorption of oxygen-containing species, which is conducive to the oxidation of formic acid and a change in electronic properties of Pd. Yu *et al.* [84] prepared carbon supported bimetallic PdPb catalysts which were found to be more resistant to deactivation in the DFAFC than Pd/C and to consistently show better long-term performance. The addition of Pb to Pd stabilized it significantly to deactivation during formic acid oxidation. Wang *et al.* [65] synthesized highly dispersed boron-doped Pd nanoparticles supported on carbon black with high Pd loadings (*ca.* 40 wt. % Pd) by using NaBH4 as the reductant. The as-prepared Pd-B/C catalyst showed extraordinary high activity toward formic acid oxidation compared with that of a commercially available Pd/C catalyst. Thermal treatment further enhanced the durability of the oxidation current on Pd-B/C. The superior performance of the Pd-B/C catalyst may arise from uniformly dispersed nanoparticles within optimal size ranges, the increase in surface-active sites, and the electronic modification effect of boron species. The Pd–Co, PdCeO$_x$/C [85] and PtRu/C [86–88] catalyst also exhibited excellent catalytic activity and stability in the oxidation of formic acid.

3.1.3. Pd Supported on Carbon Nanotubes

It remains a chanllege to load Pd nanoparticels on the surface of carbon nanotubes because of the graphitization of carbon nanotubes. However, due to the advantages of carbon nanotubes especially in the contribution to stability of the catalyst, many works are devoted to the deposition of Pd on carbon nanotubes [89,90]. Hu *et al.* [63] synthesized Pd nanoparticles supported on untreated multiwalled carbon nanotubes (MWCNTs). The Pd/MWCNT catalyst displayed superior electrocatalytic activity and stability in formic acid oxidation. Chakraborty *et al.* [91] synthesized nanosized Pd particles supported on MWCNTs. Bai *et al.* [67,92] functionalized MWCNTs with 1,10-phenanthroline (phen-MWCNTs) as a catalyst support for Pd nanoparticles. It was found that Pd nanoparticles were evenly deposited without obvious agglomeration, and the average particle size of the Pd nanoparticles was only as small as 2.3 nm. The as-prepared Pd/phen-MWCNTs catalyst had a better electrocatalytic activity and stability for the oxidation of formic acid than Pd catalyst on acid treated MWCNTs. Phen made a strong impact on the electrocatalytic activity of the catalyst through the functionalization of the MWCNTs and the formation of the active Pd-N sites. Therefore, the dispersivity and the ESA of the Pd nanoparticles were obviously enhanced in the presence of phen, resulting in better electrocatalytic activity and utilization efficiency of the catalyst. Cui *et al.* [55] loaded Pd nanopartilces on phosphomolybdic (HPMo) acid functionalized multiwalled carbon nanotubes supports. The catalysts exhibited a much higher electrocatalytic activity and stability for formic acid oxidation reaction as compared with that on traditional Pd/C. The high electrocatalytic activities were most likely related to highly dispersed and fine Pd nanoparticles as well as synergistic effect between Pd and HPMo immobilized on functionalized MWCNTs.

3.1.4. Pd Alloys Supported on Carbon Nanotubes

(1) PdPt alloys

Selvaraj *et al.* [93,94] prepared PtPd nanoparticles supported on purified singlewalled carbon nanotubes. The modified electrode exhibited significantly high electrocatalytic activity toward formic acid oxidation due to the uniform dispersion of nanoparticles on SWCNTs and the efficacy of Pd species in Pt-Pd system. Winjobi *et al.* [61] prepared Pt, Pd and Pt_xPd_y alloy nanoparticles supported on carbon nanotubes with high and uniform dispersion. With increasing Pd amount of the catalysts, the mass activity of formic acid oxidation reaction on the CNT supported catalysts increased. A direct oxidation pathway of formic acid oxidation occurred on the Pd surface, while, the formic acid oxidation was through CO_{ads} intermediate pathway on the Pt surface. The Pd/CNT demonstrated 7 times better mass activity than that of Pt/CNT at an applied potential of 0.27 V (vs. RHE) in the chronoamperometry test.

(2) PdAu alloys

Chen *et al.* [95] prepared PdAu/multiwalled carbon nanotubes (Pd–Au/MWCNTs) to increase the stability and performance of the Pd-based catalysts in DFAFCs. The catalyst was highly active in formic acid oxidation, due to the hydrogen treated catalysts have smaller metal particles and better contact with MWCNTs support. When Pd was alloyed with Au the leaching of Pd was considerably slower, which may be caused by much slower Pd leaching from Pd-Au alloy than from Pd. The Pd-Au/MWCNTs had higher current and lower onset potential for formic acid oxidation than Pd/MWCNTs. Chen *et al.* [96] prepared multiwalled carbon nanotubes (MWCNTs) supported Pd–Au catalyst for oxidation of formic acid and compared with a similarly prepared Pd/MWCNTs and a commercial Pt-Ru/C catalyst. Both the Pd-Au/MWCNTs and the Pd/MWCNTs catalysts used were more active than that of a commercial Pt-Ru/C catalyst. The specific activity of Pd in the novel Au-Pd/MWCNTs catalyst was over two times higher than that on the Pd/MWCNTs catalyst. Mikolajczuk *et al.* [97] found that the Pd-Au/MWCNTs catalyst exhibited higher activity and more stable in oxidation reaction of formic acid. The higher initial catalytic activity of Pd-Au/MWCNTs catalyst than Pd/MWCNTs catalyst in formic acid oxidation reaction was attributed to the electronic effect of gold in Pd-Au alloy [98,99].

(3) Other alloys

Morales-Acosta *et al.* [46,100] compared the Pd-Co and Pd catalysts prepared by the impregnation synthesis method on MWCNTs. The current density achieved with the PdCo/MWCNTs catalyst was 3 times higher than that of the Pd/MWCNTs catalyst. The onset potential for formic acid oxidation on PdCo/MWCNTs catalyst showed a negative shift *ca.* 50 mV compared with Pd/MWCNTs. The anodic peak current density obtained by using PdCo/MWCNTs was 3 and 4.3 times higher than Pd/MWCNTs and Pd/C, respectively. The PdCo/MWCNTs exhibited good stability in acidic media, higher current density and more negative anodic potential associated to this reaction than that of Pd/MWCNTs and Pd/C catalysts. The difference could be attributed to a better dispersion of the metallic nanoparticles with a lower particle size achieved with the MWCNTs-supported materials. Specific surface area obtained from PdCo and Pd supported on MWCNTs was higher than that obtained on Pd supported on Vulcan XC-72 carbon.

3.1.5. Pd Supported on Graphene

Graphene is a rapidly rising star on the horizon of the materials science and technology which has attracted tremendous attention and holds great promise for advanced fields. Especially in fuel cells, the emergence of graphene has opened a new avenue for utilizing two-dimensional planer carbon material as a catalytic support owing to its high conductivity, unique graphitized basal plane structure. This is propitious to not only maximize the availability of nanosized electrocatalyst surface area for electron transfer but also provide better mass transport of reactants to the electro-catalyst [101]. Fu *et al.* [59] developed an effective surfactant-free strategy for the small-sized Pd particles deposited on graphen. The Pd/GN catalyst exhibited excellent catalytic activity and stability toward formic acid oxidation compared with the Pd/C catalysts [102]. The enhanced performance of the Pd/GN catalyst was contributed to the small size and high dispersion of Pd NPs and the stabilizing effect of the graphene support. While the group has also fabricated the Pd/graphene hybrid via sacrifice template route [58]. Bong *et al.* [68] synthesized high loadings of 80 wt% Pd on graphene catalysts which showed significantly enhanced electrocatalytic activity and stability for formic acid oxidation compared with Pd/C catalysts.

3.2. Pd Supported on Oxides

As alternatives to improve the catalytic activity and stability of the Pd-based catalysts, the carbon support materials were modified with semiconducting oxides, such as TiO_2, MoO_x and SnO_2 [103–105]. The influence of TiO_2 support on the electronic effect and the bifunctional mechanism can be summarized as follows. (1) TiO_2 support imposes an electronic effect in which the hypo-*d*-electronic titanium ions promote the electrocatalytic properties of hyper-*d*-electronic noble metal surface atoms, thus lowering the adsorption energy of CO intermediates and increasing the mobility of the CO group on Pd nanostructures; (2) Adsorption of OH species (OH_{ad}) on TiO_2 can facilitate the conversion of the catalytically poisonous CO intermediates into CO_2, thereby improving the durability of Pd catalysts. Additionally, TiO_2 can facilitate the dispersion of noble metal nanoparticles to anchor them. Based on the theory that TiO_2 nanoparticles in the catalyst can adsorb OH_{ads} species and promote the oxidation reaction on the electrode, Xu *et al.* [44] prepared highly dispersed carbon supported Pd-TiO_2 catalyst with intermittent microwave irradiation. The activity of Pd-TiO_2/C catalyst for the oxidation of formic acid was higher than that of the Pd/C catalyst [106]. Wang *et al.* [54] investigated Pd nanoparticles supported on carbon-modified rutile TiO_2 (CMRT) as catalyst for formic acid oxidation. The Pd/CMRT showed three times the catalytic activity of Pd/C, as well as better catalytic stability toward formic acid oxidation. The enhanced catalytic property of Pd/CMRT mainly arised from the improved electronic conductivity of the carbon-modified rutile TiO_2, the dilated lattice constant of Pd nanoparticles, an increasing of surface steps and kinks in the microstructure of Pd nanoparticles and slightly better tolerance to the adsorption of poisonous intermediates [107]. Wang *et al.* [108] investigated the influence of the crystal structure of TiO_2 supporting material on formic acid oxidation. TiO_2 with the rutile structure improved the catalytic activity of Pd nanoparticles toward formic acid oxidation. The enhancement of Pd/TiO_2 (rutile) catalytic activity arose from uniform dispersion of Pd nanoparticles, an increase in surface-active sites, and good tolerance to the adsorption of poisonous intermediates (such as CO_{ad}, $COOH_{ad}$ and so on). This study proved that rutile TiO_2 was a better supporting material than anatase TiO_2 or composites.

3.3. Pd Supported on other Supporting Materials

At present, carbon nanotubes (CNTs) and Vulcan XC-72 carbon are the most often used supporting materials to load Pd nanoparticles. However, the rate of formic acid oxidation is lower partly due to lower Pd utilization on such conventional carbon supports, which is related to the lower electrochemically accessible surface area for the deposition of Pd particles. Tremendous efforts have been devoted to search for new catalyst supports to achieve good dispersion, utilization, activity and stability [109]. Bai *et al.* [110] supported Pd nanoparticles on polypyrrole-modified fullerene and the $Pd/ppy-C_{60}$ catalyst showed a good electrocatalytic activity and stability for the oxidation of formic acid. Pd on polyaniline (Pd-PANI) nanofiber film [56], on polypyrrole (PPy) film [111] and PMo_{12} [112] showed excellent catalytic activity in the oxidation reaction of formic acid in acidic media. Qin *et al.* [113] synthesized highly dispersed and active Pd/carbon nanofiber (Pd/CNF) catalyst which exhibited good catalytic activity and stability for the oxidation of formic acid. Cheng *et al.* [114] loaded Pd nanoparticles on the carbon nanoparticle-chitosan host, the chitosan matrix was shown to be beneficial in making nanosized catalyst convenient, effective, and reproducible. The Pd on carbon nanoparticle-chitosan host catalysts was highly active for the oxidation of formic acid.

3.4. Unsupported Pd

Based on the pratical application of Pd based catalysts in fuel cells, Pd is usually supported on other materials. However, since the nanotechnology provides the possibility to control the morphology of Pd material, and the intrinsic properties of the nanostructured material can be dramatically enhanced by shape and structural variations, there are also some research on the preparation of unsupported Pd materials for fuel cell application [115–117]. The one-dimensional Pd nanostructures like nanowires, nanobelts and nanotubes have attracted significant interest due to their exotic technological applications. Wang *et al.* [60] and Ye *et al.* [31] fabricated unsupported nanoporous Pd networks which exhibited high electrochemical active specific surface area, and high catalytic activity for oxidation of formic acid. Pd hollow nanostructures is an effeicient way to reduce the cost of catalsyt, for example Pd/Au hollow cone-like microstructures [30] and Pd clusters on highly dispersed Au nanoparticles [57] are found to show superior performance at much lower cost. Alloyed nanowires such as $PdNi_2$ [47] and $Pd_{57}Ni_{43}$ [23] alloy nanowires also shows improved performance. The size and nature of surface structures, such as crystalline planes and surface ligands is one key issue to improve the mass activity. Zhou *et al.* [49] synthesized monodispersed butylphenyl-functionalized Pd (Pd-BP) nanoparticles with unique surface functionalization and a high specific electrochemical surface area ($122 \ m^2 \ g^{-1}$), the Pd–BP nanoparticles exhibited a mass activity of 4.5 times as that of commercial Pd black for HCOOH oxidation as shown in Figure 5. The Pd-BP catalyst shows obvious improvement in both activity and stability compared with commercial Pd-black.

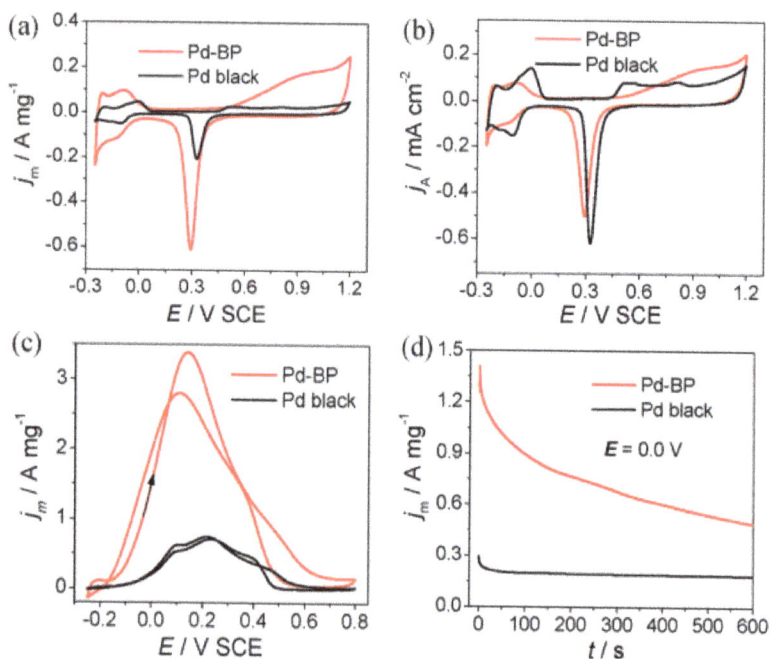

Figure 5. Cyclic voltammograms of butylphenyl-functionalized Pd (Pd-BP) nanoparticles and commer-cial Pd black in 0.1 M H_2SO_4, with the currents normalized (**a**) by the mass loadings of Pd and (**b**) by the effective electrochemical surface areas at a potential scan rate of 100 mV s^{-1}; Panels (**c**) and (**d**) depict the cyclic voltammograms and current-time curves acquired at 0.0 V for HCOOH oxidation, respectively, at the Pd-BP nanoparticles and Pd black-modified electrode in 0.1 M HCOOH + 0.1 M H_2SO_4 at room temperature. Reproduced with permission from reference [49].

3.5. Single Fuel Cell Characterization

3.5.1. Single Fuel Cell Performance

The aim of the study on the Pd based catalyst is to realize its pratical application in fuel cells. However, because of the difference in the preparation technique of the membrane electrode assembly (MEA) it is relatively difficult to compare the performance of a catalyst. The single fuel cell performance of the different catalysts was selectively summarized in Figure 6. Meng et al. [118] designed a novel MEA structure for DFAFCs where Pd nanothorns were directly electrodeposited onto the carbon paper to form the anode catalyst layer. The novel MEA provided 2.4 times higher peak power density than that of the conventional MEA. The increase in the performance was due to the improved mass transport of the formic acid in the catalyst and diffusion layers, better Pd utilization and higher electroactivity of the Pd single crystal nanothorns. Cheng et al. [119] electrodeposited Pd on graphite felt (GF) and the resulting catalyst was compared with Pt-Ru/GF for the oxidation of formic acid. The Pd/GF anode reached 852 W m^{-2} compared to 392 W m^{-2} with a commercial Pd catalyst-coated membrane (CCM). Mikołajczuk et al. [42] prepared a new carbon black supported Pd catalyst for DFAFC applications. The maximum power density of the novel 10 wt. % Pd catalyst was only 23% lower than that of the commercial 20 wt. % Pd/C. Pd-Au on multiwalled carbon nanotubes (MWCNTs) exhibited higher power density and better stability in DFAFC than that of the similar Pd/MWCNTs catalyst [79]. The 17.8 wt. % Pd/MWCNTs catalyst reached three times of the peak power density compared with that of a 20 wt. % Pt-Ru/C catalyst [120]. A MEA of the Pd/MWCNTs catalyst showed a power density of 3.3 mW cm^{-2} with 50% less Pd loadings than that of commercial Pd/C [121]. The addition of Pb and Sb into the Pd black catalyst caused a strong promotion for the formic acid oxidation in a cell [43]. Yu et al. [122] systematically evaluated and compared a number of carbon-supported Pt-based and Pd-based catalysts with commercial Pd/C, PtRu/C, and Pt/C catalysts in a multi-anode DFAFC. It is found that the PdBi/C provided higher stability than that of the commercial Pd/C catalyst, while both of the PdMo/C and PdV/C catalysts provided poor cell performances. The results provided strong evidence that both Mo and V poison Pd through electronic and/or chemical effects. Based on above analysis and Figure 6 it can be concluded that alloying with another metal is an efficient way to improve the performance of Pd based catalyst in direct formic acid fuel cell full cell performance, as is evidenced by the PdSn or Pd Sb alloy [98,123]. Next to the alloy is the Pd black with lower performance than the alloy. Pd nanothorns prepared by a novel electrodeposition technique shows higher performance than most Pd black, showing the effect of morphology control on catalytic activity. However the unexpected conclusion is Pd or Pd alloy supported on novel carbon supports such as graphene and carbon nanotube does not show much improvement in the performance. The conclusion points out futher aspects to be explored in this field: try to realize the morphology or crystallography control of Pd nanostructure aiming at high activity and stability, try to load the novel nanostructure on carbon support material aiming at reduce the loading amount of Pd, finally reducing the cost.

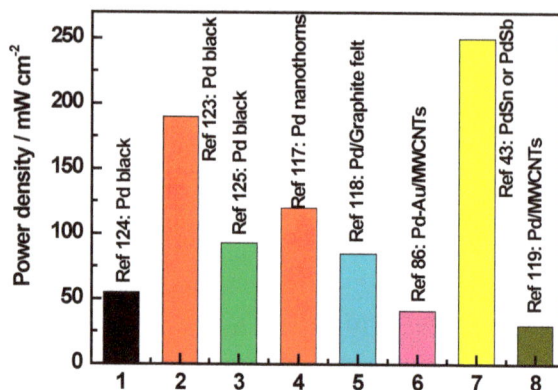

Figure 6. Comparison of the power density of the single fuel cell performance with selected catalysts.

3.5.2. Deactivation Mechanism and Reactivation

The limited durability of Pd-based anode for DFAFCs has seriously restricted their practical application. The mechanistic clarification on how to remove the poisoning species and to reactivate the Pd-based anode still remains a big challenge. The major factors causing performance degradation in the DFAFCs are an increment in the anode charge-transfer resistance and a growth in the particle size of the Pd anode catalyst [124]. The anode charge-transfer resistance, confirmed by EIS, increases with the operation time due to poisoning of the catalyst surface. The performance loss caused by surface poisoning could be completely recovered by the reactivation process. However, the increase in the catalyst size induces a reduction in active surface area and the performance loss caused by the growth in catalyst size cannot be recovered by the reactivation process. The deactivation of Pd/C increases sharply with increasing the formic acid concentration but only depends on the potential at high cell voltages. Reactivation can be achieved by driving the cell voltage to a reverse polarity of -0.2 V or higher. Although the reason for the activity loss is still unclear, it has been found that almost full activity can be recovered by applying an anodic potential of *ca.* 1.0 V *vs.* RHE or more. Ren *et al.* [125] found that the Pd oxides/hydrous oxides (POHOs) play a crucial role in promoting better performance and minimizing performance degradation of the Pd-based DFAFCs. The intrinsic presence or introduction of Pd oxides/hydrous oxides during catalysis of formic acid oxidation was found to promote elimination of poisoning species, thereby leading to a better performance of DFAFCs. Zhou *et al.* [126] found that Pd catalyst poisoned at the anode of a DFAFC under constant current discharging could be fully regenerated by a non-electrochemical method, *i.e.*, just switching pure water to DFAFC for 1 h. The voltage variation during the regeneration showed that one platform of 0.35 V was formed by the intermediate species of formic acid oxidation, which is proven to be critical for cell performance regeneration. The results indicated that the absorption of poisoning species on Pd was the main reason for the decaying of cell performance. Yu *et al.* [127] systematically studied the deactivation and electrochemical reactivation of a carbon supported Pd catalyst. The reactivation can be accomplished within a matter of seconds at $\geqslant 1.0$ V *vs.* DHE (cell voltage $\leqslant$ -0.3 V). However, reactivation at a cell voltage of 0 V or higher is required from a practical perspective. Analyzing above results it is concluded that the decaying of the cell performance is mainly caused by two factors: one is the growth of Pd particle size and the other is the absorption of poisoning species on Pd surface. The performance loss caused by surface poisoning could be recovered by the reactivation process, while the loss caused by particle growth can not be recovered. So how to limit the growth of particle size becomes one crucial problem in the stability study of the Pd based catalyst.

4. Pd-Based Catalysts for the Direct Ethanol Fuel Cells (DEFCs)

Among the alcohols, methanol and ethanol are the two most commonly used fuels for direct alcohol fuel cells (DAFCs). Direct ethanol fuel cell (DEFC) has attracted significant attention since ethanol is non-toxic, naturally available, renewable, and higher power density comparing to methanol (8030 Wh kg^{-1} for ethanol and 6100 Wh kg^{-1} for methanol). However, the sluggish reaction kinetics of ethanol oxidation is still a challenge to the commercialization of DEFC. There are many difficulties associated with ethanol oxidation and consequently in its use in fuel cell. Complete oxidation of ethanol to CO_2 involves 12 electrons and the process involves the scission of a C–C bond thus demanding high activation energies to be overcome. Many of the intermediates (mainly CO and –CHO) produced during the oxidation reaction poison the anode catalyst and in turn reduce the catalytic efficiency. Pt-based catalysts are recognized as the best catalysts for low temperature fuel cells. However, the high cost and limited resource of Pt limited its use as catalysts. At the same time, the poor utilization and the poisoning of Pt catalyst particularly in alkaline solution also limited its applications. A great deal of interest has recently been focused on the cheaper materials than platinum and the use of non-noble transition metals in alkaline media, in particular, the performance of alloys of Pd with non-noble metals for the oxidation of ethanol. In acidic environment, the complete oxidation of ethanol is difficult and the catalytic activities of the catalysts for ethanol oxidation reaction (EOR) could be significantly improved in alkaline media, where Pd-based catalyst has comparable or even better catalytic activity compared with Pt-based catalysts for ethanol oxidation [128,129]. This is why that the Pd-based materials are intensively studied for the oxidation of ethanol.

Table 2 summerized the key factors of the EOR including the specific activity, the mass activity, the onset potential and the peak potential. It was found that the mass activity of Pd for ethanol oxidation could reach as high as 2200 A g^{-1}_{Pd} [130]. Such high activity originated from the higher utilization of Pd and at the same time the reduced metal content. The mass activity could be even higher for the alloys if the current was only account by the mass of Pd. Alloying with other metals, especially, Ni could improve the kinetics of the ethanol oxidation as revealed by the negatively shifted onset potential and peak potential [131–133]. The construction of the core-shell structure was found to be an effective way to reduce the overpoential [38]. It is found that the Ru@PtPd/C [134] and Pd–Ni–Zn/C [135] show highest mass activity in the EOR, implying the advantages of multi-alloy and core-shell structure. Two components alloying of Pd with other metals have been intensively studied, while the multi-alloying still needs to be put more attention, and the muli-alloying also provides vast variations in the choice and combination of different metals. Novel nanostructure of Pd is still a popular topic in this field [136–141].

Table 2. Details of the ethanol oxidation on Pd catalysts. (SA: Specific Activity; MA: Mass Activity; OP: Onset Potential and PP: Peak Potential).

Catalysts	SA[1]/mA cm^{-2}	MA[2]/A g$^{-1}_{Pd}$	OP[3]/V RHE	PP[4]/V RHE	Conditions Ethanol/KOH/Scan Rate	References
Dendritic Pd	17	/	−0.353	−0.003	1 M/1 M/20 mV s^{-1}	[28]
Pd/CNTs	/	800	−0.353	0.007	0.5 M/0.5 M/50 mV s^{-1}	[35]
Pd shell/Au core	0.890	/	−0.582	−0.15	1 M/1 M/50 mV s^{-1}	[38]
Pd/HCHs	42	2200	−0.352	−0.072	2 M/1 M/50 mV s^{-1}	[130]
PdNPs/CFCNT	16	/	−0.24	0.14	1 M/1 M/60 mV s^{-1}	[136]
Pd–In$_2$O$_3$/CNTs	61	/	−0.347	−0.022	1 M/0.5 M/50 mV s^{-1}	[137]
Pd/CNFs	74.5	/	−0.541	−0.021	1 M/1 M/50 mV s^{-1}	[138]
Pd/CNFs	66.1	1187	−0.511	−0.057	1 M/1 M/50 mV s^{-1}	[139]
Pd/CNFs	80	1400	−0.46	−0.04	1 M/1 M/50 mV s^{-1}	[140]
Pd-Ni/CNFs	200	/	−0.602	−0.072	1 M/1 M/10 mV s^{-1}	[141]
Pd/SnO$_2$-GNS	46.1	/	−0.403	0.227	0.25 M/0.25 M/50 mV s^{-1}	[142]
Pd-PANI	/	1300	−0.46	0.07	1 M/0.5 M/100 mV s^{-1}	[56]
Pd/polyamide 6	70	/	/	−0.072	0.5 M/1.5 M/50 mV s^{-1}	[143]
Pd/Nickel foam	107.7	/	−0.442	0.078	1 M/1 M/50 mV s^{-1}	[144]
Pd-TiN	2.87	59.2	−0.474	0.022	0.5 M/1 M/20 mV s^{-1}	[145]
Pd-NiO/MgO@C	69.3	/	−0.602	−0.052	1 M/1 M/10 mV s^{-1}	[146]
Ru@PtPd/C	/	3600	−0.402	−0.052	1 M/1 M/30 mV s^{-1}	[134]
Pd particles	151	/	−0.548	/	1 M/1 M/50 mV s^{-1}	[147]
Nanoporous Pd	90.64	227.7	−0.403	0	0.5 M/1 M/10 mV s^{-1}	[148]
Pd NWs	/	7.96 *	/	−0.01	0.5 M/1 M/50 mV s^{-1}	[149]
Pd NWs	/	2.16 *	−0.566	−0.068	1 M/1 M/50 mV s^{-1}	[150]
Pd/Au NWA	199	/	−0.402	0.138	1 M/1 M/50 mV s^{-1}	[151]
PdPt	34	478	−0.523	0.077	1 M/1 M/50 mV s^{-1}	[152]
Pd-Sn/C	121.59	/	−0.46	0.135	3 M/0.5 M/50 mV s^{-1}	[153]
Pd-Ru-Sn/C	65	/	−0.536	0.097	3 M/0.5 M/50 mV s^{-1}	[153]
Pd$_{91}$Sn$_9$/C	130	/	−0.205	0.347	1 M/1 M/50 mV s^{-1}	[154]
Pd$_7$Ir/C	103	/	−0.584	0.008	1 M/1 M/50 mV s^{-1}	[155]
Pd-Ni	6	/	/	−0.03	0.5 M/1 M/50mV s^{-1}	[156]
Pd$_{40}$Ni$_{60}$	180	/	−0.712	0.098	1 M/1 M/50mV s^{-1}	[157]
Pd-Ni/CNF	199.8	/	−0.602	−0.0072	1 M/1 M/10mV s^{-1}	[131]
Pd-Ni-Zn/C	78.5	3600	−0.423	0.057	10 wt%/2 M/50mV s^{-1}	[135]

184

Table 2. *Cont.*

Catalysts	SA^1/mA cm^{-2}	MA^2/A g^{-1}$_{Pd}$	OP^3/V RHE	PP^4/V RHE	Conditions Ethanol/KOH/Scan Rate	References
Pd-Ni-Zn-P/C	108.7	3030	−0.373	0.097	10wt%/2 M/50 mV s^{-1}	[135]
Pd- Ag/C	3.7**	/	−0.602	0.018	1 M/1 M/50mV s^{-1}	[158]
Pd-Ag film	5 **	/		0.02	1 M/1 M/20 mV s^{-1}	[159]
Pd–Pb/C	4.25 **	/	−0.452	−0.0012	1 M/1 M/20 mV s^{-1}	[160]
PdAu/C	165	/	−0.402	0.0273	1 M/1 M/50 mV s^{-1}	[132]
Pd$_4$Au/C	/	/	0.5	0.88	1 M/0.25 M/10 mV s^{-1}	[161]
Pd$_{2.5}$Sn/C	/	/	0.55	0.86	1 M/0.25 M/10 mV s^{-1}	[161]
PdAu nanowire	83.7	/	−0.472	−0.001	1 M/1 M/50 mV s^{-1}	[133]
Pd@Au/C	/	800	−0.41	0.04	1 M/1 M/50 mV s^{-1}	[162]

* Unit: A· cm^{-2}· mg^{-1}; ** This value was divided by the EASA.

4.1. Pd Supported on Carbon Materials

To improve the utilization efficiency of Pd, the commonly used catalysts are Pd nanoparticles loaded on supports. Various carbon materials such as carbon nanotubes, carbon nanospheres, carbon nanowires, carbon nanofibers, porous carbon, fullerene and graphene have been used as the support. The supporting materials are required to have high surface area, low density, high chemical stability and excellent electrical conductivity.

4.1.1. Pd Supported on Carbon Spheres

In a series of work, Yan *et al.* [163–165] synthesized hollow carbon spheres/hemispheres and used them as catalyst support. The hollow carbon hemispheres (HCHs) provided high surface area (up to 1095.59 m^2 g^{-1}) at reduced volume to improve the dispersion of the nanoparticles of the noble metal. At the same time, the hemispherical structure with hollow shell resulted in the improvement in the mass transfer, which leads to greatly improved stability. The peak current density of the ethanol oxidation on the Pd/carbon spheres catalyst reached almost four times higher than that of Pd/C catalyst. Figure 7 shows the typical results of the materials for the alcohol oxidation in alkaline solution. The catalyst showed the best performance for ethanol oxidation. It was revealed that on the same carbon support, the morphology of Pd greatly influenced the activity of ethanol oxidation, the Pd nanobars on carbon have much negative oxidation peak of ethanol than Pd particle on carbon [166].

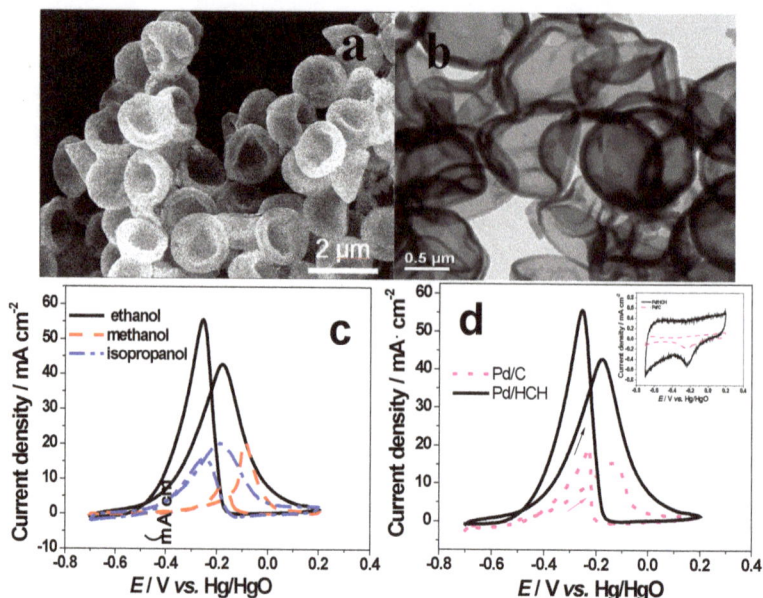

Figure 7. SEM microgram (**a**) and TEM image (**b**) of the HCHs with the mass ratio of PSs to glucose of 1:2 and the cyclic voltammograms of (**c**) different alcohol oxidation on Pd/HCH electrodes in 1.0 mol dm^{-3} KOH/1.0 mol dm^{-3} alcohol solution at 303 K, scan rate: 50 mV s^{-1} and (**d**) ethanol oxidation on Pd/HCH and Pd/C electrodes in 1.0 mol dm^{-3} KOH/1.0 mol dm^{-3} ethanol solution at 303 K, scan rate: 50 mV s^{-1}. The inset in Figure 7d is the cyclic voltammograms of Pd/HCH and Pd/C in 1.0 mol dm^{-3} KOH solution at 303 K, scan rate: 50 mV s^{-1}. Reproduced with permission from reference [130].

4.1.2. Pd Supported on Carbon Nanotubes

Carbon nanotubes (CNTs) can work as ideal substrate to modify the electrode surface used in electrochemistry. Immobilizing metal nanoparticles on CNTs has turned into an interesting field mainly due to the key role of CNTs and metal nanoparticles in the field of electrocatalysis. There are three typical methods for generating Pd nanoparticles on a CNTs surface: chemical reduction reaction, thermal decomposition and electrochemical reduction reaction. Chen *et al.* [136] synthesized Pd nanoparticles-carboxylic functional carbon nanotubes without surfactant. The material revealed high electrochemical activity and excellent catalytic characteristic for alcohol oxidation. Ding *et al.* [167,168] prepared Pd nanoparticles supported on multiwalled carbon nanotubes. Chu *et al.* [169] prepared Pd-In$_2$O$_3$/CNTs composite catalysts. The results showed that the addition of nanoparticles of In$_2$O$_3$ into Pd catalysts could significantly promote the catalytic activity for ethanol oxidation. Carbon nanotubes have higher graphitization degree than amorphous carbon which leads to higher stability in electrical environment, but higher graphitization also brings a drawback: it is not easy to load Pd nanoparticles on the surface of carbon nanotube evenly. The resolve of the dilemma will greatly accelerate the application of carbon naotube as catalyst support material in fuel cell.

4.1.3. Pd Supported on Carbon Nanofibers

Carbon nanofibers (CNFs) as a novel carbon material offered an ideal opportunity as catalyst support due to their superior electronic conductivity, anti-corrosion ability, and high surface area. Qin *et al.* [170–172] prepared Pd catalyst supported on carbon nanofibers (CNFs). The structure of the CNFs significantly affected the catalytic activity of the catalyst because the CNFs had high ratio of edge atoms to basal atoms and correspondingly faster electrode kinetics and stronger Pd-CNFs interaction. Maiyalagan *et al.* [131] prepared carbon nanofibers (CNF) supported Pd–Ni nanoparticles. The onset potential was 200 mV lower and the peak current density four times higher for ethanol oxidation for Pd–Ni/CNF compared to that for Pd/C. Cabon fibers prepared with novel nanotechnology can take the advantage of carbon nanotubes such as high graphitization and high surface area, at the same time it is also possible to construct functional groups on the surface of carbon fibers which is beneficial of the deposition of Pd nanoparticles, so carbon fiber will be a potential candidate as support material of Pd.

4.1.4. Pd Supported on Graphene

The combination of the high specific surface area (theoretical value of 2600 m^2 g^{-1}), excellent electronic conductivity, high chemical stability, unique graphitized basal plane structure and potentially low manufacturing cost, graphene nanosheets (GNS) can thus be exploited as an alternative material for catalyst support in fuel cells. Recently, graphene has received great attention as the catalyst support for fuel cell application. Wen *et al.* [173] prepared Tin oxide (SnO_2)/GNS composite as the catalyst support for direct ethanol fuel cells. Compared with Pd/GNS, the Pd/SnO_2-GNS catalyst showed superior electrocatalytic activity for ethanol oxidation. Chen *et al.* [174] prepared ultrafine Pd nanoparticles on graphene oxide (GO) surfaces. The as-made catalyst expressed high electrocatalytic ability in ethanol oxidation relative to a commercial Pd/C catalyst. Singh *et al.* [142] used graphene nanosheets (GNS) as a catalyst support of palladium nanoparticles for the electrooxidation of ethanol. The Pd nanoparticles dispersed on GNS were more active compared to those dispersed on nanocarbon particles (NC) or multiwall carbon nanotubes (MWCNTs) for electrooxidation under similar experimental conditions. The enhanced electrochemical activity of Pd/GNS toward alcohol oxidation can be ascribed to the greatly enhanced electrochemical active surface area of Pd nanoparticles on the GNS support.

4.2. Pd Supported on Non-Carbon Supports

Conventional carbon supports are prone to undergo corrosion in aggressive electrolytes that are very often encountered in fuel cells. The corroded carbon support cannot hold the catalyst on its surface, leading to aggregation or sintering of noble metal particles (reduces electrochemical active surface area) and often resulting in oxidation and subsequent leaching of the catalyst. Corrosion of the support/catalyst happens mainly because they are exposed to aggressive electrolytes, high temperature and pressure, and high humidity. Carbon is known to undergo corrosion even at open circuit voltages of the fuel cell. So there are some attempts on the application of other materials other than carbon to use as support material of Pd.

4.2.1. Pd Supported on Conducting Polymers

Polyaniline (PANI) is one of the most important conducting polymers remarkable for its high stability, solution processability and tunable electrical conductivity, which can be controlled by simple doping of the polymer. Pandey *et al.* [56] deposited a porous Pd-polyaniline (Pd-PANI) nanofiber film on conducting surfaces and the Pd-PANI showed excellent electrocatalytic activity towards the oxidation of ethanol. Su *et al.* [143] prepared Pd/polyamide 6 (Pd/PA6) nanofibers with high surface area using a simple electroless plating method. The Pd/PA6 showed excellent mechanical property, good conductivity, and high porosity. The large surface area and reduced diffusion resistance of the free-standing Pd/PA6 nanofibers led to a superior catalytic property.

4.2.2. Pd Supported on Zeolite

Zeolite has a large specific area with strongly organized microporous channel systems in which both a regular and high dispersion of metal nanoparticles can be obtained. El-Shafei *et al.* [175] prepared Pd-zeolite graphite (Pd-ZG) electrodes for ethanol oxidation. The Pd-ZG electrodes showed a better activity as well as poisoning tolerance during ethanol oxidation in alkaline medium in comparison with Pd electrode.

4.2.3. Pd Supported on Metal Supports

Nickel foam has the advantages of extinguished electronic conductivity, low weight, and 3-D cross-linked grid structure which provids high porosity and surface area. It can be used as an ideal support of catalyst. The nickel foam would not only reduce the diffusion resistance of the electrolyte but also enhance the facility of ion transportation and maintain the very smooth electron pathways in the rapid electrochemical reactions. Wang *et al.* [144] fabricated a three-dimensional, hierarchically structured Pd electrode by direct electrodepositing. The improved electrocatalytic activity and excellent stability of the Pd/Nickel foam electrode make it a favorable platform for direct ethanol fuel cell applications.

Metal nanowire or nanotube array architecture can be a potential substrate to improve noble metal utilization efficiency [176]. Besides it can act as a template, which means that no stabilizer is required, it also serves as an excellent current collector. Cherevko *et al.* [151] prepared highly ordered Pd decorated Au nanowire arrays (Pd/Au NWA). The maximum current densities were several times higher on the modified electrodes than on the unmodified Pd NWA. The highly active electrode showed almost 4-fold increase in the peak current for ethanol oxidation. The synergistic effect between substrate and deposited materials was a most important factor infecting such unusually high activity.

4.2.4. Pd Supported on Nitrides

Titanium nitride (TiN) is a very hard, conducting ceramic material used as an abrasive coating for engineering components. It possesses metal-like electronic conductivity with a very reproducible surface for electron transfer. Thotiyl *et al.* [145] studied the excellent metal-support interaction between Pd and TiN and found the efficient ethanol oxidation coupled with excellent stability of the Pd-TiN catalyst.

4.3. Performance of Pd Novel Nano Structures

4.3.1. Core-Shell Structure

Hybrid nanomaterials, particularly, the core-shell structured hybrid nanomaterials are promising due to their multi-functional and designable properties. The core/shell structured nanomaterials has the ability to improve the stability and surface chemistry of the core materials. It is possible to obtain unique structures and properties for applications via a combination of the different characteristics of the components that are not available with their single-component counterparts. The carbon-coated nanomaterials are of great interest due to their stability toward oxidation and degradation [177]. The creation of the core/shell nanostructures containing bi-metal oxides greatly enhanced the catalytic efficiency of these structures over pure single metal oxide particles [178]. Mahendiran *et al.* [146] synthesized carbon coated NiO/MgO in a core/shell nanostructure. The results indicated that the Pd-NiO/MgO@C catalyst has excellent electrocatalytic activity and stability. Gao *et al.* [134] synthesized a core-shell structured Ru@PtPd/C catalyst, with PtPd on the surface and a Ru as core. The ethanol oxidation activities of the Ru@PtPd/C catalysts were 1.3, 3, 1.4 and 2.0 times as high as that of PtPd/C, PtRu/C, Pd/C and Pt/C with same PtPd loadings, respectively. The stability of the Ru@PtPd/C was higher than that of Pt/C and PtPd/C. Ksar *et al.* [38] synthesized bimetallic Pd-Au nanostructures with a core rich in gold and a Pd porous shell. The Pd shell-Au core nanostructures synthesized in mesophases were promising for application in direct ethanol fuel cells as they exhibited a very good electrocatalytic activity and a high stability. The core-shell structure has some essential advantages such as the synergistic effect between the core material and the shell which can improve the kinetics of the reaction, the porous structure which greatly improves the mass transfer in the electrode and the reduced cost because of the less use of the noble metal. So vast efforts have been devoted to this field, but the core-shell structure is limited by the preparation technique and its application in the membrane electrode assembly of fuel cell is still a chanllenge.

4.3.2. Other Structures

Yu *et al.* [28] prepared dendritic Pd nanostructures, which exhibited high catalytic activity toward ethanol oxidation in alkaline media. Yin *et al.* [147] prepared Pd nanoparticles and the ethanol oxidation on the Pd catalysts took place at a more negative anodic potential, implying a reduced overpotential. Wang *et al.* [148] fabricated nanoporous Pd composites through chemical dealloying of the $Al_{70}Pd_{30}$ alloy. The nanoporous Pd composites had high electrochemical active surface areas and exhibited remarkable catalytic activities toward ethanol oxidation in alkaline media. Liu *et al.* [149] synthesized Pd nanowires with high catalytic activity and long-term stability toward the oxidation of alcohols. Ksar *et al.* [150] synthesized Pd nanowires with length of a few tens of nanometers. The Pd nanowires exhibited both a very important catalytic activity for ethanol oxidation and a very high stability. Liu *et al.* [31] prepared raspberry hollow Pd nanospheres (HPNs)-decorated carbon nanotube (CNT) for the oxidation of ethanol in alkaline media. The catalyst was fabricated by attaching HPNs onto the surface of the functionalized CNT. The hybrid nanostructure exhibited higher mass activity toward ethanol oxidation which increased the utilization of Pd. Pd dentric structure, Pd nanowires, hollow raspberry spheres, hollow spheres [179], nanoparticles and nanomembrane [180] are prepared and all showed superior catalytic activity in ehthanol oxidation, but they share the same problem with the core-shell structure: how to a material with high half-cell performance into a material with high full cell performance, which still needs great efforts in the preparation technique.

4.4. Pd Alloys

Pd is well known to be very active for ethanol oxidation in alkaline. Alloying Pd with another metal M (M = Au, Sn, Ru, Ag, Ni, Pb and Cu) is expected to increase the activity and at the same time the stability of the catalyst for the EOR in alkaline media [181].

4.4.1. PdNi Alloys

Qiu et al. [156] prepared bimetallic Pd-Ni thin film on glass carbon electrodes (GCEs). The high catalytic activity and the low cost of the Pd-Ni films enable them to be promising catalyst for the oxidation of methanol and ethanol in alkaline media. Qi et al. [157] fabricated $Pd_{40}Ni_{60}$ alloy catalyst with an enhanced catalytic performance toward ethanol oxidation in alkaline media comoared with nanoporous Pd. Maiyalagan et al. [131] prepared carbon nanofibers (CNF) supported Pd–Ni nanoparticles. The onset potential was 200 mV lower and the peak current density was four times higher for ethanol oxidation on Pd–Ni/CNF compared with that of Pd/C. Shen et al. [182] synthesized carbon-supported PdNi catalysts for the ethanol oxidation reaction in alkaline direct ethanol fuel cells. The Pd_2Ni_3/C catalyst exhibited higher activity and stability for the EOR in an alkaline medium than that on Pd/C catalyst. Bambagioni et al. [135] prepared Pd–(Ni–Zn)/C and Pd–(Ni–Zn–P)/C catalysts which provided excellent results in terms of the specific current and onset potential at room temperature.

4.4.2. PdAu Alloys

Xu et al. [132] prepare Pd-Au alloy catalyst for the EOR in an alkaline medium. The catalyst samples were in sequence of Pd/C > Pd_3Au/C > Pd_7Au/C > PdAu/C in terms of the peak current density. However, the stability tests demonstrated that the catalyst samples were in sequence of PdAu/C > Pd_3Au/C > Pd_7Au/C > Pd/C. Cheng et al. [133] prepared highly ordered PdAu nanowire arrays (NWAs). The onset potential of ethanol oxidation on the PdAuNWAs electrode was 123 mV more negative compared with that on the Pd NWAs due to the synergistic effect of Pd-Au bimetallic alloy. He et al. [161] prepared carbon-supported Pd_4Au and $Pd_{2.5}Sn$-alloyed nanoparticles, the results suggested that the Pd-based alloy catalysts represented promising candidates for the oxidation of ethanol. The Pd_4Au/C displayed the best catalytic activity among the series for the ethanol oxidation in alkaline media. Zhu et al. [162] decorated carbon-supported gold nanoparticles with mono- or sub-monolayer Pd atoms, the Pd@Au/C had higher specific activities than that of Pd/C for the oxidation of ethanol in alkaline media. This suggested that the Pd utilization was improved with such a surface-alloyed nanostructure. Several other alloys such as the PdPt alloy [152,183–185], PdSn alloy [153,186], PdIr alloy [155,187], PdAg alloy [158,159,188], PdTi alloy [189] and PdPb alloy [160] were also studied as the catalysts for the EOR.

4.5. Single Fuel Cell Characterizations

Although there were quite a lot reports about the Pd-based catalysts for ethanol oxidation, especially, in alkaline media, the performance of a direct ethanol fuel cell operating in alkaline membrane was rarely reported. The reason is that the anion-exchange membrane is not commercially available. The Tokuyama company in Japan is one of the pioneers in the anion-exchange membrane and most reported work were using their membranes. In recent years, Zhao's group [132,182,190,191] has done most of the work on the anion-exchange membrane direct ethanol fuel cells (AEMDEFCs). Other researchers such as Antolini [192] and Bambagioni [193] also tested the performance of the AEMDEFC, the results of these groups were compared in Figure 8. It is found that the PdNi alloy had the highest activity which was in accordance with the half cell testing. Among all catalysts, the Pd/C had the lowest activity. The change of carbon powders to carbon nanotubes improved the performance. A promising way was to alloy Pd with other metals, especially, with Ni. The highest activity of the PdNi alloy catalyst reached more than 3 times higher than that of Pd/C catalyst. The conclusion is similar with the half-cell results: alloying with other metals will improve the performance, attention should be put to multi-alloys. Another conclusion reached is the advantage of carbon nanotube *versus* carbon powder with enhanced performance.

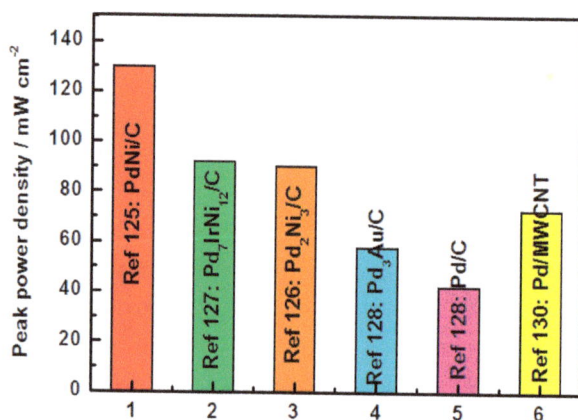

Figure 8. Peak powder density of single fuel cell performance of the Pd catalysts in anion-exchange membrane direct ethanol fuel cells (AEMDEFCs).

5. Pd-Based Catalysts for the Oxygen Reduction Reaction (ORR)

Oxygen reduction reaction (ORR) is one of the important catalytic reactions due to its role in metal corrosion and electrochemical energy converters and, in particularly, fuel cells. Pt group metal based catalysts were currently used for PEMFCs to reduce the large ORR overpotential. Unfortunately, even on the most active Pt surface, the overpotential was over 250 mV at open circuit voltage (OCV). The thermodynamic efficiency droped from 83% (1.229 V) to 66% at an OCV value of 0.98 V under standard conditions. Since the exchange current density of the

ORR is several orders of magnitude slower than that of the hydrogen oxidation reaction (HOR) (10^{-9} A cm^{-2} $vs.$ 10^{-3} A cm^{-2}), the operating voltage must be largely reduced to about 0.65 V at a reasonable current density, making the electronic efficiency of the PEMFCs only at 44% [194]. Pt is widely used as the catalyst for ORR due to its high activity and excellent chemical stability. However, Pt is expensive and the limited supply of Pt poses serious problems to widespread commercialization of the fuel cell technology. Thus, research efforts in the development of cathode catalyst have been focused on decreasing the Pt content or replacing it with less expensive materials while maintaining high ORR activity. An alternative approach is to replace Pt with less expensive, catalytically active, and relatively stable noble metals and to alloy the noble metals with base metals to enhance their stability and activity via electronic modification. Pd is the second most active metal for the ORR. Pd possesses a similar valence electronic configuration and lattice constant to Pt and highly methanol-tolerant ability. However its mass activity for the ORR is approximately five times lower than that of Pt. In an acidic electrolyte, the exchange current density of Pd for ORR is 10^{-10} A cm^{-2}, which is one magnitude lower than that of Pt (10^{-9} A cm^{-2}). Binary Pd-base metal (BM) (where BM = Co, Ni, Fe, Cu, W and Mo) systems have been identified as promising PEFC cathode catalyst with enhanced activity for ORR and stability compared to Pd alone. The origin of the enhanced activity has been linked to the modification of the electronic structure of Pd upon bonding with the alloying metal. In addition to enhancing activity, the dissolution potentials of the noble metals may be shifted to higher potentials, stabilizing the catalyst against dissolution in acidic medium.

With the rapid development of alkaline anion exchange membranes for substituting the conventional aqueous alkaline electrolyte, attention has been drawn to the study of ORR catalysts in alkaline media. More ORR catalysts are available for operation in alkaline than in acidic media due to the excessive corrosion in acidic solutions. In alkaline solution, Pd-based alloys are suitable alternative of Pt not only due to its lower costs and more abundance but also to the lower activity for the adsorption and oxidation of methanol in direct methanol fuel cells which tends to crossover to the cathode compartment and inhibits ORR. Recent reports have shown that ORR activity on the Pd alloys is comparable or slightly better than that on Pt/C.

Table 3 summarized the performance of Pd and Pd alloys as catalysts for ORR. In acidic solution, the most positive onset potential on Pd catalyst was only 0.87 V $vs.$ RHE which was almost 0.2 V negative compared with Pt catalyst. In alkaline solution, the onset potential on Pd catalyst was comparable to that of Pt catalyst. The onset potential could be improved to 1.05 V when alloyed with Pt, the highest record for alloying with non-noble metal was 0.96 V in half cell testing. Fuel cell testing gave similar conclusions. The performance of Pd catalyst could not surpass that of Pt catalyst in both PEMFC and direct alcohol fuel cells, but when Pd was alloyed the performance could be slightly better than or comparable to the commercial Pt catalyst.

Table 3. Comparison of the onset potential and mass activity of Pd for oxygen reduction reaction (ORR).

Catalysts	Mass Activity/mA mg^{-1}$_{Pd}$	Onset Potential/V vs. RHE	Conditions Solution **/Scan Rate/Rotating Speed	References
Pd/C	/	0.87	A/5 mV s^{-1}/1600 rpm	[195]
Pt/C (JM)	/	1.05	A/5 mV s^{-1}/1600 rpm	[195]
Pd–PPy/C	/	0.82 (NHE)	A/5 mV s^{-1}/1600 rpm	[196]
40% E-TEK Pt/C	/	0.98 (NHE)	A/5 mV s^{-1}/1600 rpm	[196]
Pd/Vulcan XC-72R	47 at 0.7 V *	0.82	A/5 mV s^{-1}/1600 rpm	[197]
20% E-TEK Pt/C		0.92	A/5 mV s^{-1}/1600 rpm	[197]
E-Tek 20% Pd/C	23 at 0.7 V *	0.77	A/5 mV s^{-1}/1600 rpm	[197]
PdFe-WC/C	/	0.91	A/10 mV s^{-1}/1600 rpm	[198]
Pt$_{50}$Au$_{50}$/Ce$_x$C	4.1 at 0.7 V *	/	A/5 mV s^{-1}/1600 rpm	[180]
PdCoMo/CDX975	4.1 at 0.7 V *	0.915	A/5 mV s^{-1}/1600 rpm	[199]
Pd/CDX975	1.6 at 0.7 V *	0.84	A/5 mV s^{-1}/1600 rpm	[199]
Pd$_{100-x}$W$_x$	/	0.85	A/5 mV s^{-1}/1600 rpm	[200]
Pd	/	0.7	A/5 mV s^{-1}/1600 rpm	[200]
PdPt/C	/	0.98	A/5 mV s^{-1}/1600 rpm	[201]
Pt rich-core Pd rich-shell	9 at 0.85 V * (catalyst)	/	A/5 mV s^{-1}/1600 rpm	[202]
JM20 Pt/C	5.8 at 0.85 V * (catalyst)	/	A/5 mV s^{-1}/1600 rpm	[202]
Pd/Au	0.34 at 0.8 V *	/	A/10 mV s^{-1}	[203]
	1.1 at 0.8 V *	/	B/10 mV s^{-1}	[203]
Pd/MWCNTs	/	0.76 (SHE)	B/10 mV s^{-1}	[91]
PdCu/C	23 at 0.9 V *	0.96	B/10 mV s^{-1}/1600 rpm	[204]
Pd–Cu film	/	0.86	B/5 mV s^{-1}/1000 rpm	[205]
Pd–Co/C	3.6 at 0.7 V *	/	B/5 mV s^{-1}/1600 rpm	[206]
Pt–Pd/C	/	0.92	B/5 mV s^{-1}/1600 rpm	[207]
Pt–Pd/C	114.87 at 0.9 V * (Pt)	1.05	B/5 mV s^{-1}/1600 rpm	[208]
Pt/C	73.87 at 0.9 V * (Pt)	1.0	B/5 mV s^{-1}/1600 rpm	[208]
PdFe@PdPt/C	1.92 at 0.8 V * (Pt)	/	B/20 mV s^{-1}/1600 rpm	[209]
	1.2 at 0.8 V * (Pt)	/	B/20 mV s^{-1}/1600 rpm	[209]
PdCo@PdPt/C	65 at 0.9 V * (Pt)	/	B/20 mV s^{-1}/1600 rpm	[210]
Pt/C	14 at 0.9 V * (Pt)	/	B/20 mV s^{-1}/1600 rpm	[210]
PdFe Nanorods	284 at 0.85 V *	/	B/10 mV s^{-1}/1600 rpm	[211]
Pt/C	265 at 0.85 V * (Pt)	/	B/10 mV s^{-1}/1600 rpm	[211]
Pt/Pd/Pd3Fe	1.8 at 0.9 V *	/	B/20 mV s^{-1}/1600 rpm	[212]
Pt(111)	0.8 at 0.9 V *	/	B/20 mV s^{-1}/1600 rpm	[212]
Pd/SnO2–KB	0.75 at 0.8 V *	0.88	B/5 mV s^{-1}/900 rpm	[64]
Pd/KB	0.28 at 0.8 V *	0.86	B/5 mV s^{-1}/900 rpm	[213]
Tanaka Pt/C	/	0.95	B/5 mV s^{-1}/900 rpm	[213]
Pd/GNS	280 at 0.9 V *	1.06	C/10 mV s^{-1}/1600 rpm	[188]
Pt/GNS	110 at 0.9 V *	1.0	C/10 mV s^{-1}/1600 rpm	[188]
Pd@MnO$_2$/C	450 at 0.9 V *	1.02	D/10 mV s^{-1}/2500 rpm	[214]
Pd black	180 at 0.9 V *	1.07	D/10 mV s^{-1}/2500 rpm	[214]

* The potentials were *vs.* RHE; **solution A: 0.5 M H$_2$SO$_4$, solution B: 0.1 M HClO$_4$, solution C: 0.1 M NaOH and solution D: 0.1 M KOH.

5.1. Pd Supported on Carbon Materials

5.1.1. Pd on Carbon Powder

Although Pd is the second most active metal for the ORR, its mass activity is approximately five times lower than that of Pt since the ORR exchange current density is one magnitude lower than that of Pt. Tang *et al.* [195] synthesized a carbon supported Pd/C catalyst which showed high activity for the ORR. However, the performance of the Pd/C was still much poorer than that of the commercial Pt/C.

There was a big difference in onset potential for the ORR in acidic solution. Single cell with Pd/C as the cathode displayed a maximum power density of 508 mW cm^{-2}, which was almost half of that of the commercial Pt/C catalyst. Different supporting materials such as highly ordered pyrolytic graphite and modified carbon was studied [215]. Jeyabharathi et al. [196] synthesized carbon-supported Pd-polypyrrole (Pd-PPy/C) nanocomposite. The introduction of Pd in the conducting PPy/C matrix gave higher catalytic activity toward ORR with resistance to methanol oxidation. The performance of the Pd-PPy/C catalyst was still inferior to the Pt/C catalyst. Kumar et al. [197] studied the influence of the chemical pretreatment of carbon support for ORR on Pd nanoparticles in acidic electrolyte. They found that the chemical treatment significantly changed the surface chemical properties and surface area of the carbon support. The kinetics of ORR on these catalysts predominantly involved a four-electron step. The performance of the Pd on pretreated carbon support was found to be much higher than the commercial E-Tek 20% Pd/C catalyst with the mass activity of 47 mA mg$^{-1}_{Pd}$ at 0.7 V RHE compared to 23 mA mg$^{-1}_{Pd}$ of E-Tek catalyst. It is concluded that limited by the intrinsic kinetic of ORR on different metals, the performance of the most popular Pd/C catalyst is much less than the Pt/C catalyst. However, considering the difference in the cost, it is still meaningful to study the Pd/C catalyst, but new material or structure should be put more attention.

5.1.2. Pd Supported on Carbon Nanotubes

Chakraborty et al. [91] synthesized nanosized Pd particles supported on multiwalled carbon nanotubes (MWCNTs). The authors concluded that the catalytic reduction of the oxygen followed a four electron pathway on Pd-based catalysts. Jukk et al. [216] found enhanced electrocatalytic activity of PdNP/MWCNT modified GC electrodes and the oxygen electroreduction kinetics were higher compared with those of bulk palladium electrodes. The number of electrons transferred per oxygen molecule was calculated to be 4. Kim et al. [217] studied the influence of counter ions on the oxygen reduction of Pd catalyst on functionalized carbon nanotube and found that the electrocatalytic activity is affected by the nature of anion of imidazolium salt. As is known the activity of Pd on carbon powder is less than Pt/C, not too much work is done on the attempts of loading Pd on carbon nanotubes for ORR application. However if carbon nanotubes can be helpful in the stability there will be new interest in this field.

5.1.3. Pd Supported on Graphene

Graphene is found to have vast applications in many fields where carbon powder or carbon nanotubes are used, for example as catalyst support in electrocatalysis. There was no report to prove that the activity of Pd catalyst for the ORR could surpass that of Pt catalyst in acidic solution. However, it is possible for the Pd catalyst to perform better than that of the Pt catalyst for the ORR in alkaline solution when supported on graphene. Seo *et al.* [218] studied graphene supported Pd catalyst in alkaline media. The graphene-supported Pd catalyst (Pd/GNS) showed significantly high catalytic activity for the ORR with higher mass activity and surface area for ORR in an alkaline solution. The catalyst was more favorable for ORR than that of the Pt/GNS catalyst at high metal loadings. The mass activity at 0.85 V *vs.* RHE of the Pd/GNS reached 0.84 mA μg^{-1}_{Pd}, while, the value was 0.35 mA μg^{-1}_{Pd} for the Pt/GNS. Pd metal decorated graphene oxide was found to have high ORR activity via the direct four electron pathway [219]. Gotoh *et al.* [220,221] also proved that Pd supported on graphene oxide showed better performance in ORR than Pt.

5.2. Pd Supported on Oxides

Oxides such as manganese dioxide was used as support to form $Pd@MnO_2$ catalyst [214]. The ORR onset potential on the $Pd@MnO_2$ catalyst positively shifted for more than 250 mV compared with the MnO_2 catalyst without Pd. Both the ORR onset potential and the limit current density obtained by the rotating disk electrode (RDE) measurements on the $Pd@MnO_2$ catalyst were close to those on the Pd black catalyst. The mass activity of the $Pd@MnO_2$ catalysts (normalized by Pd mass) was 2.5 times higher than that of the Pd black catalyst. The $PtPd/TiO_2$ electrocatalyst with a proper ratio of Pt/Pd showed activity comparable to that of a commercial Pt/C catalyst [222]. The interaction between Pd and highly dispersed TiO_2 is proven to improve the catalytic activity of Pd supported on TiO_2-modified carbons [223].

5.3. Pd Alloys

The intrinsic catalytic activity of Pd for the ORR is lower than that of Pt and the long-term stability at high potentials is also not as good as that of Pt. It has been proven that the ORR takes place on Pd in the same manner as that on Pt. As like Pt, the oxygen intermediate species will be covered on Pd surface at technically relevant potentials regioning from 0.7 to 0.9 V *vs.* NHE and hinders oxygen reduction. So the development of the Pd alloy catalysts that inhibit the adsorption of oxy/hydroxy species and enhance the ORR activity has been of interest. Bimetallic catalysts often exhibit notably different catalytic and chemical properties than their corresponding monometallic component. Bimetallic systems often provide enhanced selectivity, stability, and/or activity [224,225]. Several mechanisms to explain the catalytic properties of bimetallic catalysts have been proposed including geometric or ensemble effects, formation of bifunctional surfaces and electronic modification of the surface sites [226].

5.3.1. PdFe Alloys

Yin et al. [198] synthesized a PdFe-WC/C cathodic catalyst for ORR. The ORR activity of the PdFe–WC/C catalyst in acidic solution was found to be comparable to that of Pt/C catalyst. It was believed that the high catalytic activity as a Pt-free catalyst originated from the synergistic effect between PdFe and WC. The alcohol-tolerance and selectivity of the PdFe-WC/C catalyst are favorable for the ORR. Tungsten-based materials as novel supports have been intensively studied due to their chemical and/or electrochemical activities for various reactions as they exhibit a synergistic effect in many reactions. The use of WC as the catalyst support for PEMFCs and DMFCs has also been reported [227–231] PdFe/C catalyst was also found to have better performance in ORR than Pd/C catalyst [232,233].

5.3.2. PdCu Alloys

Gobal et al. [234,235] prepared CuPd alloys with different compositions on nickel. The number of transferred electrons involved in the ORR on Pd-Cu alloys is four, which is the same as Pt. A 60 mV/dec Tafel slope for the ORR was found for all the PdCu alloys. The enhancement of the activity of the alloy toward ORR was attributed to the change in geometric and electronic structures of Pd caused by the insertion of Cu. Kariuki et al. [204] prepared monodispersed PdCu alloy nanoparticles, which showed high ORR activity in acidic electrolyte. Fouda-Onana et al. [205] found $Pd_{50}Cu_{50}$ exhibited the high activity in ORR. The enhancement was attributed to an optimal d-band property that made the OOH dissociative adsorption easier, which was considered as chemical rate determining step (RDS) for the ORR [236,237].

5.3.3. PdAg Alloys

Ag catalysts have lower activity in ORR than those of Pt catalysts because of their weak interaction for binding O_2. However, the inexpensive Ag nanoparticles have been shown to have higher stability than pure Pt cathodes during long-term operation. Lee et al. [238] supported AgPd alloy on multiwalled carbon nanotubes and found that the ORR proceeded through a two-electron pathway, while according to other authors the ORR on AgPd alloy was a four-electron process [239,240]. In alkaline medium the electrode reaction kinetics is higher than that in the acidic medium, enabling the use of Pt-free catalysts. Oliveira et al. [159] evaluated PdAg alloys toward the ORR in alkaline medium and found that alloying Pd with Ag leaded to an increase in the ORR kinetics relative to Pd.

5.3.4. PdAu Alloys

Gold has been used as support material for studying the catalytic behaviour of Pd. Sarapuu et al. [203] evaluated the influence of the Pd film thickness and

Au substrate to the ORR activity of Pd. The ORR proceeded through 4-electron pathway on all PdAu electrodes. The specific activity of ORR was lower in H_2SO_4 solution and decreased slightly with decreasing the Pd film thickness. In $HClO_4$, the specific activity was higher and was not significantly dependent on the film thickness. Xu et al. [241] found Au-modified Pd catalyst exhibited increased catalytic activity for ORR in alkaline media, which was 1.4 times higher than that with the mono-Pd catalyst.

5.3.5. PdCo Alloys

There was an in-depth understanding of the various factors that ifluence the catalytic activity of the PdCo nanoalloys [242–244]. It was found that a mild annealing of the alloys at moderate temperatures (350 °C) was desirable to clean the surface and maximize the catalytic activity and durability [245–247]. Tominaka et al. [248] synthesized a mesoporous PdCo sponge-like nanostructure with a most desirable lattice contraction into a Pd catalyst for the ORR. The mesoporous PdCo catalyst had a higher specific activity than that of the Pt catalyst. Wei et al. [206] obtained PdCo/C alloy catalysts with an atomic ratio of 3:1. They found the well-formed PdCo alloy showed excellent ORR activity. Serov et al. [249] investigated PdCo catalysts for ORR in a direct methanol fuel cell. Such a non-Pt catalyst showed comparable power density with a commercial MEA prepared using Pt cathode. Rao et al. [199] prepared PdCoMo alloy nanoparticles with better catalytic activity compared with Pd.

5.3.6. PdPt Alloys

The nanosized Pd supported on carbon black will gradually grow larger during long-term operation, thus reducing the electrochemical active surface area and resulting in irreversible performance loss. Many works had been tried to improve the performance and durability of the catalyst, alloying with other metal is one important method. Lots of works had been done on the alloys of Pd with other noble metal, for example Pt [250–252]. Pd and Pt had a face-centered cubic (fcc) phase with a unit length of 3.92 Å for Pt and 3.89 Å for Pd. The small lattice mismatch meant that the epitaxial growth should be favored. Figure 9 shows a typical result of PdPt alloy for the ORR. The PdPt nanodendrites were two and a half times more active than the state-of-the-art Pt/C catalyst [253–255].

The PtPd/C showed a comparable performance and better durability than that of the Pt/C [201]. Thanasilp et al. [256] demonstrated that the different Pt:Pd atomic ratios had a significant effect on the catalyst activity. Decreasing the Pt:Pd atomic ratio led to an increase in the particle size and decrease in the electrochemical activity. Fıçıcılar et al. [257] found that when the particle size of Pd increased with the content and a lower Pd content exhibited a considerable activity and increased stability Chang et al. [202] found that the Pt_3Pd_1/C nanocatalyst has a 50% enhancement in ORR due to the synergistic effect. Ohashi *et al.* [207] prepared various PtPd/C bimetallic catalysts with a higher tolerance to ripening induced by potential cycling. Peng *et al.* [208] designed a Pt particle-on-Pd structure to address both the activity and stability issues. Other alloys of Pd–W, [200] Pd–V, [258] Pd–Ni [259,260], Pd–Sn [261,262] and Pt–Ir–Re [263] were also studied as catalysts for the ORR.

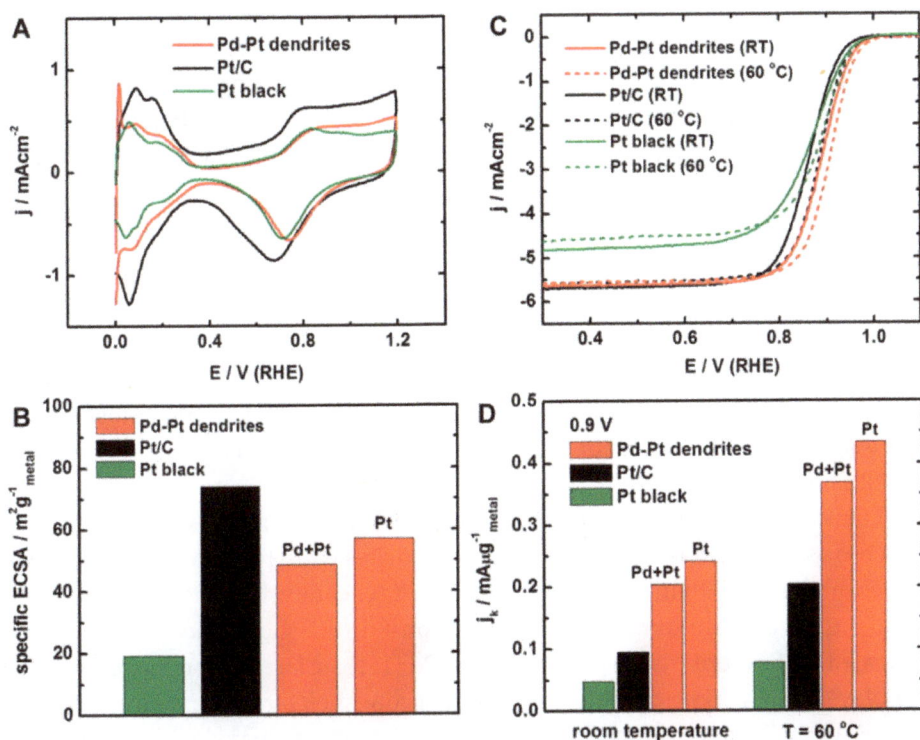

Figure 9. Comparison of the catalytic properties of the Pd-Pt nanodendrites, Pt/C catalyst (E-Tek), and Pt black (Aldrich). (**A**) The CV curves recorded at room temperature in an Ar-purged 0.1 M $HClO_4$ solution with a sweep rate of 50 mV s^{-1}; (**B**) specific ECSAs (electrochemical active surface area) for the Pd-Pt nanodendrites, Pt/C catalyst, and Pt black; (**C**) ORR polarization curves for the Pd-Pt nanodendrites, Pt/C catalyst and Pt black recorded at room temperature and 60 °C in an O_2-saturated 0.1 M $HClO_4$ solution with a sweep rate of 10 mV s^{-1} and a rotation rate of 1600 rpm and (**D**) mass activity at 0.9 V *versus* RHE (reversible hydrogen electrode) for these three catalysts. Reproduced with permission from reference [253].

5.4. Novel Nanostructures

Recent attention has been drawn to the strong dependence of the catalytic properties of Pd on their surface morphologies. For example, the specific activity of the Pd nanorods prepared by electrodeposition was found to be close to that of Pt in acidic solution and was 10 times higher than that of electrodeposited Pd nanoparticles [264]. The higher activity was attributed to the exposure of Pd (110) surface facet. The activity of the Pd with low index planes increased in the following order: Pd(110) < Pd(111) < Pd(100) [265–267]. The Pd(100) single crystal plane was even more active than Pt(110) in 0.1 M $HClO_4$. The low indexed Pd became favorable in the electrochemical reaction, controlling the morphology is one of the commonly used methods to get low indexed metal.

5.4.1. Core-Shell Structures

The heterogeneous core shell structure has many advantages. First of all, the core-shell structure can greatly reduce the cost of the catalyst. Second, the strain caused by the lattice mismatch between the surface and core components may be used to modify the electronic properties of the surface metal atoms, most notably their d-band centres, which affect the rates of one or more elementary steps in the overall catalytic reaction [268–270].

Lim et al. [253] synthesized PdPt bimetallic nanodendrites consisting of Pt branches as shell and Pd as core, the nanodendrites were two and half times more active for the ORR than the state-of-the-art Pt/C catalyst. Peng et al. [208] also found improved stability for ORR with a Pt-on-Pd core-shell nanostructure. It was calculated that the optimal coverage of Pt on Pd (111) surface was on the order of two monolayers [271]. Yang et al. [209] prepared a core-shell structure with a PdFe core and a PdPt shell, which showed four times ORR activity compared with a commercial Pt/C catalyst. Yang et al. [210] constructed a PdPt shell on a PdCo core with six fold increases in the activity and with much higher stability. Sasaki et al. [272] illustrated a core/shell catalyst with Pd and Pd_9Au_1 alloy as core and Pt monolayer as shell with high activity and very high stability. The origin of the improved activity and stability of the core-shell catalyst was studied. The Pd core not only assured the long-term stability of the monolayer Pt shell, but increased the activity of Pt by causing it to contract slightly, lowering its d-band centre energy and reducing the bond strength of the adsorbed oxygen intermediates. These effects decreased the bonding of OH and O to Pt that inhibited the ORR kinetics and also stabilize Pt against oxidation and dissolution.

5.4.2. Unsupported Pd

Besides the most frequently used supported Pd catalysts, the application of nano-structured Pd as catalyst for the ORR has also been studied. For example, PdFe nanorods [211], $PdCo_xCN_y$ [273], Pt monolayer supported on Pd/Pd_3Fe [212] and Pd/SnO_2 [213] demonstrated significantly increased ORR activity. The catalytic activity of Pd could become comparable to that of Pt upon appropriate modification of its electronic structure. The surface specific activity of Pd nanorods (Pd-NRs) toward the ORR was found to be not only 10-fold higher than that of Pd nanoparticles (Pd-NPs), but also comparable to that of Pt at operating potentials of fuel cell cathodes [264]. Zhang et al. [194] prepared PdFe-nanoleaves with Pd-rich nanowires surrounded by Fe-rich sheets. The structure demonstrated three times increased specific activity and 2.7 times increased mass activity compared witha commercial Pt/C catalyst.5.5. Singal Fuel Cell Characterizations

Two kinds of fuel cells, the PEMFC feeding with H_2/O_2 and direct alcohol (ethanol of methanol) fuel cells, were studied by using above mentioned catalysts. In PEMFC applications, Tang et al. [195] proved the performance of Pd/C was much less than that of the commercial Pt/C. Single cell with Pd/C as the cathode displayed a maximum power density of 508 mW cm^{-2} which was almost half of that of commercial Pt/C catalyst. This result was the same with the half cell test, that is, the performance of Pd/C catalyst was much inferior to that of the Pt/C catalyst for the ORR. Thanasilp et al. [256] studied the influence of Pt:Pd atomic ratios on a carbon supported upon its suitability as a cathode for a PEMFC. Although the different Pt:Pd atomic ratios had a significant effect on the performance in a H_2/O_2 fuel cell, the performance of the PdPt alloy was still much lower than that of Pt/C catalyst. Rosa et al. [274] directly sprayed Pd ink on carbon paper to form a novel oxygen diffusion electrode for the PEMFC, but the utilization efficiency of Pd was not satisfactory. With the help of nanotechnology, the performance of Pd catalyst with novel nano-structures as the PEMFC cathode was greatly improved. Li et al. [211] synthesized PdFe nanorods with tunable length which showed a better PEMFC performance than that of the commercial Pt/C due to their high intrinsic activity to ORR at reduced cell inner resistance and improved mass transport.

In direct alcohol fuel cells, Xu et al. [241] studied Au-modified Pd catalysts on carbon nanotubes which yielded a peak power density of 1.4 times higher than that with the mono-Pd cathode but was still less than Pt cathode. Pd alloys such as PdCoMo alloy [199] and PdCo alloy [199] showed comparable performance with that of commercial Pt/C, the PdNi alloy showed much higher performance than of Pd/C but still inferior to that of Pt/C catalyst [260]. It turns out that alloying Pd with other metals like Fe, Co and Ni is possible to improve the performance, reduce the cost and improve the stability of fuel cell.

6. Conclusions and Future Perspective

This article reviewed the latest advances in Pd-based catalysts for fuel cells. The review focused on Pd nanostructure, Pd catalysts for formic acid oxidation, alcohol oxidation and oxygen reduction reaction.

Different Pd morphologies were prepared. Due to the intrinsic advantages in crystallography and morphology, most of the Pd nanomaterials have high performance in fuel cell half cell characterization. Both the activity and stability of the catalysts could be improved, which would significantly reduce the usage of Pd in fuel cells. But the nano-structured Pd could not be vast applied in fuel cell, because of the limited yield and the difficulty in the MEA preparation. Futhure studies should be conducted to realize the mass production and find ways for efficient MEA preparing techniques.

As fuel cell catalysts for formic acid oxidation, methanol oxidation and oxyren reduction, Pd was loaded on carbon powders or other novel supports such as graphene and carbon nanotubes to achieve high electrochemical active surface area and improve the utilization efficiency of the Pd catalyst. The nature of the support materials also had great influence on the activity of Pd catalyst. Supported Pd nano-strucutures often had better performance than particles. Alloying with other metals could modify the electronic structure and induce tensile strain of the Pd clusters and finally infiuence their catalytic activities. There was great potential in the development of Pd alloy catalysts especially with non-noble metals to perform improved performance and stability and at the same time the reduced cost.

Considering the cost and comparable activity with Pt, the Pd-based catalysts are potential candidates as main catalysts for fuel cells to reduce the use of Pt and the cost for commercialization.

Acknowledgments: This work was supported by National Natural Science Foundation of China (21106190, 21476096, 51303217); Pearl River S&T Nova Program of Guangzhou (2013J2200040); Blue Ocean Talent Project of Nanhai, Foshan; 44th Scientific Research Foundation for the Returned Overseas Chinese Scholars, State Education Ministry; Key Laboratory of Functional Inorganic Material Chemistry (Heilongjiang University), Ministry of Education; the Fundamental Research Funds for the Central Universities.

Author Contributions: Hui Meng worte the review and designed the structure. Fangyan Xie and Dongrong Zeng collected and classified the references.

Conflicts of Interest: The authors declare no conflict of interest.

References

1.   Grove, W.R. XXIV. On voltaic series and the combination of gases by platinum. *Philos. Mag. Ser.* **1839**, *14*, 127–130.

2.   Soszko, M.; Lukaszewski, M.; Mianowska, Z.; Czerwinski, A. Electrochemical characterization of the surface and methanol electrooxidation on Pt–Rh–Pd ternary alloys. *J. Power Sources* **2011**, *196*, 3513–3522.

3.   Shao, M. Palladium-based electrocatalysts for hydrogen oxidation and oxygen reduction reactions. *J. Power Sources* **2011**, *196*, 2433–2444.

4.   Palladium Chart—Last 5 years. Available online: http://www.kitco.com/charts/popup/pd1825nyb.html (accessed on 1 May 2015).

5.   Bianchini, C.; Shen, P.K. Palladium-Based Electrocatalysts for Alcohol Oxidation in Half Cells and in Direct Alcohol Fuel Cells. *Chem. Rev.* **2009**, *109*, 4183–4206.

6.   Antolini, E. Palladium in fuel cell catalysis. *Energy Environ. Sci.* **2009**, *2*, 915–931.

7.   Mazumder, V.; Lee, Y.; Sun, S.H. Recent Development of Active Nanoparticle Catalysts for Fuel Cell Reactions. *Adv. Funct. Mater.* **2010**, *20*, 1224–1231.

8.   Antolini, E.; Gonzalez, E.R. Alkaline direct alcohol fuel cells. *J. Power Sources* **2010**, *195*, 3431–3450.

9.   Morozan, A.; Jousselme, B.; Palacin, S. Low-platinum and platinum-free catalysts for the oxygen reduction reaction at fuel cell cathodes. *Energy Environ. Sci.* **2011**, *4*, 1238–1254.

10.   Zhao, X.; Yin, M.; Ma, L.; Liang, L.; Liu, C.P.; Liao, J.H.; Lu, T.H.; Xing, W. Recent advances in catalysts for direct methanol fuel cells. *Energy Environ. Sci.* **2011**, *4*, 2736–2753.

11.   Adams, B.D.; Chen, A.C. The role of palladium in a hydrogen economy. *Mater. Today* **2011**, *14*, 282–289.

12.   Zhou, Z.Y.; Tian, N.; Li, J.T.; Broadwell, I.; Sun, S.G. Nanomaterials of high surface energy with exceptional properties in catalysis and energy storage. *Chem. Soc. Rev.* **2011**, *40*, 4167–4185.

13.   Tian, N.; Zhou, Z.Y.; Yu, N.F.; Wang, L.Y.; Sun, S.G. Direct Electrodeposition of Tetrahexahedral Pd Nanocrystals with High-Index Facets and High Catalytic Activity for Ethanol Electrooxidation. *J. Am. Chem. Soc.* **2010**, *132*, 7580–7581.

14.   Zhou, Z.Y.; Tian, N.; Huang, Z.Z.; Chen, D.J.; Sun, S.G. Nanoparticle catalysts with high energy surfaces and enhanced activity synthesized by electrochemical method. *Faraday Discuss.* **2008**, *140*, 81–92.

15.   Shen, Q.M.; Min, Q.H.; Shi, J.J.; Jiang, L.P.; Zhang, J.R.; Hou, W.H.; Zhu, J.J. Morphology-Controlled Synthesis of Palladium Nanostructures by Sonoelectrochemical Method and Their Application in Direct Alcohol Oxidation. *J. Phys. Chem. C* **2009**, *113*, 1267–1273.

16.   Ding, H.; Shi, X.Z.; Shen, C.M.; Hui, C.; Xu, Z.C.; Li, C.; Tian, Y.A.; Wang, D.K.; Gao, H.J. Synthesis of monodisperse palladium nanocubes and their catalytic activity for methanol electrooxidation. *Chin. Phys. B* **2010**, *19*.

17.   Yin, Z.; Zheng, H.J.; Ma, D.; Bao, X.H. Porous Palladium Nanoflowers that Have Enhanced Methanol Electro-Oxidation Activity. *J. Phys. Chem. C* **2009**, *113*, 1001–1005.

18.   Meng, H.; Sun, S.; Masse, J.P.; Dodelet, J.P. Electrosynthesis of Pd Single-Crystal Nanothorns and Their Application in the Oxidation of Formic Acid. *Chem. Mater.* **2008**, *20*, 6998–7002.

203

19. Meng, H.; Wang, C.X.; Shen, P.K.; Wu, G. Palladium thorn clusters as catalysts for electrooxidation of formic acid. *Energy Environ. Sci.* **2011**, *4*, 1522–1526.

20. Tian, N.; Zhou, Z.Y.; Sun, S.G. Electrochemical preparation of Pd nanorods with high-index facets. *Chem. Commun.* **2009**, 1502–1504.

21. Patra, S.; Viswanath, B.; Barai, K.; Ravishankar, N.; Munichandraiah, N. High-Surface Step Density on Dendritic Pd Leads to Exceptional Catalytic Activity for Formic Acid Oxidation. *ACS Appl. Mater. Interfaces* **2010**, *2*, 2965–2969.

22. Zhang, J.; Cheng, Y.; Lu, S.F.; Jia, L.C.; Shen, P.K.; Jiang, S.P. Significant promotion effect of carbon nanotubes on the electrocatalytic activity of supported Pd NPs for ethanol oxidation reaction of fuel cells: The role of inner tubes. *Chem. Commun.* **2014**, *50*, 13732–13734.

23. Du, C.Y.; Chen, M.; Wang, W.G.; Yin, G.P. Nanoporous PdNi Alloy Nanowires As Highly Active Catalysts for the Electro-Oxidation of Formic Acid. *ACS Appl. Mater. Interfaces* **2011**, *3*, 105–109.

24. Lee, Y.W.; Lim, M.A.; Kang, S.W.; Park, I.; Han, S.W. Facile synthesis of noble metal nanotubes by using ZnO nanowires as sacrificial scaffolds and their electrocatalytic properties. *Chem. Commun.* **2011**, *47*, 6299–6301.

25. Wen, D.; Guo, S.J.; Dong, S.J.; Wang, E.K. Ultrathin Pd nanowire as a highly active electrode material for sensitive and selective detection of ascorbic acid. *Biosens. Bioelectron.* **2010**, *26*, 1056–1061.

26. Jena, B.K.; Sahu, S.C.; Satpati, B.; Sahu, R.K.; Behera, D.; Mohanty, S. A facile approach for morphosynthesis of Pd nanoelectrocatalysts. *Chem. Commun.* **2011**, *47*, 3796–3798.

27. Zhou, R.; Zhou, W.Q.; Zhang, H.M.; Du, Y.K.; Yang, P.; Wang, C.Y.; Xu, J.K. Facile template-free synthesis of pine needle-like Pd micro/nano-leaves and their associated electro-catalytic activities toward oxidation of formic acid. *Nanoscale Res. Lett.* **2011**, *6*.

28. Yu, J.S.; Fujita, T.; Inoue, A.; Sakurai, T.; Chen, M.W. Electrochemical synthesis of palladium nanostructures with controllable morphology. *Nanotechnology* **2010**, *21*.

29. Fang, Y.X.; Guo, S.J.; Zhu, C.Z.; Dong, S.J.; Wang, E.K. Twenty Second Synthesis of Pd Nanourchins with High Electrochemical Activity through an Electrochemical Route. *Langmuir* **2010**, *26*, 17816–17820.

30. Li, L.Q.; E, Y.; Yuan, J.M.; Luo, X.Y.; Yang, Y.; Fan, L.Z. Electrosynthesis of Pd/Au hollow cone-like microstructures for electrocatalytic formic acid oxidation. *Electrochim. Acta* **2011**, *56*, 6237–6244.

31. Ye, L.; Wang, Y.; Chen, X.Y.; Yue, B.; Tsang, S.C.; He, H.Y. Three-dimensionally ordered mesoporous Pd networks templated by a silica super crystal and their application in formic acid electrooxidation. *Chem. Commun.* **2011**, *47*, 7389–7391.

32. Carrera-Cerritos, R.; Fuentes-Ramirez, R.; Cuevas-Muniz, F.M.; Ledesma-Garcia, J.; Arriaga, L.G. Performance and stability of Pd nanostructures in an alkaline direct ethanol fuel cell. *J. Power Sources* **2014**, *269*, 370–378.

33. Choi, S.I.; Shao, M.H.; Lu, N.; Ruditskiy, A.; Peng, H.C.; Park, J.; Guerrero, S.; Wang, J.G.; Kim, M.J.; Xia, Y.N. Synthesis and Characterization of Pd@Pt–Ni Core-Shell Octahedra with High Activity toward Oxygen Reduction. *ACS Nano* **2014**, *8*, 10363–10371.

34. Lim, Y.; Kim, S.K.; Lee, S.C.; Choi, J.; Nahm, K.S.; Yoo, S.J.; Kim, P. One-step synthesis of carbon-supported Pd@Pt/C core-shell nanoparticles as oxygen reduction electrocatalysts and their enhanced activity and stability. *Nanoscale* **2014**, *6*, 4038–4042.

35. Liu, Z.L.; Zhao, B.; Guo, C.L.; Sun, Y.J.; Shi, Y.; Yang, H.B.; Li, Z.A. Carbon nanotube/raspberry hollow Pd nanosphere hybrids for methanol, ethanol, and formic acid electro-oxidation in alkaline media. *J. Colloid Interface Sci.* **2010**, *351*, 233–238.

36. Bai, Z.Y.; Yang, L.; Li, L.; Lv, J.; Wang, K.; Zhang, J. A Facile Preparation of Hollow Palladium Nanosphere Catalysts for Direct Formic Acid Fuel Cell. *J. Phys. Chem. C* **2009**, *113*, 10568–10573.

37. Fang, P.P.; Duan, S.; Lin, X.D.; Anema, J.R.; Li, J.F.; Buriez, O.; Ding, Y.; Fan, F.R.; Wu, D.Y.; Ren, B.; *et al.* Tailoring Au-core Pd-shell Pt-cluster nanoparticles for enhanced electrocatalytic activity. *Chem. Sci.* **2011**, *2*, 531–539.

38. Ksar, F.; Ramos, L.; Keita, B.; Nadjo, L.; Beaunier, P.; Remita, H. Bimetallic Palladium-Gold Nanostructures: Application in Ethanol Oxidation. *Chem. Mater.* **2009**, *21*, 3677–3683.

39. Hsu, C.J.; Huang, C.W.; Hao, Y.W.; Liu, F.Q. Au/Pd core-shell nanoparticles with varied hollow Au cores for enhanced formic acid oxidation. *Nanoscale Res. Lett.* **2013**, *8*.

40. Kim, D.Y.; Kang, S.W.; Choi, K.W.; Choi, S.W.; Han, S.W.; Im, S.H.; Park, O.O. Au@Pd nanostructures with tunable morphologies and sizes and their enhanced electrocatalytic activity. *Crystengcomm* **2013**, *15*, 7113–7120.

41. Miao, F.J.; Tao, B.R.; Sun, L.; Liu, T.; You, J.C.; Wang, L.W.; Chu, P.K. Preparation and characterization of novel nickel-palladium electrodes supported by silicon microchannel plates for direct methanol fuel cells. *J. Power Sources* **2010**, *195*, 146–150.

42. Mikolajczuk, A.; Borodzinski, A.; Kedzierzawski, P.; Stobinski, L.; Mierzwa, B.; Dziura, R. Deactivation of carbon supported palladium catalyst in direct formic acid fuel cell. *Appl. Surface Sci.* **2011**, *257*, 8211–8214.

43. Haan, J.L.; Stafford, K.M.; Masel, R.I. Effects of the Addition of Antimony, Tin, and Lead to Palladium Catalyst Formulations for the Direct Formic Acid Fuel Cell. *J. Phys. Chem. C* **2010**, *114*, 11665–11672.

44. Xu, W.F.; Gao, Y.; Lu, T.H.; Tang, Y.W.; Wu, B. Kinetic Study of Formic Acid Oxidation on Highly Dispersed Carbon Supported Pd-TiO$_2$ Electrocatalyst. *Catal. Lett.* **2009**, *130*, 312–317.

45. Zhang, H.X.; Wang, C.; Wang, J.Y.; Zhai, J.J.; Cai, W.B. Carbon-Supported Pd-Pt Nanoalloy with Low Pt Content and Superior Catalysis for Formic Acid Electro-oxidation. *J. Phys. Chem. C* **2010**, *114*, 6446–6451.

46. Morales-Acosta, D.; Ledesma-Garcia, J.; Godinez, L.A.; Rodriguez, H.G.; Alvarez-Contreras, L.; Arriaga, L.G. Development of Pd and Pd–Co catalysts supported on multi-walled carbon nanotubes for formic acid oxidation. *J. Power Sources* **2010**, *195*, 461–465.

47. Du, C.Y.; Chen, M.; Wang, W.G.; Yin, G.P.; Shi, P.F. Electrodeposited PdNi$_2$ alloy with novelly enhanced catalytic activity for electrooxidation of formic acid. *Electrochem. Commun.* **2010**, *12*, 843–846.

48. Wu, Y.N.; Liao, S.J.; Su, Y.L.; Zeng, J.F.; Dang, D. Enhancement of anodic oxidation of formic acid on palladium decorated Pt/C catalyst. *J. Power Sources* **2010**, *195*, 6459–6462.

49. Zhou, Z.Y.; Kang, X.W.; Song, Y.; Chen, S.W. Butylphenyl-functionalized palladium nanoparticles as effective catalysts for the electrooxidation of formic acid. *Chem. Commun.* **2011**, *47*, 6075–6077.

50. Wang, X.M.; Wang, M.E.; Zhou, D.D.; Xia, Y.Y. Structural design and facile synthesis of a highly efficient catalyst for formic acid electrooxidation. *Phys. Chem. Chem. Phys.* **2011**, *13*, 13594–13597.

51. Ge, J.J.; Chen, X.M.; Liu, C.P.; Lu, T.H.; Liao, J.H.; Liang, L.A.; Xing, W. Promoting effect of vanadium ions on the anodic Pd/C catalyst for direct formic acid fuel cell application. *Electrochim. Acta* **2010**, *55*, 9132–9136.

52. Liang, Y.; Zhu, M.N.; Ma, J.; Tang, Y.W.; Chen, Y.; Lu, T.H. Highly dispersed carbon-supported Pd nanoparticles catalyst synthesized by novel precipitation-reduction method for formic acid electrooxidation. *Electrochim. Acta* **2011**, *56*, 4696–4702.

53. Zhang, Z.H.; Ge, J.J.; Ma, L.A.; Liao, J.H.; Lu, T.H.; Xing, W. Highly Active Carbon-supported PdSn Catalysts for Formic Acid Electrooxidation. *Fuel Cells* **2009**, *9*, 114–120.

54. Wang, X.M.; Wang, J.; Zou, Q.Q.; Xia, Y.Y. Pd nanoparticles supported on carbon-modified rutile TiO$_2$ as a highly efficient catalyst for formic acid electrooxidation. *Electrochim. Acta* **2011**, *56*, 1646–1651.

55. Cui, Z.M.; Kulesza, P.J.; Li, C.M.; Xing, W.; Jiang, S.P. Pd nanoparticles supported on HPMo-PDDA-MWCNT and their activity for formic acid oxidation reaction of fuel cells. *Int. J. Hydrogen Energy* **2011**, *36*, 8508–8517.

56. Pandey, R.K.; Lakshminarayanan, V. Electro-Oxidation of Formic Acid, Methanol, and Ethanol on Electrodeposited Pd-Polyaniline Nanofiber Films in Acidic and Alkaline Medium. *J. Phys. Chem. C* **2009**, *113*, 21596–21603.

57. Park, I.S.; Lee, K.S.; Yoo, S.J.; Cho, Y.H.; Sung, Y.E. Electrocatalytic properties of Pd clusters on Au nanoparticles in formic acid electro-oxidation. *Electrochim. Acta* **2010**, *55*, 4339–4345.

58. Zhao, H.; Yang, J.; Wang, L.; Tian, C.G.; Jiang, B.J.; Fu, H.G. Fabrication of a palladium nanoparticle/graphene nanosheet hybrid via sacrifice of a copper template and its application in catalytic oxidation of formic acid. *Chem. Commun.* **2011**, *47*, 2014–2016.

59. Yang, J.; Tian, C.G.; Wang, L.; Fu, H.G. An effective strategy for small-sized and highly-dispersed palladium nanoparticles supported on graphene with excellent performance for formic acid oxidation. *J. Mater. Chem.* **2011**, *21*, 3384–3390.

60. Wang, X.G.; Wang, W.M.; Qi, Z.; Zhao, C.C.; Ji, H.; Zhang, Z.H. High catalytic activity of ultrafine nanoporous palladium for electro-oxidation of methanol, ethanol, and formic acid. *Electrochem. Commun.* **2009**, *11*, 1896–1899.

61. Winjobi, O.; Zhang, Z.Y.; Liang, C.H.; Li, W.Z. Carbon nanotube supported platinum-palladium nanoparticles for formic acid oxidation. *Electrochim. Acta* **2010**, *55*, 4217–4221.

62. Tu, D.D.; Wu, B.; Wang, B.X.; Deng, C.; Gao, Y. A highly active carbon-supported PdSn catalyst for formic acid electrooxidation. *Appl. Catal. B* **2011**, *103*, 163–168.

63. Hu, C.G.; Bai, Z.Y.; Yang, L.; Lv, J.; Wang, K.; Guo, Y.M.; Cao, Y.X.; Zhou, J.G. Preparation of high performance Pd catalysts supported on untreated multi-walled carbon nanotubes for formic acid oxidation. *Electrochim. Acta* **2010**, *55*, 6036–6041.

64. Gao, Y.W.; Wang, G.; Wu, B.; Deng, C.; Gao, Y. Highly active carbon-supported PdNi catalyst for formic acid electrooxidation. *J. Appl. Electrochem.* **2011**, *41*, 1–6.

65. Wang, J.Y.; Kang, Y.Y.; Yang, H.; Cai, W.B. Boron-Doped Palladium Nanoparticles on Carbon Black as a Superior Catalyst for Formic Acid Electro-oxidation. *J. Phys. Chem. C* **2009**, *113*, 8366–8372.

66. Zhang, G.J.; Wang, Y.E.; Wang, X.; Chen, Y.; Zhou, Y.M.; Tang, Y.W.; Lu, L.D.; Bao, J.C.; Lu, T.H. Preparation of Pd-Au/C catalysts with different alloying degree and their electrocatalytic performance for formic acid oxidation. *Appl. Catal. B* **2011**, *102*, 614–619.

67. Bai, Z.Y.; Guo, Y.M.; Yang, L.; Li, L.; Li, W.J.; Xu, P.L.; Hu, C.G.; Wang, K. Highly dispersed Pd nanoparticles supported on 1,10-phenanthroline-functionalized multi-walled carbon nanotubes for electrooxidation of formic acid. *J. Power Sources* **2011**, *196*, 6232–6237.

68. Bong, S.; Uhm, S.; Kim, Y.R.; Lee, J.; Kim, H. Graphene Supported Pd Electrocatalysts for Formic Acid Oxidation. *Electrocatalysis* **2010**, *1*, 139–143.

69. Hu, G.Z.; Nitze, F.; Barzegar, H.R.; Sharifi, T.; Mikolajczuk, A.; Tai, C.W.; Borodzinski, A.; Wagberg, T. Palladium nanocrystals supported on helical carbon nanofibers for highly efficient electro-oxidation of formic acid, methanol and ethanol in alkaline electrolytes. *J. Power Sources* **2012**, *209*, 236–242.

70. Nitze, F.; Mazurkiewicz, M.; Malolepszy, A.; Mikolajczuk, A.; Kedzierzawski, P.; Tai, C.W.; Hu, G.Z.; Kurzydlowski, K.J.; Stobinski, L.; Borodzinski, A.; *et al.* Synthesis of palladium nanoparticles decorated helical carbon nanofiber as highly active anodic catalyst for direct formic acid fuel cells. *Electrochim. Acta* **2012**, *63*, 323–328.

71. Feng, Y.Y.; Yin, Q.Y.; Lu, G.P.; Yang, H.F.; Zhu, X.; Kong, D.S.; You, J.M. Enhanced catalytic performance of Pd catalyst for formic acid electrooxidation in ionic liquid aqueous solution. *J. Power Sources* **2014**, *272*, 606–613.

72. Cheng, N.C.; Lv, H.F.; Wang, W.; Mu, S.C.; Pan, M.; Marken, F. An ambient aqueous synthesis for highly dispersed and active Pd/C catalyst for formic acid electro-oxidation. *J. Power Sources* **2010**, *195*, 7246–7249.

73. Suo, Y.; Hsing, I.M. Size-controlled synthesis and impedance-based mechanistic understanding of Pd/C nanoparticles for formic acid oxidation. *Electrochim. Acta* **2009**, *55*, 210–217.

74. Lee, J.Y.; Kwak, D.H.; Lee, Y.W.; Lee, S.; Park, K.W. Synthesis of cubic PtPd alloy nanoparticles as anode electrocatalysts for mehtanol and formic acid oxidation reactions. *Phys. Chem. Chem. Phys.* **2015**, *17*, 8642–8648.

75. Baranova, E.A.; Miles, N.; Mercier, P.H.J.; le Page, Y.; Patarachao, B. Formic acid electro-oxidation on carbon supported Pd_xPt_{1-x} ($0 \leqslant x \leqslant 1$) nanoparticles synthesized via modified polyol method. *Electrochim. Acta* **2010**, *55*, 8182–8188.

76. Matin, M.A.; Jang, J.H.; Lee, E.; Kwon, Y.U. Sonochemical synthesis of Pt-doped Pd nanoparticles with enhanced electrocatalytic activity for formic acid oxidation reaction. *J. Appl. Electrochem.* **2012**, *42*, 827–832.

77. Liu, Z.L.; Zhang, X.H. Carbon-supported PdSn nanoparticles as catalysts for formic acid oxidation. *Electrochem. Commun.* **2009**, *11*, 1667–1670.

78. Celorrio, V.; de Oca, M.G.M.; Plana, D.; Moliner, R.; Lazaro, M.J.; Fermin, D.J. Effect of Carbon Supports on Electrocatalytic Reactivity of Au-Pd Core-Shell Nanoparticles. *J. Phys. Chem. C* **2012**, *116*, 6275–6282.

79. Celorrio, V.; de Oca, M.G.M.; Plana, D.; Moliner, R.; Fermin, D.J.; Lazaro, M.J. Electrochemical performance of Pd and Au-Pd core-shell nanoparticles on surface tailored carbon black as catalyst support. *Int. J. Hydrogen Energy* **2012**, *37*, 7152–7160.

80. Wang, H.; Ge, X.B. Facile Fabrication of Porous Pd-Au Bimetallic Nanostructures for Electrocatalysis. *Electroanalysis* **2012**, *24*, 911–916.

81. Shi, R.R.; Wang, J.S.; Cheng, N.C.; Sun, X.L.; Zhang, L.; Zhang, J.J.; Wang, L.C. Electrocatalytic activity and stability of carbon nanotubes-supported Pt-on-Au, Pd-on-Au, Pt-on-Pd-on-Au, Pt-on-Pd, and Pd-on-Pt catalysts for methanol oxidation reaction. *Electrochim. Acta* **2014**, *148*, 1–7.

82. Li, G.Q.; Feng, L.G.; Chang, J.F.; Wickman, B.; Gronbeck, H.; Liu, C.P.; Xing, W. Activity of Platinum/Carbon and Palladium/Carbon Catalysts Promoted by Ni_2P in Direct Ethanol Fuel Cells. *Chemsuschem* **2014**, *7*, 3374–3381.

83. Liu, H.Q.; Koenigsmann, C.; Adzic, R.R.; Wong, S.S. Probing Ultrathin One-Dimensional Pd–Ni Nanostructures As Oxygen Reduction Reaction Catalysts. *ACS Catal.* **2014**, *4*, 2544–2555.

84. Yu, X.W.; Pickup, P.G. Novel Pd-Pb/C bimetallic catalysts for direct formic acid fuel cells. *J. Power Sources* **2009**, *192*, 279–284.

85. Feng, L.G.; Yang, J.; Hu, Y.; Zhu, J.B.; Liu, C.P.; Xing, W. Electrocatalytic properties of $PdCeO_x$/C anodic catalyst for formic acid electrooxidation. *Int. J. Hydrogen Energy* **2012**, *37*, 4812–4818.

86. Liu, Z.L.; Zhang, X.H.; Tay, S.W. Nanostructured PdRu/C catalysts for formic acid oxidation. *J. Solid State Electrochem.* **2012**, *16*, 545–550.

87. Arikan, T.; Kannan, A.M.; Kadirgan, F. Binary Pt–Pd and ternary Pt–Pd–Ru nanoelectrocatalysts for direct methanol fuel cells. *Int. J. Hydrogen Energy* **2013**, *38*, 2900–2907.

88. Awasthi, R.; Singh, R.N. Graphene-supported Pd–Ru nanoparticles with superior methanol electrooxidation activity. *Carbon* **2013**, *51*, 282–289.

89. Qin, Y.H.; Jia, Y.B.; Jiang, Y.; Niu, D.F.; Zhang, X.S.; Zhou, X.G.; Niu, L.; Yuan, W.K. Controllable synthesis of carbon nanofiber supported Pd catalyst for formic acid electrooxidation. *Int. J. Hydrogen Energy* **2012**, *37*, 7373–7377.

90. Li, Y.H.; Xu, Q.Z.; Li, Q.Y.; Wang, H.Q.; Huang, Y.G.; Xu, C.W. Pd deposited on MWCNTs modified carbon fiber paper as high-efficient electrocatalyst for ethanol electrooxidation. *Electrochim. Acta* **2014**, *147*, 151–156.

91. Chakraborty, S.; Raj, C.R. Electrocatalytic performance of carbon nanotube-supported palladium particles in the oxidation of formic acid and the reduction of oxygen. *Carbon* **2010**, *48*, 3242–3249.

92. Liu, J.; Liu, R.; Yuan, C.L.; Wei, X.P.; Yin, J.L.; Wang, G.L.; Cao, D.X. Pd-Co/MWCNTs Catalyst for Electrooxidation of Hydrazine in Alkaline Solution. *Fuel Cells* **2013**, *13*, 903–909.

93. Selvaraj, V.; Grace, A.N.; Alagar, M. Electrocatalytic oxidation of formic acid and formaldehyde on nanoparticle decorated single walled carbon nanotubes. *J. Colloid Interface Sci.* **2009**, *333*, 254–262.

94. Limpattayanate, S.; Hunsom, M. Electrocatalytic activity of Pt-Pd electrocatalysts for the oxygen reduction reaction in proton exchange membrane fuel cells: Effect of supports. *Renew. Energy* **2014**, *63*, 205–211.

95. Chen, C.H.; Liou, W.J.; Lin, H.M.; Wu, S.H.; Mikolajczuk, A.; Stobinski, L.; Borodzinski, A.; Kedzierzawski, P.; Kurzydlowski, K. Carbon nanotube-supported bimetallic palladium-gold electrocatalysts for electro-oxidation of formic acid. *Phys. Status Solidi Appl. Mater. Sci.* **2010**, *207*, 1160–1165.

96. Chen, C.H.; Liou, W.J.; Lin, H.M.; Wu, S.H.; Borodzinski, A.; Stobinski, L.; Kedzierzawski, P. Palladium and Palladium Gold Catalysts Supported on MWCNTs for Electrooxidation of Formic Acid. *Fuel Cells* **2010**, *10*, 227–233.

97. Mikolajczuk, A.; Borodzinski, A.; Stobinski, L.; Kedzierzawski, P.; Lesiak, B.; Laszlo, K.; Jozsef, T.; Lin, H.M. Study of Pd-Au/MWCNTs formic acid electrooxidation catalysts. *Phys. Status Solidi B* **2010**, *247*, 2717–2721.

98. Qin, Y.H.; Jiang, Y.; Niu, D.F.; Zhang, X.S.; Zhou, X.G.; Niu, L.; Yuan, W.K. Carbon nanofiber supported bimetallic PdAu nanoparticles for formic acid electrooxidation. *J. Power Sources* **2012**, *215*, 130–134.

99. Wang, X.G.; Tang, B.; Huang, X.B.; Ma, Y.; Zhang, Z.H. High activity of novel nanoporous Pd-Au catalyst for methanol electro-oxidation in alkaline media. *J. Alloys Compd.* **2013**, *565*, 120–126.

100. Takenaka, S.; Tsukamoto, T.; Matsune, H.; Kishida, M. Carbon nanotube-supported Pd–Co catalysts covered with silica layers as active and stable cathode catalysts for polymer electrolyte fuel cells. *Catal. Sci. Technol.* **2013**, *3*, 2723–2731.

101. Yang, S.D.; Shen, C.M.; Lu, X.J.; Tong, H.; Zhu, J.J.; Zhang, X.G.; Gao, H.J. Preparation and electrochemistry of graphene nanosheets-multiwalled carbon nanotubes hybrid nanomaterials as Pd electrocatalyst support for formic acid oxidation. *Electrochim. Acta* **2012**, *62*, 242–249.

102. Qu, K.G.; Wu, L.; Ren, J.S.; Qu, X.G. Natural DNA-Modified Graphene/Pd Nanoparticles as Highly Active Catalyst for Formic Acid Electro-Oxidation and for the Suzuki Reaction. *ACS Appl. Mater. Interfaces* **2012**, *4*, 5001–5009.

103. Li, R.; Hao, H.; Huang, T.; Yu, A.S. Electrodeposited Pd–MoO$_x$ catalysts with enhanced catalytic activity for formic acid electrooxidation. *Electrochim. Acta* **2012**, *76*, 292–299.

104. Lu, H.T.; Fan, Y.; Huang, P.; Xu, D.L. SnO$_2$ nanospheres supported Pd catalyst with enhanced performance for formic acid oxidation. *J. Power Sources* **2012**, *215*, 48–52.

105. Kim, I.T.; Choi, M.; Lee, H.K.; Shim, J. Characterization of methanol-tolerant Pd–WO$_3$ and Pd–SnO$_2$ electrocatalysts for the oxygen reduction reaction in direct methanol fuel cells. *J. Ind. Eng. Chem.* **2013**, *19*, 813–818.

106. Maheswari, S.; Sridhar, P.; Pitchumani, S. Pd–TiO$_2$/C as a methanol tolerant catalyst for oxygen reduction reaction in alkaline medium. *Electrochem. Commun.* **2013**, *26*, 97–100.

107. Matos, J.; Borodzinski, A.; Zychora, A.M.; Kedzierzawski, P.; Mierzwa, B.; Juchniewicz, K.; Mazurkiewicz, M.; Hernandez-Garrido, J.C. Direct formic acid fuel cells on Pd catalysts supported on hybrid TiO$_2$–C materials. *Appl. Catal. B* **2015**, *163*, 167–178.

108. Wang, X.M.; Xia, Y.Y. The influence of the crystal structure of TiO$_2$ support material on Pd catalysts for formic acid electrooxidation. *Electrochim. Acta* **2010**, *55*, 851–856.

109. Liao, M.Y.; Hu, Q.; Zheng, J.B.; Li, Y.H.; Zhou, H.; Zhong, C.J.; Chen, B.H. Pd decorated Fe/C nanocatalyst for formic acid electrooxidation. *Electrochim. Acta* **2013**, *111*, 504–509.

110. Bai, Z.Y.; Yang, L.; Guo, Y.M.; Zheng, Z.; Hu, C.G.; Xu, P.L. High-efficiency palladium catalysts supported on ppy-modified C-60 for formic acid oxidation. *Chem. Commun.* **2011**, *47*, 1752–1754.

111. Ding, K.Q.; Jia, H.T.; Wei, S.Y.; Guo, Z.H. Electrocatalysis of Sandwich-Structured Pd/Polypyrrole/Pd Composites toward Formic Acid Oxidation. *Ind. Eng. Chem. Res.* **2011**, *50*, 7077–7082.

112. Zhao, X.; Zhu, J.B.; Liang, L.; Liu, C.P.; Liao, J.H.; Xing, W. Enhanced electroactivity of Pd nanocrystals supported on H$_3$PMo$_{12}$O$_{40}$/carbon for formic acid electrooxidation. *J. Power Sources* **2012**, *210*, 392–396.

113. Qin, Y.H.; Yue, J.; Yang, H.H.; Zhang, X.S.; Zhou, X.G.; Niu, L.; Yuan, W.K. Synthesis of highly dispersed and active palladium/carbon nanofiber catalyst for formic acid electrooxidation. *J. Power Sources* **2011**, *196*, 4609–4612.

114. Cheng, N.C.; Webster, R.A.; Pan, M.; Mu, S.C.; Rassaei, L.; Tsang, S.C.; Marken, F. One-step growth of 3–5 nm diameter palladium electrocatalyst in a carbon nanoparticle-chitosan host and characterization for formic acid oxidation. *Electrochim. Acta* **2010**, *55*, 6601–6610.

115. Zhang, B.A.; Ye, D.D.; Li, J.; Zhu, X.; Liao, Q. Electrodeposition of Pd catalyst layer on graphite rod electrodes for direct formic acid oxidation. *J. Power Sources* **2012**, *214*, 277–284.

116. Liu, S.L.; Han, M.; Shi, Y.; Zhang, C.Z.; Chen, Y.; Bao, J.C.; Dai, Z.H. Gram-Scale Synthesis of Multipod Pd Nanocrystals by a Simple Solid-Liquid Phase Reaction and Their Remarkable Electrocatalytic Properties. *Eur. J. Inorg. Chem.* **2012**, *2012*, 3740–3746.

117. Xu, C.X.; Liu, Y.Q.; Wang, J.P.; Geng, H.R.; Qiu, H.J. Nanoporous PdCu alloy for formic acid electro-oxidation. *J. Power Sources* **2012**, *199*, 124–131.

118. Meng, H.; Xie, F.Y.; Chen, J.; Shen, P.K. Electrodeposited palladium nanostructure as novel anode for direct formic acid fuel cell. *J. Mater. Chem.* **2011**, *21*, 11352–11358.

119. Cheng, T.T.; Gyenge, E.L. Novel catalyst-support interaction for direct formic acid fuel cell anodes: Pd electrodeposition on surface-modified graphite felt. *J. Appl. Electrochem.* **2009**, *39*, 1925–1938.

120. Mikolajczuk, A.; Borodzinski, A.; Stobinski, L.; Kedzierzawski, P.; Lesiak, B.; Kover, L.; Toth, J.; Lin, H.M. Physicochemical characterization of the Pd/MWCNTs catalysts for fuel cell applications. *Phys. Status Solidi B* **2010**, *247*, 3063–3067.

121. Morales-Acosta, D.; Rodriguez, H.; Godinez, L.A.; Arriaga, L.G. Performance increase of microfluidic formic acid fuel cell using Pd/MWCNTs as catalyst. *J. Power Sources* **2010**, *195*, 1862–1865.

122. Yu, X.W.; Pickup, P.G. Screening of PdM and PtM catalysts in a multi-anode direct formic acid fuel cell. *J. Appl. Electrochem.* **2011**, *41*, 589–597.

123. Yan, Z.X.; Xie, J.M.; Shen, P.K.; Zhang, M.M.; Zhang, Y.; Chen, M. Pd supported on 2–4 nm MoC particles with reduced particle size, synergistic effect and high stability for ethanol oxidation. *Electrochim. Acta* **2013**, *108*, 644–650.

124. Jung, W.S.; Han, J.; Yoon, S.P.; Nam, S.W.; Lim, T.H.; Hong, S.A. Performance degradation of direct formic acid fuel cell incorporating a Pd anode catalyst. *J. Power Sources* **2011**, *196*, 4573–4578.

125. Ren, M.J.; Kang, Y.Y.; He, W.; Zou, Z.Q.; Xue, X.Z.; Akins, D.L.; Yang, H.; Feng, S.L. Origin of performance degradation of palladium-based direct formic acid fuel cells. *Appl. Catal. B* **2011**, *104*, 49–53.

126. Zhou, Y.; Liu, J.G.; Ye, J.L.; Zou, Z.G.; Ye, J.H.; Gu, J.; Yu, T.; Yang, A.D. Poisoning and regeneration of Pd catalyst in direct formic acid fuel cell. *Electrochim. Acta* **2010**, *55*, 5024–5027.

127. Yu, X.W.; Pickup, P.G. Deactivation/reactivation of a Pd/C catalyst in a direct formic acid fuel cell (DFAFC): Use of array membrane electrode assemblies. *J. Power Sources* **2009**, *187*, 493–499.

128. Chen, Y.X.; Lavacchi, A.; Chen, S.P.; di Benedetto, F.; Bevilacqua, M.; Bianchini, C.; Fornasiero, P.; Innocenti, M.; Marelli, M.; Oberhauser, W.; *et al.* Electrochemical Milling and Faceting: Size Reduction and Catalytic Activation of Palladium Nanoparticles. *Angew. Chem. Int. Ed.* **2012**, *51*, 8500–8504.

129. Wu, Q.M.; Rao, Z.X.; Yuan, L.Z.; Jiang, L.H.; Sun, G.Q.; Ruan, J.M.; Zhou, Z.C.; Sang, S.B. Carbon supported PdO with improved activity and stability for oxygen reduction reaction in alkaline solution. *Electrochim. Acta* **2014**, *150*, 157–166.

130. Yan, Z.X.; Hu, Z.F.; Chen, C.; Meng, H.; Shen, P.K.; Ji, H.B.; Meng, Y.Z. Hollow carbon hemispheres supported palladium electrocatalyst at improved performance for alcohol oxidation. *J. Power Sources* **2010**, *195*, 7146–7151.

131. Maiyalagan, T.; Scott, K. Performance of carbon nanofiber supported Pd–Ni catalysts for electro-oxidation of ethanol in alkaline medium. *J. Power Sources* **2010**, *195*, 5246–5251.

132. Xu, J.B.; Zhao, T.S.; Shen, S.Y.; Li, Y.S. Stabilization of the palladium electrocatalyst with alloyed gold for ethanol oxidation. *Int. J. Hydrogen Energy* **2010**, *35*, 6490–6500.

133. Cheng, F.L.; Dai, X.C.; Wang, H.; Jiang, S.P.; Zhang, M.; Xu, C.W. Synergistic effect of Pd–Au bimetallic surfaces in Au-covered Pd nanowires studied for ethanol oxidation. *Electrochim. Acta* **2010**, *55*, 2295–2298.

134. Gao, H.L.; Liao, S.J.; Liang, Z.X.; Liang, H.G.; Luoa, F. Anodic oxidation of ethanol on core-shell structured Ru@PtPd/C catalyst in alkaline media. *J. Power Sources* **2011**, *196*, 6138–6143.

135. Bambagioni, V.; Bianchini, C.; Filippi, J.; OberhauserIal, W.; Marchionni, A.; Vizza, F.; Psaro, R.; Sordelli, L.; Foresti, M.L.; Innocenti, M. Ethanol Oxidation on Electrocatalysts Obtained by Spontaneous Deposition of Palladium onto Nickel-Zinc Materials. *Chemsuschem* **2009**, *2*, 99–112.

136. Chen, X.M.; Lin, Z.J.; Jia, T.T.; Cai, Z.M.; Huang, X.L.; Jiang, Y.Q.; Chen, X.; Chen, G.N. A facile synthesis of palladium nanoparticles supported on functional carbon nanotubes and its novel catalysis for ethanol electrooxidation. *Anal. Chim. Acta* **2009**, *650*, 54–58.

137. Fu, S.F.; Zhu, C.Z.; Du, D.; Lin, Y.H. Facile One-Step Synthesis of Three-Dimensional Pd–Ag Bimetallic Alloy Networks and Their Electrocatalytic Activity toward Ethanol Oxidation. *ACS Appl. Mater. Interfaces* **2015**, *7*, 13842–13848.

138. Lv, J.J.; Wisitruangsakul, N.; Feng, J.J.; Luo, J.; Fang, K.M.; Wang, A.J. Biomolecule-assisted synthesis of porous PtPd alloyed nanoflowers supported on reduced graphene oxide with highly electrocatalytic performance for ethanol oxidation and oxygen reduction. *Electrochim. Acta* **2015**, *160*, 100–107.

139. Wang, A.L.; He, X.J.; Lu, X.F.; Xu, H.; Tong, Y.X.; Li, G.R. Palladium-Cobalt Nanotube Arrays Supported on Carbon Fiber Cloth as High-Performance Flexible Electrocatalysts for Ethanol Oxidation. *Angew. Chem. Int. Ed.* **2015**, *54*, 3669–3673.

140. Ma, J.W.; Wang, J.; Zhang, G.H.; Fan, X.B.; Zhang, G.L.; Zhang, F.B.; Li, Y. Deoxyribonucleic acid-directed growth of well dispersed nickel-palladium-platinum nanoclusters on graphene as an efficient catalyst for ethanol electrooxidation. *J. Power Sources* **2015**, *278*, 43–49.

141. Peng, C.; Hu, Y.L.; Liu, M.R.; Zheng, Y.X. Hollow raspberry-like PdAg alloy nanospheres: High electrocatalytic activity for ethanol oxidation in alkaline media. *J. Power Sources* **2015**, *278*, 69–75.

142. Singh, R.N.; Awasthi, R. Graphene support for enhanced electrocatalytic activity of Pd for alcohol oxidation. *Catal. Sci. Technol.* **2011**, *1*, 778–783.

143. Su, L.; Jia, W.Z.; Schempf, A.; Ding, Y.; Lei, Y. Free-Standing Palladium/Polyamide 6 Nanofibers for Electrooxidation of Alcohols in Alkaline Medium. *J. Phys. Chem. C* **2009**, *113*, 16174–16180.

144. Wang, Y.L.; Zhao, Y.Q.; Xu, C.L.; Zhao, D.D.; Xu, M.W.; Su, Z.X.; Li, H.L. Improved performance of Pd electrocatalyst supported on three-dimensional nickel foam for direct ethanol fuel cells. *J. Power Sources* **2010**, *195*, 6496–6499.

145. Thotiyl, M.M.O.; Kumar, T.R.; Sampath, S. Pd Supported on Titanium Nitride for Efficient Ethanol Oxidation. *J. Phys. Chem. C* **2010**, *114*, 17934–17941.

146. Mahendiran, C.; Maiyalagan, T.; Scott, K.; Gedanken, A. Synthesis of a carbon-coated NiO/MgO core/shell nanocomposite as a Pd electro-catalyst support for ethanol oxidation. *Mater. Chem. Phys.* **2011**, *128*, 341–347.

147. Yi, Q.F.; Niu, F.J.; Sun, L.Z. Fabrication of novel porous Pd particles and their electroactivity towards ethanol oxidation in alkaline media. *Fuel* **2011**, *90*, 2617–2623.

148. Wang, X.G.; Wang, W.M.; Qi, Z.; Zhao, C.C.; Ji, H.; Zhang, Z.H. Fabrication, microstructure and electrocatalytic property of novel nanoporous palladium composites. *J. Alloys Compd.* **2010**, *508*, 463–470.

149. Liu, R.; Liu, J.F.; Jiang, G.B. Use of Triton X-114 as a weak capping agent for one-pot aqueous phase synthesis of ultrathin noble metal nanowires and a primary study of their electrocatalytic activity. *Chem. Commun.* **2010**, *46*, 7010–7012.

150. Ksar, F.; Surendran, G.; Ramos, L.; Keita, B.; Nadjo, L.; Prouzet, E.; Beaunier, P.; Hagege, A.; Audonnet, F.; Remita, H. Palladium Nanowires Synthesized in Hexagonal Mesophases: Application in Ethanol Electrooxidation. *Chem. Mater.* **2009**, *21*, 1612–1617.

151. Cherevko, S.; Xing, X.L.; Chung, C.H. Pt and Pd decorated Au nanowires: Extremely high activity of ethanol oxidation in alkaline media. *Electrochim. Acta* **2011**, *56*, 5771–5775.

152. Lin, S.C.; Chen, J.Y.; Hsieh, Y.F.; Wu, P.W. A facile route to prepare PdPt alloys for ethanol electro-oxidation in alkaline electrolyte. *Mater. Lett.* **2011**, *65*, 215–218.

153. Modibedi, R.M.; Masombuka, T.; Mathe, M.K. Carbon supported Pd-Sn and Pd-Ru-Sn nanocatalysts for ethanol electro-oxidation in alkaline medium. *Int. J. Hydrogen Energy* **2011**, *36*, 4664–4672.

154. Jou, L.H.; Chang, J.K.; Whang, T.J.; Sun, I.W. Electrodeposition of Palladium-Tin Alloys from 1-Ethyl-3-methylimidazolium Chloride-Tetrafluoroborate Ionic Liquid for Ethanol Electro-Oxidation. *J. Electrochem. Soc.* **2010**, *157*, D443–D449.

155. Shen, S.Y.; Zhao, T.S.; Xu, J.B. Carbon-supported bimetallic PdIr catalysts for ethanol oxidation in alkaline media. *Electrochim. Acta* **2010**, *55*, 9179–9184.

156. Qiu, C.C.; Shang, R.; Xie, Y.F.; Bu, Y.R.; Li, C.Y.; Ma, H.Y. Electrocatalytic activity of bimetallic Pd-Ni thin films towards the oxidation of methanol and ethanol. *Mater. Chem. Phys.* **2010**, *120*, 323–330.

157. Qi, Z.; Geng, H.R.; Wang, X.G.; Zhao, C.C.; Ji, H.; Zhang, C.; Xu, J.L.; Zhang, Z.H. Novel nanocrystalline PdNi alloy catalyst for methanol and ethanol electro-oxidation in alkaline media. *J. Power Sources* **2011**, *196*, 5823–5828.

158. Nguyen, S.T.; Law, H.M.; Nguyen, H.T.; Kristian, N.; Wang, S.Y.; Chan, S.H.; Wang, X. Enhancement effect of Ag for Pd/C towards the ethanol electro-oxidation in alkaline media. *Appl. Catal. B* **2009**, *91*, 507–515.

159. Oliveira, M.C.; Rego, R.; Fernandes, L.S.; Tavares, P.B. Evaluation of the catalytic activity of Pd–Ag alloys on ethanol oxidation and oxygen reduction reactions in alkaline medium. *J. Power Sources* **2011**, *196*, 6092–6098.

160. Wang, Y.; Nguyen, T.S.; Liu, X.W.; Wang, X. Novel palladium-lead (Pd–Pb/C) bimetallic catalysts for electrooxidation of ethanol in alkaline media. *J. Power Sources* **2010**, *195*, 2619–2622.

161. He, Q.G.; Chen, W.; Mukerjee, S.; Chen, S.W.; Laufek, F. Carbon-supported PdM (M = Au and Sn) nanocatalysts for the electrooxidation of ethanol in high pH media. *J. Power Sources* **2009**, *187*, 298–304.

162. Zhu, L.D.; Zhao, T.S.; Xu, J.B.; Liang, Z.X. Preparation and characterization of carbon-supported sub-monolayer palladium decorated gold nanoparticles for the electro-oxidation of ethanol in alkaline media. *J. Power Sources* **2009**, *187*, 80–84.

163. Hu, Z.F.; Yan, Z.X.; Shen, P.K.; Zhong, C.J. Nano-architectures of ordered hollow carbon spheres filled with carbon webs by template-free controllable synthesis. *Nanotechnology* **2012**, *23*, 48–53.

164. Yan, Z.X.; Meng, H.; Shen, P.K.; Meng, Y.Z.; Ji, H.B. Effect of the templates on the synthesis of hollow carbon materials as electrocatalyst supports for direct alcohol fuel cells. *Int. J. Hydrogen Energy* **2012**, *37*, 4728–4736.

165. Yan, Z.X.; Meng, H.; Shi, L.; Li, Z.H.; Shen, P.K. Synthesis of mesoporous hollow carbon hemispheres as highly efficient Pd electrocatalyst support for ethanol oxidation. *Electrochem. Commun.* **2010**, *12*, 689–692.

166. Cerritos, R.C.; Guerra-Balcazar, M.; Ramirez, R.F.; Ledesma-Garcia, J.; Arriaga, L.G. Morphological Effect of Pd Catalyst on Ethanol Electro-Oxidation Reaction. *Materials* **2012**, *5*, 1686–1697.

167. Ding, K.Q.; Yang, G.K. HCl-assisted pyrolysis of $PdCl_2$ to immobilize palladium nanoparticles on multi-walled carbon nanotubes. *Mater. Chem. Phys.* **2010**, *123*, 498–501.

168. Ding, K.Q.; Yang, G.K. Using RTILs of EMIBF4 as "water" to prepare palladium nanoparticles onto MWCNTs by pyrolysis of $PdCl_2$. *Electrochim. Acta* **2010**, *55*, 2319–2324.

169. Chu, D.B.; Wang, J.; Wang, S.X.; Zha, L.W.; He, J.G.; Hou, Y.Y.; Yan, Y.X.; Lin, H.S.; Tian, Z.W. High activity of Pd–In_2O_3/CNTs electrocatalyst for electro-oxidation of ethanol. *Catal. Commun.* **2009**, *10*, 955–958.

170. Qin, Y.H.; Yang, H.H.; Zhang, X.S.; Li, P.; Ma, C.A. Effect of carbon nanofibers microstructure on electrocatalytic activities of Pd electrocatalysts for ethanol oxidation in alkaline medium. *Int. J. Hydrogen Energy* **2010**, *35*, 7667–7674.

171. Qin, Y.H.; Li, H.C.; Yang, H.H.; Zhang, X.S.; Zhou, X.G.; Niu, L.; Yuan, W.K. Effect of electrode fabrication methods on the electrode performance for ethanol oxidation. *J. Power Sources* **2011**, *196*, 159–163.

172. Qin, Y.H.; Yang, H.H.; Zhang, X.S.; Li, P.; Zhou, X.G.; Niu, L.; Yuan, W.K. Electrophoretic deposition of network-like carbon nanofibers as a palladium catalyst support for ethanol oxidation in alkaline media. *Carbon* **2010**, *48*, 3323–3329.

173. Wen, Z.L.; Yang, S.D.; Liang, Y.Y.; He, W.; Tong, H.; Hao, L.A.; Zhang, X.G.; Song, Q.J. The improved electrocatalytic activity of palladium/graphene nanosheets towards ethanol oxidation by tin oxide. *Electrochim. Acta* **2010**, *56*, 139–144.

174. Chen, X.M.; Wu, G.H.; Chen, J.M.; Chen, X.; Xie, Z.X.; Wang, X.R. Synthesis of "Clean" and Well-Dispersive Pd Nanoparticles with Excellent Electrocatalytic Property on Graphene Oxide. *J. Am. Chem. Soc.* **2011**, *133*, 3693–3695.

175. El-Shafei, A.A.; Elhafeez, A.M.A.; Mostafa, H.A. Ethanol oxidation at metal-zeolite-modified electrodes in alkaline medium. Part 2: Palladium-zeolite-modified graphite electrode. *J. Solid State Electrochem.* **2010**, *14*, 185–190.

176. Hasan, M.; Newcomb, S.B.; Rohan, J.F.; Razeeb, K.M. Ni nanowire supported 3D flower-like Pd nanostructures as an efficient electrocatalyst for electrooxidation of ethanol in alkaline media. *J. Power Sources* **2012**, *218*, 148–156.

177. Wang, W.J.; Zhang, J.; Yang, S.C.; Ding, B.J.; Song, X.P. Au@Pd Core-Shell Nanobricks with Concave Structures and Their Catalysis of Ethanol Oxidation. *Chemsuschem* **2013**, *6*, 1945–1951.

178. Li, S.J.; Cheng, D.J.; Qiu, X.G.; Cao, D.P. Synthesis of Cu@Pd core-shell nanowires with enhanced activity and stability for formic acid oxidation. *Electrochim. Acta* **2014**, *143*, 44–48.

179. Li, C.L.; Su, Y.; Lv, X.Y.; Shi, H.J.; Yang, X.G.; Wang, Y.J. Enhanced ethanol electrooxidation of hollow Pd nanospheres prepared by galvanic exchange reactions. *Mater. Lett.* **2012**, *69*, 92–95.

180. Wu, H.X.; Li, H.J.; Zhai, Y.J.; Xu, X.L.; Jin, Y.D. Facile Synthesis of Free-Standing Pd-Based Nanomembranes with Enhanced Catalytic Performance for Methanol/Ethanol Oxidation. *Adv. Mater.* **2012**, *24*, 1594–1597.

181. Ozturk, Z.; Sen, F.; Sen, S.; Gokagac, G. The preparation and characterization of nano-sized Pt-Pd/C catalysts and comparison of their superior catalytic activities for methanol and ethanol oxidation. *J. Mater. Sci.* **2012**, *47*, 8134–8144.

182. Shen, S.Y.; Zhao, T.S.; Xu, J.B.; Li, Y.S. Synthesis of PdNi catalysts for the oxidation of ethanol in alkaline direct ethanol fuel cells. *J. Power Sources* **2010**, *195*, 1001–1006.

183. Zhu, C.Z.; Guo, S.J.; Dong, S.J. PdM (M = Pt, Au) Bimetallic Alloy Nanowires with Enhanced Electrocatalytic Activity for Electro-oxidation of Small Molecules. *Adv. Mater.* **2012**, *24*, 2326–2331.

184. Seweryn, J.; Lewera, A. Electrooxidation of ethanol on carbon-supported Pt–Pd nanoparticles. *J. Power Sources* **2012**, *205*, 264–271.

185. Lv, J.J.; Zheng, J.N.; Zhang, H.B.; Lin, M.; Wang, A.J.; Chen, J.R.; Feng, J.J. Simple synthesis of platinum-palladium nanoflowers on reduced graphene oxide and their enhanced catalytic activity for oxygen reduction reaction. *J. Power Sources* **2014**, *269*, 136–143.

186. Ding, L.X.; Wang, A.L.; Ou, Y.N.; Li, Q.; Guo, R.; Zhao, W.X.; Tong, Y.X.; Li, G.R. Hierarchical Pd–Sn Alloy Nanosheet Dendrites: An Economical and Highly Active Catalyst for Ethanol Electrooxidation. *Sci. Rep.* **2013**, *3*.

187. Assumpcao, M.; da Silva, S.G.; de Souza, R.F.B.; Buzzo, G.S.; Spinace, E.V.; Santos, M.C.; Neto, A.O.; Silva, J.C.M. Investigation of PdIr/C electrocatalysts as anode on the performance of direct ammonia fuel cell. *J. Power Sources* **2014**, *268*, 129–136.

188. Nguyen, S.T.; Yang, Y.H.; Wang, X. Ethanol electro-oxidation activity of Nb-doped-TiO_2 supported PdAg catalysts in alkaline media. *Appl. Catal. B* **2012**, *113*, 261–270.

189. Liu, Y.Q.; Xu, C.X. Nanoporous PdTi Alloys as Non-Platinum Oxygen-Reduction Reaction Electrocatalysts with Enhanced Activity and Durability. *Chemsuschem* **2013**, *6*, 78–84.

190. Li, Y.S.; Zhao, T.S. A high-performance integrated electrode for anion-exchange membrane direct ethanol fuel cells. *Int. J. Hydrogen Energy* **2011**, *36*, 7707–7713.

191. Shen, S.Y.; Zhao, T.S.; Xu, J.B.; Li, Y.S. High performance of a carbon supported ternary PdIrNi catalyst for ethanol electro-oxidation in anion-exchange membrane direct ethanol fuel cells. *Energy Environ. Sci.* **2011**, *4*, 1428–1433.

192. Antolini, E.; Colmati, F.; Gonzalez, E.R. Ethanol oxidation on carbon supported (PtSn)(alloy)/SnO$_2$ and (PtSnPd)(alloy)/SnO$_2$ catalysts with a fixed Pt/SnO$_2$ atomic ratio: Effect of the alloy phase characteristics. *J. Power Sources* **2009**, *193*, 555–561.

193. Bambagioni, V.; Bianchini, C.; Marchionni, A.; Filippi, J.; Vizza, F.; Teddy, J.; Serp, P.; Zhiani, M. Pd and Pt-Ru anode electrocatalysts supported on multi-walled carbon nanotubes and their use in passive and active direct alcohol fuel cells with an anion-exchange membrane (alcohol = methanol, ethanol, glycerol). *J. Power Sources* **2009**, *190*, 241–251.

194. Zhang, Z.Y.; More, K.L.; Sun, K.; Wu, Z.L.; Li, W.Z. Preparation and Characterization of PdFe Nanoleaves as Electrocatalysts for Oxygen Reduction Reaction. *Chem. Mater.* **2011**, *23*, 1570–1577.

195. Tang, Y.F.; Zhang, H.M.; Zhong, H.X.; Ma, Y.W. A facile synthesis of Pd/C cathode electrocatalyst for proton exchange membrane fuel cells. *Int. J. Hydrogen Energy* **2011**, *36*, 725–731.

196. Jeyabharathi, C.; Venkateshkumar, P.; Mathiyarasu, J.; Phani, K.L.N. Carbon-Supported Palladium-Polypyrrole Nanocomposite for Oxygen Reduction and Its Tolerance to Methanol. *J. Electrochem. Soc.* **2010**, *157*, B1740–B1745.

197. Kumar, S.M.S.; Herrero, J.S.; Irusta, S.; Scott, K. The effect of pretreatment of Vulcan XC-72R carbon on morphology and electrochemical oxygen reduction kinetics of supported Pd nano-particle in acidic electrolyte. *J. Electroanal. Chem.* **2010**, *647*, 211–221.

198. Yin, S.B.; Cai, M.; Wang, C.X.; Shen, P.K. Tungsten carbide promoted Pd–Fe as alcohol-tolerant electrocatalysts for oxygen reduction reactions. *Energy Environ. Sci.* **2011**, *4*, 558–563.

199. Rao, C.V.; Viswanathan, B. Carbon supported Pd–Co–Mo alloy as an alternative to Pt for oxygen reduction in direct ethanol fuel cells. *Electrochim. Acta* **2010**, *55*, 3002–3007.

200. Sarkar, A.; Murugan, A.V.; Manthiram, A. Low cost Pd-W nanoalloy electrocatalysts for oxygen reduction reaction in fuel cells. *J. Mater. Chem.* **2009**, *19*, 159–165.

201. Zhou, Z.M.; Shao, Z.G.; Qin, X.P.; Chen, X.G.; Wei, Z.D.; Yi, B.L. Durability study of Pt–Pd/C as PEMFC cathode catalyst. *Int. J. Hydrogen Energy* **2010**, *35*, 1719–1726.

202. Chang, S.H.; Su, W.N.; Yeh, M.H.; Pan, C.J.; Yu, K.L.; Liu, D.G.; Lee, J.F.; Hwang, B.J. Structural and Electronic Effects of Carbon-Supported Pt$_x$Pd$_{1-x}$ Nanoparticles on the Electrocatalytic Activity of the Oxygen-Reduction Reaction and on Methanol Tolerance. *Chem. Eur. J.* **2010**, *16*, 11064–11071.

203. Sarapuu, A.; Kasikov, A.; Wong, N.; Lucas, C.A.; Sedghi, G.; Nichols, R.J.; Tammeveski, K. Electroreduction of oxygen on gold-supported nanostructured palladium films in acid solutions. *Electrochim. Acta* **2010**, *55*, 6768–6774.

204. Kariuki, N.N.; Wang, X.P.; Mawdsley, J.R.; Ferrandon, M.S.; Niyogi, S.G.; Vaughey, J.T.; Myers, D.J. Colloidal Synthesis and Characterization of Carbon-Supported Pd–Cu Nanoparticle Oxygen Reduction Electrocatalysts. *Chem. Mater.* **2010**, *22*, 4144–4152.

216

205. Fouda-Onana, F.; Bah, S.; Savadogo, O. Palladium-copper alloys as catalysts for the oxygen reduction reaction in an acidic media I: Correlation between the ORR kinetic parameters and intrinsic physical properties of the alloys. *J. Electroanal. Chem.* **2009**, *636*, 1–9.

206. Wei, Y.C.; Liu, C.W.; Chang, Y.W.; Lai, C.M.; Lim, P.Y.; Tsai, L.D.; Wang, K.W. The structure-activity relationship of Pd–Co/C electrocatalysts for oxygen reduction reaction. *Int. J. Hydrogen Energy* **2010**, *35*, 1864–1871.

207. Ohashi, M.; Beard, K.D.; Ma, S.G.; Blom, D.A.; St-Pierre, J.; van Zee, J.W.; Monnier, J.R. Electrochemical and structural characterization of carbon-supported Pt–Pd bimetallic electrocatalysts prepared by electroless deposition. *Electrochim. Acta* **2010**, *55*, 7376–7384.

208. Peng, Z.M.; Yang, H. Synthesis and Oxygen Reduction Electrocatalytic Property of Pt-on-Pd Bimetallic Heteronanostructures. *J. Am. Chem. Soc.* **2009**, *131*, 7542–7543.

209. Yang, J.H.; Zhou, W.J.; Cheng, C.H.; Lee, J.Y.; Liu, Z.L. Pt-Decorated PdFe Nanoparticles as Methanol-Tolerant Oxygen Reduction Electrocatalyst. *ACS Appl. Mater. Interfaces* **2010**, *2*, 119–126.

210. Yang, J.; Cheng, C.H.; Zhou, W.; Lee, J.Y.; Liu, Z. Methanol-Tolerant Heterogeneous PdCo@PdPt/C Electrocatalyst for the Oxygen Reduction Reaction. *Fuel Cells* **2010**, *10*, 907–913.

211. Li, W.Z.; Haldar, P. Supportless PdFe nanorods as highly active electrocatalyst for proton exchange membrane fuel cell. *Electrochem. Commun.* **2009**, *11*, 1195–1198.

212. Zhou, W.P.; Yang, X.F.; Vukmirovic, M.B.; Koel, B.E.; Jiao, J.; Peng, G.W.; Mavrikakis, M.; Adzic, R.R. Improving Electrocatalysts for O_2 Reduction by Fine-Tuning the Pt-Support Interaction: Pt Monolayer on the Surfaces of a $Pd_3Fe(111)$ Single-Crystal Alloy. *J. Am. Chem. Soc.* **2009**, *131*, 12755–12762.

213. Jin, S.A.; Kwon, K.; Pak, C.; Chang, H. The oxygen reduction electrocatalytic activity of intermetallic compound of palladium-tin supported on tin oxide-carbon composite. *Catal. Today* **2011**, *164*, 176–180.

214. Sun, W.; Hsu, A.; Chen, R.R. Palladium-coated manganese dioxide catalysts for oxygen reduction reaction in alkaline media. *J. Power Sources* **2011**, *196*, 4491–4498.

215. Arroyo-Ramirez, L.; Rodriguez, D.; Otano, W.; Cabrera, C.R. Palladium Nanoshell Catalysts Synthesis on Highly Ordered Pyrolytic Graphite for Oxygen Reduction Reaction. *Acs Appl. Mater. Interfaces* **2012**, *4*, 2018–2024.

216. Jukk, K.; Alexeyeva, N.; Johans, C.; Kontturi, K.; Tammeveski, K. Oxygen reduction on Pd nanoparticle/multi-walled carbon nanotube composites. *J. Electroanal. Chem.* **2012**, *666*, 67–75.

217. Kim, Y.S.; Shin, J.Y.; Chun, Y.S.; Lee, C.; Lee, S.G. Anion-Dependent Electrocatalytic Activity of Supported Palladium Catalysts onto Imidazolium Salt-Functionalized Carbon Nanotubes in Oxygen Reduction Reaction. *Bull. Korean Chem. Soc.* **2011**, *32*, 3209–3210.

218. Seo, M.H.; Choi, S.M.; Kim, H.J.; Kim, W.B. The graphene-supported Pd and Pt catalysts for highly active oxygen reduction reaction in an alkaline condition. *Electrochem. Commun.* **2011**, *13*, 182–185.

219. Kim, D.; Ahmed, M.S.; Jeon, S. Different length linkages of graphene modified with metal nanoparticles for oxygen reduction in acidic media. *J. Mater. Chem.* **2012**, *22*, 16353–16360.

220. Gotoh, K.; Kawabata, K.; Fujii, E.; Morishige, K.; Kinumoto, T.; Miyazaki, Y.; Ishida, H. The use of graphite oxide to produce mesoporous carbon supporting Pt, Ru, or Pd nanoparticles. *Carbon* **2009**, *47*, 2120–2124.

221. Du, S.F.; Lu, Y.X.; Steinberger-Wilckens, R. PtPd nanowire arrays supported on reduced graphene oxide as advanced electrocatalysts for methanol oxidation. *Carbon* **2014**, *79*, 346–353.

222. Huang, S.Y.; Ganesan, P.; Popov, B.N. Electrocatalytic Activity and Stability of Titania-Supported Platinum-Palladium Electrocatalysts for Polymer Electrolyte Membrane Fuel Cell. *ACS Catal.* **2012**, *2*, 825–831.

223. Bae, S.J.; Nahm, K.S.; Kim, P. Electroreduction of oxygen on Pd catalysts supported on Ti-modified carbon. *Curr. Appl. Phys.* **2012**, *12*, 1476–1480.

224. Ramanathan, M.; Ramani, V.; Prakash, J. Kinetics of the oxygen reduction reaction on Pd_3M (M = Cu, Ni, Fe) electrocatalysts synthesized at elevated annealing temperatures. *Electrochim. Acta* **2012**, *75*, 254–261.

225. Wen, M.; Zhou, B.; Fang, H.; Wu, Q.S.; Chen, S.P. Novel-Phase Structural High-Efficiency Anode Catalyst for Methanol Fuel Cells: α-$(NiCu)_3Pd$ Nanoalloy. *J. Phys. Chem. C* **2014**, *118*, 26713–26720.

226. Lee, K.R.; Jung, Y.; Woo, S.I. Combinatorial Screening of Highly Active Pd Binary Catalysts for Electrochemical Oxygen Reduction. *ACS Comb. Sci.* **2012**, *14*, 10–16.

227. Meng, H.; Shen, P.K. The beneficial effect of the addition of tungsten carbides to Pt catalysts on the oxygen electroreduction. *Chem. Commun.* **2005**.

228. Meng, H.; Shen, P.K. Tungsten carbide nanocrystal promoted Pt/C electrocatalysts for oxygen reduction. *J. Phys. Chem. B* **2005**, *109*, 22705–22709.

229. Nie, M.; Shen, P.K.; Wu, M.; Wei, Z.D.; Meng, H. A study of oxygen reduction on improved Pt–WC/C electrocatalysts. *J. Power Sources* **2006**, *162*, 173–176.

230. Cui, G.F.; Shen, P.K.; Meng, H.; Zhao, J.; Wu, G. Tungsten carbide as supports for Pt electrocatalysts with improved CO tolerance in methanol oxidation. *J. Power Sources* **2011**, *196*, 6125–6130.

231. He, G.Q.; Yan, Z.X.; Ma, X.M.; Meng, H.; Shen, P.K.; Wang, C.X. A universal method to synthesize nanoscale carbides as electrocatalyst supports towards oxygen reduction reaction. *Nanoscale* **2011**, *3*, 3578–3582.

232. Pan, Y.; Zhang, F.; Wu, K.; Lu, Z.Y.; Chen, Y.; Zhou, Y.M.; Tang, Y.W.; Lu, T.H. Carbon supported Palladium-Iron nanoparticles with uniform alloy structure as methanol-tolerant electrocatalyst for oxygen reduction reaction. *Int. J. Hydrogen Energy* **2012**, *37*, 2993–3000.

233. Han, B.H.; Xu, C.X. Nanoporous PdFe alloy as highly active and durable electrocatalyst for oxygen reduction reaction. *Int. J. Hydrogen Energy* **2014**, *39*, 18247–18255.

234. Gobal, F.; Arab, R. A preliminary study of the electro-catalytic reduction of oxygen on Cu–Pd alloys in alkaline solution. *J. Electroanal. Chem.* **2010**, *647*, 66–73.

235. Yang, R.Z.; Bian, W.Y.; Strasser, P.; Toney, M.F. Dealloyed PdCu$_3$ thin film electrocatalysts for oxygen reduction reaction. *J. Power Sources* **2013**, *222*, 169–176.

236. You, D.J.; Jin, S.A.; Lee, K.H.; Pak, C.; Choi, K.H.; Chang, H. Improvement of activity for oxygen reduction reaction by decoration of Ir on PdCu/C catalyst. *Catal. Today* **2012**, *185*, 138–142.

237. Xiong, L.; Huang, Y.X.; Liu, X.W.; Sheng, G.P.; Li, W.W.; Yu, H.Q. Three-dimensional bimetallic Pd-Cu nanodendrites with superior electrochemical performance for oxygen reduction reaction. *Electrochim. Acta* **2013**, *89*, 24–28.

238. Lee, C.L.; Chiou, H.P.; Chang, K.C.; Huang, C.H. Carbon nanotubes-supported colloidal Ag–Pd nanoparticles as electrocatalysts toward oxygen reduction reaction in alkaline electrolyte. *Int. J. Hydrogen Energy* **2011**, *36*, 2759–2764.

239. Godinez-Garcia, A.; Perez-Robles, J.F.; Martinez-Tejada, H.V.; Solorza-Feria, O. Characterization and electrocatalytic properties of sonochemical synthesized PdAg nanoparticles. *Mater. Chem. Phys.* **2012**, *134*, 1013–1019.

240. Xu, L.; Luo, Z.M.; Fan, Z.X.; Zhang, X.; Tan, C.L.; Li, H.; Zhang, H.; Xue, C. Triangular Ag–Pd alloy nanoprisms: Rational synthesis with high-efficiency for electrocatalytic oxygen reduction. *Nanoscale* **2014**, *6*, 11738–11743.

241. Xu, J.B.; Zhao, T.S.; Li, Y.S.; Yang, W.W. Synthesis and characterization of the Au-modified Pd cathode catalyst for alkaline direct ethanol fuel cells. *Int. J. Hydrogen Energy* **2010**, *35*, 9693–9700.

242. Kim, D.S.; Kim, J.H.; Jeong, I.K.; Choi, J.K.; Kim, Y.T. Phase change of bimetallic PdCo electrocatalysts caused by different heat-treatment temperatures: Effect on oxygen reduction reaction activity. *J. Catal.* **2012**, *290*, 65–78.

243. Son, D.N.; Takahashi, K. Selectivity of Palladium-Cobalt Surface Alloy toward Oxygen Reduction Reaction. *J. Phys. Chem. C* **2012**, *116*, 6200–6207.

244. Ren, Y.B.; Zhang, S.C.; Fang, H.; Wei, X.; Yang, P.H. Investigation of Co$_3$O$_4$ nanorods supported Pd anode catalyst for methanol oxidation in alkaline solution. *J. Energy Chem.* **2014**, *23*, 801–808.

245. Liu, H.; Li, W.; Manthiram, A. Factors influencing the electrocatalytic activity of Pd$_{100-x}$Co$_x$ ($0 \leqslant x \leqslant 50$) nanoalloys for oxygen reduction reaction in fuel cells. *Appl. Catal. B-Environ.* **2009**, *90*, 184–194.

246. Oishi, K.; Savadogo, O. Electrochemical investigation of Pd-Co thin films binary alloy for the oxygen reduction reaction in acid medium. *J. Electroanal. Chem.* **2013**, *703*, 108–116.

247. Barakat, N.A.M.; Abdelkareem, M.A.; Shin, G.; Kim, H.Y. Pd-doped Co nanofibers immobilized on a chemically stable metallic bipolar plate as novel strategy for direct formic acid fuel cells. *Int. J. Hydrogen Energy* **2013**, *38*, 7438–7447.

248. Tominaka, S.; Hayashi, T.; Nakamura, Y.; Osaka, T. Mesoporous PdCo sponge-like nanostructure synthesized by electrodeposition and dealloying for oxygen reduction reaction. *J. Mater. Chem.* **2010**, *20*, 7175–7182.

249. Serov, A.; Nedoseykina, T.; Shvachko, O.; Kwak, C. Effect of precursor nature on the performance of palladium-cobalt electrocatalysts for direct methanol fuel cells. *J. Power Sources* **2010**, *195*, 175–180.

250. Alia, S.M.; Jensen, K.O.; Pivovar, B.S.; Yan, Y.S. Platinum-Coated Palladium Nanotubes as Oxygen Reduction Reaction Electrocatalysts. *ACS Catal.* **2012**, *2*, 858–863.

251. Slanac, D.A.; Li, L.; Mayoral, A.; Yacaman, M.J.; Manthiram, A.; Stevenson, K.J.; Johnston, K.P. Atomic resolution structural insights into PdPt nanoparticle-carbon interactions for the design of highly active and stable electrocatalysts. *Electrochim. Acta* **2012**, *64*, 35–45.

252. Chen, X.T.; Jiang, Y.Y.; Sun, J.Z.; Jin, C.H.; Zhang, Z.H. Highly active nanoporous Pt-based alloy as anode and cathode catalyst for direct methanol fuel cells. *J. Power Sources* **2014**, *267*, 212–218.

253. Lim, B.; Jiang, M.J.; Camargo, P.H.C.; Cho, E.C.; Tao, J.; Lu, X.M.; Zhu, Y.M.; Xia, Y.N. Pd–Pt Bimetallic Nanodendrites with High Activity for Oxygen Reduction. *Science* **2009**, *324*, 1302–1305.

254. Tang, Y.F.; Gao, F.M.; Yu, S.X.; Li, Z.P.; Zhao, Y.F. Surfactant-free synthesis of highly methanol-tolerant, polyhedral Pd–Pt nanocrystallines for oxygen reduction reaction. *J. Power Sources* **2013**, *239*, 374–381.

255. Limpattayanate, S.; Hunsom, M. Effect of supports on activity and stability of Pt–Pd catalysts for oxygen reduction reaction in proton exchange membrane fuel cells. *J. Solid State Electrochem.* **2013**, *17*, 1221–1231.

256. Thanasilp, S.; Hunsom, M. Effect of Pt: Pd atomic ratio in Pt–Pd/C electrocatalyst-coated membrane on the electrocatalytic activity of ORR in PEM fuel cells. *Renew. Energy* **2011**, *36*, 1795–1801.

257. Ficicilar, B.; Bayrakceken, A.; Eroglu, I. Effect of Pd loading in Pd-Pt bimetallic catalysts doped into hollow core mesoporous shell carbon on performance of proton exchange membrane fuel cells. *J. Power Sources* **2009**, *193*, 17–23.

258. Ang, S.Y.; Walsh, D.A. Palladium-vanadium alloy electrocatalysts for oxygen reduction: Effect of heat treatment on electrocatalytic activity and stability. *Appl. Catal. B* **2010**, *98*, 49–56.

259. Li, B.; Prakash, J. Oxygen reduction reaction on carbon supported Palladium-Nickel alloys in alkaline media. *Electrochem. Commun.* **2009**, *11*, 1162–1165.

260. Zhao, J.; Sarkar, A.; Manthiram, A. Synthesis and characterization of Pd–Ni nanoalloy electrocatalysts for oxygen reduction reaction in fuel cells. *Electrochim. Acta* **2010**, *55*, 1756–1765.

261. Miah, M.R.; Masud, J.; Ohsaka, T. Kinetics of oxygen reduction reaction at electrochemically fabricated tin-palladium bimetallic electrocatalyst in acidic media. *Electrochim. Acta* **2010**, *56*, 285–290.

262. Kim, J.; Park, J.E.; Momma, T.; Osaka, T. Synthesis of Pd-Sn nanoparticles by ultrasonic irradiation and their electrocatalytic activity for oxygen reduction. *Electrochim. Acta* **2009**, *54*, 3412–3418.

263. Karan, H.I.; Sasaki, K.; Kuttiyiel, K.; Farberow, C.A.; Mavrikakis, M.; Adzic, R.R. Catalytic Activity of Platinum Mono layer on Iridium and Rhenium Alloy Nanoparticles for the Oxygen Reduction Reaction. *ACS Catal.* **2012**, *2*, 817–824.

264. Xiao, L.; Zhuang, L.; Liu, Y.; Lu, J.T.; Abruna, H.D. Activating Pd by Morphology Tailoring for Oxygen Reduction. *J. Am. Chem. Soc.* **2009**, *131*, 602–608.

265. Kondo, S.; Nakamura, M.; Maki, N.; Hoshi, N. Active Sites for the Oxygen Reduction Reaction on the Low and High Index Planes of Palladium. *J. Phys. Chem. C* **2009**, *113*, 12625–12628.

266. Erikson, H.; Sarapuu, A.; Alexeyeva, N.; Tammeveski, K.; Solla-Gullon, J.; Feliu, J.M. Electrochemical reduction of oxygen on palladium nanocubes in acid and alkaline solutions. *Electrochim. Acta* **2012**, *59*, 329–335.

267. Lee, C.L.; Chiou, H.P. Methanol-tolerant Pd nanocubes for catalyzing oxygen reduction reaction in H_2SO_4 electrolyte. *Appl. Catal. B* **2012**, *117*, 204–211.

268. Wang, D.L.; Xin, H.L.; Wang, H.S.; Yu, Y.C.; Rus, E.; Muller, D.A.; DiSalvo, F.J.; Abruna, H.D. Facile Synthesis of Carbon-Supported Pd–Co Core-Shell Nanoparticles as Oxygen Reduction Electrocatalysts and Their Enhanced Activity and Stability with Monolayer Pt Decoration. *Chem. Mater.* **2012**, *24*, 2274–2281.

269. Jang, J.H.; Pak, C.; Kwon, Y.U. Ultrasound-assisted polyol synthesis and electrocatalytic characterization of Pd_xCo alloy and core-shell nanoparticles. *J. Power Sources* **2012**, *201*, 179–183.

270. Koenigsmann, C.; Sutter, E.; Adzic, R.R.; Wong, S.S. Size- and Composition-Dependent Enhancement of Electrocatalytic Oxygen Reduction Performance in Ultrathin Palladium-Gold ($Pd_{1-x}Au_x$) Nanowires. *J. Phys. Chem. C* **2012**, *116*, 15297–15306.

271. Wang, J.X.; Inada, H.; Wu, L.J.; Zhu, Y.M.; Choi, Y.M.; Liu, P.; Zhou, W.P.; Adzic, R.R. Oxygen Reduction on Well-Defined Core-Shell Nanocatalysts: Particle Size, Facet, and Pt Shell Thickness Effects. *J. Am. Chem. Soc.* **2009**, *131*, 17298–17302.

272. Sasaki, K.; Naohara, H.; Cai, Y.; Choi, Y.M.; Liu, P.; Vukmirovic, M.B.; Wang, J.X.; Adzic, R.R. Core-Protected Platinum Monolayer Shell High-Stability Electrocatalysts for Fuel-Cell Cathodes. *Angew. Chem.-Int. Ed.* **2010**, *49*, 8602–8607.

273. Di Noto, V.; Negro, E. Synthesis, characterization and electrochemical performance of tri-metal Pt-free carbon nitride electrocatalysts for the oxygen reduction reaction. *Electrochim. Acta* **2010**, *55*, 1407–1418.

274. Rego, R.; Oliveira, M.C.; Alcaide, F.; Alvarez, G. Development of a carbon paper-supported Pd catalyst for PEMFC application. *Int. J. Hydrogen Energy* **2012**, *37*, 7192–7199.

Recent Advances on Electro-Oxidation of Ethanol on Pt- and Pd-Based Catalysts: From Reaction Mechanisms to Catalytic Materials

Ye Wang and Wen-Bin Cai

Abstract: The ethanol oxidation reaction (EOR) has drawn increasing interest in electrocatalysis and fuel cells by considering that ethanol as a biomass fuel has advantages of low toxicity, renewability, and a high theoretical energy density compared to methanol. Since EOR is a complex multiple-electron process involving various intermediates and products, the mechanistic investigation as well as the rational design of electrocatalysts are challenging yet essential for the desired complete oxidation to CO_2. This mini review is aimed at presenting an overview of the advances in the study of reaction mechanisms and electrocatalytic materials for EOR over the past two decades with a focus on Pt- and Pd-based catalysts. We start with discussion on the mechanistic understanding of EOR on Pt and Pd surfaces using selected publications as examples. Consensuses from the mechanistic studies are that sufficient active surface sites to facilitate the cleavage of the C–C bond and the adsorption of water or its residue are critical for obtaining a higher electro-oxidation activity. We then show how this understanding has been applied to achieve improved performance on various Pt- and Pd-based catalysts through optimizing electronic and bifunctional effects, as well as by tuning their surface composition and structure. Finally we point out the remaining key problems in the development of anode electrocatalysts for EOR.

Reprinted from *Catalysts*. Cite as: Wang, Y.; Zou, S.; Cai, W.-B. Recent Advances on Electro-Oxidation of Ethanol on Pt- and Pd-Based Catalysts: From Reaction Mechanisms to Catalytic Materials. *Catalysts* **2015**, *5*, 1507–1534.

1. Introduction

Rising demands for energy coupled with concerns over ecosystem damage and growing consumption of non-regenerative fossil energy pose a great need for clean and efficient power sources [1–4]. Fuel cells are widely considered as sustainable energy conversion devices. Low-temperature fuel cells are undergoing rapid development for mobile applications and in particular for the transport sector. Among different fuels that have been used for fuel cells, hydrogen, methanol, and ethanol have been the most explored and each has its advantages and disadvantages. The choice of the fuel depends on the applications. Proton exchange membrane fuel cells (PEMFCs) using hydrogen as the fuel have the advantages of low operating temperature, sustained operation at high current densities, low weight, compactness, and suitability for discontinuous operation, but face challenges in the production, storage and transport of hydrogen. As an alternative fuel, ethanol which can be produced on a massive scale from biomass feed stocks originating from agriculture (first-generation bioethanol), forestry, and urban residues (second-generation bioethanol), is attracting increasing interest [5–8]. Compared to another common fuel, methanol, ethanol complements the shortcomings of methanol owing to its non-toxicity, higher boiling point, and most importantly, renewability. In addition, ethanol has a high specific energy of 8.01 kWh·kg^{-1}, which is comparable to that of gasoline [9,10]. Nevertheless, the relatively sluggish kinetics for the ethanol oxidation reaction (EOR) presents a major roadblock for the development of direct ethanol fuel cells (DEFCs) [3,7]. Higher performance catalysts are needed to overcome this bottleneck. A detailed understanding of the reaction mechanism and in particular of the rate-limiting step(s) in EOR under continuous reaction conditions is of critical importance for the design of highly active catalysts [11,12]. Although numerous experimental studies using Fourier transform infrared spectroscopy (FTIR) [13–26] or differential electrochemical mass spectrometry (DEMS) [27–37], as well as theoretical studies [38–45] have been conducted to understand the EOR process, a detailed mechanism of EOR remains unclear or even contradictory. Nevertheless, a so-called dual-pathway (C1 and C2) mechanism has been largely agreed upon: the C1 pathway proceeds via adsorbed carbon monoxide (CO_{ads}) intermediate to form CO_2 (or carbonate in alkaline solutions) by delivering 12 electrons, and the C2 pathway mainly leads to the formation of acetic acid (or acetate in alkaline solutions) by delivering four electrons and/or acetaldehyde by delivering two electrons. Though a higher electro-efficiency can be achieved by the C1 pathway, the C2 pathway is generally dominant in the overall EOR [46–49]. Therefore selectively enhancing the C1 pathway by rational design of high performance catalysts is an effective way to increase the DEFC efficiency.

Pt is the most commonly used catalytic metal in the anode of DEFCs because of its excellent properties in the adsorption and dissociation of ethanol. However, the cost of Pt is a major impediment in the commercialization of fuel cell technology, because it alone accounts for approximately 54% of the total fuel cell stack cost [3]. On the other hand, Pd has similar catalytic properties to Pt (in the same group of the periodic table, having the same face centered cubic (fcc) crystal structure and a similar atomic size) [12], but is much lower in material cost. Moreover, the abundance of Pd on the Earth's crust is 200 times higher than that of Pt (0.6 part per billion (ppb) vs. 0.003 ppb), making it very attractive for long-term industrial applications [3]. Though Pt and Pd show relatively good activity, a complete oxidation of ethanol in both acidic and/or basic media remains virtually impossible. A large number of studies have reported the enhancement of the electrocatalytic performance of Pt-M and Pd-M binary or ternary catalysts by adding additional elements such as metallic elements Ru, Sn, Ir, Bi, Rh, Mo, Fe, Co, Cu, Ni, Au, Ag and nonmetallic elements, oxides *etc.* Yet, there are many unanswered questions regarding the role of these foreign materials in improving the electrocatalytic activities. To rationally design Pt- or Pd-based materials as anode catalysts and to develop DEFC technology, a better understanding of the structure-electrocatalytic activity relationships in the EOR is a pre-requisite.

In this review, by discussing selected publications on mechanism studies and the development of advanced catalysts, we present an overview of how the achievements in mechanism studies have been used to guide the rational design of catalysts. Recent advancements in fundamental studies as well as in developing promising new anode EOR catalysts are briefly surveyed. Finally, we summarize the key problems in the investigation on catalysts for EOR and provide outlooks for their future development. Because EOR is under active research, it is impossible to cover every aspect of the new developments. We therefore focus our discussion on Pt- and Pd-based catalysts. Selected examples are only used to facilitate the discussion and inevitably we may have omitted other significant contributions in the field.

2. Reaction Mechanism of EOR

Activity, selectivity, and stability are critical issues that need to be addressed for any catalysts. Comprehensive fundamental studies of EOR form the basis of design rules for high efficiency catalysts [2,21,26,47,50]. A great deal of work on the mechanisms of Pt- or Pd-based catalysts have been devoted to solve the long-standing puzzle concerning the intermediates and the products from EOR. The pioneering work on the mechanism of EOR can be traced back to the 1950s [51] and now has been evolved into a commonly accepted dual-pathway mechanism on Pt- or Pd-based catalysts in either acidic or alkaline media as shown in Figure 1. [44,47,52–55].

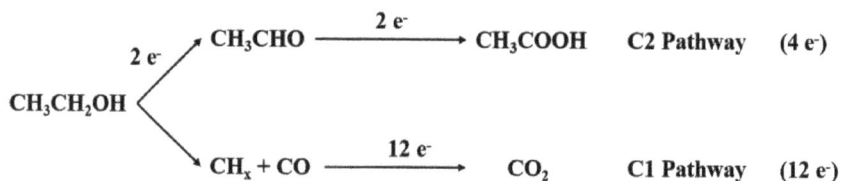

$$CH_3CH_2OH \begin{cases} \xrightarrow{2\ e^-} CH_3CHO \xrightarrow{2\ e^-} CH_3COOH & \text{C2 Pathway} \quad (4\ e^-) \\ \xrightarrow{} CH_x + CO \xrightarrow{12\ e^-} CO_2 & \text{C1 Pathway} \quad (12\ e^-) \end{cases}$$

Figure 1. Schematic representation of the parallel pathways for ethanol oxidation on Pt electrodes in acidic media.

The C1 pathway is the complete oxidation of ethanol to CO_2 or carbonates via CO_{ads} intermediate by delivering 12 electrons and the C2 pathway is the partial oxidation of ethanol to acetate by delivering four electrons or to acetaldehyde by delivering two electrons without the breaking of the C–C bond as shown in the following equations:

C1 pathway:

$$CH_3 - CH_2OH + 3H_2O \rightarrow 2CO_2 + 12H^+ + 12e^- \qquad (1)$$

$$CH_3 - CH_2OH + 5H_2O \rightarrow 2HCO_3^- + 14H^+ + 12e^- \qquad (2)$$

$$CH_3 - CH_2OH + 5H_2O \rightarrow 2CO_3^{2-} + 16H^+ + 12e^- \qquad (3)$$

C2 pathway:

$$CH_3 - CH_2OH + H_2O \rightarrow CH_3 - COOH + 4H^+ + 4e^- \qquad (4)$$

$$CH_3 - CH_2OH \rightarrow CH_3 - CHO + 2H^+ + 2e^- \qquad (5)$$

Adsorbed CO, C1 and C2 hydrocarbon residues have been identified as the major adsorbed intermediates on Pt- or Pd-based catalysts, while acetaldehyde and acetic acid have been detected as the main by-products using techniques such as infrared spectroscopy [13–26], online DEMS [27–37], ion chromatography [56,57], and liquid chromatography [14]. However, EOR has been shown to occur via a series of complex reactions involving a number of sequential and parallel reaction steps, thus resulting in more than 40 possible volatile and adsorbed intermediates or oxidative derivatives [43]. Previous studies agree that CO is a dominant adsorbed species formed during EOR, however, they disagree on details such as the adsorbed state of other intermediates and on the question of the rate limiting steps: the adsorption of intermediate or the cleavage of C–C bond or the formation of OH or oxides [20,23,26,31,38,45,47,54]. In the following we summarize EOR mechanisms developed in the last two decades from experimental as well as DFT calculation studies and discuss the unsolved issues in understanding the EOR mechanism.

2.1. Experimental Detection and Quantification of Reaction Intermediates and Products

As mentioned above, the reaction mechanism is complex involving several adsorbed intermediates and numerous products and by-products. Determining product distribution and identifying reactive intermediates are the keys for solving the EOR mechanism puzzle and therefore are always hot topics under active debate. To address these two issues, many efforts have been made to combine traditional electrochemical methods (cyclic voltammetry, chronoamperometry, rotating disc electrodes, *etc.*) with other physicochemical methods, such as *in situ* FTIR [9,13–24], broadband sum-frequency generation (BB-SFG) spectroscopy [58], DEMS [16,27–36], HPLC [56,57], GC [14], electrochemical quartz crystal microbalance (EQCM) [59] and *in situ* NMR [60] to probe the adsorbed intermediates and/or quantify the reaction products and by-products [56]. We devote the following section to summarize and discuss how these techniques were used for product quantification as well as the clarification of intermediates, starting with FTIR studies and followed by mass spectrometric results.

In the early days Weaver and co-workers [25,61,62] adopted real-time FTIR spectroscopy to study EOR on Pt surfaces in acidic media, and found that the final reaction products included acetic acid, acetaldehyde along with a smaller amount of CO_2. Their work was the first quantification of specific oxidation products, and provides values of the effective absorption coefficient, ε_{eff} of CO_2, acetic acid, and acetaldehyde which are 3.5×10^4, 5.8×10^3, and 2.2×10^3 $M^{-1} \cdot cm^{-2}$, respectively. The yields of oxidation products were calculated using respective integrated band intensities (A_i), and the amount of a given species Q (mol$\cdot$cm^{-2}) trapped inside the thin layer between the electrode surface and the optical window followed the relationship:

$$Q = \frac{A_i}{\varepsilon_{eff}} \tag{6}$$

To better compare the selectivity and activity of EOR on catalysts, Adzic's [9] group applied *in situ* infrared reflection-absorption spectroscopy (*in situ* IRRAS) to quantify the ratio of C1 pathway to C2 pathway on the ternary Pd–Rh–SnO$_2$/C electrocatalysts using the following Equation (7):

$$\frac{C_{CO_2}}{C_{CH_3COOH} + C_{CH_3CHO}} = \frac{6 \times Q_{CO_2}}{4 \times Q_{CH_3COOH} + 2 \times Q_{CH_3CHO}} \tag{7}$$

where C_{CO_2} and $C_{CH_3COOH} + C_{CH_3CHO}$ represent the charges associated with the total oxidation pathway (C1) and the partial oxidation pathway (C2), respectively.

As can be seen from reactions 1–5, water, or its adsorption residue (adsorbed OH) is involved in EOR and therefore the product distribution strongly depends on the nature of the electrolyte, such as the concentration of ethanol, pH, or the anion [18,63,64]. Camara and Iwasita [55] systematically investigated the effects of ethanol concentration on the product distribution on polycrystalline Pt by FTIR. They found that the C1 pathway is more pronounced at low ethanol concentration (below 0.1 M) with negligible acetaldehyde. When the ethanol concentration was higher than 0.2 M, the formation of CO_2 and acetic acid was inhibited and acetaldehyde was the main product.

Recently the fast development of anion exchange membranes has renewed the interest in the development of alkaline polymer electrolyte fuel cells. In alkaline media, a facile EOR can be achieved less costly with relatively abundant non-Pt metal catalysts. However, it is not clear whether or to what extent the mechanism proposed for EOR on Pt in acidic media can be extended to basic media. By combining electrochemical and spectroscopic techniques (SERS and *in situ* FTIR), Koper's group [18,54,65] showed that the activity of the reaction on Pt electrodes increases significantly when the pH of the electrolyte was higher than 10. Detailed mechanisms were proposed for EOR at low electrolyte pH (<6) and at high electrolyte pH (>11) as shown in Figure 2:

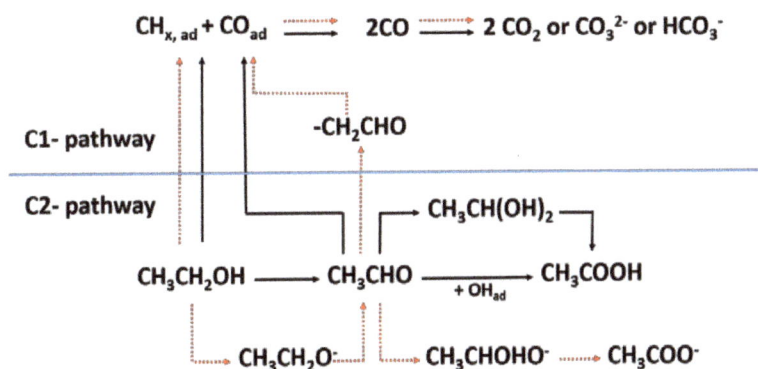

Figure 2. Proposed reaction mechanism for electro-oxidation of ethanol on Pt electrodes. Solid arrows denote the mechanism at low electrolyte pH, while dashed arrows denote the mechanism at high electrolyte pH. Adapted from Reference [18].

On the other hand, based on their *in situ* FTIR spectroscopic results in studying EOR on Pt in alkaline media, Christensen and his co-workers [20] proposed that in contrast to the acidic solutions, under alkaline conditions, the intermediates interact with the surface through O rather than C as shown in Figure 3. $Pt_s–CH_2–C(=O)–O–Pt_s$ was speculated as a new intermediate, and the solution acetate species was the predominant product.

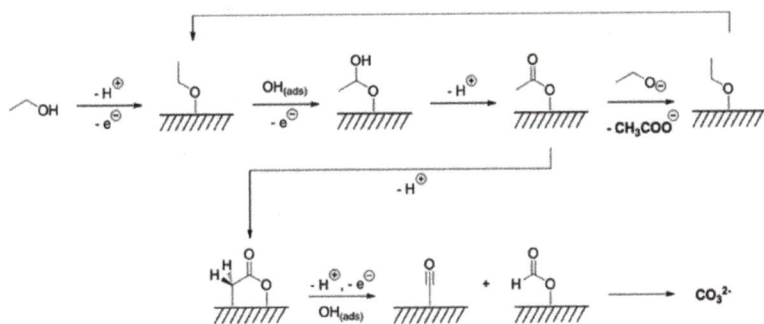

Figure 3. The mechanism of ethanol oxidation at polycrystalline Pt in alkaline solutions proposed by Christensen. Reproduced with permission from Reference [20] Copyright 2012, American Chemical Society.

Pd exhibits a much higher electrocatalytic activity in alkaline media compared to Pt due to its higher oxophilicity and relatively inert nature [17,47,66]. Ethanol oxidation on Pd is dramatically affected by the pH of the solution: virtually no reaction occurs in acidic solutions, while the reaction is fast in alkaline solutions. Through IRRAS studies [19], it was demonstrated that the oxidation of ethanol is incomplete on Pd electrodes and the main product is acetate. In addition, a quantitative FTIR study showed the selectivity for ethanol oxidation to CO_2 is less than 2.5% on Pd in the potential region of -0.60 to 0 V [67], but it is still slightly higher than that of Pt in alkaline media.

Though the external infrared reflection absorption spectroscopy with a thin-layer configuration enables the evaluation of the selectivity of the reaction products (*i.e.*, CO_2, acetate, and acetaldehyde) as a function of the applied potential, it is not sufficiently sensitive to probe low-coverage or weakly adsorbed intermediates and therefore may not provide complete information for understanding the reaction mechanism [68]. In addition the limited mass transport to and from the thin layer can skew or even alter the product distribution [20]. In contrast, attenuated total reflection surface enhanced infrared absorption spectroscopy (ATR-SEIRAS) provides high surface sensitivity and unobstructed mass transport and thus is promising for the complete disclosure of the EOR mechanism [26,69]. Yang *et al.* [23] recently investigated the surface reaction of ethanol on Pd in alkaline media using ATR-SEIRAS and H–D isotope replacement on α-C to shed new light on the self-dissociation and oxidation processes. As illustrated in Figure 4, ethanol may undergo dehydrogenation at α-C to form adsorbed acetyl rather than acetaldehyde, followed by successive decomposition to form C1 species, including CO_{ad} and CH_x at open circuit potential or lower potentials. Moreover, ATR-FTIR was also adopted to clarify the rate limiting steps on Pd thin film electrodes. It was found that at higher potential the subsequent dissociation (C–C bond breaking) of the adsorbed acetyl species is the rate limiting step rather than the formation of adsorbed acetyl [70]. Thus, despite controversy over the details, the reaction pathways for EOR with Pd electrodes in alkaline media are more or less similar to that with Pt electrodes in basic and acidic media.

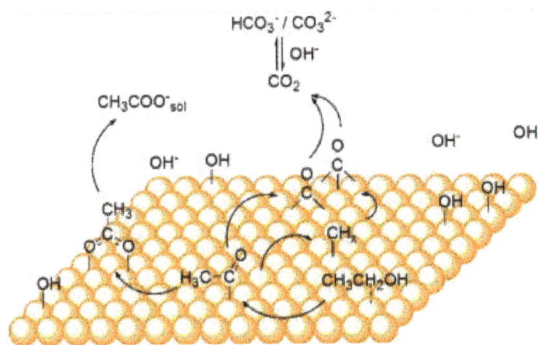

Figure 4. Reaction pathways for interfacial CH_3CH_2OH at Pd electrodes in alkaline media. Reproduced with permission from Reference [23]. Copyright 2014, American Chemical Society.

Despite the fact that external IRRAS has been frequently used in identifying reaction intermediates and products, the thin-layer configuration limits its capability for quantification. First, the thin layer structure is not reproducible from experiment to experiment. Second, product accumulation and diffusion out of the thin layer take place at the same time with species-dependent rates, *i.e.*, the diffusion rate varies with different species in the thin layer. CO_2 is the most volatile among the three major products and diffuses out faster than CH_3COOH and CH_3CHO, resulting in its lower estimation. Complimentary to IR spectroscopy, on-line DEMS can provide accurate quantitative information on ethanol oxidation products. In fact, mass spectrometry has often been used to study EOR in conjunction with IR spectroscopy. Studies using DEMS [28–37], electrochemical thermal desorption mass spectroscopy (ECTDMS) [71], and multipurpose electrochemical mass spectrometry studies (MPEMS) [29] have drawn the conclusion that all mono- or bi-aliphatic alcohols, except tertiary ones, yield minor amounts of CO_2 with the corresponding aldehydes or keto-compounds as major products during their electro-oxidation. Typically, on-line DEMS, especially under well-defined transport and diffusion conditions, has the ability to quantitatively determine kinetic parameters (reaction orders, activation energies, steady-state rates) for the overall EOR, and also for the partial reactions leading to the individual reaction products such as CO_2, acetic acid, and acetaldehyde [16,30–33,35,72].

By combining cyclic voltammetry and potential step measurements of the reaction transients with DEMS, Behm and co-workers performed a thorough investigation on the EOR products on a carbon-supported Pt nanoparticle catalyst at reaction temperature (23–60 °C) [30], and on PtRu and Pt_3Sn catalysts [33]. Absolute rates for CO_2 and acetaldehyde formation were determined via the doubly ionized carbon dioxide at $m/z = 22$ and the CHO^+ fragment at $m/z = 29$ from the calibrated mass spectrometric currents, whereas acetic acid yields were determined indirectly by calculating the difference between the measured Faradaic current and the partial currents of ethanol oxidation to CO_2 and acetaldehyde. More importantly, they convincingly showed that Pt-based catalysts exhibit selectivity towards CO_2 ranging from 0.5%–7.5%, which is far below the selectivity needed for economic implementation of the DEFC technology. They further explored the reaction in a wider temperature range (up to 100 °C) and found that the current efficiency for CO_2 formation increased significantly with the temperature while it decreased with increasing potential [37]. The latter observation suggests that the rate limiting step was changed from CO_{ad} oxidation at lower potential to C–C bond breaking at higher potentials. This transition is reasonable because the cleavage of the C–C bond becomes vital when the surface has the ability to easily remove poison species at higher potentials.

2.2. Theoretical Studies

Over the past decade the theoretical description of surface reactions has undergone a radical development [38,39]. Advances in density functional theory make it now possible to describe catalytic reactions on surfaces with the detail and accuracy required, so that computational results compare favorably with experiments. Simulations and theoretical studies have helped to advance our understanding of the EOR including predictions of vital intermediates, the preferable pathway, or the underlying electron transport process [40–42]. Some of these theoretical works corroborate with works that have been proposed by experimentalists, including the confirmation of some adsorbed intermediates in the step-wise mechanism, which are not detectable, probably due to the limited time resolution of the current experimental techniques. Therefore, theoretical studies may help to reconcile controversies and to understand variations in catalytic activity from one catalyst to another [73].

A higher percentage of C1 pathway is desirable for high efficiency DEFCs, but the production of CO_2 can be as low as 0.5% as shown by DEMS [31]. DFT was used to elucidate the reasons for the low efficiency of EOR on Pt [38] and the study found that a higher percentage of C1-pathway requires a careful control of oxidant surface coverage to allow facile C–C bond cleavage. Ethanol oxidation shows significant structure sensitivity in that the defect sites activate both the O–H and the C–C bonds [74]. The presence of OH or O species will considerably increase the energy barrier of the C–C bond cleavage as shown by DFT calculations, therefore the C1 pathway will be largely reduced compared to the corresponding clean surface. On a clean Pt surface with defects at low applied potentials, and thus low oxidant coverages, the formation of acetic acid and CO/CO_2 are energetically favorable and, interestingly, comparable. This finding is consistent with the experimental observations [18,20,23,66] that CO formation is indeed reasonably facile at low applied potentials on clean Pt surfaces. Increasing the applied potential, results in increasing oxidant surface coverage and leads to a large reduction in the rate of C–C bond cleavage. However, surface oxidants are required for conversion of CO to CO_2. These two competing processes explain the inability of pure platinum catalysts to act as efficient DEFC catalysts.

There has been some debate over whether C–C bond breaking is the rate limiting step in EOR. Very recently Anderson's group calculated the reversible potentials for the reaction intermediates of EOR on Pt (111) [45] using DFT. They found that surface potentials for the path to CO_2 were low and close to the calculated 0.004 V reversible potential for the 12 electron oxidation of ethanol. The main activation energy in the total oxidation of ethanol to CO_2 comes from the formation of OH_{ads} from H_2O with a reversible potential of 0.49 V, the highest potential as shown in Figure 5. The favorable path to CO_2 takes the right hand branch to $OCCH_3$ and then to $OCCH_{ads}$ [45]. OH_{ads} is essential for the oxidation of CO_{ads} and CH_{ads}, which leaves OH_{ads} formation to be the rate limiting step. Accordingly, an ideal catalyst would have the ability to adsorb most intermediates weakly but OH more strongly. This conclusion agrees with the observation that the maximum rate of adsorbed acetyl decomposition into CO_{ads} and $CH_{x, ads}$ appeared at 0.3–0.4 V $vs.$ RHE [70], which is significantly lower than the onset potential of OH_{ads} formation (~0.5 V) [75].

The role of water and hydroxyls during EOR on Pd electrodes in alkaline media was further investigated by Lin's group [39] by acquiring first principle calculations. The possible pathways for the formation of acetate from acetaldehyde were evaluated by comparing the reaction barriers (E_a) as well as thermodynamic (ΔE) and structural parameters. Their results suggested that acetaldehyde is first hydrated in water to form germinal diol, and then the dehydrogenation of germinal diol produces acetate. According to this study, the OH^- anion acts as the center in the concerted-like dehydrogenation path as shown in Figure 6 confirming what has been found in experiments.

The total oxidation current at lower potentials was found to be rather structure sensitive where the presence of steps enhances the rupture of the C–C bond and the complete oxidation to CO_2 [74]. Theoretical studies have identified the platinum monoatomic steps as the most likely sites for full ethanol oxidation and concluded that the close-packed surfaces are unsuitable [76]. Liu's group [43,44] clarified the location of the transition state and saddle points for most surface reactions during EOR on different Pt surfaces based on gradient-corrected DFT as shown in Figure 7. Their results suggest the EOR is a structure-sensitive reaction that is influenced by two key reaction steps: (i) the initial dehydrogenation of ethanol and (ii) the oxidation of acetyl (CH_3CO). By simulating three typical Pt surfaces, namely close-packed Pt (111), monatomic stepped Pt (211), and open Pt (100), these authors demonstrated for the first time that the selectivity of ethanol oxidation on Pt is highly structure sensitive among which Pt (100) is the best surface to fully oxidize ethanol to CO_2 at low coverages. It shows that CO_2 and acetic acid originate from the same surface intermediate *i.e.*, CH_3CO as experimentally evidenced by our group [23], but acetaldehyde is from ethanol directly. The cleavage of the C–C bond occurs through the strongly chemisorbed precursor CH_2CO or $CHCO$ only at low-coordinated surface sites, not from CH_3CO as proposed by FTIR study [13,14]. Acetaldehyde is produced via the one-step concerted dehydrogenation of ethanol, which occurs mainly on close-packed (111), and is enhanced by increased CH_x coverage. Acetic acid is the dominant oxidation product on Pt(111) at oxidative conditions, but its formation is significantly inhibited by the monoatomic steps.

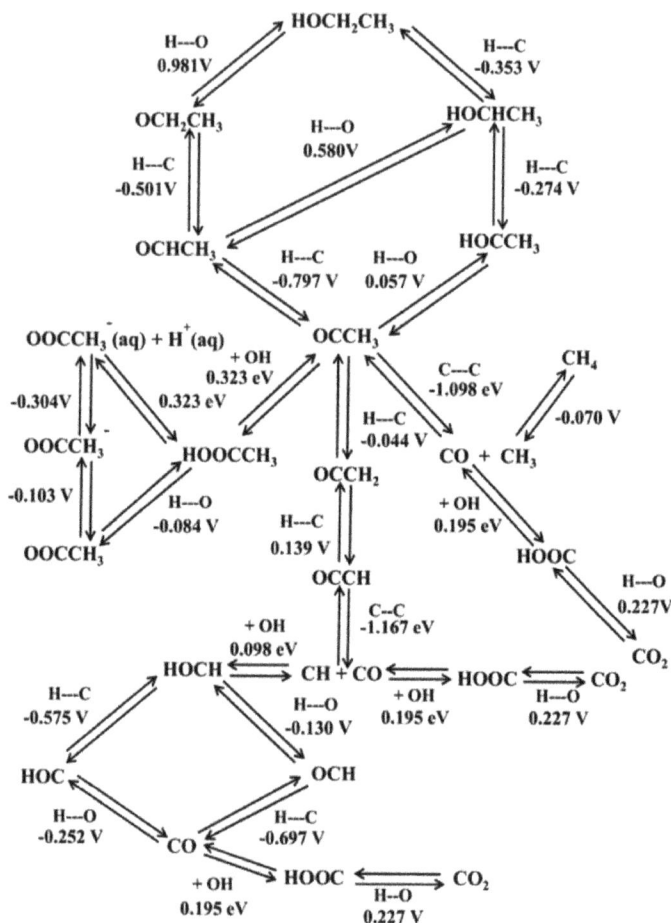

Figure 5. Reaction network calculated for ethanol electro-oxidation to CO_2, methane, and acetate. Reproduced with permission from Reference [45]. Copyright 2015, the Electrochemical Society.

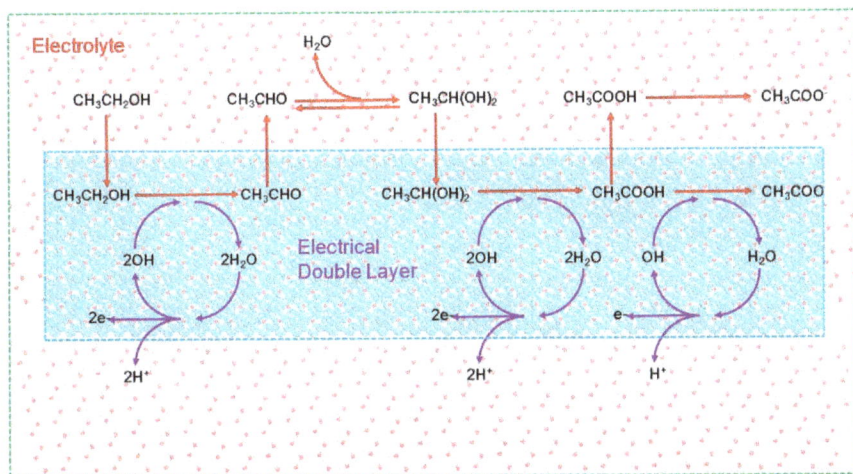

Figure 6. General scheme describing the ethanol oxidation reaction on Pd electrodes in the presence of an electrical double layer proposed by first principle calculations. Reproduced with permission from Reference [39]. Copyright 2014, American Chemical Society.

In summary, EOR is a very complex reaction that can proceed via different pathways. The product distribution and the dominant reaction pathway depend on many factors including temperature, catalyst material and structure, applied potential, ethanol concentration, reaction media, *etc.* [20,30–32,37,45–47]. The interplay of these factors affects EOR greatly. Typically, the adsorption of hydroxyl is vital in both acidic and alkaline media [18,39,40,45–47]. In acidic media, at low potentials C–C bond cleavage occurs readily to form CO [30,43]. Owing to the unavailability of oxidants to remove CO_{ads}, the surface is poisoned and CO_2 production is limited. At higher potentials, there are abundant oxidants, but C–C bond cleavage is inhibited by the high coverage of oxidants, thus leading again to small CO/CO_2 production [45]. Similarly in alkaline media, the dissociative adsorption of ethanol proceeds rather quickly and the rate-determining step is the removal of the adsorbed species, which varies among literature, by the adsorbed hydroxyl. At higher potentials the kinetics is not only affected by the electro-adsorption of OH^- ions, but also by the formation of the inactive surface oxide layer [20,23,39]. Therefore EOR on Pt- or Pd-based catalysts proceeds predominantly with the C2 pathway and acetic acid or acetate is the main product. Carbon dioxide or carbonate is relatively low in the product distribution [3,30,37,57]. In general EOR proceeds through similar reaction pathways but some differences have been discussed. In acidic media, the initial bond breaking step is dehydrogenation at α-carbon, while in strong alkaline media it is the O–H bond cleavage [18,20,43,45].

A

CH_3CH_2OH

$\overset{2.98}{-\!-}\ CH_3+CH_2OH$

$\overset{0.89}{-\!-}\ CH_2CH_2OH$

$\overset{0.88(0.70)}{-\!-\!-}\ CH_3CH_2O$

$\overset{0.64(0.46)}{-\!-\!-}\ CH_3CHOH\ \overset{0.11}{-\!-}\ CH_3CO$

$\overset{0.65(0.40)}{-\!-\!-}\ CH_3CHO$

$CH_3COH\ \overset{0.84}{-\!-}\ CH_2COH$

$\overset{1.68}{-\!-}\ CH_3+COH$

$0.18\ \|\ 0.13$

$\overset{0.31}{\underset{+OH}{-\!-}}\ CH_3COOH\ \overset{0.60}{-\!-}\ CH_3COO$

$\overset{0.96}{-\!-}\ CH_2CO$

$\overset{1.12}{-\!-}\ CH_2CHO\ \overset{1.36}{-\!-}\ CH_3+CO$

$\overset{1.31}{-\!-}\ CH_3+CHO$

$\overset{1.18}{\underset{+OH}{-\!-}}\ CH_2COO+H_2O$

$\overset{1.55}{-\!-}\ CH_2COO$

$\overset{2.16}{-\!-}\ CH_3+CO_2$

B

CH_3CO

$\overset{0.58/0.71}{-\!-}\ CH_2CO$

$\overset{0.70/0.86}{\underset{+OH}{-\!-}}\ CH_3COOH\ \longrightarrow\ CH_3COO$

$\overset{1.02/1.19}{-\!-}\ CH_3+CO$

$\overset{0.57/0.36}{-\!-}\ CHCO$

$\overset{0.66/0.65}{-\!-}\ CH_2+CO$

$\overset{0.93/0.53}{-\!-}\ CH+CO\ \overset{-/\leq0.7}{-\!-}\ CO_2 + O$

$\overset{0.94/1.20}{-\!-}\ CCO$

$\overset{0.98/-}{\underset{+OH}{-\!-}}\ CH_2COO+H_2O$

$\overset{1.62/-}{-\!-}\ CH_2COO$

$\overset{2.77/-}{-\!-}\ CH_3+CO_2$

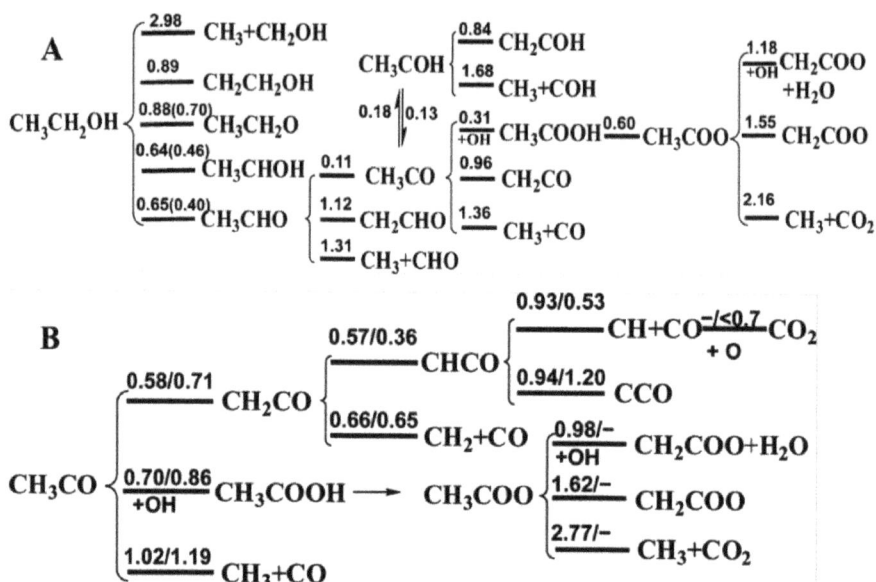

Figure 7. Calculated reaction network and reaction barriers (units, eV) for ethanol oxidation (**A**) on Pt (111) and acetyl oxidation (**B**) on Pt (211) (data on the left) and Pt (100) (data on the right). Reprinted and adapted with permission from Ref. [43]. Copyright 2008, American Chemical Society.

Despite the diverse reaction mechanisms for EOR presented in the literature, some characteristics of a highly efficient catalyst for complete oxidation of ethanol to CO_2 have emerged. The rationally designed catalysts would have: (i) suitable surface sites for C–C bond breaking; (ii) a suitable surface composition to increase selectivity for CO_2 formation; (iii) a bifunctional effect to facilitate the adsorption and activation of water to form OH_{ads} for the removal of CO and -CH_x species [45,76]. Utilizing these design guidelines, researchers have developed many new catalysts with higher selectivity and activity, and longer durability as demonstrated in the following studies.

3. Catalytic Role of the Electrode Materials

As discussed above, both experimental and theoretical studies suggest that the lower electro-efficiency C2 pathway is the dominant for EOR on Pt and Pd surfaces. Therefore a good tactic to obtain a better electrocatalytic performance is to increase the ratio of C1 to C2 pathway to achieve a more complete ethanol oxidation, which as shown above requires active surface sites for C–C bond breaking, CO and –CH_x species removal together with a suitable surface structure to increase selectivity for CO_2 formation [77]. Moreover, C1 pathway always involves the participation of water or its adsorption residue, a good electrocatalyst must be able to activate both ethanol and water adsorption, which can be achieved by varying the composition and structure of the rationally designed catalysts. In the following we first summarize some design principles and then use selected examples to elucidate the applications of these principles to obtain high performance catalysts.

3.1. Principles in Rational Design

Electronic effect: when Pt or Pd is alloyed or modified with another metal, the electronic interaction between Pt or Pd and the other metal results in the changes in their valence electronic structure through ligand effect and strain effect which can be described as the shift in the d-band center (ε_d) as proposed by Norskov and co-workers [50,73,78,79] and reviewed by Demirci [10]. The d-band center directly relates to the binding energy of surface poison or reactive intermediates. A higher-lying ε_d suggests a more reactive surface that tends to bind adsorbates more strongly while a surface with a lower-lying ε_d tends to bind adsorbates more weakly and facilitates the formation of bonds among them. In the EOR mechanism mentioned above, a suitable ε_d with a moderate binding energy of CO, CH_x, and acetyl or acetaldehyde, but higher energy for OH binding is needed by adjusting the electronic effect.

A bifunctional effect originates from Pt-Ru alloys and is extended to Pt/Sn and Pt/SnO_x [1]. In this mechanism, the presence of Ru, Sn, or SnO_x aids in the activation of water dissociation to form surface hydroxides, which can more readily oxidize CO and CH_x intermediates and therefore exhibit relatively higher EOR performance [80]. These alloys also tend to promote the partial oxidation of acetaldehyde to acetic acid. However, catalysts with a bifunctional effect do not particularly enhance C–C bond cleavage during EOR [81].

Surface-structure effect is another important parameter that significantly changes the activity of catalysts. EOR is a surface sensitive reaction and its efficiency largely depends on the crystal orientation of the catalyst surface [48]. In alkaline media Pt(111) electrodes display the highest current and lowest onset oxidation potential, however with little CO_2 production. On the other hand, Pt(100) electrodes are considered to be more active in the breaking of C–C bond and the formation of CO [43,48] no matter whether acidic or alkaline media. Construction of Pt or Pd based catalysts with well-defined morphology and a tunable surface is therefore another way to gain higher electrocatalytic performance catalysts.

3.2. Pt and Pd Based Electrocatalysts

3.2.1. Pt Based Catalysts

Previous studies have shown a monometallic catalyst such as Pt exhibits a selectivity of oxidation of ethanol to CO_2 of 0.5%–7.5% in acidic media [30], which falls short for the commercialization of DEFCs. Pure Pt can easily be poisoned by intermediates (including CO) generated during EOR. Regardless of the media, Pt-based materials represent the benchmark catalysts for ethanol oxidation. Elements such as Ru [32,33,35,60,82,83], Sn [15,32,33,35,80,84–86], Pb [21,87–91], Bi [36,89,90,92,93], Re [86], Sb [93], Ir [94], Au [95], Ce [96], Rh [81,97], Pd [81,98,99], Fe [100], Ni [95,101], P [102], Mo [83,103] have been widely employed to enhance the activity and selectivity. In addition, metal oxides, such as MgO, CeO_2, ZrO_2, SnO_x [76,94,104] RuO_2, PbO_x [88] have also been investigated for facilitating the removal of CO. The enhancement has been attributed to the steric hindrance of the surface, or electronic and/or bifunctional effects when the adatoms are directly involved in the catalytic process.

Up to now, PtRu [32,33,35,60,82,83] and PtSn [15,32,33,35,80,84–86] have been regarded as some of the most efficient catalysts for EOR according to the bifunctional effects. The distribution of products is well-studied by FTIR [35] and DEMS [33,35,55]. The addition of Sn or Ru, though beneficial for the overall activity of EOR, and the partial oxidation of acetaldehyde to acetic acid, does not enhance the activity for C–C breaking [35]. The higher current is mainly contributed from the higher yields of C2 products. These alloys aid in the oxidation of CO but actually lower the total conversion to CO_2 because they slow down the breaking of the C–C bond. Because Pt is the active metal for C–C bond activation, alloying decreases the amount of Pt and its ability to activate the C–C bond.

To further promote selectivity and activity of EOR, Adzic's group developed Pt monolayer (Pt_{ML}) electrocatalysts comprising a one atom thick layer of Pt placed on selected extended or nanoparticle surfaces. They observed a correlation between substrate-induced lateral strains in the Pt monolayer and its activity/selectivity towards EOR. In agreement with previous theories [10,73], a positive- or tensile-surface strain in the metal overlayer tends to upshift ε_d and therefore facilitates OH_{ads} formation resulting in an enhanced EOR as shown in Figure 8. The IRRAS spectra showed that acetic acid is the predominant product [105]. This work demonstrates nicely the importance of electronic effect in tuning electrocatalytic activity.

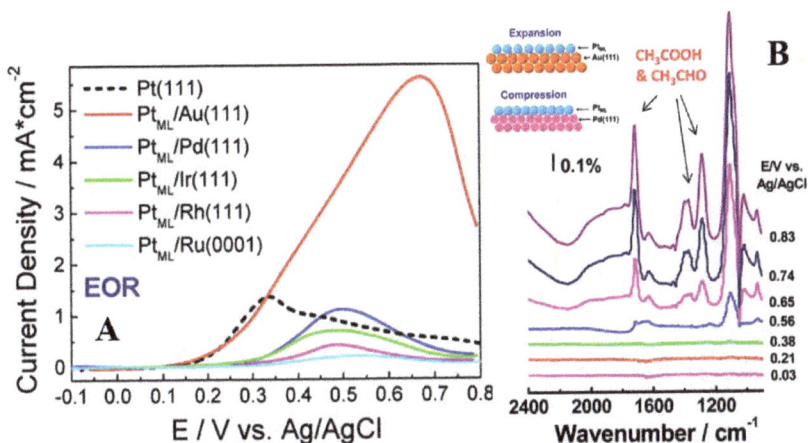

Figure 8. (A) Positive voltammetric scans for Pt(111) and Pt_{ML} supported on five different substrates in 0.1 M $HClO_4$ containing 0.5 M ethanol; (B) *In situ* infrared reflection-absorption spectroscopy (IRRAS) spectra recorded during EOR on the Pt_{ML} / Au(111) electrode in 0.1 M $HClO_4$ containing 0.5 M ethanol. Inserted are models of pseudomorphic monolayers of Pt on two different substrates of Au(111) and Pd(111). Reprinted and adapted with permission from Reference [105]. Copyright 2012, American Chemical Society.

Along this line, various Pt-Au alloys have been proposed and have shown relatively high EOR performance [41,106]. Recently Pt-Au hetero-nanostructures were synthesized by varying the reduction kinetics of a gold precursor to obtain dimer Pt-Au or core-satellite (Pt@Au) structures. These catalysts show high selectivity and improved efficiency in alkaline media compared to their monometallic counterparts [107].

Ternary nanoalloys incorporating suitable metal and metalloid components are expected to exhibit more flexibility in tuning the geometric and electronic properties of Pt surfaces, thus are promising to achieve a higher electrocatalytic performance. Adzic's group found the ternary-electrocatalysts, Pt-Rh-SnO_2/C [9,76] can effectively split the C–C bond in ethanol at room temperature in acidic solutions and the highest activity was obtained with a composition of Pt:Rh:Sn = 3:1:4 [9]. As shown in Figure 9, the integrated band intensities of CO_2 (2343 cm^{-1}), CH_3CHO (933 cm^{-1}), and CH_3COOH (1280 cm^{-1}) for both Pt-Rh–SnO_2/C and Pt–SnO_2/C samples proved the enhanced cleavage of the C–C bond in ethanol and all three constituents Pt, Rh, and SnO_2 are needed to gain the synergistic effect in facilitating the total oxidation of ethanol.

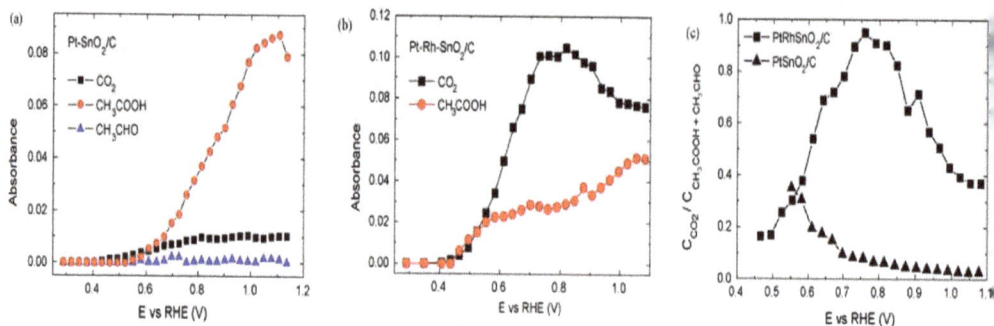

Figure 9. Integrated band intensities of CO_2, CH_3CHO and CH_3COOH in IRRAS spectra from (**a**). Pt–SnO$_2$/C with the atomic ratio Pt:Sn = 3:4; (**b**). Pt–Rh–SnO$_2$/C with the atomic ratio Pt:Rh:Sn = 3:1:4; (**c**) The charge ratio of the total oxidation pathway (C_{CO_2}) over the partial oxidation pathway ($C_{CH_3COOH} + C_{CH_3CHO}$) as a function of electrode potential for both electrocatalysts in 0.1 M HClO$_4$ and 0.1 M ethanol. Reprinted and adapted with permission from Reference [9]. Copyright 2010, Elsevier.

Electrode arrays of 91 combinations of Pt–Sn–M (M = Fe, Ni, Pd, and Ru) were prepared and screened by a fluorescence assay (Figure 10) to optimize the catalysts with the highest electrocatalytic activity by Abruña and co-workers [101]. They found that Fe-containing catalysts exhibited the highest activity followed by Ni- and Pd-containing materials with similar results. This work shows that the variation and combination of different components can exert better electro-activity performance with different electronic effects.

EOR strongly depends on the electronic and surface structures of the catalysts. The different EOR activities observed on Pt surfaces with different crystallographic orientations offer the possibility of optimizing activities of nanoscale practical catalysts by controlling the particle shape [48,49,74]. By using cubic Pt nanoparticles, on which (100) surface sites are predominant, the performance of DEFCs can be increased from 14–24 mW per mg of Pt when compared with cuboctahedral nanoparticles. Moreover, the open circuit potential shifts about 50 mV toward more positive potentials [108]. Pt nanoparticles with 24 high-index facets such as (730), (210), and/or (520) surfaces were synthesized and showed enhancement on EOR compared with commercial Pt/C as well as a higher selectivity for the cleavage of the C–C bond [109].

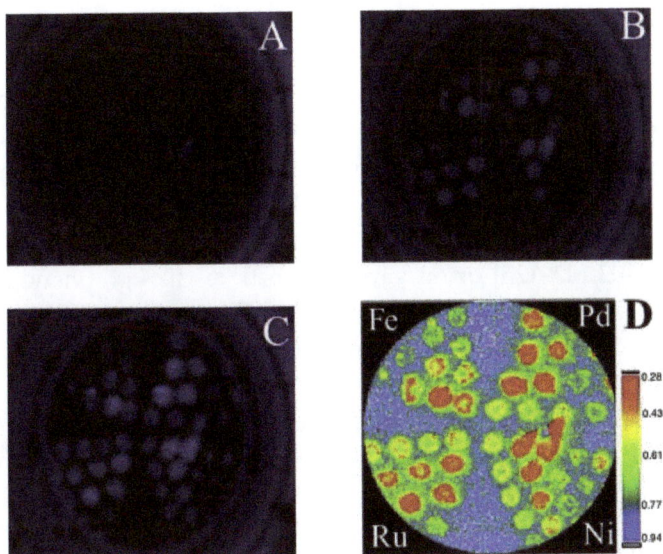

Figure 10. Combinatorial array and screening results by fluorescence imaging of PtSnM electrode arrays in 6.0 M ethanol and quinine fluorescent indicator. Active compositions for ethanol electro-oxidation are shown as bright spots: fluorescence image at (**A**) lower overpotential (~0.27V *vs.* RHE); (**B**) intermediate overpotential (~0.46 V *vs.* RHE) and (**C**) high overpotential (~0.93 V *vs.* RHE); (**D**) Fluorescence onset potential for ethanol electro-oxidation on PtM and PtSnM library. Reproduced with permission from Reference [101]. Copyright 2015, Elsevier.

Bimetallic or ternary nanocrystals have also been studied to explore the optimal combination of higher activity surface facets and electronic effect. The composition-varied (100)-terminated Pt–Pd–Rh nanocubes (NCs) and (111)-terminated Pt–Pd–Rh nanotruncated-oxtahedrons (NTOs) were synthesized with the help of halides [81]. Owing to the synergistic effects resulted from appropriate surface composition and exposed facets, the Pt–Pd–Rh NTOs exhibited the highest selectivity to CO_2 and PtPdRh NCs possessed the best durability. With the help of *in situ* FTIR and DFT calculations, the influence of the exposed facet and surface composition on the capability of C–C bond cleavage was examined. It was found that (100)-bounded surface is favorable to the cleavage of C–C bond while (111)-bounded surface tends to oxidize CO more easily.

3.2.2. Pd Based Catalysts

There has been a surge of interest in developing Pd-based catalysts mainly because facile EOR kinetics are expected in alkaline media on the less costly and more abundant Pd [17,47,66]. However, Pd itself cannot meet the practical demand because of shortage of activity and durability. In particular, a great deal of interest has been focused on the use of Pd-based alloy catalysts as an alternative for EOR where the catalytic activity may be further increased by the addition of a second metal or metal oxide promoters through the effects mentioned above. Various Pd-based catalysts have been synthesized with the addition of one or more elements including Ni [110,111], Ag [112,113], Au [114–116], P [117], Co [118], Sn [80,110,119,120], Ru [53], Zn [121] as well as metal oxides like SnO_2, CeO_x, Co_3O_4, Mn_3O_4, NiO [102,122–125] or on various substrates like graphene [126], carbon microspheres [127], nanowire [128], carbon fiber [118], *etc.*

Sn can promote the catalytic activity of Pd toward EOR by providing oxygen-containing species at more negative potentials due to the bifunctional effect and electronic effects, and has therefore been widely studied [80,119,120]. Teng's group synthesized carbon supported Pd–Sn electrocatalysts with different amounts of Sn, and determined that the optimum Sn content in Pd–Sn for EOR was 14%. The promotional effect of Sn on EOR activity was confirmed by kinetic study and DFT calculations [80]. The reaction energies of the initial steps of EOR (H removal) were studied with several catalysts using DFT (a Pd surface, a Pd–Sn_1, and a Pd–Sn_5 surface) in this work. As shown in Figure 11, Pd–Sn alloy structures resulted in lower reaction energies for the dehydrogenation of ethanol compared to pure Pd. Despite that the DFT calculations in reference [80] only involve the initial steps of EOR, their experimental results support the premise that Pd-Sn may be a better catalyst than Pd for EOR.

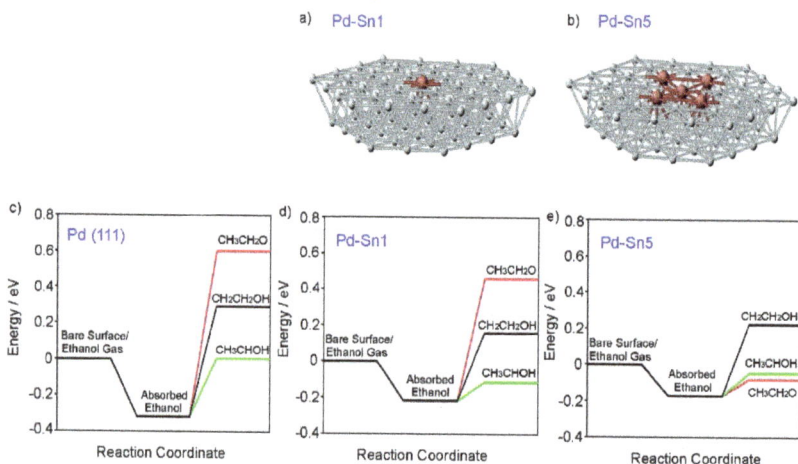

Figure 11. Surface cluster models of (**a**) Pd–Sn_1 and (**b**) Pd–Sn_5 used for modeling ethanol dehydrogenation, and reaction energies for H removal from ethanol over (**c**) Pd (111); (**d**) Pd–Sn_1, (**e**) Pd–Sn_5 surfaces. Reproduced with permission from Reference [80]. Copyright 2012, American Chemical Society.

Au is another promising candidate for promoting EOR on Pd catalysts. Zhao's group [114] synthesized a monolayer or a sub-monolayer of Pd adatoms decorated on Au/C with different Pd:Au atomic ratios in the precursors via a chemical epitaxial growth method and found Pd_1Au_4 shows the highest specific activity due to the electronic effect between the Au support and the Pd decoration, and the enhanced poison resistance. To further tune the electronic and geometric effect of the catalysts, ternary catalysts PdNiAu [116] were synthesized which possessed a peak power density about three times that of the monometallic Pd catalyst, and twice that of the bimetallic PdNi catalyst. A relatively preferable C2 pathway on the Pd–Au–Ni catalyst compared to its single and binary counterparts in alkaline media was estimated by ion chromatography.

Doping of various oxides to Pd catalysts is another strategy to enhance the efficiency of EOR. Xu *et al.* [124] demonstrated that the addition of oxides such as CeO_2, NiO, Co_3O_4, and Mn_3O_4 significantly promoted catalytic activity and stability. Most importantly these oxides have the ability to lower the onset potential for EOR. A recent study showed a maximum energy efficiency of about 7% from room temperature air breathing DEFCs with Pd on a mixed CeO_2–C support as the anode catalysts [125]. The use of a mixed carbon-CeO_2 support extends the stability of the Pd catalyst under working conditions by promoting ethanol electro-oxidation at lower anode potentials.

Nonmetal component can also play a role in the rational design of catalysts [129–132]. Recently, our group discovered that Pd–Ni–P/C ternary nanocatalysts showed a remarkable enhancement towards EOR compared to Pd–Ni/C, Pd–P/C and Pd/C [117]. After a careful analysis of the structure and oxidation state of the catalysts, we found apparently opposite effects of alloying the elements P and Ni. P expands the Pd lattice with the incorporation of P atoms in the interstice of Pd lattice, while alloying with Ni partially replaced Pd sites with smaller Ni atoms, resulting in a contraction of the Pd lattice. In addition, P accepts electrons from the surrounding Pd atoms while Ni donates, resulting in a slight downshift of the Pd d-band center. Alloying Pd with Ni and P creates appropriate electronic and geometric modifications to Pd, leading to a modestly weakened adsorption of intermediates on Pd sites. In addition, the oxophilic nature of Ni provides OH_{ads} at lower potentials which facilitates the removal of surface poisons. (Figure 12).

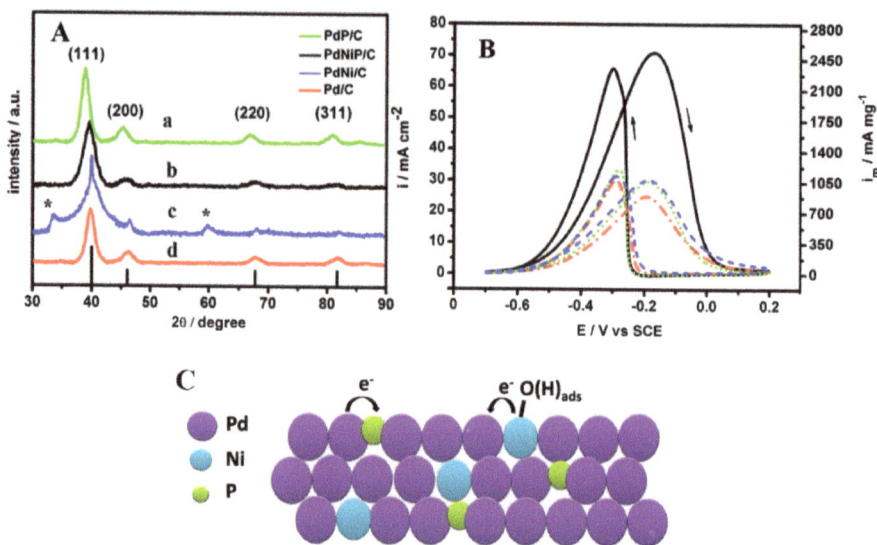

Figure 12. (**A**) XRD patterns of the Pd–P/C (curve a), Pd–Ni–P/C (curve b), Pd–Ni/C (curve c) and Pd/C (curve d). (**B**) Cyclic voltammograms for Pd-based catalysts in 0.5 M NaOH and 1 M C_2H_5OH at 50 mV·s^{-1}; (**C**) Scheme of Pd–Ni–P atomic arrangement. The asterisks at 33.5° and 59.2° mark the peaks from Ni(OH)$_2$ (100) and (110) facets. Reprinted and adapted with permission from Reference [117]. Copyright 2013, Elsevier.

Furthermore, by de-alloying some of the Ni and P in an electrodeposited Pd–Ni–P film, we investigated the de-alloying effect on EOR performance in alkaline media. We found that the enhancement of electrocatalytic activity does not simply originate from the increase of active surface area, but is due to the variation of relative contributions of the two pathways as evidenced by *in situ* infrared spectroscopic results shown in Figure 13 [24]. CO and acetate band intensities (Figure 13B,D) can be used to approximate the relative contributions of the C1 and C2 pathways. Notably, the more pronounced CO$_{ad}$ and acetate bands observed on the de-alloyed Pd–Ni–P film provide molecular spectral evidence supporting the assumption that suitable de-alloying can enhance both C–C cleavage in the C1 pathway with CO formation and production of acetate in the C2 pathway, correlating well with the observed higher EOR current on the de-alloyed film. Furthermore, the relative intensities of the CO bands *versus* the acetate bands appeared to be much higher on the de-alloyed film as compared to those on the as-deposited film, indicating that the C1 pathway is relatively more favorable after the de-alloying treatment.

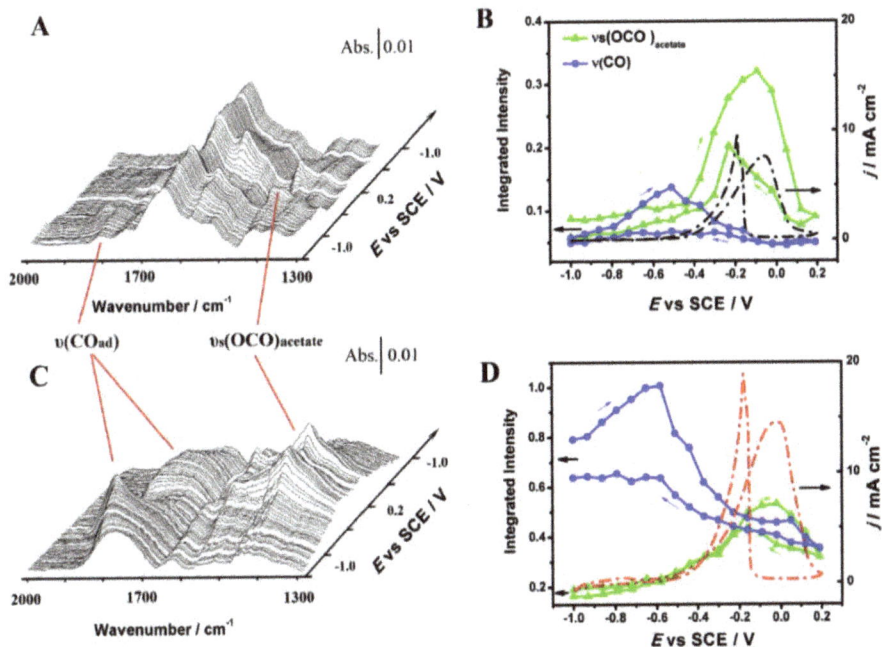

Figure 13. Potentiodynamic ATR-SEIRAS spectra on the as-deposited Pd–Ni–P film (**A**) and the de-alloyed film (**C**) in 0.1 M NaOH + 0.5 M ethanol; Potential-dependent band intensities for $\nu(CO_{ad})$ (blue) and $\nu_{s(OCO)}$ of adsorbed acetate (green) with corresponding CVs recorded at 5 mV·s^{-1} on the as-deposited Pd–Ni–P film (**B**) and the de-alloyed film (**D**) in 0.1 M NaOH + 0.5 M ethanol. Reproduced with permission from Reference [24]. Copyright 2014, Elsevier.

4. Conclusions and Outlook

From the survey of experimental and theoretical studies, some basic principles for rational design of high performance EOR catalysts can be obtained. The catalysts should have (i) active surface sites for C–C bond breaking; (ii) a suitable surface composition to increase selectivity for CO_2 formation; (iii) the ability to facilitate the adsorption and activation of water for the removal of CO and –CH_x species. Utilizing these design guidelines, researchers have developed many new catalysts with higher selectivity, activity, and longer durability. These advances have significantly propelled the development of DEFCs.

Despite this progress, several fundamental and practical issues of catalysts remain to be addressed. From a fundamental point of view, the EOR mechanism is far from solidified. Some key aspects need to be clarified. These include finding the key factors in determining whether EOR goes through the C1 pathway or the C2 pathway; identifying intermediates during the C–C cleavage step in the C1 pathway;

and understanding the nature of the intermediates, especially their adsorption mode on Pt and Pd surfaces. Until these issues are satisfactorily addressed, the rational design of high performance catalysts for DEFCs will remain in its infancy. From a practical point of view, the reported methods for synthesizing high performance catalysts are not suitable for large scale commercial production. We believe efforts in these directions are essential for the further development and deployment of commercially viable DEFCs.

Acknowledgments: This work is supported by the 973 Program (No. 2015CB932303) of MOST and NSFC (Nos. 21473039 and 21273046) to W.-B.C and the US-NSF (CHE 1156425) to SZ.

Author Contributions: Wen-Bin Cai conceived the project. Ye Wang wrote the first draft of the manuscript which was then revised by Shouzhong Zou and Wen-Bin Cai. All the authors researched the literature.

Conflicts of Interest: The authors declare no conflict of interest.

References

1. Antolini, E.; Gonzalez, E.R. Alkaline direct alcohol fuel cells. *J. Power Sources* **2010**, *195*, 3431–3450.
2. Kamarudin, M.Z.F.; Kamarudin, S.K.; Masdar, M.S.; Daud, W.R.W. Review: Direct ethanol fuel cells. *Int. J. Hydrogen Energy* **2013**, *38*, 9438–9453.
3. Teng, X. Anodic Catalyst Design for the Ethanol Oxidation Fuel Cell. Available online: http://www.formatex.info/energymaterialsbook/book/473-484.pdf (accessed on 17 August 2015).
4. Yao, L.X.; Chang, Y.H. Shaping china's energy security: The impact of domestic reforms. *Energy Policy* **2015**, *77*, 131–139.
5. Xuan, J.; Leung, M.K.; Leung, D.Y.; Ni, M. A review of biomass-derived fuel processors for fuel cell systems. *Renew. Sustain. Energy Rev.* **2009**, *13*, 1301–1313.
6. Wee, J.-H. Applications of proton exchange membrane fuel cell systems. *Renew. Sustain. Energy Rev.* **2007**, *11*, 1720–1738.
7. Antolini, E. Catalysts for direct ethanol fuel cells. *J. Power Sources* **2007**, *170*, 1–12.
8. Rao, L.; Jiang, Y.; Zhang, B.; You, L.; Li, Z.; Sun, S. Electrocatalytic oxidation of ethanol. *Prog. Chem.* **2014**, *26*, 727–736.
9. Li, M.; Kowal, A.; Sasaki, K.; Marinkovic, N.; Su, D.; Korach, E.; Liu, P.; Adzic, R.R. Ethanol oxidation on the ternary Pt–Rh–SnO$_2$/C electrocatalysts with varied Pt:Rh:Sn ratios. *Electrochim. Acta* **2010**, *55*, 4331–4338.
10. Demirci, U.B. Theoretical means for searching bimetallic alloys as anode electrocatalysts for direct liquid-feed fuel cells. *J. Power Sources* **2007**, *173*, 11–18.
11. Bianchini, C.; Shen, P.K. Palladium-based electrocatalysts for alcohol oxidation in half cells and in direct alcohol fuel cells. *Chem. Rev.* **2009**, *109*, 4183–4206.
12. Antolini, E. Palladium in fuel cell catalysis. *Energy Environ. Sci.* **2009**, *2*, 915–931.

13. Chang, S.C.; Leung, L.W.H.; Weaver, M.J. Metal crystallinity effects in electrocatalysis as probed by real-time FTIR spectroscopy: Electrooxidation of formic acid, methanol, and ethanol on ordered low-index platinum surfaces. *J. Phys. Chem.* **1990**, *94*, 6013–6021.

14. Hitmi, H.; Belgsir, E.; Léger, J.-M.; Lamy, C.; Lezna, R. A kinetic analysis of the electro-oxidation of ethanol at a platinum electrode in acid medium. *Electrochim. Acta* **1994**, *39*, 407–415.

15. Vigier, F.; Coutanceau, C.; Hahn, F.; Belgsir, E.; Lamy, C. On the mechanism of ethanol electro-oxidation on Pt and PtSn catalysts: Electrochemical and *in situ* IR reflectance spectroscopy studies. *J. Electroanal. Chem.* **2004**, *563*, 81–89.

16. Raskó, J.; Dömök, M.; Baán, K.; Erdőhelyi, A. FTIR and mass spectrometric study of the interaction of ethanol and ethanol-water with oxide-supported platinum catalysts. *Appl. Catal. A* **2006**, *299*, 202–211.

17. Fang, X.; Wang, L.; Shen, P.K.; Cui, G.; Bianchini, C. An *in situ* fourier transform infrared spectroelectrochemical study on ethanol electrooxidation on Pd in alkaline solution. *J. Power Sources* **2010**, *195*, 1375–1378.

18. Lai, S.C.S.; Kleijn, S.E.F.; Ozturk, F.T.Z.; Vellinga, V.C.V.; Koning, J.; Rodriguez, P.; Koper, M.T.M. Effects of electrolyte pH and composition on the ethanol electro-oxidation reaction. *Catal. Today* **2010**, *154*, 92–104.

19. Zhou, Z.Y.; Wang, Q.A.; Lin, J.L.; Tian, N.; Sun, S.G. *In situ* FTIR spectroscopic studies of electrooxidation of ethanol on Pd electrode in alkaline media. *Electrochim. Acta* **2010**, *55*, 7995–7999.

20. Christensen, P.A.; Jones, S.W.M.; Hamnett, A. *In situ* FTIR studies of ethanol oxidation at polycrystalline Pt in alkaline solution. *J. Phys. Chem. C* **2012**, *116*, 26109–26109.

21. Christensen, P.A.; Jones, S.W.; Hamnett, A. An *in situ* FTIR spectroscopic study of the electrochemical oxidation of ethanol at a Pb-modified polycrystalline Pt electrode immersed in aqueous KOH. *Phys. Chem. Chem. Phys.* **2013**, *15*, 17268–17276.

22. Anjos, D.M.; Hahn, F.; Leger, J.M.; Kokoh, K.B.; Tremiliosi, G. *In situ* FTIRS studies of the electrocatalytic oxidation of ethanol on Pt alloy electrodes. *J. Solid State Electrochem.* **2007**, *11*, 1567–1573.

23. Yang, Y.-Y.; Ren, J.; Li, Q.-X.; Zhou, Z.-Y.; Sun, S.-G.; Cai, W.-B. Electrocatalysis of ethanol on a Pd electrode in alkaline media: An *in situ* attenuated total reflection surface-enhanced infrared absorption spectroscopy study. *ACS Catal.* **2014**, *4*, 798–803.

24. Wang, Y.; Jiang, K.; Cai, W.-B. Enhanced electrocatalysis of ethanol on dealloyed Pd–Ni–P film in alkaline media: An infrared spectroelectrochemical investigation. *Electrochim. Acta* **2015**, *162*, 100–107.

25. Leung, L.W.H.; Chang, S.C.; Weaver, M.J. Real-time FTIR spectroscopy as an electrochemical mechanistic probe—Electrooxidation of ethanol and related species on well-defined Pt(111) surfaces. *J. Electroanal. Chem.* **1989**, *266*, 317–336.

26. Shao, M.H.; Adzic, R.R. Electrooxidation of ethanol on a Pt electrode in acid solutions: *In situ* ATR-SEIRAS study. *Electrochim. Acta* **2005**, *50*, 2415–2422.

27. Willsau, J.; Heitbaum, J. Elementary steps of ethanol oxidation on Pt in sulfuric acid as evidenced by isotope labelling. *J. Electroanal. Chem. Interface* **1985**, *194*, 27–35.

245

28. Iwasita, T.; Pastor, E. A DEMS and FTIR spectroscopic investigation of adsorbed ethanol on polycrystalline platinum. *Electrochim. Acta* **1994**, *39*, 531–537.

29. Wang, J.; Wasmus, S.; Savinell, R. Evaluation of ethanol, 1-propanol, and 2-propanol in a direct oxidation polymer-electrolyte fuel cell a real-time mass spectrometry study. *J. Electrochem. Soc.* **1995**, *142*, 4218–4224.

30. Wang, H.; Jusys, Z.; Behm, R.J. Ethanol electrooxidation on a carbon-supported Pt catalyst: Reaction kinetics and product yields. *J. Phys. Chem. B* **2004**, *108*, 19413–19424.

31. Wang, H.; Jusys, Z.; Behm, R.J. Ethanol and acetaldehyde adsorption on a carbon-supported Pt catalyst: A comparative DEMS study. *Fuel Cells* **2004**, *4*, 113–125.

32. Colmenares, L.; Wang, H.; Jusys, Z.; Jiang, L.; Yan, S.; Sun, G.Q.; Behm, R.J. Ethanol oxidation on novel, carbon supported Pt alloy catalysts-model studies under defined diffusion conditions. *Electrochim. Acta* **2006**, *52*, 221–233.

33. Wang, H.; Jusys, Z.; Behm, R. Ethanol electro-oxidation on carbon-supported Pt, PtRu and Pt₃Sn catalysts: A quantitative DEMS study. *J. Power Sources* **2006**, *154*, 351–359.

34. Rao, V.; Cremers, C.; Stimming, U. Investigation of the ethanol electro-oxidation in alkaline membrane electrode assembly by differential electrochemical mass spectrometry. *Fuel Cells* **2007**, *7*, 417–423.

35. Wang, Q.; Sun, G.Q.; Jiang, L.H.; Xin, Q.; Sun, S.G.; Jiang, Y.X.; Chen, S.P.; Jusys, Z.; Behm, R.J. Adsorption and oxidation of ethanol on colloid-based Pt/C, PtRu/C and Pt₃Sn/C catalysts: *In situ* FTIR spectroscopy and on-line DEMS studies. *Phys. Chem. Chem. Phys.* **2007**, *9*, 2686–2696.

36. Figueiredo, M.C.; Aran-Ais, R.M.; Feliu, J.M.; Kontturi, K.; Kallio, T. Pt catalysts modified with Bi: Enhancement of the catalytic activity for alcohol oxidation in alkaline media. *J. Catal.* **2014**, *312*, 78–86.

37. Sun, S.; Halseid, M.C.; Heinen, M.; Jusys, Z.; Behm, R.J. Ethanol electrooxidation on a carbon-supported Pt catalyst at elevated temperature and pressure: A high-temperature/high-pressure DEMS study. *J. Power Sources* **2009**, *190*, 2–13.

38. Kavanagh, R.; Cao, X.M.; Lin, W.F.; Hardacre, C.; Hu, P. Origin of low CO₂ selectivity on platinum in the direct ethanol fuel cell. *Angew. Chem. Int. Ed. Engl.* **2012**, *51*, 1572–1575.

39. Sheng, T.; Lin, W.F.; Hardacre, C.; Hu, P. Role of water and adsorbed hydroxyls on ethanol electrochemistry on Pd: New mechanism, active centers, and energetics for direct ethanol fuel cell running in alkaline medium. *J. Phys. Chem. C* **2014**, *118*, 5762–5772.

40. Sheng, T.; Lin, W.-F.; Hardacre, C.; Hu, P. Significance of β-dehydrogenation in ethanol electro-oxidation on platinum doped with Ru, Rh, Pd, Os and Ir. *Phys. Chem. Chem. Phys.* **2014**, *16*, 13248–13254.

41. Neurock, M. First-Principles Modeling for the Electro-Oxidation of Small Molecules. In *Handbook of Fuel Cells*; Vielstich, W.G., Lamm, A., Yokokawa, H., Eds.; John Wiley & Sons, Ltd.: Hoboken, NJ, USA, 2010.

42. Cui, G.F.; Song, S.Q.; Shen, P.K.; Kowal, A.; Bianchini, C. First-principles considerations on catalytic activity of Pd toward ethanol oxidation. *J. Phys. Chem. C* **2009**, *113*, 15639–15642.

43. Wang, H.F.; Liu, Z.P. Comprehensive mechanism and structure-sensitivity of ethanol oxidation on platinum: New transition-state searching method for resolving the complex reaction network. *J. Am. Chem. Soc.* **2008**, *130*, 10996–11004.

44. Wang, H.F.; Liu, Z.P. Selectivity of direct ethanol fuel cell dictated by a unique partial oxidation channel. *J. Phys. Chem. C* **2007**, *111*, 12157–12160.

45. Asiri, H.A.; Anderson, A.B. Mechanisms for ethanol electrooxidation on Pt(111) and adsorption bond strengths defining an ideal catalyst. *J. Electrochem. Soc.* **2015**, *162*, F115–F122.

46. Zhiani, M.; Majidi, S.; Rostami, H.; Taghiabadi, M.M. Comparative study of aliphatic alcohols electrooxidation on zero-valent palladium complex for direct alcohol fuel cells. *Int. J. Hydrogen Energy* **2015**, *40*, 568–576.

47. Liang, Z.X.; Zhao, T.S.; Xu, J.B.; Zhu, L.D. Mechanism study of the ethanol oxidation reaction on palladium in alkaline media. *Electrochim. Acta* **2009**, *54*, 2203–2208.

48. Buso-Rogero, C.; Herrero, E.; Feliu, J.M. Ethanol oxidation on Pt single-crystal electrodes: Surface-structure effects in alkaline medium. *Chemphyschem* **2014**, *15*, 2019–2028.

49. Zhou, W.J.; Li, M.; Zhang, L.; Chan, S.H. Supported PtAu catalysts with different nano-structures for ethanol electrooxidation. *Electrochim. Acta* **2014**, *123*, 233–239.

50. Hammer, B.; Nørskov, J.K. Theoretical Surface Science and Catalysis-Calculations and Concepts. *Adv. Catal.* **2000**, *45*, 71–129.

51. Srinivasan, S.; Dave, B.B.; Murugesamoorthi, K.A.; Parthasarathy, A.; Appleby, A.J. Overview of Fuel Cell Technology. In *Fuel cell Systems*; Plenum Press: New York, NY, USA, 1993.

52. Lamy, C.; Belgsir, E.; Leger, J. Electrocatalytic oxidation of aliphatic alcohols: Application to the direct alcohol fuel cell (DAFC). *J. Appl. Electrochem.* **2001**, *31*, 799–809.

53. Zhou, W.; Zhou, Z.; Song, S.; Li, W.; Sun, G.; Tsiakaras, P.; Xin, Q. Pt based anode catalysts for direct ethanol fuel cells. *Appl. Catal. B* **2003**, *46*, 273–285.

54. Lai, S.C.S.; Koper, M.T.M. Ethanol electro-oxidation on platinum in alkaline media. *Phys. Chem. Chem. Phys.* **2009**, *11*, 10446–10456.

55. Camara, G.A.; Iwasita, T. Parallel pathways of ethanol oxidation: The effect of ethanol concentration. *J. Electroanal. Chem.* **2005**, *578*, 315–321.

56. Belgsir, E.M.; Bouhier, E.; Yei, H.E.; Kokoh, K.B.; Beden, B.; Huser, H.; Leger, J.M.; Lamy, C. Electrosynthesis in aqueous medium: A kinetic study of the electrocatalytic oxidation of oxygenated organic molecules. *Electrochim. Acta* **1991**, *36*, 1157–1164.

57. Tarnowski, D.J.; Korzeniewski, C. Effects of surface step density on the electrochemical oxidation of ethanol to acetic acid. *J. Phys. Chem. B* **1997**, *101*, 253–258.

58. Kutz, R.B.; Braunschweig, B.; Mukherjee, P.; Dlott, D.D.; Wieckowski, A. Study of ethanol electrooxidation in alkaline electrolytes with isotope labels and sum-frequency generation. *J. Phys. Chem. Lett.* **2011**, *2*, 2236–2240.

59. Ke, X.; Deng, L.L.; Shen, P.K.; Cu, G.F. Electrochemical quartz crystal microbalance (EQCM) characterization of electrodeposition and catalytic activity of Pd-based electrocatalysts for ethanol oxidation. *Chem. Res. Chin. Univ.* **2010**, *26*, 443–448.

60. Huang, L.; Sorte, E.; Sun, S.-G.; Tong, Y.Y.J. A straightforward implementation of *in situ* solution electrochemical ^{13}C NMR spectroscopy for studying reactions on commercial electrocatalysts: Ethanol oxidation. *Chem. Commun.* **2015**, *51*, 8086–8088.

61. Leung, L.W.H.; Weaver, M.J. Real-time FTIR spectroscopy as a quantitative kinetic probe of competing electrooxidation pathways of small organic molecules. *J. Phys. Chem.* **1988**, *92*, 4019–4022.

62. Gao, P.; Chang, S.C.; Zhou, Z.H.; Weaver, M.J. Electrooxidation pathways of simple alcohols at platinum in pure nonaqueous and concentrated aqueous environments as studied by real-time FTIR spectroscopy. *J. Electroanal. Chem.* **1989**, *272*, 161–178.

63. Buso-Rogero, C.; Grozovski, V.; Vidal-Iglesias, F.J.; Solla-Gullon, J.; Herrero, E.; Feliu, J.M. Surface structure and anion effects in the oxidation of ethanol on platinum nanoparticles. *J. Mater. Chem. A* **2013**, *1*, 7068–7076.

64. Paulino, M.E.; Nunes, L.M.; Gonzalez, E.R.; Tremiliosi-Filho, G. *In situ* FTIR spectroscopic study of ethanol oxidation on Pt (111)/Rh/Sn surface: The anion effect. *Electrochem. Commun.* **2015**, *52*, 85–88.

65. Lai, S.C.; Koper, M.T. Electro-oxidation of ethanol and acetaldehyde on platinum single-crystal electrodes. *Faraday Discuss.* **2009**, *140*, 399–416.

66. Ma, L.; Chu, D.; Chen, R. Comparison of ethanol electro-oxidation on Pt/C and Pd/C catalysts in alkaline media. *Int. J. Hydrogen Energy* **2012**, *37*, 11185–11194.

67. Zhou, Z.-Y.; Chen, D.-J.; Li, H.; Wang, Q.; Sun, S.-G. Electrooxidation of dimethoxymethane on a platinum electrode in acidic solutions studied by *in situ* FTIR spectroscopy. *J. Phys. Chem. C* **2008**, *112*, 19012–19017.

68. Osawa, M. Surface-Enhanced Infrared Absorption Spectroscopy. In *Handbook of Vibrational Spectroscopy*; Osawa, M.C., Griffiths, P.R., Eds.; Wiley: New York, NY, USA; Chichester, UK, 2002; Volume 1, pp. 785–799.

69. Yang, Y.Y.; Zhang, H.X.; Cai, W.B. Recent experimental progresses on electrochemical ATR-SEIRAS. *J. Electrochem.* **2013**, *19*, 6–16.

70. Heinen, M.; Jusys, Z.; Behm, R.J. Ethanol, acetaldehyde and acetic acid adsorption/electrooxidation on a Pt thin film electrode under continuous electrolyte flow: An *in situ* ATR-FTIRs flow cell study. *J. Phys. Chem. C* **2010**, *114*, 9850–9864.

71. Bittins-Cattaneo, B.; Wilhelm, S.; Cattaneo, E.; Buschmann, H.W.; Vielstich, W. Intermediates and products of ethanol oxidation on platinum in acid solution. *Ber. Bunsenges. Phys. Chem.* **1988**, *92*, 1210–1218.

72. Lai, S.C.; Koper, M.T. The influence of surface structure on selectivity in the ethanol electro-oxidation reaction on platinum. *J. Phys. Chem. Lett.* **2010**, *1*, 1122–1125.

73. Norskov, J.K.; Bligaard, T.; Rossmeisl, J.; Christensen, C.H. Towards the computational design of solid catalysts. *Nat. Chem.* **2009**, *1*, 37–46.

74. Da Silva, S.G.; Silva, J.C.M.; Buzzo, G.S.; de Souza, R.F.B.; Spinace, E.V.; Neto, A.O.; Assumpcao, M.H.M.T. Electrochemical and fuel cell evaluation of PtAu/C electrocatalysts for ethanol electro-oxidation in alkaline media. *Int. J. Hydrogen Energy* **2014**, *39*, 10121–10127.

75. Climent, V.; Gómez, R.; Orts, J.M.; Feliu, J.M. Thermodynamic analysis of the temperature dependence of oh adsorption on Pt(111) and Pt(100) electrodes in acidic media in the absence of specific anion adsorption. *J. Phys. Chem. B* **2006**, *110*, 11344–11351.

76. Kowal, A.; Li, M.; Shao, M.; Sasaki, K.; Vukmirovic, M.; Zhang, J.; Marinkovic, N.; Liu, P.; Frenkel, A.; Adzic, R. Ternary Pt/Rh/SnO₂ electrocatalysts for oxidizing ethanol to CO₂. *Nat. Mater.* **2009**, *8*, 325–330.

77. Assumpção, M.; Nandenha, J.; Buzzo, G.; Silva, J.; Spinacé, E.; Neto, A.; de Souza, R. The effect of ethanol concentration on the direct ethanol fuel cell performance and products distribution: A study using a single fuel cell/attenuated total reflectance-fourier transform infrared spectroscopy. *J. Power Sources* **2014**, *253*, 392–396.

78. Greeley, J.; Norskov, J.K.; Mavrikakis, M. Electronic structure and catalysis on metal surfaces. *Annu. Rev. Phys. Chem.* **2002**, *53*, 319–348.

79. Rabis, A.; Rodriguez, P.; Schmidt, T.J. Electrocatalysis for polymer electrolyte fuel cells: Recent achievements and future challenges. *ACS Catal.* **2012**, *2*, 864–890.

80. Du, W.X.; Mackenzie, K.E.; Milano, D.F.; Deskins, N.A.; Su, D.; Teng, X.W. Palladium-Tin alloyed catalysts for the ethanol oxidation reaction in an alkaline medium. *ACS Catal.* **2012**, *2*, 287–297.

81. Zhu, W.; Ke, J.; Wang, S.-B.; Ren, J.; Wang, H.-H.; Zhou, Z.-Y.; Si, R.; Zhang, Y.-W.; Yan, C.-H. Shaping single-crystalline trimetallic Pt–Pd–Rh nanocrystals toward high-efficiency C–C splitting of ethanol in conversion to CO₂. *ACS Catal.* **2015**, *5*, 1995–2008.

82. Camara, G.A.; de Lima, R.B.; Iwasita, T. Catalysis of ethanol electrooxidation by PtRu: The influence of catalyst composition. *Electrochem. Commun.* **2004**, *6*, 812–815.

83. Neto, A.O.; Giz, M.J.; Perez, J.; Ticianelli, E.A.; Gonzalez, E.R. The electro-oxidation of ethanol on Pt–Ru and Pt–Mo particles supported on high-surface-area carbon. *J. Electrochem. Soc.* **2002**, *149*, A272–A279.

84. Lamy, C.; Rousseau, S.; Belgsir, E.M.; Coutanceau, C.; Leger, J.M. Recent progress in the direct ethanol fuel cell: Development of new platinum-tin electrocatalysts. *Electrochim. Acta* **2004**, *49*, 3901–3908.

85. Zhou, W.J.; Song, S.Q.; Li, W.Z.; Zhou, Z.H.; Sun, G.Q.; Xin, Q.; Douvartzides, S.; Tsiakaras, P. Direct ethanol fuel cells based on PtSn anodes: The effect of Sn content on the fuel cell performance. *J. Power Sources* **2005**, *140*, 50–58.

86. Tayal, J.; Rawat, B.; Basu, S. Effect of addition of rhenium to Pt-based anode catalysts in electro-oxidation of ethanol in direct ethanol PEM fuel cell. *Int. J. Hydrogen Energy* **2012**, *37*, 4597–4605.

87. He, Q.G.; Shyam, B.; Macounova, K.; Krtil, P.; Ramaker, D.; Mukerjee, S. Dramatically enhanced cleavage of the C–C bond using an electrocatalytically coupled reaction. *J. Am. Chem. Soc.* **2012**, *134*, 8655–8661.

88. Suffredini, H.B.; Salazar-Banda, G.R.; Avaca, L.A. Enhanced ethanol oxidation on PbO_x-containing electrode materials for fuel cell applications. *J. Power Sources* **2007**, *171*, 355–362.

89. Huang, Y.; Cai, J.; Liu, M.; Guo, Y. Fabrication of a novel PtPbBi/C catalyst for ethanol electro-oxidation in alkaline medium. *Electrochim. Acta* **2012**, *83*, 1–6.

90. Matsumoto, F. Ethanol and methanol oxidation activity of PtPb, PtBi, and PtBi$_2$ intermetallic compounds in alkaline media. *Electrochemistry* **2012**, *80*, 132–138.

91. Gunji, T.; Tanabe, T.; Jeevagan, A.J.; Usui, S.; Tsuda, T.; Kaneko, S.; Saravanan, G.; Abe, H.; Matsumoto, F. Facile route for the preparation of ordered intermetallic Pt$_3$Pb–PtPb core-shell nanoparticles and its enhanced activity for alkaline methanol and ethanol oxidation. *J. Power Sources* **2015**, *273*, 990–998.

92. Tusi, M.M.; Polanco, N.S.; da Silva, S.G.; Spinacé, E.V.; Neto, A.O. The high activity of PtBi/C electrocatalysts for ethanol electro-oxidation in alkaline medium. *Electrochem. Commun.* **2011**, *13*, 143–146.

93. Figueiredo, M.C.; Santasalo-Aarnio, A.; Vidal-Iglesias, F.J.; Solla-Gullon, J.; Feliu, J.M.; Kontturi, K.; Kallio, T. Tailoring properties of platinum supported catalysts by irreversible adsorbed adatoms toward ethanol oxidation for direct ethanol fuel cells. *Appl Catal. B* **2013**, *140*, 378–385.

94. Li, M.; Cullen, D.A.; Sasaki, K.; Marinkovic, N.S.; More, K.; Adzic, R.R. Ternary electrocatalysts for oxidizing ethanol to carbon dioxide: Making Ir capable of splitting C–C bond. *J. Am. Chem. Soc.* **2013**, *135*, 132–141.

95. Dutta, A.; Ouyang, J.Y. Ternary niaupt nanoparticles on reduced graphene oxide as catalysts toward the electrochemical oxidation reaction of ethanol. *ACS Catal.* **2015**, *5*, 1371–1380.

96. Jacob, J.M.; Corradini, P.G.; Antolini, E.; Santos, N.A.; Perez, J. Electro-oxidation of ethanol on ternary Pt–Sn–Ce/C catalysts. *Appl. Catal. B* **2015**, *165*, 176–184.

97. Zhao, Y.H.; Wang, R.Y.; Han, Z.X.; Li, C.Y.; Wang, Y.S.; Chi, B.; Li, J.Q.; Wang, X.J. Electrooxidation of methanol and ethanol in acidic medium using a platinum electrode modified with lanthanum-doped tantalum oxide film. *Electrochim. Acta* **2015**, *151*, 544–551.

98. Datta, J.; Dutta, A.; Mukherjee, S. The beneficial role of the cometals Pd and Au in the carbon-supported PtPdAu catalyst toward promoting ethanol oxidation kinetics in alkaline fuel cells: Temperature effect and reaction mechanism. *J. Phys. Chem. C* **2011**, *115*, 15324–15334.

99. Yang, X.; Yang, Q.; Xu, J.; Lee, C.-S. Bimetallic PtPd nanoparticles on nafion-graphene film as catalyst for ethanol electro-oxidation. *J. Mater. Chem.* **2012**, *22*, 8057–8062.

100. Liao, F.L.; Lo, T.W.B.; Sexton, D.; Qu, J.; Wu, C.T.; Tsang, S.C.E. PdFe nanoparticles as selective catalysts for C–C cleavage in hydrogenolysis of vicinal diol units in biomass-derived chemicals. *Catal. Sci. Technol.* **2015**, *5*, 887–896.

101. Almeida, T.S.; van Wassen, A.R.; van Dover, R.B.; de Andrade, A.R.; Abruña, H.D. Combinatorial PtSnM (M = Fe, Ni, Ru and Pd) nanoparticle catalyst library toward ethanol electrooxidation. *J. Power Sources* **2015**, *284*, 623–630.

102. Carmo, M.; Sekol, R.C.; Ding, S.Y.; Kumar, G.; Schroers, J.; Taylor, A.D. Bulk metallic glass nanowire architecture for electrochemical applications. *ACS Nano* **2011**, *5*, 2979–2983.

103. Li, L.; Yuan, X.X.; Xia, X.Y.; Du, J.; Ma, Z.; Ma, Z.F. Effects of Mo doping on properties of Pt/C as catalyst towards electro-oxidation of ethanol. *J. Inorg. Mater.* **2014**, *29*, 1044–1048.

104. Jiang, L.; Colmenares, L.; Jusys, Z.; Sun, G.Q.; Behm, R.J. Ethanol electrooxidation on novel carbon supported Pt/SnO$_x$/C catalysts with varied Pt:Sn ratio. *Electrochim. Acta* **2007**, *53*, 377–389.

105. Li, M.; Liu, P.; Adzic, R.R. Platinum monolayer electrocatalysts for anodic oxidation of alcohols. *J. Phys. Chem. Lett.* **2012**, *3*, 3480–3485.

106. Cheng, F.L.; Dai, X.C.; Wang, H.; Jiang, S.P.; Zhang, M.; Xu, C.W. Synergistic effect of Pd-Au bimetallic surfaces in Au-covered Pd nanowires studied for ethanol oxidation. *Electrochim. Acta* **2010**, *55*, 2295–2298.

107. Mourdikoudis, S.; Chirea, M.; Zanaga, D.; Altantzis, T.; Mitrakas, M.; Bals, S.; Liz-Marzán, L.M.; Perez-Juste, J.; Pastoriza-Santos, I. Governing the morphology of Pt–Au heteronanocrystals with improved electrocatalytic performance. *Nanoscale* **2015**, *7*, 8739–8747.

108. Figueiredo, M.C.; Solla-Gullón, J.; Vidal-Iglesias, F.J.; Nisula, M.; Feliu, J.M.; Kallio, T. Carbon-supported shape-controlled Pt nanoparticle electrocatalysts for direct alcohol fuel cells. *Electrochem. Commun.* **2015**, *55*, 47–50.

109. Tian, N.; Zhou, Z.-Y.; Sun, S.-G.; Ding, Y.; Wang, Z.L. Synthesis of tetrahexahedral platinum nanocrystals with high-index facets and high electro-oxidation activity. *Science* **2007**, *316*, 732–735.

110. Sheikh, A.; Silva, E.; Moares, L.; Antonini, L.; Abellah, M.Y.; Malfatti, C. Pd-based catalysts for ethanol oxidation in alkaline electrolyte. *Am. J. Min. Metal.* **2014**, *2*, 64–69.

111. Kumar, K.S.; Haridoss, P.; Seshadri, S.K. Synthesis and characterization of electrodeposited Ni–Pd alloy electrodes for methanol oxidation. *Surf. Coat. Technol.* **2008**, *202*, 1764–1770.

112. Hosseini-Sarvari, M.; Khanivar, A.; Moeini, F. Magnetically recoverable nano Pd/Fe$_3$O$_4$/ZnO catalyst: Preparation, characterization, and application for the synthesis of 2-oxazolines and benzoxazoles. *J. Mater. Sci.* **2015**, *50*, 3065–3074.

113. Li, G.; Jiang, L.; Jiang, Q.; Wang, S.; Sun, G. Preparation and characterization of Pd$_x$Ag$_y$/C electrocatalysts for ethanol electrooxidation reaction in alkaline media. *Electrochim. Acta* **2011**, *56*, 7703–7711.

114. Zhu, L.D.; Zhao, T.S.; Xu, J.B.; Liang, Z.X. Preparation and characterization of carbon-supported sub-monolayer palladium decorated gold nanoparticles for the electro-oxidation of ethanol in alkaline media. *J. Power Sources* **2009**, *187*, 80–84.

115. Wang, H.; Xu, C.W.; Cheng, F.L.; Jiang, S.P. Pd nanowire arrays as electrocatalysts for ethanol electrooxidation. *Electrochem. Commun.* **2007**, *9*, 1212–1216.

116. Dutta, A.; Datta, J. Outstanding catalyst performance of pdauni nanoparticles for the anodic reaction in an alkaline direct ethanol (with anion-exchange membrane) fuel cell. *J. Phys. Chem. C* **2012**, *116*, 25677–25688.

117. Wang, Y.; Shi, F.F.; Yang, Y.Y.; Cai, W.B. Carbon supported Pd–Ni–P nanoalloy as an efficient catalyst for ethanol electro-oxidation in alkaline media. *J. Power Sources* **2013**, *243*, 369–373.

118. Bahemmat, S.; Ghassemzadeh, M.; Afsharpour, M.; Harms, K. Synthesis, characterization and crystal structure of a Pd(ii) complex containing a new bis-1,2,4-triazole ligand: A new precursor for the preparation of Pd(0) nanoparticles. *Polyhedron* **2015**, *89*, 196–202.

119. Da Silva, S.G.; Assumpcao, M.H.M.T.; Silva, J.C.M.; de Souza, R.F.B.; Spinace, E.V.; Neto, A.O.; Buzzo, G.S. PdSn/C electrocatalysts with different atomic ratios for ethanol electro-oxidation in alkaline media. *Int. J. Electrochem. Sci.* **2014**, *9*, 5416–5424.

120. Mao, H.M.; Wang, L.L.; Zhu, P.P.; Xu, Q.J.; Li, Q.X. Carbon-supported PdSn SnO$_2$ catalyst for ethanol electro-oxidation in alkaline media. *Int. J. Hydrogen Energy* **2014**, *39*, 17583–17588.

121. Bambagioni, V.; Bianchini, C.; Filippi, J.; Oberhauser, W.; Marchionni, A.; Vizza, F.; Psaro, R.; Sordelli, L.; Foresti, M.L.; Innocenti, M. Ethanol oxidation on electrocatalysts obtained by spontaneous deposition of palladium onto nickel-zinc materials. *ChemSusChem* **2009**, *2*, 99–112.

122. Takeguchi, T.; Anzai, Y.; Kikuchi, R.; Eguchi, K.; Ueda, W. Preparation and characterization of CO-tolerant Pt and Pd anodes modified with SnO$_2$ nanoparticles for PEFC. *J. Electrochem. Soc.* **2007**, *154*, B1132–B1137.

123. Mao, H.; Huang, T.; Yu, A.-S. Facile synthesis of trimetallic Cu$_1$Au$_{0.15}$Pd$_{1.5}$/C catalyst for ethanol oxidation with superior activity and stability. *J. Mater. Chem. A* **2014**, *2*, 16378–16380.

124. Shen, P.K.; Xu, C. Alcohol oxidation on nanocrystalline oxide Pd/C promoted electrocatalysts. *Electrochem. Commun.* **2006**, *8*, 184–188.

125. Wang, L.; Lavacchi, A.; Bevilacqua, M.; Bellini, M.; Fornasiero, P.; Filippi, J.; Innocenti, M.; Marchionni, A.; Miller, H.A.; Vizza, F. Energy efficiency of alkaline direct ethanol fuel cells employing nanostructured palladium electrocatalysts. *Chemcatchem* **2015**, *7*, 2214–2221.

126. Wang, Y.; Wang, X.M.; Wang, Y.Z.; Li, J.P. Acid-treatment-assisted synthesis of Pt–Sn/graphene catalysts and their enhanced ethanol electro-catalytic activity. *Int. J. Hydrogen Energy* **2015**, *40*, 990–997.

127. Xu, C.; Cheng, L.; Shen, P.; Liu, Y. Methanol and ethanol electrooxidation on Pt and Pd supported on carbon microspheres in alkaline media. *Electrochem. Commun.* **2007**, *9*, 997–1001.

128. Wang, A.L.; He, X.J.; Lu, X.F.; Xu, H.; Tong, Y.X.; Li, G.R. Palladium-Cobalt nanotube arrays supported on carbon fiber cloth as high-performance flexible electrocatalysts for ethanol oxidation. *Angew. Chem. Int. Ed.* **2015**, *54*, 3669–3673.

129. Zhao, M.; Abe, K.; Yamaura, S.-I.; Yamamoto, Y.; Asao, N. Fabrication of Pd–Ni–P metallic glass nanoparticles and their application as highly durable catalysts in methanol electro-oxidation. *Chem. Mater.* **2014**, *26*, 1056–1061.

130. Jiang, R.Z.; Tran, D.T.; McClure, J.P.; Chu, D. A class of (Pd–Ni–P) electrocatalysts for the ethanol oxidation reaction in alkaline media. *ACS Catal.* **2014**, *4*, 2577–2586.

131. Shao, A.F.; Wang, Z.B.; Chu, Y.Y.; Jiang, Z.Z.; Yin, G.P.; Liu, Y. Evaluation of the performance of carbon supported Pt–Ru–Ni–P as anode catalyst for methanol electrooxidation. *Fuel Cells* **2010**, *10*, 472–477.

132. Lo, Y.L.; Hwang, B.J. Kinetics of ethanol oxidation on electroless Ni–P/SnO$_2$/Ti electrodes in KOH solutions. *J. Electrochem. Soc.* **1995**, *142*, 445–450.

Highly Active Non-PGM Catalysts Prepared from Metal Organic Frameworks

Heather M. Barkholtz, Lina Chong, Zachary B. Kaiser, Tao Xu and Di-Jia Liu

Abstract: Finding inexpensive alternatives to platinum group metals (PGMs) is essential for reducing the cost of proton exchange membrane fuel cells (PEMFCs). Numerous materials have been investigated as potential replacements of Pt, of which the transition metal and nitrogen-doped carbon composites ($TM/N_x/C$) prepared from iron doped zeolitic imidazolate frameworks (ZIFs) are among the most active ones in catalyzing the oxygen reduction reaction based on recent studies. In this report, we demonstrate that the catalytic activity of ZIF-based $TM/N_x/C$ composites can be substantially improved through optimization of synthesis and post-treatment processing conditions. Ultimately, oxygen reduction reaction (ORR) electrocatalytic activity must be demonstrated in membrane-electrode assemblies (MEAs) of fuel cells. The process of preparing MEAs using ZIF-based non-PGM electrocatalysts involves many additional factors which may influence the overall catalytic activity at the fuel cell level. Evaluation of parameters such as catalyst loading and perfluorosulfonic acid ionomer to catalyst ratio were optimized. Our overall efforts to optimize both the catalyst and MEA construction process have yielded impressive ORR activity when tested in a fuel cell system.

Reprinted from *Catalysts*. Cite as: Barkholtz, H.M.; Chong, L.; Kaiser, Z.B.; Xu, T.; Liu, D.-J. Highly Active Non-PGM Catalysts Prepared from Metal Organic Frameworks. *Catalysts* **2015**, *5*, 955–965.

1. Introduction

Polymer electrolyte membrane fuel cells are the future powertrain for automotive applications due to their high power density, relatively quick start-up, high efficiency, and emission of only water from the vehicle [1–5]. However, the cathodic oxygen reduction reaction (ORR) kinetics are significantly slower than the anodic hydrogen oxidation reaction [6] therefore requiring more catalyst. To date, the preferred electrocatalysts are platinum or platinum group metals (PGMs), which contribute a significant fraction to the overall fuel cell stack cost [5]. In order for widespread commercialization to take place, technological advancements in cathode catalysts are required, including decreased cost while increasing performance and durability.

The discovery of ORR activity in cobalt phthalocyanine [7] nearly 50 years ago inspired the search for non-PGM ORR electrocatalysts using nitrogen group chelated transition metal complexes, such as metallated phthalocyanines, porphyrins and their analogues, as the precursors of preparing transition metal/nitrogen/carbon ($TM/N_x/C$) catalysts activated via pyrolysis. These precursors typically have square-planar configuration and frequently require high surface area substrates such as amorphous carbons onto which they are supported and dispersed. Using an inert support dilutes the volumetric and gravimetric densities of the possible active sites, limiting the potential of producing highly efficient catalyst. At Argonne National Laboratory, we pioneered the use of zeolitic imidazolate frameworks (ZIFs), a subclass of metal-organic frameworks (MOFs), as precursor templates to prepare ORR electroactive catalysts [8]. MOFs are porous materials comprised organic ligands coordinated to transition metal ions [9,10]. Recently, research surrounding the chemistry and diverse applications of MOF-based materials has experienced tremendous growth [11–14]. For instance, Proietti *et al.* demonstrated that the addition of iron precursors with a Zn-ZIF can give superior fuel cell performance [15]. Since then, several binary ZIF or transition metal doped ZIF systems have been explored as precursors for highly active ORR electrocatalysts [16–19]. For example, we reported an all solid-state one-pot synthesis technique to prepare ZIF based electrocatalysts that demonstrated impressive ORR activity measured by both rotating disk electrode (RDE) and membrane electrode assembly (MEA) in a fuel cell [17].

Our recent study suggests that the catalytic performance of ZIF-based catalysts is very sensitive to the synthesis and processing conditions. In this report, we describe the impact of electrocatalyst processing and MEA fabrication conditions on the overall ORR activity when tested in a single cell fuel cell. We will also discuss how different iron additives could change the fuel cell performance while keeping all other processing parameters the same. Additionally, the influence to fuel cell performance by the addition of small amount of carbon black at different steps of the electrocatalyst synthesis process was investigated. Finally, the weight ratio of Nafion$^{®}$ ionomer to catalyst was optimized in the MEA fabrication process. Overall, an impressive current density of 221.9 mA cm^{-2} at 0.8 V was achieved.

2. Results and Discussion

2.1. Influence of Catalyst Activity by Fe Complex in Precursor

The general procedures for "one-pot" synthesis have been published previously in detail [17]. As a continuation of this study, we investigated the influence of iron precursor and its chelating chemistry to the solid state synthesis and the resulting catalyst performance. In this report, three combinations of iron complexes and organic ligand were applied. They were iron (II) acetate ($Fe(Ac)_2$), tris-1,10-phenanthroline iron (II) perchlorate (TPI) and a combination of iron (II) acetate and 1,10-phenanthroline (Phen) with molar ratio 1:6 ($Fe(Ac)_2(Phen)_6$). For

ZIF-8, we used its molecular formula as $Zn(mIm)_2$. The catalyst samples were labeled based on their iron complex and ligand combinations in the $Zn(mIm)_2$ precursor synthesis before the thermal activation. For example, $Zn(mIm)_2Fe(Ac)_2$, represents adding iron acetate directly into mixture for solid state synthesis for $Zn(mIm)_2$. Similarly, $Zn(mIm)_2TPI$ and $Zn(mIm)_2Fe(Ac)_2(Phen)_6$ represent adding TPI and $Fe(Ac)_2$/Phen at 1/6 in the mixtures for solid state synthesis for $Zn(mIm)_2$, respectively. All electrocatalysts underwent normal thermal activation and were fabricated into the cathode of MEAs. The current-voltage polarization and power density curves of the fuel cell tests are shown in Figure 1a,b, respectively. Several key catalyst performance parameters in MEA/fuel cell tests are given by Table 1. It is easily seen that the addition of TPI gave the best overall fuel cell performance. Replacing TPI with $Fe(Ac)_2$ resulted in a 2.5 fold decrease in power density as well as a 3.4 fold decrease in the current density at 0.8 V. Adding in 1,10-phenanthroline as well as iron (II) acetate improved the performance from $Fe(Ac)_2$ but was not equivalent to TPI.

Table 1. Brunauer-Emmett-Teller (BET) Surface area and Activity with Different Iron Additives. BET surface area reported on autoclaved samples before any further processing was performed. Fuel cell data is reported for all three samples, including current density at 0.8 $V_{iR-free}$, limiting current, and maximum power density. Fuel cell conditions: $P_{O_2} = P_{H_2} = 1$ bar (back pressure = 7.3 psig) fully humidified; $T = 80\,°C$; N211 membrane; 5 cm^2 membrane-electrode assemblies (MEA); cathode catalyst = 3.5–4 mg/cm^2, anode catalyst = 0.4 mg_{Pt}/cm^2.

Sample	BET Surface Area ($m^2\,g^{-1}$)	Current Density at 0.8 V (mA cm^{-2})	Limiting Current (A cm^{-2})	Power Density (mW cm^{-2})
$Zn(mIm)_2Fe(Ac)_2$	264.8	64.7	0.883	241.5
$Zn(mIm)_2Fe(Ac)_2(Phen)_6$	702.2	136.9	1.57	441.9
$Zn(mIm)_2TPI$	859.3	221.9	1.86	603.3

We speculate that the Fe-ligand coordination strength plays a very important role in overall electrocatalytic performance. If the iron precursor is able to undergo metal ion exchange with ZIF-8, previous studies have shown that there will be a decrease in fuel cell performance [20]. During the one-pot synthesis, $Fe(Ac)_2$ could be readily dissolved into the liquefied imidazole and water (produced through reaction with zinc oxide) into Fe^{2+} and Ac^-. The ionic iron(II) could be incorporated into ZIF framework in the place of Zn^{2+}. On the other hand, Fe^{2+} is tightly chelated by six nitrogens from three phenanthroline in an octahedral direction to form a propeller-shaped configuration and will not readily undergo ion exchange with Zn in $Zn(mIm)_2$ [20]. With significant amounts of Phen present in the iron acetate, imidazole and zinc oxide mixture, it is possible that a certain fraction of Fe^{2+} could ligate with Phen through iron (II)–N bond, as suggested by the previous study [15]. It is difficult, however, to expect that Fe^{2+} will react exclusively with Phen in the presence of excessive amount of imidazole. The consumption of a fraction of available iron (II) to form a Fe-Phen intermediate while the other Fe^{2+} ions chelate with imidazoles as part of ZIF framework may explain the trend seen in Figure 1. Additionally, it is generally accepted that the surface area is an important factor determining fuel cell performance [21]. The Brunauer-Emmett-Teller (BET) surface areas of all three autoclaved catalyst precursors are also given in Table 1. As is shown by the Table, there is a direct correlation between the surface area of the starting material and the overall fuel cell performance. Iron additives in the form of $Fe(Ac)_2$ produced the lowest surface area electrocatalysts therefore the lowest activity, among all three catalysts studied. When Phen was added in addition to $Fe(Ac)_2$, there was a 62.3% increase in precursor surface area, corresponding to a 52.7% increase in the current density at 0.8 $V_{iR\text{-free}}$.

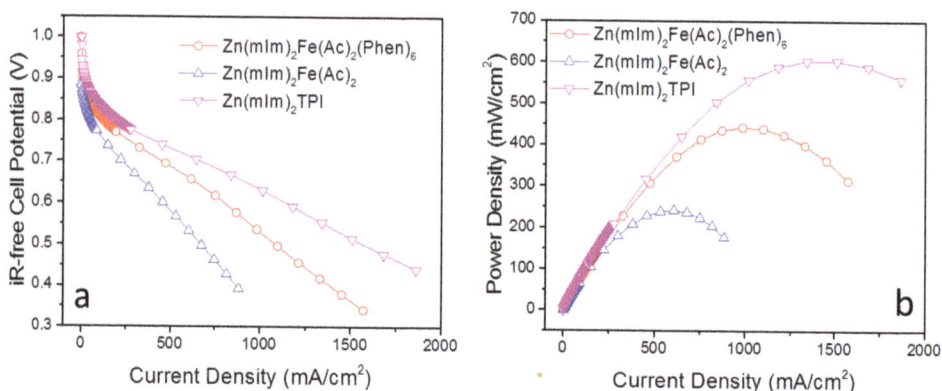

Figure 1. (a) Polarization curves from different iron additive electrocatalysts. The current densities at 0.8 $V_{iR\text{-free}}$ for $Zn(mIm)_2Fe(Ac)_2$ = 64.7 mA cm^{-2}, $Zn(mIm)_2Fe(Ac)_2(Phen)_6$ = 136.9 mA cm^{-2}, and $Zn(mIm)_2TPI$ = 221.9 mA cm^{-2}. (b) Power density curves corresponding to polarization curves shown in (a). Maximum power density for $Zn(mIm)_2Fe(Ac)_2$ = 241.5 mW cm^{-2}, $Zn(mIm)_2Fe(Ac)_2(Phen)_6$ = 441.9 mW cm^{-2}, and $Zn(mIm)_2TPI$ = 603.3 mW cm^{-2}. Conditions: P_{O_2} = P_{H_2} = 1 bar (back pressure = 7.3 psig) fully humidified; T = 80 °C; N211 membrane; 5 cm^2 MEA; cathode catalyst = 3.5–4 mg/cm^2, anode catalyst = 0.4 mg$_{Pt}$/cm^2.

2.2. Influence of Adding Carbon Black to MEA Performance

High temperature activation of the ZIF-based precursor converts individual imidazole molecules to graphitic carbon. Such a process often generates incomplete conversion [22], resulting in low electro-conductance of the catalyst and subsequent MEA formed [22]. Mixing a small quantity of carbon black into the heat-activated catalyst could potentially mitigate such an impedance issue. On the other hand, addition of catalytically inert carbon will dilute the ZIF-based catalyst density and therefore could cause a loss of volumetric and areal specific activities. In this experiment, we selected Ketjen Black as the carbon additive, which is a frequently used carbon support for fuel cell catalysts [23–26]. Two samples were prepared with 10 wt % Ketjen Black EC-300J added at different steps of the electrocatalyst synthesis process. For the first sample, Ketjen Black was added during the ball milling step before thermal activation had occurred. The sample was labeled Zn(mIm)$_2$TPI-10-BM. The second sample was prepared wherein Ketjen Black was added after the thermal activation and acid wash. The sample was labeled Zn(mIm)$_2$TPI-10-AW.

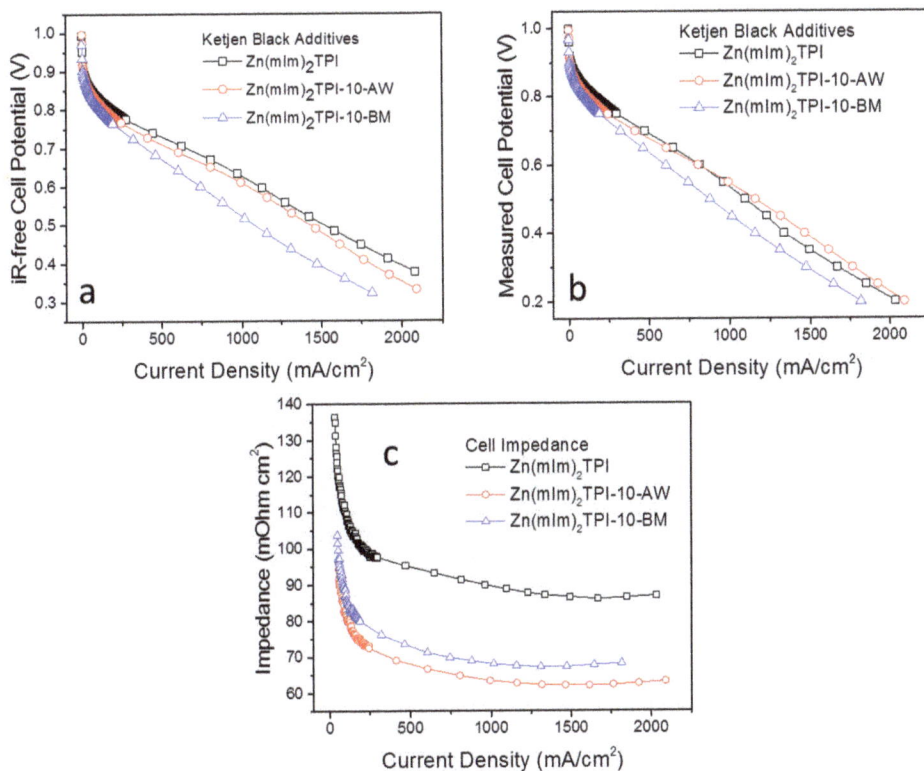

Figure 2. (a) Impedance corrected polarization curves from Ketjen Black additives and a control sample without any Ketjen Black. (b) Polarization curved without impedance correction from Ketjen Black additives and a control sample without any Ketjen Black. (c) Cell impedance curves corresponding to polarization curves shown in (a) and (b). Conditions: $P_{O_2} = P_{H_2} = 1$ bar (back pressure = 7.3 psig) fully humidified; $T = 80$ °C; N211 membrane; 5 cm^2 MEA; cathode catalyst = 3.5–4 mg/cm^2, anode catalyst = 0.4 mg$_{Pt}$/cm^2.

Figure 2a shows the *iR*-free current-voltage polarizations of MEAs with Zn(mIm)$_2$TPI-10-BM and Zn(mIm)$_2$TPI-10-AW as the cathode catalysts, and a MEA without Ketjen Black additive as a benchmark; measured values (without impedance correction) are displayed in Figure 2b. As expected, the cell impedance was reduced by about ~20% for both catalysts containing Ketjen Black additive compared to the benchmark, as is shown in Figure 2c, with Zn(mIm)$_2$TPI-10-AW having the lowest overall impedance. However, the internal impedance corrected (*iR*-free) polarizations (Figure 2a) show that Zn(mIm)$_2$TPI has the highest fuel cell specific current density among all the samples, suggesting that addition of Ketjen Black causes the dilution of the active sites thereby reducing the overall activity even at a mere 10 wt % loading. Between Zn(mIm)$_2$TPI-10-BM and Zn(mIm)$_2$TPI-10-AW, the latter demonstrates better specific current density at any given cell voltage. Even though both samples contain same amount of Ketjen Black, the amorphous carbon in Zn(mIm)$_2$TPI-10-BM had been subjected to a heat treatment at >1000 °C. Under such temperature, the carbon black would undergo the phase transformation from amorphous to graphitic carbon, accompanied by loss of porosity, volume, and surface area. In contrast, Ketjen Black in Zn(mIm)$_2$TPI-10-AW was added after the high temperature treatment but before the ammonia treatment at 750 °C. The carbon black was still highly porous and more reactive with NH$_3$ for additional active site formation compared to that in Zn(mIm)$_2$TPI-10-BM. Higher catalytic activity is therefore expected. For practical application, the current-voltage polarization without impedance correction represents actual fuel cell performance. Such polarization embodies the balance between the electrode catalyst activity and electron/proton conductivities. To this point, the fuel cell with Zn(mIm)$_2$TPI-10-AW at the cathode offers slightly less desirable performance than Zn(mIm)$_2$TPI at $V > 0.6$ V and slightly better current density at $V < 0.6$ V, see Figure 2b. The cell with Zn(mIm)$_2$TPI-10-BM, however, was inferior over the entire operating region.

2.3. MEA Fabrication Optimization

Although the rotating disk electrode (RDE) is the most commonly used method for activity measurements, the electrode catalyst ultimately needs to be measured at MEA/single fuel cell levels since they mostly resemble the operating conditions in a commercial fuel cell stack. Some of the crucial catalyst performance attributes, such as available catalyst area, mass/charge transport efficiencies, triple-phase boundary exposure, *etc.*, will not be properly measured by RDE even though they play extremely important roles in controlling the fuel cell current and power densities. One of the key process parameters for MEA fabrication is the weight ratio between the Nafion® ionomer and the catalyst. Correct balance between the use and intermixing of ionomer and catalyst could result in optimal exposure of the catalytic active sites and effective proton/electron transfers as well as oxygen and water transports to and from the catalytic sites. This is, in fact, particularly important for our ZIF-derived non-PGM catalysts since their active sites are believed to be uniformly decorated throughout the catalyst surface. We have conducted a series of experiments in MEA optimization by preparing the catalytic ink containing different amount of ionomer and catalyst. The preparation of a cathode catalyst ink includes mixing 5 wt % Nafion®, isopropyl alcohol, water, and thermally activated catalyst. In this experiment, three dry weight ratios between ionomer to the catalyst were applied, from 0.5:1 to 0.9:1.

When optimizing ionomer to catalyst ratios for the ink, it was demonstrated by Figure 3 that 0.7:1 was the optimal ratio to producing the best performing MEA tested in fuel cell. Also noted is that when the ionomer to catalyst ratio was decreased below 0.7:1 to 0.5:1, there was a more discernible drop of cell voltage in both ohmic and limiting-current regions, presumably due to the lack of ion transport caused by insufficient Nafion®. When the ionomer to catalyst ratio was increased to 0.9:1, there was also an 18.4% decrease in single cell performance at 0.8 $V_{iR-free}$. This decrease in fuel cell performance may be attributed to excess Nafion® present, clogging the electrocatalyst pores and the electrochemically active surface area, preventing the access of the gas phase oxygen.

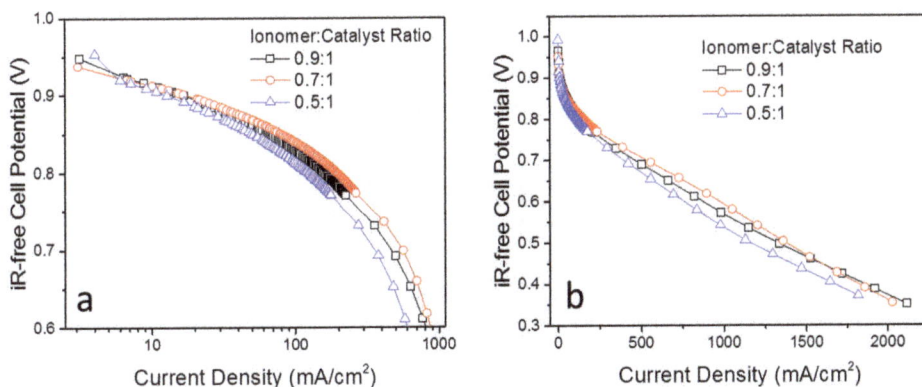

Figure 3. (a) Tafel plots from samples prepared with different ionomer to catalyst ratios. (b) Polarization curved corresponding to the Tafel plots shown in (a). The current density at 0.8 V for 0.9:1 = 158.4 mA cm^{-2}, 0.7:1 = 194.0 mA cm^{-2}, and 0.5:1 = 123.5 mA cm^{-2}. Conditions: P_{O_2} = P_{H_2} = 1 bar (back pressure = 7.3 psig) fully humidified; T = 80 °C; N211 membrane; 5 cm^2 MEA; cathode catalyst = 3.5–4 mg/cm^2, anode catalyst = 0.4 mg$_{Pt}$/cm^2.

3. Experimental Section

3.1. Materials and Methods

Commercially available reagents were used as received without further purification. Ball milling was carried out with a Retsch PM 100 planetary ball mill (Haan, Germany). Fuel cell test stand measurements were carried out on a Scribner 850e fuel cell test stand station (Southern Pines, NC, USA) using a Poco graphite blocks single cell with serpentine flow channels and a geometric electrode surface area of 5 cm^2 (Fuel Cell Technology, Albuquerque, NM, USA). The single cell test conditions are given individually under each figure.

3.2. One-Pot Synthesis of ZIF-Based Electrocatalyst Precursor

The one-pot synthesis of the ZIF-based electrocatalyst precursor was carried out according to previously reported procedures [17,27]. Preparation of Zn(mIm)$_2$TPI was carried out by adding 2-methylimidazole (668.2 mg, 8 mmol, Aldrich, St. Louis, MO, USA), ZnO (323.5 mg, 4 mmol, Aldrich), and 0.022 mol % iron in the form of 1,10-phenanthroline iron (II) perchlorate (70.5 mg, 0.09 mmol, Aldrich) together, grinding, and sealing in an autoclave under an Ar atmosphere. The autoclave was heated to 180 °C and held for 18 h. A pink powder was obtained and used as described below.

When performing the comparative study, the iron precursor was changed while keeping all other reagents the same as well as the molar percentage of iron added. Two other catalysts were prepared, Zn(mIm)$_2$Fe(Ac)$_2$, and Zn(mIm)$_2$Fe(Ac)$_2$(Phen)$_6$. Using Zn(mIm)$_2$Fe(Ac)$_2$(Phen)$_6$ as an example, the catalyst was prepared by adding 2-methylimidazole (668.2 mg, 8 mmol, Aldrich), ZnO (323.5 mg, 4 mmol, Aldrich), iron (II) acetate (15.6 mg, 0.09 mmol, Aldrich), and 1, 10-phenanthroline (97.3 mg, 0.05 mmol, Aldrich) together, grinding, and sealing in an autoclave under Ar atmosphere. Heating profiles were not changed from what is described above.

3.3. Preparation of Electrocatalyst

Pyrolysis of ZIF-based electrocatalyst precursor was carried out by placing about 250 mg of ball milled Zn(mIm)$_2$TPI into a ceramic boat and inserting it into a quartz tube (2 inch diameter). The tube was sealed and purged with Ar for one hour before it was heated to 1050 °C and held for one hour under flowing Ar atmosphere. The pyrolyzed sample was then acid washed in 0.5 M H$_2$SO$_4$ via sonication for 30 min then continuously agitated for 16 h at room temperature. The sample was then washed with water until neutral and dried in a vacuum oven at 40 °C. The sample was then pyrolyzed again at 750 °C for 30 min under flowing NH$_3$ atmosphere to give the final product.

When electrocatalysts with Ketjen Black additive were prepared, 10 wt % Ketjen Black was added during either at the ball mill step or after the acid wash and before NH_3 thermal activation. As an example, the preparation is given below for $Zn(mIm)_2TPI$-10-AW, where Ketjen Black (4 mg, EC-300J, Azko Nobel, Willowbrook, IL, USA) was added to the electrocatalyst (33.1 mg) after acid wash but before NH_3 thermal activation. The sample was dispersed in isopropyl alcohol via sonication for 30 min the dried in a vacuum oven. Once dry, the prepared $Zn(mIm)_2TPI$-10-AW sample was thermally activated at 750 °C for 30 min under NH_3 flow per usual.

3.4. Single Fuel Cell Test

3.4.1. Preparation of Cathode

An ink solution was prepared for the molar ratio of 0.7:1 ionomer to catalyst by combining $Zn(mIm)_2TPI$ (25 mg) with Nafion® (350 mg, 5 wt. % solution, Aldrich), isopropyl alcohol (890 µL), and water (350 µL). The ink solution was sonicated for 30 min and constantly agitated for 16 h at room temperature before it was painted onto carbon paper (5 cm^2, Fuel Cells Etc., Sigracet 25 BC, College Station, TX, USA) to create the cathode, which was dried in a vacuum oven at 40 °C for two hours. The catalyst loading for all the tests was approximately 3.5–4.0 mg cm^{-2}.

3.4.2. Preparation of Anode

An ink solution was prepared by combining Pt/C (10 mg, 40 wt. % of Pt, E-TEK, Somerset, NJ, USA), Nafion® (80 mg, 5 wt. % solution, Aldrich), isopropyl alcohol (205 µL), and water (80 µL). This ink solution was solicited for 30 min and constantly agitated for 16 h at room temperature before it was painted onto carbon paper (5 cm^2, Fuel Cells Etc., Sigracet 25 BC) to create the anode, which was dried in a vacuum oven at 40 °C for 2 h. The Pt loading for all tests was approximately 0.4 mg_{Pt}/cm^2.

3.4.3. Preparation of Membrane Electrode Assembly

The prepared cathode and anode were hot pressed on either side of a Nafion® N211 membrane (DuPont, New Castle, DE, USA) at 120 °C for 30 s using a pressure of 5.4×10^6 Pa, the pressure was then increased to 1.1×10^7 Pa and held for an additional 30 s. Pressure values were calculated assuming that the load is evenly applied to the 5 cm^2 electrode.

3.4.4. Fuel Cell Activity Test

Fuel cell activity tests were carried out by placing the prepared MEA into a single cell. Data was gathered by a Scribner 850e fuel cell test stand. A polarization curve was recorded by scanning from open circuit potential (OCV) to 750 mV at 10 mA s^{-1} then from 750 mV to 200 mV at 50 mV s^{-1}. The area current density,

I_A, was recorded directly from the polarization measurement. Cell impedance was measured by the current interrupt method installed in the fuel cell test stand.

4. Conclusions

Our study shows that significant improvements in cathodic electrocatalyst performance can be achieved when synthesis, processing, and fabrication parameters are optimized. We found that iron complex with six-N coordination such as 1,10-phenanthroline iron (II) perchlorate could provide excellent catalytic activity and overall single cell performance. Amorphous carbon such as Ketjen Black was also added at different steps of catalyst preparation to reduce the cell impedance. However, such addition also dilutes the catalytically active sites and thereby reduced the fuel cell performance. Finally, different Nafion® to catalyst ratios were used to improve the cathode ink formulation before MEA fabrication was performed. These improvements at both catalyst and MEA levels have yielded impressive ORR activity when tested in a fuel cell system, moving towards the performance targets set by the U.S. DOE for the automotive application.

Acknowledgments: This work is supported by the U.S. Department of Energy's Office of Science and the Office of Energy Efficiency and Renewable Energy, Fuel Cell Technologies Program. The authors wish to thank Deborah J. Myers for her assistance in the single cell test.

Author Contributions: HMB prepared the samples, performed the experiment and wrote manuscript. LC and ZBK supported the experiments. TX supervised HMB and DJL designed and oversaw the experiment and manuscript preparation.

Conflicts of Interest: The authors declare no conflict of interest.

References

1. Costamagna, P.; Srinivasan, S. Quantum jumps in the pemfc science and technology from the 1960s to the year 2000 part ii. Engineering, technology development and application aspects. *J. Power Sources* **2001**, *102*, 253–269.
2. Debe, M.K. Electrocatalyst approaches and challenges for automotive fuel cells. *Nature* **2012**, *486*, 43–51.
3. Mehta, V.; Cooper, J.S. Review and analysis of pem fuel cell design and manufacturing. *J. Power Sources* **2003**, *114*, 32–53.
4. Wagner, F.T.; Lakshmanan, B.; Mathias, M.F. Electrochemistry and the future of the automobile. *J. Phys. Chem. Lett.* **2010**, *1*, 2204–2219.
5. Wong, W.Y.; Daud, W.R.W.; Mohamad, A.B.; Kadhum, A.A.H.; Loh, K.S.; Majlan, E.H. Recent progress in nitrogen-doped carbon and its composites as electrocatalysts for fuel cell applications. *Int. J. Hydrogen Energ.* **2013**, *38*, 9370–9386.
6. Gewirth, A.A.; Thorum, M.S. Electroreduction of dioxygen for fuel-cell applications: Materials and challenges. *Inorg. Chem.* **2010**, *49*, 3557–3566.
7. Jasinski, R. A new fuel cell cathode catalyst. *Nature* **1964**, *201*, 1212–1213.

8. Ma, S.Q.; Goenaga, G.A.; Call, A.V.; Liu, D.J. Cobalt imidazolate framework as precursor for oxygen reduction reaction electrocatalysts. *Chem. Eur. J.* **2011**, *17*, 2063–2067.

9. Carne, A.; Carbonell, C.; Imaz, I.; Maspoch, D. Nanoscale metal-organic materials. *Chem. Soc. Rev.* **2011**, *40*, 291–305.

10. James, S.L. Metal-organic frameworks. *Chem. Soc. Rev.* **2003**, *32*, 276–288.

11. Allendorf, M.D.; Schwartzberg, A.; Stavila, V.; Talin, A.A. A roadmap to implementing metal-organic frameworks in electronic devices: Challenges and critical directions. *Chem. Eur. J.* **2011**, *17*, 11372–11388.

12. Ariga, K.; Yamauchi, Y.; Rydzek, G.; Ji, Q.M.; Yonamine, Y.; Wu, K.C.W.; Hill, J.P. Layer-by-layer nanoarchitectonics: Invention, innovation, and evolution. *Chem. Lett.* **2014**, *43*, 36–68.

13. Gascon, J.; Corma, A.; Kapteijn, F.; Xamena, F.X.L.I. Metal organic framework catalysis: Quo vadis? *ACS Catal.* **2014**, *4*, 361–378.

14. Ryder, M.R.; Tan, J.C. Nanoporous metal organic framework materials for smart applications. *Mater. Sci. Technol.* **2014**, *30*, 1598–1612.

15. Proietti, E.; Jaouen, F.; Lefevre, M.; Larouche, N.; Tian, J.; Herranz, J.; Dodelet, J.P. Iron-based cathode catalyst with enhanced power density in polymer electrolyte membrane fuel cells. *Nat. Commun.* **2011**, *2*.

16. Zhao, D.; Shui, J.L.; Chen, C.; Chen, X.Q.; Reprogle, B.M.; Wang, D.P.; Liu, D.J. Iron imidazolate framework as precursor for electrocatalysts in polymer electrolyte membrane fuel cells. *Chem. Sci.* **2012**, *3*, 3200–3205.

17. Zhao, D.; Shui, J.L.; Grabstanowicz, L.R.; Chen, C.; Commet, S.M.; Xu, T.; Lu, J.; Liu, D.J. Highly efficient non-precious metal electrocatalysts prepared from one-pot synthesized zeolitic imidazolate frameworks. *Adv. Mater.* **2014**, *26*, 1093–1097.

18. Xia, W.; Zhu, J.H.; Guo, W.H.; An, L.; Xia, D.G.; Zou, R.Q. Well-defined carbon polyhedrons prepared from nano metal-organic frameworks for oxygen reduction. *J. Mater. Chem. A* **2014**, *2*, 11606–11613.

19. Zhao, S.L.; Yin, H.J.; Du, L.; He, L.C.; Zhao, K.; Chang, L.; Yin, G.P.; Zhao, H.J.; Liu, S.Q.; Tang, Z.Y. Carbonized nanoscale metal-organic frameworks as high performance electrocatalyst for oxygen reduction reaction. *ACS Nano* **2014**, *8*, 12660–12668.

20. Tian, J.; Morozan, A.; Sougrati, M.T.; Lefevre, M.; Chenitz, R.; Dodelet, J.P.; Jones, D.; Jaouen, F. Optimized synthesis of Fe/N/C cathode catalysts for pem fuel cells: A matter of iron-ligand coordination strength. *Angew. Chem. Int. Edit.* **2013**, *52*, 6867–6870.

21. Jaouen, F.; Herranz, J.; Lefevre, M.; Dodelet, J.P.; Kramm, U.I.; Herrmann, I.; Bogdanoff, P.; Maruyama, J.; Nagaoka, T.; Garsuch, A.; *et al.* Cross-laboratory experimental study of non-noble-metal electrocatalysts for the oxygen reduction reaction. *ACS Appl. Mater. Inter.* **2009**, *1*, 1623–1639.

22. Mikhailenko, S.D.; Afsahi, F.; Kaliaguine, S. Complex impedance spectroscopy study of the thermolysis products of metal-organic frameworks. *J. Phys. Chem. C* **2014**, *118*, 9165–9175.

23. Inoue, H.; Hosoya, K.; Kannari, N.; Ozaki, J. Influence of heat-treatment of ketjen black on the oxygen reduction reaction of Pt/C catalysts. *J. Power Sources* **2012**, *220*, 173–179.

24. Lee, J.S.; Park, G.S.; Kim, S.T.; Liu, M.L.; Cho, J. A highly efficient electrocatalyst for the oxygen reduction reaction: N-doped ketjenblack incorporated into Fe/Fe$_3$C-functionalized melamine foam. *Angew. Chem. Int. Edit.* **2013**, *52*, 1026–1030.

25. Nam, G.; Park, J.; Kim, S.T.; Shin, D.B.; Park, N.; Kim, Y.; Lee, J.S.; Cho, J. Metal-free ketjenblack incorporated nitrogen-doped carbon sheets derived from gelatin as oxygen reduction catalysts. *Nano Lett.* **2014**, *14*, 1870–1876.

26. Yu, J.R.; Islam, M.N.; Matsuura, T.; Tamano, M.; Hayashi, Y.; Hori, M. Improving the performance of a PEMFC with Ketjenblack EC-600JD carbon black as the material of the microporous layer. *Electrochem. Solid-State Lett.* **2005**, *8*, A320–A323.

27. Lin, J.B.; Lin, R.B.; Cheng, X.N.; Zhang, J.P.; Chen, X.M. Solvent/additive-free synthesis of porous/zeolitic metal azolate frameworks from metal oxide/hydroxide. *Chem. Commun.* **2011**, *47*, 9185–9187.

Effect of ZIF-8 Crystal Size on the O_2 Electro-Reduction Performance of Pyrolyzed Fe–N–C Catalysts

Vanessa Armel, Julien Hannauer and Frédéric Jaouen

Abstract: The effect of ZIF-8 crystal size on the morphology and performance of Fe–N–C catalysts synthesized via the pyrolysis of a ferrous salt, phenanthroline and the metal-organic framework ZIF-8 is investigated in detail. Various ZIF-8 samples with average crystal size ranging from 100 to 1600 nm were prepared. The process parameters allowing a templating effect after argon pyrolysis were investigated. It is shown that the milling speed, used to prepare catalyst precursors, and the heating mode, used for pyrolysis, are critical factors for templating nano-ZIFs into nano-sized Fe–N–C particles with open porosity. Templating could be achieved when combining a reduced milling speed with a ramped heating mode. For templated Fe–N–C materials, the performance and activity improved with decreased ZIF-8 crystal size. With the Fe–N–C catalyst templated from the smallest ZIF-8 crystals, the current densities in H_2/O_2 polymer electrolyte fuel cell at 0.5 V reached *ca.* 900 mA cm^{-2}, compared to only *ca.* 450 mA cm^{-2} with our previous approach. This templating process opens the path to a morphological control of Fe–N–C catalysts derived from metal-organic frameworks which, when combined with the versatility of the coordination chemistry of such materials, offers a platform for the rational design of optimized Metal–N–C catalysts.

Reprinted from *Catalysts*. Cite as: Armel, V.; Hannauer, J.; Jaouen, F. Effect of ZIF-8 Crystal Size on the O_2 Electro-Reduction Performance of Pyrolyzed Fe–N–C Catalysts. *Catalysts* **2015**, *5*, 1333–1351.

1. Introduction

Fuel cells offer a combination of efficiency and power density that meets the requests of the most demanding applications, and, in particular, those of the transportation sector [1,2]. Among the fuel cells allowing fast start-up and shut-down operation, an imperative criterion for the automotive industry, the polymer electrolyte membrane fuel cell (PEMFC) is today the most advanced technology due to the existence of proton conductive polymer membranes [3,4]. While PEMFCs have reached the commercialisation level for materials handling vehicles and, since recently, for personal automobiles that are being released in Japan and California, its long term success is bound to cost and sustainability. These two aspects are intimately linked to the usage of platinum for catalyzing the anodic and cathodic reactions. Ultimately, the accessibility to enough platinum is questionable if the PEMFC technology is to replace internal combustion engines. Since 80%–90% of platinum in a PEMFC is needed to catalyze the sluggish oxygen reduction reaction (ORR) [5], reducing the Pt content at the cathode or replacing Pt-based catalysts by catalysts based on Earth-abundant elements is a topic of intense research since 2003 [6].

Catalysts based on iron, cobalt, nitrogen and carbon that feature Metal–N_xC_y moieties covalently integrated in N-doped carbons have emerged as a promising alternative to Pt-based catalysts [6–12]. Rational approaches for the selection of metal, nitrogen and carbon precursors, for the synthetic conditions and for more durable Metal–N–C catalysts, are, however, still needed. In 2011, the use of metal organic frameworks (MOFs) as sacrificial N and C precursors for the synthesis of Co– and Fe–N–C catalysts was first reported [8,13]. In the approach by Liu's group, the cobalt ions were engaged in the MOF structure and therefore ideally dispersed and coordinated to nitrogen atoms [13]. The inherent disadvantage is the high content of cobalt in Co-based MOFs, 30–40 wt. %. This is well above the optimum content for Me–N–C catalyst precursors before pyrolysis, typically below 2–3 wt. % [7,14–16]. Too large Fe or Co contents lead to the formation of highly graphitized carbon structures during pyrolysis [13,17]. In such graphitized structures, the number of MeN_xC_y active sites is low, which leads to a low ORR activity [18,19]. In order to maximize the activity, it is necessary to reach a high specific area, and especially a high microporous area [7,20–22]. In the approach by Dodelet's group, a Zn-based MOF was combined with Fe(II) and phenanthroline in order to prepare a catalyst precursor which, after pyrolysis in Ar and then NH_3, resulted in a Fe–N–C catalyst with unprecedented initial power performance in PEMFC [8]. The Zn-based MOF used in 2011 by Dodelet's group was ZIF-8, a well-known zeolitic imidazolate framework (ZIF) [23,24], commercially available under the trade name Basolite® Z1200 (produced by BASF, purchased from Sigma Aldrich, St. Louis, MO, USA), referred to hereafter as Basolite®. ZIF-8 has a sodalite topology and is entirely microporous with a BET area of *ca.* 1600 m^2 g^{-1} [23]. Fe–N–C catalysts derived with various approaches from Fe(II) acetate, 1,10-phenanthroline and ZIF-8 but sharing a common pyrolytic step in NH_3, still represent the state-of-the-art in terms of initial power performance in PEMFC [8,10,25–27]. They however suffer from a poor durability, with *ca.* halved power performance after 50 h of operation, characteristic for NH_3-pyrolyzed catalysts [8,10,28]. In contrast, Fe–N–C catalysts derived from Fe(II), 1,10-phenanthroline and ZIF-8 but pyrolyzed in Argon are initially less active but more stable [8,29,30]. Other Fe– and Fe–Co–N–C catalysts pyrolyzed in inert atmosphere have resulted in a constant current density at 0.4–0.5 V over several hundred hours of operation in PEMFC, a promising achievement toward more durable Me–N–C catalysts [9,31].

266

Since the first reports on the synthesis of Metal–N–C materials via the sacrificial pyrolysis of microporous ZIFs [8,13], modifications of the preparation of the catalyst precursor from Fe(II) acetate, N-ligands and ZIF-8 have minimized the formation of undesired iron particles during pyrolysis. Low energy milling of the dry precursor powders has been shown to optimise the dispersion of ferrous ions and to favour the formation of FeN_xC_y moieties during pyrolysis [25,32]. The replacement of Fe(II) and 1,10-phenanthroline (phen) by an iron porphyrin was also shown to preferentially result in the formation of FeN_xC_y moieties after pyrolysis [33]. Alternative Zn-based ZIFs have also started being investigated, using a one-pot synthesis approach [10,26]. The impregnation of ZIF-8 with furfuryl alcohol introduced mesoporosity in the resulting Fe–N–C catalysts [34]. However, the increased mesoporous volume did not increase the fuel cell performance.

To a large extent, the transport properties of Metal–N–C cathodes today limit the fuel cell performance at high load [6]. Identifying rational approaches to optimize transport properties without negatively affecting the ORR activity of the most active or most stable Me–N–C catalysts reported to date is therefore important. Hence, controlling the size of ZIF crystals before pyrolysis and controlling all process parameters in order to template nanosized ZIF crystals into nano-sized catalytic Me–N–C particles is a promising approach. It could significantly reduce the average diffusion length for O_2 molecules from the electrode macropores to the active FeN_4 sites located in intra-particle pores. In the commercial ZIF-8, Basolite®, the crystal size ranges from 200 to 500 nm, which is not optimum [8].

Hitherto, the size effect of ZIF crystals on the activity and accessibility of ORR active sites in Me–N–C catalysts derived from the pyrolysis of ZIFs has not been investigated in depth. Recent attempts include Co–N–C catalysts derived from ZIF-67 and metal-free N–C catalysts derived from ZIF-8 [35,36]. ZIF-67, a ZIF comprising Co(II) ion ligated with 2-MeIm organised in sodalite topology, was synthesized in three average sizes of 300, 800 and 1700 nm. ZIF-67 powders were then pyrolyzed in Ar at 600–900 °C to form Co–N–C catalysts. The Co–N–C catalyst derived from ZIF-67 with the smallest crystal size of 300 nm showed a higher activity in rotating disk electrode measurements in 0.1 M $HClO_4$ [35]. SEM and TEM images showed a remarkable templating of the rhombic dodecahedron shape of ZIF-67 nanocrystals into Co–N–C catalytic particles with similar size. The second study reported the synthesis of ZIF-8 crystals with average size of 60 nm [36]. ZIF-8 was pyrolyzed in N_2 at 700–1000 °C. A templating effect was similarly observed, even after pyrolysis at 1000 °C. A high ORR activity of the resulting N-doped carbons was measured in 0.1 M KOH. The effect of smaller catalytic Fe(Co)–N–C particles has however not yet been demonstrated in fuel cell measurements at high current density, where a short diffusion path for O_2 is expected to be most beneficial. The restricted current densities of a few mA·cm^{-2} in rotating disk electrode measurements cannot inform whether such templated catalysts perform better in a fuel cell.

The development of synthetic methods to control the size of MOF or ZIF crystals is relatively recent by itself [37–43]. Of particular interest, the rapid crystallization at room temperature of Zn(II) and 2-MeIm into ZIF-8 in either methanol or aqueous solution is now well established [36,42,44,45]. Control of the size of ZIF-8 crystals has been gained with the approaches of (i) modulating-ligand [38,44] (ii) surfactants [40] and (iii) excess-ligands [37,42,45]. The first approach relies on the competition for Zn(II) ions between 2-MeIm and a second modulating ligand added in the reagent solution. For example, large ZIF-8 crystals of *ca.* 1 μm were obtained from Zn(II) and 2-MeIm in the presence of the modulating ligand 1-MeIm [44]. In the absence of 1-MeIm, ZIF-8 crystals of 60–70 nm were obtained. In the presence of another modulating ligand, n-butylamine, ultra small ZIF-8 crystals of 9–10 nm could be obtained. The effects were rationalized on the basis of the ability of the modulating ligand to deprotonate 2-MeIm during the nucleation of ZIF-8. The second approach relies on the use of surfactants that stabilize nanosized ZIF-8 crystals [40]. With surfactants, phase pure ZIF-8 crystals of sub-100 nm size could be obtained at nearly stoichiometric ratio 2:1 for 2-MeIm:Zn(II). The third approach relies on overstoichiometric ratios of 2-MeIm:Zn(II) in the reagent solution. The excess of 2-MeIm results in a high density of nucleation sites for the crystallization of ZIF-8, and hence in smaller crystals [37,42,45]. Molar ratios of 2-MeIm to Zn(II) from four to 200 have been investigated resulting in ZIF-8 crystals with size ranging from 1900 to 250 nm, respectively [42].

In the present work, we synthesized ZIF-8 materials with a wide range of crystal size through the ligand-excess approach. These materials were then milled with Fe(II) acetate and 1,10-phenanthroline to form catalyst precursors. The latter were pyrolyzed in Ar at 1050 °C, with a heating either in ramp or flash mode, to form Fe–N–C electrocatalysts. The structure and morphology of ZIF-8 powders, catalyst precursors and catalysts were investigated with SEM, X-ray diffraction and N$_2$ sorption while the Fe–N–C catalysts were electrochemically characterized in a single cell PEMFC.

2. Results and Discussion

2.1. Morphology and Size of ZIF-8 Nanocrystals

Figure 1 shows the SEM images for the five synthetic conditions of ZIF-8 (labelled hereafter as Z8) with ratios of 2-MeIm to Zn(II) (denoted as X hereafter) varying from 40 to 140 in the initial reagent solution. It is seen that the morphology of these Z8-X samples did not change with the molar ratio X, but the crystal size drastically decreased with increasing molar ratio. Crystals 1200–1800 nm in size are seen for Z8-40, 250–700 nm for Z8-60, 160–400 nm for Z8-80, 100–280 nm for Z8-100, and 80–200 nm for Z8-140. This agrees with the results reported by Kida *et al.* [42]. The vast majority of the crystals show a well-defined truncated rhombic dodecahedron morphology [42,44]. As a comparison, Figure 1 also shows a typical SEM image for Basolite®. A large dispersion size from 280–640 nm is observed, but most of the crystals are larger than 400 nm. The average size observed with Basolite® is similar to that of Z8-60 (Figure 1). Hence, the set of Z8 materials synthesized here forms an interesting basis to investigate the size effect of Z8 on the activity and performance of Fe–N–C catalysts that will be derived from them. Figure S1 shows the powder XRD patterns of the various Z8-X materials. All materials are well crystallized, even at the highest ratio of 2-MeIm to Zn(II) corresponding to the smallest crystals. All diffraction lines can be assigned to the sodalite topology of ZIF-8. Kida *et al.*, observed XRD peaks assigned to zinc hydroxides by-products at ratios of 2-MeIm to Zn(II) $\leqslant$ 20, peaks that were absent at molar ratios $\geqslant$40 [42]. In this work, we restricted ourselves to ratios $\geqslant$40 and the materials are thus phase-pure Z8 crystals. The accessibility of the cavities in these materials was then investigated with N_2 sorption (Figure S2). No hysteresis is observed and the isotherms have a type I shape, characteristic for microporous materials. The specific surface area slightly decreases with decreasing Z8 crystal size (Table 1, 2nd column). For the smallest crystals (Z8-100 and 140), the isotherms show a sharp rise at high P/P_0, assigned to the filling of mesopores existing between ZIF crystals [42].

2.2. Characterization of Fe–N–C Catalysts Obtained from Various ZIF-8 Materials without Templating

A first series of Fe–N–C catalysts was synthesized via flash pyrolysis in Ar of catalyst precursors prepared from Fe(II) acetate, phen and Z8-X materials. All catalyst precursors discussed in this sub-section were prepared by planetary milling the dry precursor powders at 400 rpm (see Section 3.2). These process parameters are those used by us previously [8,46]. An example of a catalyst label is Z8-40-400 rpm-F, indicating that Z8-40 was used, the milling speed was 400 rpm and the pyrolysis was carried out in flash mode. The SEM images of this first series of Fe–N–C catalysts are shown in Figure 2. Clearly, the morphology and size of the Fe–N–C catalytic particles are totally different from those of the starting Z8-X crystals. Even when starting from the largest Z8 crystals (Z8-40), most particles in Z8-40-400-F lost their original size (Figures 1 and 2). The visual impression is that the pristine Z8 batches with different average crystal size (100–1600 nm) resulted after milling at 400 rpm and flash pyrolysis in Fe–N–C agglomerates with a common size of *ca.* 300–600 nm, independent of the initial Z8 crystal size. It therefore seems that the largest Z8 crystals (Z8-40) are transformed into catalytic particles of a smaller size, while the smallest Z8 crystals (Z8-140) are transformed into larger catalytic particles.

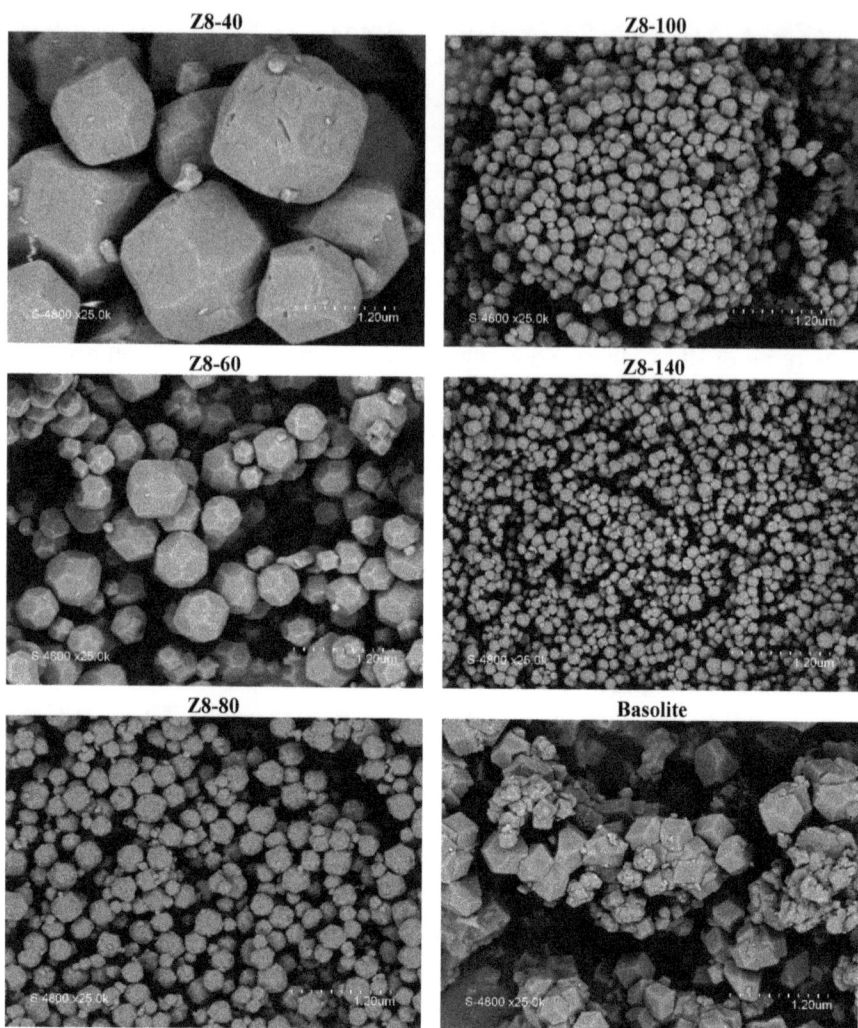

Figure 1. SEM micrographs of Z8-X crystals synthesized with the excess-ligand approach in aqueous solution at different molar ratios X of 2-MeIm to Zn(II), and SEM of the commercial Z8, Basolite®. The scale bar is 1.20 μm for all micrographs.

Table 1. BET specific areas of non-pyrolyzed and pyrolyzed materials.

Z8 Material	BET Specific Area/m^2 g^{-1}			
	as-synthesized Z8-X	Z8-X-400 rpm-F	Z8-X-100 rpm-F	Z8-X-100 rpm-R
Z8-40	1798	412	-	726
Z8-60	1767	372	-	745
Z8-80	1654	439	-	738
Z8-100	1543	412	367	673
Z8-140	1524	292	-	722
Basolite®	1618	357	-	719

Figure 2. SEM micrographs of Fe–N–C catalysts derived from Fe(II), phen and Z8-X powders or Basolite®, ballmilled at 400 rpm and pyrolyzed at 1050 °C in Ar in flash mode. The scale bar is 1.20 μm for all SEM micrographs.

Figure S3 shows the N_2 isotherms for this series of catalysts, and Table 1 (3rd column) reports their BET areas. No clear trend is observed for the BET area. Figure 3 shows the PEMFC polarization curves recorded for a cathode loading of 4 mg cm^{-2} for this series of Fe–N–C catalysts. As seen in the inset of Figure 3, the ORR activity at 0.9 V iR-free is similar for all Z8-X-400 rpm-F samples (0.4–0.8 mA cm^{-2}) but lower than that of the Fe–N–C catalyst derived from Basolite® (1.5 mA cm^{-2}). The similar ORR activities and fuel cell performance obtained within this series agree with their similar morphology and specific surface area. The higher ORR activity obtained with Basolite® is possibly assigned to the presence of 100 ppm of iron atomically-dispersed in Basolite®. We will report on this in the near future.

In view of this first set of morphological and electrochemical results, we then investigated the effect of two process parameters that might explain the lack of templating observed on this first series of Fe–N–C catalysts. Those parameters are (i) the rotation rate applied during the ball milling of Fe(II), phen and Z8-X, and (ii) the heating mode. Previous works have recently reported on the templating of nano-ZIFs after pyrolysis in inert gas, but the heating mode was a ramp at 5 °C min^{-1} [35,36]. Our flash pyrolysis mode is unusual but was found to be crucial in order to precisely control the pyrolysis duration, which is in turn important when pyrolyzing in reactive NH$_3$ atmosphere [7,20]. Precisely controlling the pyrolysis duration is however less important when the pyrolysis is carried out in inert gas since no continuous chemical reaction occurs between the formed carbonaceous material and inert gas.

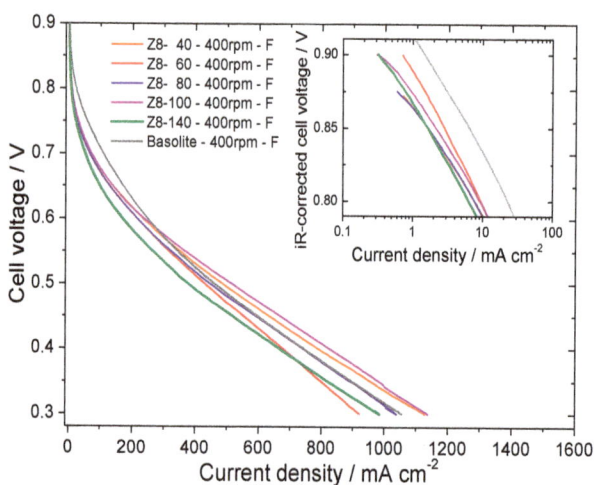

Figure 3. Non iR-corrected PEMFC polarization curves with cathodes comprising 4 mg cm^{-2} of Fe–N–C catalysts. Ballmilling was carried out at 400 rpm for forming catalyst precursors from Fe(II) acetate, phen and Z8-X powders. The catalyst precursors were subsequently pyrolyzed in flash mode. Inset: Tafel plots at high potential of the iR-corrected polarization curves.

2.3. Effects of Milling Speed and Heating Mode on the Morphology & Performance of Fe–N–C Catalysts

In this section, we selected Z8-100 to investigate the effects of milling speed (400 or 100 rpm) and heating mode (flash or ramp) on the morphology of Fe–N–C catalysts. Z8-100 was selected because its average crystal size is smaller than that of Basolite® and smaller than that of the agglomerates in the first series of Fe–N–C catalysts. The morphological changes of the materials were investigated with SEM at different stages of the synthesis. Fixing the milling speed at 400 rpm, the SEM images of (a) Z8-100, (b) the corresponding catalyst precursor ballmilled at 400 rpm and (c,d) the corresponding Fe–N–C catalysts after pyrolysis in flash or ramp mode are shown in Figure 4. Milling at 400 rpm leads to the formation of agglomerates that are larger than the pristine Z8-100 crystals (Figure 4b) and also leads to an amorphization of the catalyst precursor, as shown by XRD (Figure S4).

Following the milling step, the pyrolysis does not significantly modify the macroscopic morphology of those materials, regardless of whether the pyrolysis is carried out in flash or ramp mode (Figure 4c,d).

Figure 4. SEM micrographs of (**a**) Z8-100 crystals; (**b**) the catalyst precursor derived from Fe(II), phen and Z8-100 mixed via ballmilling at 400 rpm, and (**c,d**) Fe–N–C catalysts obtained by pyrolyzing that catalyst precursor at 1050 °C in Argon, either in flash (**c**) or ramp mode (**d**). The scale bar is 1.20 μm for all SEM micrographs.

The same investigation of morphological changes at different stages of synthesis was then carried out with a lowered milling speed of 100 rpm (Figure 5). The lowered milling speed not only preserved the original shape of Z8-100 crystals in the catalyst precursor but also prevented the formation of aggregates (Figure 5a,b). Moreover, powder X-ray diffraction confirmed that the lowered milling speed resulted in an X-ray diffractogram now superimposed with that of Z8-100 (Figure S4). It therefore seems that a correlation exists between crystallographic amorphization at high milling speed (XRD pattern) and the agglomeration observed on the SEM images of catalyst precursors. For various ZIFs and MOFs, it can thus be expected that the milling speed should be minimized to avoid amorphization during the milling stage. The threshold milling speed at which amorphization starts will likely depend on the mechanical properties of each specific MOF structure. Following the milling stage at 100 rpm, the macroscopic morphology of Z8-100 crystals is also maintained in the Fe–N–C catalyst after pyrolysis in flash or ramp mode (Figure 5c,d). With the latter pyrolysis mode, a well dispersed structure with spherical entities is observed, while in the case of the flash pyrolysis mode, the Fe–N–C catalytic particles seem to be elongated and slightly more interconnected or fused together (Figure 5c). Higher resolution SEM images better highlight such microscopic morphological and surface-roughness differences between Z8-100-100 rpm-F and Z8-100-100 rpm-R (Figure S5).

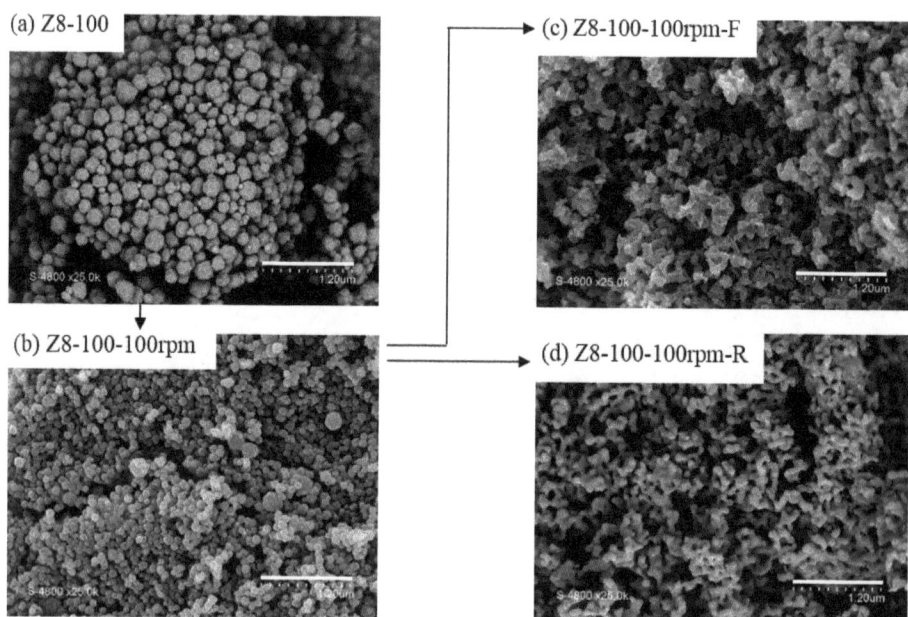

Figure 5. SEM micrographs of (**a**) Z8-100 crystals; (**b**) the catalyst precursor derived from Fe(II), phen and Z8-100 mixed via ballmilling at 100 rpm, and (**c,d**) Fe–N–C catalysts obtained by pyrolyzing that catalyst precursor at 1050 °C in Argon, either in flash (**c**) or ramp mode (**d**). The scale bar is 1.20 μm for all SEM micrographs.

A much more obvious difference is however observed in the BET areas of the two materials, 367 and 673 m^2 g^{-1} for Z8-100-100 rpm-F and Z8-100-100 rpm-R, respectively (Table 1). The twice higher uptake of N$_2$ at the initial stage of the isotherm ($P/P_0 < 0.02$) also indicates a much higher microporous surface area after a ramp pyrolysis (Figure S6). This major difference is interpreted as an increased accessibility of the intra-particular pores after a ramp pyrolysis, compared to the case with flash pyrolysis. It is proposed that the flash mode, implying a high gasification rate of Z8 (about 66% of the initial Z8 mass is gasified, probably during the first minute of pyrolysis), not only leads to the deformation of the templating Z8 crystals but also to the formation of closed pores within carbonized crystals. If the outer surface of Z8 crystals fuses or becomes less porous before the carbonization is completed in the depth of the crystals, this will impede or slow down the outgassing of volatile products formed within the crystals. Pressure increase within Z8 crystals under carbonization at the initial stage of a flash pyrolysis may also ensue, distorting the shape of the crystals and modifying the interconnections between the pores of the final pyrolysis product.

Figure 6 shows the PEMFC polarization curves recorded with a cathode loading of 4 mg cm^{-2} for the four Fe–N–C catalysts derived from Z8-100, and ballmilled at 400 or 100 rpm and pyrolyzed either in flash or ramp mode. As observed previously, the ORR activity at 0.9 V iR-free is low for the Fe–N–C catalyst prepared via milling at 400 rpm and pyrolyzed in flash mode, when compared to that of the catalyst derived from Basolite® (insets of Figures 3 and 6).

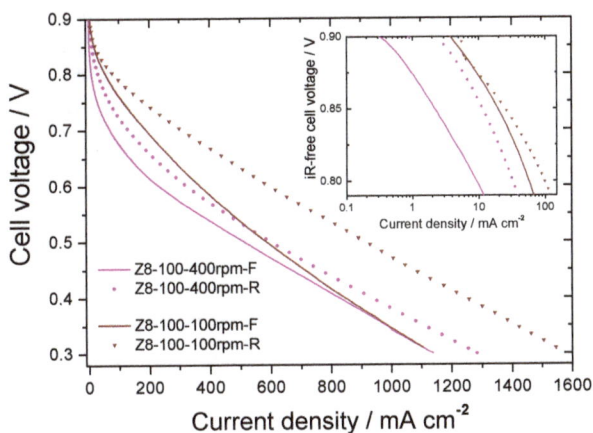

Figure 6. PEMFC polarization curves with cathodes comprising 4 mg cm^{-2} of Fe–N–C catalysts. Ballmilling was carried out at 400 or 100 rpm for forming catalyst precursors from Fe(II) acetate, phen and Z8-100, catalyst precursors that were subsequently pyrolyzed in flash (F) or ramp mode (R). Inset: Tafel plots at high potential of the iR-corrected polarization curves.

The ORR activity at 0.9 V iR-free is however enhanced by a factor of seven when the heating mode is switched to ramp (magenta solid to magenta dotted curve in the inset of Figure 6). The corresponding enhancement of the current density at 0.5 V is however minor (magenta curves in Figure 6), probably due to the formation of large agglomerates during the milling at 400 rpm. Even higher ORR activities are observed for the two Fe–N–C catalysts derived from Z8-100 and prepared via milling at 100 rpm, then pyrolyzed in flash or ramp mode (inset of Figure 6, brown curves). The ORR activities at 0.9 V iR-free of those two catalysts are now significantly higher than that observed with the Fe–N–C catalyst derived from Basolite® via 400 rpm milling and flash pyrolysis (3.7 and 5.3 mA cm^{-2} vs. 1.5 mA cm^{-2}). Even more interesting, the combined low milling speed of 100 rpm with the ramped pyrolysis mode results in a much improved performance at lower cell voltage, reaching 910 mA cm^{-2} at 0.5 V (Figure 6), instead of 374–515 mA cm^{-2} for the first series of Fe–N–C catalysts synthesized with 400 rpm milling speed and flash pyrolysis (Figure 3). The beneficial effect of a combined low milling speed (100 rpm) with a ramp pyrolysis mode is obvious at 0.5 V, compared to the 100 rpm milling speed combined with a flash pyrolysis (910 vs. 585 mA cm^{-2}, brown curves in Figure 6). The high ORR activity at 0.9 V but medium performance at 0.5 V of Z8-100-100 rpm-F was reproducible. Hence, the catalytic sites are more accessible by O_2 in Z8-100-100 rpm-R and this can be correlated to its higher BET area. While Z8-100-100 rpm-F shows a particle size similar to that in Z8-100-100 rpm-R (Figure 5c,d), the intra-particle pore network is less developed (lower BET area, Table 1) and less microporous (Figure S6).

2.4. Characterization of Fe–N–C Catalysts Templated from Various ZIF-8 Materials

In view of the templating effect and positive electrochemical result obtained with Z8-100-100 rpm-R, the synthesis parameters were then fixed (milling speed 100 rpm and ramp pyrolysis) and the investigation of the templating effect for the different Z8 materials shown in Figure 1 is now possible. The BET areas for this second series of Fe–N–C materials are reported in Table 1 (5th column). In this second series of catalysts, all BET areas are significantly higher than those measured for the first series of Fe–N–C catalysts (Table 1, 3rd column). High BET and microporous area is required for reaching a high ORR activity at high potential. This is demonstrated by the high ORR activity observed for all materials within this second series of Fe–N–C catalysts (inset of Figure 7).

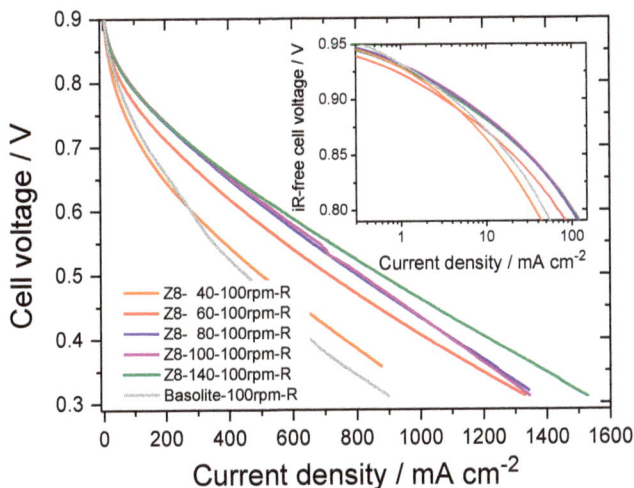

Figure 7. Non-iR corrected PEMFC polarization curves with cathodes comprising 4 mg cm^{-2} of Fe–N–C catalysts. Ballmilling was carried out at 100 rpm for forming catalyst precursors from Fe(II) acetate, phen and Z8-X powders. The catalyst precursors were heated in ramp mode. Inset: Tafel plots at high potential of iR-free polarization curves (the green curve is superimposed below the blue and magenta curves).

Beyond the fact that the average BET areas of these two series of catalysts differ, the N$_2$ sorption isotherms also reveal the presence of a hysteresis for the materials pyrolyzed with flash mode, but little or no hysteresis for the materials pyrolyzed with ramp mode (Figures S7 and S8). This is interpreted as a less tortuous path and the absence of bottlenecks in the intra-particle pore network of materials pyrolyzed in ramp mode. The ORR activity at 0.9 V iR-free is high and quite homogeneous within this second series of Fe–N–C catalysts (3.2–5.3 mA cm^{-2}, see inset of Figure 7), in agreement with their similar BET areas and N$_2$ isotherms (Table 1, Figure S7). Hence, the size of Z8 crystals before pyrolysis does not play a major role regarding the ORR activity after pyrolysis. However, the effect of Z8 particle size (and thus of Fe–N–C catalytic particle size after pyrolysis, due to the templating observed earlier for Z8-100) is obviously demonstrated in the linear *E vs. I* plots, showing a continuous increase of the current density at e.g., 0.5 V with decreased Z8 particle size (Figure 7). The clear assignment of the increased performance at 0.5 V to the reduced catalytic particle size is made possible due to the very similar ORR activities observed at 0.9 V for this second series of catalysts. This synthesis approach therefore allows a control of the particle size and quality of the intra-particle porous network in Fe–N–C catalysts derived from the pyrolysis of metal-organic frameworks.

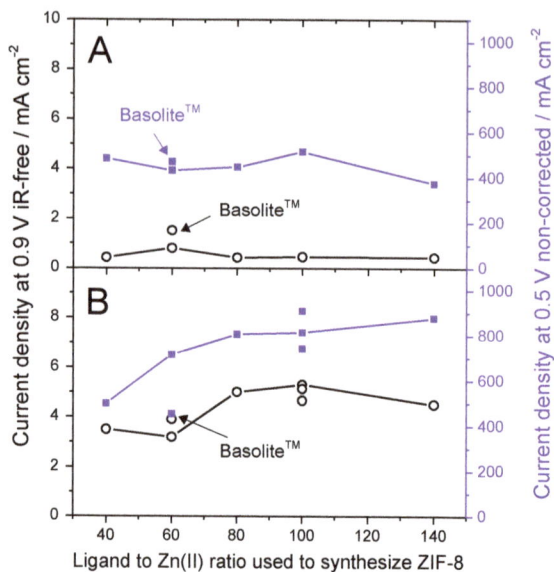

Figure 8. (**a**) Current density at 0.9 V iR-free and 0.5 V against the ligand-to-Zn(II) ratio used for the synthesis of Z8 samples. The catalysts were synthesized with 400 rpm milling and flash pyrolysis; (**b**) the same figure for catalysts synthesized with 100 rpm milling and ramp pyrolysis. Basolite® is artificially positioned at a ligand-to-Zn(II) ratio of 60.

Figure 8 summarizes the ORR activity at 0.9 V iR-free cell voltage (left handside Y-axis) and current density at 0.5 V uncorrected voltage (right handside Y-axis) for the two series of Fe–N–C catalysts. The X-axis reports the ligand to Zn(II) ratio used for synthesizing Z8 materials, and is thus inversely correlated with the average crystal size of Z8 materials. In this figure, the results obtained with Basolite® are also shown. In order to do this, Basolite® was artificially attributed a ratio of ligand-to-Zn(II) of 60, on the basis of similar Z8 crystals size observed for Z8-60 and Basolite® (Figure 1). For Z8-100-100 rpm-R, the reproducibility of the results was verified on three different MEAs (Figure 8B). The best MEA reached 910 mA cm^{-2} at 0.5 V cell voltage and a peak power density of 500 mW cm^{-2} at 0.39 V (Figure 6, brown dotted curve). This result surpasses by a factor of two the current density at 0.5 V previously obtained with Basolite® and with our typical process parameters (400 rpm, flash pyrolysis in Ar, Figure 8A) [46]. The initial peak power density of 500 mW cm^{-2} is among the highest reported for Fe–N–C catalysts, including those pyrolyzed in NH$_3$ [8–10,12,26,27,30,47]. This is particularly interesting since such Fe–N–C catalysts pyrolyzed in inert gas are intrinsically more stable than NH$_3$-pyrolyzed Fe–N–C catalysts [8]. While the crystal size of ZIF-8 may be further decreased, this might not necessarily lead to further increased performance since other transport phenomena may then become limiting (proton conduction and O_2 diffusion across the cathode layer, rather than O_2 diffusion in single catalytic particles). The present work also clearly demonstrates that the desired pore network for a well-performing Fe–N–C catalyst is highly microporous, and with no hysteresis attributed to small mesopores or bottlenecks (Figure S8). The durability of such templated Ar-pyrolyzed Fe–N–C catalysts with improved initial performance will be investigated in the near future. The effect of even higher pyrolysis temperature in inert gas (1080–1150 °C) will also be investigated with this approach, since a beneficial effect on the durability was recently reported for non-templated Fe–N–C catalysts derived from Basolite® [30].

3. Experimental Section

3.1. Synthesis of Nanosized ZIF-8 Crystals

Phase pure ZIF-8 crystals were synthesized at room temperature in aqueous solution according to the report by Kida and coworkers [42]. The size of ZIF-8 crystals was controlled with the molar ratio of 2-methylimidazole to Zn(II) nitrate hexahydrate. For example, to reach a ratio of 60, a mass of 0.744 g of Zn(II) nitrate hexahydrate (2.5 mmol) was dissolved in 10 mL of deionized water and added to a solution consisting of 12.3 g of 2-MeIm (150 mmol) previously dissolved in 90 mL of deionized water. The final molar composition of this reagent solution is 1:60:2228 for Zn(II):2-MeIm:water. The solution was constantly stirred and quickly turned cloudy and a suspension was obtained. Twenty-four hours later, the suspension was centrifuged at 11,000 rpm for 15 min, washed with methanol and re-dispersed with ultrasounds. The centrifugation and re-dispersion step was repeated three times. The product was then vacuum-dried for 24 h at 80 °C. In order to obtain ZIF-8 crystals with different sizes, the ratio 2-MeIm:Zn was adjusted to different values (40, 60, 80, 100, 140) by adjusting the amount of 2-MeIm while keeping the amount of the Zinc salt and of water constant. The resulting ZIF-8 materials are labelled Z8-X, with X being the ratio of 2-MeIm:Zn.

3.2. Synthesis of Fe–N–C Catalysts

All catalyst precursors were prepared via the dry ballmilling of ZIF-8 nanocrystals, Fe(II) acetate and 1,10-phenanthroline [32]. Then, 32.45 mg Fe(II)Ac, 100 mg of 1,10-phenanthroline and 800 mg of ZIF-8 were weighed and poured into a ZrO_2 crucible. This corresponds to 1 wt. % Fe in the catalyst precursor before pyrolysis. Hundred zirconium-oxide balls of 5 mm diameter were then added. The ZrO_2 crucible was then sealed under air and placed in a planetary ball-miller (Fritsch Pulverisette 7 Premium, Fritsch, Idar-Oberstein, Germany). The powders were milled during 4 cycles of 30 min, at either 400 or 100 rpm milling speed. The catalyst precursors resulting from the milling were pyrolyzed at 1050 °C in flowing Ar for

1 h, via either a heat-chock procedure previously described by us [20], or via a ramp mode with a heating rate of 5 °C min^{-1}. For the latter, the catalyst dwell time at 1050 °C in Ar was also 1 h. Last, the obtained powders were ground in an agate mortar. The mass loss during pyrolysis due to Zn, C and N volatile compounds from ZIF-8 and phenanthroline was *ca.* 60–65 wt. %, leading to Fe contents of 2.5–2.9 wt. % in the final catalysts. The catalysts are labelled Z8-X-Yrpm-Z, where X is the ratio of ligand to Zn(II) used during ZIF-8 synthesis, Y is the rotation rate during milling of the catalyst precursor (100 or 400 rpm) and Z is the heating mode (F for flash pyrolysis and R for ramp mode). For example, Z8-60-400 rpm-F corresponds to a catalyst precursor made from Fe, phen and Z8-60, milled at 400 rpm, and pyrolyzed in argon for 1 h with flash mode.

3.3. Material Characterization

The crystalline structure of ZIF-8 materials was investigated with X-ray diffraction using a PANanalytical X'Pert Pro powder X-ray diffractometer (Almelo, The Netherlands). The Brunauer–Emmett–Teller (BET) surface area and pore volume were measured with N_2 sorption at liquid nitrogen temperature (77 K) using a Micromeritics ASAP 2020 instrument (Norcross, GA, USA). Samples were degassed at 200 °C for 5 h in flowing nitrogen prior to measurements to remove guest molecules. The microstructure of ZIF-8 and Fe–N–C materials was investigated with SEM (Hitachi S-4800, Hitachi, Tokyo, Japan) after gold metallization.

3.4. Electrochemical Characterization

Electrochemical activity towards the ORR and initial power performance of the catalysts was determined in a single-cell laboratory fuel cell. The inks for the cathode electrode were prepared using the formulation of 20 mg of catalyst, 652 µL of an alcohol-based 5 wt. % Nafion® solution that also contains 15%–20% water, 326 µL of ethanol and 272 µL of de-ionized water. The inks were first sonicated, then agitated with a vortex mixer. These sonication-agitation steps were repeated every 15 min for a total duration of 1 h. Then, three aliquots of 405 µL of the cathode catalyst ink were deposited on the microporous layer of a commercial gas diffusion layer (Sigracet S10-BC, 4.48 cm^2, SGL Group The Carbon Company, Wiesbaden, Germany). The gas diffusion layer was heated on a heating plate to facilitate solvent evaporation, and the second and third aliquots were deposited only when the first and second aliquots had dried, respectively, in order to avoid layer cracking. This resulted in a total cathode catalyst loading of 4 mg cm^{-2}. The remaining solvents and water were then completely evaporated at 80 °C. The anode was a 0.5 mg$_{Pt}$·cm^{-2} electrode pre-deposited on Sigracet S10-BC. Membrane-electrode-assemblies were fabricated by hot-pressing the anode and cathode with geometric areas of 4.84 cm^2 against a Nafion® NRE-211 membrane at 135 °C for 2 min. The membrane electrode assemblies were installed in a single-cell PEMFC with serpentine flow fields (Fuel Cell Technologies Inc., Albuquerque, NM, USA). The fuel cell bench was an in-house bench connected to a Biologic Potentiostat with a 50 amperes booster. The experiments were controlled with the EC-Lab software (Bio-Logic Science Instruments, Claix, France). For all fuel cell tests reported in the present work, the cell temperature was 80 °C, the humidifier's temperature was 85 °C, and the inlet gas pressures were 1 bar gauge at both the anode and the cathode. The humidified H_2 and O_2 flow rates were *ca.* 50–70 sccm, as controlled downstream of the fuel cell. The fuel cell polarization curves were recorded with EC lab software using the cycling voltammetry experiment and scanning the cell at 0.5 mV·s^{-1}.

4. Conclusions

The templating of nano-sized ZIF-8 crystals into catalytic Fe–N–C particles with open porosity is shown to result in improved power performance at high current density in PEM fuel cells. In order to achieve a templating effect in the pyrolyzed catalysts, the milling speed used to mix the iron salt, phenanthroline and ZIF-8 precursors before pyrolysis is lowered to avoid agglomeration. In a second stage, the heating mode used to pyrolyze the catalyst precursors under inert gas is crucial in order to reach high BET and microporous areas and an intra-particle pore-network free of bottlenecks. A flash pyrolysis mode (room temperature to 1050 °C in circa 1.5 min) resulted in a low BET area and large hysteresis in the N_2 sorption isotherms while a ramp heating mode at 5 °C min^{-1} resulted in a high BET area and little or no hysteresis in the N_2 sorption isotherms. The use of ZIF-8 nanocrystals of average size 100 nm combined with a low milling speed for preparing the catalyst precursor that was subsequently pyrolyzed in argon in ramp mode resulted in a much improved ORR activity at high potential and also improved power performance at 0.5 V. The synthesis of Fe–N–C catalysts pyrolyzed in inert atmosphere and demonstrating improved power performance is important due to their known better durability in PEM fuel cells compared to NH_3-pyrolyzed Fe–N–C catalysts. The templating and open porosity effects resulting in improved power performance reported here for ZIF-8 will most likely be applicable to other MOFs as well and also open the door to the design of advanced composite materials comprising MOFs and corrosion-resistant supports such as carbon nanotubes or fibers.

Acknowledgments: We acknowledge funding from Agence Nationale de la Recherche, ANR, contract 2011 CHEX 004 01. The authors thank Didier Cot (Institut Européen des Membranes, Montpellier) for his contribution to the SEM measurements.

Author Contributions: Julien Hannauer synthesized and characterized nano ZIFs. Vanessa Armel synthesized and characterized Fe–N–C catalysts. Frédéric Jaouen supervised the research. Vanessa Armel and Frédéric Jaouen wrote the manuscript.

Conflicts of Interest: The authors declare no conflict of interest.

References

1. Debe, M. Electrocatalyst approaches and challenges for automotive fuel cells. *Nature* **2012**, *486*, 43–51.
2. Wagner, F.T.; Lakshmanan, B.; Mathias, M.F. Electrochemistry and the future of the automobile. *J. Phys. Chem. Lett.* **2010**, *1*, 2204–2219.
3. Resnick, P.R. A short history of Nafion. *Actual. Chim.* **2006**, *301*, 144–147.
4. Banerjee, S.; Curtin, D.E. Nafion® perfluorinated membranes in fuel cells. *J. Fluor. Chem.* **2004**, *125*, 1211.
5. Gasteiger, H.A.; Panels, J.E.; Yan, S.G. Dependence of PEM fuel cell performance on catalyst loading. *J. Power Sources* **2004**, *127*, 162–171.
6. Jaouen, F.; Proietti, E.; Lefèvre, M.; Chenitz, R.; Dodelet, J.-P.; Wu, G.; Chung, H.T.; Johnston, C.M.; Zelenay, P. Recent advances in non-precious metal catalysis for oxygen reduction reaction in polymer electrolyte fuel cells. *Energy Environ. Sci.* **2011**, *4*, 114–130.
7. Lefèvre, M.; Proietti, E.; Jaouen, F.; Dodelet, J.P. Iron-based catalysts with improved oxygen reduction activity in polymer electrolyte fuel cells. *Science* **2009**, *324*, 71–74.
8. Proietti, E.; Jaouen, F.; Lefèvre, M.; Larouche, N.; Tian, J.; Herranz, J.; Dodelet, J.-P. Iron-based cathode catalyst with enhanced power density in polymer electrolyte membrane fuel cells. *Nat. Commun.* **2011**, *2*, 416.
9. Wu, G.; More, K.L.; Johnston, C.M.; Zelenay, P. High-performance electrocatalysts for oxygen reduction derived from polyaniline, iron, and cobalt. *Science* **2011**, *332*, 443–447.
10. Zhao, D.; Shui, J.-L.; Grabstanowicz, L.R.; Chen, C.; Commet, S.M.; Xu, T.; Lu, J.; Liu, D.-J. Highly Efficient Non-Precious Metal Electrocatalysts Prepared from One-Pot Synthesized Zeolitic Imidazolate Frameworks. *Adv. Mater.* **2014**, *26*, 1093–1097.
11. Chang, S.-T.; Wang, C.-H.; Du, H.-Y.; Hsu, H.-C.; Kang, C.-M.; Chen, C.-C.; Wu, J.C.S.; Yen, S.-C.; Huang, W.-F.; Chen, L.-C.; *et al.* Vitalizing fuel cells with vitamins: Pyrolyzed vitamin B12 as a non-precious catalyst for enhanced oxygen reduction reaction of polymer electrolyte fuel cells. *Energy Environ. Sci.* **2012**, *5*, 5305–5314.
12. Serov, A.; Artyushkova, K.; Atanassov, P. Fe–N–C Oxygen Reduction Fuel Cell Catalyst Derived from Carbendazim: Synthesis, Structure, and Reactivity. *Adv. Energy Mater.* **2014**, *4*, 1301735.
13. Ma, S.; Goenaga, G.A.; Call, A.V.; Liu, D.J. Cobalt imidazolate framework as precursor for oxygen reduction electrocatalysts. *Chem. Eur. J.* **2011**, *17*, 2063–2067.
14. Jaouen, F.; Dodelet, J.P. Average turn-over frequency of O_2 electro-reduction for Fe/N/C and Co/N/C catalysts in PEFCs. *Electrochim. Acta* **2007**, *52*, 5975–5984.
15. He, P.; Lefèvre, M.; Faubert, G.; Dodelet, J.P. Oxygen reduction catalysts for polymer electrolyte fuel cells from the pyrolysis of various transition metal acetates adsorbed on 3,4,9,10-perylenetetracarboxylic dianhydride. *J. New Mater. Electrochem. Syst.* **1999**, *2*, 243–251.
16. Wu, G.; Johnston, C.M.; Mack, N.H.; Artyushkova, K.; Ferrandon, M.; Nelson, M.; Lezama-Pacheco, J.S.; Conradson, S.D.; More, K.L.; Myers, D.J.; *et al.* Synthesis-structure-performance correlation for polyaniline-Me-C non-precious metal cathode catalysts for oxygen reduction in fuel cells. *J. Mater. Chem.* **2011**, *21*, 11392–11405.

17. Lefèvre, M.; Dodelet, J.P. Recent advances in non-precious metal electrocatalysts for oxygen reduction in PEM fuel cells. *Electrochem. Soc. Trans.* **2012**, *45*, 35–44.

18. Charreteur, F.; Jaouen, F.; Dodelet, J.P. Iron porphyrin-based cathode catalysts for PEM fuel cells: Influence of pyrolysis gas on activity and stability. *Electrochim. Acta* **2009**, *54*, 6622–6630.

19. Kramm, U.I.; Herrmann-Geppert, I.; Fiechter, S.; Zehl, G.; Zizak, I.; Dorbandt, I.; Schmeißer, D.; Bogdanoff, P. Effect of iron-carbide formation on the number of active sites in Fe–N–C catalysts for the oxygen reduction reaction in acidic media. *J. Mater. Chem. A* **2014**, *2*, 2663–2670.

20. Jaouen, F.; Lefèvre, M.; Dodelet, J.P.; Cai, M. Heat-Treated Fe/N/C Catalysts for O_2 Electroreduction: Are Active Sites Hosted in Micropores? *J. Phys. Chem. B* **2006**, *110*, 5553–5558.

21. Jaouen, F.; Herranz, J.; Lefèvre, M.; Dodelet, J.P.; Kramm, U.I.; Herrmann, I.; Bogdanoff, P.; Maruyama, J.; Nagaoka, T.; Garsuch, A.; *et al.* Cross-laboratory experimental study of non-noble-metal electrocatalysts for the oxygen reduction reaction. *Appl. Mater. Interf.* **2009**, *1*, 1623–1639.

22. Ferrandon, M.; Kropf, A.J.; Myers, D.J.; Artyushkova, K.; Kramm, U.; Bogdanoff, P.; Wu, G.; Johnston, C.M.; Zelenay, P. Multitechnique characterisation of a polyaniline-iron-carbon oxygen reduction catalyst. *J. Phys. Chem. C* **2012**, *116*, 16001–16013.

23. Park, K.S.; Ni, Z.; Côté, A.P.; Choi, J.Y.; Huang, R.; Uribe-Romo, F.J.; Chae, H.K.; O'Keeffe, M.; Yaghi, O.M. Exceptional chemical and thermal stability of zeolitic imidazolate frameworks. *Proc. Natl. Acad. Sci. USA* **2006**, *103*, 10186–10191.

24. Wu, H.; Zhou, W.; Yildirim, T. Hydrogen Storage in a Prototypical Zeolitic Imidazolate Framework-8. *J. Am. Chem. Soc.* **2007**, *129*, 5314–5315.

25. Tian, J.; Morozan, A.; Sougrati, M.T.; Lefèvre, M.; Chenitz, R.; Dodelet, J.-P.; Jones, D.; Jaouen, F. Optimized Synthesis of Fe/N/C Cathode Catalysts for PEM Fuel Cells: A Matter of Iron—Ligand Coordination Strength. *Angew. Chem. Int. Ed.* **2013**, *52*, 6867–6870.

26. Strickland, K.; Miner, E.; Jia, Q.; Tylus, U.; Ramaswamy, N.; Liang, W.; Sougrati, M.-T.; Jaouen, F.; Mukerjee, S. Highly active oxygen reduction non-platinum group metal electrocatalyst without direct metal-nitrogen coordination. *Nat. Commun.* **2015**, *6*, 7343.

27. Barkholtz, H.M.; Chong, L.; Kaiser, Z.B.; Xu, T.; Liu, D.J. Highly Active Non-PGM Catalysts Prepared from Metal Organic Frameworks. *Catalysts* **2015**, *5*, 955–965.

28. Herranz, J.; Jaouen, F.; Lefèvre, M.; Kramm, U.I.; Proietti, E.; Dodelet, J.-P.; Bogdanoff, P.; Fiechter, S.; Abs-Wurmbach, I.; Bertrand, P.; *et al.* Unveiling N-protonation and anion-binding effects on Fe/N/C catalysts for O_2 reduction in proton-exchange-membrane fuel cells. *J. Phys. Chem. C* **2011**, *115*, 16087–16097.

29. Larouche, N.; Chenitz, R.; Lefèvre, M.; Proietti, E.; Dodelet, J.P. Activity and stability in proton exchange membrane fuel cells of iron-based cathode catalysts synthesized with addition of carbon fibers. *Electrochim. Acta* **2014**, *115*, 170–182.

30. Yang, L.; Larouche, N.; Chenitz, R.; Zhang, G.; Lefèvre, M.; Dodelet, J.-P. Activity, performance, and durability of for the reduction of oxygen in PEM fuel cells, of Fe/N/C electrocatalysts obtained from the pyrolysis of metal-organic-framework and iron porphyrin precursors. *Electrochim. Acta* **2015**, *159*, 184–197.

31. Wu, G.; Artyushkova, K.; Ferrandon, M.; Kropf, A.J.; Myers, D.; Zelenay, P. Performance durability of polyaniline-derived non-precious cathode catalysts. *Electrochem. Soc. Trans.* **2009**, *25*, 1299–1311.

32. Goellner, V.; Baldizzone, C.; Schuppert, A.; Sougrati, M.T.; Mayrhofer, K.; Jaouen, F. Degradation of Fe/N/C catalysts upon high polarization in acid medium. *Phys. Chem. Chem. Phys.* **2014**, *16*, 18454–18462.

33. Kramm, U.I.; Lefèvre, M.; Larouche, N.; Schmeisser, D.; Dodelet, J.P. Correlations between mass activity and physicochemical properties of Fe/N/C catalysts for the ORR in PEM fuel cell via ^{57}Fe Mössbauer spectroscopy and other techniques. *J. Am. Chem. Soc.* **2013**, *136*, 978–985.

34. Morozan, A.; Sougrati, M.T.; Goellner, V.; Jones, D.; Stievano, L.; Jaouen, F. Effect of Furfuryl Alcohol on Metal Organic Framework-based Fe/N/C Electrocatalysts for Polymer Electrolyte Membrane Fuel Cells. *Electrochim. Acta* **2014**, *119*, 192–205.

35. Xia, W.; Zhu, J.; Guo, W.; An, L.; Xia, D.; Zou, R. Well-defined carbon polyhedrons prepared from nano metal-organic frameworks for oxygen reduction. *J. Mater. Chem. A* **2014**, *2*, 11606–11613.

36. Zhang, L.; Su, Z.; Jiang, F.; Yang, L.; Qian, J.; Zhou, Y.; Li, W.; Hong, M. Highly graphitized nitrogen-doped porous carbon nanopolyhedra derived from ZIF-8 nanocrystals as efficient electrocatalysts for oxygen reduction reactions. *Nanoscale* **2014**, *6*, 6590–6602.

37. Cravillon, J.; Münzer, S.; Lohmeier, S.-J.; Feldhoff, A.; Huber, K.; Wiebcke, M. Rapid room-temperature synthesis and characterization of nanocrystals of a prototypical zeolitic imidazolate framework. *Chem. Mater.* **2009**, *21*, 1410–1412.

38. Zacher, D.; Nayuk, R.; Schweins, R.; Fischer, R.A.; Huber, K. Monitoring the Coordination Modulator Shell at MOF Nanocrystals. *Crystal Growth Design* **2014**, *14*, 4859–4863.

39. Sindoro, M.; Yanai, N.; Jee, A.-J.; Granick, S. Colloidal-Sized Metal Organic Frameworks: Synthesis and Applications. *Acc. Chem. Res.* **2014**, *47*, 459–469.

40. Fan, X.; Wang, W.; Li, W.; Zhou, J.; Wang, B.; Zheng, J.; Li, X. Highly Porous ZIF-8 Nanocrystals Prepared by a Surfactant Mediated Method in Aqueous Solution with Enhanced Adsorption kinetics. *ACS Appl. Mater. Interf.* **2014**, *6*, 14994–14999.

41. Diring, S.; Furukawa, S.; Takashima, Y.; Tsuruoka, T.; Kitagawa, S. Controlled Multiscale Synthesis of Porous Coordination Polymer in Nano/Micro Regimes. *Chem. Mater.* **2010**, *22*, 4531–4538.

42. Kida, K.; Okita, M.; Fujita, K.; Tanaka, S.; Miyake, Y. Formation of high crystalline ZIF-8 in an aqueous solution. *CrystEngComm* **2013**, *15*, 1794–1801.

43. Tsuruoka, T.; Furukawa, S.; Takashima, Y.; Yoshida, K.; Isoda, S.; Kitagawa, S. Nanoporous Nanorods Fabricated by Coordination Modulation and Oriented Attachment Growth. *Angew. Chem. Int. Ed.* **2009**, *48*, 4739–4743.

44. Cravillon, J.; Nayuk, R.; Springer, S.; Feldhoff, A.; Huber, K.; Wiebcke, M. Controlling zeolitic imidazolate framework nano- and microcrystal formation: Insight into crystal growth by time-resolved *in situ* static light scattering. *Chem. Mater.* **2011**, *23*, 2130–2141.

45. Pan, Y.; Liu, Y.; Zeng, G.; Zhao, L.; Lai, Z. Rapid synthesis of zeolitic imidazolate framework-8 (ZIF-8) nanocrystals in an aqueous system. *Chem. Commun.* **2011**, *47*, 2071–2073.

46. Goellner, V.; Armel, V.; Zitolo, A.; Fonda, E.; Jaouen, F. Degradation by Hydrogen Peroxide of Metal-Nitrogen-Carbon Catalysts for Oxygen Reduction. *J. Electrochem. Soc.* **2015**, *162*, H403–H414.

47. Yuan, S.; Shui, J.-L.; Grabstanowicz, L.; Chen, C.; Commet, S.; Reprogle, B.; Xu, T.; Yu, L.; Liu, D.-J. A Highly Active and Support-Free Oxygen Reduction Catalyst Prepared from Ultrahigh-Surface-Area Porous Polyporphyrin. *Angew. Chem. Int. Ed.* **2013**, *52*, 1–6.

Surfactant-Template Preparation of Polyaniline Semi-Tubes for Oxygen Reduction

Shiming Zhang and Shengli Chen

Abstract: Nitrogen and metal doped nanocarbons derived from polyaniline (PANI) have been widely explored as electrocatalysts for the oxygen reduction reaction (ORR) in fuel cells. In this work, we report surfactant-template synthesis of PANI nanostructures and the ORR electrocatalysts derived from them. By using cationic surfactant such as the cetyl trimethyl ammonium bromide (CTAB) as the template and the negatively charged persulfate ions as the oxidative agent to stimulate the aniline polymerization in the micelles of CTAB, PANI with a unique 1-D semi-tubular structure can be obtained. The semi-tubular structure can be maintained even after high-temperature treatment at 900 °C, which yields materials exhibiting promising ORR activity.

Reprinted from *Catalysts*. Cite as: Zhang, S.; Chen, S. Surfactant-Template Preparation of Polyaniline Semi-Tubes for Oxygen Reduction. *Catalysts* **2015**, *5*, 1202–1210.

1. Introduction

Seeking the highly-active electrocatalysts for oxygen reduction reaction (ORR) has become the one of the urgent demands for fuel cells, which would take a key role in the "hydrogen energy economy" [1]. In recent years, non-precious metal and/or metal-free materials based on nitrogen (N)-doped nanocarbons have shown great promise in substituting Pt and its alloys for catalyzing the ORR [2,3].

Polyaniline (PANI), a low-cost and easy-making conjugate conducting polymer containing rich content of nitrogen, has received extensive research interest [4–6]. Very recently, a variety of N-doped carbon catalysts based on PANI have been constructed which showed superior electrocatalytic activities for the ORR [7–12]. The multi-technique characterization have suggested that the enviable performance should be ascribed to the formation of metal-N complexion structures as well as the carbon nanostructures such as thin graphene sheets and nanofibers [10–12]. It has been generally accepted that the formation of uniform and ordered carbon nanostructures is very important in enhancing the catalytic activity [10–19].

Up to now, doped carbon electrocatalysts of different morphologies, such as nanoparticles [20], nanowires [21], nanotubes [22], nanorods [23], hollow nanospheres [24] and amorphous carbons [25], have been constructed by using various methods. The soft-template synthesis through self-assembly processes is among the most straightforward methods for nanostructure formation. In this work, we use assembly architectures of a variety of surfactants as the soft-templates to synthesize PANI nanostructures. In particular, PANI semi-tubes with uniform diameters of ~80 nm are obtained by using cationic surfactant. The electrocatalysts derived from these PANI semi-tubes show good ORR catalytic activity in alkaline.

2. Results and Discussion

Figure 1 shows the morphologies of PANI materials obtained by using CTAB (80 mM) as the template and APS as the oxidative agent. It can be seen that uniform 1-D nanostructures with diameters of ~80 nm and lengths of a few micrometers were obtained under this condition. Careful inspection revealed that these 1-D nanostructures possessed semi-cannular structures. As seen from the TEM images (Figure 1b and its insert), the walls of the individual tubes were highly rugged and full of cone-shaped protuberances of ~10 nm lengths, exhibiting centipede-like morphologies. We denoted this sample as $PANI_{s-tubes}$.

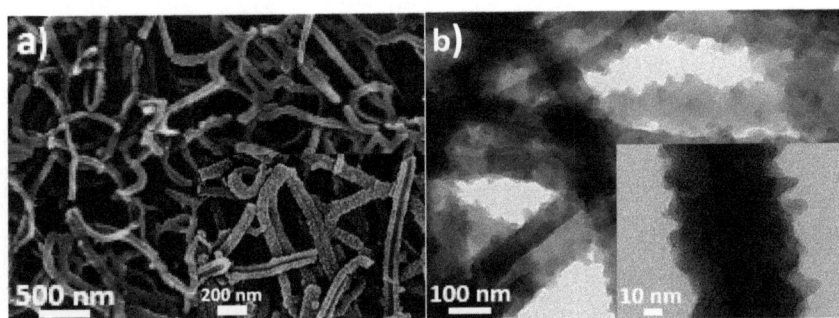

Figure 1. (a) SEM and (b) TEM images for polyaniline (PANI) obtained by using cetyl trimethyl ammonium bromide (CTAB) as template and ammonium persulfate (APS) as oxidative agent.

In the case when the preparation was conducted using the same procedure as that giving $PANI_{s-tubes}$ but the CTAB was absent, highly agglomerated PANI particles were obtained (Figure 2a). When the $FeCl_3$ was used to replace APS as the oxidative agent to stimulate the polymerization of aniline in CTAB solution, irregular PANI nanosheets were obtained (Figure 2b). These results indicated that the semi-cannular structured PANIs can be uniquely formed through the oxidative polymerization of anilines by APS in the assemblies of CTAB.

Figure 2. Morphologies of PANI materials obtained when (**a**) the CTAB was absent, or (**b**) FeCl₃ instead of APS was used as the oxidative reagent. The other conditions are the same as that for Figure 1.

We have also explored the effects of the surfactant types on the morphologies of the formed PANI materials. For anionic surfactants, e.g., SDBS, and non-ionic surfactants, e.g., X-100 and Span 40, mixtures of PANI nanoparticles and nanorods were obtained (Figure 3), which indicated the uniqueness of CTAB in producing the tubular structures of PANI.

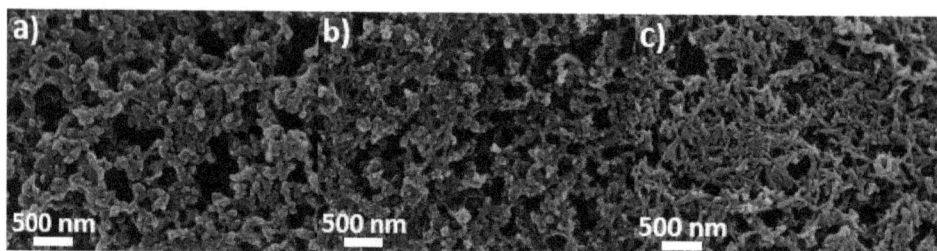

Figure 3. Morphologies of PANI materials obtained when the CTAB was replaced by (**a**) sodium dodecyl benzene sulfonate (SDBS), (**b**) Triton X-100 or, (**c**) Span 40. The other conditions are the same as that for Figure 1.

We believed that the formation of the PANI semi-tubes was related to the rod-like CTAB micelles and the opposite charges between the oxidative persulfate ions and the protonated aniline. Scheme 1 depicts the possible growth mechanism. The aniline molecules should be dissolved into the rod-like CTAB micelles. In the presence of HCl, the aniline molecules should be protonated. Therefore, they would be mainly located in the outer region of the micelles. The electrostatic attraction made the negatively charged persulfate ions approach the outer surface of micelles and oxidize the aniline molecules, which stimulate the polymerization in the outer region of cylindrical micelles. The formation of half instead of full PANI tubes was probably due to that there were only limited amounts of aniline molecules dissolved in the micelles. When the CTAB micelles were present or positively charged Fe(III) ions were used as oxidative agent, the mechanism shown in Scheme 1 could be altered. A deeper understanding of the effects of the oxidants and surfactants on the resultant PANI morphologies requires much more detailed investigation.

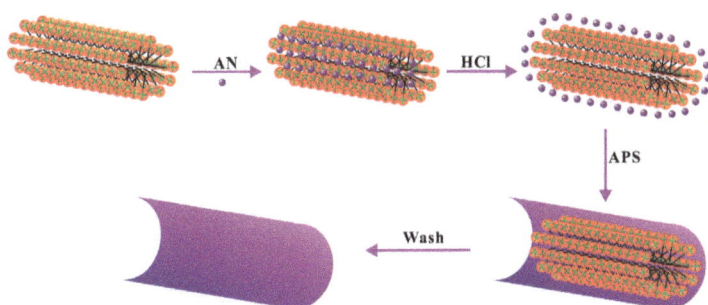

Scheme 1. The possible growth process of 1-D PANI$_{s\text{-tubes}}$ by CTAB micelle as the soft-template.

The obtained PANI samples were converted into N-doped carbon materials by heat-treating at 900 °C under Ar atmosphere. As shown in Figure 4, semi-tubular morphologies were maintained after the heat-treatment.

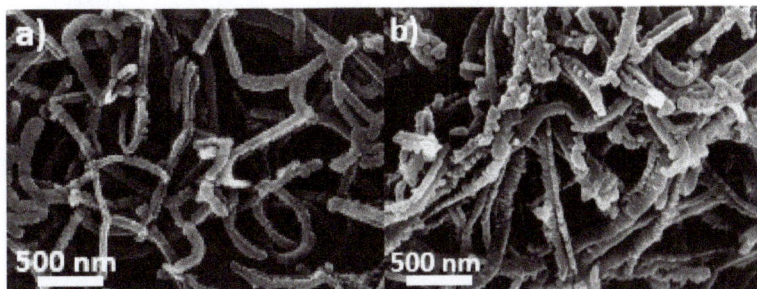

Figure 4. Comparison between the morphologies of PANI$_{s\text{-tubes}}$ (**a**) before and (**b**) after heat-treatment at 900 °C.

We investigated the physical properties and chemical composition of the PANI$_{s\text{-tubes}}$ materials before and after the heat-treatment. It was found that the heat-treatment resulted in materials exhibiting BET surface area (*ca.* 351.8 m^2/g) and pore volume (0.45 m^3/g) which were much higher than that before heat-treatment (*ca.* 54.8 m^2/g and 0.29 m^3/g respectively). This was probably due to the volatilizing release of some components during the heat-treatment, making the resulted materials more porous. XPS characterization results indicated that the contents of N and O decreased significantly after heat-treatment (Table 1), which confirmed the volatilization of some components.

Table 1. The content values of C, N and O estimated from XPS results for PANI$_{s\text{-tubes}}$ before and after heat-treatment.

PANI$_{s\text{-tubes}}$	C (at. %)	N (at. %)	O (at. %)
After Heat-Treatment	88.96	5.79	5.25
Before Heat-Treatment	78.85	10.88	10.27

Figure 5 compares the ORR polarization curves of the materials obtained through heat-treating the PANI prepared without using surfactant (see Figure 2a), the PANI$_{s\text{-tubes}}$ prepared by using CTAB micelles as templates, and the PANI$_{s\text{-tubes}}$/GS composite. For comparison, the ORR polarization curve for the pure GS is also given. It can be seen that the heat-treated PANI$_{s\text{-tubes}}$ exhibited significantly enhanced ORR activity as compared with the material derived from the PANI that was prepared without using surfactant template. This should be due to the open semi-tubular structure, which gave higher specific surface areas. As compared with the heat-treated PANI$_{s\text{-tubes}}$, the GS exhibited slightly more positive ORR onset potential, but slower current rising rate and lower limiting current. The heat-treated PANI$_{s\text{-tubes}}$/GS composite showed much higher ORR activity than that exhibited by the heat-treated PANI$_{s\text{-tubes}}$ and the GS alone.

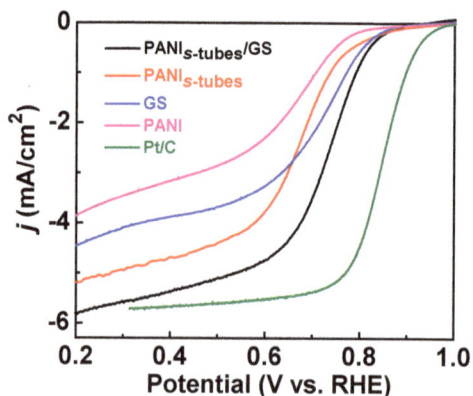

Figure 5. ORR polarization curves for different catalyst samples in O_2-saturated 0.1 M KOH at an electrode rotation speed of 1600 rpm. The PANI refers to the sample obtained by heat-treatment of PANI prepared without using surfactant; the PANI$_{s\text{-tubes}}$/GS refers to the sample obtained by heat-treatment of PANI$_{s\text{-tubes}}$ and graphene sheets together. The catalyst loadings were 0.3 mg cm^{-2} for the non-precious metal catalysts and 0.1 mg cm^{-2} for Pt/C (20 µg cm^{-2} for Pt).

Since the ORR polarization curves in Figure 5 were obtained with same total mass loading of 0.3 mg cm^{-2} for the heat-treated PANI$_{s\text{-tubes}}$, GS and PANI$_{s\text{-tubes}}$/GS, one may expect that the ORR activity of the heat-treated PANI$_{s\text{-tubes}}$/GS composite is between that of the heat-treated PANI$_{s\text{-tubes}}$ and GS. The actually higher ORR activity of the composite thus indicated that there were synergetic interaction between PANI$_{s\text{-tubes}}$ and GS. The introduction of GS may increase the electrical conductivity of the composite. On the other hand, the formation of composite between the 2-D GS and the 1-D PANI$_{s\text{-tubes}}$ would prevent the GS and PANI$_{s\text{-tubes}}$ from agglomerating and stacking during the heat treatment and electrode preparation. There have been numerous studies showing that the electrochemical performance of nanomaterials can be enhanced by forming composites with GS, due to the good electric conductivity of GS and the capability of GS to improve the dispersion of the electroactive materials [26,27].

It can be seen that the limiting current of the composite was very similar to that of the Pt/C electrocatalyst. The value of the limited current is directly related to the electron transfer number of the reaction. It is known that the Pt-based electrocatalysts catalyze the ORR through a 4-electron pathway. Therefore, we have reason to believe that ORR proceeded on the heat-treated composite mainly through a 4-electron process. It is noted that the present PANI$_{s\text{-tubes}}$ materials were still less efficient for the ORR than the Pt/C catalyst. Further optimization on the material preparation is necessary to promote the electrocatalytic activity.

3. Experimental Section

3.1. Chemicals and Materials

Various surfactants, such as cetyl trimethyl ammonium bromide (CTAB), sodium dodecyl benzene sulfonate (SDBS), and octoxinol (Triton X-10), and sorbitan monopalmitate (Span 40), and other chemicals were purchased from Sinopharm Chemical Reagent Co., Ltd. (Shanghai, China). The 20 wt. % Pt/C from Johnson Matthey (JM, London, UK) was used as reference catalyst.

3.2. Materials Synthesis

In a typical synthesis, a desired amount of CTAB was mixed with 15 mL of ultrapure water under ultrasonication for more than 30 min. After fully dissolution of CTAB, 200 µL of aniline (AN) and 20 mL of 1 M HCl were successively added under ultrasonication for another 30 min and then the solution was allow to stay for 24 h. 5 mL of 0.4 M ammonium persulfate (APS) solution was then added into the solution to stimulate the polymerization of aniline in the CTAB assemblies. Different CTAB concentrations (10–80 mM) were explored and no substantial difference in the morphology was seen for the obtained PANI. During the reaction progress, the color gradually changed to blue and cloudy precipitates were formed, which were collected through ultrafiltration and then alternately washed by ethanol and water followed by freeze-drying. Finally, dark blue product was obtained. For comparison, the preparation was also conducted by replacing CTAB with other surfactants.

We also prepared the composite of PANIs with graphene nanosheets (GS) that were prepared by high temperature thermal reduction of graphene oxide [17,18]. In this case, 30 mg of GS were added into the CTAB solution to prepare the PANI, which corresponded to a PANI/GS ratio of 1/8. Different PANI/GS ratios were explored and the 1/8 was found to be the most optimized value. To obtain electrocatalysts from the obtained nanostructured PANI, they were pyrolyzed under a flow of Ar at 900 °C for 1 h.

3.3. Characterization

Scanning electron microscopy (SEM) images were obtained by Hitachi S-4800 Scanning Electron Microsope (Tokyo, Japan). Transmission electron microscope (TEM) images were obtained at JEM-2100F (JEOL Ltd., Tokyo, Japan). The values of Brunner-Emmet-Teller (BET) surface area and total pore volume (TPV) were from N_2 adsorption isotherms using an ASAP2020 Surface Area and Porosity Analyzer (Micromeritics, Atlanta, GA, USA). X-ray photoelectron spectroscopy (XPS) measurements were carried out using a Kratos Ltd. XSAM-800 spectrometer (Kratos Analytical Ltd., Manchester, UK) with Mg Kα radiator. The data were fitted by using Gaussian/Lorentzian fitting in the software XPSPEAK41 (Kratos Analytical Ltd., Manchester, UK) with Shirley function as baseline.

3.4. Electrochemical Measurement

The three-electrode configuration were used for electrochemical measurements using Pt foil counter electrode and saturated calomel reference electrode. To prepare the working electrodes, catalyst samples as a thin film were coated onto a glass carbon (GC) RDE substrate (diameter: 5 mm) with Nafion as the binding agent. For the PANI nanostructures, 5 mg catalysts were dispersed in 1 mL Nafion solution (0.5 wt. % Nafion in isopropyl alcohol) to form the catalyst inks and 12 μL ink suspension was pipetted onto the GC RDE. For the Pt/C catalyst, 5 mg catalyst was dispersed ultrasonically in 1 mL Nafion-isopropyl alcohol solution and 4 μL of the resulted suspension was then pipetted onto the GC RDE. The catalyst loadings were respectively 0.3 mg cm^{-2} for the non-precious metal catalysts and 0.1 mg cm^{-2} for the Pt/C (20 μg cm^{-2} for Pt).

4. Conclusions

In this work, a unique 1-D semi-tubular structure of PANI has been obtained by using self-assemblies of CTAB molecules as soft-templates and APS as oxidative agent in aqueous solution. The obtained PANI nanostructure can be maintained in the course of high-temperature treatment. The materials derived from heat-treating the composite of PANI semi-tubes and GS show significantly enhanced ORR performance.

Acknowledgments: This work was supported by the Ministry of Science and Technology of China under the National Basic Research Program (Grant nos. 2012CB215500 and 2012CB932800).

Author Contributions: S.M.Z. performed the experiments. S.L.C. and S.M.Z. analyzed the data and wrote the paper.

Conflicts of Interest: The authors declare no conflict of interest.

References

1. *Handbook of Fuel Cells: Fundamentals, Technology and Application*; Vielstich, W., Lamm, A., Gasteiger, H.A., Eds.; Wiley: West Sussex, UK, 2003.

2. Mazumder, V.; Lee, Y.; Sun, S. Recent Development of Active Nanoparticle Catalysts for Fuel Cell Reactions. *Adv. Funct. Mater.* **2010**, *20*, 1224–1231.

3. Chen, Z.; Higgins, D.; Yu, A.; Zhang, L.; Zhang, J. A review on non-precious metal electrocatalysts for PEM fuel cells. *Energy Environ. Sci.* **2011**, *4*, 3167–3192.

4. Zhang, D.; Wang, Y. Synthesis and applications of one-dimensional nano-structured polyaniline: An overview. *Mater. Sci. Eng. B* **2006**, *134*, 9–19.

5. Li, D.; Huang, J.; Kaner, R.B. Polyaniline Nanofibers: A Unique Polymer Nanostructure for Versatile Applications. *Acc. Chem. Res.* **2008**, *42*, 135–145.

6. Wang, L.; Lu, X.; Lei, S.; Song, Y. Graphene-Based Polyaniline Nanocomposites: Preparation, Properties and Applications. *J. Mater. Chem. A* **2014**, *2*, 4491–4509.

7. Zhong, H.; Zhang, H.; Xu, Z.; Tang, Y.; Mao, J. A Nitrogen-Doped Polyaniline Carbon with High Electrocatalytic Activity and Stability for the Oxygen Reduction Reaction in Fuel Cells. *ChemSusChem* **2012**, *5*, 1698–1702.

8. Hu, Y.; Zhao, X.; Huang, Y.; Li, Q.; Bjerrum, N.J.; Liu, C.; Xing, W. Synthesis of Self-Supported Non-Precious Metal Catalysts for Oxygen Reduction Reaction with Preserved Nanostructures from the Polyaniline Nanofiber Precursor. *J. Power Sources* **2013**, *225*, 129–136.

9. Gavrilov, N.; Pašti, I.A.; Mitrić, M.; Travas-Sejdić, J.; Ćirić-Marjanović, G.; Mentus, S.V. Electrocatalysis of Oxygen Reduction Reaction on Polyaniline-Derived Nitrogen-Doped Carbon Nanoparticle Surfaces in Alkaline Media. *J. Power Sources* **2012**, *220*, 306–316.

10. Wu, G.; More, K.L.; Johnston, C.M.; Zelenay, P. High-Performance Electrocatalysts for Oxygen Reduction Derived from Polyaniline, Iron, and Cobalt. *Science* **2011**, *332*, 443–447.

11. Ferrandon, M.; Kropf, A.J.; Myers, D.J.; Artyushkova, K.; Kramm, U.; Bogdanoff, P.; Wu, G.; Johnston, C.M.; Zelenay, P. Multitechnique Characterization of a Polyaniline-Iron-Carbon Oxygen Reduction Catalyst. *J. Phys. Chem. C* **2012**, *116*, 16001–16013.

12. Wu, G.; Zelenay, P. Nanostructured Nonprecious Metal Catalysts for Oxygen Reduction Reaction. *Acc. Chem. Res.* **2013**, *46*, 1878–1889.

13. Gong, K.; Du, F.; Xia, Z.; Durstock, M.; Dai, L. Nitrogen-doped carbon nanotube arrays with high electrocatalytic activity for oxygen reduction. *Science* **2009**, *323*, 760–764.

14. Cheon, J.Y.; Kim, T.; Choi, Y.; Jeong, H.Y.; Kim, M.G.; Sa, Y.J.; Kim, J.; Lee, Z.; Yang, T.H.; Kwon, K.; *et al.* Ordered Mesoporous Porphyrinic Carbons with Very High Electrocatalytic Activity for the Oxygen Reduction Reaction. *Sci. Rep.* **2013**, *3*, 2715.

15. Ding, W.; Li, L.; Xiong, K.; Wang, Y.; Li, W.; Nie, Y.; Chen, S.; Qi, X.; Wei, Z. Shape Fixing via Salt Recrystallization: A Morphology-Controlled Approach To Convert Nanostructured Polymer to Carbon Nanomaterial as a Highly Active Catalyst for Oxygen Reduction Reaction. *J. Am. Chem. Soc.* **2015**, *137*, 5414–5420.

16. Zhang, S.; Zhang, H.; Hua, X.; Chen, S. Tailoring Molecular Architectures of Fe Phthalocyanine on Nanocarbon Supports for High Oxygen Reduction Performance. *J. Mater. Chem. A* **2015**, *3*, 10013–10019.

17. Zhang, S.; Zhang, H.; Liu, Q.; Chen, S. Fe–N doped carbon nanotube/graphene composite: Facile synthesis and superior electrocatalytic activity. *J. Mater. Chem. A* **2013**, *1*, 3302–3308.

18. Zhang, S.; Liu, B.; Chen, S. Synergistic increase of oxygen reduction favourable Fe–N coordination structures in a ternary hybrid of carbon nanospheres/carbon nanotubes/graphene sheets. *Phys. Chem. Chem. Phys.* **2013**, *15*, 18482–18490.

19. Liu, Q.; Zhang, H.; Zhong, H.; Zhang, S.; Chen, S. N-doped graphene/carbon composite as non-precious metal electrocatalyst for oxygen reduction reaction. *Electrochim. Acta* **2012**, *81*, 313–320.

20. Yan, J.; Wei, T.; Fan, Z.; Qian, W.; Zhang, M.; Shen, X.; Wei, F. Preparation of Graphene Nanosheet/Carbon Nanotube/Polyaniline Composite as Electrode Material for Supercapacitors. *J. Power Sources* **2010**, *195*, 3041–3045.

21. Xu, J.; Wang, K.; Zu, S.Z.; Han, B.H.; Wei, Z. Hierarchical Nanocomposites of Polyaniline Nanowire Arrays on Graphene Oxide Sheets with Synergistic Effect for Energy Storage. *ACS Nano* **2010**, *4*, 5019–5026.

22. Huang, Y.; Lin, C. Facile Synthesis and Morphology Control of Graphene Oxide/Polyaniline Nanocomposites via *in situ* Polymerization Process. *Polymer* **2012**, *53*, 2574–2582.

23. Hu, L.; Tu, J.; Jiao, S.; Hou, J.; Zhu, H.; Fray, D.J. *In situ* Electrochemical Polymerization of a Nanorod-PANI-Graphene Composite in a Reverse Micelle Electrolyte and Its Application in a Supercapacitor. *Phys. Chem. Chem. Phys.* **2012**, *14*, 15652–15656.

24. Fan, W.; Zhang, C.; Tjiu, W.W.; Pramoda, K.P.; He, C.; Liu, T. Graphene-Wrapped Polyaniline Hollow Spheres as Novel Hybrid Electrode Materials for Supercapacitor Applications. *ACS Appl. Mater. Interfaces* **2013**, *5*, 3382–3391.

25. Lai, L.; Potts, J.R.; Zhan, D.; Wang, L.; Poh, C.K.; Tang, C.; Gong, H.; Shen, Z.; Linc, J.; Ruoff, R.S. Exploration of the Active Center Structure of Nitrogen-Doped Graphene-Based Catalysts for Oxygen Reduction Reaction. *Energy Environ. Sci.* **2012**, *5*, 7936–7942.

26. Lee, S.H.; Lee, D.H.; Lee, W.J.; Kim, S.O. Tailored Assembly of Carbon Nanotubes and Graphene. *Adv. Funct. Mater.* **2011**, *21*, 1338–1354.

27. Fan, Z.; Yan, J.; Zhi, L.; Zhang, Q.; Wei, T.; Feng, J.; Zhang, M.; Qian, W.; Wei, F. A Three-Dimensional Carbon Nanotube/Graphene Sandwich and Its Application as Electrode in Supercapacitors. *Adv. Mater.* **2010**, *22*, 3723.

Polyaniline-Derived Ordered Mesoporous Carbon as an Efficient Electrocatalyst for Oxygen Reduction Reaction

Kai Wan, Zhi-Peng Yu and Zhen-Xing Liang

Abstract: Nitrogen-doped ordered mesoporous carbon was synthesized by using polyaniline as the carbon source and SBA-15 as the template. The microstructure, composition and electrochemical behavior were extensively investigated by the nitrogen sorption isotherm, X-ray photoelectron spectroscopy, cyclic voltammetry and rotating ring-disk electrode. It is found that the pyrolysis temperature yielded a considerable effect on the pore structure, elemental composition and chemical configuration. The pyrolysis temperature from 800 to 1100 °C yielded a volcano-shape relationship with both the specific surface area and the content of the nitrogen-activated carbon. Electrochemical tests showed that the electrocatalytic activity followed a similar volcano-shape relationship, and the carbon catalyst synthesized at 1000 °C yielded the best performance. The post-treatment in NH_3 was found to further increase the specific surface area and to enhance the nitrogen doping, especially the edge-type nitrogen, which favored the oxygen reduction reaction in both acid and alkaline media. The above findings shed light on electrocatalysis and offer more strategies for the controllable synthesis of the doped carbon catalyst.

Reprinted from *Catalysts*. Cite as: Wan, K.; Yu, Z.-P.; Liang, Z.-X. Polyaniline-Derived Ordered Mesoporous Carbon as an Efficient Electrocatalyst for Oxygen Reduction Reaction. *Catalysts* **2015**, *5*, 1034–1045.

1. Introduction

The oxygen reduction reaction (ORR) is one key electrochemical process for the energy conversion devices, like fuel cells and metal-air batteries. Pt-based materials have been so far acknowledged to be the most effective catalysts for the ORR at low temperatures [1,2]; however, the source scarcity and high cost pose great challenges to the large-scale applications to fuel cells [3–5]. Hence, enormous effort has been devoted to search for greater efficiency, durability and less cost [6,7].

In recent years, nanostructured carbon materials have attracted increasing attention as the Pt-alternative electrocatalysts. Dai [8] synthesized 1D nitrogen-doped carbon nanotubes by the chemical vapor deposition (CVD) method with iron (II) phthalocyanine as the precursor, which featured high charge transfer and, thus, facilitated the ORR. Feng [9] synthesized 2D graphene-based carbon nitride nanosheets, of which the high specific surface favored the dense assembling of the active sites on the surface. Lu [10] synthesized 3D hierarchically porous nitrogen-doped carbons with a hieratical porous structure, which enabled low mass transfer resistance and improved accessibility of catalytic sites for the ORR.

Among them, nitrogen-doped ordered mesoporous carbon (NOMC) features high specific surface area and a uniform pore structure, which respectively facilitate the reaction kinetics and mass transfer to the electrode [11,12]. The hard-template method is a universal strategy to synthesize such materials, and SBA-15 is one of the most often used templates to synthesize NOMCs. Asefa [13] synthesized NOMC catalysts by pyrolyzing polyaniline in the framework of SBA-15. It was found that the NOMC catalyst pyrolyzed at 800 °C showed the best ORR performance. Guo [14] synthesized a series of NOMC catalysts with honey as the precursor, which showed a high specific surface area ranging from 1050 to 1273 $m^2 \cdot g^{-1}$. Mullen [15] synthesized high-performance NOMC catalysts by pyrolyzing ionic liquid within SBA-15. All of the above-mentioned catalysts showed a decent electrocatalytic activity to the ORR in alkaline media. Other mesoporous silicas have been also used for the synthesis. For example, Popov [16] synthesized nitrogen-doped ordered porous carbon with the aid of SBA-12, which also showed a superior activity for the ORR. Joo [17] synthesized a series of carbon catalysts by using various templates, like SBA-15, MSU-F, KIT-6 and fumed Carb-O-sil M-5. It was found that the template SBA-15 yielded an extremely high surface area of 1500 $m^2 \cdot g^{-1}$ and the best electrocatalytic activity.

Beside the template, the carbon precursor has a considerable effect on the microstructure and composition of the final carbon materials. A variety of nitrogen-containing organic chemicals, like phthalocyanine [18], porphyrin [19,20] and ionic liquid [21,22], have been used as the carbon precursor to synthesize the nitrogen-doped carbon catalyst. Polyaniline represents an aromatic ring connected via nitrogen-containing groups and, thus, facilitates the incorporation of nitrogen-containing active sites into the carbon matrix during the heat treatment [23]. Zelenay [23] found that the polyaniline-derived carbon electrocatalyst showed a superior performance to the ORR, which exhibited the highest maximum power density of 0.55 $W \cdot cm^{-2}$ at 0.38 V.

It has been well acknowledged that the template, carbon precursor and pyrolysis conditions have significant and complicated effects on the composition, structure and electrocatalytic activity of the ORR. In our previous work, we developed a method to synthesize the nitrogen-doped ordered mesoporous carbon featuring a high specific surface area [24,25]. Additionally, the active sites for the ORR were claimed to be the nitrogen-activated carbon atoms, on which the ORR proceeded by a surface-confined redox-mediation mechanism in both acid and alkaline media [24,26]. In this work, we will use polyaniline as the carbon source to synthesize the NOMC catalysts and to optimize the pyrolysis conditions. In order to improve the electrocatalytic activity, the as-prepared catalyst is further subjected to the NH_3-activation at high temperatures.

Then, the nitrogen sorption isotherm, electron microscopy and X-ray photoelectron spectroscopy are used to study the microstructure and composition. The cyclic voltammetry and rotating-ring-disk electrode methods are used to investigate the electrochemical behavior for the ORR.

2. Results and Discussion

Figure 1 shows the nitrogen sorption isotherm of the as-synthesized carbon materials. It is seen that all curves show a similar shape of the typical Type-IV isotherm, which indicates the mesoporous nature of the synthesized carbon (see Figure S1). Pore parameters are then extracted from the isotherms and listed in Table 1. It is seen that the specific surface area increases from 470 to 629 $m^2 \cdot g^{-1}$ with increasing the pyrolysis temperature from 800 to 1000 °C, which can be ascribed to the deepened decomposition in this course. Then, a further increase in the pyrolysis temperature to 1100 °C results in a decrease in the specific surface area, which may be linked to the collapse of the carbon framework. A similar trend is also seen in the specific pore volume, which reaches the highest value of 0.81 $cm^{-3} \cdot g^{-1}$ at 1000 °C. It is noted that the post-treatment in NH_3 yields a dramatic increase in the specific surface area (1312 $m^2 \cdot g^{-1}$), which can be attributed to the gasification of amorphous carbon and the consequent formation of micro-/meso-pores in the active atmosphere [27–29].

Table 1. Pore features of the synthesized carbon materials.

Samples	$A_{BET}/m^2 \cdot g^{-1}$	$A_{MP}/m^2 \cdot g^{-1}$	D_{BJH}/nm	$V/cm^3 \cdot g^{-1}$
C-PA-800	470	67	5.4	0.58
C-PA-900	569	41	5.9	0.80
C-PA-1000	629	132	6.1	0.81
C-PA-1100	517	61	5.9	0.67
C-PA-1000-NH$_3$	1312	229	5.9	1.73

The surface composition is characterized by XPS, for which the survey spectra are shown in Figure S2. The content of the main elements are qualified and listed in Table 2. It is found that both the pyrolysis temperature and NH_3-activation yield a significant effect on the nitrogen content. First, the nitrogen content shows a monotonic decrease from 5.07 to 1.25 at. % with increasing the pyrolysis temperature from 800 to 1100 °C, which should be attributed to the enhanced decomposition of the nitrogen-containing functional groups at higher temperatures. Second, the nitrogen content is 2.20 at. % for C-PA-1000 and 3.15 at. % for C-PA-1000-NH$_3$, indicating that the nitrogen doping can be enhanced by the pyrolysis in the nitrogen-containing active gases. The change in the surface composition is expected to yield effects on the electrocatalysis, as discussed below.

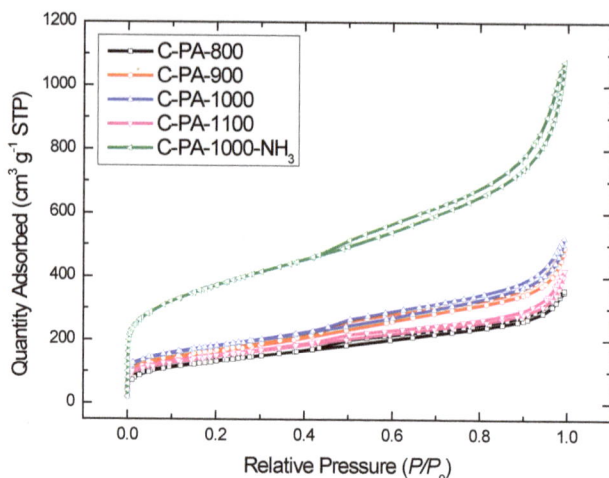

Figure 1. Nitrogen sorption isotherms of the synthesized carbon materials (STP: standard temperature and pressure).

Table 2. Elemental composition (at. %) of the synthesized carbon materials.

Samples	C	N	O	N:C
C-PA-800	86.63	5.07	5.33	0.057
C-PA-900	92.03	3.13	4.40	0.034
C-PA-1000	93.45	2.20	3.91	0.024
C-PA-1100	94.88	1.25	3.57	0.013
C-PA-1000-NH3	94.32	3.15	2.33	0.033

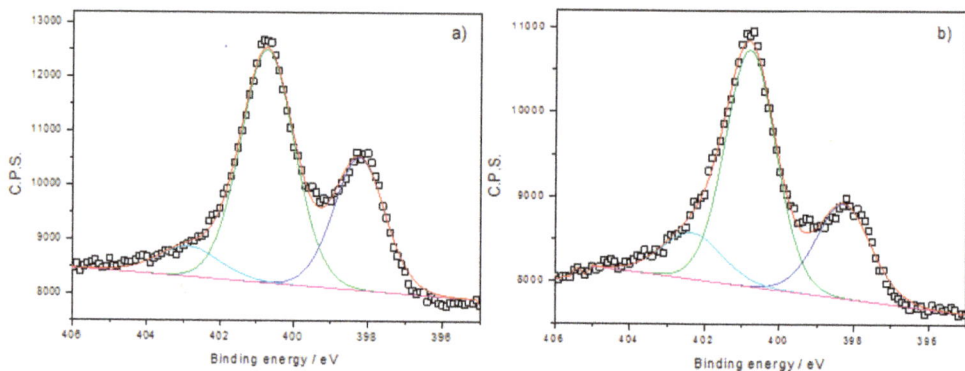

Figure 2. *Cont.*

298

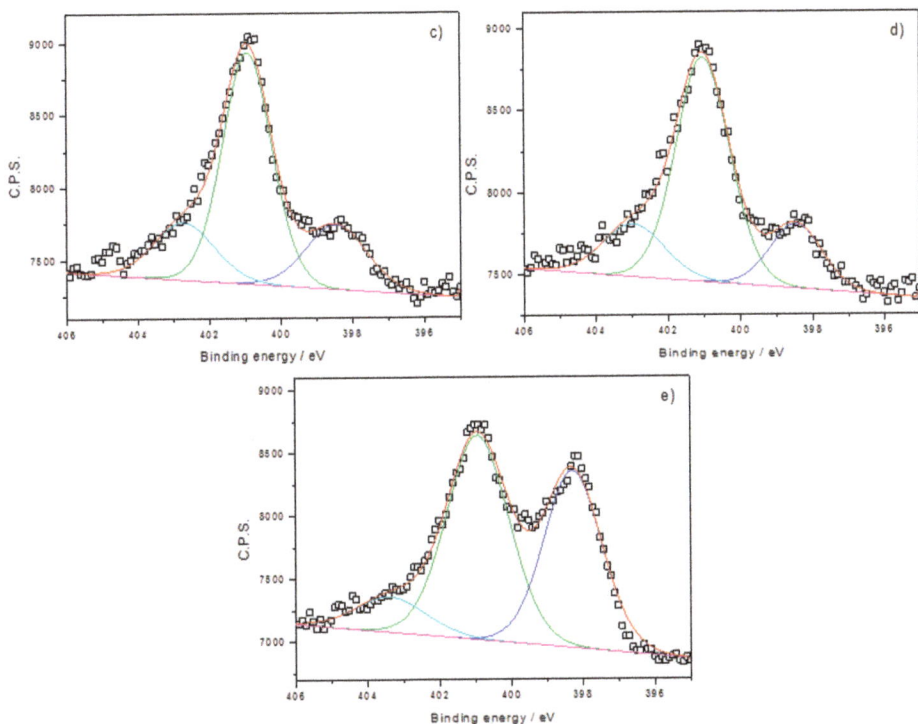

Figure 2. N 1s peak and the peak fitting results of the following materials: (a) C-PA-800; (b) C-PA-900; (c) C-PA-1000; (d) C-PA-1100; and (e) C-PA-1000-NH$_3$. The unit of y-axis unit is counts per second (C.P.S.).

Figure 2 shows the high-resolution XPS spectra of N 1s. These spectra can be deconvoluted to three peaks based on the binding energy: 398.4 ± 0.2, 401.0 ± 0.1 and 401.5–404 eV, which respectively correspond to pyridinic-nitrogen, graphitic-nitrogen and nitrogen-oxide [30,31]. Then, the content of each species is seen in Table 3. It is seen that the graphitic nitrogen is the predominant component among the three species, of which the content slightly increases with the pyrolysis temperature. In comparison, the content of the pyridinic-nitrogen decreases from 31.88% to 17.90% with increasing the pyrolysis temperature from 800 to 1100 °C. These results are consistent with the previous findings that the graphitic nitrogen is the most stable nitrogen species at high temperatures [32,33]. Finally, C-PA-1000-NH$_3$ shows an extraordinarily high content of pyridinic-nitrogen (40.31%), which should be attributed to the nitrogen doping at the edge of the graphite plane during the NH$_3$-etching process. It seems that the content of the graphitic nitrogen decreases to 49.88% in the etching process; however, it should be noted that the "absolute" content of this species remains unchanged, as compared with the un-etched one, by considering the total nitrogen content (see Table 2).

Table 3. Content of each nitrogen component (%) of the synthesized carbon materials.

Samples	Pyridinic-N	Graphitic-N	O-N
C-PA-800	31.88	59.10	9.02
C-PA-900	26.42	59.55	14.04
C-PA-1000	20.90	60.62	18.48
C-PA-1100	17.90	64.72	17.38
C-PA-1000-NH$_3$	40.31	49.88	9.81

In our previous work, the relationship has been well established between the content of the nitrogen-activated carbon and the electrocatalytic activity for ORR [24]. In line with this understanding, the curve fitting of high-resolution C1s peak is performed (see Figure S3), and the results are listed in Table 4. It is found that the content of the nitrogen-activated carbon slightly increases with increasing pyrolysis temperature from 800 to 1000 °C and then decreases at 1100 °C. Additionally, C-PA-1000-NH$_3$ shows the highest content of nitrogen-activated carbon among all of the carbon materials. The ORR performance is expected to follow this change in content, as seen below.

Table 4. Content of each carbon component (%) of the synthesized carbon materials.

Samples	C–C=C	C–N	C–O/C=N	C=O	COOH
C-PA-800	65.58	13.19	14.39	5.52	1.33
C-PA-900	68.29	15.60	8.81	5.46	1.84
C-PA-1000	68.84	18.51	7.53	3.85	1.28
C-PA-1100	73.60	15.11	6.10	3.71	1.48
C-PA-1000-NH$_3$	67.98	20.49	7.84	1.65	2.04

Figure 3 shows the CV curves in Ar-saturated 0.10 M KOH. It is seen that all of the curves are similar in shape with large capacitance currents, and broad symmetrical redox peaks are found in the potential range of 0 to 0.9 V. It is understandable that a large capacitance current should result from the high specific surface area, and the pseudocapacitance current is associated with the chemical adsorption of OH$^-$ onto the enriched redox couples. Basically, the capacitance current shows a volcano-shape relationship with the pyrolysis temperature, which first increases and then dramatically decreases at temperatures of 1100 °C. Notably, C-PA-1000-NH$_3$ shows the largest capacitance current. Such a change can be rationalized as a result of the specific surface area (*vide supra*). In comparison, the pseudocapacitance current shows a monotonic decrease with increasing pyrolysis temperature. Additionally, this result should be associated with the deepened decomposition of the electrochemically-active functional groups (like nitrogen-containing species) on the surface at higher temperatures. In line with the above analysis, C-PA-1000-NH$_3$ should yield the largest pseudocapacitance/capacitance currents due to its extraordinarily high specific surface area and nitrogen content. However, the increase in the capacitance current is not that large, and the pseudocapacitance current does not increase. This seeming contradiction may be understandable by considering the pore structure. The NH$_3$ post-treatment can effectively gasify/etch the amorphous carbon, leaving enriched micropores in the bulk. Additionally, these pores may not be fully utilized due to the lack of contact with the liquid electrolyte.

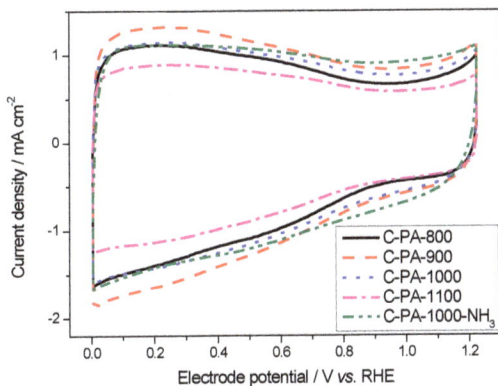

Figure 3. Cyclic voltammograms of the carbon materials in Ar-saturated 0.10 M KOH solution. RHE, reversible hydrogen electrode.

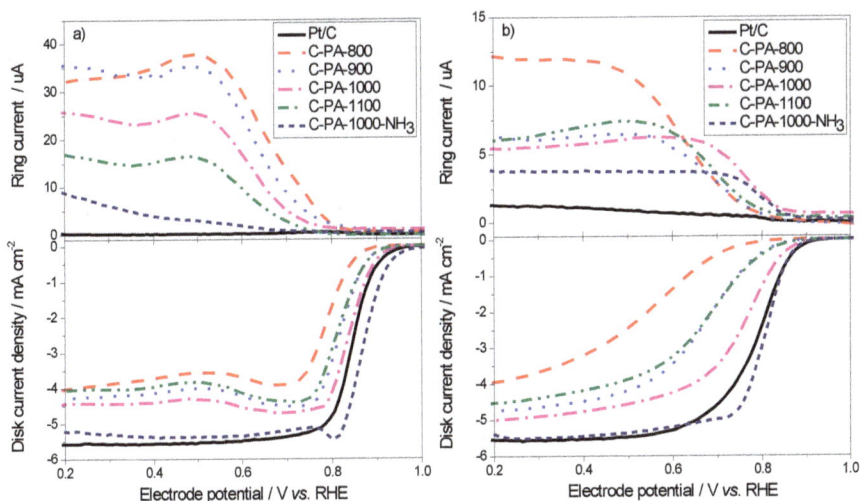

Figure 4. Ring (top) and disk (bottom) currents for the RRDE tests of the carbon materials: (**a**) O$_2$-saturated 0.10 M KOH solution; (**b**) O$_2$-saturated 0.10 M HClO$_4$ solution.

Figure 4 shows the polarization curves of the carbon materials in O_2-saturated 0.10 M KOH and 0.10 M $HClO_4$ solutions. Figure 4a shows that in alkaline media, the electrocatalytic activity shows a volcano-shape relationship, which increases with the pyrolysis temperature from 800 to 1000 °C and then decreases at 1100 °C. The same trend is further seen in the change of the yield of hydrogen peroxide and electron transfer number (see Figure S4a). Increasing the pyrolysis temperature lowers the yield of hydrogen peroxide, and thereby, selectively favors the 4-e reduction of oxygen. Such a trend in the electrocatalytic activity and selectivity correlates well with the change in the content of nitrogen-activated carbon and the specific surface area (*vide supra*). These findings further confirm that the active sites for the ORR should be the nitrogen-activated carbon atoms [24,26]. In acid media, the change in the electrocatalytic activity is similar to that in alkaline media (Figure 4b), revealing that the active site should remain substantially the same in a wide range of pH. Finally, it is noted that for C-PA-1000-NH_3, both the activity and the selectivity are considerably improved upon NH_3-activation. This result quantitatively agrees with the change in the specific surface area and the content of nitrogen-activated carbon, as discussed above. However, it seems irrational to directly correlate the small increase in the nitrogen-activated carbon (Table 4) with the extremely high activity and low H_2O_2 yield. The mechanism has not been fully understood yet, but the reason may be associated with high content of the pyridinic-nitrogen after the NH_3 treatment.

3. Experimental Section

3.1. Materials Preparation

Nitrogen-doped ordered mesoporous carbon (NOMC) was synthesized by a modified nanocasting method [24]. The process is briefly described as follows. (i) Synthesis of the template SBA-15 [34]: An aqueous mixture, consisting of Pluronic P123, HCl and tetraethoxysilane, was stirred for 20 h at 35 °C and then hydrothermally treated at 100 °C for 24 h. The resultant powders were calcined in air at 550 °C for 6 h, and SBA-15 was finally obtained. (ii) Impregnation of the carbon precursor: Aniline was impregnated into SBA-15 by the vaporization-capillary condensation method [24]. Then, the monomer was polymerized upon adding $FeCl_3$ to form polyaniline (PA). (iii) Pyrolysis and template removal: The resultant powders were then subjected to the pyrolysis at high temperatures (800, 900, 1000, 1100 °C) for 3 h in argon, respectively. Finally, the carbon catalyst was obtained by removing the silicate template by boiling in 10 M NaOH solution for 24 h. The samples were referred to as C-PA-X. Here, X refers to the pyrolysis temperature, *viz.* 800, 900, 1000 and 1100. In addition, the sample C-PA-1000 was heat-treated again at 1000 °C in ammonia for 30 min, which was labeled as C-PA-1000-NH_3.

3.2. Physical Characterizations

X-ray photoelectron spectroscopy (XPS, Physical Electronics PHI 5600, Chanhassen, MN, USA) measurement was carried out with a multi-technique system using an Al monochromatic X-ray at a power of 350 W. Transmission electron microscopy (TEM) was performed on an FEI Tecnai G2 F20 S-TWIN (Hillsboro, OR, USA) operated at 200 kV. Nitrogen adsorption/desorption isotherms were measured at 77 K using a Micromeritics TriStar II 3020 analyzer (Norcross, GA, USA). The total surface area was analyzed with the well-established Brunauer–Emmett–Teller (BET) method; the microporous surface area was obtained with the MP (micropore) method (*t*-plot method); and the pore size distribution was analyzed by the Barrett–Joyner–Halenda (BJH) method.

3.3. Electrochemical Characterization

The electrochemical behavior of the catalyst was characterized by the cyclic voltammetry (CV) and linear sweeping voltammetry (LSV) using a three-electrode cell with an electrochemical work station Zennium (Zahner, Germany) at room temperature (25 °C). A platinum wire and a double-junction Ag/AgCl reference electrode were used as the counter and reference electrodes, respectively. The working electrode was a rotating ring-disk electrode (RRDE, glassy carbon disk: 5.0 mm in diameter; platinum ring: 6.5 mm inner diameter and 7.5 mm outer diameter). The thin-film electrode on the disk was prepared as follows. Ten milligrams of the catalyst were dispersed in 1.0 mL Nafion/ethanol (0.84 wt. % Nafion) by sonication for 120 min. Then, 10 µL of the dispersion were transferred onto the glassy carbon disk by using a pipette, yielding the catalyst loading of $0.50 \text{ mg} \cdot \text{cm}^{-2}$. The ORR activity of the Pt/C catalyst (HiSPEC4000, Johnson Matthey, London, UK) with the metal loading of $20 \text{ µg} \cdot \text{cm}^{-2}$ was collected for comparison.

The electrolyte solution, 0.10 M KOH, was first bubbled with argon for 60 min. Then, a CV test was conducted at $20 \text{ mV} \cdot \text{s}^{-1}$ in the potential range between 0 and 1.23 V (*vs.* reversible hydrogen electrode, RHE) for 20 cycles. If not specified, the LSV curve was collected by scanning the disk potential from 1.2 down to 0 V at $5 \text{ mV} \cdot \text{s}^{-1}$ in the oxygen-saturated electrolyte solution under 1600 rpm, from which the ORR polarization curve was extracted by subtracting the capacitive current. During the collection, the potential of the ring was set to be 0.5 V (*vs.* RHE) to determine the yield of hydrogen peroxide.

The electron-transfer number (n) and hydrogen peroxide yield ($H_2O_2\%$) in the ORR were calculated from the following equations:

$$n = \frac{4|i_d|}{|i_d| + i_r/N} \tag{1}$$

$$H_2O_2\,(\%) = \frac{2i_r/N}{|i_d| + i_r/N} \times 100 \tag{2}$$

where i_d is the disk current, i_r is the ring current and N is the collection efficiency (=20.50%).

4. Conclusions

Nitrogen-doped ordered mesoporous carbon was synthesized by the modified nanocasting method with high electrocatalytic activities to the ORR in both acid and alkaline media. The results revealed that both the pyrolysis temperature and the NH_3-activation yielded significant effects on the specific surface area, nitrogen doping and, thus, the electrocatalytic activity, as well. First, the pyrolysis temperature yielded a volcano-shape relationship with the specific surface area and the content of the nitrogen-activated carbon. Additionally, it was found that such a change could be correlated with the electrocatalytic activity to the ORR, revealing the importance of the specific surface area and the chemical nature of the active sites. Second, the post-treatment in NH_3 could further increase the specific surface and enhance the nitrogen doping, which thereby improved the electrocatalytic activity and selectivity to the ORR. Additionally, the C-PA-1000-NH_3 catalyst outperformed the Pt/C one in both acid and alkaline media, which make it promising to be applied in fuel cells.

Acknowledgments: The work described in this paper was jointly supported by the National Natural Science Foundation of China (No. 21476087), the Pearl River S&T Nova Program of Guangzhou (No. 2013J2200041), the Science & Technology Research Project of Guangdong Province (No. 2014A010105041), the Guangdong Natural Science Foundation (No. S2013010012469) and the Innovation Project of Guangdong Department of Education (No. 2014KTSCX016).

Author Contributions: K.W. and Z.P.Y. performed the experiment and analyzed the data; K.W. and Z.X.L. wrote the paper.

Conflicts of Interest: The authors declare no conflicts of interest.

References

1. Job, N.; Lambert, S.; Zubiaur, A.; Cao, C.; Pirard, J.P. Design of Pt/carbon xerogel catalysts for pem fuel cells. *Catalysts* **2015**, *5*, 40–57.

2. Jia, Q.; Liang, W.; Bates, M.K.; Mani, P.; Lee, W.; Mukerjee, S. Activity descriptor identification for oxygen reduction on platinum-based bimetallic nanoparticles: *In situ* observation of the linear composition-strain-activity relationship. *ACS Nano* **2015**, *9*, 387–400.

3. Ye, T.N.; Lv, L.B.; Li, X.H.; Xu, M.; Chen, J.S. Strongly veined carbon nanoleaves as a highly efficient metal-free electrocatalyst. *Angew. Chem. Int. Ed.* **2014**, *53*, 6905–6909.

4. Wu, G.; Zelenay, P. Nanostructured nonprecious metal catalysts for oxygen reduction reaction. *Acc. Chem. Res.* **2013**, *46*, 1878–1889.

5. Zhang, Y.; Zhuang, X.D.; Su, Y.Z.; Zhang, F.; Feng, X.L. Polyaniline nanosheet derived B/N co-doped carbon nanosheets as efficient metal-free catalysts for oxygen reduction reaction. *J. Mater. Chem. A* **2014**, *2*, 7742–7746.

6. Su, D.S.; Perathoner, S.; Centi, G. Nanocarbons for the development of advanced catalysts. *Chem. Rev.* **2013**, *113*, 5782–5816.

7. Wang, H.L.; Dai, H.J. Strongly coupled inorganic-nano-carbon hybrid materials for energy storage. *Chem. Soc. Rev.* **2013**, *42*, 3088–3113.

8. Gong, K.; Du, F.; Xia, Z.; Durstock, M.; Dai, L. Nitrogen-doped carbon nanotube arrays with high electrocatalytic activity for oxygen reduction. *Science* **2009**, *323*, 760–764.

9. Yang, S.; Feng, X.; Wang, X.; Müllen, K. Graphene-based carbon nitride nanosheets as efficient metal-free electrocatalysts for oxygen reduction reactions. *Angew. Chem. Int. Ed.* **2011**, *50*, 5339–5343.

10. He, W.; Jiang, C.; Wang, J.; Lu, L. High-rate oxygen electroreduction over graphitic-n species exposed on 3D hierarchically porous nitrogen-doped carbons. *Angew. Chem. Int. Ed.* **2014**, *53*, 9503–9507.

11. Zhu, Y.; Zhang, B.; Liu, X.; Wang, D.W.; Su, D.S. Unravelling the structure of electrocatalytically active Fe–N complexes in carbon for the oxygen reduction reaction. *Angew. Chem. Int. Ed.* **2014**, *53*, 10673–10677.

12. Sun, X.; Song, P.; Zhang, Y.; Liu, C.; Xu, W.; Xing, W. A class of high performance metal-free oxygen reduction electrocatalysts based on cheap carbon blacks. *Sci. Rep.* **2013**, *3*, 2505.

13. Silva, R.; Voiry, D.; Chhowalla, M.; Asefa, T. Efficient metal-free electrocatalysts for oxygen reduction: Polyaniline-derived N- and O-doped mesoporous carbons. *J. Am. Chem. Soc.* **2013**, *135*, 7823–7826.

14. Ramaswamy, N.; Tylus, U.; Jia, Q.; Mukerjee, S. Activity descriptor identification for oxygen reduction on nonprecious electrocatalysts: Linking surface science to coordination chemistry. *J. Am. Chem. Soc.* **2013**, *135*, 15443–15449.

15. Liang, H.W.; Wei, W.; Wu, Z.S.; Feng, X.; Mullen, K. Mesoporous metal-nitrogen-doped carbon electrocatalysts for highly efficient oxygen reduction reaction. *J. Am. Chem. Soc.* **2013**, *135*, 16002–16005.

16. Liu, G.; Li, X.G.; Ganesan, P.; Popov, B.N. Development of non-precious metal oxygen-reduction catalysts for pem fuel cells based on N-doped ordered porous carbon. *Appl. Catal. B* **2009**, *93*, 156–165.

17. Cheon, J.Y.; Kim, T.; Choi, Y.; Jeong, H.Y.; Joo, S.H. Ordered mesoporous porphyrinic carbons with very high electrocatalytic activity for the oxygen reduction reaction. *Sci. Rep.* **2013**, *3*, 2715.

18. Yin, H.; Zhang, C.; Liu, F.; Hou, Y. Hybrid of iron nitride and nitrogen-doped graphene aerogel as synergistic catalyst for oxygen reduction reaction. *Adv. Funct. Mater.* **2014**, *24*, 2930–2937.

19. Xi, P.B.; Liang, Z.X.; Liao, S.J. Stability of hemin/C electrocatalyst for oxygen reduction reaction. *Int. J. Hydrogen Energy* **2012**, *37*, 4606–4611.

20. Liang, Z.X.; Song, H.Y.; Liao, S.J. Hemin: A highly effective electrocatalyst mediating the oxygen reduction reaction. *J. Phys. Chem. C* **2011**, *115*, 2604–2610.

21. Fellinger, T.P.; Hasche, F.; Strasser, P.; Antonietti, M. Mesoporous nitrogen-doped carbon for the electrocatalytic synthesis of hydrogen peroxide. *J. Am. Chem. Soc.* **2012**, *134*, 4072–4075.

22. Sa, Y.J.; Park, C.; Jeong, H.Y.; Park, S.H.; Lee, Z.; Kim, K.T.; Park, G.G.; Joo, S.H. Carbon nanotubes/heteroatom-doped carbon core-sheath nanostructures as highly active, metal-free oxygen reduction electrocatalysts for alkaline fuel cells. *Angew. Chem. Int. Ed.* **2014**, *53*, 4102–4106.

23. Wu, G.; More, K.L.; Johnston, C.M.; Zelenay, P. High-performance electrocatalysts for oxygen reduction derived from polyaniline, iron, and cobalt. *Science* **2011**, *332*, 443–447.

24. Wan, K.; Long, G.F.; Liu, M.Y.; Du, L.; Liang, Z.X.; Tsiakaras, P. Nitrogen-doped ordered mesoporous carbon: Synthesis and active sites for electrocatalysis of oxygen reduction reaction. *Appl. Catal. B* **2015**, *165*, 566–571.

25. Long, G.F.; Wan, K.; Liu, M.Y.; Li, X.H.; Liang, Z.X.; Piao, J.H. Effect of pyrolysis conditions on nitrogen-doped ordered mesoporous carbon electrocatalysts. *Chin. J. Catal.* **2015**.

26. Wan, K.; Yu, Z.P.; Li, X.H.; Liu, M.Y.; Yang, G.; Piao, J.H.; Liang, Z.X. PH effect on electrochemistry of nitrogen-doped carbon catalyst for oxygen reduction reaction. *ACS Catal.* **2015**, *5*, 4325–4332.

27. Meng, H.; Larouche, N.; Lefèvre, M.; Jaouen, F.; Stansfield, B.; Dodelet, J.-P. Iron porphyrin-based cathode catalysts for polymer electrolyte membrane fuel cells: Effect of NH$_3$ and Ar mixtures as pyrolysis gases on catalytic activity and stability. *Electrochim. Acta* **2010**, *55*, 6450–6461.

28. Zhao, Y.; Watanabe, K.; Hashimoto, K. Self-supporting oxygen reduction electrocatalysts made from a nitrogen-rich network polymer. *J. Am. Chem. Soc.* **2012**, *134*, 19528–19531.

29. Liang, H.W.; Zhuang, X.; Bruller, S.; Feng, X.; Mullen, K. Hierarchically porous carbons with optimized nitrogen doping as highly active electrocatalysts for oxygen reduction. *Nat. Commun.* **2014**, *5*, 4973.

30. Chen, S.; Bi, J.Y.; Zhao, Y.; Yang, L.J.; Zhang, C.; Ma, Y.W.; Wu, Q.; Wang, X.Z.; Hu, Z. Nitrogen-doped carbon nanocages as efficient metal-free electrocatalysts for oxygen reduction reaction. *Adv. Mater.* **2012**, *24*, 5593–5597.

31. Hu, Y.; Jensen, J.O.; Zhang, W.; Cleemann, L.N.; Xing, W.; Bjerrum, N.J.; Li, Q. Hollow spheres of iron carbide nanoparticles encased in graphitic layers as oxygen reduction catalysts. *Angew. Chem. Int. Ed.* **2014**, *53*, 3675–3679.

32. Geng, D.; Chen, Y.; Chen, Y.; Li, Y.; Li, R.; Sun, X.; Ye, S.; Knights, S. High oxygen-reduction activity and durability of nitrogen-doped graphene. *Energ. Environ. Sci.* **2011**, *4*, 760–764.

33. Favaro, M.; Perini, L.; Agnoli, S.; Durante, C.; Granozzi, G.; Gennaro, A. Electrochemical behavior of n and ar implanted highly oriented pyrolytic graphite substrates and activity toward oxygen reduction reaction. *Electrochim. Acta* **2013**, *88*, 477–487.

34. Zhao, D.Y.; Feng, J.L.; Huo, Q.S.; Melosh, N.; Fredrickson, G.H.; Chmelka, B.F.; Stucky, G.D. Triblock copolymer syntheses of mesoporous silica with periodic 50 to 300 angstrom pores. *Science* **1998**, *279*, 548–552.

Phosphorus and Nitrogen Dual Doped and Simultaneously Reduced Graphene Oxide with High Surface Area as Efficient Metal-Free Electrocatalyst for Oxygen Reduction

Xiaochang Qiao, Shijun Liao, Chenghang You and Rong Chen

Abstract: A P, N dual doped reduced graphene oxide (PN-rGO) catalyst with high surface area (376.20 $m^2 \cdot g^{-1}$), relatively high P-doping level (1.02 at. %) and a trace amount of N (0.35 at. %) was successfully prepared using a one-step method by directly pyrolyzing a homogenous mixture of graphite oxide (GO) and diammonium hydrogen phosphate ($(NH_4)_2HPO_4$) in an argon atmosphere, during which the thermal expansion, deoxidization of GO and P, N co-doping were realized simultaneously. The catalyst exhibited enhanced catalytic performances for oxygen reduction reaction (ORR) via a dominated four-electron reduction pathway, as well as superior long-term stability, better tolerance to methanol crossover than that of commercial Pt/C catalyst in an alkaline solution.

Reprinted from *Catalysts*. Cite as: Qiao, X.; Liao, S.; You, C.; Chen, R. Phosphorus and Nitrogen Dual Doped and Simultaneously Reduced Graphene Oxide with High Surface Area as Efficient Metal-Free Electrocatalyst for Oxygen Reduction. *Catalysts* **2015**, *5*, 981–991.

1. Introduction

A crucial component of a fuel cell is the electrocatalyst for the cathodic oxygen reduction reaction (ORR) [1]. Pt-based precious metals are regarded as the most effective ORR electrocatalysts developed to date. However, they suffer from a number of drawbacks including the scarcity and consequent high cost of Pt, as well as their poor durability and low tolerance to methanol crossover [2]. Accordingly, considerable effort has been devoted to developing nonprecious-metal [3–10] and metal-free [11–14] ORR catalysts. Among such candidates, carbon materials doped with heteroatoms have attracted a great deal of attention due to their relative cost-effectiveness, good long-term durability, and excellent tolerance to methanol crossover.

Graphene, a two-dimensional monolayer of sp^2-hybridized carbon atoms packed in a honeycomb lattice, has recently become an attractive candidate, due to its superior electrical conductivity, high surface area and excellent mechanical properties. Both theoretical calculations and experimental studies reveal that incorporating foreign atoms into the graphene structure can effectively tailor the material's electronic and chemical properties [15–17]. Recently, graphene doped with heteroatoms such as nitrogen, sulfur, boron, and iodine has yielded metal-free ORR electrocatalysts with enhanced electrochemical performance [13,18–22]. This performance boost is attributed to the heteroatoms, because their electronegativity (N: 3.04; S: 2.58; B: 2.04; I: 2.66) differs from that of carbon (2.55), they break carbon's electroneutrality, creating charged sites and consequently favoring O_2 adsorption during the ORR process. Since phosphorus has a lower electronegativity (2.19) than carbon, it is well worth exploring the unique properties of P-doped graphene. Liu *et al.* prepared P-doped graphene by pyrolyzing graphene oxide with 1-butyl-3-methlylimidazolium hexafluorophosphate, and achieved, in an alkaline solution, an ORR catalytic performance comparable to that of commercial Pt/C [23]. Zhang *et al.* synthesized P-doped graphene by thermally annealing a mixture of graphite oxide (GO) and triphenylphosphine (TPP), and the resultant catalyst showed remarkable catalytic activity toward the ORR [24]. However, while exciting results have been obtained with P-doped graphene, just a few investigations into this type of catalyst have been reported to date. Furthermore, it has been reported that the co-doping of P and N can further improve the carbon materials' ORR catalytic activity, due to the synergistic effect [25].

Herein, we propose a one-step method for preparing a P, N dual-doped reduced graphene oxide (PN-rGO) catalyst, using diammonium hydrogen phosphate ($(NH_4)_2HPO_4$) as both phosphorus and nitrogen sources. In an alkaline medium, the as-prepared PN-rGO exhibited enhanced ORR electrocatalytic activity, good long-term stability, high tolerance to methanol crossover, and high selectivity for the four-electron reduction pathway.

2. Results and Discussion

Figure 1 shows typical SEM and TEM images of the PN-rGO catalyst. As can be seen in Figure 1a,b ultrathin, crumpled PN-rGO nanosheets are randomly arranged and overlapped with each other, these could easily have formed a slit-shaped porous structure, and indeed, such a structure was confirmed by Brunauer-Emmett-Teller (BET) testing (Figure 3). In Figure 1c, the PN-rGO nanosheets are transparent and wrinkled, like wavy silk veils. The high-resolution TEM image (Figure 1d) shows well-defined graphitic lattice fringes, indicating the good crystallization of the PN-rGO nanosheets. Actually, the morphology of our product is quite consistent with those reported previously [18,19].

Figure 2a shows the X-ray diffraction (XRD) patterns of GO, rGO and PN-rGO. GO exhibited a peak at $2\theta = 11°$ with an interlayer distance of 0.8 nm, which is larger than the interlayer distance of graphite (0.34 nm), revealing that many different oxygen-containing groups were intercalated within the interlayer space. The peak at 11° completely disappeared after annealing, replaced by a broad peak at 2θ = 22° for rGO and PN-rGO, with a d-spacing of 0.4 nm, implying that the successful reduction of GO to reduced graphene oxide.

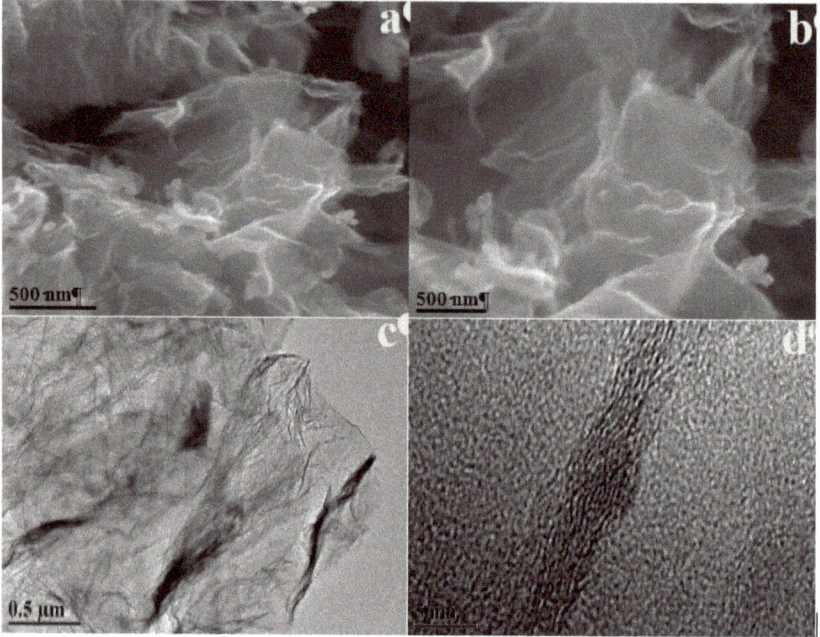

Figure 1. Scanning electron microscopy (SEM) images (**a**, **b**) and transmission electron microscopy (TEM) images (**c**, **d**) of P, N dual doped reduced graphene oxide (PN-rGO).

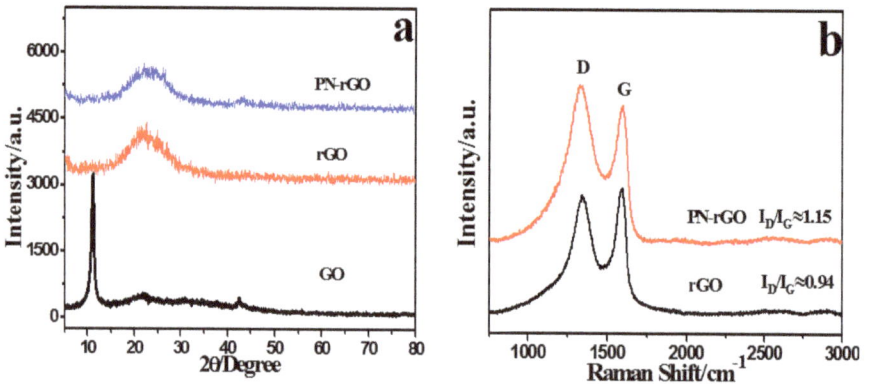

Figure 2. X-ray diffraction (XRD) patterns of GO, rGO and PN-rGO (**a**), Raman spectra of rGO and PN-rGO (**b**).

Further structural information about PN-rGO was obtained from Raman spectroscopy. As shown in Figure 2b, similar to all sp^2-carbons, two distinct peaks appeared near 1350 cm^{-1} and 1580 cm^{-1}, corresponding to the D band and G band, respectively. The D band is resulting from the disordered carbon atoms, whereas the G band from sp^2-hybridized graphitic carbon atoms. The intensity ratio of I_D/I_G generally provides a gauge for the lever of disorder. Evidently, the I_D/I_G value of PN-rGO (1.15) was relatively higher than that of rGO (0.94) due to the incorporated phosphorus atoms.

The N_2 adsorption-desorption isotherms and the corresponding pore size distribution curves of PN-rGO and rGO are shown in Figure 3. According to the International Union of Pure and Applied Chemistry classification, the N_2 adsorption-desorption isotherms of the two samples were type IV, with hysteresis loops type H_3. A type IV adsorption-desorption isotherm indicates the presence of mesopores, while a type H_3 hysteresis loop of is correlated with slit-shaped pores, possibly between parallel layers. This result is consistent with the SEM observations. Surface area and pore volume for PN-rGO were 376.2 $m^2 \cdot g^{-1}$ and 1.50 $cm^3 \cdot g^{-1}$, and for rGO 260.2 $m^2 \cdot g^{-1}$ and 1.17 $cm^3 \cdot g^{-1}$, respectively. The greatly increased BET surface area and pore volume of PN-rGO may have been due to the activation effect of $(NH_4)_2HPO_4$ on carbon [26,27]. The high surface area and large pore volume of PN-rGO could have (i) exposed more active sites and (ii) favored the mass transport of reactants and products.

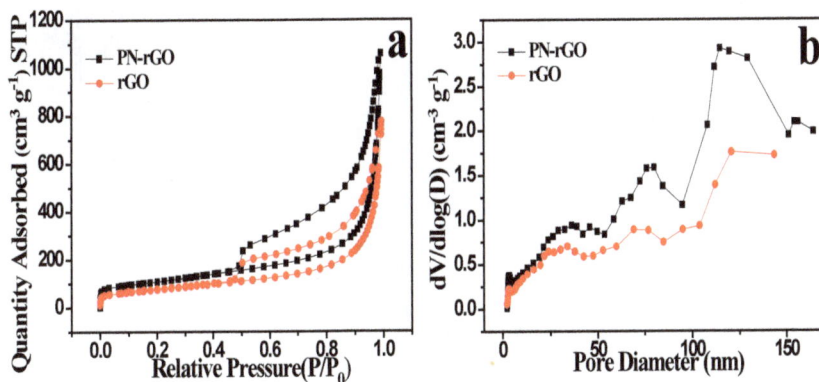

Figure 3. Nitrogen adsorption-desorption isotherms (**a**) and the corresponding pore size distribution curve of PN-rGO and rGO (**b**).

To further investigate the elemental composition of PN-rGO, we carried out XPS measurement. As shown in Figure 4a, the XPS survey spectrum of PN-rGO presented a dominant C1s peak (~284.5 eV), a O1s peak (~532.0 eV), a P2p peak (~132.8 eV), and a N1s peak (~400.0 eV), confirming successful P and N co-doping [28]; the corresponding atomic percentages were 92.47, 6.02, 1.16, and 0.35 at. %, respectively.

High-resolution spectra were then obtained to gain more insight into the phosphorus and nitrogen doping.

As shown in Figure 4b, the high-resolution P2p spectrum can be deconvoluted into two main component peaks located at 131.7 and 133.1 eV, corresponding to P–C and P–O bonding, respectively [23]. In addition, the peak area ratio of P–C to P–O is close to 2:3. Doping phosphorus atoms into the carbon lattice (forming a P–C covalent bond) can induce negatively delocalized C atoms adjacent to P atoms; meanwhile, in P–O bonding (where an oxygen bridge is formed between C and P), the oxygen atoms can enhance electron poverty in the carbon atoms. These two kinds of structure have been reported to be advantageous for the ORR [29]. It should be pointed out that, there is always the debate if P can access to the honeycomb crystal lattice of graphene like what N or B atom does, due to the big difference of carbon and phosphorous in radius [30].

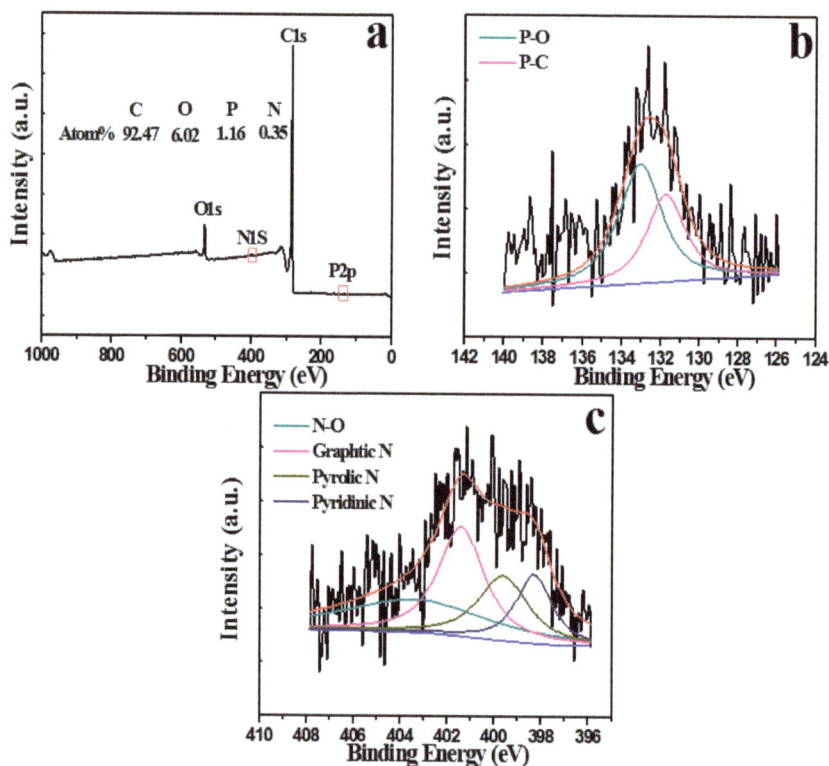

Figure 4. X-ray photoelectron spectroscopy (XPS) survey (**a**) and high resolution P2p (**b**), and N1s (**c**) spectra of PN-rGO.

312

The deconvolution results of the high-resolution N1s spectrum were shown in Figure 4c. It's shown the prepared PN-rGO catalyst had four types of N species, corresponding to oxidized N (~403.3 eV), graphitic N (~401.4 eV), pyrrolic N (~399.6 eV), and pyridinic N (~398.3 eV) [31,32], with compositions of 31.0, 32.3, 20.8, and 15.9 at. %, respectively [33]. The total amount of active N species (graphitic N, pyrrolic N, and pyridinic N) reached 69.0 at. % [34].

To explore the electrocatalytic activity of PN-rGO for the ORR, cyclic voltammetry (CV) experiments were carried out in an O_2-saturated 0.1 M KOH solution. The CV curves of a bare GCE and rGO were also measured for comparison. As shown in Figure 5a, for all the electrodes, the CV curves displayed distinct oxygen reduction cathodic peaks. The ORR peak potential positively shifted from −0.39 V for the GCE to −0.26 V for rGO and −0.21 V for PN-rGO. In addition, the PN-rGO had the highest peak current density, at −0.96 mA cm^{-2}, which was about four times higher than that of the GCE. The most positive ORR peak potential and the highest peak current density of PN-rGO, suggest that phosphorus and trace nitrogen co-doping can greatly enhance the ORR catalytic activity of graphene. The CV area of the PN-rGO was also much greater than that of the rGO, indicating the former had a much greater electroactive area, as CV area is closely related to a sample's capacitance, which is proportional to its specific surface area. This result is in good agreement with the BET results.

To gain further insight into the role of P, N co-doping in the ORR, the linear sweep voltammetry (LSV) curve of PN-rGO was recorded in an O_2-saturated 0.1 M KOH solution; for comparison, analogous LSV curves were also obtained for GCE, rGO, and commercial 20 wt. % Pt/C. As can be seen in Figure 5b, PN-rGO had a much more positive ORR onset potential and a much higher limiting current density than GCE or rGO, indicating that doping graphene with phosphorus and trace nitrogen can facilitate the ORR. The LSV results are consistent with the CV results.

To gain more information on the ORR kinetics of the PN-rGO catalyst, we recorded LSV curves in an O_2-saturated 0.1 M KOH solution at various rotation rates, from 1600 to 3600 rpm (Figure 5c). The diffusion current density increased rapidly as the rotation rate increased. In addition, the K-L plots at different electrode potentials displayed good linearity, We used the K-L equation to calculate the electron transfer number (n) of PN-rGO in the potential range of −0.40 to −0.60 V and obtained an average n value of 3.66, indicating that the ORR proceeded via a dominated four-electron pathway.

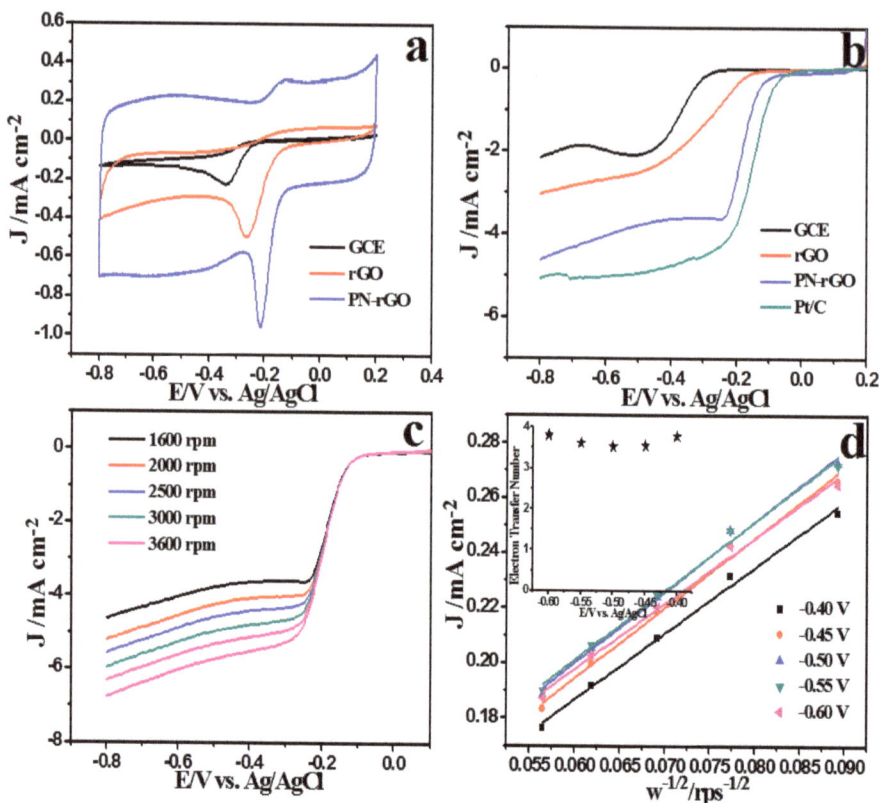

Figure 5. Cyclic voltammetry (CV) curves (**a**) and linear sweep voltammetry (LSV) curves at 1600 rpm (**b**) for different samples, LSV curves at different rotation rates (**c**) and the corresponding K-L plots (**d**) of PN-rGO.

For practical application in fuel cells, the fuel crossover effect should be considered because fuel molecules (e.g., methanol) may pass from anode to cathode through the membrane and poison the cathode catalyst. Thus, we recorded the chronoamperometric responses of PN-rGO and Pt/C upon the addition of 3 M methanol (Figure 6a). After the methanol was introduced into an O_2-saturated 0.1 M KOH solution at about 200 s, no noticeable change was observed in the ORR current for PN-rGO; in contrast, Pt/C showed a significant drop in ORR current. These results indicated that PN-rGO possessed a high immunity to methanol crossover.

As durability is also of great importance in practical applications of fuel-cell technology, the chronoamperometric durabilities of PN-rGO and Pt/C were measured at −0.3 V for 20,000 s in an O_2-saturated 0.1 M KOH solution. As can be seen in Figure 6b, slight performance attenuation with high current retention

(96%) was achieved with our PN-rGO catalyst. However, commercial Pt/C suffered a current loss of 12% under the same conditions, indicating that the PN-rGO electrocatalyst was much more stable in an alkaline medium.

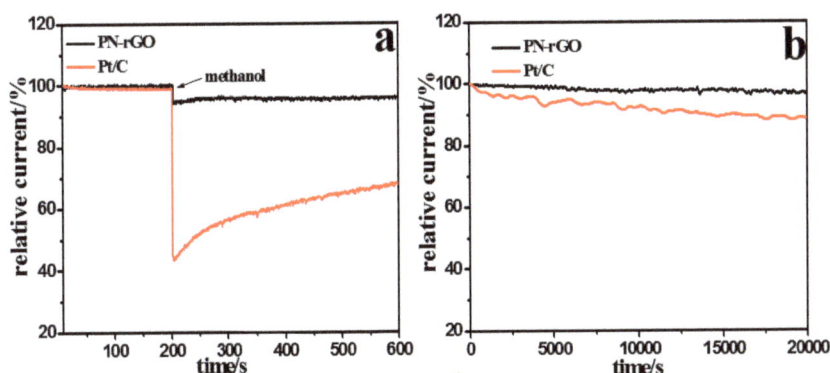

Figure 6. chronoamperometric responses of PN-rGO and Pt/C at −0.3 V upon the addition of methanol (**a**), durability testing curves of PN-rGO and Pt/C for 20,000 s at 1600 rpm (**b**).

3. Experimental Section

3.1. Catalysts Preparation

Graphite oxide (GO) was prepared from 10,000 mesh graphite powder using a modified Hummers' Method. PN-rGO was synthesized by the thermal annealing of GO and $(NH_4)_2HPO_4$. In a typical procedure, 50 mg of GO was mixed with 15 mg of $(NH_4)_2HPO_4$ in 50 mL of deionized water, at room temperature, under stirring in an open beaker. After the water was completely removed using a rotary evaporator at 50 °C, the resulting mixture was transferred into a quartz boat in the center of a tube furnace and annealed at 900 °C for 1 h, with high-purity argon as the protective atmosphere. For comparison, reduced graphene oxide without P, N doping (rGO) was also prepared using the same procedure but in the absence of $(NH_4)_2HPO_4$.

3.2. Physical Characterization

Scanning electron microscopy (SEM) was performed on a Nova Nano 430 field emission scanning electron microscope (FEI, Hillsboro, OR, USA). Transmission electron microscopy (TEM) images were recorded on JEM-2100HR transmission electron microscope (JEOL, Tokyo, Japan). X-ray diffraction (XRD) patterns were conducted on a TD-3500 powder diffractometer (Tongda, Liaoning, China). Raman spectroscopy measurements were carried out on a Lab RAM Aramis Raman spectrometer (HORIBA Jobin Yvon, Edison, NJ, USA) with a laser wave length

315

of 632.8 nm. Surface area and pore characteristics were determined by recording nitrogen adsorption-desorption isotherms using a Tristar II 3020 gas adsorption analyzer (Micromeritics, Norcross, GA, USA). X-ray photoelectron spectroscopy (XPS) was performed with an ESCALAB 250 X-ray photoelectron spectrometer (Thermo-VG Scientific, Waltham, MA, USA).

3.3. Electrochemical Measurements

Electrochemical measurements were carried out on an electrochemical workstation (Ivium, Eindhoven, The Netherlands) with a standard three-electrode system at room temperature. A glassy carbon rotating disk electrode (GC-RDE) (5 mm diameter, 0.196 cm^2 geometric area) was used as the working electrode, while a Pt wire and an Ag/AgCl (3 M NaCl) electrode were the counter and reference electrodes, respectively. The electrolyte was 0.1 M aqueous KOH solution. For each sample, a catalyst ink was prepared by dispersing 5 mg of the corresponding catalyst in 1 mL Nafion ethanol solution (0.25· wt. %). Then 20 μL of the dispersed catalyst ink was pipetted onto the GC-RDE and dried under an infrared lamp. The mass loading of the catalyst was 0.5 mg· cm^{-2}. Before testing, the electrolyte solution was purged with high-purity nitrogen or oxygen gas for at least 30 min. Unless otherwise specified, the scanning rate was 10 mV· s^{-1}. The electron transfer number (n) per oxygen molecule involved was calculated on the basis of the Koutecky-Levich (K-L) equation:

$$J^{-1} = J_L^{-1} + J_K^{-1} = B^{-1}\omega^{-1/2} + J_K^{-1}$$
$$B = 0.62nFC_0D_0^{2/3}\gamma^{-1/6} \tag{1}$$
$$J_K = nF\kappa C_0$$

where J is the measured current density; J_K and J_L are the kinetic current density and the diffusion limiting current density, respectively; ω is the angular velocity of the disk ($\omega = 2 \pi N$, where N is the linear rotation rate); n is the number of electrons transferred for the ORR; F is the Faraday constant ($F = 96485$ C· mol^{-1}); C_0 is the bulk concentration of O_2 (1.2×10^{-3} mol· L^{-1}); D_0 is the diffusion coefficient of O_2 in 0.1 M KOH (1.9×10^{-5} cm^2· s^{-1}); γ is the kinetic viscosity of the electrolyte (0.01 cm^2· s^{-1}); and κ is the electron transfer rate constant.

4. Conclusions

A metal-free phosphorus and nitrogen dual-doped reduced graphene oxide (PN-rGO) catalyst was successfully synthesized using a one-step thermal annealing method by directly pyrolyzing a homogenous mixture of graphite oxide (GO) and diammonium hydrogen phosphate (($NH_4)_2HPO_4$). The specific surface area of PN-rGO, 376.2 m^2· g^{-1}, was much higher than that of rGO (260.2 m^2· g^{-1}). The catalyst exhibited enhanced ORR activity via a dominant four-electron reduction pathway and showed outstanding selectivity and stability in an alkaline solution. Certainly, the details of the ORR mechanism and active sites of this new catalyst require further investigation.

Acknowledgments: The authors would like to acknowledge the financial support of the National Science Foundation of China (NSFC Project No. 21076089, 21276098, 11132004, and U1301245), the Ministry of Science and Technology of China (Project No. 2012AA053402), the Guangdong Natural Science Foundation (Project No. S2012020011061), the Doctoral Fund of the Ministry of Education of China (20110172110012), and the Basic Scientific Foundation of the Central Universities of China (No. 2013ZP0013).

Author Contributions: X. Q. performed the expermient and analyzed the date; X. Q. and S. L. wrote the paper; C. Y. and R. C analyzed the date.

Conflicts of Interest: The authors declare no conflict of interest.

References

1.   Debe, M.K. Electrocatalyst approaches and challenges for automotive fuel cells. *Nature* **2012**, *486*, 43–51.

2.   Winter, M.; Brodd, R.J. What are batteries, fuel cells, and supercapacitors? *Chem Rev.* **2004**, *104*, 4245–4270.

3.   Lefèvre, M.; Proietti, E.; Jaouen, F.; Dodelet, J.-P. Iron-based catalysts with improved oxygen reduction activity in polymer electrolyte fuel cells. *Science* **2009**, *324*, 71–74.

4.   Wu, G.; More, K.L.; Xu, P.; Wang, H.L.; Ferrandon, M.; Kropf, A.J.; Myers, D.J.; Ma, S.; Johnston, C.M.; Zelenay, P. A carbon-nanotube-supported graphene-rich non-precious metal oxygen reduction catalyst with enhanced performance durability. *Chem. Commun.* **2013**, *49*, 3291–3293.

5.   Liang, Y.; Wang, H.; Diao, P.; Chang, W.; Hong, G.; Li, Y.; Gong, M.; Xie, L.; Zhou, J.; Wang, J.; *et al.* Oxygen reduction electrocatalyst based on strongly coupled cobalt oxide nanocrystals and carbon nanotubes. *J. Am. Chem. Soc.* **2012**, *134*, 15849–15857.

6.   Duan, J.; Zheng, Y.; Chen, S.; Tang, Y.; Jaroniec, M.; Qiao, S. Mesoporous hybrid material composed of Mn_3O_4 nanoparticles on nitrogen-doped graphene for highly efficient oxygen reduction reaction. *Chem. Commun.* **2013**, *49*, 7705–7707.

7.   Andersen, N.I.; Serov, A.; Atanassov, P. Metal oxides/CNT nano-composite catalysts for oxygen reduction/oxygen evolution in alkaline media. *Appl. Catal. B* **2015**, *163*, 623–627.

8.   Qiao, X.; You, C.; Shu, T.; Fu, Z.; Zheng, R.; Zeng, X.; Li, X.; Liao, S. A one-pot method to synthesize high performance multielement co-doped reduced graphene oxide catalysts for oxygen reduction. *Electrochem. Commun.* **2014**, *47*, 49–53.

9.   Holade, Y.; Sahin, N.E.; Servat, K.; Napporn, T.W.; Kokoh, K.B. Recent advances in carbon supported metal nanoparticles preparation for oxygen reduction reaction in low temperature fuel cells. *Catalysts* **2015**, *5*, 310–348.

10. Niu, W.; Li, L.; Liu, X.; Wang, N.; Liu, J.; Zhou, W.; Tang, Z.; Chen, S. Mesoporous N-doped carbons prepared with thermally removable nanoparticle templates: An efficient electrocatalyst for oxygen reduction reaction. *J. Am. Chem. Soc.* **2015**, *137*, 5555–5562.

11. Xia, W.; Masa, J.; Bron, M.; Schuhmann, W.; Muhler, M. Highly active metal-free nitrogen-containing carbon catalysts for oxygen reduction synthesized by thermal treatment of polypyridine-carbon black mixtures. *Electrochem. Commun.* **2011**, *13*, 593–596.

12. Gong, K.; Du, F.; Xia, Z.; Durstock, M.; Dai, L. Nitrogen-doped carbon nanotube arrays with high electrocatalytic activity for oxygen reduction. *Science* **2009**, *323*, 760–764.

13. Yang, Z.; Yao, Z.; Li, G.; Fang, G.; Nie, H.; Liu, Z.; Zhou, X.; Chen, X.A.; Huang, S. Sulfur-doped graphene as an efficient metal-free cathode catalyst for oxygen reduction. *ACS Nano* **2011**, *6*, 205–211.

14. Bo, X.; Guo, L. Ordered mesoporous boron-doped carbons as metal-free electrocatalysts for the oxygen reduction reaction in alkaline solution. *Phys. Chem. Chem. Phys.* **2013**, *15*, 2459–2465.

15. Lherbier, A.; Blase, X.; Niquet, Y.-M.; Triozon, F.; Roche, S. Charge transport in chemically doped 2D graphene. *Phys. Rev. Lett.* **2008**, *101*, 036808.

16. Liu, H.; Liu, Y.; Zhu, D. Chemical doping of graphene. *J. Mater. Chem.* **2011**, *21*, 3335–3345.

17. Zhou, X.; Qiao, J.; Yang, L.; Zhang, J. A review of graphene-based nanostructural materials for both catalyst supports and metal-free catalysts in PEM fuel cell oxygen reduction reactions. *Adv. Energy Mater.* **2014**, *4*.

18. Sheng, Z.-H.; Gao, H.-L.; Bao, W.-J.; Wang, F.-B.; Xia, X.-H. Synthesis of boron doped graphene for oxygen reduction reaction in fuel cells. *J. Mater. Chem.* **2012**, *22*, 390–395.

19. Qu, L.; Liu, Y.; Baek, J.-B.; Dai, L. Nitrogen-doped graphene as efficient metal-free electrocatalyst for oxygen reduction in fuel cells. *ACS Nano* **2010**, *4*, 1321–1326.

20. Yao, Z.; Nie, H.; Yang, Z.; Zhou, X.; Liu, Z.; Huang, S. Catalyst-free synthesis of iodine-doped graphene via a facile thermal annealing process and its use for electrocatalytic oxygen reduction in an alkaline medium. *Chem. Commun.* **2012**, *48*, 1027–1029.

21. Sfaelou, S.; Zhuang, X.; Feng, X.; Lianos, P. Sulfur-doped porous carbon nanosheets as high performance electrocatalysts for photofuelcells. *RSC Adv.* **2015**, *5*, 27953–27963.

22. Geng, D.; Chen, Y.; Chen, Y.; Li, Y.; Li, R.; Sun, X.; Ye, S.; Knights, S. High oxygen-reduction activity and durability of nitrogen-doped graphene. *Energy Environ. Sci.* **2011**, *4*, 760–764.

23. Li, R.; Wei, Z.; Gou, X.; Xu, W. Phosphorus-doped graphene nanosheets as efficient metal-free oxygen reduction electrocatalysts. *RSC Adv.* **2013**, *3*, 9978.

24. Zhang, C.; Mahmood, N.; Yin, H.; Liu, F.; Hou, Y. Synthesis of phosphorus-doped graphene and its multifunctional applications for oxygen reduction reaction and lithium ion batteries. *Adv. Mater.* **2013**, *25*, 4932–4937.

25. Choi, C.H.; Chung, M.W.; Kwon, H.C.; Park, S.H.; Woo, S.I. B, N-and P, N-doped graphene as highly active catalysts for oxygen reduction reactions in acidic media. *J. Mater. Chem. A* **2013**, *1*, 3694–3699.

26. Kyotani, T. Control of pore structure in carbon. *Carbon* **2000**, *38*, 269–286.

27. Sun, J.; Wu, L.; Wang, Q. Comparison about the structure and properties of pan-based activated carbon hollow fibers pretreated with different compounds containing phosphorus. *J. Appl. Polym. Sci.* **2005**, *96*, 294–300.

28. Liu, Z.W.; Peng, F.; Wang, H.J.; Yu, H.; Zheng, W.X.; Yang, J. Phosphorus-doped graphite layers with high electrocatalytic activity for the O_2 reduction in an alkaline medium. *Angew. Chem.* **2011**, *123*, 3315–3319.

29. Choi, C.H.; Park, S.H.; Woo, S.I. Binary and ternary doping of nitrogen, boron, and phosphorus into carbon for enhancing electrochemical oxygen reduction activity. *ACS Nano* **2012**, *6*, 7084–7091.

30. Geng, D.; Ding, N.; Hor, T.S.A.; Liu, Z.; Sun, X.; Zong, Y. Potential of metal-free "graphene alloy" as electrocatalysts for oxygen reduction reaction. *J. Mater. Chem. A* **2015**, *3*, 1795–1810.

31. Wohlgemuth, S.-A.; White, R.J.; Willinger, M.-G.; Titirici, M.-M.; Antonietti, M. A one-pot hydrothermal synthesis of sulfur and nitrogen doped carbon aerogels with enhanced electrocatalytic activity in the oxygen reduction reaction. *Green Chem.* **2012**, *14*, 1515–1523.

32. Nagaiah, T.C.; Kundu, S.; Bron, M.; Muhler, M.; Schuhmann, W. Nitrogen-doped carbon nanotubes as a cathode catalyst for the oxygen reduction reaction in alkaline medium. *Electrochem. Commun.* **2010**, *12*, 338–341.

33. Vikkisk, M.; Kruusenberg, I.; Joost, U.; Shulga, E.; Kink, I.; Tammeveski, K. Electrocatalytic oxygen reduction on nitrogen-doped graphene in alkaline media. *Appl. Catal. B* **2014**, *147*, 369–376.

34. Qiao, X.; Peng, H.; You, C.; Liu, F.; Zheng, R.; Xu, D.; Li, X.; Liao, S. Nitrogen, phosphorus and iron doped carbon nanospheres with high surface area and hierarchical porous structure for oxygen reduction. *J. Power Sources* **2015**, *288*, 253–260.

Titanium-Niobium Oxides as Non-Noble Metal Cathodes for Polymer Electrolyte Fuel Cells

Akimitsu Ishihara, Yuko Tamura, Mitsuharu Chisaka, Yoshiro Ohgi, Yuji Kohno, Koichi Matsuzawa, Shigenori Mitsushima and Ken-ichiro Ota

Abstract: In order to develop noble-metal- and carbon-free cathodes, titanium-niobium oxides were prepared as active materials for oxide-based cathodes and the factors affecting the oxygen reduction reaction (ORR) activity were evaluated. The high concentration sol-gel method was employed to prepare the precursor. Heat treatment in Ar containing 4% H_2 at 700–900 °C was effective for conferring ORR activity to the oxide. Notably, the onset potential for the ORR of the catalyst prepared at 700 °C was approximately 1.0 V *vs.* RHE, resulting in high quality active sites for the ORR. X-ray (diffraction and photoelectron spectroscopic) analyses and ionization potential measurements suggested that localized electronic energy levels were produced via heat treatment under reductive atmosphere. Adsorption of oxygen molecules on the oxide may be governed by the localized electronic energy levels produced by the valence changes induced by substitutional metal ions and/or oxygen vacancies.

Reprinted from *Catalysts*. Cite as: Ishihara, A.; Tamura, Y.; Chisaka, M.; Ohgi, Y.; Kohno, Y.; Matsuzawa, K.; Mitsushima, S.; Ota, K.-I. Titanium-Niobium Oxides as Non-Noble Metal Cathodes for Polymer Electrolyte Fuel Cells. *Catalysts* **2015**, *5*, 1289–1303.

1. Introduction

Polymer electrolyte fuel cells (PEFCs) offer many advantages, including high power density, high energy conversion efficiency, and lower operating temperatures. PEFCs are therefore suitable as power sources for vehicles and residential co-generation power systems. However, the use of Pt as a cathode electrocatalyst for PEFCs is problematic due to the high cost and limited availability of Pt, and insufficient stability of these catalysts. To successfully commercialize PEFCs, low-cost non-platinum cathode catalysts with high stability must be developed.

Since Jasinski discovered the oxygen reduction reaction (ORR) activity of cobalt phthalocyanine [1], the search for promising non-platinum ORR catalysts has led to the development of several cobalt- and iron-containing catalysts [2,3]. Approaches to enhance the activity of these catalysts include the use and optimization of carbon supports and heat treatment conditions. Heat treatment of iron salts adsorbed on carbon supports under ammonia gas is a recent breakthrough that produces catalysts with high ORR activities comparable to that of platinum-based catalysts [4]. Despite significant improvement of the ORR activity of non-platinum catalysts, issues regarding their long-term durability remain unresolved.

Based on the high stability of Group 4 and 5 metal oxide-based compounds in acidic media, low cost, [5,6] and lower solubility in acid solution compared to platinum-based catalysts, these compounds have piqued our interest as they are expected to be stable even under the conditions encountered at the PEFC cathode. Recently, we successfully synthesized oxide-based nanoparticles using oxy-metal phthalocyanines (MeOPc; Me = Ta, Zr, and Ti) as the starting material and multi-walled carbon nanotubes (MWCNTs) as the support as well as the electro-conductive material [7,8]. However, carbon materials are easily oxidized at high potentials with a consequent decrease of the ORR activity due to degradation of the electron conduction paths [8]. Thus, carbon-free electrocatalysts are required to achieve high durability of the oxide-based cathodes. To prepare noble-metal- and carbon-free cathodes, the basic approach is to combine electro-conductive oxides with oxides that possess ORR active sites.

Previously, we prepared noble-metal- and carbon-free cathodes comprising niobium-titanium oxides with active sites and titanium oxides with magneli phase Ti_4O_7 as the electro-conductive material (*i.e.*, $Ti_xNb_yO_z$ + Ti_4O_7) [9]. The highest onset potential of $Ti_xNb_yO_z$ + Ti_4O_7 was *ca.* 1.1 V *versus* the reversible hydrogen electrode (RHE). No degradation of the ORR performance of $Ti_xNb_yO_z$ + Ti_4O_7 was observed during the start-stop and load cycle tests in 0.1 mol·dm^{-3} H_2SO_4 at 80 °C, where these conditions are close to the operating conditions of the existing PEFC [10]. Therefore, we successfully demonstrated superior durability of noble-metal- and carbon-free oxide-based cathodes under the cathode conditions of the PEFC.

However, the ORR activities of the $Ti_xNb_yO_z$ + Ti_4O_7 catalysts were still low because these catalysts were prepared under argon containing 4% hydrogen at high temperature, 1050 °C, where Ti_4O_7 was generated by the reduction of TiO_2. That is, the preparation conditions encouraged the formation of Ti_4O_7 but were not optimal for the formation of niobium-titanium oxides with active sites. Domen *et al.* demonstrated that Nb-doped TiO_2 synthesized by the oxidation of Nb-doped TiN nanoparticles exhibited definite ORR activity and high long-term stability in acidic solutions [11]. However, these catalysts contained carbon residues that functioned to improve the conductivity between the particle aggregates. The preparation conditions used in that study were thus not suitable for the formation of ORR active titanium-niobium oxides without carbon. Consequently, it is necessary to separately optimize the conditions for the formation of titanium-niobium oxides with active sites and the formation of electro-conductive oxides. In this study, we focus on the formation of active sites on titanium-niobium oxides using a high concentration sol-gel method. The factors that influence the ORR activity in the absence of a carbon support are evaluated. However, it is necessary to obtain sufficient electro-conductivity to evaluate the ORR activity of the titanium-niobium oxides. Even a glassy carbon (GC) rod is heat-treated in air, an insulating oxide film is not formed on the surface. Therefore, the GC rod is superior to use as a substrate for the working electrode. The present strategy utilizes pre-heat-treatment (600 °C in air for 10 min) to achieve sufficient electrical contact between the titanium-niobium oxides and the GC substrate. It is necessary to secure the sufficient electro-conductivity between oxide-based catalysts and conductive oxide support when carbon-free cathodes are prepared. For example, the electro-conductive oxide network is made preparations in advance. Then, after oxide-based precursor is supported on the network it is heat-treated to create the ORR active sites and to obtain sufficient electro-conductivity. In this study, the effects of the preparation conditions, such as the gas atmosphere and heat treatment temperatures, on the ORR activity of the titanium-niobium oxides employing a GC rod are evaluated.

2. Results and Discussion

2.1. Characterization of Catalysts

We prepared the titanium-niobium oxide samples with the charged total composition of $Ti_{0.841}Nb_{0.126}O_2$. Figure 1 shows the X-ray diffraction (XRD) patterns of the titanium-niobium oxide samples prepared at 600, 700, 800, 900, and 1050 °C (a) in air and (b) in Ar containing 4% H_2. The crystalline phase of the catalysts prepared by heat treatment in air at temperatures between 600 and 900 °C was identified as anatase TiO_2 (JCPDS no. 00-021-1272), indicating that the niobium atoms were incorporated into the TiO_2 anatase structure. According to phase diagram of TiO_2–Nb_2O_5 [12], Nb(V) ions dissolve into TiO_2 rutile structure only below ca. 10 atomic % in this temperature range. On the other hand, quasi-stable phase, TiO_2 anatase structure, can dissolve more Nb(V) ions. The phase transition from anatase to rutile occurred at temperatures above 900 °C. For samples prepared at higher temperatures, peaks corresponding to the rutile TiO_2 (JCPDS no. 00-021-1276) and $TiNb_2O_7$ (JCPDS no. 1001270) phases were observed. This is because the Nb(V) ions that cannot dissolve in the TiO_2 rutile structure forms complex oxides $TiNb_2O_7$ that is solid solution of TiO_2 and Nb_2O_5. Simultaneously, Nb-containing phases such as $TiNb_2O_7$ appeared at 1050 °C. These results are consistent with previous observations [13].

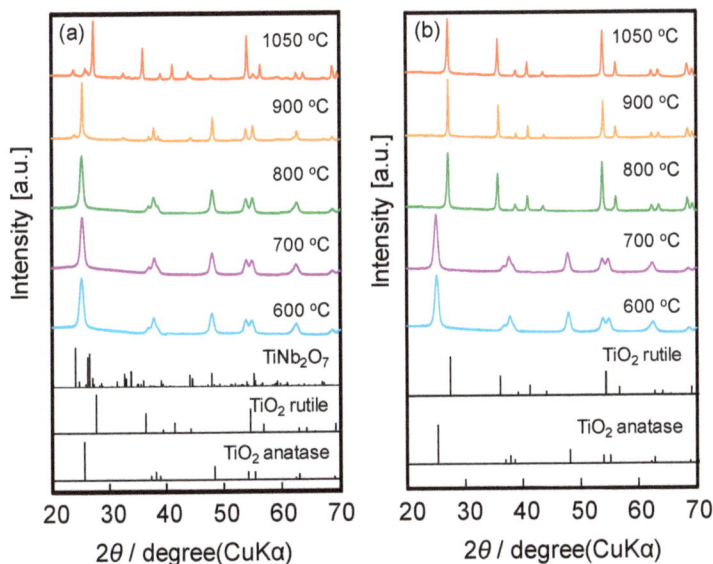

Figure 1. XRD patterns of titanium-niobium oxides prepared at 600, 700, 800, 900, and 1050 °C (**a**) in air and (**b**) in Ar containing 4% H_2.

The crystalline phase of the samples subjected to heat treatment at 600 and 700 °C in Ar containing 4% H_2 could be indexed to the TiO_2 anatase structure. However, the samples prepared at temperatures above 800 °C under this reductive atmosphere could be indexed to rutile TiO_2 with no Nb_2O_5 peaks. The shift of the XRD peaks to lower angles (Figure S1) with increasing treatment temperature suggested that the catalysts are substitutional solid solutions in which the niobium ions substitute titanium ions in the rutile TiO_2 lattice. Compared to formation of the rutile phase above 900 °C for the samples heat-treated in air, the rutile was phase formed at 800 °C under reductive atmosphere. Thus, the transformation from the anatase to rutile phase occurred at lower temperature under reductive atmosphere. In addition, the substitutional solid solution (rutile phase) was stable up to 1050 °C under reductive atmosphere. The XRD analysis clearly demonstrated that the TiO_2 rutile-based structure was more stable under reductive atmosphere than in air. This stabilization of the TiO_2 rutile-based structure is not predicted from the viewpoint of thermochemistry. The role of heat-treatment under reductive atmosphere and doped niobium ions must be elucidated.

Figure 2 and Figure S2 show scanning electron microscopy (SEM) images of the titanium-niobium oxides prepared at 600 °C in air, and 600, 700, 800, 900, and 1050 °C in Ar containing 4% H_2. The SEM images demonstrate that the surface morphology of the titanium-niobium oxides depends on the heat treatment temperature. Very little difference in the surface morphology was observed for the samples prepared by heat treatment at 600 °C under different atmospheres. The particle size of the catalysts prepared at 600 °C was *ca.* several tens of nanometers. A significant change in the morphologies of the catalysts was observed with treatment at 800 °C, indicative of particle aggregation above 800 °C. Aggregation became progressive with increasing heat treatment temperatures. Thus, the surface area of the catalysts decreased with temperature, especially above 800 °C.

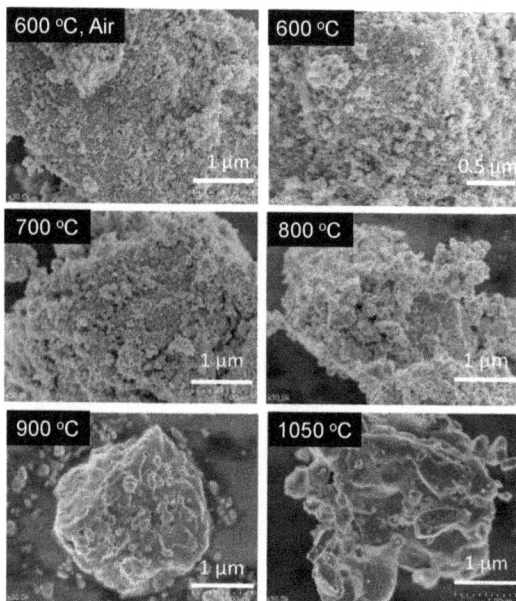

Figure 2. SEM images of the titanium-niobium oxides prepared at 600 °C in air, and 600, 700, 800, 900, and 1050 °C in Ar containing 4% H_2.

Figure 3 shows photographs of the catalysts prepared at 600 °C in air, and 600, 700, 800, 900, and 1050 °C in Ar containing 4% H_2. The powder heat-treated at 600 °C was white, as expected from the wide bandgap of TiO_2 (all samples treated in air were white). On the other hand, the samples heat-treated at 600 °C under reductive atmosphere had a light-blue color and the color deepened with increasing temperature. This color change suggests that there is some difference in the electronic energy levels of the samples prepared under reductive atmosphere relative to those prepared in air. Namely, the difference between the highest occupied and lowest unoccupied electronic energy levels decreases with increasing temperature. This color change suggests the development of a localized energy level of electrons in the bandgap of TiO_2.

Figure 3. Photographs of the catalysts prepared at 600 °C in air, and 600, 700, 800, 900, and 1050 °C in Ar containing 4% H_2.

Figure 4a shows the Ti 2p XPS spectra of the catalysts prepared at 800 °C in air and in Ar containing 4% H_2. As anticipated, the Ti 2p XPS spectra revealed that Ti adopted the tetravalent state for the specimen heat-treated in air based on the $2p_{3/2}$ peak (TiO_2; 458.8 eV [14]). On the other hand, a low valence state, *i.e.*, Ti^{3+} (Ti_2O_3; 456.8 eV [15]), was observed for the catalyst heat-treated at 800 °C under reductive atmosphere. The ratios of Ti^{3+}/Ti^{4+} calculated from areas of the XPS spectra of the specimens heat-treated at 800 °C in air and in Ar containing 4% H_2 were 5.0% and 10%, respectively. The ratio of the specimen prepared under reductive atmosphere was twice as large as that prepared in air. In addition, the total atomic ratio of Nb/Ti is 0.15 according to the charged total composition of $Ti_{0.841}Nb_{0.126}O_2$. The atomic ratios of Nb/Ti calculated from areas of the XPS spectra of the specimens heat-treated at 800 °C in air and in Ar containing 4% H_2 were 0.43 and 0.23, respectively. Both ratios are larger than the total atomic ratio, suggested that the niobium ions accumulate the surface of the oxide particles. In particular, the Nb/Ti ratio of the specimen heat-treated in air was about three times larger than the total atomic ratio. As mentioned in XRD patterns, because the rutile TiO_2 phase cannot dissolve the Nb(V) ions, the dissolved Nb(V) ions in the anatase TiO_2 phase began to accumulate near the surface of the particles at higher temperature heat treatment.

Figure 4b shows the Ti 2p XPS spectra of the catalysts prepared at 600, 700, 800, 900, and 1050 °C in Ar containing 4% H_2. Low valence states of Ti were observed for the catalyst heat-treated at 600 °C under reductive atmosphere, suggesting that the oxides underwent little reduction at 600 °C in Ar containing 4% H_2 upon treatment for 10 min. Heat-treatment above 700 °C under reductive atmosphere resulted in the formation of low valence state Ti as shown in Figure 4.

325

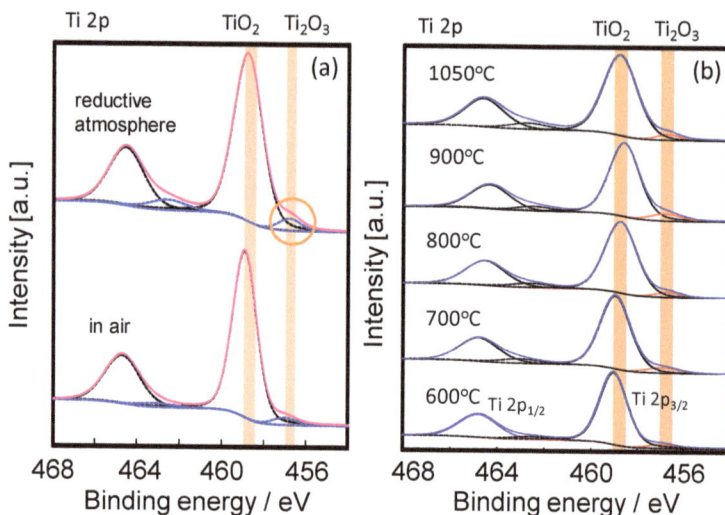

Figure 4. Ti 2p XPS spectra of the catalysts prepared at 800 °C in air and in Ar containing 4% H$_2$ (**a**) and prepared at 600, 700, 800, 900, and 1050 °C in Ar containing 4% H$_2$ (**b**).

Figure 5a shows the dependence of the ratios of Ti^{3+}/Ti^{4+} (expressed as $S_{Ti(III)}/S_{Ti(IV)}$) calculated from areas of the XPS spectra of the specimens heat-treated under reductive atmosphere on the temperature. The ratio of Ti^{3+}/Ti^{4+} of the specimen prepared at 600 °C is 6.7%. Ti^{3+} ions are produced by the substitution of the Nb^{5+} ions with Ti^{4+} ions of the TiO$_2$ lattice. Figure 5b shows the Nb 3d XPS spectra of the catalysts prepared at 600, 700, 800, 900, and 1050 °C in Ar containing 4% H$_2$. The peak in the Nb 3d spectra shifted to higher binding energy (NbO$_2$; 205.3 eV [16], Nb$_2$O$_5$; 207.1 eV [17]) with increasing heat treatment temperatures, in contrast with the Ti 2p peak. Therefore, the Nb 3d XPS spectra revealed that most of Nb ions were highest oxidation state, 5+. Thus, the state of the specimens can be expressed as Ti(IV)$_{1-2x}$Ti(III)$_x$Nb(V)$_x$O$_2$. If all Nb ions substitute Ti^{4+} ions of the TiO$_2$ lattice as Nb(V) ions, the composition is Ti(IV)$_{0.74}$Ti(III)$_{0.13}$Nb(V)$_{0.13}$O$_2$. Therefore, in that case, the ratio of Ti^{3+}/Ti^{4+} is calculated to be *ca.* 18%. The ratio of Ti^{3+}/Ti^{4+} at 600 °C, *ca.* 6.7%, was smaller than 18%, indicating that the Nb(V) ions did not sufficiently incorporate into the TiO$_2$ lattice at 600 °C. As shown in Figure 5a, the ratio of Ti^{3+}/Ti^{4+} increased with increasing temperature from 600 °C to 700 °C and saturated around 10%. These results deduced that reductive heat-treatment above 700 °C induced the formation of low valence state Ti.

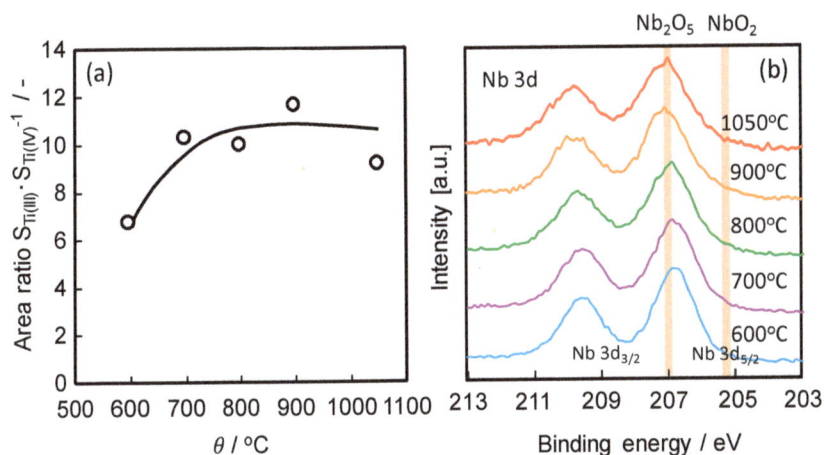

Figure 5. (a) Dependence of the ratios of Ti^{3+}/Ti^{4+}, $S_{Ti(III)}/S_{Ti(IV)}$, calculated from areas of the XPS spectra of the specimens heat-treated under reductive atmosphere on the temperature. **(b)** Nb 3d XPS spectra of the catalysts prepared at 600, 700, 800, 900, and 1050 °C in Ar containing 4% H_2.

Figure 6 shows the dependence of the atomic ratio of Nb/Ti calculated from XPS spectra of the specimens prepared under reductive atmosphere on the heat treatment temperature. The atomic ratio of Nb/Ti decreased with increasing temperature above 700 °C and approached the bulk value at 1050 °C. The XRD patterns revealed that the bulk phase transition occurred between 700 and 800 °C under reductive atmosphere. The XPS spectra indicated that the titanium ions near the surface were reduced and the Nb(V) ions near the surface incorporated into the TiO_2 lattice at *ca.* 700 °C. Therefore, the phase transition was probably caused by a change in the valence of titanium. We previously demonstrated that tantalum and zirconium oxide-based catalysts had some oxygen vacancies that acted as active sites for the ORR [6]. In case of the titanium-niobium oxide system, the low valence state of the metal ions does not always indicate the presence of oxygen vacancies. The low valence state of the metal ions can be achieved even in the absence of oxygen vacancies because the highest valence states of titanium and niobium are different. The relationship between the presence of oxygen vacancies and the active sites remains a topic for further study.

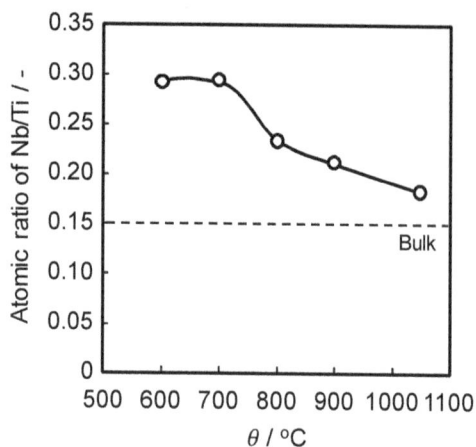

Figure 6. Dependence of the atomic ratio of Nb/Ti calculated from XPS spectra of the specimens prepared under reductive atmosphere on the heat treatment temperature.

It was difficult to evaluate the differences in the electronic state of the catalysts heat-treated under reductive atmosphere at temperatures between 700 and 1050 °C based on the XPS spectra, as shown in Figures 4b and 5a. Thus, the ionization potential of the specimens was used as a parameter to evaluate these differences. The ionization potentials of the specimens were measured using a photoelectron spectrometer surface analyzer in order to investigate the differences in the surfaces of the specimens heat-treated in reductive atmosphere at different temperatures. Figure 7a shows the relationship between the square root of the photoelectric quantum yield and the photon energy (that is, the photoelectron spectra of the specimens heat-treated at 800 °C in air or in Ar containing 4% H_2). The square root of the photoelectric quantum yield increased linearly with an increase in the photon energy applied to each specimen. The slope of the straight line reflects the tendency of the photoelectron emission of the specimens, that is, the density of state of the electrons near the Fermi level. Fewer photoelectrons were emitted in the case of the catalyst prepared in air. The slope of the straight line for the specimen heat-treated in air, where TiO_2 was identified on the sample surface by XPS, was apparently lower than that of the congener prepared under reductive atmosphere. It is remarkable that the slope of this plot was steeper for the specimen prepared in Ar containing 4% H_2. The intersection between the straight line and the background line in the photoelectron spectra provides the threshold energy corresponding to the photoelectric ionization potential. The photoelectric ionization potential corresponds to the highest energy level of the electrons in the materials. The ionization potential is directly affected by the localized electronic levels of the lattice defects and impurities in the metal oxides, such as valence changes due to substitutional metal ions, oxygen vacancies, and donor impurities.

Figure 7b shows the dependence of the ionization potential of the catalysts prepared at 600, 800, and 1050 °C in air, and 600, 700, 800, 900, and 1050 °C in Ar containing 4% H_2 on the heat treatment temperature, θ. The ionization potential of commercial rutile and anatase TiO_2 is 5.8 eV. The ionization potential was the same (*i.e., ca.* 5.8 eV) for the catalysts prepared at 600, 800, and 1050 °C in air, suggesting that the surface of the catalysts prepared in air had few localized electronic levels from lattice defects and impurities in the metal oxides, similar to commercial TiO_2. On the other hand, the ionization potentials of the catalysts prepared under reductive atmosphere decreased with increasing temperature. The decrease in the ionization potential reflects an increase in the localized electronic levels. In other words, the valence changes due to substitutional metal ions, oxygen vacancies, and donor impurities increase with increasing temperature.

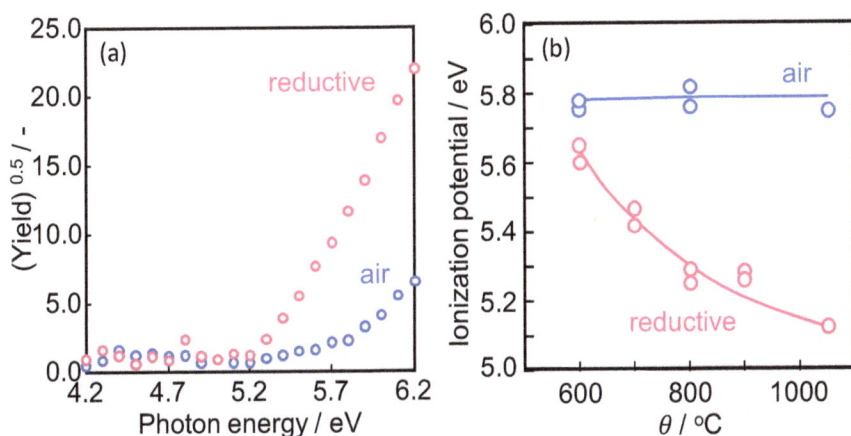

Figure 7. (a) Relationship between the square root of the photoelectric quantum yield ($Y^{1/2}$) and the photon energy of the specimens heat-treated at 800 °C in air or in Ar containing 4% H_2. **(b)** Dependence of the ionization potential of the catalysts prepared at 600, 800, and 1050 °C in air, and 600, 700, 800, 900, and 1050 °C in Ar containing 4% H_2 on the heat treatment temperature, θ.

2.2. Oxygen Reduction Activity in Acidic Media

Figure 8a shows the potential-i_{ORR} curves for the catalysts prepared at 600, 700, and 1050 °C in Ar containing 4% H_2. The heat treatment temperature apparently affected the ORR activity. We focused on the ORR activity in the higher potential region. Figure 8b shows the potential-i_{ORR} curves for the catalysts prepared at 600, 700, 800, 900, and 1050 °C in Ar containing 4% H_2. All samples prepared in air had a low ORR current in the potential range above 0.6 V, indicating that these catalysts have low ORR activity. On the other hand, although the ORR current was low, the catalysts prepared under reductive atmosphere exhibited some ORR activity. In particular, the onset potential of the ORR for the catalyst prepared at 700 °C was approximately 1.0 V *vs.* RHE. This high onset potential indicates the good suitability of the active sites for the ORR. Therefore, high quality active sites were created by heat treatment at 700 °C under reductive atmosphere.

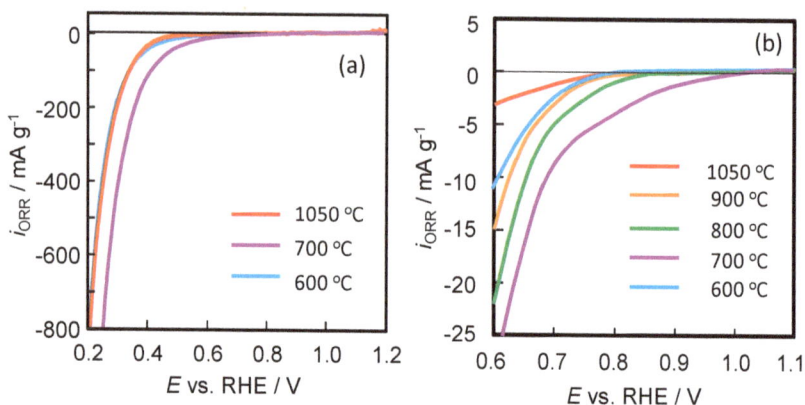

Figure 8. (a) Potential-i_{ORR} curves for the catalysts prepared 600, 700, and 1050 °C in Ar containing 4% H_2 and (b) potential-i_{ORR} curves in the higher potential region for the catalysts prepared 600, 700, 800, 900, and 1050 °C in Ar containing 4% H_2.

Figure 9 shows the dependence of the i_{ORR} @ 0.7 V on the heat-treatment temperature for the samples prepared under reductive atmosphere. The i_{ORR} @ 0.7 V reached a maximum around 700 °C. The i_{ORR} presented in Figure 9 is based on the mass of the catalysts loaded on the GC rod. As shown in Figure 2 and Figure S2, the surface area of the catalysts declined precipitously above 800 °C. Thus, the decrease in the i_{ORR} @ 0.7 V above 800 °C seems to be due to the decrease in the surface area. To evaluate the specific activity (*i.e.*, the ORR current density based on surface area) the actual surface area of the oxides must be estimated. However, it is difficult to estimate the surface area of the oxides because neither hydrogen nor CO is adsorbed by the oxides. Therefore, the electrical charges of the double layer of the catalysts calculated from the cyclic voltammogram (CV) in N_2 atmosphere were used to evaluate the apparent specific activity of the catalysts. Figure S3 shows the cyclic voltammograms of the GC rod only and of titanium-niobium oxide supported on the GC rod ($Ti_xNb_yO_z$/GC) heat-treated at 800 °C under reductive atmosphere. Because the amount of oxide catalyst loaded on the rod was small (*ca.* 1 mg), the charge/discharge current was mainly derived from that due to the GC substrate. The electrical charge due to the oxide was estimated from the difference between the CV of $Ti_xNb_yO_z$/GC and that of GC only. Figure S4 shows the dependence of the electrical charge of the double layer of the oxides on the catalyst loading. The SEM images showed that the surface area of the catalysts decreased above 800 °C due to aggregation of the particles. However, a linear relationship was obtained, suggesting that the electrical charge was determined not by the heat treatment temperature but by the catalyst loading. Therefore, the trend in the apparent specific activity (ORR current density based on electrical charge) is similar to that of the mass activity. It is anomalous that the electrical charge is independent of the heat treatment temperature. The surface area estimated using the electrical charge may be different from that predicted from the SEM images. Because the electrical conductivity of even the catalysts prepared under reductive atmosphere is low, the surface area of the electrochemical active region in contact with the GC rod might be small. Thus, a more accurate estimation of the actual surface area of the catalysts is necessary.

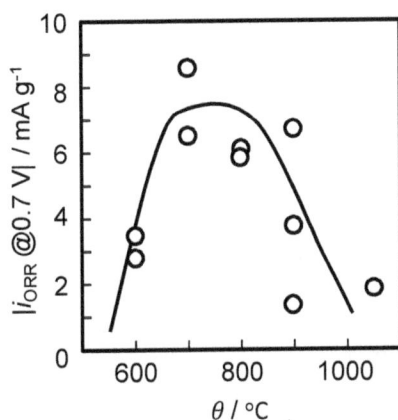

Figure 9. Dependence of i_{ORR} @ 0.7 V on temperature used for heat-treatment of the samples under reductive atmosphere.

2.3. Relationship between ORR Activity and Physico-Chemical Properties

The ORR activity was enhanced by reductive heat treatment in the region of 700 to 900 °C. The XRD patterns indicated that the crystalline structure of the catalysts prepared under reductive atmosphere changed from anatase to rutile TiO_2 around 800 °C. On the other hand, the XPS spectra revealed that low valence state Ti is generated by heat treatment above 700 °C under reductive atmosphere. Therefore, reduction of the sample surface occurs around 700 °C. The ionization potential is more sensitive to the surface state as shown in Figure 7b. Henrich *et al.* found that the work function (*i.e.*, ionization potential in this study) of TiO_2 decreased as the density of oxygen vacancies increased [18]. Therefore, the low ionization potential suggested that the catalysts heat-treated under reductive atmosphere at higher temperature had more surface defects. In this study, the oxygen vacancies as well as the valence changes induced by substitutional metal ions were found to produce localized electronic energy levels in the bandgap.

Figure 10 shows the relationship between the ionization potential and the i_{ORR} @ 0.7 V of the catalysts prepared under reductive atmosphere. A "volcano plot" with a maximum at 5.4 eV was obtained, suggesting that the electronic state of the sample surface is suitable for the ORR.

Adsorption of oxygen molecules on the surface is required as the first step for the ORR to proceed. Many studies have demonstrated that surface defect sites are required for adsorption of oxygen molecules on the surface of the oxides [19]. Therefore, a larger number of surface defects furnishes more sites for adsorption of oxygen molecules. In addition, the interaction of oxygen with the catalyst surface is essential because adsorption of oxygen and desorption of water from the surface are both necessary for robust progress of the ORR. When the interaction of oxygen with the catalyst surface is strong, desorption of water does not proceed readily. On the other hand, when the interaction of oxygen with the catalyst surface is weak, less adsorption of oxygen molecules occurs. Therefore, there is an optimal strength for the interaction between oxygen and the catalyst surface. Metallic Ti adsorbs oxygen strongly because of the large adsorption energy of oxygen (759 kJ·mol^{-1}) and the strong energy of the Ti-oxygen bonds (calculated: 625 kJ·mol^{-1}) [20]. In the case of Pt, the energy for adsorption of oxygen and the calculated Pt-oxygen bond energy are 272 kJ·mol^{-1} and 385 kJ·mol^{-1}, respectively [20]. Therefore, the corresponding values for Ti are much larger than those of Pt. As the degree of oxidization of metallic Ti increases, the interaction of oxygen with Ti on the catalyst surface is weakened because the oxide ions attract the electrons in the highest occupied molecular orbital of Ti thereby conferring a positive charge on Ti, $i.e.$, higher valence state. Because the ionization potential is related to the strength of the interaction between the surface of the specimen and oxygen, the volcano plot shown in Figure 10 suggests that there is a suitable interaction between the surface of the specimen and oxygen. Consequently, the strength of the interaction between oxygen and the oxide surface could be manipulated by controlling the local energy level of the electrons, $i.e.$, by controlling the valence changes induced by the substitutional ions and/or oxygen vacancies.

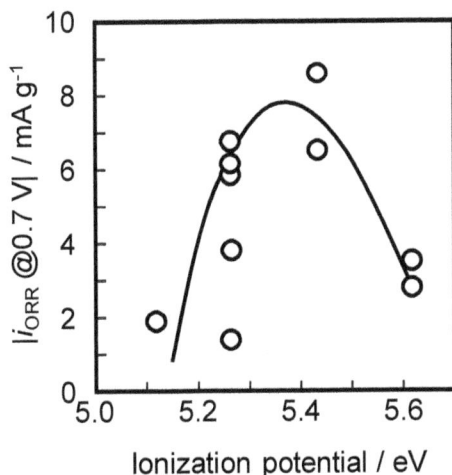

Figure 10. Relationship between the ionization potential and the i_{ORR} @ 0.7 V of the catalysts prepared under reductive atmosphere.

3. Experimental Section

The high concentration sol-gel method [21,22] was used for preparation of the precursor. A 30 cm^3 aliquot of titanium(IV) tetraisopropoxide ($C_{12}H_{28}O_4Ti$, 99.99%, Sigma-Aldrich Japan Co. LLC, Tokyo, Japan) and 4 cm^3 of niobium(V) ethoxide ($C_{10}H_{25}NbO_5$, 99.95%, Aldrich) were dissolved in 200 cm^3 of 2-methoxyethanol with a TiO_2:Nb_2O_5 weight ratio of 8:2. The mixed solution was maintained at -50 °C, and 15 cm^3 of 2-methoxyethanol in 15 cm^3 of pure water was added to the mixed solution dropwise. The temperature of the solution was raised to 80 °C and maintained for 3 weeks as an aging treatment, resulting in the formation of nano-sized complex oxides. The precipitates were dispersed in 2-methoxyethanol to obtain a dispersion of nano-sized titanium-niobium oxide.

A 3-mm^3 aliquot of the dispersion was dropped onto a GC rod ($\varphi = 5.0$ mm; TOKAI CARBON CO., LTD., Tokyo, Japan) followed by drying at room temperature. The coated rod was heat-treated at 600 °C for 10 min in air as a pre-heat-treatment step to remove organic species and carbon residue and to provide sufficient electrical contact between the titanium-niobium oxide and the GC substrate. Subsequently, samples of titanium-niobium oxide supported on the GC rods were heat-treated at 600, 700, 800, 900, and 1050 °C in air or in Ar containing 4% H_2 to prepare the working electrodes. For the powder XRD and ionization potential measurements, 3 cm^3 of the dispersion of nano-sized titanium-niobium oxide was dried on a hot plate at 160 °C to obtain the powder samples. The powders were then heat-treated at 600 °C for 10 min in air to remove organic species and carbon residue. The powders were subsequently heat-treated at 600, 700, 800, 900, or 1050 °C in air or in Ar containing 4% H_2 for powder XRD and ionization potential measurements.

The morphologies, crystalline structures, and chemical states of the synthesized catalysts were investigated by transmission electron microscopy (TEM; JEOL Ltd., JEM-2100F, Akishima, Japan, X-ray diffraction (XRD; Rigaku Corporation, Ultima IV, X-ray source: Cu-Kα, Akishima, Japan) and X-ray photoelectron spectroscopy (XPS; ULVAC-PHI, Inc. Quantum-2000, X-ray source: monochromated Al-Kα radiation, Chigasaki, Japan). The peak of the C–C bond attributed to free carbon at 284.6 eV in the C 1s spectrum was used to compensate for surface charging.

The ionization potential of the specimens was measured using a photoelectron spectrometer surface analyzer (Model AC-2, RIKEN KEIKI Co., Ltd., Tokyo, Japan) [23,24].

All electrochemical measurements were performed in 0.1 mol·dm^{-3} H_2SO_4 at 30 °C with a 3-electrode cell. A reversible hydrogen electrode (RHE) and a glassy carbon plate were used as the reference and counter electrodes, respectively. As a pre-treatment, 300 CV cycles were performed in O_2 atmosphere in the range of 0.05 to 1.2 V with respect to the RHE at a scan rate of 150 mV·s^{-1}. Slow scan voltammetry was performed under O_2 and N_2 atmosphere in the range of 0.2 to 1.2 V with respect to RHE at a scan rate of 5 mV·s^{-1}. The ORR current density, i_{ORR}, based on the mass of the catalyst (mass activity), was determined by calculating the difference between the current density under O_2 and N_2 atmosphere.

4. Conclusions

In order to develop noble-metal- and carbon-free cathodes, titanium-niobium oxides were prepared for use as oxide-based cathodes and the factors affecting the ORR activity and active sites were evaluated. The high concentration sol-gel method was employed for preparation of the precursor. Secure adhesion between the oxide catalysts and the substrate was achieved by heating the precursor supported GC rod at 600 °C in air to maintain the electrical contact as a pretreatment step. To create ORR active sites, the precursor supported GC rod was heat-treated in the temperature range of 600 to 1050 °C in air or in Ar containing 4% H_2. Heat treatment in reductive atmosphere at 700–900 °C was effective for conferring ORR activity to the catalysts. Notably, the onset potential for the ORR was approximately 1.0 V *vs.* RHE for the catalyst prepared at 700 °C. This high onset potential indicates the high quality of the active sites for the ORR. XRD, XPS and ionization potential measurements suggested that localized electronic energy levels were produced by heat treatment under reductive atmosphere. The electronic energy levels produced by the valence changes of Ti induced by substitutional metal ions and/or oxygen vacancies might govern adsorption of the oxygen molecules. Therefore, the strength of the interaction between oxygen and the oxide surface can be manipulated by controlling the valence changes induced by the substitutional ions and/or oxygen vacancies.

Acknowledgments: The authors acknowledge financial support from the New Energy and Industrial Technology Development Organization (NEDO). This work was conducted under the auspices of the Ministry of Education, Culture, Sports, Science and Technology (MEXT) Program for Promoting the Reform of National Universities.

Author Contributions: A.I. designed the research; Y.T. performed research; M. C. analyzed XPS data; Y.O. analyzed data; and all authors provided feedback during preparation of the manuscript.

Conflicts of Interest: The authors declare no conflict of interest.

References

1. Jasinski, R. A new fuel cell cathode catalyst. *Nature* **1964**, *201*, 1212–1213.

2. Jaouen, F.; Goellner, V.; Lefevre, M.; Herranz, J.; Proietti, E.; Dodelet, J.P. Oxygen reduction activities compared in rotating-disk electrode and proton exchange membrane fuel cells for highly active Fe–N–C catalysts. *Electrochim. Acta* **2013**, *87*, 619–628.

3. Wu, G.; More, K.L.; Johnston, C.M.; Zelenay, P. High-Performance Electrocatalysts for Oxygen Reduction Derived from Polyaniline, Iron, and Cobalt. *Science* **2011**, *332*, 443–447.

4. Proietti, E.; Jaouen, F.; Lefevre, M.; Larouche, N.; Tian, J.; Herranz, J.; Dodelet, J.P. Iron-based cathode catalyst with enhanced power density in polymer electrolyte membrane fuel cells. *Nat. Commun.* **2011**, *2*, 416.

5. Ishihara, A.; Lee, K.; Doi, S.; Mitsushima, S.; Kamiya, N.; Hara, M.; Domen, K.; Fukuda, K.; Ota, K. Tantalum Oxynitride for a Novel Cathode of PEFC. *Electrochem. Solid-State Lett.* **2005**, *8*, A201–A203.

6. Ishihara, A.; Tamura, M.; Ohgi, Y.; Matsumoto, M.; Matsuzawa, K.; Mitsushima, S.; Imai, H.; Ota, K. Emergence of Oxygen Reduction Activity in Partially Oxidized Tantalum Carbonitrides: Roles of Deposited Carbon for Oxygen-Reduction-Reaction-Site Creation and Surface Electron Conduction. *J. Phys. Chem. C* **2013**, *117*, 18837–18844.

7. Ishihara, A.; Chisaka, M.; Ohgi, Y.; Matsuzawa, K.; Mitsushima, S.; Ota, K. Synthesis of nano-TaO_x oxygen reduction reaction catalysts on multi-walled carbon nanotubes connected via a decomposition of oxy-tantalum phthalocyanine. *Phys. Chem. Chem. Phys.* **2015**, *17*, 7643–7647.

8. Okada, Y.; Ishihara, A.; Matsumoto, M.; Imai, H.; Kohno, Y.; Matsuzawa, K.; Mitsushima, S.; Ota, K. Electrochemical stability of zirconium oxide-based electrocatalysts made from oxy-zirconium phthalocyanines. *J. Electrochem. Soc.* **2015**, *162*, F959–F964.

9. Ishihara, A.; Hamazaki, M.; Kohno, Y.; Matsuzawa, K.; Mitsushima, S.; Ota, K. Titanium-niobium oxides mixed with Ti_4O_7 as noble-metal- and carbon-free cathodes for polymer electrolyte fuel cells. *Electrochim. Acta.* submitted.

10. Hamazaki, M.; Ishihara, A.; Kohno, Y.; Matsuzawa, K.; Mitsushima, S.; Ota, K. Evaluation of durability of titanium-niobium oxides mixed with Ti_4O_7 as non-noble and carbon-free cathodes for PEFC in H_2SO_4 at 80 °C. *Electrochemistry*, in press.

11. Arashi, T.; Seo, J.; Takanabe, K.; Kubota, J.; Domen, K. Nb-doped TiO_2 cathode catalysts for oxygen reduction of polymer electrolyte fuel cells. *Catal. Today* **2014**, *233*, 181–186.

12. The American Ceramic Society. *Phase Equilibria Diagrams Volume XII Oxides*; McHale, A., Roth, R., Gen, S., Eds.; The American Ceramic Society, United States of America: Westerville, OH, USA, 1996; p. 119.

13. Sedneva, T.A.; Lokshin, E.P.; Belikov, M.L.; Belyaevskii, A.T. TiO_2-and Nb_2O_5-Based Photocatalytic Composites. *Inorg. Mater.* **2013**, *49*, 382–389.

14. Haukka, S.; Lakomaa, E.-L.; Jylha, O.; Vilhunen, J.; Hornytzkyj, S. Dispersion and Distribution of Titanium Species Bound to Silica from $TiCl_4$. *Langmuir* **1993**, *9*, 3497–3506.

15. González-Elipe, A.R.; Munuera, G.; Espinos, J.P.; Sanz, J.M. Compositional changes induced by 3.5 keV Ar^+ ion bombardment in Ni–Ti oxide systems: A comparative study. *Surf. Sci.* **1989**, *220*, 368–380.

16. Bahr, M.K. ESCA studies of some niobium compounds. *J. Phys. Chem. Solids* **1975**, *36*, 485–491.

17. Olejniczak, M.; Ziolek, M. Comparative study of Zr, Nb, Mo containing SBA-15 grafted with amino-organosilanes. *Microporous Mesoporous Mater.* **2014**, *196*, 243–253.

18. Henrich, V.E.; Dresselhaus, G.; Zeiger, H.J. Observation of Two-Dimensional Phases Associated with Defect States on the Surface of TiO_2. *Phys. Rev. Lett.* **1976**, *36*, 1335–1339.

19. Bourgeois, S.; Domenichini, B.; Jupille, J. Excess Electrons at Oxide Surfaces. In *Defects at Oxide Surfaces*; Jupille, J., Thornton, G., Eds.; Springer Series in Surface Sciences 58; Springer International Publishing: Basel, Switzerland, 2015; pp. 123–148.

20. Miyazaki, E.; Yasumori, I. Heats of chemisorption of gases. *Surf. Sci.* **1976**, *55*, 747–753.

21. Matsuda, H.; Mizushima, T.; Kuwabara, M. Low-Temperature Synthesis and Electrical Properties of Semiconducting $BaTiO_3$ Ceramics by the Sol-Gel Method with High Concentration Alkoxide Solutions. *J. Ceram. Soc. Jpn.* **1999**, *107*, 290–292.

22. Matsuda, H.; Kobayashi, N.; Kobayashi, T.; Miyazawa, K.; Kuwabara, M. Room-temperature synthesis of crystalline barium titanate thin films by high-concentration sol-gel method. *J. Non-Cryst. Solids* **2000**, *271*, 162–166.

23. Kirihata, H.; Uda, M. Externally quenched air counter for low-energy electron emission measurements. *Rev. Sci. Instrum.* **1981**, *52*, 68–70.

24. Uda, M.; Nakagawa, Y.; Yamamoto, T.; Kawasaki, M.; Nakamura, A.; Saito, T.; Hirose, K. Successive change in work function of Al exposed to air. *J. Electron. Spectrosc. Relat. Phenom.* **1998**, *88–91*, 767–771.

Positive Effect of Heat Treatment on Carbon-Supported CoS Nanocatalysts for Oxygen Reduction Reaction

Haihong Zhong, Jingmin Xi, Pinggui Tang, Dianqing Li and Yongjun Feng

Abstract: It is of increasing interest and an important challenge to develop highly efficient less-expensive cathode catalysts for anion-exchange membrane fuel cells (AEMFCs). In this work, we have directly prepared a carbon-supported CoS nanocatalyst in a solvothermal route and investigated the effect of heat-treatment on electrocatalytic activity and long-term stability using rotating ring-disk electrode (RRDE). The results show that the heat-treatment below 400 °C under nitrogen atmosphere significantly enhanced the electrocatalytic performance of CoS catalyst as a function of annealed temperature in terms of the cathodic current density, the half-wave potential, the HO_2^- product and the number of electrons transferred. The CoS catalyst that annealed at 400 °C (CoS-400) has exhibited a promising performance with the half-wave potential of 0.71 V *vs.* RHE (the highest one for non-precious metal chalcogenides), the minimum HO_2^- product of 4.3% at 0.60 V *vs.* RHE and close to the 4-electron pathway during the oxygen reduction reaction in 0.1 M KOH. Also, the CoS-400 catalyst has comparable durability to the Pt/C catalyst.

Reprinted from *Catalysts*. Cite as: Zhong, H.; Xi, J.; Tang, P.; Li, D.; Feng, Y. Positive Effect of Heat Treatment on Carbon-Supported CoS Nanocatalysts for Oxygen Reduction Reaction. *Catalysts* **2015**, *5*, 1211–1220.

1. Introduction

The polymer electrolyte membrane fuel cells system is one of the best alternative candidates to fossil power systems because of high power density, high efficiency and near zero emission [1–3]. Compared with proton-exchange membrane fuel cells (PEMFCs), anion-exchange membrane fuel cells (AEMFCs) are attracting more interest because cathode electrocatalysts have higher oxygen reduction kinetics, longer durability and more choices among non-Pt metals in alkaline media [4]. However, it still remains a big challenge and it is of great interest to develop highly efficient less-expensive electrocatalysts for oxygen reduction reaction (ORR) in alkaline media [5–8].

Recently, non-precious metal chalcogenides, e.g., MS_2 {M = Fe [9], Co and (Co, Ni) [10]} thin film, CoS_2 [11], $CoSe_2$ [12,13], $Co_{1-x}S$ [14] nanoparticles have attracted extensive attention due to promising catalytic activity for ORR, low cost and abundant reserve in the earth's crust. In Ref. [13,14], one notes that heat treatment plays an important role to produce electrocatalysts with high performance. However, Dai and his coauthors did not explain the reason for annealing treatment in Ref. [14]. As reviewed by Zhang *et al.* [15], heat treatment can significantly affect the ORR catalytic activity and stability of the supported catalysts. However, few researchers have focused on the effect of heat treatment on the ORR electrocatalytic performance of non-precious metal chalcogenides [16,17].

In this work, we directly prepared the carbon-supported CoS nanocatalyst as one of the non-precious metal cathodic catalysts in a solvothermal route and specially investigated the effect of heat-treatment temperature on electrocatalytic performance of the prepared nanocatalysts towards ORR in alkaline medium.

2. Results and Discussion

2.1. Structure and Morphology

Figure 1 shows powder X-ray diffraction (XRD) patterns of five prepared 20 wt. % CoS/C samples: those that were as-prepared and those that were annealed at 250, 300, 400 and 450 °C under nitrogen atmosphere, respectively. Compared with the ICDD-PDF2-2004 card of CoS No. 75-0605 in space group P63/mmc (No. 194), the first four samples exhibit four characteristic Bragg reflection peaks of CoS phase, *i.e.*, (100), (101), (102) and (110) as marked in the graph while CoS-450 reveals reflections from another phase as denoted with an asterisk besides CoS phase. After careful comparison and analysis, the impurity is possibly attributed to Co_9S_8 (PDF No. 86-2273) with two typical characterization peaks (2θ) located at $29.84°/311$ and $53.10°/440$. Furthermore, the corresponding crystallite size was evaluated based on the Scherrer equation: $D_{hkl} = K\lambda/\beta \cdot \cos\theta$, where D_{hkl}, K, λ, β and θ are the crystallite size in the hkl direction (nm), the shape factor ($K = 1.0$), the X-ray wavelength ($\lambda = 0.15406$ nm), the full-width at half maximum (rad) and the Bragg reflection angle (°), respectively. Based on (100), (101) and (102) Bragg reflection peaks, the average crystallite size was individually 6.6 nm, 8.5 nm, 10.1 nm, 11.6 nm and 12.7 nm for the five samples with an increase in the annealed temperature. These results suggest that this mild synthesis route is available for CoS nanoparticles when the S/Co molar ratio in the initial reaction solution is equal to 3.0. Furthermore, the thermal stability of the prepared CoS nanoparticles is below 450 °C, which is similar to that of $CoSe_2$ nanoparticles [13], and lower than that of $Co_{1-x}S$ [14]. In this work, therefore, our following investigation will mainly concentrate on four samples: as-prepared, CoS-250, -300 and -400.

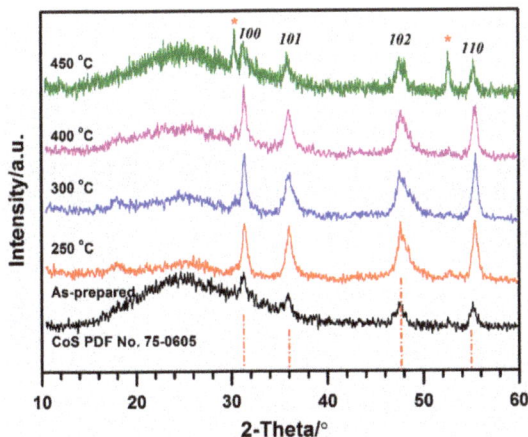

Figure 1. Powder X-ray diffraction patterns of the as-prepared 20 wt. % CoS/C sample and four annealed samples at four different temperatures: CoS-250, -300, -400 and -450, respectively. Vertical dot-dashed lines represent the ICDD-PDF2-2004 card of CoS No.75-0605.

In order to further investigate the influence of CoS nanoparticles on carbon substrate during the annealed process, the Raman spectra of the prepared CoS/C at different temperatures were examined as shown in Figure 2. Raman spectroscopy as one of non-destructive techniques is a powerful tool to detect ordered and disordered crystal structures of carbon materials [18]. The intensity of the D band centered at *ca.* 1350 cm^{-1} and the G band centered at *ca.* 1600 cm^{-1} is generally used to characterize the degree of disorder structure based on $R = I_D/I_G = 4.4\,La$ (nm), where *La* is the microcrystalline planar size in products. The *R* value was calculated to be 2.62, 2.76, 2.46, and 2.75 for CoS-250, -300, -400, and -450, respectively. The corresponding *La* value was obtained to be 0.60, 0.63, 0.56, and 0.63 nm, which are little different. These results suggest that the disordered degree almost has no change during the annealed process.

Figure 3 demonstrates TEM images of four samples: as-prepared (a), CoS-250 (b), CoS-300 (c), and CoS-400 (d). It is difficult to distinguish the CoS nanoparticles because of very little color contrast between CoS and carbon. Also, the magnetic property of the Co-containing samples obviously reduces the clarity of the TEM images, although the samples were pretreated by degaussing. The average particle size was approximately evaluated in the range of 5–20 nm. With an increase in the annealed temperature, the nanoparticles were aggregated to form hierarchical structure (Figure 3d).

In order to further confirm the CoS phase, the XPS spectra from the Co and S regions of the CoS/C samples were determined as presented in Figure 4. The

Co2p spectrum in Figure 4A has two peaks: one located at 778.1 eV for $Co2p_{2/3}$, and the other one at 793.0 eV for $Co2p_{1/2}$, which are attributed to Co–S. Besides, a small emission peak at *ca.* 781.0 eV could result from Co–O due to the strong affinity between cobalt ions and atmospheric oxygen. The peak at 162.1 eV in Figure 4B corresponds to the binding energy of Co–S [19]. In comparison, the intensities of Co2p and S2p peaks increase with the increase of the annealed temperature and no position shift is detected.

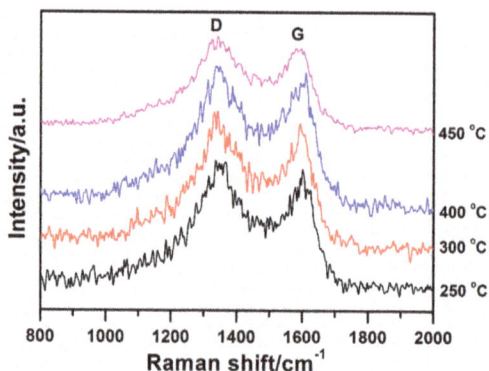

Figure 2. Comparative Raman spectra of CoS/C annealed at temperatures from 250 to 450 °C.

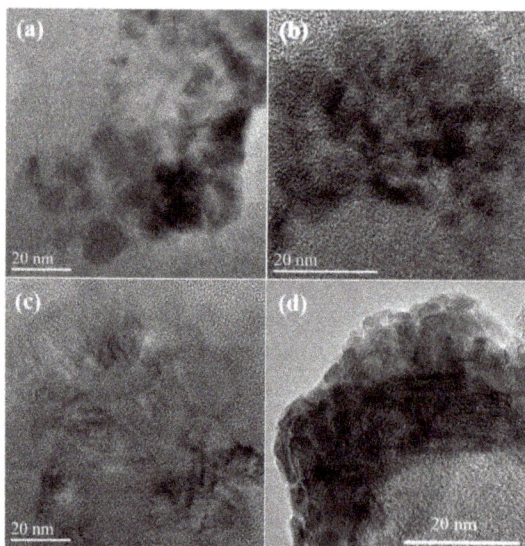

Figure 3. Transmission electron microscopy (TEM) images of as-prepared (**a**); CoS-250 (**b**); CoS-300 (**c**); and CoS-400 (**d**).

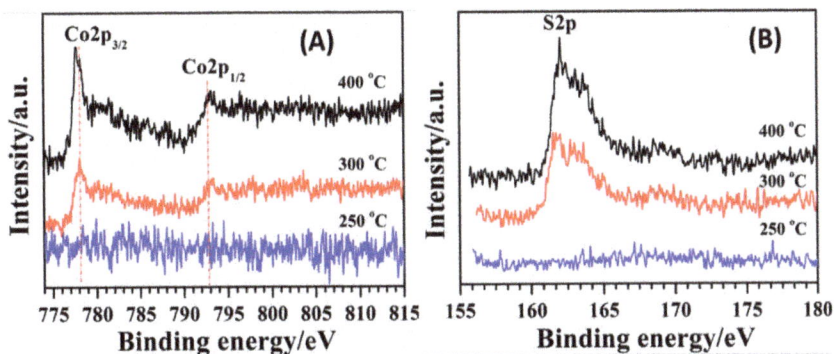

Figure 4. X-ray Photoelectron Spectroscopy (XPS) spectra of (**A**) Co2p and (**B**) S2p for CoS samples obtained at different annealed temperatures from 250 to 400 °C.

2.2. Influence of Heat-Treatment on Electrocatalytic Activity towards ORR

Figure 5 demonstrates the effect of heat-treatment temperature on the ORR activity of 20 wt. % CoS/C nanoparticles in O_2-saturated 0.1 M KOH at a rotating speed of 2500 rpm as measured by the RRDE technique. The mass loading of CoS catalyst is 80 $\mu g \cdot cm^{-2}$ on the working disk. In the disk current density (j_D) curves (down), the heat-treatment remarkably improves the ORR activity of 20 wt. % CoS/C nanoparticles in terms of the current density and the half-wave potential ($E_{1/2}$, see the inset). The disk current density at 0.70 V is increased from 0.60 mA$\cdot cm^{-2}$ for the as-prepared sample to 2.22 mA$\cdot cm^{-2}$ for CoS-400 whereas only 0.33 mA$\cdot cm^{-2}$ is observed for CoS-450. The CoS-400 shows the plateau-like current density of 5.0 mA$\cdot cm^{-2}$ at a potential of lower than 0.52 V, which is close to the diffusion-limited current density expected for 4-electron pathway in 0.1 M KOH [14]. The half-wave potential is also enhanced from 0.64 V for the as-prepared sample to 0.71 V for CoS-400, and then reduced to 0.51 V for CoS-450. The value of 0.71 V is the highest value among non-precious metal chalcogenides [3,11] and very close to 0.77 V for RuSe$_x$/C [20]. At 2500 rpm, in comparison, j_R (HO$_2^-$ oxidation) is much smaller than j_D (oxygen reduction), meaning that H_2O was the main product of ORR catalyzed by the prepared catalysts. Additionally, the ring current density (j_R) is decreased with increase of the annealing temperature, further suggesting the positive effect of the heat-treatment on the electrocatalytic activity of CoS/C catalyst under the investigated conditions.

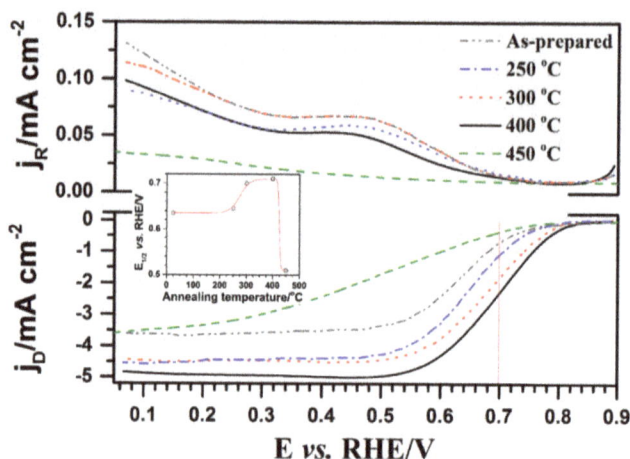

Figure 5. Ring (top) and disk (down) current density from RRDE measurements of 20 wt. % CoS/C samples after annealing at different temperature in O_2-saturated 0.1 M KOH at 25 °C with a sweep rate of 5 mV s^{-1} at a rotating speed of 2500 rpm. Inset shows the half-wave potential ($E_{1/2}$) as a function of the annealing temperature. The CoS mass loading on the disk electrode is 80 µg·cm^{-2}.

Furthermore, the electrocatalytic activity towards ORR has been further evaluated based on the RRDE measurements: the electron reduction pathway and the percentage of H_2O^- product as a function of annealing temperature, see Figure 6. The HO_2^-% and the number of electrons transferred (n) was calculated by the following two equations:

$$HO_2^-\% = 200 \times \frac{I_R/N}{I_D + I_R/N} \tag{1}$$

$$n = 4 \times \frac{I_D}{I_D + I_R/N} \tag{2}$$

where I_D, I_R and N is disk current, ring current and collection efficiency (0.26), respectively. With increase of the annealing temperature from room temperature to 400 °C, the H_2O^- product is decreased from 11.0% to 4.3% at the potential of 0.60 V with a mass loading of 80 µg·cm^{-2}. The H_2O^- product of 4.3% for CoS-400 at 0.60 V is much less than that of *ca.* 8% for Co_3O_4/rmGO and comparable with Co_3O_4/N-rmGO with a mass loading of 100 µg·cm^{-2} [21]. Also, the number of electrons transferred is increased from 3.5 for the as-prepared sample to 3.9 for CoS-400 in the range from 0.10 V to 0.72 V, which is comparable with that for Co_3O_4/rmGo and Co_3O_4/N-rmGO [21]. The results suggest that CoS-400 has promising electrocatalytic activity for ORR in 0.1 M KOH.

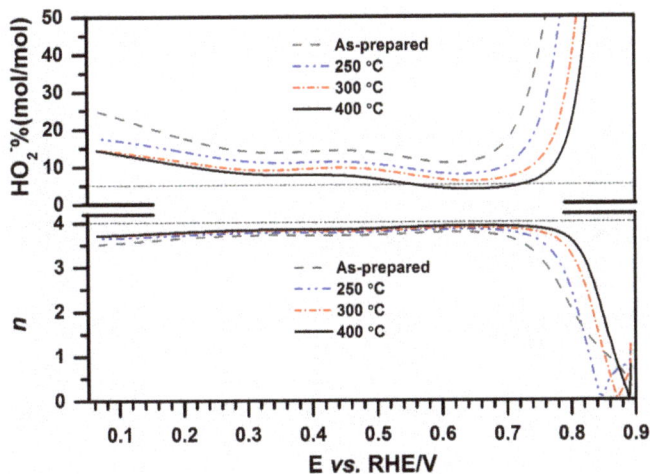

Figure 6. Molar fraction of HO_2^- formation and electron transfer number n at different potentials calculated from rotating ring-disk electrode (RRDE) curves in Figure 3. The fraction of HO_2^- formation was calculated according to Equation (1) and the electron transfer number (n) was evaluated from Equation (2). The collection efficiency N of 0.26 was used for all the calculations.

2.3. Electrocatalytic Stability

The long-term stability is the other important parameter for non-precious metal catalysts. Commercial 20 wt. % Pt/C (E-TEK) is usually used as the reference to evaluate the catalytic activity and durability of non-Pt catalysts in alkaline medium [4,8,21–23]. Figure 7 depicts the effect of the annealing temperature on the electrocatalytic stability of 20 wt. % CoS/C. The normalized current ($I/I_0\%$) means a percentage of the determined current (I) at operating time over the initial current (I_0) at a fixed potential E. In our case, the potential value of 0.66 V was chosen for the accelerated stability test based on Dai's work [14,21]. After 18,000 s of continuous operation, the decrease in activity is observed, e.g., 60% for the as-prepared sample and 35% for CoS-400, suggesting that the annealed treatment markedly improves the electrocatalytic stability of CoS/C catalyst. The decrease value in current for CoS-400 is very close to that of 30% for Pt/C as reported in Ref. [21], where a 20%–48% decrease in current for Pt/C was determined in 0.1–6 M KOH with a high mass loading of 240 µg·cm^{-2}.

343

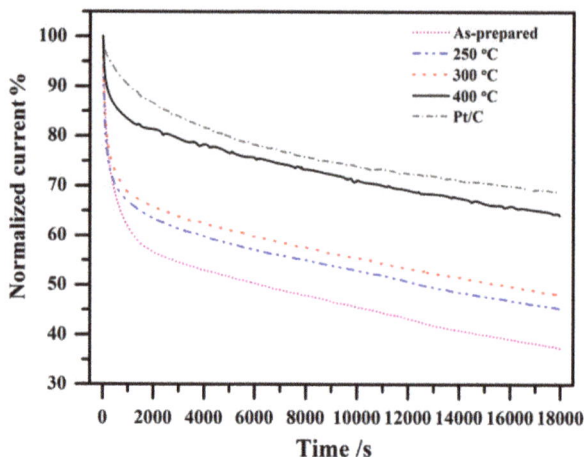

Figure 7. Chronoamperometric response (percentage of current retained *vs.* operation time) of 20 wt. % CoS/C after annealed at different temperature and 20 wt. % Pt/C (E-TEK) on a glassy carbon electrode kept at 0.66 V *vs.* RHE in O_2-saturated 0.1 M KOH at 25 °C. Mass loading on the disk electrode is 80 $\mu g \cdot cm^{-2}$ for CoS and Pt.

3. Materials and Methods

3.1. Chemicals

All the chemicals with analytical grade were used as received without further purification. Vulcan XC-72R carbon, received from CABOT Co. (Shanghai, China), was activated at 400 °C under a high purity nitrogen atmosphere for 4 h before use.

3.2. Synthesis of Carbon-Supported CoS Nanocatalyst

Carbon-supported CoS nanocatalyst was directly synthesized in a solvothermal route using cobalt nitrate $(Co(NO_3)_2 \cdot 6H_2O)$ and thiourea (CH_4N_2S) as the source of Co and S, which followed the similar procedure for CoS_2 nanoparticles as described in [24]. For 20 wt. % CoS/C, typically, 0.8 mmol $Co(NO_3)_2 \cdot 6H_2O$, 2.4 mmol CH_4N_2S, and 0.3 g carbon (Vulcan XC-72R) were mixed in 80 mL absolute ethanol and then removed into 100 mL Teflon-lined stainless steel autoclave. The sealed autoclave was kept at 200 °C for 24 h and then cooled down to room temperature. The final product was collected after six cycles of centrifugation and resuspension with distilled water and ethanol and drying at 60 °C for 6 h. Finally, the obtained sample was further annealed at 250, 300, 400 and 450 °C for 3 h under high purity nitrogen atmosphere and denoted as as-prepared, CoS-250, -300, -400 and -450, respectively.

3.3. Structural Characterization and Electrochemical Measurements

Powder X-ray diffraction (XRD) patterns were recorded on a Shimadzu XRD-6000 X-ray diffractometer (Kyoto, Japan) (Cu Kα radiation, λ = 0.15406 nm) at tube current of 30 mA and tube potential of 40 kV from 10° to 60°/2θ with a scan speed of $2°·min^{-1}$.

Transmission electron microscopy (TEM) and high-resolution transmission electron microscopy (a resolution of 0.19 nm) were carried out on a JEOL JEM-2010 electron microscope (Tokyo, Japan) at 200 kV.

The rotating ring-disk electrode (RRDE) measurements were carried out at 25 °C using an interchangeable Pine ring-disk electrodes with a bi-potentiostat (CHI760C, ChenHua Instruments Co., Shanghai, China) and a rotation control system (Pine Instruments, Grove, PA, USA). The Pt ring electrode with a geometric area of 0.152 cm^2 was potentiostated at 1.2 V, where the detection of peroxide is diffusion limited. The disk electrode was glassy carbon with a geometric area of 0.162 cm^2. The counter and the reference electrodes were a Pt ring and a saturated calomel electrode. All potentials reported in this paper are referred to a reference hydrogen electrode (RHE). The current density on the disk and the ring was calculated based on the corresponding geometric area, respectively. Before use, the glassy carbon disk electrode was polished with a 5 A alumina powder, and washed in water and ethanol by ultrasound. The working electrode was prepared by depositing homogeneous catalyst ink on the glassy carbon disk, which was formed by dispersing 9.0 mg 20 wt. % CoS/C powder in a mixture solvent of 250 mL Nafion® solution (5 wt. %, DuPont, Shanghai, China) and 1250 mL water in an ultrasonic bath for 2 h. Newly prepared 0.1 M KOH solution (pH = 12.97) was used as the electrolyte. Prior to linear-sweep voltammograms (LSV), the electrode was subjected to 50 cycles of cyclic voltammetry under high purity argon atmosphere to clean the surface. After this cleaning, the LSVs under saturated oxygen were recorded by scanning the disk potential *vs.* RHE at 5 mV·s^{-1} at the rotating speed of 2500 rpm.

4. Conclusions

In summary, carbon-supported CoS electrocatalyst, as one of non-precious metal chalcogenides, has been directly prepared in a facile solvothermal route. The annealed treatment has significantly improved the electrocatalytic activity and long-term stability of CoS catalyst towards ORR in alkaline medium with increase of annealed temperature from room temperature to 400 °C. The CoS catalyst that annealed at 400 °C has the maximum electrocatalytic performance among the annealed CoS samples and shows promising applications for alkaline fuel cells as one of non-precious metal cathodes.

Acknowledgments: This work was supported by the National Basic Research Program of China (Grant No. 2011CBA00508) and Beijing Engineering Center for Hierarchical Catalysts.

Author Contributions: Y.J. F, P.G. T. and D.Q. L conceived and designed the experiments; H.H. Z and J.M. X performed the experiments; H.H. Z and J.M. X analyzed the data; J.M. X contributed reagents/materials/analysis tools; Y.J. F and H.H. Z wrote the paper.

Conflicts of Interest: The authors declare no conflict of interest. The founding sponsors had no role in the design of the study; in the collection, analyses, or interpretation of data; in the writing of the manuscript, and in the decision to publish the results.

References

1. Wang, Y.; Chen, K.S.; Mishler, J.; Cho, S.C.; Adroher, X.C. A review of polymer electrolyte membrane fuel cells: Technology, applications, and needs on fundamental research. *Appl. Energy* **2011**, *88*, 981–1007.
2. Miller, M.; Bazylak, A. A review of polymer electrolyte membrane fuel cell stack testing. *J. Power Sources* **2011**, *196*, 601–613.
3. Feng, Y.; Gago, A.; Timperman, L.; Alonso-Vante, N. Chalcogenide metal centers for oxygen reduction reaction: Activity and tolerance. *Electrochim. Acta* **2011**, *56*, 1009–1022.
4. Guo, J.; Li, H.; He, H.; Chu, D.; Chen, R. CoPc- and CoPcF16-modified Ag nanoparticles as novel catalysts with tunable oxygen reduction activity in alkaline media. *J. Phys. Chem. C* **2011**, *115*, 8494–8502.
5. Robertson, N.J.; Kostalik, H.A.; Clark, T.J.; Mutolo, P.F.; Abruña, H.D.; Coates, G.W. Tunable High Performance Cross-Linked Alkaline Anion Exchange Membranes for Fuel Cell Applications. *J. Am. Chem. Soc.* **2010**, *132*, 3400–3404.
6. Tamain, C.; Poynton, S.D.; Slade, R.C.T.; Carroll, B.; Varcoe, J.R. Development of cathode architectures customized for H_2/O_2 metal-cation-free alkaline membrane fuel cells. *J. Phys. Chem. C* **2007**, *111*, 18423–18430.
7. Wu, H.; Chen, W. Copper nitride nanocubes: Size-controlled synthesis and application as cathode catalyst in alkaline fuel cells. *J. Am. Chem. Soc.* **2011**, *133*, 15236–15239.
8. Yang, D.S.; Bhattacharjya, D.; Inamdar, S.; Park, J.; Yu, J.S. Phosphorus-doped ordered mesoporous carbons with different lengths as efficient metal-free electrocatalysts for oxygen reduction reaction in alkaline media. *J. Am. Chem. Soc.* **2012**, *134*, 16127–16130.
9. Susac, D.; Zhu, L.; Teo, M.; Sode, A.; Wong, K.C.; Wong, P.C.; Parsons, R.R.; Bizzotto, D.; Mitchell, K.A.R.; Campbell, S.A. Characterization of FeS_2-Based Thin Films as Model Catalysts for the oxygen reduction reaction. *J. Phys. Chem. C* **2007**, *111*, 18715–18723.
10. Zhu, L.; Susac, D.; Teo, M.; Wong, K.C.; Wong, P.C.; Parsons, R.R.; Bizzotto, D.; Mitchell, K.A.R.; Campbell, S.A. Investigation of CoS_2-based Thin Films as Model Catalysts for the Oxygen Reduction Reaction. *J. Catal.* **2008**, *258*, 235–242.
11. Zhao, C.; Li, D.; Feng, Y. Size-controlled hydrothermal synthesis and high electrocatalytic performance of CoS_2 nanocatalysts as non-precious metal cathode materials for fuel cells. *J. Mater. Chem. A* **2013**, *1*, 5741–5746.

12. Feng, Y.J.; He, T.; Alonso-Vante, N. *In situ* Surfactant-Free Synthesis and ORR-Electrochemistry of Carbon-Supported Co_3S_4 and $CoSe_2$ Nanoparticles. *Chem. Mater.* **2008**, *20*, 26–28.

13. Feng, Y.J.; He, T.; Alonso-Vante, N. Carbon-Supported $CoSe_2$ Nanoparticles for Oxygen Reduction Reaction in Acid Medium. *Fuel Cells* **2010**, *10*, 77–83.

14. Wang, H.; Liang, Y.; Li, Y.; Dai, H. $Co_{1-x}S$-Graphene Hybrid: A High-Performance Metal Chalcogenide Electrocatalyst for Oxygen Reduction. *Angew. Chem. Int. Ed.* **2011**, *50*, 10969–10972.

15. Bezerra, C.W.B.; Zhang, L.; Liu, H.; Lee, K.; Marques, A.L.B.; Marques, E.P.; Wang, H.; Zhang, J. A review of heat-treatment effects on activity and stability of PEM fuel cell catalysts for oxygen reduction reaction. *J. Power Sources* **2007**, *173*, 891–908.

16. Cheng, H.; Yuan, W.; Scott, K. Influence of Thermal Treatment on RuSe Cathode Materials for Direct Methanol Fuel Cells. *Fuel Cells* **2007**, *7*, 16–20.

17. Cheng, H.; Yuan, W.; Scott, K. The influence of a new fabrication procedure on the catalytic of ruthenium-selenium catalysts. *Electrochim. Acta* **2006**, *52*, 466–473.

18. Dresselhaus, M.S.; Jorio, A.; Hofmann, M.; Dresselhaus, G.; Saito, R. Perspectives on carbon nanotubes and graphene Raman spectroscopy. *Nano Lett.* **2010**, *10*, 751–758.

19. Bao, S.-J.; Li, Y.B.; Li, C.M.; Bao, Q.L.; Lu, Q.; Guo, J. Shape evolution and magnetic properties of cobalt sulfide. *Cryst. Growth Des.* **2008**, *8*, 3745–3749.

20. Delacôte, C.; Bonakdarpour, A.; Johnston, C.M.; Zelenay, P.; Wieckowski, A. Aqueous-based synthesis of ruthenium-selenium catalyst for oxygen reduction reaction. *Faraday Discuss.* **2008**, *140*, 269–281.

21. Liang, Y.; Li, Y.; Wang, H.; Zhou, J.; Wang, J.; Regier, T.; Dai, H. Co_3O_4 nanocrystals on graphene as a synergistic catalyst for oxygen reduction reaction. *Nat. Mater.* **2011**, *10*, 1780–1786.

22. Tuci, G.; Zafferoni, C.; D'Ambrosio, P.; Caporali, S.; Ceppatelli, M.; Rossin, A.; Tsoufis, T.; Innocenti, M.; Giambastiani, G. Tailoring carbon nanotube N-dopants while designing metal-free electrocatalys for the oxygen reduction reaction in alkaline medium. *ACS Catal.* **2013**, *3*, 2108–2111.

23. Lee, J.S.; Park, G.S.; Lee, H.I.; Kim, S.T.; Cao, R.; Liu, M.; Cho, J. Ketjenblack carbon supported amorphous manganese oxides nanowires as highly efficient electrocatalyst for oxygen reduction reaction in alkaline solution. *Nano Lett.* **2011**, *11*, 5362–5366.

24. Han, J.-T.; Huang, Y.-H.; Huang, W. Solvothermal synthesis and magnetic properties of pyrite $Co_{1-x}Fe_xS_2$ with various morphologies. *Mater. Lett.* **2006**, *60*, 1805–1808.

Microwave Synthesis of High Activity FeSe$_2$/C Catalyst toward Oxygen Reduction Reaction

Qiaoling Zheng, Xuan Cheng and Hengyi Li

Abstract: The carbon supported iron selenide catalysts (FeSe$_2$/C) were prepared with various selenium to iron ratios (Se/Fe), namely, Se/Fe = 2.0, 2.5, 3.0, 3.5 and 4.0, through facile microwave route by using ferrous oxalate (FeC$_2$O$_4 \cdot$ 2H$_2$O) and selenium dioxide (SeO$_2$) as precursors. Accordingly, effects of Se/Fe ratio on the crystal structure, crystallite size, microstructure, surface composition and electrocatalytic activity for oxygen reduction reaction (ORR) of FeSe$_2$/C in an alkaline medium were systematically investigated. The results revealed that all the FeSe$_2$/C catalysts obtained with the Se/Fe ratios of 2.0–4.0 exhibited almost pure orthogonal FeSe$_2$ structure with the estimated mean crystallite sizes of 32.9–36.2 nm. The electrocatalytic activities in potassium hydroxide solutions were higher than those in perchloric acid solutions, and two peak potentials or two plateaus responded to ORR were observed from cyclic voltammograms and polarization curves, respectively. The ORR potentials of 0.781–0.814 V with the electron transfer numbers of 3.3–3.9 at 0.3 V could be achieved as the Se/Fe ratios varied from 2.0 to 4.0. The Fe and Se were presented at the surface of FeSe$_2$/C upon further reduction on FeSe$_2$. The Se/Fe ratios slightly influenced the degree of graphitization in carbon support and the amount of active sites for ORR.

Reprinted from *Catalysts*. Cite as: Zheng, Q.; Cheng, X.; Li, H. Microwave Synthesis of High Activity FeSe$_2$/C Catalyst toward Oxygen Reduction Reaction. *Catalysts* **2015**, *5*, 1079–1091.

1. Introduction

Alkaline fuel cells have been attracting extensive attention due to their great advantages in cathode dynamics and with the reduction of ohmic polarization [1]. Oxygen reduction reaction (ORR) is an important process in an electrochemical energy conversion and a four-electron reaction is desirable to take place for a given catalyst in order to achieve good electrocatalytic performance. In recent years, the transition metals such as Mn [2], Fe [3,4], Co [5], Ni [6], Cu [7] and the heteroatom dopants B [8], N [9,10], P [11,12], S [13,14], Se [15] have been reported to modify the catalytic properties of various carbon materials including amorphous carbon, carbon nanotubes, and graphene, which arouse a great deal of interest for the research and development of non-noble catalysts.

Chalcogenides are promising for the potential replacement Pt based cathode catalysts because of their good electrocatalytic activity and high selectivity toward ORR in both acidic and basic media. The onset potential of 0.823 V for ORR could be attained in H_2SO_4 solutions with CoSe/C synthesized through microwave assisted routes [16], while the ORR potentials of 0.6–0.7 V and the electron transfer numbers of 3.1 Oxygen reduction reaction 4.0 could be obtained for $CoSe_2$/C prepared with the Se/Co ratios of 2.5–4.0 [17]. In addition, $CoSe_2$ nanoparticles showed a higher ORR activity in KOH than in H_2SO_4 solutions and a higher tolerance to methanol as compared with a commercial 20 wt % Pt/C catalyst [18]. The tetragonal and cubic Cu_2Se nanowires were found to have the four-electron mechanism, while the cubic nanowires were a dual-path mode in KOH solution [7]. Although FeSe and $FeSe_2$ have been reported for the applications in superconductors or magnetic semiconductors [19–21], their ORR activities have not been investigated so far.

In this work, a series of carbon supported $FeSe_2$ nanoparticles were synthesized using the microwave method with different molar ratios of Se/Fe. The crystal phases, microstructures, chemical compositions and electrocatalytic activities of the as-prepared $FeSe_2$/C catalysts were explored by X-ray diffraction (XRD), transmission electron microscopy (TEM), selected area electron diffraction (SAED), Raman spectroscopy, energy dispersive X-ray spectroscopy (EDS), X-ray photoelectron spectroscopy (XPS), cyclic voltammetry and rotating disk electrode (RDE) techniques. The effect of Se/Fe ratio on ORR activity in an alkaline medium is discussed in terms of ORR active site and carbon graphitization.

2. Results and Discussion

Typical powder XRD patterns of $FeSe_2$/C catalysts prepared with different Se/Fe ratios are shown in Figure 1. Compared with the standard lines of orthogonal $FeSe_2$ phase (PDF#65-2570) included in the bottom of Figure 1a, the major characteristic diffraction peaks appeared at $2\theta \approx 35.0°$, $36.4°$ and $48.4°$ belonged to (111), (120) and (211) planes, while a pair of twin peaks near $31°$ and $50°$ to (101)/(020) and (031)/(130) planes. A closer examination in the range of $25°–50°$ revealed that the diffraction peak corresponded to (120) shifted to smaller Bragg angles as evident in Figure 1b. Despite this, the formation of orthogonal $FeSe_2$ structure is strongly indicated. The average crystallite sizes were evaluated using Scherrer equation described below:

$$d = \frac{K\lambda}{\beta\cos\theta} \tag{1}$$

where d is the mean crystallite size; K is a dimensionless shape factor and has a typical value of 0.89, λ is the X-ray wavelength, $\lambda = 0.1546$ nm; β is the full width at half maximum (FWHM); θ is the Bragg scattering angle. The results are given in Table 1. Apparently, the Se/Fe ratios did not significantly influence the average crystallite sizes of the as-prepared FeSe$_2$/C catalysts, which ranged 36.2–32.9 nm with Se/Fe ratios of 2.0–4.0. The empirical Se/Fe ratios, also included in Table 1, were roughly evaluated from EDS data and agreed reasonably well with those nominal ones.

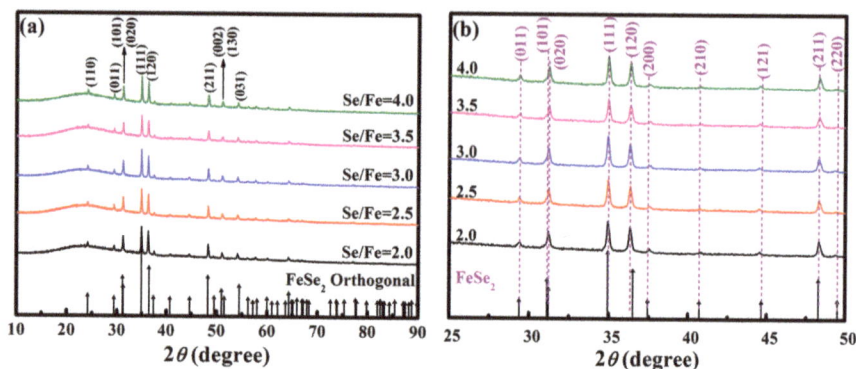

Figure 1. Typical powder XRD patterns of FeSe$_2$/C prepared with different Se/Fe ratios. The standard lines of orthogonal FeSe$_2$ phase are included for comparison. (a) Full range in 10°–90°; (b) Enlarged in 25°–50°.

Table 1. Parameters of iron selenide catalysts (FeSe$_2$/C) prepared with different Se/Fe ratios.

Parameter		Se/Fe ratio				
Nominal		2.0	2.5	3.0	3.5	4.0
Evaluated by EDS		2.1	2.5	3.3	3.6	4.2
Crystallite Size (nm)		36.2	35.6	32.9	33.1	35.4
E_P (V, vs.	(I)	0.733	0.704	0.733	0.699	0.727
RHE)	(II)	0.511	0.514	0.509	0.499	0.478
n at 0.3 V (vs. RHE)		3.7	3.9	3.5	3.3	3.6
E_{ORR} (V, vs. RHE)		0.814	0.781	0.809	0.795	0.814
I_D/I_G		1.64	1.71	1.76	1.90	1.74
A_{sp}^3/A_{sp}^2		0.37	0.48	0.44	0.51	0.46

The electrocatalytic activities of FeSe$_2$/C prepared with different Se/Fe ratios toward ORR were studied in both acidic and alkaline media. Figure 2 presents the cyclic voltammograms obtained in N$_2$ (dashed lines) and O$_2$ (solid lines) saturated 0.1 mol·L^{-1} HClO$_4$ solutions at 50 mV·s^{-1}. No apparent peaks were observed in N$_2$ atmosphere, while one or two reduction peaks in O$_2$ atmosphere. Two peak potentials of 0.211 V, −0.045 V and 0.204 V, −0.028 V were obtained only for FeSe$_2$/C prepared with Se/Fe = 2.5 and Se/Fe = 4.0. Obviously, the ORR activity of FeSe$_2$/C in an acidic medium was poor. The cyclic voltammograms and RDE polarization curves for FeSe$_2$/C prepared with different Se/Fe ratios measured in N$_2$ (dashed lines) and O$_2$ (solid lines) saturated 0.1 mol·L^{-1} KOH solutions are illustrated in Figure 3. A large reduction peak I was observed near 0.7 V, and followed by a small reduction peak II around 0.5 V in O_2 saturated KOH solutions (solid lines in Figure 3a). This phenomenon was also observed for vertically aligned carbon nanotubes in a KOH solution [22]. The peak potentials (E_P) could be obtained from Figure 3a and are also summarized in Table 1. The E_P values ranged from 0.699–0.773 V for Peak I and 0.499–0.514 V for Peak II, suggesting an enhanced ORR activity in an alkaline medium. As can be seen in Figure 3b, the plateaus observed for FeSe$_2$/C were not well defined in KOH solutions, which significantly differed from those observed for CoSe$_2$/C in H$_2$SO$_4$ solutions [17]. Similarly, the electron transfer numbers (n) of FeSe$_2$/C prepared for different Se/Fe ratios could be determined from the slops of Koutecky-Levich plots at 0.3 V as shown in Figure 4a. The dashed lines indicated the slopes corresponding to two-electron and four-electron reactions. The calculated results are also included in Table 1. The n values varied from 3.3–3.9 at 0.3 V in KOH solutions for FeSe$_2$/C prepared with Se/Fe ratios of 2.0–4.0, while they were 3.1–4.0 in H$_2$SO$_4$ solutions for CoSe$_2$/C prepared with Se/Co ratios of 2.0–4.0 [17].

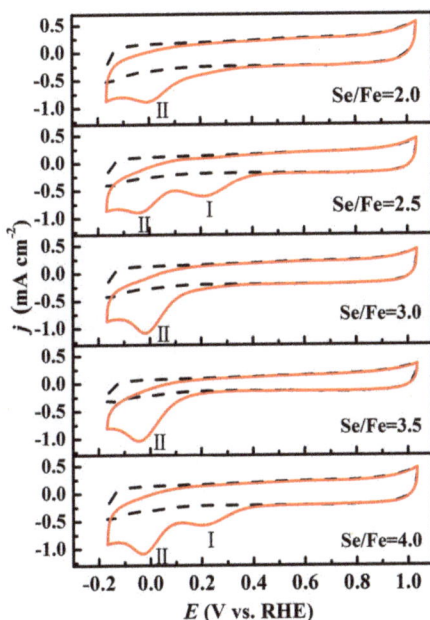

Figure 2. Cyclic voltammograms of FeSe$_2$/C prepared with different Se/Fe ratios in N$_2$ (dashed lines) and O$_2$ (solid lines) saturated 0.1 mol· L^{-1} HClO$_4$ solutions at 50 mV· s^{-1}.

The polarization curves of FeSe$_2$/C prepared with different Se/Fe ratios at 1600 rpm are compared with that of a commercial 20% Pt/C in Figure 4b. The potential at the current density of -0.5 mA· cm^{-2} is defined as E_{ORR} (the inset in

Figure 4b) and the values are also provided in Table 1. The E_{ORR} value for 20% Pt/C was 0.992 V, while those for FeSe$_2$/C ranged 0.781–0.814 V with the Se/Fe ratios varying 2.0–4.0. However, two platforms observed with FeSe$_2$/C resulted in slightly larger limiting current densities than 20% Pt/C.

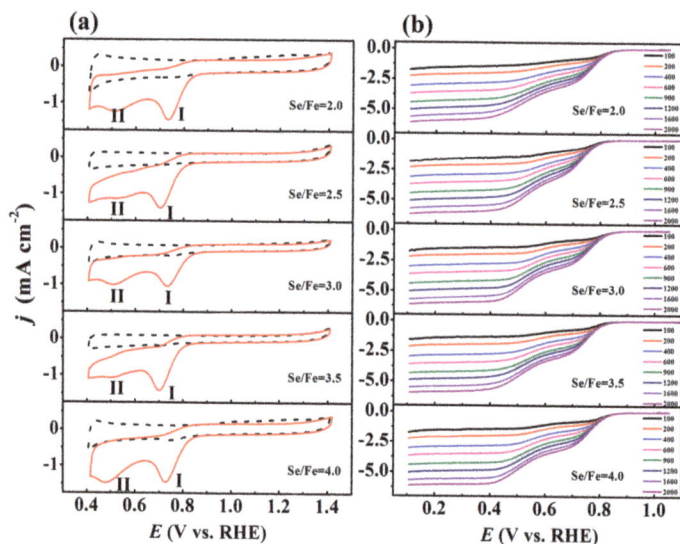

Figure 3. Cyclic voltammograms at 50 mV·s^{-1} (**a**) and RDE polarization curves at 5 mV·s^{-1}; (**b**) of FeSe$_2$/C prepared with different Se/Fe ratios in N$_2$ (dashed lines) and O$_2$ (solid lines) saturated 0.1 mol·L^{-1} KOH solutions.

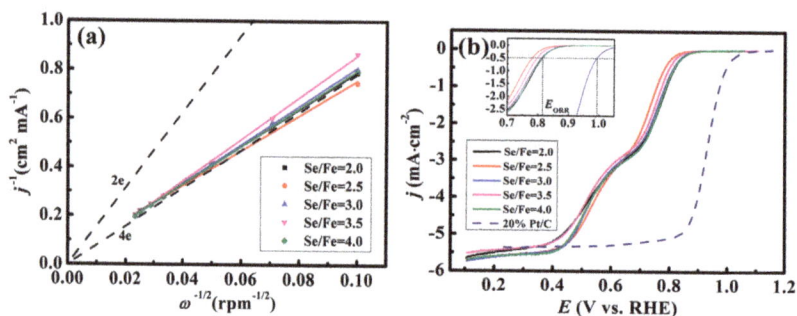

Figure 4. (**a**) Koutecky-Levich plots at 0.3 V; (**b**) RDE polarization curves of Pt/C (dashed line) and FeSe$_2$/C (solid lines) prepared with different Se/Fe ratios in O$_2$ saturated 0.1 mol·L^{-1} KOH solutions at 1600 rpm. The inset in (**b**) illustrates the potential corresponding to ORR at the current density of -0.5 mA·cm^{-2} (E_{ORR}).

The information in surface species and carbon support could be further studied by obtaining Raman spectra using Ar ion laser excitation of 532 nm as shown in

Figure 5. The presences of Fe–Se near 219 cm^{-1}, 284 cm^{-1} and 597 cm^{-1}, as well as Fe–O at 400 cm^{-1} [23,24] are identified in Figure 5a for all the Se/Fe ratios. In Figure 5b, the Raman bands appeared at 1336 cm^{-1} and 1593 cm^{-1} corresponded to D-band of sp^2 type carbon ascribed to the finite-sized crystals of graphite due to the reduction in symmetry and G-band of all sp^2 bonds in an ideal graphitic layer, respectively. Figure 5c illustrates the curve fitting plots for the Raman data given in Figure 5b. Two additional weak bands at 1191 cm^{-1} and 1499 cm^{-1} belonged to sp^3 type carbon. The relative intensity of the D band over G band (I_D/I_G) and the relative ratio under the areas of sp^3 and sp^2 types of carbon (A_{sp}^3/A_{sp}^2) were calculated, and the results are compared in Table 1. The I_D/I_G and A_{sp}^3/A_{sp}^2 values ranged 1.64–1.90 and 0.37–0.51, respectively, for the Se/Fe ratios of 2.0–4.0. The small difference of I_D/I_G valves might mean that the carbon surface was partially oxygenated without significant structural deformation [25]. The least I_D/I_G and A_{sp}^3/A_{sp}^2 values were obtained for the FeSe$_2$/C prepared with Se/Fe = 2.0, implying the presence of less defect and higher degree graphitization in carbon support. Contrarily, FeSe$_2$/C prepared with Se/Fe = 3.5 showed the largest I_D/I_G and A_{sp}^3/A_{sp}^2 values, and resulted in more defect and lower degree graphitization. Similar I_D/I_G and A_{sp}^3/A_{sp}^2 values were observed for FeSe$_2$/C prepared with Se/Fe = 2.5, 3.0 and 4.0 as evident in Table 1.

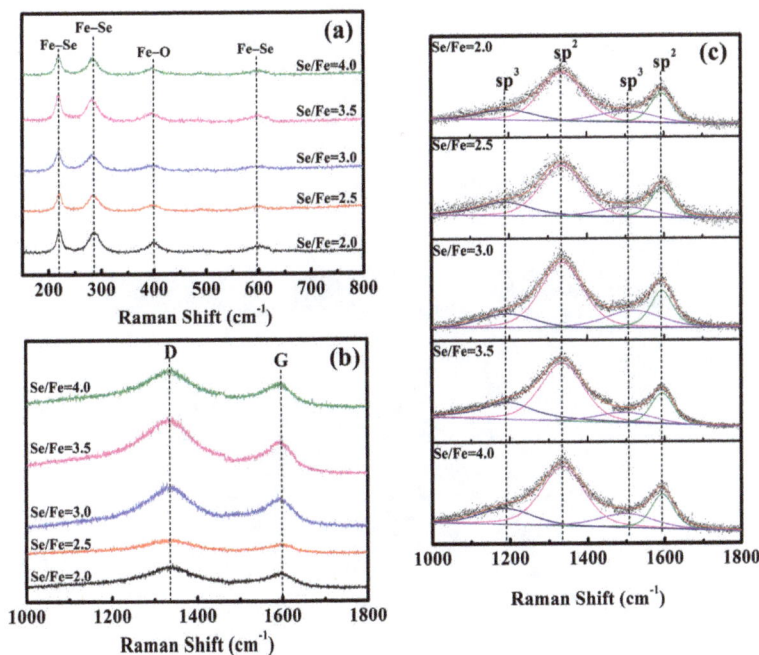

Figure 5. (a,b) Raman spectra of FeSe$_2$/C prepared with different Se/Fe ratios; (c) Fitting curves for Raman spectra (b).

The high resolution TEM images and SAED patterns of $FeSe_2/C$ prepared with different Se/Fe ratios are supplied in Figures 6 and 7 respectively. The particles sized about 3–12 nm were observed in Figure 6, which are much smaller than those calculated from XRD data (32.9–36.2 nm in T – 1) because XRD gives volume-weighted measurements that tend to overestimate the geometric particle size [26]. The formation of orthogonal $FeSe_2$ nanoparticles by microwave synthesis was verified by combining both TEM and SAED data, which is consistent with the XRD results in Figure 1.

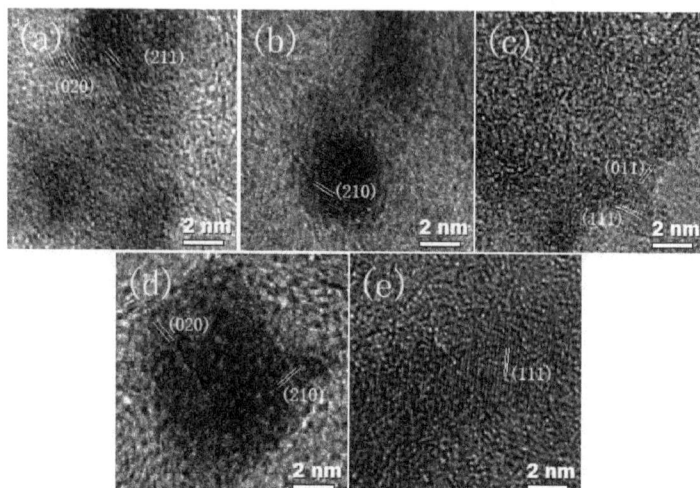

Figure 6. High resolution TEM images of the $FeSe_2/C$ prepared with different Se/Fe ratios. (a) Se/Fe = 2.0; (b) Se/Fe = 2.5; (c) Se/Fe = 3.0; (d) Se/Fe = 3.5; (e) Se/Fe = 4.0.

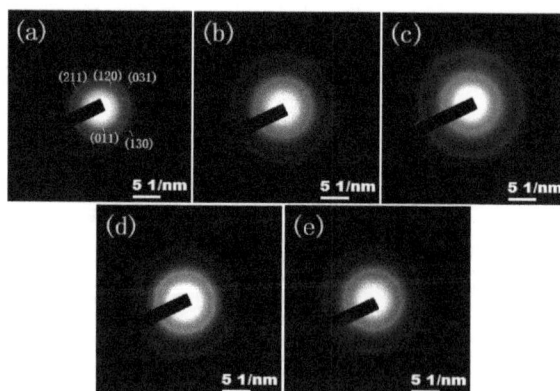

Figure 7. SAED patterns of $FeSe_2/C$ prepared with different Se/Fe ratios. (a) Se/Fe = 2.0; (b) Se/Fe = 2.5; (c) Se/Fe = 3.0; (d) Se/Fe = 3.5; (e) Se/Fe = 4.0.

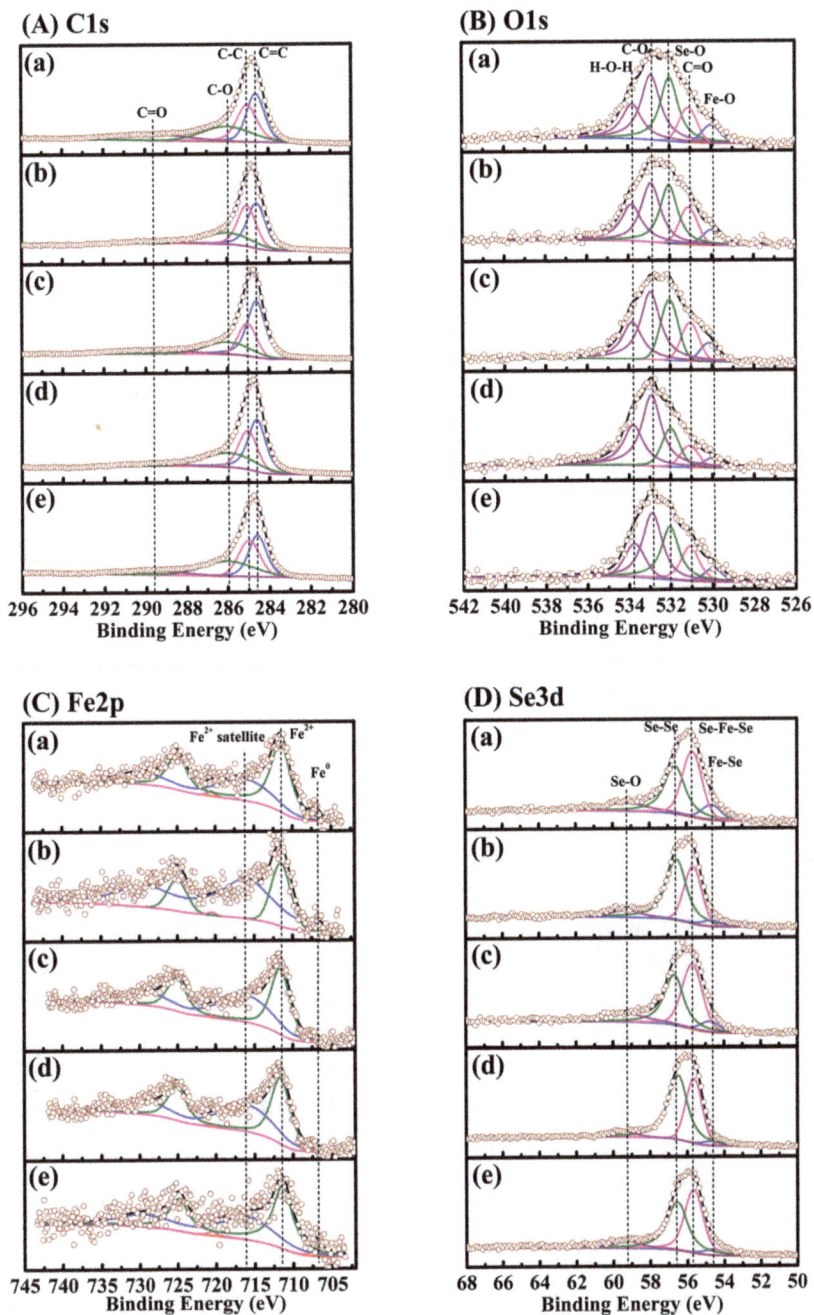

Figure 8. Deconvoluted XPS spectra of (**A**) C1s; (**B**) O1s; (**C**) Fe2p and (**D**) Se3d for $FeSe_2$/C prepared with different Se/Fe ratios. (**a**) Se/Fe = 2.0; (**b**) Se/Fe = 2.5; (**c**) Se/Fe = 3.0; (**d**) Se/Fe = 3.5; (**e**) Se/Fe = 4.0.

Table 2. Surface compositions of $FeSe_2/C$ determined based on XPS analyses.

Parameter		Fe/Se ratio								
Nominal		2.0		2.5		3.0		3.5		4.0
Calculated		2.0		2.5		3.0		3.5		4.0
Species	E_B (eV)	R.A. (%)	E_B (eV)	R.A. (%).	E_B (eV)	R.A. (%).	E_B (eV)	R.A. (%)	E_B (eV)	R.A. (%)
C1s C=C	284.6	31.8	284.6	39.8	284.6	47.1	284.6	41.2	284.6	31.3
C–C	285.0	29.5	285.0	34.3	285.0	22.3	285.0	28.6	285.0	28.7
C–O	286.0	24.3	286.0	22.4	286.0	21.3	286.0	26.5	286.0	31.0
C=O	289.6	14.4	289.6	3.5	289.6	9.3	289.6	3.7	289.6	9.0
O1s Fe–O	530.0	5.9	530.0	6.4	530.1	6.5	530.0	3.7	530.0	5.1
C=O	531.1	11.2	531.1	13.0	531.1	13.2	531.1	8.3	531.1	15.8
Se–O	532.0	30.0	532.0	28.2	532.0	20.9	532.0	16.5	532.0	27.9
C–O	532.9	33.8	532.9	30.6	532.9	37.8	532.9	44.8	532.9	35.5
H–O–H	533.8	19.0	533.8	21.7	533.8	21.5	533.8	26.6	533.8	15.7
Fe2p Fe°	707.0	3.7	706.9	3.3	707.0	1.5	707.0	1.4	707.0	5.3
Fe^{2+}	711.5	96.3	711.5	96.7	711.6	98.5	711.6	98.6	711.4	94.7
Satellite	716.0	/	715.9	/	716.0	/	716.0	/	716.0	/
Se3d Fe–Se	54.5	8.8	54.5	5.1	54.6	6.4	54.5	2.7	54.5	4.5
Se–Fe–Se	55.5	38.4	55.5	32.0	55.6	40.7	55.5	35.5	55.5	43.7
Se–Se	56.4	38.4	56.4	53.5	56.5	38.5	56.4	57.0	56.4	43.4
Se–O	59.1	14.3	59.1	9.3	59.1	14.4	59.2	4.8	59.1	8.4

High resolution XPS spectra of C1s, O1s, Fe2p and Se3d were obtained for $FeSe_2/C$ prepared with different Se/Fe ratios and are presented in Figure 8. The surface compositions could be evaluated by performing multi-peak fitting analysis of every spectrum, and the deconvoluted XPS spectra are also illustrated in Figure 8. The C1 peaks in Figure 8A were fitted with four components centered at 284.6, 285.0, 286.0 and 290.5 eV, which were attributed to C=C, C–C, C–O and C=O, respectively. The surfaces of carbon support (BP2000) consisted of C=C (sp^2 type) and C–C (sp^3 type) bonding. The origins of C=O and C–O might possibly be from oxygenation of the carbon surface during the preparation [27]. The O1s spectra in Figure 8B were deconvoluted into oxide oxygen species mainly associated with Fe oxide (Fe–O) at 530.0 eV and Se oxide (Se–O) at 532.0 eV, as well as C=O at 531.1 eV, C–O at 532.9 eV, and adsorbed water molecule (H–O–H) at 533.8 eV. The presence of Fe–O was also indicated by Raman spectra in Figure 5a, and the Se–O mainly came from the non-reacted raw material. The Fe2p spectra in Figure 8C suggested the existences of Fe^{2+} related to the formation of major compound of $FeSe_2$ with possible formations of FeSe and Fe oxide (Fe–O). Possible formation of Fe^0 could be resulted from further reduction in $FeSe_2$ (or FeSe) during the microwave preparation. Furthermore, the surface of $FeSe_2$ might become oxidized during the preparation and characterization, which was clearly indicated as Fe–O in Raman spectra (Figure 5a). However, the Fe2p spectra in Figure 8C could not directly differentiated the Fe_3O_4 (Fe^{2+} and Fe^{3+}), Fe_2O_3 (Fe^{3+}) and FeSe, because the binding energies of Fe_3O_4 (711.4 eV), Fe_2O_3 (711.0 eV) and FeSe (711.5 eV) are very close. Similarly, they could not be readily differentiated in O1s based on the binding energy O–Fe–O (530.1 eV) and Fe–O (530.0 eV). The formation of FeSe$_2$ (Se–Fe–Se) could be accompanied by over reduction of FeSe$_2$ since the strong reducing environment was created by using ethylene glycol and glycerol during microwave preparation. Further reduction on FeSe$_2$ occurred for all the Se/Fe ratios at the surface of FeSe$_2$ and led to formations of FeSe, Fe and Se. In addition, other species such as FeSe$_4$ could also exist for the excess Se, but the exact verification required more detailed study. It has been found that the appropriate excess amounts of SeO$_2$ could prevent the CoSe$_2$/C nanoparticles from agglomeration and dissolution, which contributed to the improved ORR activity and good stability [17]. The Se3d peaks in Figure 8D indicated the presences of Fe–Se, Se–Fe–Se, Se–Se and Se–O. The relative amounts (R. A.) of surface species could be obtained by multi-peak fitting the XPS data and are summarized in Table 2. The FeSe$_2$ and FeSe were ORR active sites, and the total amounts of FeSe$_2$ and FeSe were 47.2%, 37.1%, 47.1%, 38.2% and 48.2% for Se/Fe = 2.0, 2.5, 3.0, 3.5, and 4.0, respectively, which were consistent with the ORR activities indicated by E_{ORR} values in Table 1. The FeSe$_2$/C catalysts prepared with Se/Fe = 2.5 and Se/Fe = 3.5 showed relatively smaller E_{ORR} values, while those with Se/Fe = 2.0, Se/Fe = 3.0, and Se/Fe = 4.0 had larger E_{ORR} values.

3. Experimental Section

3.1. Materials

The chemical reagents of ferrous oxalate (FeC$_2$O$_4 \cdot$ 2H$_2$O), selenium dioxide (SeO$_2$), ethylene glycol and glycerol of analytic grade were purchased from Sinopharm Chemical Reagent Co. Ltd. in China. The carbon support material of Black Pearls 2000 (BP2000) was purchased from Cabot Co. The mean grain size and specific surface area (Brunauer–Emmet–Teller, BET) were 12 nm and 1500 m$^2 \cdot$ g^{-1}, respectively.

3.2. Catalyst Synthesis

The amounts of 40.0 mg FeC$_2$O$_4 \cdot$ 2H$_2$O were dissolved in 2 mL ethylene glycol and a certain amount of 0.161 mmol/mL SeO$_2$ aqueous solution with the different molar ratios of Se/Fe, namely, 2.0, 2.5, 3.0, 3.5, 4.0. Then, a certain amount of glycerol was added, and the mixed solution was agitated with a glass rod and homogenized in an ultrasonic bath for 30 min. The BP2000 was continually added during the ultrasonic processing. The loading amount of FeSe$_2$ on carbon was about 36% according to the weight ratio at the start of the feeding. The homogeneous solution was placed in a microwave oven by using 800 W for 180 s while the solution cooled to room temperature, which was sonicated and stirred for 4 h. The product was finally centrifuged, washed with ethanol and deionized water, and dried in a vacuum oven at 338 K for 12 h.

3.3. Electrochemical Characterization

The electrochemical measurements were carried out by using the electrochemical test station (Autolab-PGSTAT30) with rotation disc electrode (RDE) system (Pine Research Instrument) in a conventional three-electrode cell. A glassy carbon RDE was a working electrode, a Pt mesh (2 cm × 2 cm) a counter electrode and a Ag/AgCl a reference electrode. Catalyst ink was prepared by homogeneously dispersing 2 mg of the as-prepared FeSe$_2$/C powder ultrasonically in a solution mixture containing 0.5 mL isopropanol and 10 μL 5 wt. % Nafion solution. Then, 10 μL of the mixture was transferred onto the 0.196 cm^{-2} polished glassy carbon electrode surface and dried at room temperature. The catalyst loadings on the electrodes were evaluated to be 0.2 mg· cm^{-2} (including the support).

The cyclic voltammetry (CV) and RDE measurements were done at 20 °C in either nitrogen purged or oxygen saturated 0.1 mol· L^{-1} HClO$_4$ and 0.1 mol· L^{-1} KOH solutions. Prior to the measurement, the electrolyte was deaerated by nitrogen or oxygen throughout the 30 min. The scanning potentials started from -0.160 V to 1.034 V in HClO$_4$ or 0.413 V to 1.410 V in KOH at a sweep rate of 50 mV· s^{-1}. The linear sweep voltammetry (LSV) curves were recorded in the potential range of 0.107–1.105 V with 5 mV· s^{-1} over a rotation rate of 0–2000 rpm in oxygen saturated electrolyte. All the potentials in this work were reported with respect to reversible hydrogen electrode (RHE).

3.4. Physicochemical Characterization

XRD analysis of the catalyst nanoparticles was performed with a powder diffractometer (Rigaku Ultima IV XRD) using Cu K$_\alpha$ radiation (λ = 0.1546 nm). Raman spectra were acquired using the 532 nm laser on a Princeton TriVista CRS557 Raman spectrometer. A high resolution transmission electron microscope (TEM) (JEOL JEM-2100), field emission scanning electron microscope (SEM) with built-in energy dispersive X-ray spectroscope (EDS) (Zeiss Sigma SEM) and X-ray photoelectron spectroscope (XPS) (PHI Quantum 2000) using Al K$_\alpha$ radiation were used to examine the microstructures and chemical compositions of the as-prepared catalyst nanoparticles, respectively.

4. Conclusions

The FeSe$_2$/C catalysts could be rapidly prepared through a simple microwave method by using various Se/Fe ratios. The formation of the orthogonal FeSe$_2$ structure was confirmed by XRD, TEM and SAED analyses. The estimated average crystallite sizes were 32.9–36.2 nm for the Se/Fe ratios of 2.0–4.0. The catalysts exhibited the enhanced ORR activities in alkaline media rather than in acidic media. The ORR potentials of 0.781–0.814 V with the electron transfer numbers of 3.3–3.9 at 0.3 V could be achieved in KOH solutions as the Se/Fe ratios varied from 2.0 to 4.0. The Se/Fe ratios slightly influenced the amounts of ORR active sites and the defects of carbon support, as well as the degrees of graphitization, which together affected the ORR activities.

Acknowledgments: The authors wish to thank the financial support provided by the National Natural Science Foundation of China (11372263).

Author Contributions: Q.Z. prepared the samples and performed all the measurements; X.C. prepared the manuscript; H.L. assisted for experimental design.

Conflicts of Interest: The authors declare no conflict of interest.

References

1. McLean, G.; Niet, T.; Prince-Richard, S.; Djilali, N. An assessment of alkaline fuel cell technology. *Int. J. Hydrogen Energy* **2002**, *27*, 507–526.
2. Liang, Y.; Wang, H.; Zhou, J.; Li, Y.; Wang, J.; Regier, T.; Dai, H. Covalent hybrid of spinel manganese–cobalt oxide and graphene as advanced oxygen reduction electrocatalysts. *J. Am. Chem. Soc.* **2012**, *134*, 3517–3523.
3. Lefèvre, M.; Proietti, E.; Jaouen, F.; Dodelet, J.P. Iron-based catalysts with improved oxygen reduction activity in polymer electrolyte fuel cells. *Science* **2009**, *324*, 71–74.
4. Cao, R.; Thapa, R.; Kim, H.; Xu, X.; Kim, M.G.; Li, Q.; Park, N.; Liu, M.; Cho, J. Promotion of oxygen reduction by a bio-inspired tethered iron phthalocyanine carbon nanotube-based catalyst. *Nat. Commun.* **2013**, *4*, 2076.
5. Shin, D.; Jeong, B.; Mun, B.S.; Jeon, H.; Shin, H.J.; Baik, J.; Lee, J. On the Origin of Electrocatalytic Oxygen Reduction Reaction on Electrospun Nitrogen–Carbon Species. *J. Phys. Chem. C* **2013**, *117*, 11619–11624.
6. Ding, L.; Xin, Q.; Zhou, X.; Qiao, J.; Li, H.; Wang, H. Electrochemical behavior of nanostructured nickel phthalocyanine (NiPc/C) for oxygen reduction reaction in alkaline media. *J. Appl. Electrochem.* **2013**, *43*, 43–51.
7. Liu, S.; Zhang, Z.; Bao, J.; Lan, Y.; Tu, W.; Han, M.; Dai, Z. Controllable Synthesis of Tetragonal and Cubic Phase Cu_2Se Nanowires Assembled by Small Nanocubes and Their Electrocatalytic Performance for Oxygen Reduction Reaction. *J. Phys. Chem. C* **2013**, *117*, 15164–15173.
8. Yang, L.; Jiang, S.; Zhao, Y.; Zhu, L.; Chen, S.; Wang, X.; Wu, Q.; Ma, J.; Ma, Y.; Hu, Z. Boron-Doped Carbon Nanotubes as Metal-Free Electrocatalysts for the Oxygen Reduction Reaction. *Angew. Chem.* **2011**, *123*, 7270–7273.
9. Gong, K.; Du, F.; Xia, Z.; Durstock, M.; Dai, L. Nitrogen-doped carbon nanotube arrays with high electrocatalytic activity for oxygen reduction. *Science* **2009**, *323*, 760–764.
10. Hibino, T.; Kobayashi, K.; Heo, P. Oxygen reduction reaction over nitrogen-doped graphene oxide cathodes in acid and alkaline fuel cells at intermediate temperatures. *Electrochim. Acta* **2013**, *112*, 82–89.

11. Wu, J.; Yang, Z.; Li, X.; Sun, Q.; Jin, C.; Strasser, P.; Yang, R. Phosphorus-doped porous carbons as efficient electrocatalysts for oxygen reduction. *J. Mater. Chem. A* **2013**, *1*, 9889–9896.

12. Zhu, J.; He, G.; Liang, L.; Wan, Q.; Shen, P.K. Direct anchoring of platinum nanoparticles on nitrogen and phosphorus-dual-doped carbon nanotube arrays for oxygen reduction reaction. *Electrochim. Acta* **2015**, *158*, 374–382.

13. Liang, J.; Jiao, Y.; Jaroniec, M.; Qiao, S.Z. Sulfur and Nitrogen Dual-Doped Mesoporous Graphene Electrocatalyst for Oxygen Reduction with Synergistically Enhanced Performance. *Angew. Chem. Int. Ed.* **2012**, *51*, 11496–11500.

14. Susac, D.; Zhu, L.; Teo, M.; Sode, A.; Wong, K.C.; Wong, P.C.; Parsons, R.R.; Bizzotto, D.; Mitchell, K.A.R.; Campbell, S.A. Characterization of FeS_2-based thin films as model catalysts for the oxygen reduction reaction. *J. Phys. Chem. C* **2007**, *111*, 18715–18723.

15. Jin, Z.; Nie, H.; Yang, Z.; Zhang, J.; Liu, Z.; Xu, X.; Huang, S. Metal-free selenium doped carbon nanotube/graphene networks as a synergistically improved cathode catalyst for oxygen reduction reaction. *Nanoscale* **2012**, *4*, 6455–6460.

16. Nekooi, P.; Akbari, M.; Amini, M.K. CoSe nanoparticles prepared by the microwave-assisted polyol method as an alcohol and formic acid tolerant oxygen reduction catalyst. *Int. J. Hydrogen Energy* **2010**, *35*, 6392–6398.

17. Li, H.; Gao, D.; Cheng, X. Simple microwave preparation of high activity Se-rich $CoSe_2$/C for oxygen reduction reaction. *Electrochim. Acta* **2014**, *138*, 232–239.

18. Feng, Y.; Alonso-Vante, N. Carbon-supported cubic $CoSe_2$ catalysts for oxygen reduction reaction in alkaline medium. *Electrochim. Acta* **2012**, *72*, 129–133.

19. Oyler, K.D.; Ke, X.; Sines, I.T.; Schiffer, P.; Schaak, R.E. Chemical Synthesis of Two-Dimensional Iron Chalcogenide Nanosheets: FeSe, FeTe, Fe (Se, Te), and $FeTe_2$. *Chem. Mater.* **2009**, *21*, 3655–3661.

20. Han, D.S.; Batchelor, B.; Abdel-Wahab, A. Sorption of selenium (IV) and selenium (VI) onto synthetic pyrite (FeS_2): Spectroscopic and microscopic analyses. *J. Colloid Interface Sci.* **2012**, *368*, 496–504.

21. Burrard-Lucas, M.; Free, D.G.; Sedlmaier, S.J.; Wright, J.D.; Cassidy, S.J.; Hara, Y.; Corkett, A.J.; Lancaster, T.; Baker, P.J.; Blundell, S.J. Enhancement of the superconducting transition temperature of FeSe by intercalation of a molecular spacer layer. *Nat. Mater.* **2013**, *12*, 15–19.

22. Wang, S.; Iyyamperumal, E.; Roy, A.; Xue, Y.; Yu, D.; Dai, L. Vertically Aligned BCN Nanotubes as Efficient Metal-Free Electrocatalysts for the Oxygen Reduction Reaction: A Synergetic Effect by Co-Doping with Boron and Nitrogen. *Angew. Chem. Int. Ed.* **2011**, *50*, 11756–11760.

23. Campos, C.; de Lima, J.; Grandi, T.; Machado, K.; Pizani, P. Structural studies of iron selenides prepared by mechanical alloying. *Solid State Commun.* **2002**, *123*, 179–184.

24. Frost, R.L.; Xi, Y.; López, A.; Scholz, R.; de Carvalho Lana, C.; e Souza, B.F. Vibrational spectroscopic characterization of the phosphate mineral barbosalite $Fe^{2+}Fe^{3+}_2(PO_4)_2(OH)_2$ – Implications for the molecular structure. *J. Mol. Struct.* **2013**, *1051*, 292–298.

25. Hibino, T.; Kobayashi, K.; Nagao, M.; Kawasaki, S. High-temperature supercapacitor with a proton-conducting metal pyrophosphate electrolyte. *Sci. Rep.* **2015**, *5*, 7903.

26. Warren, B.E. *X-ray Diffraction*; Dover Publications: Mineola, NY, USA, 1969; Volume II, pp. 251–257.

27. Kobayashi, K.; Nagao, M.; Yamamoto, Y.; Heo, P.; Hibino, T. Rechargeable PEM fuel-cell batteries using porous carbon modified with carbonyl groups as anode materials. *J. Electrochem. Soc.* **2015**, *162*, F868–F877.

Pt Monolayer Shell on Nitrided Alloy Core—A Path to Highly Stable Oxygen Reduction Catalyst

Jue Hu, Kurian A. Kuttiyiel, Kotaro Sasaki, Dong Su, Tae-Hyun Yang, Gu-Gon Park, Chengxu Zhang, Guangyu Chen and Radoslav R. Adzic

Abstract: The inadequate activity and stability of Pt as a cathode catalyst under the severe operation conditions are the critical problems facing the application of the proton exchange membrane fuel cell (PEMFC). Here we report on a novel route to synthesize highly active and stable oxygen reduction catalysts by depositing Pt monolayer on a nitrided alloy core. The prepared $Pt_{ML}PdNiN/C$ catalyst retains 89% of the initial electrochemical surface area after 50,000 cycles between potentials 0.6 and 1.0 V. By correlating electron energy-loss spectroscopy and X-ray absorption spectroscopy analyses with electrochemical measurements, we found that the significant improvement of stability of the $Pt_{ML}PdNiN/C$ catalyst is caused by nitrogen doping while reducing the total precious metal loading.

Reprinted from *Catalysts*. Cite as: Hu, J.; Kuttiyiel, K.A.; Sasaki, K.; Su, D.; Yang, T.-H.; Park, G.-G.; Zhang, C.; Chen, G.; Adzic, R.R. Pt Monolayer Shell on Nitrided Alloy Core—A Path to Highly Stable Oxygen Reduction Catalyst. *Catalysts* **2015**, 5, 1321–1332.

1. Introduction

Proton exchange membrane fuel cell (PEMFC) is expected to be an alternative power-generation for vehicles, stationary, and portable power applications because of its high energy density, low operation temperature, low air pollution and the use of renewable fuels, such as hydrogen and some alcohol [1,2]. Although the PEMFC power source technique has been really influent in the last decade, the slow kinetics of the oxygen reduction reaction (ORR) is still one of the main obstacles hampering the large scale applications of PEMFC [3]. Platinum (Pt) as the most effective catalyst for ORR has been the general choice. However, high Pt loading at the cathode as well as inadequate activity and stability of Pt under severe operation conditions are still unresolved problems facing the PEMFC [4,5]. To overcome these problems, it is essential to decrease the Pt amount in electrocatalysts, and at the same time, improve the performance of the Pt-based cathode catalyst both in terms of activity and stability. To this end, one of the strategies is to develop the metal@Pt core-shell structure catalysts in which a non-Pt core is employed and covered by atomically

362

thin layers of Pt. This core-shell structure allows efficient use of Pt, and thereby can reduce the demands on Pt while enhancing the catalyst performance [6–9].

Significant progress has been made through the combination of experimental and theoretical studies [10,11]. We developed a new class of catalysts consisting of a Pt monolayer on different metals and alloy supporting cores, including Pd, Ru, Ir, Rh, Au, PdAu, IrNi, IrRe, and AuNiFe [12–18]. The ORR activity of Pt monolayer on different metal surfaces shows a volcano-type dependence on the d-band center of Pt [18]. The strain-induced d-band center shifts and electronic ligand effects between the substrate and the overlayer are the two main factors determining the activity of these core-shell catalysts [6]. Nevertheless, the improving the electrocatalytic activity and stability of Pt-based cathode catalysts simultaneously is still a challenge. Great efforts have been made to modify the Pt surface with other elements such as Au [19]. The oxidation of Pt on Au-modified Pt surfaces requires much higher potentials than that on unmodified Pt surface, resulting in the enhancement of the catalyst stability [19]. Another strategy is to modify the metal core. Gong *et al.* synthesized highly stable $Pt_{ML}AuNi_{0.5}Fe$ catalysts and found that the Au shell in the core precluded the exposure of NiFe to the electrolyte leading to the high electrochemical stability [20]. Kuttiyiel *et al.* also developed a highly stable ORR catalyst by Au-stabilized PdNi [21]. More recently, we have reported a new approach to develop Pt-M (Ni, Co, and Fe) core-shell catalysts with high stability and activity by nitriding core metals [22,23]. The synchrotron XRD analysis proved the generation of the highly stable Fe_4N, Co_4N, and Ni_4N nitride cores. Since the Pt monolayer on Pd core catalyst is on the top of the volcano plot as mentioned above, and also the price of Pd is considerably lower than that of Pt [24], we selected Pt monolayer on nitride stabilized PdNi core ($Pt_{ML}PdNiN$) for studying its synthesis and structure in detail with the possibility to simultaneously improve its stability and activity, while reducing the PGM metal content.

2. Results and Discussion

PdNi alloy nanoparticles were first synthesized by chemical reduction (see experimental section), followed by thermal annealing in N_2 at 250 °C for 1 h, and subsequent annealing at 510 °C for 2 h in NH_3 as the nitrogen precursor. As illustrated in Figure 1, the PdNiN nanoparticles have a core-shell structure with Ni in the core and Pd on the surface. Figure 1a shows a high angle annular dark field scanning transmission electron microscope (HAADF-STEM) image of a representative single PdNiN nanoparticle. Elementary characterization of the PdNiN nanoparticle was performed by the electron energy-loss spectroscopy (EELS) mapping for Pd (M-edge, 2122 eV) and Ni (L-edge, 855 eV) from the nanoparticle shown in Figure 1a. As shown in Figure 1b, overlapping the mapping of Pd and Ni EELS signal validates an obvious Ni-core and Pd-shell structure. However, the outside of the particle is decorated by a trace amount of Ni/Ni oxides. The Ni/Ni oxides would not affect the electrocatalytic activity of these particles because they quickly dissolve in acid conditions during the Pt monolayer deposition. Figure 1c and Figure S1 (Supplementary Information) shows a line profile analysis by STEM-EELS illustrating the distribution of the Pd and Ni components in a single representative nanoparticle. It is evident that the Pd atoms are distributed uniformly over the Ni; the Pd shell thickness is determined to be around 0.6–1.5 nm by examining a number of particles. From the TEM images, the average particle size of the PdNiN nanoparticles was determined to be around 11 nm (Figure S2).

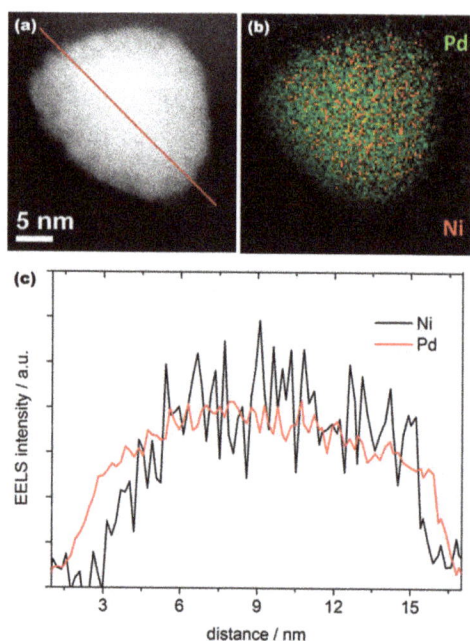

Figure 1. (a) HAADF-STEM image of PdNiN core-shell nanoparticle; (b) Two dimensional EELS mapping of Ni L signal (red) and Pd M signal (green) from a single nanoparticle; (c) EELS line scan profile for Pd M-edge and Ni L-edge along the scanned line indicated in (a).

To verify the formation of NiN$_x$ core in PdNiN nanoparticles we carried out X-ray absorption spectroscopy (XAS) measurements and compared the obtained spectra with those of reference metal foils, as shown in Figure 2. X-ray absorption near-edge structure (XANES) of Ni K edge from PdNiN nanoparticles shows that the electronic state of Ni has been changed due to the presence of N forming NiN$_x$ species. The Fourier transform (FT) magnitudes of the extended x-ray absorption fine structure (EXAFS) data for Ni-K edge (Figure 2c) for PdNiN presents a decrease in Ni bonding distance due to the formation of Ni nitrides. Also previous studies have shown that EXAFS for NiO or $Ni(OH)_2$ species demonstrate a peak at 1.6 Å corresponding to the Ni–O bond, accompanied by small peak at around 2.4 Å corresponding to the Ni–Ni bond [25]. The absence of these peaks along with the changes in the bonding distance compared to Ni metal verifies the presences of N_x species in the PdNiN. The alloying effect of $PdNiN_x$ has changed the electronic states of Pd as well, and these distinctions are clearly observed in the XANES and EXAFS regions when compared to those from a Pd foil (Figure 2b,d). The appearance of a peak around 2.0 Å in FT EXAFS of Pd K edge for PdNiN is likely caused by Pd–Ni bond. Although the exact species of NiN_x could not be determined, the XAS results along with the STEM-EELS analysis indicate that Ni in the core-shell structured PdNiN nanoparticles is nitrided. Our previous studies on nitrided Pt–M (M = Ni, Fe or Co) core-shell nanoparticles have indicated the presences of M_4N species [22,23]. As the synthesis parameters are similar to the previous study we presume the presence of Ni_4N species in our PdNiN core-shell nanoparticles.

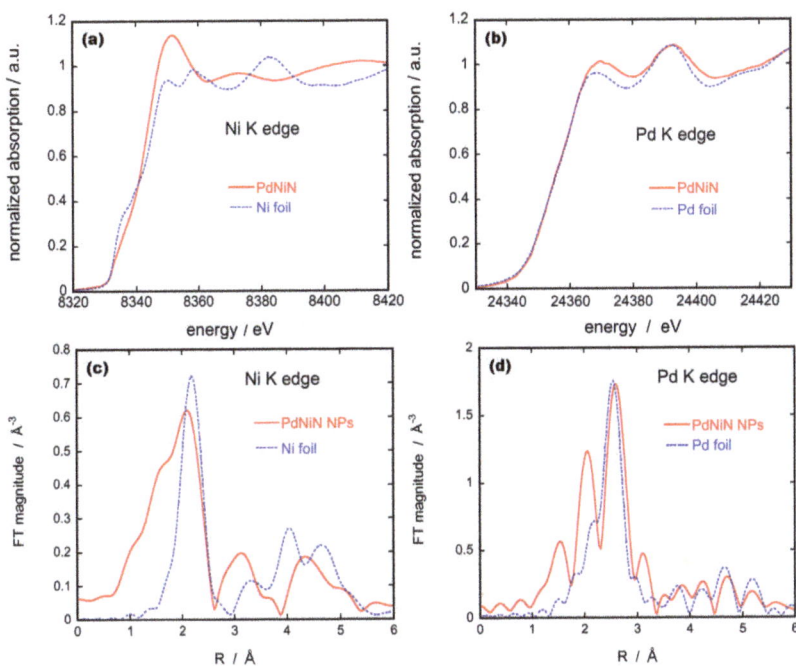

Figure 2. (a,b) Normalized XANES spectra for Ni and Pd K edges respectively along with Ni and Pd reference foil; (c,d) FT EXAFS spectra for Ni and Pd K edges respectively along with their reference foils.

The cyclic voltammetry (CV) curves obtained on the PdNiN/C and $Pt_{ML}PdNiN/C$ catalysts in Ar-saturated 0.1 M $HClO_4$ solution are shown in Figure 3a. It is observed that the curves in the hydrogen adsorption/desorption region of the $Pt_{ML}PdNiN/C$ resembled those of a typical Pt/C surface although the peaks from (110) and (100) planes are suppressed due to the interaction of the substrate materials. Moreover, after a Pt_{ML} depositing on the PdNiN/C surface, the oxide adsorption/desorption potentials shift more positively. The surface area of *i-E* plot associated with the hydrogen desorption can be used to estimate the electrochemical surface area (ECSA) of Pt catalysts. The ECSA of catalyst can be calculated according to Equation (1) [26]:

$$S_{ECSA} = \frac{Q_H}{L_{Pt} \times 0.21} \tag{1}$$

in which L_{Pt} represents the Pt loading (1.13 $\mu g \cdot cm^{-2}$ derived from the Cu under-potential deposition charge), Q_H ($mC \cdot cm^{-2}$) is the charge exchanged during the electro-desorption of hydrogen on Pt surface and 0.21 ($mC \cdot cm^{-2}$) is the charge required to oxidize a monolayer of hydrogen on a smooth Pt [27]. The ECSA value of the catalyst is 90 $m^2 \cdot g^{-1}_{Pt}$. Comparison of CVs from the commercial Pt/C (E-TEK, 10 wt. %), Pt_{ML} deposited commercial Pd/C (E-TEK, 10 wt. %, 3.5 nm Pd particle size) and $Pt_{ML}PdNiN/C$ catalysts (Figure 3b) showed that the oxide adsorption/desorption wave of $Pt_{ML}PdNiN/C$ occurred 37 mV and 60 mV positive compared to the $Pt_{ML}Pd/C$ and commercial Pt/C catalyst, respectively. The elevation of Pt oxidation potential on the $Pt_{ML}PdNiN/C$ catalyst indicates stabilization of the Pt_{ML} on the PdNiN/C substrate [19].

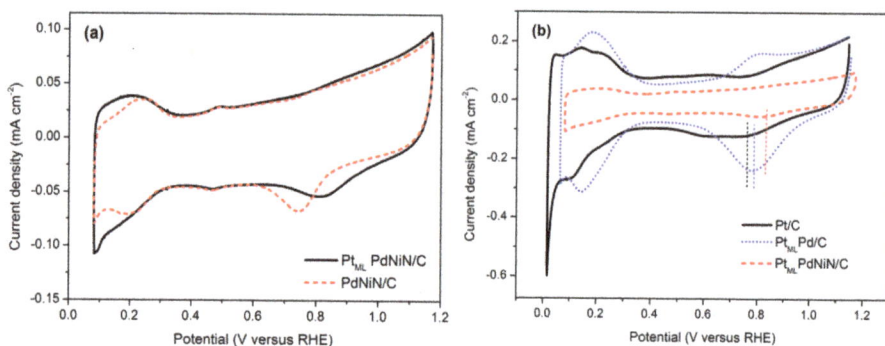

Figure 3. Cyclic voltammograms for (a) obtained PdNiN/C and $Pt_{ML}PdNiN/C$; and (b) commercial Pt/C, $Pt_{ML}Pd/C$ and $Pt_{ML}PdNiN/C$ nanoparticles in 0.1 M $HClO_4$ solution at a scan rate of 20 $mV \cdot s^{-1}$.

Figure 4a shows rotating disk electrode (RDE) measurements of the ORR on the $Pt_{ML}PdNiN/C$ catalyst in O_2 saturated 0.1 M $HClO_4$ solution at a sweep rate of

10 mV·s^{-1} and the rotation speeds from 100 to 3025 rpm. The high onset potential (*ca.* 1.0 V) and half-wave potential (850 mV at the rotation rate of 1600 rpm) of O$_2$ reduction at an ultra-low Pt loading (1.13 µg·cm^{-2}) indicate a good ORR activity of a Pt$_{ML}$PdNiN/C catalyst. The kinetic current density j_k was calculated from these ORR polarization curves (Figure 4a) using the Koutecky-Levich equation [22]:

$$\frac{1}{j} = \frac{1}{j_k} + \frac{1}{B\omega^{1/2}} \tag{2}$$

where j is the measured current density, B and ω are the constant and rotation rate, respectively. As can be seen from the Koutecky-Levich plot ($1/j$ plotted as a function of $\omega^{-1/2}$), shown in Figure 4b, the linearity and parallelism of the plots at 0.8 V, 0.85 V, and 0.9 V indicate the first-order kinetics with respect to molecular oxygen [28]. The intercept with the y-axis gives the inverse kinetic current density. The specific activity was determined from the normalization of kinetic current density to the ECSA while the kinetic current density was normalized to the loading of Pt or platinum group metal (PGM) to calculate the mass activity. The specific activity of the Pt$_{ML}$PdNiN/C catalyst is 1.17 mA·cm^{-2} at 0.9 V, which is more than four times higher than that of commercial Pt/C catalyst (0.24 mA·cm^{-2}), and 2.5 times higher than that of the commercial Pd/C with Pt monolayer (0.42 mA·cm^{-2}). However, higher ORR activities for commercial Pt/C catalyst were observed in some literature [29,30]. The Pt mass activity of Pt$_{ML}$PdNiN/C catalyst (1.05 A·mg^{-1}) is more than five times higher than the commercial Pt/C catalyst (0.2 A·mg^{-1}) and is also greater than the Pt$_{ML}$Pd/C catalyst (0.95 A·mg^{-1}) [21].

In addition to the high electrochemical activities, the Pt$_{ML}$PdNiN/C catalyst also exhibited excellent stability. The stability of the electrocatalyst was evaluated by an accelerated durability test involving potential cycling between 0.6 V and 1.0 V at the sweep rate of 50 mV·s^{-1} using a RDE in an air-saturated 0.1 M HClO$_4$ solution at room temperature. Figure 5a shows the ORR polarization curves of the Pt$_{ML}$PdNiN/C catalyst at 1600 rpm before and after 30,000 and 50,000 potential cycles. After 30,000 cycles, the half-wave potential of the ORR polarization curve remained at almost the initial value. After 50,000 cycles, the ORR measurements showed only 10 mV loss in the half-wave potential. This observation is similar to the previous results of Pt$_{ML}$Pd/C nanoparticles that retained their ORR activity even after losing their electrochemical surface area (ECSA) [31]. This can be explained by the concept that the Pd dissolution in the catalyst induces contraction to the Pt bonds and thereby increases the ORR activity [12,32]. Such a mechanism may be operative in the present system. However, as shown below, the loss in ECSA of Pt$_{ML}$PdNiN/C is much smaller than that of Pt$_{ML}$Pd/C, presumably because the presence of nitride phase retards the dissolution rate.

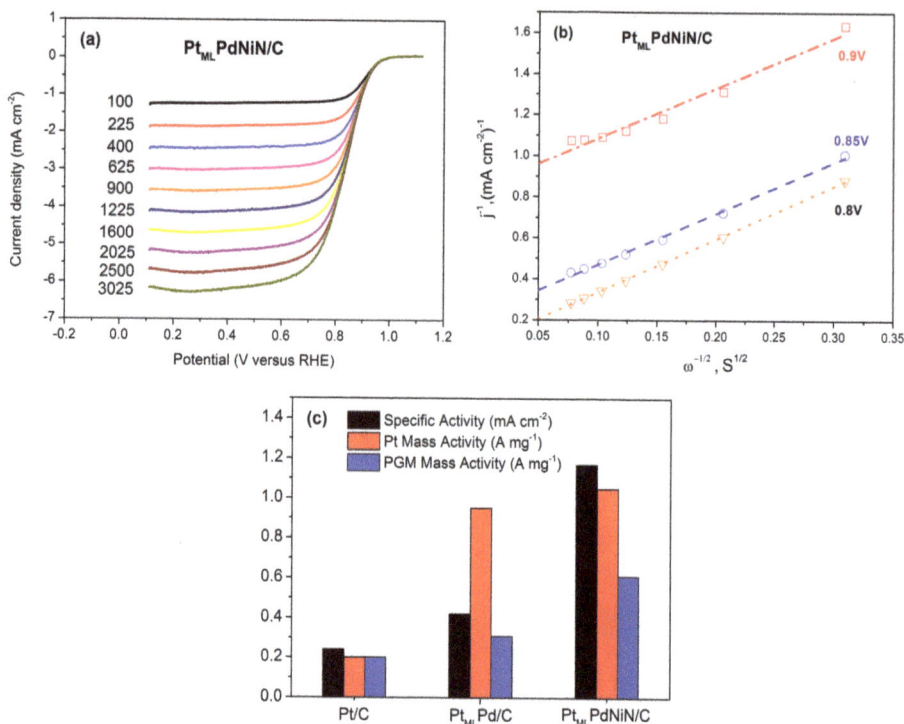

Figure 4. ORR polarization curves for the $Pt_{ML}PdNiN/C$ nanoparticles in 0.1 M $HClO_4$ solution at a scan rate of 10 mV·s^{-1} at various rpm. (**b**) The Koutechy-Levich plots at 0.8 V, 0.85 V and 0.9 V obtained from the ORR polarization curves as shown in (**a**). (**c**) Specific and mass activities for the commercial Pt/C, $Pt_{ML}Pd/C$ and $Pt_{ML}PdNiN/C$ catalysts at 0.9 V.

Figure 5b shows the CV curves of the $Pt_{ML}PdNiN/C$ catalyst in 0.1 M $HClO_4$ solution before and after cycling indicating a negligible loss of Pt surface area. The ECSA losses of the $Pt_{ML}PdNiN/C$ catalyst after different cycles are shown in Figure 5c. As reported in our previous paper, the $Pt_{ML}Pd/C$ catalyst exhibited a drastic decrease in ECSA after electrochemical cycling (27% after 5000 cycles and 34% after 15,000 cycles) due to the dissolution of Pd from the core [21]. Incorporation of Ni in the Pd core can slow down the Pd dissolution and as a result decrease the ECSA loss to 11.5% after 5000 cycles. But further cycling of the PdNi core leads to an ECSA loss of 28% after 15,000 cycles. Nitriding the PdNi core restrains the dissolution process, the ECSA loss of the $Pt_{ML}PdNiN/C$ catalyst, as shown in Figure 5c, is only 11% after 50,000 cycles. By further comparing to the commercial Pt/C catalyst which only retains 55% of its initial ECSA after 30,000 cycles, the less ECSA loss of the obtained $Pt_{ML}PdNiN/C$ catalyst indicates that stabilization in the metal core by nitrogen modification exhibits a significant improvement in Pt stability [19]. ORR activities of the $Pt_{ML}PdNiN/C$ catalyst before and after an accelerated durability test are listed in Table 1.

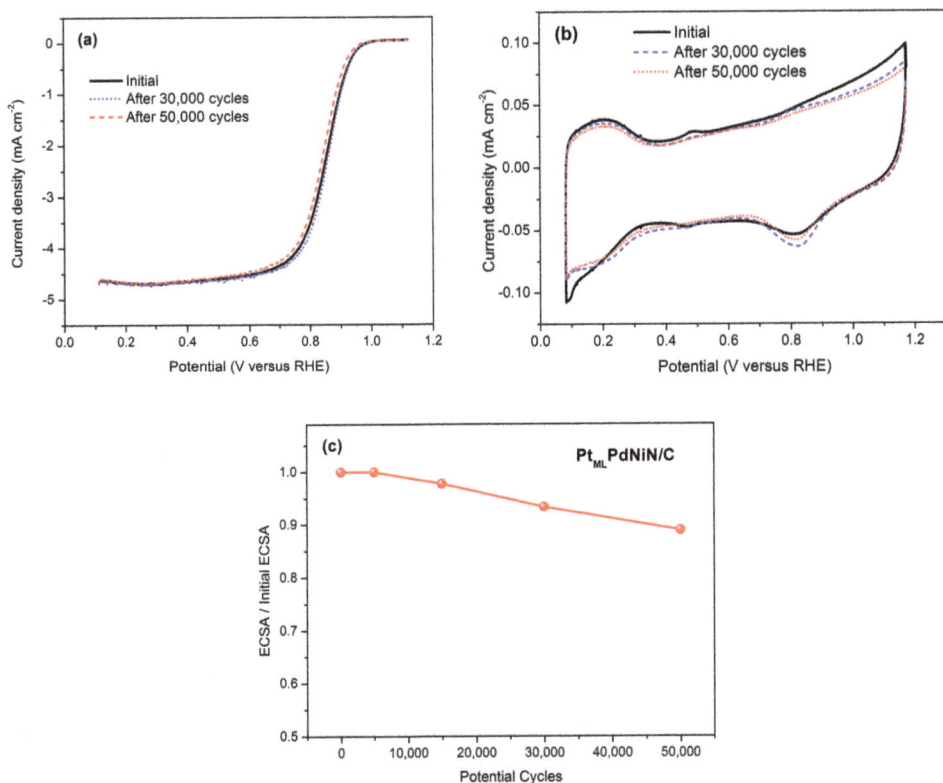

Figure 5. (a) ORR polarization curves and (b) cyclic voltammograms of the obtained $Pt_{ML}PdNiN/C$ catalyst before and after 50,000 cycles test between 0.6 and 1.0 V in 0.1 M $HClO_4$ solution; (c) ECSA degradation for $Pt_{ML}PdNiN/C$ catalyst plotted as a function of the number of after potential cycles between 0.6 and 1.0 V.

Table 1. Catalytic activities of the $Pt_{ML}PdNiN/C$ catalyst before and after the accelerated durability test.

$Pt_{ML}PdNiN/C$	ECSA ($m^2 \cdot g^{-1}Pt$)	$E_{1/2}$ (mV)	Specific activity ($mA \cdot cm^{-2}$)	Pt mass activity ($A \cdot mg^{-1}$)
Initial	90	850	1.17	1.05
After 30,000 cycles	84	854	1.35	1.13
After 50,000 cycles	80	840	0.84	0.67

3. Experimental Section

3.1. Preparation of PdNiN/C Nanoparticles

PdNi nanoparticles were synthesized by mixing a 1:1 molar ratio of $Pd(NO_3)_2 \cdot H_2O$ (Sigma-Aldrich, St. Louis, MO, USA) and $Ni(HCO_2)_2 \cdot 2H_2O$ (Sigma-Aldrich) salts with high area Vulcan XC-72R carbon black in MiliQ UV-plus water (Millipore corporation, Billerica, MA, USA) to obtain a total metal loading of 20 wt %. After sonicating the mixture for an hour under continuous Ar flow, $NaBH_4$ (Sigma-Aldrich) was added into the mixture and was then kept under sonication for 1 h. The mixture was filtered and rinsed with MilliQ UV-plus water (Millipore corporation), and then dried. The obtained PdNi/C nanoparticles were annealed in N_2 at 250 °C for 1 h followed by annealing at 510 °C for 2 h using NH_3 as the nitrogen precursor to get the PdNiN/C nanoparticles.

3.2. Characterization

The microstructure of the synthesized PdNiN/C nanoparticles was characterized by HD-2700C aberration-corrected STEM (Hitachi, Clarksburg, MD, USA) using a 1.4 Å electron probe with probe current ~50 pA and an energy resolution of 0.35 eV, at the Center for Functional Nanomaterials (CFN), Brookhaven National Laboratory (BNL). Elementary sensitive EELS line scan and mapping were carried out for Pd M-edge (2122 eV), Ni L-edge (855 eV) across various single PdNiN/C nanoparticle. The XAS measurements were undertaken at the National Synchrotron Light Source, BNL (Upton, NY, USA) using Beam Line X19A. The content of Pd and Ni in the PdNiN/C, measured by inductively coupled plasma-optical emission spectrometry (ICP-OES), were 8.2 wt % and 7.0 wt % respectively.

3.3. Electrochemical Measurements

Electrochemical testing was carried out in a three-electrode test cell by using a potentiostat (CHI 700B, CH Instruments, Austin, TX, USA). Before testing, catalyst ink was prepared by ultrasonic mixing of 5 mg of catalyst with 5 mL Millipore water until a dark and uniform aqueous dispersion was achieved. A thin film of the catalyst was prepared on a glassy carbon RDE with the area of 0.196 cm^2 by placing 10–15 µL of the obtained dispersion and then covered by a 10 µL dilute Nafion solution (2 µg·µL^{-1}). We deposited Pt monolayer both on the prepared PdNiN/C nanoparticle and commercial Pd/C nanoparticle surfaces using the galvanic displacement of Cu monolayer formed by Cu under-potential deposition (UPD) [6,16]. The Pt loadings on the RDE for the $Pt_{ML}PdNiN/C$ and $Pt_{ML}Pd/C$ catalysts were 1.13 and 3.75 µg·cm^{-2} respectively whereas their Pd loadings were 0.82 and 2.0 µg·cm^{-2} respectively. However, we note that a catalyst with higher loadings would be required for MEA preparation (future work) to replicate the ORR activity as that of RDE. The Pt loading on RDE for the commercial Pt/C catalyst was 7.65 $\mu g \cdot cm^{-2}$. The electrochemical measurements were all performed at room temperature, and the potentials were referenced to that of the reversible hydrogen electrode (RHE).

4. Conclusions

We described a promising route to develop nitride-stabilized substrates for Pt monolayer catalyst with substantial reduction in platinum group metal loading while retaining high ORR activity and stability. Using STEM-EELS mapping techniques we have investigated the core-shell structure of the catalyst while XAS measurement emphasized the NiN_x species in the core of the nanoparticles providing a stable support for Pt monolayer electrocatalysts.

Acknowledgments: This manuscript has been authored by employees of Brookhaven Science Associates, LLC under Contract No. DE-SC0012704 with the U.S. Department of Energy. The publisher by accepting the manuscript for publication acknowledges that the United States Government retains a non-exclusive, paid-up, irrevocable, world-wide license to publish or reproduce the published form of this manuscript, or allow others to do so, for United States Government purposes. This research used electron microscopy facility of the Center for Functional Nanomaterials, which is a U.S. DOE Office of Science Facility, at Brookhaven National Laboratory. Beam lines X19A at the National Synchrotron Light Source are supported in part by the Synchrotron Catalysis Consortium, U.S. Department of Energy Grant No DE-FG02-05ER15688. This work was also conducted under the framework of KIER's (Korea Institute of Energy Research) Research and Development Program (B5-2425).

Author Contributions: R.R.A., K.S. conceived and designed the experiments; K.A.K. performed the experiments; K.A.K. and J.H. analyzed the data and co-wrote the manuscript; D.S. performed the STEM-EELS analysis; K.S. and K.A.K. performed the XAS analysis; All authors contributed to the manuscript and the interpretation of the results.

Conflicts of Interest: The authors declare no conflict of interest.

References

1. Borup, R.; Meyers, J.; Pivovar, B.; Kim, Y.S.; Mukundan, R.; Garland, N.; Myers, D.; Wilson, M.; Garzon, F.; Wood, D.; *et al.* Scientific Aspects of Polymer Electrolyte Fuel Cell Durability and Degradation. *Chem. Rev.* **2007**, *107*, 3904–3951.

2. Jacobson, M.Z.; Colella, W.G.; Golden, D.M. Cleaning the Air and Improving Health with Hydrogen Fuel-Cell vehicles. *Science* **2005**, *308*, 1901–1905.

3. Gasteiger, H.A.; Markovic, N.M. Just a Dream-or Future Reality? *Science* **2009**, *324*, 48–49.

4. Debe, M.K. Electrocatalyst Approaches and Challenges for Automotive Fuel Cells. *Nature* **2012**, *486*, 43–51.

5. Yasuda, K.; Taniguchi, A.; Akita, T.; Ioroi, T.; Siroma, Z. Platinum Dissolution and Depostion in the Polymer Electrolyte Membrane of a PEM Fuel Cell as Studied by Potential Cycling. *Phys. Chem. Chem. Phys.* **2006**, *8*, 746–752.

6. Adzic, R.R.; Zhang, J.; Sasaki, K.; Vukmirovic, M.B.; Shao, M.; Wang, J.X.; Nilekar, A.U.; Mavrikakis, M.; Valerio, J.A.; Uribe, F. Platinum Monolayer Fuel Cell Electrocatalysts. *Top. Catal.* **2007**, *46*, 249–262.

7. Shao, M.; Shoemaker, K.; Peles, A.; Kaneko, K.; Protsailo, L. Pt Monolayer on Porous Pd-Cu Alloys as Oxygen Reduction Electrocatalysts. *J. Am. Chem. Soc.* **2010**, *132*, 9253–9255.

8. Strasser, P.; Koh, S.; Anniyev, T.; Greeley, J.; More, K.; Yu, C.F.; Liu, Z.C.; Kaya, S.; Nordlund, D.; Ogasawara, H.; *et al.* Lattice-strain control of the activity in dealloyed core-shell fuel cell catalysts. *Nat. Chem.* **2010**, *2*, 454–460.

9. Wang, D.L.; Xin, H.L.L.; Hovden, R.; Wang, H.S.; Yu, Y.C.; Muller, D.A.; DiSalvo, F.J.; Abruna, H.D. Structurally ordered intermetallic platinum-cobalt core-shell nanoparticles with enhanced activity and stability as oxygen reduction electrocatalysts. *Nat. Mater.* **2013**, *12*, 81–87.

10. Brimaud, S.; Behm, R.J. Electrodeposition of a Pt Monolayer Film: Using Kinetic Limitations for Atomic Layer Epitaxy. *J. Am. Chem. Soc.* **2013**, *135*, 11716–11719.

11. Zhao, X.; Chen, S.; Fang, Z.; Ding, J.; Sang, W.; Wang, Y.; Zhao, J.; Peng, Z.; Zeng, J. Octahedral Pd@Pt1.8Ni Core-Shell Nanocrystals with Ultrathin PtNi Alloy Shells as Active Catalysts for Oxygen Reduction Reaction. *J. Am. Chem. Soc.* **2015**, *137*, 2804–2807.

12. Wang, J.X.; Inada, H.; Wu, L.J.; Zhu, Y.M.; Choi, Y.M.; Liu, P.; Zhou, W.P.; Adzic, R.R. Oxygen Reduction on Well-Defined Core-Shell Nanocatalysts: Particle Size, Facet, and Pt Shell Thickness Effects. *J. Am. Chem. Soc.* **2009**, *131*, 17298–17302.

13. Karan, H.I.; Sasaki, K.; Kuttiyiel, K.; Farberow, C.A.; Mavrikakis, M.; Adzic, R.R. Catalytic Activity of Platinum Mono layer on Iridium and Rhenium Alloy Nanoparticles for the Oxygen Reduction Reaction. *ACS Catal.* **2012**, *2*, 817–824.

14. Zhang, Y.; Hsieh, Y.C.; Volkov, V.; Su, D.; An, W.; Si, R.; Zhu, Y.M.; Liu, P.; Wang, J.X.; Adzic, R.R. High Performance Pt Mono layer Catalysts Produced via Core-Catalyzed Coating in Ethanol. *ACS Catal.* **2014**, *4*, 738–742.

15. Zhang, Y.; Ma, C.; Zhu, Y.M.; Si, R.; Cai, Y.; Wang, J.X.; Adzic, R.R. Hollow core supported Pt monolayer catalysts for oxygen reduction. *Catal. Today* **2013**, *202*, 50–54.

16. Kuttiyiel, K.A.; Sasaki, K.; Choi, Y.; Su, D.; Liu, P.; Adzic, R.R. Bimetallic IrNi Core Platinum Monolayer Shell Electrocatalysts for the Oxygen Reduction Reaction. *Energy Environ. Sci.* **2012**, *5*, 5297–5304.

17. Hsieh, Y.-C.; Zhang, Y.; Su, D.; Volkov, V.; Si, R.; Wu, L.; Zhu, Y.; An, W.; Liu, P.; He, P.; *et al.* Ordered Bilayer Ruthenium-Platinum Core-Shell Nanoparticles as Carbon Monoxide-Tolerant Fuel Cell Catalysts. *Nat. Commun.* **2013**, *4*, 1–9.

18. Zhang, J.; Vukmirovic, M.B.; Xu, Y.; Mavrikakis, M.; Adzic, R.R. Controlling the Catalytic Activity of Platinum-Monolayer Electrocatalysts for Oxygen Reduction with Different Substrates. *Angew. Chem. Int. Ed.* **2005**, *44*, 2132–2135.

19. Zhang, J.; Sasaki, K.; Sutter, E.; Adzic, R.R. Stabilization of Platinum Oxygen-Reduction Electrocatalysts Using Gold Clusters. *Science* **2007**, *315*, 220–222.

20. Gong, K.; Su, D.; Adzic, R.R. Platinum-Monolayer Shell on AuNi0.5Fe Nanoparticle Core Electrocatalyst with High Activity and Stability for the Oxygen Reduction Reaction. *J. Am. Chem. Soc.* **2010**, *132*, 14364–14366.

21. Kuttiyiel, K.A.; Sasaki, K.; Su, D.; Vukmirovic, M.B.; Marinkovic, N.S.; Adzic, R.R. Pt monolayer on Au-Stabilized PdNi Core-Shell Nanoparticles for Oxygen Reduction Reaction. *Electrochim. Acta* **2013**, *110*, 267–272.

22. Kuttiyiel, K.A.; Sasaki, K.; Choi, Y.M.; Su, D.; Liu, P.; Adzic, R.R. Nitride Stabilized PtNi Core-Shell Nanocatalyst for High Oxygen Reduction Activity. *Nano Lett.* **2012**, *12*, 6266–6271.

23. Kuttiyiel, K.A.; Choi, Y.; Hwang, S.-M.; Park, G.-G.; Yang, T.-H.; Su, D.; Sasaki, K.; Liu, P.; Adzic, R.R. Enhancement of the Oxygen Reduction on Nitride Stabilized Pt-M (M = Fe, Co, and Ni) Core-Shell Nanoparticle Electrocatalysts. *Nano Energy* **2015**, *13*, 442–449.

24. Liu, H.; Koenigsmann, C.; Adzic, R.R.; Wong, S.S. Probing Ultrathin One-Dimensional Pd–Ni Nanostructures As Oxygen Reduction Reaction Catalysts. *ACS Catal.* **2014**, *4*, 2544–2555.

25. Subbaraman, R.; Tripkovic, D.; Strmcnik, D.; Chang, K.-C.; Uchimura, M.; Paulikas, A.P.; Stamenkovic, V.; Markovic, N.M. Enhancing Hydrogen Evolution Activity in Water Splitting by Tailoring Li+-Ni(OH)$_2$-Pt Interfaces. *Science* **2011**, *334*, 1256–1260.

26. Chu, Y.-Y.; Wang, Z.-B.; Gu, D.-M.; Yin, G.-P. Performance of Pt/C catalysts prepared by microwave-assisted polyol process for methanol electrooxidation. *J. Power Sources* **2010**, *195*, 1799–1804.

27. Schmidt, T.J.; Gasteiger, H.A.; Stäb, G.D.; Urban, P.M.; Kolb, D.M.; Behm, R.J. Characterization of High-Surface-Area Electrocatalysts Using a Rotating Disk Electrode Configuration. *J. Electrochem. Soc.* **1998**, *145*, 2354–2358.

28. Sasaki, K.; Wang, J.X.; Naohara, H.; Marinkovic, N.; More, K.; Inada, H.; Adzic, R.R. Recent Advances in Platinum Monolayer Electrocatalysts for Oxygen Reduction Reaction: Scale-up Synthesis, Structure and Activity of Pt Shells on Pd Cores. *Electrochim. Acta* **2010**, *55*, 2645–2652.

29. Garsany, Y.; Singer, I.L.; Swider-Lyons, K.E. Impact of film drying procedures on RDE characterization of Pt/VC electrocatalysts. *J. Electroanal. Chem.* **2011**, *662*, 396–406.

30. Takahashi, I.; Kocha, S.S. Examination of the activity and durability of PEMFC catalysts in liquid electrolytes. *J. Power Sources* **2010**, *195*, 6312–6322.

31. Sasaki, K.; Naohara, H.; Cai, Y.; Choi, Y.M.; Liu, P.; Vukmirovic, M.B.; Wang, J.X.; Adzic, R.R. Core-Protected Platinum Monolayer Shell High-Stability Electrocatalysts for Fuel-Cell Cathodes. *Angew. Chem. Int. Ed.* **2010**, *49*, 8602–8607.

32. Wang, J.X.; Ma, C.; Choi, Y.M.; Su, D.; Zhu, Y.M.; Liu, P.; Si, R.; Vukmirovic, M.B.; Zhang, Y.; Adzic, R.R. Kirkendall Effect and Lattice Contraction in Nanocatalysts: A New Strategy to Enhance Sustainable Activity. *J. Am. Chem. Soc.* **2011**, *133*, 13551–13557.

Simple Preparation of Pd Core Nanoparticles for Pd Core/Pt Shell Catalyst and Evaluation of Activity and Durability for Oxygen Reduction Reaction

Hiroshi Inoue, Ryotaro Sakai, Taiki Kuwahara, Masanobu Chiku and Eiji Higuchi

Abstract: Pd core nanoparticles less than 5 nm in mean size were prepared on carbon black (CB) without any stabilizer by using palladium acetate as a precursor and CO as a reducing agent, and then used for preparing Pd core/Pt shell nanoparticles-loaded CB (Pt/Pd/CB). The mean size of Pd nanoparticles could be controlled by the concentration of palladium acetate and the CO bubbling time. The cyclic voltammograms of two Pd nanoparticles-loaded CB ($Pd_{4.2}$/CB, $Pd_{3.3}$/CB) electrodes whose mean size was 4.2 and 3.3 nm, respectively, had characteristics similar to a Pt electrode after the formation of a Pt monolayer shell, suggesting that the Pd core nanoparticles were almost covered with the Pt monolayer shell. The oxygen reduction reaction (ORR) on both Pt/Pd/CB proceeded in 4-electron reduction mechanism. Both Pt/Pd/CB electrodes was *ca.* 1.5 times higher in ORR activity per electrochemical surface area of Pt (specific activity, SA) than the commercial Pt nanoparticles-loaded CB (Tanaka Kikinzoku Kogyo, Pt/CB-TKK) electrode, and the $Pt/Pd_{3.3}$/CB electrode had higher SA than the $Pt/Pd_{4.2}$/CB electrode. The ORR activity per unit mass of Pt for both Pt/Pd/CB electrodes was 5.0 and 5.5 times as high as that for the Pt/CB-TKK electrode, respectively. The durability of both Pt/Pd/CB electrodes was comparable to that of Pt/CB-TKK.

Reprinted from *Catalysts*. Cite as: Inoue, H.; Sakai, R.; Kuwahara, T.; Chiku, M.; Higuchi, E. Simple Preparation of Pd Core Nanoparticles for Pd Core/Pt Shell Catalyst and Evaluation of Activity and Durability for Oxygen Reduction Reaction. *Catalysts* **2015**, *5*, 1375–1387.

1. Introduction

Polymer electrolyte fuel cells (PEFCs) attract great attention as a clean power source for habitations and electric vehicles due to their high energy conversion efficiency, low emission of pollutants and low operating temperature. However, a serious issue for their practical use is high price of Pt, which has the highest activity for oxygen reduction reaction (ORR) at the cathode, so reducing the consumption of Pt is an urgent mission. For this purpose, various strategies including the preparation of nanoparticles of various Pt-based alloys and bimetals have been attempted so far [1–8].

It is well-known that only a few surface atomic layers of catalyst participate in heterogeneous catalysis, suggesting that covering foreign metal core nanoparticles with a Pt monolayer shell is the most effective way to reduce the Pt consumption with keeping the ORR activity because the Pt utilization is ultimately enhanced. This can be another desirable strategy. For realizing the formation of a Pt monolayer shell on Pd core nanoparticles, a Cu monolayer was formed on Pd core particles by underpotential deposition (upd), followed by galvanic displacement with Pt to prepare Pd core/Pt shell nanoparticles-loaded carbon black (Pt/Pd/CB) catalysts [9]. The prepared Pt/Pd/CB exhibited high ORR activity per unit mass of Pt (mass activity, MA) [9–11], suggesting that the Pt consumption was highly reduced.

To prepare Pd nanoparticles, various stabilizers like polyvinylpyrrolidone, tetra(*n*-octylammonium) bromide, cetyltrimethylammonium bromide, sodium citrate, oleylamine *etc.* [12–15], have been used so far. However, the removal of the stabilizers is so tiresome that the preparation of Pd nanoparticles without any stabilizer is desirable in terms of low cost and low environmental load. Quite recently, the specific preparation method of Pt/Pd nanoparticles without any stabilizer has been reported although they were not loaded on CB [16]. Recently, we have succeeded in the simple preparation of Pt and Au nanoparticle-loaded CB with relatively narrow size distribution by using CO as a reducing agent [17,18]. In this study, we applied this method to the synthesis of Pd core nanoparticle-loaded CB (Pd/CB), and, consequently, we successfully prepared CB loaded Pd core nanoparticles less than 5 nm in mean size without any stabilizer by bubbling CO in acetonitrile solutions containing palladium acetate. Moreover, after a Pt monolayer shell was formed on the Pd core nanoparticles by upd of Cu and the following galvanic displacement with Pt, the ORR activity was greatly enhanced.

2. Results and Discussion

2.1. Structural Properties of Pd/CB and Pt Monolayer-Modified Pd/CB

Figure 1 shows X-ray diffraction (XRD) patterns of Pd/CB prepared under different conditions. When the concentrations of palladium acetate in acetonitrile was changed from 0.05 to 1.0 mM, CO bubbling time was fixed to 5 min. On the other hand, when CO was bubbled for 3–60 min, the concentration of palladium acetate in acetonitrile was fixed to 1 mM. In all XRD patterns of Figure 1 a broad reflection peak assigned to Pd(111) was distinctly observed at $2\theta = ca.$ 40°, suggesting the production of Pd nanoparticles [19], while there were not any peaks assigned to Pd oxides. Figure 2 shows a Pd3d core level spectrum of Pd/CB as the concentration of palladium acetate in acetonitrile was 1 mM and the CO bubbling time was 5 min.

The spectrum indicates that only metallic Pd was produced, supporting the XRD data. In Figure 1a, the intensity of the (111) peak was significantly increased as the concentration of palladium acetate in acetonitrile was increased, suggesting the increase in the amount of produced Pd nanoparticles. In contrast, the increase in the CO bubbling time did not contribute to the increase in the peak intensity, as shown in Figure 1b.

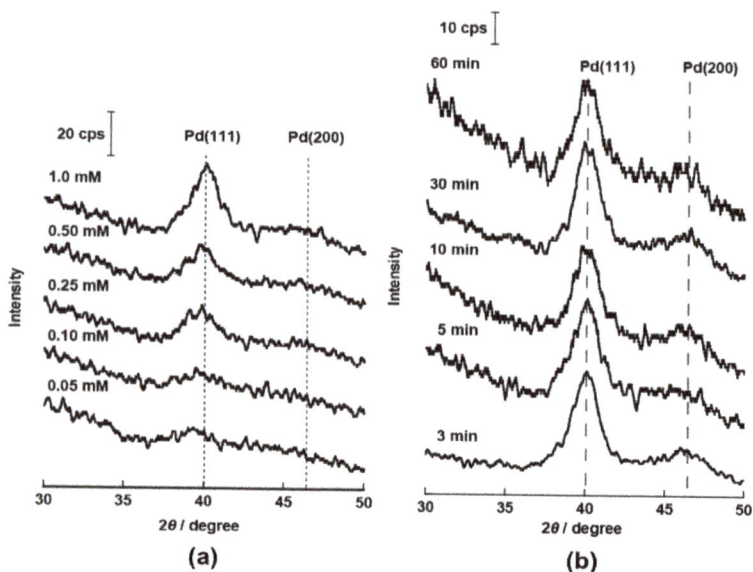

Figure 1. X-ray diffraction patterns of Pd/CB prepared with (**a**) 0.05–1.0 mM palladium acetate in acetonitrile when CO was bubbled for 5 min and (**b**) 1.0 mM palladium acetate in acetonitrile when CO was bubbled for 3–30 min.

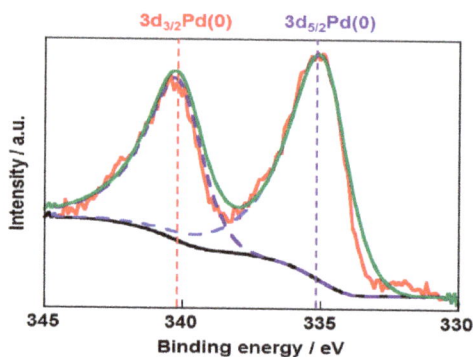

Figure 2. A Pd3d core level spectrum of Pd/CB prepared by bubbling CO in an acetonitrile solution containing 1 mM palladium acetate for 5 min.

The mean size of Pd nanoparticles in the Pd/CB prepared with a variety of the concentration of palladium acetate in acetonitrile and CO bubbling time was estimated by applying the Scherrer's Equation to the (111) peak in each XRD pattern. The results are summarized in Figure 3. In Figure 3a, when the concentrations of palladium acetate in acetonitrile was 0.1 mM or less, the mean size of the prepared Pd nanoparticles was around 3.1 nm, while at more than 0.1 mM it was increased with the concentration of palladium acetate in acetonitrile. This can be ascribed to the growth of the Pd nanoparticles due to the progress of Pd deposition. In Figure 3b, when the CO bubbling time was 10 min or less, the mean particle size was maintained at *ca.* 4.2 nm, while at more than 10 min it was increased with the CO bubbling time. This can be ascribed to the agglomeration of Pd nanoparticles before loading on the CB powder. From these results, it is concluded that the mean size of Pd nanoparticles can be controlled by the concentration of palladium acetate in acetonitrile and the CO bubbling time, and the former is effective in the preparation of smaller particles. The Pd/CB catalysts prepared by bubbling CO in acetonitrile solutions containing 1.0 and 0.25 mM palladium acetate for 5 min are used hereafter, which are named $Pd_{4.2}/CB$ and $Pd_{3.3}/CB$, respectively.

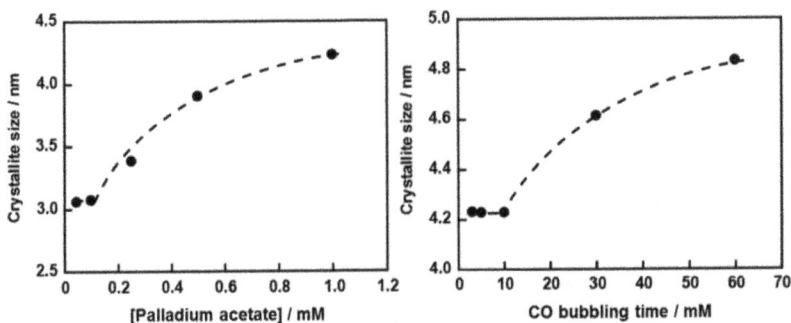

Figure 3. Crystalline size of Pd nanoparticles as a function of (**a**) the concentration of palladium acetate in acetonitrile and (**b**) the CO bubbling time.

Figure 4 shows TEM images and the histograms of the size of the Pd nanoparticles in $Pd_{4.2}/CB$ and $Pd_{3.3}/CB$. The TEM images exhibited that in both cases the Pd nanoparticles were well dispersed on CB. The mean size and standard deviation of the Pd nanoparticles were evaluated to be 3.5 ± 0.9 nm and 4.2 ± 0.7 nm for $Pd_{3.3}/CB$ and $Pd_{4.2}/CB$, respectively. In the former, the particles less than 2.5 nm in diameter were included. This seems to be because smaller nuclei are formed at lower precursor concentrations and their growth rate is slow. In both cases the mean size of the Pd nanoparticles evaluated from XRD and TEM was almost equivalent to each other. In this way, we succeeded in preparing Pd nanoparticles less than 5 nm in diameter without any stabilizer by a simple method in which palladium acetate and CO was used as a precursor and a reducing agent, respectively.

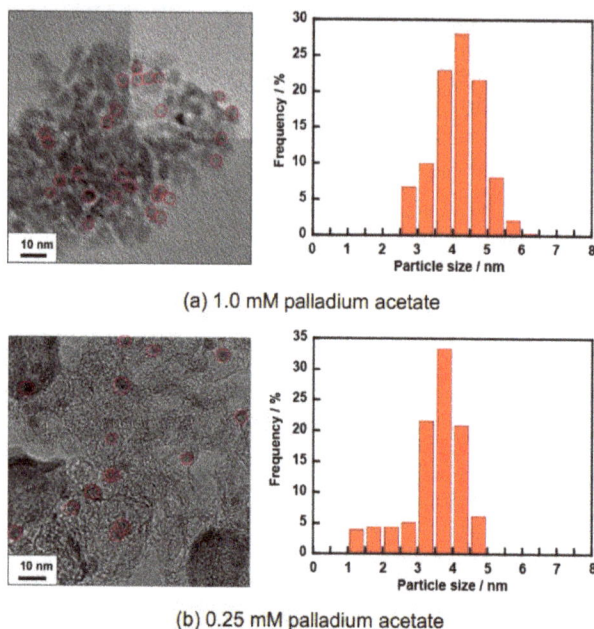

(a) 1.0 mM palladium acetate

(b) 0.25 mM palladium acetate

Figure 4. TEM images and histograms of size of Pd nanoparticles loaded on CB prepared by bubbling CO in acetonitrile containing (**a**) 0.25 mM and (**b**) 1 mM palladium acetate for 5 min.

Figure 5 shows the Pt4f and Pd3d core level spectra for the Pt/Pd$_{4.2}$/CB catalyst. Both spectra clearly indicate that both Pd and Pt have the metallic state. With the peak area for Pd and Pt in both spectra, atomic ratio of Pd and Pt (Pd:Pt) was estimated as 0.80:0.20. If a Pd core nanoparticle with the size of 4.2 nm is covered with a Pt monolayer, Pd:Pt = 0.69:0.31. However, since the Pd core nanoparticles are loaded on CB, the contact surface of Pd with CB is dead space, so the fraction of the exposed Pt surface should be less than 0.31. The utilization of Pd for the Pd$_{4.2}$/CB, which is defined by the ratio of electrochemical surface area of 4.2 nm Pd nanoparticles evaluated by CO stripping to the calculated surface area of the same Pd nanoparticles which are assumed to be spherical, was evaluated as 63%. So the ratio of the exposed Pt is estimated as 0.31 × 0.63 = 0.20, and Pd:Pt = 0.80:0.20, which is equivalent to the experimental value. Moreover, the atomic ratio was also equivalent to that evaluated from inductively coupled plasma spectroscopy. This suggests that the Pt shell monolayer almost covered the Pd core nanoparticles.

378

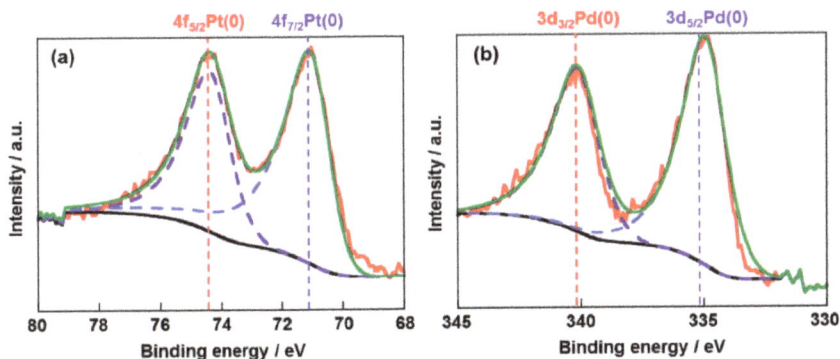

Figure 5. (a) Pt4f and (b) Pd3d core level spectra for Pt/Pd$_{4.2}$/CB.

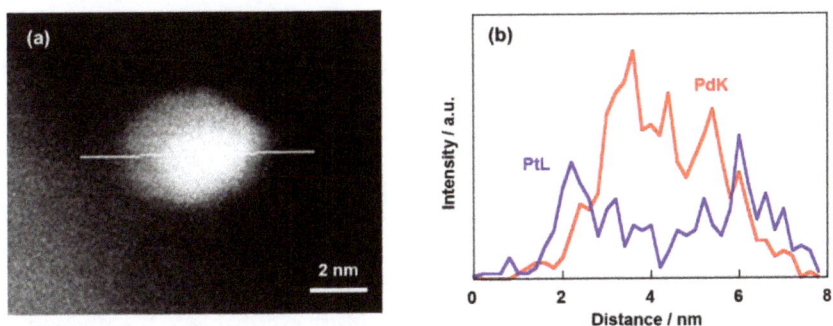

Figure 6. (a) A transmission electron micrograph of a Pt$_{4.2}$ core/Pt shell nanoparticle loaded on CB and (b) EDX line scan spectra of Pd and Pt.

Figure 6 shows a transmission electron micrograph and its EDX line profile for a 4.2 nm Pd core/Pt shell nanoparticle loaded on CB. Figure 6b exhibited that Pd and Pt were observed over the whole nanoparticle. In particular, the intensity of Pd was high in the center of the nanoparticle, whereas that of Pt was high at both edges of the nanoparticle, suggesting that the Pd core nanoparticle was covered with Pt.

2.2. Electrochemical Properties of Pt/Pd/CB Electrodes

Figure 7 shows cyclic voltammograms (CVs) of the Pt monolayer-modified Pd$_{4.2}$/CB (Pt/Pd$_{4.2}$/CB), Pt monolayer-modified Pd$_{3.3}$/CB (Pt/Pd$_{3.3}$/CB), Pd$_{4.2}$/CB and Pd$_{3.3}$/CB electrodes in an Ar-saturated 0.1 M HClO$_4$ aqueous solution. The CVs of the Pd$_{4.2}$/CB and Pd$_{3.3}$/CB electrodes had two pairs of redox peaks in the potential range less than 0.4 V in addition to a distinct reduction peak of palladium oxide at *ca.* 0.75 V, which agrees with CVs of the Pd thin layers and Pd nanoparticles [9,20,21]. A couple of large peaks at *ca.* 0.1 V can be assigned to the hydrogen absorption/desorption process in the Pd nanoparticles, while the two small peaks at *ca.* 0.27 V are assigned to the hydrogen adsorption/desorption on the Pd nanoparticles [20].

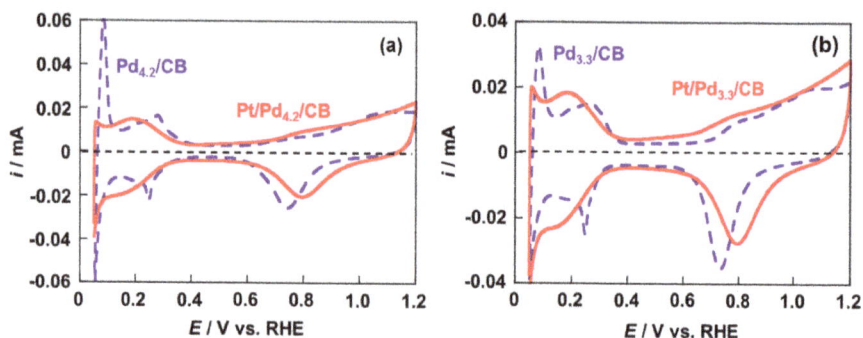

Figure 7. Cyclic voltammograms of (a) $Pt_x/Pd_{4.2}/CB$ ($x = 0, 1$) and (b) $Pt_x/Pd_{3.3}/CB$ ($x = 0, 1$) electrodes in Ar-saturated 0.1 M $HClO_4$ aqueous solution. Scan rate: 20 mV s^{-1}.

For the $Pt/Pd_{4.2}/CB$ and $Pt/Pd_{3.3}/CB$ electrodes, two pairs of peaks in the potential range less than 0.4 V almost disappeared, and the peaks assigned to the hydrogen adsorption/desorption (*ca.* 0.2 V) and hydrogen evolution/oxidation (*ca.* 0.05 V) appeared instead. Moreover, the potential of the reduction peak of oxide was shifted more positively [9]. These modifications are ascribable to the phenomena on Pt, suggesting that the Pd nanoparticles were almost covered with a Pt monolayer.

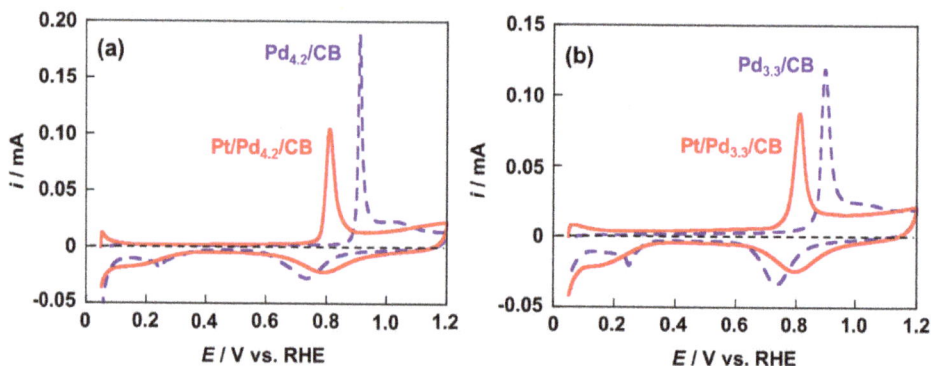

Figure 8. CO stripping voltammograms of (a) $Pt_x/Pd_{4.2}/CB$ ($x = 0, 1$) and (b) $Pt_x/Pd_{3.3}/CB$ ($x = 0, 1$) electrodes in Ar-saturated 0.1 M $HClO_4$ aqueous solution. Scan rate: 20 mV s^{-1}.

Figure 8 shows CO stripping voltammograms of the $Pt_x/Pd_{4.2}/CB$ ($x = 0, 1$) and $Pt_x/Pd_{3.3}/CB$ ($x = 0, 1$) electrodes in Ar-saturated 0.1 M $HClO_4$ aqueous solution.

For the $Pd_{4.2}/CB$ and $Pd_{3.3}/CB$ electrodes, the CO stripping peak was observed at *ca.* 0.92 V. Irrespective of the size of Pd nanoparticles, after the deposition of a Pt monolayer, the CO stripping peak significantly shifted in the negative direction, and its potential was quite close to that of Pt. These results also strongly suggest that the Pt monolayer shell almost covers the Pd core nanoparticles.

2.3. Activity and Durability for ORR of $Pt/Pd_{4.2}/CB$ and $Pt/Pd_{3.3}/CB$ Electrodes

Figure 9 shows hydrodynamic voltammograms at various rotating speeds and the Koutecky-Levich plots at various potentials for the $Pt/Pd_{4.2}/CB$ and $Pt/Pd_{3.3}/CB$ electrodes in O_2-saturated 0.1 M $HClO_4$ aqueous solution. As shown in Figure 9b,d, in both cases, there was a linear relationship between the reciprocal of square root of rotating speed ($\omega^{-1/2}$) and the reciprocal of measured current density (i^{-1}) irrespective of potential. With a slope of each straight line and the following Equation (1) [4], the number of electrons in the ORR was evaluated.

$$i^{-1} = i_k{}^{-1} + (0.62nFc_{O2}D_{O2}{}^{2/3}v^{-1/6})^{-1}\omega^{-1/2} \tag{1}$$

where i_k is the kinetic current density in mA cm^{-2}, n is the number of electrons in the ORR, F is the Faraday constant, c_{O2} is the dissolved O_2 concentration (1.18×10^{-3} mol L^{-1} [4]), D_{O2} is the diffusion coefficient of O_2 (1.9×10^{-5} cm^2 s^{-1} [4]) and v is the viscosity of the solution (0.0893×10^{-2} cm^2 s^{-1} [4]). In both cases the number of electrons in ORR was evaluated to be *ca.* 4 irrespective of potential, indicating that direct 4-electron reduction reaction to water proceeded on these electrodes, which was the same as the Pt electrode. Moreover, the mechanism of ORR was not influenced by the size of Pd core nanoparticles.

i_k can be evaluated using the following Equation (2) [4].

$$i_k = i_1 i/(i_1 - i) \tag{2}$$

where i_1 is the diffusion-limited current density and can be determined from Figure 9a,c. Using hydrodynamic voltammograms at 1600 rpm, the log i_k—electrode potential (E) plot or Tafel plot for the $Pt/Pd_{4.2}/CB$ and $Pt/Pd_{3.3}/CB$ electrodes was made, as shown in Figure 10. The Tafel slopes at the higher ($E > 0.85$ V) and lower ($E < 0.85$ V) potential regions was -61 and -120 mV dec^{-1} for the $Pt/Pd_{4.2}/CB$ electrode and -63 and -119 mV dec^{-1} for the $Pt/Pd_{3.3}/CB$ electrode. The Tafel slope for polycrystalline Pt electrodes was *ca.* -60 and *ca.* -120 mV dec^{-1} at higher and lower potential regions due to the ORR on a Pt surface covered with oxides and a clean Pt surface, respectively [22,23]. Therefore, the ORR mechanism on both Pt/Pd/CB catalysts is the same as that on Pt, and, in particular, the rate-determining step in the lower potential region was the first one-electron reduction reaction of O_2 molecules adsorbed on the Pt surface [22,23].

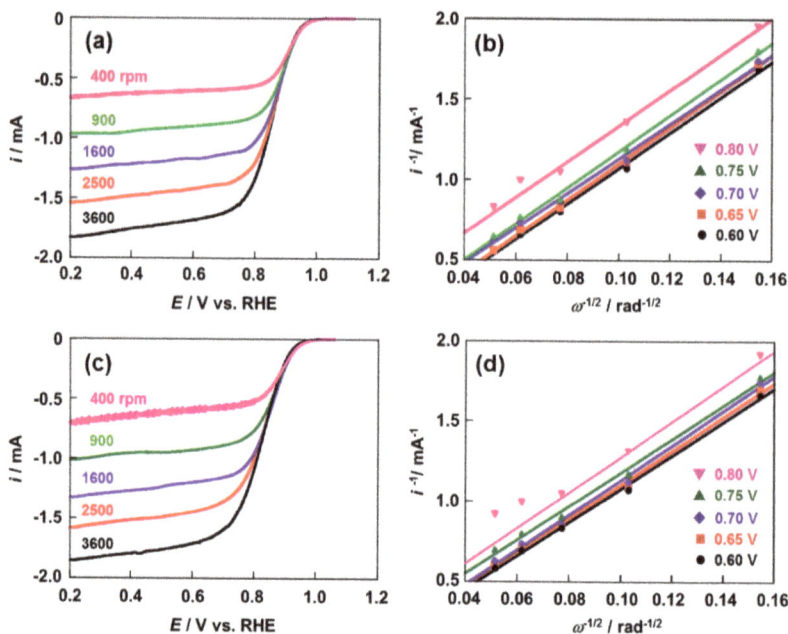

Figure 9. Hydrodynamic voltammograms at various rotating speeds and the Koutecky-Levich plots at various potentials for (**a,b**) Pt/Pd$_{4.2}$/CB and (**c,d**) Pt/Pd$_{3.3}$/CB electrodes in O$_2$-saturated 0.1 M HClO$_4$ aqueous solution. Scan rate: 10 mV s^{-1}.

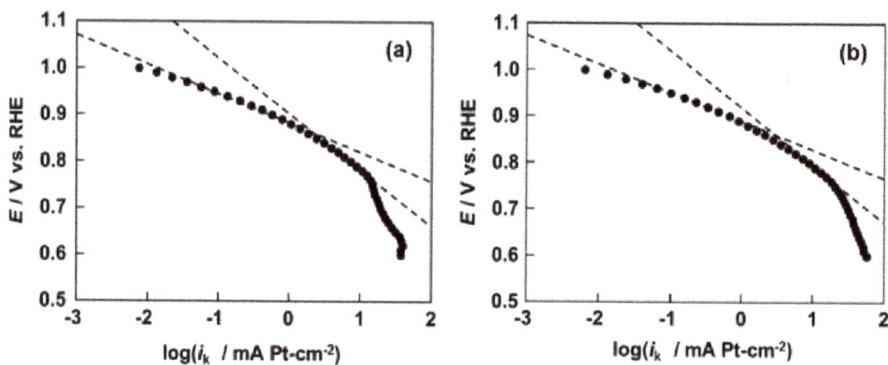

Figure 10. Tafel plots for (**a**) Pt/Pd$_{4.2}$/CB and (**b**) Pt/Pd$_{3.3}$/CB electrodes.

Figure 11 summarizes the ORR activity per electrochemical surface area of Pt (specific activity, SA) and MA, ORR current per unit mass of Pt at 0.90 V for the Pt/Pd$_{4.2}$/CB, Pt/Pd$_{3.3}$/CB and commercial Pt/CB (TEC10E50E, Tanaka Kikinzoku Kogyo, Oshu-city, Iwate, Japan; Pt/CB-TKK) electrodes. For SA, both Pt/Pd/CB electrodes was about 1.5 times as high as the Pt/CB-TKK electrode, and the $Pt/Pd_{3.3}/CB$ electrode whose Pd core size was smaller had higher SA than the $Pt/Pd_{4.2}/CB$ electrode. These results can be ascribed to the compressive strain effect of Pd core nanoparticles, which depends on core size and shape, leading to the compression of Pt-Pt distance of the Pt monolayer shell [24–26]. The strain effect induced d-band shift regulates the adsorption properties of rate-determining intermediates in catalytic processes, so the maximal compression of the Pt-Pt distance must give the highest SA [24–26]. For MA, the $Pt/Pd_{4.2}/CB$ and $Pt/Pd_{3.3}/CB$ electrodes was 5.0 and 5.5 times as high as the Pt/CB-TKK electrode, respectively. Since the specific surface areas of both Pt/Pd/CB electrodes were similar to each other, the increase in MA is ascribable to the increase in SA.

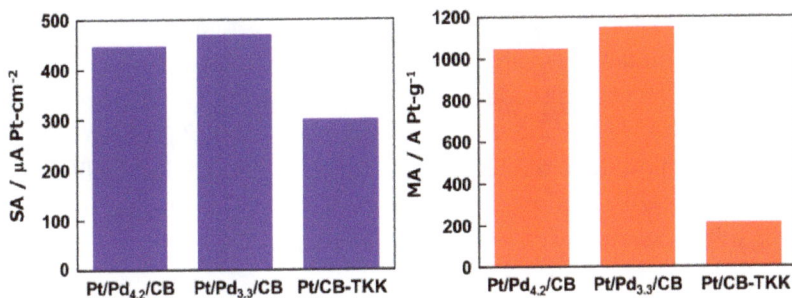

Figure 11. SA and MA for the $Pt/Pd_{4.2}/CB$, $Pt/Pd_{3.3}/CB$ and Pt/CB-TKK electrodes.

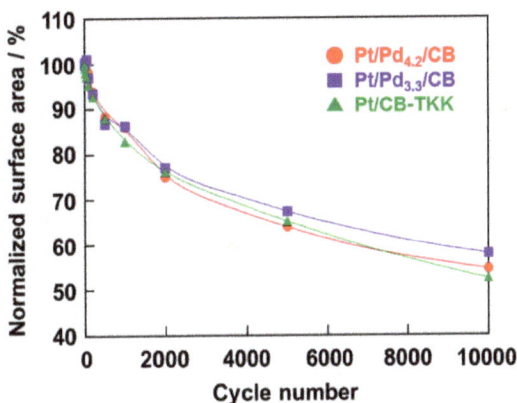

Figure 12. Change in normalized electrochemical surface area of Pt with cycle number for the $Pt/Pd_{4.2}/CB$, $Pt/Pd_{3.3}/CB$ and Pt/CB-TKK electrodes.

Figure 12 shows time courses of the normalized ECSA during durability tests at 60 °C for the $Pt/Pd_{4.2}/CB$, $Pt/Pd_{3.3}/CB$ and Pt/CB-TKK electrodes. The durability of the $Pt/Pd_{4.2}/CB$ and $Pt/Pd_{3.3}/CB$ electrodes was comparable to that of Pt/CB-TKK, suggesting that the surface of Pd core nanoparticles was almost covered by Pt atoms for both Pt/Pd/CB electrodes.

3. Experimental Section

3.1. Preparation and Characterization of Pd/CB

The Pd/CB without stabilizer was prepared as follows; CO was bubbled in acetonitrile solutions containing 0.050–1.0 M palladium acetate at 4 °C for 3–60 min, followed by adding Ketjen Black powder and sonicating for 30 min. Then the Pd/CB powder was separated by suction filtration. The loading of Pd on CB was evaluated to be 28 wt% from the mass of the residue by thermogravimetry.

Crystal structure and valence state of Pd/CB were analyzed with an X-ray diffractometer (Shimadzu, Kyoto, Japan; 50 kV, 30 mA) using CuK$_\alpha$ radiation and X-ray photoelectron spectroscope (8 kV, 30 mA) using MgK$_\alpha$ radiation (1253.6 eV). The mean size and its distribution of Pd nanoparticles in Pd/CB were evaluated with a field-emission transmission electron microscope (FE-TEM; Hitachi, Tokyo, Japan). The size distribution profiles were obtained by measuring more than 300 nanoparticles randomly chosen from TEM images.

3.2. Modification of Pd Core Nanoparticles Loaded on CB with Pt Shell

The Pd/CB (Amount of CB: 3.1 μg) was cast on a glassy carbon (GC) disk electrode (5 mmφ). The Pd/CB-modified GC disk electrode was immersed in a 0.5 M H$_2$SO$_4$ aqueous solution containing 2 mM CuSO$_4$ with a Pt plate counter electrode and a reversible hydrogen electrode (RHE). The disk electrode was polarized at 0.30 V for 10 min to deposit a Cu adlayer on the Pd nanoparticles, followed by being immersed in 5.0 mM K$_2$PtCl$_4$ aqueous solution in Ar atmosphere for 30 min. Consequently, a Pt monolayer shell was formed by galvanic displacement of the Cu adlayer atoms with Pt atoms [9]. Each Pd/CB or Pt/Pd/CB electrode was covered with a thin Nafion film by dropping 20 μL of 0.05 wt% Nafion ethanol solution and then drying.

3.3. Electrochemical Measurements

The ORR activity of the Pt/Pd/CB electrodes was measured by hydrodynamic voltammetry using the rotating disk electrode (RDE) technique. The commercial Pt/CB (TEC10E50E, Tanaka Kikinzoku Kogyo, Japan; Pt/CB-TKK) electrode was used for comparison. The counter and reference electrodes were a Pt plate and an RHE, respectively. The electrolyte solution was 0.1 M HClO$_4$. Cyclic voltammograms were recorded at a scan rate of 20 mV s^{-1} in an Ar atmosphere at 25 °C. The hydrodynamic voltammograms were measured at rotating speeds of 3600, 2500, 1600, 900, and 400 rpm in an O$_2$-saturated 0.1 M HClO$_4$ aqueous solution at 25 °C.

384

The potential of the GC disk was swept at 10 mV s^{-1} from 0.05 to 1.2 V. The SA and MA for ORR were evaluated from the kinetic current at 0.90 V $vs.$ RHE of each hydrodynamic voltammogram at 1600 rpm, respectively. The mass of Pt deposited on the Pd core nanoparticles for each catalyst was evaluated by integrating the electric charge of the amount of Cu-upd in the stripping voltammogram of the Cu adlayer, assuming that all the Cu adlayer atoms were completely displaced by Pt atoms as follows:

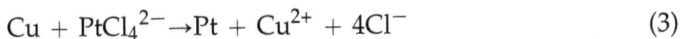

$$Cu + PtCl_4{}^{2-} \rightarrow Pt + Cu^{2+} + 4Cl^- \tag{3}$$

CO stripping voltammograms were recorded at 20 mV s^{-1} in the positive direction from 0.05 V $vs.$ RHE after CO was adsorbed on each electrode at 0.05 V $vs.$ RHE in a CO-saturated 0.1 M HClO$_4$ aqueous solution for 15 min.

The durability of Pt/Pd/CB and Pt/CB electrodes was investigated by repeating square-wave potential cycling between 0.6 V for 3 s and 1.0 V for 3 s in an Ar-saturated 0.1 M HClO$_4$ aqueous solution at 60 °C [18]. The ECSA of Pt was periodically measured during each durability test, and the loss of the ECSA was used as a measure of degradation.

4. Conclusions

Pd core nanoparticles less than 5 nm in size were successfully prepared without any stabilizer by using palladium acetate as a precursor and CO as a reducing agent. The mean size of Pd nanoparticles was controllable by the concentration of palladium acetate and the CO bubbling time, and the former was superior to the latter in the preparation of smaller particles. The CVs of the Pd/CB electrodes had two pairs of redox peaks in the hydrogen region in addition to a reduction peak of palladium oxide at $ca.$ 0.75 V. The deposition of a Pt monolayer shell on CB-loaded Pd core nanoparticles led to the disappearance of the peaks and the appearance of peaks assigned to Pt. The ORR on the Pt/Pd$_{4.2}$/CB and Pt/Pd$_{3.3}$/CB proceeded in 4-electron reduction mechanism like the Pt/CB electrode. The Pt/Pd$_{4.2}$/CB and Pt/Pd$_{3.3}$/CB electrodes was $ca.$ 1.5 times higher in SA than the Pt/CB-TKK electrode, and the Pt/Pd$_{3.3}$/CB electrode had higher SA than the Pt/Pd$_{4.2}$/CB electrode. The increase in SA was ascribed to the compression strain effect of the Pd core nanoparticles to tune a Pt-Pt distance which influenced the adsorption properties of rate-determining intermediates. The MA for the Pt/Pd$_{4.2}$/CB and Pt/Pd$_{3.3}$/CB electrodes was 5.0 and 5.5 times as high as that for the Pt/CB-TKK electrode, respectively. The durability of the Pt/Pd$_{4.2}$/CB and Pt/Pd$_{3.3}$/CB electrodes was comparable to that of Pt/CB-TKK, suggesting that the surface of Pd core nanoparticles was almost covered by Pt atoms for both Pt/Pd/CB electrodes.

Acknowledgments: This work was supported by the New Energy and Industrial Technology Development Organization (NEDO) through the Industrial Technology Research Grant Program (08002049-0).

Author Contributions: Ryotaro Sakai and Taiki Kuwahara performed the synthesis of catalysts, and the evaluation of loading of metals on CB. Masanobu Chiku carried out the preparation of catalysts-modified electrodes and their electrochemical measurements. Eiji Higuchi performed structural characterization by XRD, XPS and FE-TEM. All the authors contributed equally to the data interpretation and discussion. Hiroshi Inoue coordinated and wrote the manuscript.

Conflicts of Interest: The authors declare no conflict of interest.

References

1. Toda, T.; Igarashi, H.; Uchida, H.; Watanabe, M. Enhancement of the electroreduction of oxygen on Pt alloys with Fe, Ni, and Co. *J. Electrochem. Soc.* **1999**, *146*, 3750–3756.

2. Mukerjee, S.; Srinivasan, S.; Soriaga, M.P.; McBreen, J. Role of Structural and electronic properties of Pt and Pt alloys on electrocatalysis of oxygen reduction: An *in situ* XANES and EXAFS investigation. *J. Electrochem. Soc.* **1995**, *142*, 1409–1422.

3. Greeley, J.; Stephens, I.E.L.; Bondarenko, A.S.; Johansson, T.P.; Hansen, H.A.; Jaramillo, T.F.; Rossmeisl, J.; Chorkendorff, I.; Norskov, J.K. Alloys of platinum and early transition metals as oxygen reduction electrocatalysts. *Nat. Chem.* **2009**, *1*, 552–556.

4. Paulus, U.A.; Wokaun, A.; Scherer, G.G.; Schmidt, T.J.; Stamenkovic, V.; Radmilovic, V.; Markovic, N.M.; Ross, P.N. Oxygen reduction on carbon-supported Pt-Ni and Pt-Co alloy catalysts. *J. Phys. Chem. B* **2002**, *106*, 4181–4191.

5. Paffett, M.T.; Berry, J.G.; Gottesfeld, S. Oxygen reduction at $Pt_{0.65}Cr_{0.35}$, $Pt_{0.2}Cr_{0.8}$ and roughened platinum. *J. Electrochem. Soc.* **1998**, *135*, 1431–1436.

6. Smith, M.C.; Gilbert, J.A.; Mawdsley, J.R.; Seifert, S.; Myers, D.J. *In situ* small-angle X-ray scattering observation of Pt catalyst particle growth during potential cycling. *J. Am. Chem. Soc.* **2008**, *130*, 8112–8113.

7. Salgado, J.R.C.; Antolini, E.; Gonzalez, E.R. Structure and activity of carbon-supported Pt-Co electrocatalysts for oxygen reduction. *J. Phys. Chem. B* **2004**, *108*, 17767–17774.

8. Yano, H.; Kataoka, M.; Yamashita, H.; Uchida, H.; Watanabe, M. Oxygen reduction activity of carbon-supported Pt-M (M = V, Ni, Cr, Co, and Fe) alloys prepared by nanocapsule method. *Langmuir* **2007**, *23*, 6438–6445.

9. Zhang, J.; Mo, Y.; Vukmirovic, M.B.; Klie, R.; Sasaki, K.; Adzic, R.R. Platinum monolayer electrocatalysts for O_2 reduction: Pt monolayer on Pd(111) and on carbon-supported Pd nanoparticles. *J. Phys. Chem. B* **2004**, *108*, 10955–10964.

10. Vukmirovic, M.B.; Zhang, J.; Sasaki, K.; Nilekar, A.U.; Uribe, F.; Mavrikakis, M.; Adzic, R.R. Platinum monolayer electrocatalysts for oxygen reduction. *Electrochim. Acta* **2007**, *52*, 2257–2263.

11. Inaba, M.; Ito, H.; Tuji, H.; Wada, T.; Banno, M.; Yamada, H.; Saito, M.; Tasaka, A. Effect of core size on activity and durability of Pt core-shell catalysts for PEFCs. *ECS Trans.* **2010**, *33*, 231–238.

12. Li, Y.; El-Sayed, M.A. The effect of stabilizers on the catalytic activity and stability of Pd colloidal nanoparticles in the suzuki reactions in aqueous solution. *J. Phys. Chem. B* **2001**, *105*, 8938–8943.

13. Shao, M.; Odell, J.; Humbert, M.; Yu, T.; Xia, Y. Electrocatalysis on shape-controlled palladium nanocrystals: Oxygen reduction reaction and formic acid oxidation. *J. Phys. Chem. C* **2013**, *117*, 4172–4180.

14. Vidal-Iglesias, F.J.; Aran-Ais, R.M.; Solla-Gullon, J.; Garnier, E.; Herrero, E.; Aldaz, A.; Feliu, J.M. Shape-dependent electrocatalysis: Formic acid electrooxidation on cubic Pd nanoparticles. *Phys. Chem. Chem. Phys.* **2012**, *14*, 10258–10265.

15. Mazumdar, V.; Sun, S. Oleylamine-mediated synthesis of Pd nanoparticles for catalytic formic acid oxidation. *J. Am. Chem. Soc.* **2009**, *131*, 4588–4589.

16. Cao, K.; Zhu, Q.; Chen, R. Controlled synthesis of Pd/Pt core shell nanoparticles using area-selective atomic layer deposition. *Sci. Rep.* **2015**, *5*.

17. Higuchi, E.; Taguchi, A.; Hayashi, K.; Inoue, H. Electrocatalytic activity for oxygen reduction reaction of Pt nanoparticle catalysts with narrow size distribution prepared from $[Pt_3(CO)_3(\mu\text{-}CO)_3]_n^{2-}$ ($n = 3$–8) complexes. *J. Electroanal. Chem.* **2011**, *663*, 84–89.

18. Higuchi, E.; Hayashi, K.; Chiku, M.; Inoue, H. Simple preparation of Au nanoparticles and their application to Au core/Pt shell catalysts for oxygen reduction reaction. *Electrocatalysis* **2012**, *3*, 274–283.

19. Erdogan, H.; Metin, O.; Ozkar, S. *In situ*-generated PVP-stabilized palladium(0) nanocluster catalyst in hydrogen generation from the methanolysis of ammonia-borane. *Phys. Chem. Chem. Phys.* **2009**, *11*, 10519–10525.

20. Zhang, J.; Qiu, C.; Ma, H.; Liu, X. Facile fabrication and unexpected electrocatalytic activity of palladium thin films with hierarchical architectures. *J. Phys. Chem. C* **2008**, *112*, 13970–13975.

21. Gabrielli, C.; Grand, P.P.; Lasia, A.; Perrot, H. Investigation of hydrogen adsorption and absorption in palladium thin films: II. Cyclic voltammetry. *J. Electrochem. Soc.* **2004**, *151*, A1937–A1942.

22. Damjanovic, A.; Genshaw, M.A. Dependence of the kinetics of O_2 dissolution at Pt on the conditions for adsorption of reaction intermediates. *Electrochim. Acta* **1970**, *15*, 1281–1283.

23. Markovic, N.M.; Ross, P.N. Electrocatalysis as well-defined surfaces: Kinetics of oxygen reduction and hydrogen oxidation/evolution on Pt(*hkl*) electrodes; mechanism of methanol electro-oxidation. In *Interfacial Electrochemistry, Theory, Experiment, and Applications*; Wieckowski, A., Ed.; Marcel Dekker: New York, NY, USA, 1999; pp. 821–841.

24. Hammer, B.; Nørskov, J.K. Theoretical Surface Science and Catalysis—Calculations and Concepts. *Adv. Catal.* **2000**, *45*, 71–129.

25. Wang, J.X.; Inada, H.; Wu, L.; Zhu, Y.; Choi, Y.; Liu, P.; Zhou, W.–P.; Adzic, R.R. Oxygen reduction on well-defined core-shell nanocatalysts: Particle size, facet, and Pt shell thickness effects. *J. Am. Chem. Soc.* **2009**, *131*, 17298–17302.

26. Wang, X.; Orikasa, Y.; Takesue, Y.; Inoue, H.; Nakamura, M.; Minato, T.; Hoshi, N.; Uchimoto, Y. Quantitating the lattice strain dependence of monolayer Pt shell activity toward oxygen reduction. *J. Am. Chem. Soc.* **2013**, *135*, 5938–5941.

Electrochemical Oxidation of the Carbon Support to Synthesize Pt(Cu) and Pt-Ru(Cu) Core-Shell Electrocatalysts for Low-Temperature Fuel Cells

Griselda Caballero-Manrique, Enric Brillas, Francesc Centellas,
José Antonio Garrido, Rosa María Rodríguez and Pere-Lluís Cabot

Abstract: The synthesis of core-shell Pt(Cu) and Pt-Ru(Cu) electrocatalysts allows for a reduction in the amount of precious metal and, as was previously shown, a better CO oxidation performance can be achieved when compared to the nanoparticulated Pt and Pt-Ru ones. In this paper, the carbon black used as the support was previously submitted to electrochemical oxidation and characterized by XPS. The new catalysts thus prepared were characterized by HRTEM, FFT, EDX, and electrochemical techniques. Cu nanoparticles were generated by electrodeposition and were further transformed into Pt(Cu) and Pt-Ru(Cu) core-shell nanoparticles by successive galvanic exchange with Pt and spontaneous deposition of Ru species, the smallest ones being 3.3 nm in mean size. The onset potential for CO oxidation was as good as that obtained for the untreated carbon, with CO stripping peak potentials about 0.1 and 0.2 V more negative than those corresponding to Pt/C and Ru-decorated Pt/C, respectively. Carbon oxidation yielded an additional improvement in the catalyst performance, because the ECSA values for hydrogen adsorption/desorption were much higher than those obtained for the non-oxidized carbon. This suggested a higher accessibility of the Pt sites in spite of having the same nanoparticle structure and mean size.

Reprinted from *Catalysts*. Cite as: Caballero-Manrique, G.; Brillas, E.; Centellas, F.; Garrido, J.A.; Rodríguez, R.M.; Cabot, P.-L. Electrochemical Oxidation of the Carbon Support to Synthesize Pt(Cu) and Pt-Ru(Cu) Core-Shell Electrocatalysts for Low-Temperature Fuel Cells. *Catalysts* **2015**, *5*, 815–837.

1. Introduction

The Proton Exchange Membrane Fuel Cells (PEMFCs) are considered good environmentally friendly alternatives to the use of fossil fuel engines as power generation systems for transport applications. They have better energy efficiency, lower operation temperature, and much lower emission of pollutants [1–4]. One of the problems appears when using hydrogen obtained from reforming, because it contains CO, which is strongly adsorbed on Pt. This produces the metal poisoning and decreases the anode performance. On the other hand, Direct Methanol Fuel

Cells (DMFCs) are envisaged for low-weight portable applications such as laptops, cellular phones, sensors, and medical devices, in which the methanol fuel can be easily managed and recharged [4–8]. In this case, the anodic oxidation of methanol produces CO-type intermediates, which also lead to the Pt metal poisoning. In order to solve this problem, Pt-containing binary and ternary alloys have alternatively been studied [3,5,9–15]. The amount of Pt used could also be decreased in this form, because it is expensive and has a limited abundance in the Earth [3,7,8,11,16–18]. In this way, Pt-Ru and Pt alloy nanoparticles containing Au, Ni, Cu, Co, Pd, and/or other transition metals have been tested on different carbon substrates as anodic catalysts [1,2,6,9–11,13,16,17,19–29].

Many different allotropic forms of carbon have been used as substrates to synthesize Pt-based catalysts. Pt(IV) and Ru(III) species or their carbonyl complexes can be reduced in ethylenglycol or formate solutions [11,13,30], also using in some cases oil in water microemulsions [31], microwave radiation [15], or galvanic exchange [32–34]. However, carbon blacks and active carbons have been the mostly employed supports for low-temperature fuel cells due to their unique characteristics of high surface area, electric conductivity, porosity, stability, and low cost [1,3,10,16,22,30,35,36]. The morphology and size distribution of carbon black particles depend on the raw material and on the thermal decomposition process utilized in the synthesis procedure. Vulcan carbons XC-72 and XC-72R [37] are carbon blacks obtained from the pyrolysis of natural gas or oil fractions. They can be considered rather graphitic amorphous forms of carbon with spherical shape about 50 nm in diameter that can be aggregated in spherules of about 250 nm in size [30,36]. Vulcan XC-72R has suitable surface and micropore areas of 218 and 65.2 m^2g^{-1}, respectively, and pore, mesopore, and micropore volumes of 0.41, 0.3,7 and 0.036 cm^3g^{-1}, respectively [19,37,38].

It has been noted that carbon blacks have a large number of diverse structural defects that affect their reproducibility as substrates [1,8,19,39]. For this reason, these carbons are normally functionalized by means of surface oxidation treatments to create oxygenated organic groups that can serve as nucleation points for the metallic precursors. The oxygenated groups help to diminish the carbon hydrophobicity, thereby favoring the accessibility of the aqueous metallic precursors. On the other hand, the less acidic groups increase the interaction between the metal precursor and the carbon support, thus avoiding the agglomeration tendency of the metal on the carbon [1,40]. The functionalization of the carbon surface can be achieved by treatment with strong oxidizing acids such as nitric, sulfuric, phosphoric, sulphonic, or their mixtures, or by means of sodium hydroxide, ammonia, or hydrogen peroxide [18,27–29,35,36,38,40–46]. Oxygen or oxygen plus nitrogen mixtures, thermal oxidation [18,38], and electrochemical oxidation at constant current or potential, potential pulses, or potential cycling have also been applied [28,34,42,45,47].

All these techniques were able to introduce oxygenated surface compounds via the consecutive formation of hydroxyl, carbonyl, and carboxyl groups according to the following reaction sequence [47]:

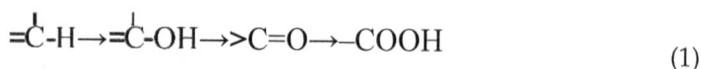

$$=\overset{|}{C}-H \rightarrow =\overset{|}{C}-OH \rightarrow >C=O \rightarrow -COOH \tag{1}$$

Hydroquinone (HQ) $C_6H_4(OH)_2$ groups, which can be oxidized to quinone (Q) $C_6H_4O_2$ groups, can also improve the electrocatalytic properties and the stability of the catalysts in the operation conditions of the fuel cells [18,19,35,36,38,42–45]:

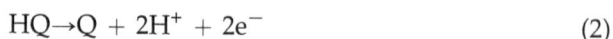

$$HQ \rightarrow Q + 2H^+ + 2e^- \tag{2}$$

However, extreme oxidizing conditions under high temperature, humidity, and low pHs can negatively affect the performance and durability of the PEM anodes because carbon can be oxidized to CO and CO_2 according to the following reactions [19]:

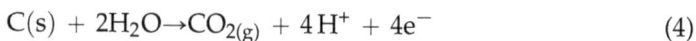

$$C(s) + H_2O \rightarrow CO_{(g)} + 2H^+ + 2e^- \tag{3}$$

$$C(s) + 2H_2O \rightarrow CO_{2(g)} + 4H^+ + 4e^- \tag{4}$$

To the best of the authors' knowledge, studies about the effect of carbon oxidation as a substrate for copper electrodeposition are scarcely found in the literature. Li et al. [29] reported that when the amount of carbonyl groups on carbon fibers was increased by thermal treatment, the amount of Cu nuclei on carbon during the Cu electrodeposition also increased. This could be also a way of increasing the dispersion of Cu nuclei on carbon as smaller nanoparticles.

In previous work by these authors [48], core-shell carbon-supported Pt(Cu) and Pt-Ru(Cu) nanoparticles were synthesized in three steps: 1) Cu electrodeposition on Vulcan carbon XC-72R; 2) Pt deposition on Cu by galvanic exchange; and 3) spontaneous deposition of Ru species on Pt. It was shown that this way allowed us to significantly reduce the Pt content of the catalyst together with increasing the CO tolerance. Thus, the CO stripping peak potentials were about 0.1 and 0.2 V more negative than those corresponding to the Pt/C and the Ru-decorated Pt/C catalysts, respectively. In addition, the efficiency of methanol oxidation per unit mass of Pt was much higher. In this paper, the effect of the electrochemical carbon oxidation on the synthesis of the core-shell Pt(Cu) and Pt-Ru(Cu) nanoparticles has been explored in order to try to effect a further improvement in the catalyst performance. Carbon Vulcan XC-72R was electrochemically oxidized under different conditions to increase the amount of oxygenated carbon groups and then the core-shell catalysts were deposited on it, using the best conditions reported in our previous work. The oxidized carbons and the catalysts thus prepared were characterized by means of structural and electrochemical techniques. The hydrogen adsorption/desorption behavior and the CO oxidation performance of these new catalysts were compared to those previously reported for the untreated carbon support.

2. Results and Discussion

2.1. XPS Analyses and Electrochemical Testing of Carbon Oxidation

Cyclic voltammograms of Vulcan carbon XC72R in 0.5 M H_2SO_4 were recorded between the initial potential of 0.0 V vs. RHE and different anodic limits up to 2.2 V to select the potentials to check its activation performance. Figure 1 shows that the anodic current in the anodic sweep significantly increased from about 0.9 V and passed through a maximum at about 2.0 V (curve a). According to this behavior, potentials of 1.6, 1.8, 2.0, and 2.2 V were tentatively selected as characteristic potentials for the anodic treatment of carbon.

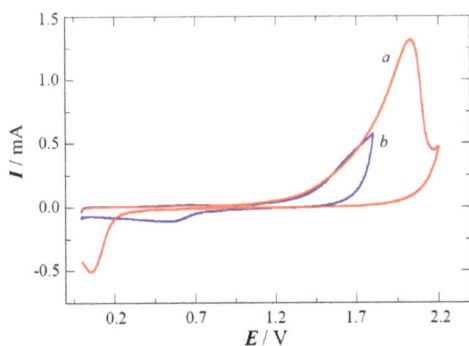

Figure 1. Cyclic voltammograms corresponding to the oxidation of Vulcan Carbon XC72R at 10 mV s^{-1} in 0.5 M H_2SO_4. The initial potential was 0.0 V and the reversal potential was (a) 2.2 V and (b) 1.8 V.

Figure 1 also highlights that the cathodic profile depended on the reversal potential. For the anodic limit of 1.8 V (curve b), the current in the cathodic sweep started to grow from about 0.7 V, passing through a rather flat cathodic peak at about 0.4–0.6 V. When the reversal potential was 2.2 V, the cathodic peak was close to 0.1 V. These curves then showed that the carbon oxidation was a function of the anodic limit. Note that in both cases, the charge of the anodic sweep largely exceeded the cathodic charge invested in the reduction of the species generated, thereby indicating the irreversibility of the oxidation process. This irreversible nature was further confirmed when performing consecutive cyclic voltammograms, in which a significant current decrease was apparent cycle by cycle (not shown here). It has been reported in the literature that the cathodic peak currents in the range

0.4–0.6 V are due to the reduction of the quinone phenolic groups of carbon to hydroquinone [18,19,35,36,38,42–45]. When the anodic and the cathodic limits were limited to the range 0.4–0.7 V, the transformation between quinone and hydroquinone groups appeared to be reversible. However, when the anodic potential exceeded 0.7 V in the positive direction, an increasing degree of carbon oxidation took place, as expected, along with the production of aldehyde and carboxylic groups and, finally, CO_2 evolution. For an anodic limit of 2.2 V, practically no cathodic current appeared between 0.4 and 0.6 V, suggesting the irreversible oxidation of most of the phenolic quinone groups to higher oxidation states.

XPS analyses of the oxidized carbons were performed in order to more precisely define the nature of the processes taking place at different potentials. The general spectra of the different carbons, including the non-oxidized one, showed only the presence of carbon and oxygen, although very residual amounts of S and Cl were also identified for the non-oxidized carbon. The main difference between the latter and the oxidized carbons was the oxygen content. The corresponding atomic ratio O:C was 0.62, 22.8, 22.0, and 22.3 atom % for the non-oxidized carbon and for carbons oxidized for 300s at 1.8, 2.0, and 2.2 V, respectively. The C1s and the O1s binding energy region of the XPS spectra are depicted in Figure 2a,b, respectively.

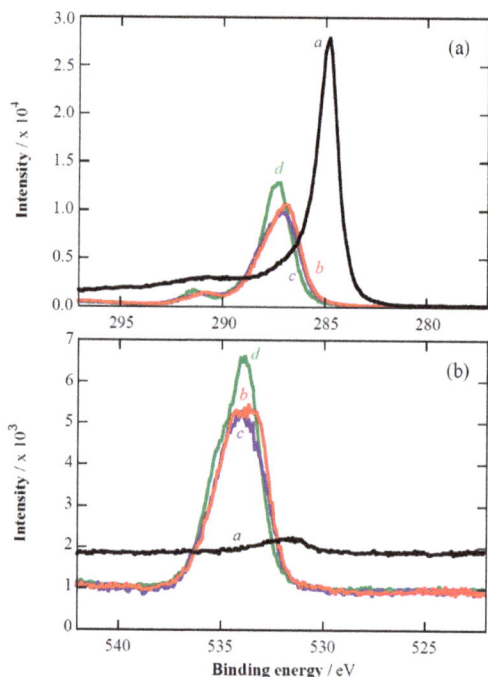

Figure 2. XPS spectra in the (**a**) C1s and (**b**) O1s binding energy region for the non-oxidized carbon (*a*) and the carbons oxidized at 1.8 V (*b*), 2.0 V (*c*), and 2.2 V(*d*).

In the C1s binding energy region (Figure 2a), it is clear that the surface carbon appears mainly in the form of the C–C state (284.5 eV) [19,49,50] for the non-oxidized carbon (curve a), whereas for the oxidized carbons, the binding energies are consistent with the states C–O (286.0 eV), C=O (287.8 eV) and O–C=O (290.2 eV) [19,51,52]. All the surface carbon appears to be oxidized. The chemical states of O1s (Figure 2b) are consistent with C–OH (531.8 eV) for the non-oxidized carbon (curve a), whereas for the oxidized carbons, O1s is mainly in the states O–C–O (532.2 eV) and O–C=O (534.3 eV) [19,51,52]. Note that when the oxidation potential of the carbon is increased, there is a shift toward higher binding energies, thus indicating that the final oxidation form as CO_2 is approached. Moreover, the band located in the range 290–292 eV is due to the π electrons of the aromatic rings, in agreement with the presence of quinonic and hydroquinonic structures [53].

Further confirmation of the above behavior was found from the chronoamperograms depicted in Figure 3a, where the steady current depended on the applied potential. For potentials up to 2.0 V, small steady currents were obtained (curves a–c), whereas for 2.2 V (curve d) the current was one order of magnitude greater. Figure 3b shows that the profile of the cathodic sweep after the constant oxidation potential for 300 s of Figure 3a was consistent with the cyclic voltammograms shown in Figure 1. Reduction of phenolic quinone groups was apparent at about 0.4–0.6 V when the applied potentials were 1.6 and 1.8 V. The absence of significant reduction peaks at 0.4–0.6 V together with the formation of new cathodic peaks at more negative potentials, curves c and d in Figure 3b, clearly indicate that higher oxidation states of carbon were produced when the carbon was previously oxidized at 2.0 and 2.2 V, in agreement with the XPS analyses. The reduction of higher oxidized states was also apparent at constant potential oxidation of 1.8 V, as indicated by the big reduction peak at about 0.2 V. Oxidation at 2.0 and 2.2 V clearly led to deeper carbon oxidation including the formation of CO_2, species that were not reduced during the cathodic sweep.

The different carbon activation treatments described in the experimental part are associated to Figures 1 and 3: Figure 3a corresponds to treatment (i); Figure 3a followed by 3b, to treatment (iii); and 10 consecutive cycles following the first one shown in Figure 1, to treatment (ii). Figure 4 depicts the cyclic voltammograms of the oxidized carbons in the potential region from 0.0 to 1.0 V, which is the region of interest when analyzing the performance of the catalysts prepared in this paper. The main feature shown in the curves of Figure 4 is the Q/HQ couple (oxidation and reduction in the ranges 0.5–0.7 and 0.4–0.6 V, respectively), which is not apparent in curve a for the non-oxidized carbon. The symmetry of the peaks and charge of the anodic and the cathodic profiles corroborates the reversibility of the couple from 0.0 to 1.0 V. The cycles were also repetitive, thus proving the stability of the resulting carbon in these conditions. As can be seen in Figure 4a, the anodic and cathodic charges of the cyclic voltammograms increased when carbon was previously oxidized following the treatment (i) (curves b–d). The corresponding currents grew when changing the treatment potential from 1.6 to 2.0 V (curves b and c, respectively). However, the current decreased when the applied potential was 2.2 V (curve d), probably due to the more intense carbon degradation favoring its loss at this potential. The anodic and cathodic currents after treatment (iii) were similar to those obtained after treatment (i). Qualitatively similar features were found for treatment (ii) (Figure 4b), where the currents increased with the anodic limit of the potential cycling (curves b and c). In this case, the currents for an anodic limit of 2.2 V were higher than those obtained for treatment (i) at the same potential (curve d in Figure 4a). This can be explained assuming that the application of 2.2 V for 300 s produced higher carbon oxidation than potentiodynamic cycling 10 times up to the same potential. The formation of a higher amount of superior oxidation states of carbon, even with CO_2 evolution, involving a carbon loss and a decrease in the content of quinone groups, would explain the smaller currents of curve d in Figure 4a with respect to curve c in Figure 4b. Note that the highest peak currents reached in the cyclic voltammograms after all the carbon oxidation treatments were about 65 µA.

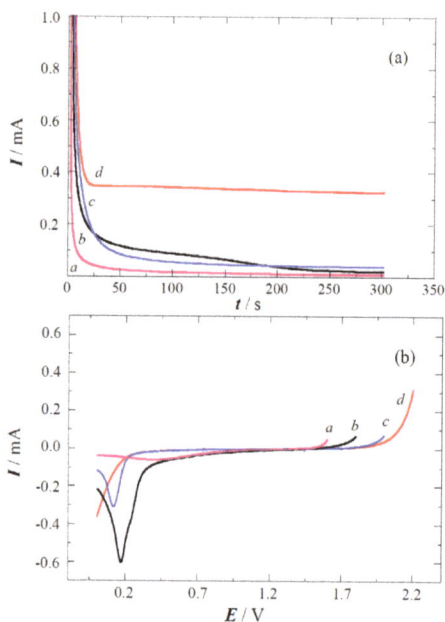

Figure 3. (a) Chronoamperograms obtained for the oxidation of Vulcan Carbon XC72R in 0.5 M H_2SO_4 at potentials: (a) 1.6 V, (b) 1.8 V, (c) 2.0 V, and (d) 2.2 V; (b) Cathodic sweep voltammograms at 10 mV s^{-1} up to 0.0 V after the potentiostatic oxidation of carbon for 300 s shown in (a), the initial potential being: (a) 1.6 V, (b) 1.8 V, (c) 2.0 V, and (d) 2.2 V.

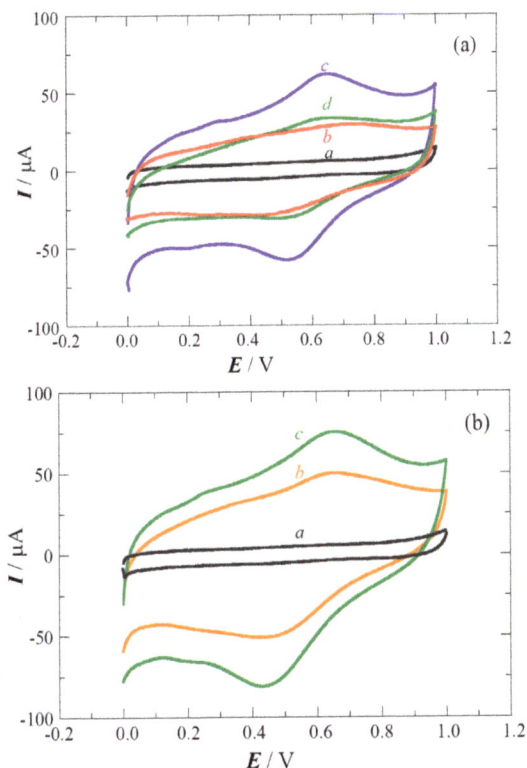

Figure 4. Cyclic voltammograms of carbon XC72R in deaerated 0.5 M H_2SO_4 at 20 mV s^{-1} after the different oxidation treatments. (**a**) Treatment (i), 300 s at (*b*) 1.6 V, (*c*) 2.0 V, and (*d*) 2.2 V; (**b**) Treatment (ii), anodic limit of (*b*) 1.8 V and (*c*) 2.2 V. Curves *a* in both graphics are the cyclic voltammograms of XC72R without previous oxidation treatment.

2.2. Copper Electrodeposition on the Oxidized Carbon

After the different activation treatments of carbon, 40 mC of Cu were electrodeposited at –0.1 V, as explained in the experimental part, and reoxidized in the same electrolyte to determine the copper electrodeposition efficiency. The corresponding voltammograms thus obtained are presented in Figure 5, where it can be observed that the copper oxidation profiles depended on the previous carbon oxidation. The Cu oxidation voltammograms on treated carbons presented anodic peaks that are wider and have much smaller peak currents than the reference curve *a* for the non-oxidized carbon. Treatments (i) (curves *b* and *c*) and (iii) (curves *d* and *e*) led to similar Cu oxidation profiles, with a peak and a shoulder at about 0.50 and 0.65 V, respectively. For treatment (ii) (curves *f* and *g*), the voltammograms were even more depressed, with still significant anodic currents at 1.0 V. This behavior was more pronounced at higher anodic limits and the treatment (ii) was not then found to be suitable for further examination and testing.

Peaks with an anodic maximum and a shoulder were also found during the reoxidation of electrodeposited copper on non-treated carbon for scan rates over 20 mV s^{-1} [48]. They were assigned to the formation of Cu^{2+} complexes and to the oxidation of the Cu(I) species generated by a disproportionation reaction, respectively. This double peak structure is not so apparent in curve a of Figure 5 because of the smaller scan rate. However, both, peak and shoulder, which were the main peak at about 0.50 V and the noticeable peak at about 0.65 V, were observed in the voltammograms of copper oxidation on the previously oxidized carbons, which could then be tentatively related to the same species as the untreated one. The shift of the anodic peaks and shoulder in the anodic direction could be explained considering that the oxidized carbons have a more open structure, with the possibility to nucleate copper in the inner part of the carbon spherules with a higher bonding energy because of the new oxygen-containing functional groups. The oxidation of the Cu nuclei could then be more difficult and demand the application of higher potentials. In fact, it is reasonable to assume that the higher currents found after carbon oxidation in the cyclic voltammograms of Figure 4 were in part faradaic (Q/HQ couple) and also capacitive (expansion of the carbon structure). This expansion of the carbon structure would justify the formation of less accessible Cu nuclei, which, together with a higher bonding energy, would demand higher oxidation potentials for Cu oxidation.

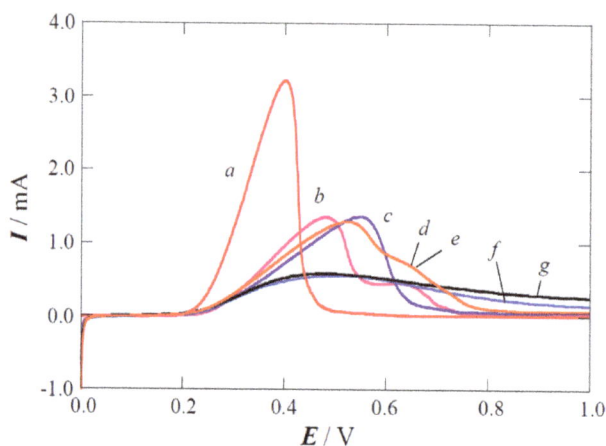

Figure 5. Cu oxidation voltammograms at 10 mV s^{-1} after 40 mC electrodeposition at −0.1 V in the same electrolyte (1.0 mM CuSO$_4$ + 0.1 M Na$_2$SO$_4$ + 0.01 M H$_2$SO$_4$). Experiments performed with carbon submitted to treatment (i): (b) 1.8 V, 300 s and (c) 2.2 V, 300 s; treatment (iii): (d) 1.8 V, 300 s and (e) 2.2 V, 300 s; and treatment (ii): cycling up to (f) 1.8 V and (g) cycling up to 2.2 V. Curve a is the reference one, obtained without previous oxidation of the carbon.

The anodic charges of the different voltammograms were measured to determine the current efficiencies of copper electrodeposition. In contrast to the untreated carbon, in which they were over 99% [48], the current efficiencies for oxidized carbons were somewhat smaller. For treatment (i) at potentials in the range 1.6–2.0 V for 300 s, the efficiency was about 86%. This can be explained by a small contribution of the reduction of oxidized carbon groups previously generated during the anodic treatment of carbon, which would take place in parallel with the Cu electrodeposition. This is in agreement with other electrodeposition experiments using carbon oxidized for shorter times. Thus, in the carbon oxidation at 1.6 and 1.8 V for 100s, the electrodeposition efficiency was about 95%.

The TEM images corresponding to the copper electrodeposition on the carbon oxidized following the treatment (i) at 1.6 V for 200 s and 2.2 V for 300 s, are depicted in Figure 6, where it is highlighted that Cu nuclei were obtained in the nanoparticle size. Figure 6b corresponds to 1.6 V, showing a HRTEM image with the FFT analyses of the squared marked area in the same figure. From the FFT analyses, the interplanar space d obtained was 0.2065 nm, which can be assigned to the planes Cu(100) ($d = 0.2088$ nm) [54], with a relative error of 1.1%. The size distributions of the nanoparticles are also shown in the insets, with respective mean sizes of 6.6 ± 3.1 and 4.4 ± 1.3 nm for 1.6 and 2.2 V, respectively. These nanoparticle sizes were comparable but somewhat higher than the mean value of 3.9 nm obtained in our previous work under the same Cu electrodeposition conditions, except for the carbon oxidation treatment [48].

In order to examine how carbon oxidation can affect the size of the nanoparticles, one can suppose that the new oxygen-containing functional groups produced by carbon oxidation acted as additional nucleation centers for Cu electrodeposition. In this case, the electrodeposition of a given amount of Cu on the oxidized carbon would give a higher number of Cu nanoparticles, which should have a smaller size when compared to the non-oxidized one. As long as the nanoparticle sizes were not smaller for the oxidized carbons, one can infer that carbon oxidation did not lead to an increased number of nucleation centers for Cu electrodeposition. In the present case, it seems that the new oxygen-containing functional groups, behaving as nucleation centers, even yielded somewhat bigger nanoparticles. This interesting point undoubtedly merits more attention but is outside the scope of the present paper.

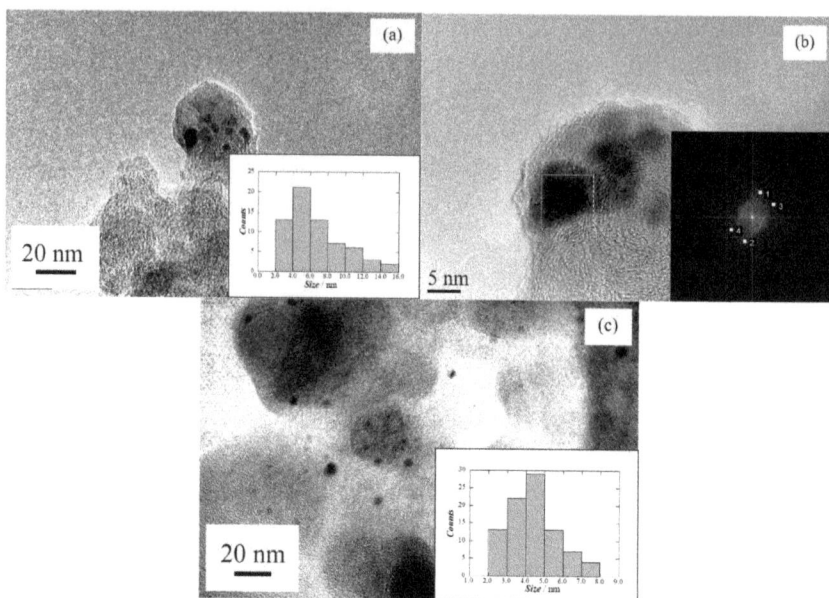

Figure 6. TEM micrographs of electrodeposited Cu on XC72R previously oxidized following treatment (i) at (**a**) and (**b**) 1.6 V, 200 s; (**c**) 2.2 V, 300 s. The size distributions are shown in the inset panels. Picture (**b**) is a HRTEM image from the sample shown in (**a**).

2.3. Performance of the Pt(Cu) and Pt-Ru(Cu) Catalysts

The prepared carbon-supported Pt(Cu) and Pt-Ru(Cu) catalysts were tested by cyclic voltammetry in 0.5 M H$_2$SO$_4$. Some examples are shown in Figure 7, where curves *a* and *c* correspond to Pt(Cu) with previous carbon oxidation according to treatment (i) (1.6 V for 200 s) and without oxidation treatment, respectively. Note that the surface Cu, if present, should be oxidized from a potential of about 0.2 V [48]. Since there is no evidence about it in the quasistationary curves shown in this figure, one can conclude that the cyclic voltammograms resulted only from the surface Pt, which is consistent with the expected core-shell structure.

It can also be observed that the currents were much higher with previous carbon oxidation. However, it is apparent that there is an important contribution of the capacitive charge effect due to the increase in area by the carbon oxidation together with the Q/HQ couple in the region of 0.4–0.7 V (see Figure 4). The respective mean charges of the hydrogen adsorption/desorption regions led to ECSA values of 1.58 × 10^3 and 0.66 × 10^3 m^2 mol$_{Cu}^{-1}$ (see Table 1). The much higher ECSA value for previously oxidized carbon is indicative of a higher ability for the hydrogen adsorption/desorption process. This effect was general because in all cases, as shown in Table 1, carbon oxidation always gave higher ECSA values for hydrogen adsorption/desorption. In any case, the highest ones were obtained for treatment (i) at 1.6 V. Note that similar profiles can be observed in the cyclic voltammograms *a* and *c* of Figure 7 for the oxidized and the non-oxidized carbon, respectively, thus suggesting that carbon oxidation leads to a higher efficiency in the use of Pt and not to a different Pt structure on the surface of the nanoparticles. This is not surprising because carbon appears to be expanded with carbon oxidation and, therefore, protons could easily reach the Pt sites to be reduced.

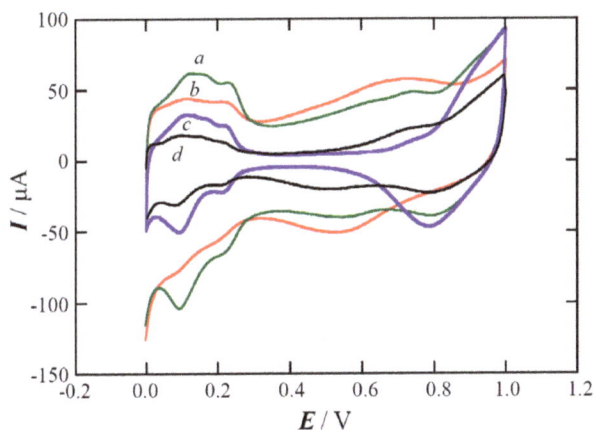

Figure 7. Cyclic voltammograms in 0.5 M H_2SO_4 at 20 mV s^{-1} for Pt(Cu)/C (curves *a* and *c*) and Pt-Ru(Cu)/C (curves *b* and *d*). Curves *a* and *b* correspond to the carbon submitted to treatment (i) at 1.6 V for 200 s, whereas curves *c* and *d* were obtained for carbon without previous oxidation.

The CO stripping curves corresponding to the same Pt(Cu) specimens as in Figure 6 are depicted in curves *a* and *c* of Figure 8, also for the carbon oxidized according to treatment (i) at 1.6 V for 200 s and for the non-oxidized carbon, respectively. Note that there was a capacitive shift of curve *a* toward higher currents, but the onset potential for CO oxidation was comparable to that obtained for the non-oxidized carbon. Table 1 highlights that the ECSA values for CO stripping were in general comparable to those obtained for the non-oxidized carbon. However, for carbon oxidation, they were also similar to those found for hydrogen adsorption/desorption, which is not the case for the non-oxidized carbon. The highest value was again obtained for 1.6 V (1.77×10^3 m^2 mol$_{Cu}^{-1}$), approximately equal to that measured for the non-oxidized carbon (1.79×10^3 m^2 mol$_{Cu}^{-1}$). Increasing the potential and time for carbon oxidation generally caused a decrease in the ECSA values. Moreover, no further improvement was found when changing treatment (i) by treatment (iii).

Table 1. Electrochemical active surface areas for hydrogen adsorption/desorption ($ECSA_{Hads/des}$) and for CO stripping ($ECSA_{CO}$), determined from the cyclic voltammograms in 0.5 M H_2SO_4 for the different catalysts and carbon oxidation treatments (i) and (iii). Results relative to the non-oxidized carbon have been taken from Ref. [48].

Carbon treatment	Catalyst	$ECSA_{Hads/des}/10^3$ m^2 mol$_{Cu}^{-1}$	$ECSA_{CO}/10^3$ m^2 mol$_{Cu}^{-1}$
no oxidation		0.66	1.79
(i) 1.6 V, 100 s		1.56	1.55
(i) 1.6 V, 200 s		1.58	1.77
(iii) 1.6 V, 200 s		1.33	1.47
(i) 1.8 V, 300 s	Pt(Cu)	1.27	1.50
(i) 2.0 V, 300 s		1.19	1.64
(iii) 2.0 V, 300 s		1.11	1.04
(i) 2.2 V, 300 s		1.05	1.16
(iii) 2.2 V, 300 s		1.05	1.18
no oxidation		0.24	1.74
(i) 1.6 V, 200 s	Pt-Ru(Cu)	0.62	1.73
(i) 2.2 V, 300 s		0.57	1.32

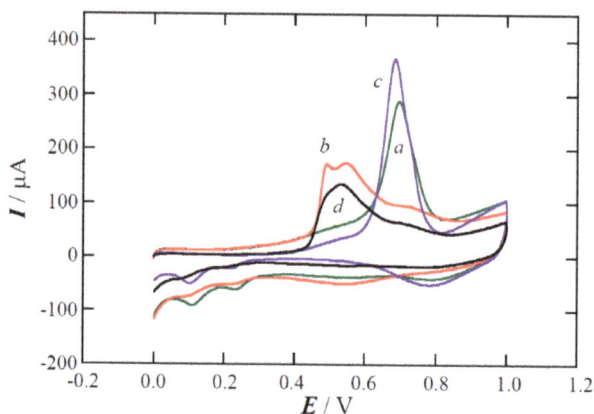

Figure 8. CO stripping voltammograms in 0.5 M H_2SO_4 at 20 mV s^{-1} for Pt(Cu)/C (curves *a* and *c*) and Pt-Ru(Cu)/C (curves *b* and *d*). Curves *a* and *b* correspond to the carbon submitted to treatment (i) at 1.6 V for 200 s, whereas curves *c* and *d* were recorded for carbon without previous oxidation.

The ECSA values for hydrogen adsorption/desorption and for CO oxidation were also measured after the spontaneous deposition of the Ru species. Cyclic voltammograms corresponding to hydrogen adsorption/desorption for the carbon oxidized according to treatment (i) at 1.6 V for 200 s and for the non-oxidized carbon, respectively, are shown in Figure 7. Curves *b* and *d* for Pt-Ru(Cu) presented a significant drop of the hydrogen adsorption/desorption currents with respect to curves *a* and *c* of Pt(Cu), because Ru species covering Pt sites were not suitable for hydrogen adsorption/desorption (see the ECSA values given in Table 1) [48,55]. The

CO stripping curves for Pt-Ru(Cu) are depicted in Figure 8, where curves b and d correspond to the carbon oxidized from the treatment (i) at 1.6 V for 200 s and to the non-oxidized carbon, respectively. The higher currents for the oxidized carbon can again be interpreted as due to the capacitive effect discussed above. However, no further improvement in the onset potential or in the ECSA value for CO oxidation (see Table 1) was found when compared to the non-oxidized carbon.

Micrographs of the Pt(Cu) and Pt-Ru(Cu) catalysts prepared on the carbon oxidized following treatment (i) at 1.6 V for 200 s are shown in Figure 9. The size distributions of the Pt(Cu) and the Pt-Ru(Cu) nanoparticles are given in the insets of Figure 9a,c, with mean particle sizes of 4.2 ± 1.3 and 3.3 ± 1.0 nm, respectively. Figure 9b depicts the FFT analyses of these Pt(Cu) nanoparticles, which gave interplanar spaces d of 0.2154 nm, which can be assigned to Pt(111) (d = 0.2265 nm) [54,55], with a relative error less than 5%. This interplanar space of Pt can be explained by the effect of the remaining Cu core, which has a smaller interplanar distance and can condition the structure of the Pt shell, in agreement with the expected core-shell structure.

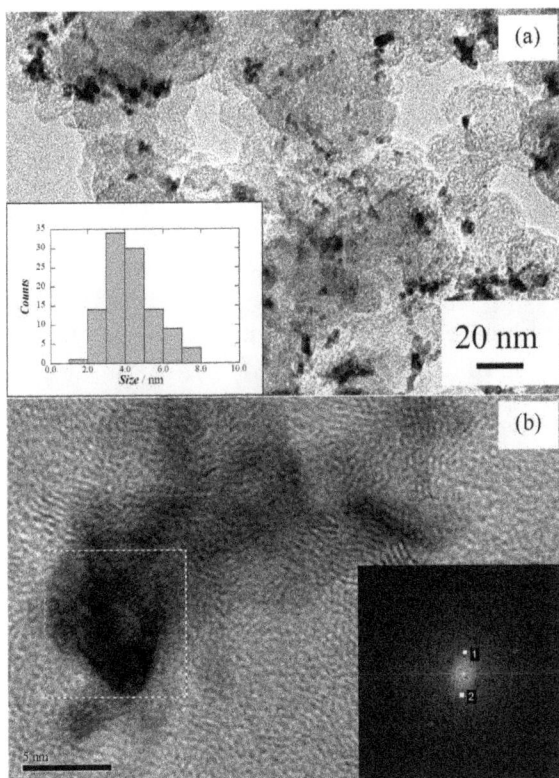

Figure 9. *Cont.*

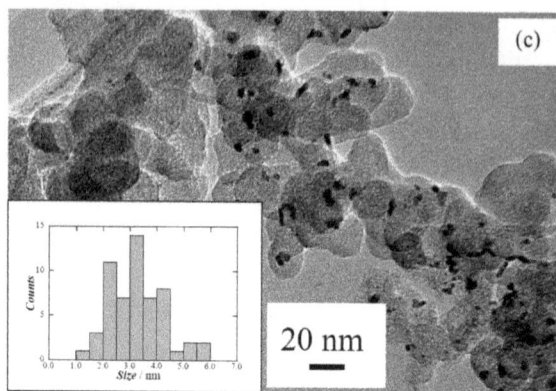

Figure 9. TEM micrographs of (**a**) and (**b**) Pt(Cu), and (**c**) Pt-Ru(Cu) electrocatalysts deposited on carbon oxidized following treatment (i) at 1.6 V, 200 s. The size distributions are shown in the inset panels. Picture (**b**) is a HRTEM image of the same sample shown in (a), which includes the corresponding FFT analysis of the marked zone.

Comparative images of the Pt(Cu) and the Pt-Ru(Cu) catalysts obtained on carbons oxidized following treatment (i) at 2.2 V for 300 s are presented in Figure 10. The corresponding nanoparticle size distributions are highlighted in the insets of Figure 10a,b, with values of 6.6 ± 1.2 and 4.8 ± 1.7 nm, respectively. These values were somewhat higher than those obtained for the same treatment at 1.6 V for 200 s, thus suggesting again that an increased carbon oxidation favored the nucleation of the metal particles around the oxidized points in the carbon and, therefore, the nanoparticle dispersion was somewhat smaller. The HRTEM image of Figure 10c allowed us to obtain the FFT analysis shown in the inset, with a mean interplanar space of $d = 0.2208$ nm, which can be assigned to Pt(111) ($d = 0.2265$ nm) [54,55] with a relative error of 2.5%.

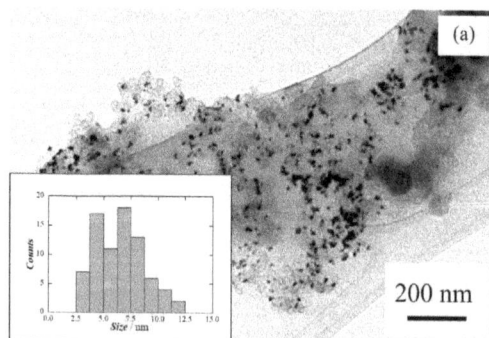

Figure 10. *Cont.*

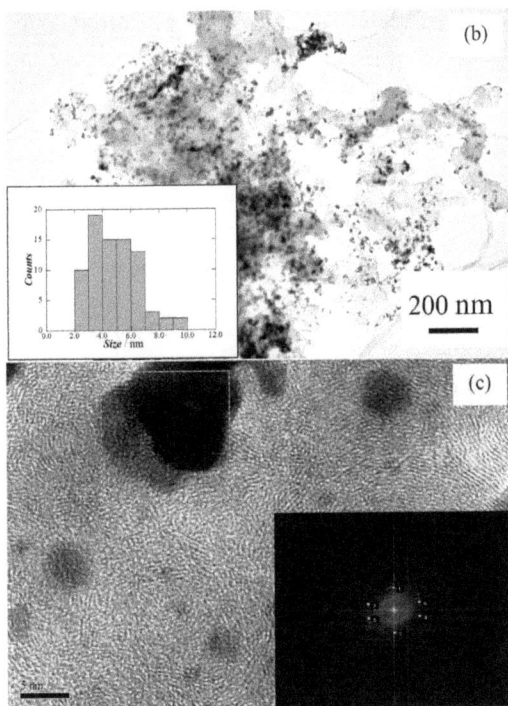

Figure 10. TEM micrographs of (**a**) Pt(Cu) and (**b**) Pt-Ru(Cu) electrocatalysts deposited on carbon oxidized following treatment (i) at 2.2 V, 300 s. The size distributions are shown in the inset panels. Picture (**c**) is a HRTEM image of the same sample shown in (**b**), which includes the corresponding FFT analysis of the marked zone.

The mean particle sizes and EDS analyses of the Pt-Ru(Cu) catalysts are summarized in Table 2. Apart from the nanoparticle size being somewhat higher when applying 2.2 V for 300s, in agreement with the discussion pointed out above, the EDS analyses showed the presence of the three elements, with a significant amount of Cu and a small quantity of Ru. The amount of Cu in all the specimens, together with the cyclic voltammograms shown in Figures 7 and 8 which do not give any evidence of Cu oxidation [48], were consistent with the core-shell structure. On the other hand, the poor quantity of Ru resulted from a slight surface deposition on Pt. This feature, together with the probable amorphous character of the deposited Ru species, could justify the fact that no Ru species were identified by the FFT analyses presented in Figure 10b [22,56].

403

Table 2. Mean particle sizes and EDS analyses of the Pt-Ru(Cu) specimens obtained on carbon oxidized according to treatment (i), compared to the non-oxidized carbon reported in Ref. [48].

Carbon treatment	Particle size/nm	Pt:Ru:Cu/at%
no oxidation	3.6	61.2:0.3:38.5
1.6 V, 200 s	3.3 ± 1.0	63.2:4.3:32.5
2.2 V, 300 s	4.8 ± 1.7	49.4:7.4:43.2

As can be seen in Table 2, the mean nanoparticle sizes and the EDS analyses were comparable to the results obtained for the non-oxidized carbon previously reported by us [48] and therefore, no further improvement with respect to them was obtained by means of the previous carbon oxidation treatment. One can then conclude that the oxidation treatments of carbon do not lead to smaller particle size, nor to a different structure or composition being able to further improve the CO tolerance and ECSA for CO oxidation with respect to the non-oxidized carbon. Note, however, that the CO tolerance continued to be as good as for the non-oxidized carbon, the onset potential for CO oxidation of Pt(Cu)/C and Pt-Ru(Cu)/C catalysts still being about 0.1 and 0.2 V more negative than those corresponding to Pt/C and Ru-decorated Pt/C ones, respectively [48]. Moreover, the improved results of the hydrogen adsorption/desorption ECSAs for the oxidized carbons when compared to the non-oxidized ones indicated a better electrolyte accessibility for the former, thus encouraging us to further test their catalyst performance in a real fuel cell.

3. Materials and Methods

3.1. Materials and Reagents

The test electrodes were prepared from E-Tek Vulcan XC72R carbon (mean particle size *ca.* 30 nm and specific surface area of about 250 m^2 g^{-1} [37]), which was deposited onto a Metrohm glassy carbon (GC) tip 3 mm in diameter. The GC was polished by means of Micropolish II deagglomerated α-alumina (0.3 μm) and γ-alumina (0.05 μm) on a Buehler PSA-backed White Felt polishing cloth. The solutions were prepared using Millipore Milli Q high-purity water (resistivity > 18 MΩ cm at 25 °C), analytical grade 96 wt.% H$_2$SO$_4$ from Acros Organics (Geel, Belgium), HClO$_4$, hydrated RuCl$_3$, and H$_2$PtCl$_6$ from Merck (Darmstadt, Germany), and CuSO$_4$.5H$_2$O, and Na$_2$SO$_4$ from Panreac Química S.A. (Barcelona, Spain). N$_2$ and CO gases were Abelló Linde 3.0 (purity $\geqslant$ 99.9%, Barcelona, Spain).

3.2. Working Electrodes and Electrochemical Testing

The electrochemical cells for the preparation of the working electrodes and testing were Metrohm 200 mL in capacity with a double-wall to control the temperature at 25.0 ± 0.1 °C by means of a Julabo MP-5 thermostat. The reference and auxiliary electrodes were a double junction Ag | AgCl | KCl(sat) (0.199 V *vs.* SHE at 25 °C) and a Pt rod, respectively. All the potentials given in this paper are referred to the Reversible Hydrogen Electrode (RHE). The working electrode was the carbon-supported catalyst, prepared using different electrolytes on the GC tip, which was coupled to an Ecochemie Autolab rotating disk electrode (RDE). The electrochemical experiments were conducted by means of an Ecochemie Autolab PGSTAT100 potentiostat-galvanostat, commanded by a NOVA 1.5 software (Metrohm Autolab, Utrecht, The Netherlands). Before the electroless deposition of Pt and Ru species and the electrochemical tests using cyclic voltammetry (CV), a N_2 flow was bubbled through the electrolyte. This gas flow passed over the electrolyte during such deposition processes and measurements.

The working electrodes were prepared as previously described [48], except that activation was introduced here. In short, 4 mg of carbon were dispersed and sonicated in 4 mL of water for at least 45 min. Then, 20 μL of this suspension were deposited onto the polished GC tip (0.28 mg$_C$ cm^{-2}) and dried under the heat of a lamp. Afterwards, the carbon was cleaned on the RDE in deaerated 0.5 M H_2SO_4 by CV scans between 0.0 and 1.0 V at 100, 50, and 20 mV s^{-1} for 10, 5, and 3 cycles, respectively (cleaning protocol). At this point and in the same electrolyte, three different activation procedures were applied: (i) potentiostatic oxidation of carbon for 100 s, 200 s, and 300 s up to 2.2 V; (ii) 10 consecutive cycles at 10 mV s^{-1} between 0.0 V and different anodic limits up to 2.2 V; and (iii) potentiostatic oxidation for 300 s at different potentials up to 2.2 V, followed by a potentiodynamic sweep at 10 mV s^{-1} to the cathodic limit of 0.0 V. After all of these activation treatments, the cleaning protocol was always applied.

After activation, the core-shell Pt(Cu)/C and Pt-Ru(Cu)/C catalysts were prepared according to the test results reported elsewhere [48], by the following consecutive steps: (a) potentiostatic deposition of Cu nuclei at −0.1 V and 100 rpm in 1 mM $CuSO_4$ + 0.1 M Na_2SO_4 + 0.01 M H_2SO_4 for 40 mC, determining the deposition efficiency of this Cu/C electrode through the Cu oxidation charge in the same solution after sweeping the potential from 0.0 to 1.0 V at 10 mV s^{-1}; (b) Pt deposition on the Cu nuclei by galvanic exchange in 1 mM H_2PtCl_6 + 0.1 M $HClO_4$ for 30 min at 100 rpm (Pt(Cu)/C electrode); and (c) spontaneous deposition of Ru species on the Pt(Cu)/C electrode in 8.0 mM $RuCl_3$ + 0.1 M $HClO_4$ (aged for at least one week) for 30 min without the electrolyte stirring (Pt-Ru(Cu)/C electrode). After the Cu deposition, the Cu/C electrode was carefully cleaned in water and, after steps (ii) and (iii), the Pt(Cu)/C and Pt-Ru(Cu)/C electrodes were also submitted to the cleaning protocol described above. It has to be noted that the cyclic voltammograms obtained from this protocol were always practically stationary after the second sweep, thus confirming the stability and cleanness of the electrodes.

The CO stripping curves for testing the CO oxidation activity and tolerance were performed in 0.5 M H_2SO_4, where CO gas was bubbled through the solution for 15 min, setting the electrode potential at 0.1 V. After removing the dissolved CO by N_2 bubbling through the solution for 30 min, CO was oxidized by sweeping the potential from 0.0 to 1.0 V at 20 mV s^{-1} without stirring. The electrochemically active area (ECSA) was estimated, taking into account that the oxidation of a CO monolayer on polycrystalline Pt needs 420 μC cm^{-2} [16,57]. After CO stripping, the activity of the Pt(Cu)/C and the Pt-Ru(Cu)/C catalysts was recovered, as shown by consecutive cyclic voltammograms, which retraced those obtained before the CO adsorption.

3.3. Microscopic Examination

The transmission electron microscopy (TEM) and high-resolution TEM (HRTEM) analyses were performed by means of a Hitachi H-800 MT, furnished with an Energy Dispersive X-ray (EDX) detector, and of a 200 kV JEOL JEM 2100 F, respectively. These analyses allowed for electron diffraction analyses and determining the size distribution, nanoparticle dispersion, and crystallographic phases. Prior to the observation, the catalyst on the GC tip was dispersed in 3 mL of n-hexane for 10 min by ultrasonication. Then, a drop of the suspension was placed on a Holley-carbon nickel grid with further evaporation of the solvent under the heat of a 40 W lamp for 5 min. The images were recorded in a Gatan MultiScan 794 charge-coupled device (CCD) camera and the Fast Fourier Transform (FFT) analyses of selected areas were obtained by means of the Gatan Digital Micrograph 3.7.0 software. The MinCryst database was used to assign the crystallographic data corresponding to the electron diffraction and FFT. Different images from different zones allowed us to count more than 100 nanoparticles to determine their size distribution.

3.4. XPS Analyses

X-ray Photoelectron Spectroscopy analyses were performed using a Physical Electronics PHI 5500 Multitechnique System spectrometer with a monochromatic X-ray source (Al Kα line of 1486.6 eV, powered at 350 W). This X-ray source was placed perpendicular to the axis of the analyzer. The energy was calibrated using the 3d5/2 line of Ag with a full width at half maximum (FWHM) of 0.8 eV. The oxidized carbons, prepared as indicated above on the GC electrode, were carefully moved by scratching to the support, after careful cleaning in water and drying. The section for the surface analyses was a circular area of 0.8 mm in diameter. A survey spectrum (187.85 eV of Pass Energy and 0.8 eV/step) was first obtained and, afterwards, the high-resolution spectra (23.5 eV of Pass Energy and 0.1 eV/step) were recorded. A low energy electron gun less than 10 eV was used in order to discharge the surface when necessary. All the measurements were made in an ultra-high vacuum chamber pressure in the range 5.0×10^{-9}–2.0×10^{-8} torr. The resulting XPS spectra were analyzed using Ulvac-phi MultiPak V8.2B software.

4. Conclusions

This paper has explored the possibility of increasing the performance of electrodeposited Pt(Cu)/C and Pt-Ru(Cu)/C core-shell catalysts by previously oxidizing the carbon support. Different XC72R oxidation treatments were applied: (i) potentiostatic oxidation for 100–300 s up to 2.2 V; (ii) cycling 10 times at 10 mV s^{-1} between 0.0 and different anodic limits up to 2.2 V; and (iii) treatment (i) followed by a potentiodynamic sweep at 10 mV s^{-1} up to 0.0 V. The oxidation treatment led to a capacitive current increase in the potential range of interest between 0.0 and 1.0 V, together with the formation of Q/HQ couples. The XPS analyses of the oxidized carbons indicated an increase in the oxidation states of carbon with the anodic potential, tending to CO_2 formation. Cu electrodeposition as well as galvanic exchange with Pt and spontaneous deposition of Ru species was performed in the best conditions reported before for the non-oxidized carbon. The Cu reoxidation after its electrodeposition indicated that Cu nuclei presented a deeper penetration into a more open carbon structure, probably with a higher bonding energy when compared to the non-oxidized carbon. Treatment (ii) was not found to be suitable because Cu oxidation took place even after 1.0 V.

The Pt(Cu)/C and Pt-Ru(Cu)/C catalysts prepared following treatment (i) at 1.6 V for 200 s led to the best results of ECSA for the hydrogen adsorption/desorption and CO oxidation reactions. The EDX analyses of the latter gave 63.2, 32.5, and 4.3 atom % for Pt, Cu, and Ru, respectively. The HRTEM and FFT analyses of these catalysts showed smaller interplanar spaces for Pt due to the effect of the Cu core. The onset potential and the ECSA values for CO oxidation as well as the mean size of the catalyst particles were comparably good to those obtained when using the non-oxidized carbon, and then behaved in a similar manner in front of the CO oxidation. However, the ECSA values for the hydrogen adsorption/desorption were much higher when carbon was previously oxidized. This was assigned not to a structural difference between the catalysts obtained with and without carbon oxidation, but to a better accessibility of the Pt sites. According to this, carbon oxidation appears to be useful to ensure a better catalyst performance.

Acknowledgments: The authors thank the financial support received from the *Generalitat de Catalunya* under the project 2014SGR83 as a consolidated research group and also that received from SENACYT (Republic of Panama) by Griselda Caballero-Manrique through the Scholarship Program for Professional Excellence. The authors also thank the CCiT-UB (Scientific and Technological Centers of the Universitat de Barcelona) for the electron microscope and the XPS analyses facilities.

Author Contributions: G.C.-M. and P.-L.C conceived and designed the experiments; G.C.-M. performed the experiments; G.C.-M, E.B., J.A.G., and P.-L.C. analyzed the data; F.C. and R.M.R. contributed reagents/materials/analysis tools; and G.C.-M., P.-L.C, and E.B. wrote the paper.

Conflicts of Interest: The authors declare no conflict of interest. The founding sponsors had no role in the design of the study; in the collection, analyses, or interpretation of data; in the writing of the manuscript; or in the decision to publish the results.

References

1. Antolini, E. Carbon supports for low-temperature fuel cell catalysts. *Appl. Catal. B* **2009**, *88*, 1–24.
2. Álvarez, G.; Alcaide, F.; Cabot, P.L.; Lázaro, M.J.; Pastor, E.; Sollá-Gullón, J. Electrochemical Performance of low temperature PEMFC with Surface Tailored Carbon Nanofibers as Catalyst Support. *Int. J. Hydrogen Energy* **2012**, *37*, 393–404.
3. Antolini, E. Platinum-based ternary catalysts for low temperature fuel cells. Part II. Electrochemical properties. *Appl. Catal. B* **2007**, *74*, 337–350.
4. Wang, Y.; Chen, K.; Mishler, J.; Cho, S.; Cordobes Adroher, X. A review of polymer electrolyte membrane fuel cells: Technology, applications, and needs on fundamental research. *Appl. Energy* **2011**, *88*, 981–1007.
5. Guo, J.W.; Zhao, T.S.; Prabhuram, J.; Chen, R.; Wong, C.W. Preparation and characterization of Pt-Ru/C nanocatalyst for direct methanol fuell cells. *Electrochim. Acta* **2005**, *51*, 754–763.
6. Salgado, J.R.C.; Alcaide, F.; Álvarez, G.; Calvillo, L.; Lázaro, M.J.; Pastor, E. Pt–Ru electrocatalysts supported on ordered mesoporous carbon for direct methanol fuel cell. *J. Power Sources* **2010**, *195*, 4022–4029.
7. Zainoodin, A.M.; Kamarudin, S.K.; Daud, W.R.W. Review: Electrode in direct methanol fuel cells. *Int. J. Hydrogen Energy* **2010**, *35*, 4606–4621.
8. Alcaide, F.; Álvarez, G.; Cabot, P.L.; Grande, H.J.; Miguel, O.; Querejeta, A. Testing of carbon supported Pd-Pt electrocatalysts for methanol electrooxidation in direct methanol fuel cells. *Int. J. Hydrogen Energy* **2011**, *36*, 4432–4439.
9. Yasuka, Y.; Fujiwara, T.; Murakami, Y.; Saaki, K.; Oguri, M.; Asaki, T.; Sugimoto, W. Effect of structure of carbon-supported PtRu electrocatalysts on the electrochemical oxidation of methanol. *J. Electrochem. Soc.* **2000**, *147*, 4421–4427.
10. Steigertwalt, E.; Deluga, G.; Cliffel, D.; Lukehart, C. A Pt-Ru/Graphitic carbon nanofiber nanocomposite exhibiting high relative performance as a direct-methanol fuel cell anode catalyst. *J. Phys. Chem.* **2001**, *105*, 8097–8101.
11. Dickinson, A.J.; Carrette, L.P.L.; Collins, J.A.; Friedrich, K.A.; Stimming, U. Preparation of Pt-Ru/C catalyst from carbonyl complexes for fuel cell applications. *Electrochim. Acta* **2002**, *47*, 3733–3739.
12. Friedrich, K.A.; Geiyzers, L.P.; Dickinson, A.J.; Stimming, U. Fundamental aspects in electrocatalysis: from the reactivity of single-crystals to fuel cell electrocatalysts. *J. Electroanal. Chem.* **2002**, *524–525*, 261–272.

13. Baena-Moncada, A.M.; Coneo-Rodríguez, R.; Calderón, J.C.; Flórez-Montaño, J.; Barbero, C.A.; Planes, G.A.; Rodríguez, J.L.; Pastor, E. Macroporous carbon as support for PtRu catalysts. *Int. J. Hydrogen Energy* **2014**, *39*, 3964–3969.

14. Velázquez-Palenzuela, A.; Centellas, F.; Garrido, J.A.; Arias, C.; Rodríguez, R.M.; Brillas, E.; Cabot, P.L. Kinetic analysis of carbon monoxide and methanol oxidation on high performance carbon- supported Pt-Ru electrocatalyst for direct methanol fuel cells. *J. Power Sources* **2011**, *196*, 3503–3512.

15. Boxall, D.L.; Deluga, G.A.; Kenik, E.A.; King, W.D.; Lukehart, C.M. Rapid synthesis of a Pt1Ru1/C nanocomposite using microwave irradiation: A DMFC anode catalyst of high relative performance. *Chem. Mater.* **2001**, *13*, 891–900.

16. Esparbé, I.; Brillas, E.; Centellas, F.; Garrido, J.A.; Rodríguez, R.M.; Arias, C.; Cabot, P.L. Structure and electrocatalytic activity of carbon-supported Pt nanoparticles for polymer electrolyte fuel cells. *J. Power Sources* **2009**, *190*, 201–209.

17. Min, M.; Cho, J.; Cho, K.; Kim, H. Particle size and alloying effects of Pt-based alloy catalysts for fuel cell applications. *Electrochim. Acta* **2010**, *45*, 4211–4217.

18. Tokarz, W.; Lota, G.; Frackowiak, E.; Czerwinski, A.; Piela, P. Fuel cell testing of Pt–Ru catalysts supported on differently prepared and pretreated carbon nanotubes. *Electrochim. Acta* **2013**, *98*, 94–103.

19. Álvarez, G.; Alcaide, F.; Miguel, O.; Cabot, P.L.; Martinez-Huerta, M.V.; Fierro, J.L.G. Electrochemical stability of carbon nanofibers in proton exchange membrane fuel cells. *Electrochim. Acta* **2011**, *56*, 9370–9377.

20. Hsieh, C.T.; Lin, J.Y.; Wei, J.L. Deposition and electrochemical activity of Pt-based bimetallic nanocatalysts on carbon nanotube electrodes. *Int. J. Hydrogen Energy* **2009**, *34*, 685–693.

21. Hwang, J.Y.; Chatterjee, A.; Shen, C.H.; Wang, J.H.; Sun, C.L.; Chya, O.; Chen, C.W.; Chen, K.H.; Chen, L.C. Mesoporous active carbon dispersed with ultra-fine platinum nanoparticles and their electrochemical properties. *Diam. Relat. Mater.* **2009**, *18*, 303–306.

22. Velázquez-Palenzuela, A.; Centellas, F.; Garrido, J.A.; Arias, C.; Rodríguez, R.M.; Brillas, E.; Cabot, P.L. Structural characterization of Ru-modified carbon-supported Pt nanoparticles using spontaneous deposition with CO oxidation activity. *J. Phys. Chem. C* **2012**, *116*, 18469–18478.

23. Tegou, A.; Papadimitriou, S.; Pavlidou, E.; Kokkinidis, G.; Sotiropoulos, S. Oxygen reduction at platinum- and gold-coated copper deposits on glassy carbon substrates. *J. Electroanal. Chem.* **2007**, *608*, 67–77.

24. Papadimitriou, S.; Tegou, A.; Pavlidou, E.; Armyanov, S.; Valova, E.; Kokkinidis, G.; Sotiropoulos, S. Preparation and characterization of platinum- and gold-coated copper, iron, cobalt and nickel deposits on glassy carbon substrates. *Electrochim. Acta* **2008**, *53*, 6559–6567.

25. Rahsepar, M.; Pakshir, M.; Piao, Y.; Kim, H. Synthesis and electrocatalytic performance of high loading active PtRu multiwalled carbon nanotube catalyst for methanol oxidation. *Electrochim. Acta* **2012**, *71*, 246–251.

26. García, G.; Flórez-Montaño, J.; Hernández-Creus, A.; Pastor, E.; Planes, G.A. Methanol electrooxidation at mesoporous Pt and Pt-Ru electrodes: A comparative study with carbon supported materials. *J. Power Sources* **2011**, *196*, 2979–2986.

27. Ding, Y.; Liu, Y.; Rao, G.; Wang, G.; Zhong, Q.; Ren, B.; Tian, Z. Electrooxidation Mechanism of Methanol at Pt-Ru Catalyst Modified GC Electrode in Electrolytes with Different pH Using Electrochemical and SERS Techniques. *Chin. J. Chem.* **2007**, *25*, 1617–1621.

28. Yue, Z.R.; Jiang, W.; Wang, L.; Gardner, S.D.; Pittman, C.U., Jr. Surface characterization of electrochemically oxidized carbon fibers. *Carbon* **1999**, *37*, 1785–1796.

29. Li, W.; Liu, L.; Zhong, Ch.; Shen, B.; Hu, W. Effects of Carbon Fiber surface treatment on Cu electrodeposition: The electrochemical behavior and the morphology of Cu deposits. *J. Alloy Compd.* **2011**, *509*, 3532–3536.

30. Carmo, M.; dos Santos, A.R.; Rocha Poco, J.G.; Linardi, M. Physical and electrochemical evaluation of commercial carbon black as electrocatalysts supports for DMFC applications. *J. Power Sources* **2007**, *173*, 860–866.

31. Zhang, X.; Chan, K. Water in Oil microemulsion synthesis of platinum-ruthenium nanoparticles, their characterization and electrocatalytic properties. *Chem. Mater.* **2003**, *15*, 451–459.

32. Podlovchenko, B.I.; Krivchenk, V.A.; Maksimov, Y.M.; Gladysheva, T.D.; Yashina, L.V.; Evlashin, S.A.; Pilevsky, A. Specific features of the formation of Pt (Cu) catalysts by galvanic displacement with carbon nanowalls used as support. *Electrochim. Acta* **2012**, *76*, 137–144.

33. Podlovchenko, B.I.; Zhumaev, U.E.; Maksimov, Y.M. Galvanic displacement of copper adatoms on platinum in PtCl$_4{}^{2-}$ solutions. *J. Electroanal. Chem.* **2011**, *651*, 30–37.

34. Podlovchenko, B.I.; Gladysheva, T.D.; Filatov, A.; Yashina, L.V. The Use of Galvanic Displacement in Synthesizing Pt(Cu) Catalysts with the Core-Shell Structure. *Russ. J. Electrochem.* **2010**, *46*, 1189–1197.

35. Wang, J.; Yin, G.; Shao, Y.; Zhang, S.; Wang, Z.; Gao, Y. Effect of carbon black support corrosion on the durability of Pt/C catalyst. *J. Power Sources* **2007**, *171*, 331–339.

36. Weissmann, M.; Baranton, S.; Clacens, J.M.; Coutanceau, C. Modification of hydrophobic /hydrophilic properties of Vulcan XC72 carbon powder by grafting of trifluoromethylphenyl and phenylsulfonic acid groups. *Carbon* **2010**, *48*, 2755–2764.

37. Cabot Corporation. Specialty Chemicals and Performance Materials. Available online: http://www.cabotcorp.com (accessed on 20 January 2014).

38. Kumar, S.; Soler Herrero, J.; Irusta, S.; Scott, K. The effect of pretreatment of Vulcan XC-72R carbon on morphology and electrochemical oxygen reduction kinetics of supported Pd nano-particle in acidic electrolyte. *J. Electroanal. Chem.* **2010**, *647*, 211–221.

39. Ghodbane, O.; Roué, L.; Bélanger, D. Copper electrodeposition on pyrolitic graphite electrodes: Effect of the copper salt on the electrodeposition process. *Electrochim. Acta* **2007**, *52*, 5843–5855.

40. Kumar, S.; Hidyatai, N.; Soler Herrero, J.; Irusta, S.; Scott, K. Efficient tuning of the Pt nano-particle mono-dispersion on Vulcan XC-72R by selective pre-treatment and electrochemical evaluation of hydrogen oxidation and oxygen reduction reactions. *Int. J. Hydrogen Energy* **2011**, *36*, 5453–5465.

41. Bae, G.; Youn, D.; Han, S.; Lee, J. The role of nitrogen in a carbon support on the increased activity and stability of a Pt catalyst in electrochemical hydrogen oxidation. *Carbon* **2013**, *51*, 274–281.

42. Yoon, Ch.; Long, D.; Jang, S.; Qiao, W.; Ling, L.; Miyawaki, J.; Rhee, Ch.; Mochida, I.; Yoon, S. Electrochemical surface oxidation of carbon nanofibers. *Carbon* **2011**, *49*, 96–105.

43. Gómez de la Fuente, J.L.; Martínez-Huerta, M.V.; Rojas, S.; Terreros, P.; Fierro, J.L.G.; Peña, M.A. Methanol electrooxidation on PtRu nanoparticles supported on functionalized carbon black. *Catal. Today* **2006**, *116*, 422–432.

44. Cao, J.; Song, L.; Tang, J.; Xu, J.; Wang, W.; Chen, Z. Enhanced activity of Pd nanoparticles supported on Vulcan XC72R carbon pretreated via a modified Hummers method for formic acid electrooxidation. *Appl. Surf. Sci.* **2013**, *274*, 138–143.

45. Yue, Z.R.; Jiang, W.; Wang, L.; Toghiani, H.; Gardner, S.D.; Pittman, C.U., Jr. Adsorption of precious metal ions onto electrochemically oxidized carbon fibers. *Carbon* **1999**, *37*, 1607–1618.

46. Carmo, M.; Linardi, M.; Rocha Poco, J.G. Characterization of nitric acid functionalized carbon black and its evaluation as electrocatalyst support for direct methanol fuel cell applications. *Appl. Catal. A* **2009**, *355*, 132–138.

47. Rueffer, M.; Bejan, D.; Bunce, N.J. Graphite: An active or an inactive anode? *Electrochim. Acta* **2011**, *56*, 2246–2253.

48. Caballero-Manrique, G.; Velázquez-Palenzuela, A.; Centellas, F.; Garrido, J.A.; Arias, C.; Rodríguez, R.M.; Brillas, E.; Cabot, P.L. Electrochemical synthesis and characterization of carbon-supported Pt and Pt-Ru nanoparticles with Cu cores for CO and methanol oxidation in polymer electrolyte fuel cells. *Int. J. Hydrogen Energy* **2014**, *39*, 12859–12869.

49. Shao, Y.; Yin, G.; Zhang, J.; Gao, Y. Comparative investigation of the resistance to electrochemical oxidation of carbon black and carbon nanotubes in aqueous sulfuric acid solution. *Electrochim. Acta* **2006**, *51*, 5853–5857.

50. Yumitori, S. Correlation of C1s chemical state intensities with the O1s intensity in the XPS analysis of anodically oxidized glass-like carbon samples. *J. Mater. Sci.* **2000**, *35*, 139–146.

51. Stankovich, S.; Dikin, D.; Piner, R.; Kohlhaas, K.; Kleinhammes, A.; Jia, Y.; Wu, Y.; Nguyen, S.; Ruoff, R. Synthesis of graphene-based nanosheets via chemical reduction of exfoliated graphite oxide. *Carbon* **2007**, *45*, 1558–1565.

52. Yang, D.; Velamakannia, A.; Bozoklu, G.; Park, S.; Stoller, M.; Piner, R.; Stankovich, S.; Jung, I.; Field, D.; Ventrice, C., Jr.; *et al.* Chemical analysis of graphene oxide films after heat and chemical treatments by X-ray photoelectron and Micro-Raman spectroscopy. *Carbon* **2009**, *47*, 145–152.

53. Terzyk, A. The influence of activated carbon surface chemical composition on the adsorption of acetaminophen (paracetamol) *in vitro*. Part II. TG, FTIR, and XPS analysis of carbons and the temperature dependence of adsorption kinetics at the neutral pH. *Colloids Surf. A* **2001**, *177*, 23–45.

54. WWW-MINCRYST. Crystallographic and Crystallochemical Database for Minerals and their Structural Analogues. Available online: http://database.iem.ac.ru/mincrysst (accessed on 30 June 2014).

55. Ruth, K.; Vogt, M.; Zuber, R. Development of CO-tolerant catalysts. In *Handbook of Fuel Cells—Fundamentals, Technology and Applications*; Vielstich, W., Gasteiger, H.A., Lamm, A., Eds.; John Wiley & Sons: New York, NY, USA, 2003; Volume 3, pp. 489–496.

56. Velázquez-Palenzuela, A.; Brillas, E.; Arias, C.; Centellas, F.; Garrido, J.A.; Rodríguez, R.M.; Cabot, P.L. Structural analysis of carbon-supported Ru-decorated Pt nanoparticles synthesized using forced deposition and catalytic performance toward CO, methanol, and ethanol electro-oxidation. *J. Catal.* **2013**, *298*, 112–121.

57. Chen, Z.; Xu, L.; Li, W.; Waje, M.; Yan, Y. Polianiline nanofibre supported platinum nanoelectrocatalysts for direct methanol fuel cells. *Nanotechnol.* **2006**, *17*, 5254–5259.

Pt Monolayer Electrocatalyst for Oxygen Reduction Reaction on Pd-Cu Alloy: First-Principles Investigation

Amra Peles, Minhua Shao and Lesia Protsailo

Abstract: First principles approach is used to examine geometric and electronic structure of the catalyst concept aimed to improve activity and utilization of precious Pt metal for oxygen reduction reaction in fuel cells. The Pt monolayers on Pd skin and $Pd_{1-x}Cu_x$ inner core for various compositions x were examined by building the appropriate models starting from Pd-Cu solid solution. We provided a detailed description of changes in the descriptors of catalytic behavior, d-band energy and binding energies of reaction intermediates, giving an insight into the underlying mechanism of catalytic activity enhancement based on the first principles density functional theory (DFT) calculations. Structural properties of the Pd-Cu bimetallic were determined for bulk and surfaces, including the segregation profile of Cu under different environment on the surface.

Reprinted from *Catalysts*. Cite as: Peles, A.; Shao, M.; Protsailo, L. Pt Monolayer Electrocatalyst for Oxygen Reduction Reaction on Pd-Cu Alloy: First-Principles Investigation. *Catalysts* **2015**, *5*, 1193–1201.

1. Introduction

The slow kinetics of the oxygen reduction reaction (ORR) and high cost of platinum electrocatalysts, in the proton exchange membrane fuel cells (PEMFC) are recognized as significant limitations toward the large scale implementation as a clean energy alternative [1–4]. Many research efforts have focused on the search for alternative catalysts with high Pt utilization and improved activity. Pt mono-layer catalyst supported on metal or metal-alloy core is a promising alternative to the traditional catalysts [5–7]. Enhanced catalytic activity for Pt mono-layer supported on Pd core was demonstrated in Adzic's group [5]. The lateral compressive strain in the Pt surface layer due to the lattice constant mismatch with Pd substrate was suggested as major driving force for the better catalytic activity [8,9]. An alternative view is based on the X-ray photo-electron spectroscopy (XPS) results combined with electrochemical cell. This study concluded that increased coverage of oxygenated reaction intermediates is driving force for enhanced activity for Pt skin like catalyst on bimetallic core [10].

Here we focus on Pt monolayer catalyst deposited on the core that itself has a core-shell structure consisting of Pd shell and $Pd_{1-x}Cu_x$ core with $x = 0.125$,

0.25, 0.5 and 0.75. We provide an insight into the underlying mechanism of catalytic activity enhancement based on the first principles density functional theory (DFT) calculations. In our approach, we examine theoretical descriptors of catalyst behavior: Structural and compositional parameters that correlate with catalytic activity changes including electronic structure effects, described by the mean energy of d-band electrons; the electron occupancy of the d-band and strain effects that modify electronic structure via lateral strain in the catalyst's top layer. We build comprehensive models, bottom-up, with the aim to identify optimal compositions of core-shell structure with promise of better catalytic activity solely based on inherent electronic structure features. In principal, catalyst with exposed base metal Cu is not stable in PEMFC, and Cu is expected to be removed over time due to dissolution in the acidic environment causing the accelerated fuel cell degradation. To ensure stability in acid, PdCu bimetallic must be protected by more stable noble metals. Selective dissolution of Cu in PdCu alloy can lead a core-shell structure where core consists of the PdCu alloy and shell of Pd protective skin [11]. The selective dissolution of alloying components provide for compositional and structural changes that play an important role in manipulating catalytic activity [6,8,9,12,13].

Here, we discuss structural properties and surface segregation profile of Pd-Cu bimetallic; the electronic structure properties and surface reactivity effects of pseudo-morphic Pt and Pd over-layers supported on Pd-Cu alloy. The comparison of the simulation results with experiments will be discussed as well.

2. Computational Method

The first-principles calculations are based on spin-polarized density functional theory (DFT) using a Generalized Gradient Approximation (GGA) [14] and projector augmented wave (PAW) method [15] as implemented in Vienna Ab-Initio Simulation Package (VASP) [16,17]. The cut-off energy for plane wave basis set was 400 eV and Brillouin zone was sampled using a Monkhorst-Pack sampling technique [18] with k-space interval Δk not larger then 0.3 Å^{-1}. Surfaces are modeled by 8-layer slabs with 2×2 surface cell separated by a 16 Å vacuum layer perpendicular to the surface. The top six layers were fully relaxed until Hellmann-Feynman forces were 0.01 eV/Å. The Pt and Pd monolayer surfaces are modeled as pseudo-morphic layers placed on top of the Pd$_{1-x}$Cu$_x$ (111) surface.

3. Results and Discussion

3.1. Structural Properties of Pd$_{1-x}$Cu$_x$

Pd and Cu, both crystallize in the face centered cubic (fcc) geometry with $Fm\bar{3}m$ space group. Binary phase diagram [19–21] of Pd-Cu bimetallic, shows formation of the solid state solution along the entire range, x, of Pd$_{1-x}$Cu$_x$ compositions above 600 °C. The ordered phases have been reported at lower temperatures. Here we investigate solid solution phases as our theoretical results were compared to the samples which were synthesized at 700 °C and subsequently subjected to dealloying [11]. Bimetallic solid solution structures were optimized by minimizing forces on atoms for the range of lattice parameters between those of Pd and Cu in $4 \times 4 \times 4$ super-cell geometry for each composition $x = 0, 0.125, 0.25, 0.5, 0.75$ and 1 in $Pd_{1-x}Cu_x$. Super-cell is illustrated in Figure 1a. Two random atomic distributions in solid state solution were considered for each x and final lattice parameter represents an average over the two. Equilibrium lattice parameters were obtained by fitting energy *versus* volume curves, shown in Figure 2, to the third order Birch-Murnaghan equation of state as follows:

$$E = E_0 + \frac{9V_0 B_0}{16} \left\{ \left[\left(\frac{V_0}{V} \right)^{\frac{2}{3}} - 1 \right]^3 B_0' + \left[\left(\frac{V_0}{V} \right)^{\frac{2}{3}} - 1 \right]^2 \left[6 - 4 \left(\frac{V_0}{V} \right)^{\frac{2}{3}} \right] \right\} \quad (1)$$

We also generated and optimized the special quasi random structures [22] at these compositions that have lead to the lattice parameters within 0.1 Å of those found by fitting equations of state. Calculated lattice parameters of Pd and Cu are 3.96 and 3.64 Å, respectively. Figure 1b displays the dependence of lattice parameter of disordered structures to the composition constant x. The computed linear changes follow empirically observed Vegard's law.

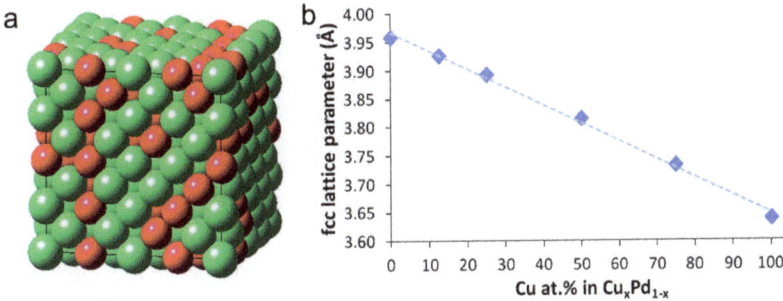

Figure 1. (a) Super-cell illustration. **(b)** Calculated lattice parameters for solid state solution of $Pd_{1-x}Cu_x$ for various compositions x.

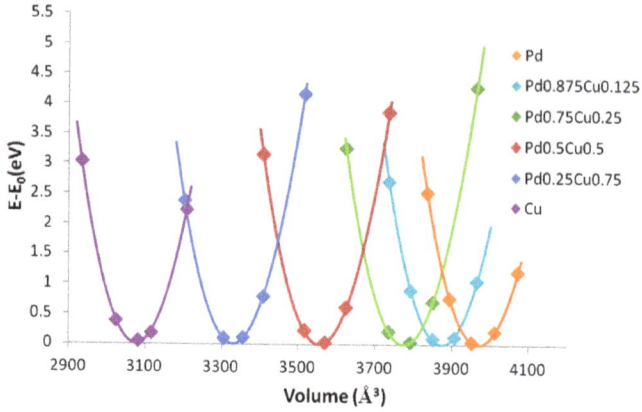

Figure 2. The total energy as a function of the super cell volumes and a Birch-Murnaghan fits to the calculated points for $Pd_{1-x}Cu_x$; $x = 0, 0.125, 0.25, 0.5.0.75$ and 1. The zero of the energy is set to the minimum of each curve, while volumes are given in Å^3.

3.2. Chemical Stability and Segregation Profile

Chemical stability of $Pd_{1-x}Cu_x$ solid state solutions is examined by calculating formation enthalpy as follows:

$$\Delta H = E_{Pd_{1-x}Cu_x} - [(1-x)E_{Pd_{1-x}} + xE_{Cu_x}] \tag{2}$$

where $E_{Pd_{1-x}}$, E_{Cu_x} and $E_{Pd_{1-x}Cu_x}$ are the free energies per atom of pure Pd, pure Cu, and the PdCu alloy at concentration x, respectively.

Chemical compositions of alloys at surfaces often differ from the bulk composition and depend on the chemical environment. In bimetallics, surface composition may become enriched by one of the alloying components. To understand which alloying components enriches the surface under the neutral and strongly interacting environment, we have computed Cu segregation energies for the (111) slabs in vacuum and with oxygen adsorbed on the surface. Segregation energy is computed as energy difference for total slab energies with Cu enriched and Pd depleted surface $E_{Cu\uparrow Pd\downarrow}$ and stoichiometric surface E_{stoich} according to:

$$E_{seg}^{Cu} = E_{Cu\uparrow Pd\downarrow} - E_{stoich} \tag{3}$$

where negative (positive) energy favors (disfavors) surface segregation of copper. Results are summarized in Table 1. Formation enthalpy is negative and increases with increased amount of Pd, whith the exception of $x = 0.5$ which presents an ordered alloy. Negative enthalpy indicate chemically stable solid solutions with stability increasing with increased amount of Pd. The trends in copper segregation are different in vacuum and in oxygen. Cu tends to segregate to the surface of solid solutions in strongly reacting oxygen environment while in inert environment, as in vacuum, Pd enriches the surface. This tendency have important consequences for PEMFC application. Exposure to the acidic conditions leads to strong interaction of more abundant copper species on the surface and their subsequent dissolution. With careful de-alloying strategy, catalyst with surface layer covered by noble Pd metal skin and bimetallic PdCu core can be engineered [11].

Table 1. Formation enthalpy ΔH (meV/atom) and surface segregation energy of Cu E_{seg}^{Cu} (meV).

Composition	ΔH	E_{seg}^{Cu} in vacuum	E_{seg}^{Cu} in oxygen
$Pd_{0.875}Cu_{0.125}$	-29.68	-	-
$Pd_{0.75}Cu_{0.25}$	-54.84	61	-100
$Pd_{0.5}Cu_{0.5}$	-95.47	72	-136
$Pd_{0.25}Cu_{0.75}$	-79.88	70	-178

3.3. Electronic Structure

Oxygen reduction reaction (ORR) in PEMFC require a catalyst that can speed up interaction of oxygen from the air with recombined electron and proton to form water as a product:

$$\frac{1}{2}O_2 + 2(H^+ + e^-) \rightarrow H_2O \tag{4}$$

The potency of precious metal catalyst has been correlated with the position of electronic d-band center at catalyst surfaces, since d electrons are involved in making and breaking of inter atomic bonds of ORR reactants and products. We have examined changes in the position of d-band center with respect to Fermi level, associated with Pd atoms on the Pd monolayer surface commensurate with $Pd_{1-x}Cu_x$ substrate and compare them to the pure Pt. Figure 3 illustrate these changes, showing the shift in the positions of d-band center which start decreasing with the amount of copper in the Pd-Cu core and asymptomatically approaches position of d-band center of Pt, illustrating a modification of physical properties at atomic scale that make Pd on substrates to behave in between pure Pd and Pt becoming more like Pt metal with increasing amount of Cu in the substrate.

The changes in the d electrons density of state of surface Pt, modeled as pseudo-morphic layer on $Pd/Pd_{1-x}Cu_x$ are shown in Figure 4. Pt 5d projected electron density of states becomes broader with increase in Cu content. The marked positions of d-band center shift away from the Fermi level when Cu content increases. The shifts of d-band ceneter value away from the Fermi level indicate a surface of lower adsortion affinity to the ORR intermediates. The ability of

417

surface to interact with adsorbate molecules can be measured by the strength of the adsorbate-surface bond. Binding energies for O and OH reaction intermediates were computed with respect to respective gas phase.

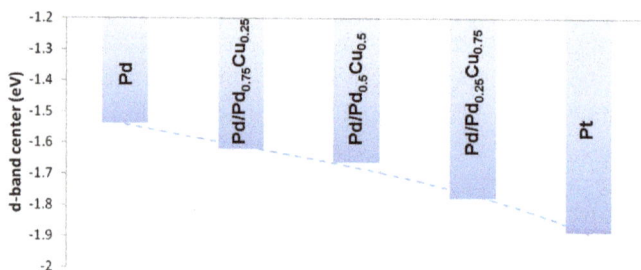

Figure 3. Changes in the position of Pd d-band center with respect to Fermi level on $Pd_{1-x}Cu_x$ substrates.

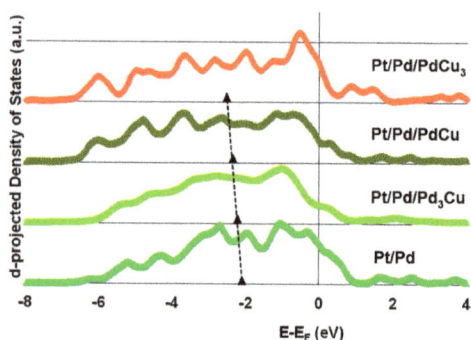

Figure 4. Calculated d-projected density of states of Pt mono-layer for $Pd/Pd_{1-x}Cu_x$ substrates and Pt metal. The Fermi level is set at zero. The position of d-band center, marked by black triangles, shift away from Fermi level for Pt on mixed $Pd_{1-x}Cu_x$ substrates.

$$E_O = E_{slab+O} - E_{slab} - \frac{1}{2}E_{gas}(O_2) \tag{5}$$

$$E_{OH} = E_{slab+OH} - E_{slab} - E_{gas}(OH) \tag{6}$$

The correlation of changes in the position of d-band center and surface reactivity, the strength of adsorbate-surface interaction are presented in Figure 5 for intermediates of ORR reaction. The OH intermediates bind on the top of Pt sites, while atomic oxygen intermediates bind on the hollow sites at (111) and more favorably on sites with fcc symmetry. The bonding strength of ORR intermediates decreases on surfaces with d-band center shifted away from the Fermi level.

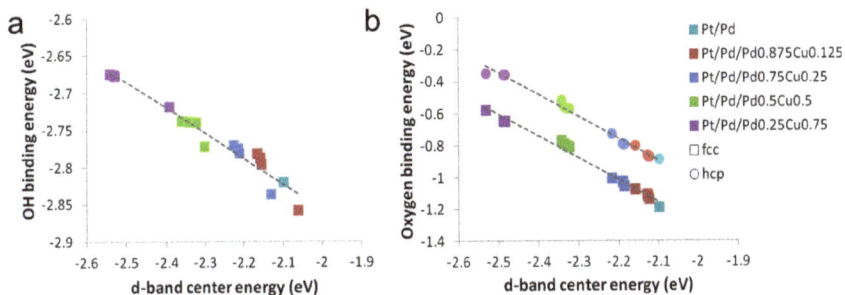

Figure 5. The correlation between surface binding strength and position of the d-band center for oxygen reduction reaction (ORR) intermediates: (**a**) OH species on the top sites and (**b**) oxygen atoms on fcc and hcp hollow sites.

The optimal catalytic activity per active site is a trade-off between a not too strong and a not too weak binding of reaction intermediates on the catalyst surface. Empirical observations show the volcano type dependence between catalytic activity and surface reactivity [7,23–25]. The maximum of the volcano plot defines most optimal catalyst. According to this principle, the rate limiting step for ORR on Pt metal is its too strong binding of oxygen in the multi-step ORR reaction. The catalysts that bind oxygen less strongly than Pt but still not too weakly, that this would become rate limiting step, hold the promise of better catalytic activity. The change toward lower binding energy of 0.2 eV have been suggested as optimal in literature [6]. Such shift is accomplished in the case of $Pt/Pd/Pd_{1-x}Cu_x$; $\times 0.125$; shown in the Figure 5. Indeed the experimental evidence for the activity of Pt mono-layer on the electrochemically de-alloyed Pd-Cu alloy, characterized to have Pd skin on the core corresponding to the $Pd_{0.85}Cu_{0.125}$ resulted in the catalyst with way superior catalytic activity compared to Pt metal [11].

4. Conclusions

We presented the comprehensive build-up of models accounting for various compositions of the Pt and Pd monolayers on $Pd_{1-x}Cu_x$ alloys with x = 0, 0.125, 0.25, 0.5, 0.75 and 1. Surface segregation profile shows different segregation tendency and surface compositions for models in vacuum and with adsorbed oxygen, pointing to the limiting stability factors for the PEMFC applications. Detailed study of the d-electrons projected electronic density of states and position of d-band center energy is presented. The correlation between the d-band center energy of surface atoms and surface reactivity is linear indicating less reactive surfaces for d-band center energy further away from the Fermi level. Relaying on the empirical observations based on Sabatier principle, $Pt/Pd/Pd_{0.875}Cu_{0.125}$

was identified as the most optimal composition and geometry that correlates with better catalytic activity compared to pure Pt and Pt/Pd core-shell catalyst. These predictions were corroborated by the experimental results [11].

Author Contributions: A. P. conceived and designed computational models, performed the computations, analyzed the data, predicted the optimal catalyst composition and wrote the paper; M. S. and L. P. conceived the concept of Pd-Cu core shell catalyst and provided the data for the experimental validation.

Conflicts of Interest: The authors declare no conflict of interest.

References

1. Gasteiger, H.A.; Kocha, S.S.; Sompalli, B.; Wagner, F.T. Activity benchmarks and requirements for Pt, Pt-alloy, and non-Pt oxygen reduction catalysts for PEMFCs. *Appl. Catal. B* **2005**, *56*, 9.
2. Adzic, R.R. Frontiers in electrochemistry. In *Electrocatalysis*; Lipkowski, J., Ross, P.N., Eds.; Wiley: New York, NY, USA, 1998; p. 197.
3. Gasteiger, H.A.; Markovic, N.M. Just a dream-or future reality? *Science* **2009**, *324*, 48–49.
4. Bashyam, R.; Zelenay, P. A class of non-precious metal composite catalysts for fuel cells. *Nature* **2006**, *443*, 63–66.
5. Adzic, R.R.; Zhang, J.; Sasaki, K.; Vukmirovic, M.B.; Shao, M.; Wang J.X.; Nilekar, A.U.; Mavrikakis,M.; Valerio, J.A.; Uribe, F. Platinum monolayer fuel cell electrocatalysts. *Top. Catal.* **2007**, *46*, 249–262.
6. Stamenkovic, V.R.; Mun, B.S.; Arenz, M.; Mayrhofer, K.J.J.; Lucas, C.A.; Wang, G.; Ross, P.N.; Markovic, N.M. Trends in electrocatalysis on extended and nanoscale Pt-bimetallic alloy surfaces *Nat. Mater.* **2007** *6* 241–247.
7. Stamenkovic, V.R.; Fowler, B.; Mun, B.S.; Wang, G.; Ross, P.N.; Lucas, C.A.; Markovic, N.M. Improved oxygen reduction activity on Pt3Ni (111) via increased surface site availability. *Science* **2007**, *315*, 493–497.
8. Hammer, B.; Nørskov, J.K. Electronic factors determining the reactivity of metal surfaces. *Surf. Sci.* **1995**, *343*, 211.
9. Greeley, J.; Mavrikakis, M. Alloy catalysts designed from first principles. *Nat. Mater.* **2004** *3*, 810–815.
10. Watanabe, M.; Wakisaka, M.; Yano, H.; Uchida, H. Analyses of oxygen reduction reaction at Pt-based electrocatalysts. *ECS Trans.* **2008**, *16*, 199.
11. Shao, M.; Shoemaker, K.; Peles, A.; Kaneko, K.; Protsailo, L. Pt monolayer on porous Pd-Cu alloys as oxygen reduction electrocatalysts. *J. Am. Chem. Soc.* **2010**, *132*, 9253–9255.
12. Koh, S.; Strasser, P. Electrocatalysis on bimetallic surfaces: Modifying catalytic reactivity for oxygen reduction by voltammetric surface dealloying. *J. Am. Chem. Soc.* **2007**, *129*, 12624–12625.
13. Zeis, R.; Mathur, A.; Fritz, G.; Lee, J.; Erlebacher, J. Platinum-plated nanoporous gold: An efficient, low Pt loading electrocatalyst for PEM fuel cells. *J. Power Sour.* **2007**, *165*, 65.

14. Pedrew, J.P.; Wang, Y. Accurate and simple analytic representation of the electron-gas correlation energy. *Phys. Rev. B* **1992**, *45*, 13244.

15. Blöch, P.E. Projector augmented-wave method. *Phys. Rev. B* **1994**, *50*, 17953–17979.

16. Kresse, G.; Furthmuller, J. Efficient iterative schemes for ab initio total-energy calculations using a plane-wave basis set. *Phys. Rev. B* **1996**, *54*, 11169.

17. Kresse, G.; Joubert, D. From ultrasoft pseudopotentials to the projector augmented-wave method. *Phys. Rev. B* **1999**, *59*, 1758.

18. Monkhorst, H.J.; Pack, J.D. Special points for Brillouin-zone integrations. *Phys. Rev. B* **1976**, *13*, 5188.

19. Subramanian, P.R.; Laughlin D.E. *Cu-Pd (Copper-Palladium), Binary Alloy Phase Diagrams*, 2nd ed., Massalski, T.B., Ed.; Springer-Verlag: Berlin, Germany, 1990; Volume 2, pp. 1454–1456.

20. Straumanis, M.E.; Yu, L.S. Lattice parameters, densities, expansion coefficients and perfection of structure of Cu and of Cu-In (α) phase. *Acta Crystallogr.* **1969**, *25*, 676.

21. Rao, C.N.; Rao, K.K. Effect of temperature on the lattice parameters of some silver-palladium alloys.*Can. J. Phys.* **1964**, *42*, 1336–1342.

22. Alex Zunger, A.; Wei, S.H.; Ferreira, L.G.; Bernard, J.E. Special quasirandom structures. *Phys. Rev. Lett.* **1995**, *65*, 353–356.

23. Stamenkovic, V.; Moon, B.S.; Mayrhofer, K.J.J.; Ross, P.N.; Markovic, N.M.; Rossmeisl, J.; Greeley, J.; Nørskov, J.K. Changing the activity of electrocatalysts for oxygen reduction by tuning the surface electronic structure. *Angew. Chem.* **2006**, *45*, 2897–2901.

24. Zhang, J.L.; Vukmirovic, M.B.; Xu, Y.; Mavrikakis, M.; Adzic, R.R. Controlling the catalytic activity of platinum-monolayer electrocatalysts for oxygen reduction with different substrates. *Angew. Chem.* **2005**, *44*, 2132–2125.

25. Greeley, J.; Stephens, I.E.L.; Bondarenko, A.S.; Johansson, T.P.; Hansen, H.A.; Jaramillo, T.F.; Rossmeisl, J.; Chorkendorff, I.; Nørskov, J.K. Alloys of platinum and early transition metals as oxygen reduction electrocatalysts. *Nat. Chem.* **2009**, *1*, 552–556.

Effect of Particle Size and Operating Conditions on Pt₃Co PEMFC Cathode Catalyst Durability

Mallika Gummalla, Sarah C. Ball, David A. Condit, Somaye Rasouli, Kang Yu, Paulo J. Ferreira, Deborah J. Myers and Zhiwei Yang

Abstract: The initial performance and decay trends of polymer electrolyte membrane fuel cells (PEMFC) cathodes with Pt₃Co catalysts of three mean particle sizes (4.9 nm, 8.1 nm, and 14.8 nm) with identical Pt loadings are compared. Even though the cathode based on 4.9 nm catalyst exhibited the highest initial electrochemical surface area (ECA) and mass activity, the cathode based on 8.1 nm catalyst showed better initial performance at high currents. Owing to the low mass activity of the large particles, the initial performance of the 14.8 nm Pt₃Co-based electrode was the lowest. The performance decay rate of the electrodes with the smallest Pt₃Co particle size was the highest and that of the largest Pt₃Co particle size was lowest. Interestingly, with increasing number of decay cycles (0.6 to 1.0 V, 50 mV/s), the relative improvement in performance of the cathode based on 8.1 nm Pt₃Co over the 4.9 nm Pt₃Co increased, owing to better stability of the 8.1 nm catalyst. The electron microprobe analysis (EMPA) of the decayed membrane-electrode assembly (MEA) showed that the amount of Co in the membrane was lower for the larger particles, and the platinum loss into the membrane also decreased with increasing particle size. This suggests that the higher initial performance at high currents with 8.1 nm Pt₃Co could be due to lower contamination of the ionomer in the electrode. Furthermore, lower loss of Co from the catalyst with increased particle size could be one of the factors contributing to the stability of ECA and mass activity of electrodes with larger cathode catalyst particles. To delineate the impact of particle size and alloy effects, these results are compared with prior work from our research group on size effects of pure platinum catalysts. The impact of PEMFC operating conditions, including upper potential, relative humidity, and temperature on the alloy catalyst decay trends, along with the EMPA analysis of the decayed MEAs, are reported.

Reprinted from *Catalysts*. Cite as: Gummalla, M.; Ball, S.C.; Condit, D.A.; Rasouli, S.; Yu, K.; Ferreira, P.J.; Myers, D.J.; Yang, Z. Effect of Particle Size and Operating Conditions on Pt₃Co PEMFC Cathode Catalyst Durability. *Catalysts* **2015**, *5*, 926–948.

1. Introduction

To reduce the electrocatalyst cost for polymer electrolyte membrane fuel cells (PEMFCs), platinum alloys with higher oxygen reduction reaction (ORR) mass activity are being developed. In particular, alloys such as Pt_3Co, Pt_3Fe and Pt_3Ni have been shown to provide higher ORR mass activity (A/g-Pt) and specific activity (mA/cm^2-Pt) than pure Pt catalysts [1–3]. In aqueous environments, the rotating disk electrode (RDE) measurements found that polycrystalline Pt_3Co with a Pt skin has ORR specific activity three times greater than pure Pt, and $Pt_3Ni(111)$ has the highest ORR specific activity recorded to date with an enhancement factor of nearly twenty *versus* polycrystalline Pt [2,4,5]. Additionally, in RDE measurements, carbon-supported nanoparticles of Pt_3Co with a mean diameter of 6 nm showed an ORR specific activity enhancement factor of three *versus* 6 nm mean diameter Pt [6]. Paulus *et al.* found that Pt based alloy catalysts exhibit higher ORR specific activity than pure Pt by an enhancement factor of approximately 1.5 for bulk electrodes, 1.5 to 2 for supported Pt_3X catalysts, and 2 to 3 for PtCo catalyst [7]. There are relatively few measurements of ORR activity for Pt alloys in the fuel cell environment. In a study where Co, Ni and Fe-based Pt alloy catalysts were evaluated in a fuel cell, the initial mass activity was found to be the highest for the Pt–Fe catalyst and lowest for pure Pt catalyst [2,3]. Recently, Huang *et al.* demonstrated an ORR specific activity enhancement factor of 3.5 in the fuel cell environment for Pt_3Co/Ketjen carbon *versus* Pt/Vulcan carbon, albeit for different particle sizes [8].

The enhancement in ORR activity observed for Pt alloys *versus* Pt has been attributed to modification of the electronic/atomic structure of the alloy catalyst surface [2,4,9–12]. Using X-ray absorption near edge structure (XANES), Min *et al.* showed that the structure sensitivity is associated with the adsorption strength of oxygen intermediates on the Pt surface [13,14]. Furthermore, they reported that the reduced Pt–Pt neighbor distance on the surface of the alloy catalysts is favorable for the adsorption of oxygen. Through experimental and theoretical studies, Mukerjee *et al.* suggested that this improvement can be attributed to a positive shift of the onset potential for forming OH_{ads} on the alloy relative to the Pt catalyst, thereby allowing O_2 to adsorb at higher potentials and reducing the overpotential for O_2 reduction [14,15].

Degradation studies of Pt_3Co alloys in a 16 cell stack showed that Co leaches out, at the nanometer scale, resulting in a "Pt skeleton" structure at the topmost surface layer of Pt_3Co particles within the first hour of operation. With longer operational time, the particles slowly evolve toward "Pt-shell/Pt–Co alloy core" structures with depleted Co content and a Pt-enriched shell (of the order of two atomic monolayers after 1124 h of operation) due to Co surface segregation/leaching and Ostwald ripening [16,17]. Chen *et al.* showed that the acid treated Pt_xCo resulted in Co dissolution, which increased the thickness of the Pt-enriched surface layer. This structural change was identified as a contributor to the reduction in the specific activity of Pt_xCo nanoparticles after potential cycling [18]. Popov *et al.* reported that pure Pt catalyst showed higher ECA decay rate than that of Pt_3Co catalyst during potential cycling in acidic aqueous environment as well as during constant current holding in-cell test, however the initial particle sizes of Pt is lower than that of Pt_3Co in this comparison [3].

Shao *et al.* have shown through systematic synthesis of mono-dispersed Pt catalysts that the catalytic activity depends on the shape and size of the nanoparticles [19]. Furthermore it is reported that the edge sites, which increase in fraction of total surface sites with decreasing particle size, have lower specific activity due to very strong oxygen binding energies [19–22]. Even with the same alloy metal combinations, the particle size effects are difficult to delineate as the heat treatment used to increase particle size results in a varied degree of alloying, and, thus, different surface activity [23,24]. Pt-alone and Pt-based alloy catalysts both show increasing ORR specific activities with decreasing specific surface area (*i.e.*, increasing particle size) [25–29].

As the heat treatment temperature is increased the degree of alloying increases along with the particle size. Min *et al.* showed that the particle size and alloying effects are the two most important factors affecting the catalytic activity towards ORR, with lowered Pt–Pt bond distance resulting in favorable adsorption of oxygen [13]. The recent work by Matsutani *et al.* reported particle size effects of the Pt and PtCo catalyst performance in MEAs and suggested that the MEAs with cathodes containing 4–5 nm Pt particles and 7–8 nm PtCo particles were the most stable. These two sizes were the largest of the catalyst particles studied [30]. These catalysts were found to be more stable when heat treated to a higher temperature. However, the heat treatment conditions changed the PtCo composition along with particle size. This leaves an unanswered question about the factor impacting durability, namely whether it is the size, the composition, or a combination of these two factors. The impact of Pt_3Co particle size with nearly identical metal ratio is needed to delineate the effects of composition and particle size on the initial performance and stability.

The catalyst structures, such as the core-shell type, offer the benefits of low platinum content in the electrodes and high activity [31–34]. However, the core-shell catalysts are large and often in the size range of 10 nm. To elucidate the role of these new catalyst structures, the contribution of the alloy and size effects needs to be understood. Recently, our research group has shown that the activity and stability of the Pt catalysts in an MEA is strongly dependent on the particle size, and an optimum particle size between 3.2 and 7.1 nm is suggested for maximized life averaged performance per mg Pt [35]. The focus of this paper is to report a similar systematic analysis of Pt_3Co catalysts that have nearly identical metal ratio but different mean particle sizes, and contrast these with Pt catalysts to delineate the impact of alloying and particle size.

2. Results and Discussion

Pt$_3$Co catalysts with nearly identical Pt:Co:carbon ratio (shown in Table 1) but different mean particle sizes were prepared by heat treatment, in order to distinguish the effects of composition from particle size on the initial performance and stability. While, in spite of nearly identical Pt:Co metal ratio, as verified by ICP-OES and X-ray fluorescence measurements, the Pt$_3$Co catalysts with different particle sizes are ineluctably associated with variance, to some extent, in the degree of alloy ordering, which is shown in the X-ray diffraction patterns (Figure 1). The superlattice reflections appear to be clearer for larger particle sizes, indicating a higher degree of ordering.

Table 1. Characterization of Pt$_3$Co/Ketjen EC 300J catalysts tested.

Sample	wt.% Pt	wt.% Co	XRD crystallite Size (nm)	XRD L.P (Å)	CO area (m^2/g-Pt)	TEM mean diameter (nm)	St error	N
1	37.9	3.88	3.9	3.848	35	4.9	0.2	200
2	38.5	3.94	5.6	3.850	23	8.1	0.3	200
3	38.4	3.93	9.5	3.850	11.3	14.8	0.6	200

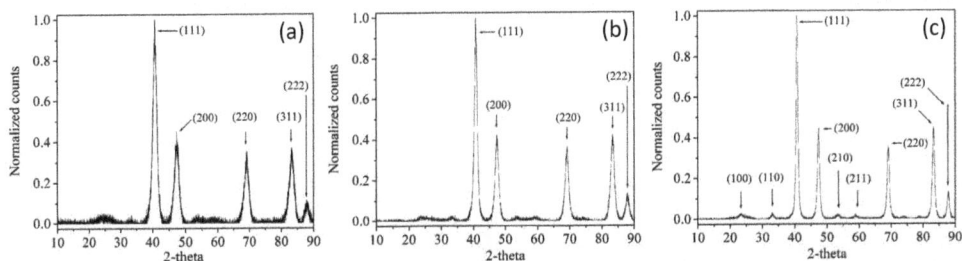

Figure 1. X-ray diffraction patterns of as prepared (a) 4.9 nm (b) 8.1 nm and (c) 14.8 nm Pt$_3$Co on carbon powders.

Systematic tests carried out on the Pt$_3$Co catalysts for particle size study are listed in Table 2, while Table 3 shows the tests performed at different operating conditions for the parametric study. The electrodes with small Pt$_3$Co particle size (4.9 nm) were used for the parametric study in order to better distinguish the difference of the operating condition impact on cells' performance degradation. The ECAs of the cathodes reported in Table 2 are the peak ECAs observed between 0–1000 potential cycles. All ECAs in this study were characterized using the hydrogen absorption area in the CV. In general, these calculated ECA values may underestimate the catalysts' real electrochemical surface areas to some extent, due to the possible disruptive influences such as hydrogen evolution and the difference

425

of catalyst surface structures [36–38]. However, in this study, since the MEA tests were carried out with the same protocol and the CVs were performed under the identical temperature, RH, and gas flow rate conditions, these calculated ECA values are meaningful for the comparison purpose and revealing the ECA evolution trend along the potential decay cycling. As expected, the values showed the trend of decreasing ECAs with increasing catalyst mean particle size.

Table 2. Pt_3Co catalyst particle sizes used in this study and cycling conditions used to test their electrochemical stability. Note that the ECA values reported are the peak ECA's.

Cell #	Mean diameter (nm) (ECA, m²/g-Pt)	Aonde/cathode Pt loading mg-Pt/cm²	Potential cycling conditions
1	4.9 (37)	0.19/0.23	Triangle wave potential cycle:
1a *	4.9 (40)	0.22/0.18	0.6 V to 1.0 V (50 mV/s ramp rate) Cell Temperature: 80 °C
2	8.1 (27)	0.20/0.22	Humidity: Anode = Cathode = 100% RH
3	14.8 (23)	0.20/0.22	Fuel/Oxidant: H_2 at 100 sccm/N_2 at 50 sccm Pressure: Atmospheric pressure

* The cathode Pt_3Co catalyst was pre-leached.

Table 3. Operating conditions used for testing the electrochemical stability of the 4.9 nm Pt_3Co catalysts.

Cell #	Description	Potential cycling conditions
4	Baseline	Square wave potential cycle: 10 s at 0.4 V, 10 s at 0.95 V (20 s/cycle) Cell Temperature: 80 °C Humidity: Anode = Cathode = 100% RH Fuel/Oxidant: 0.5 SPLM 4% H_2/0.5 SPLM N_2 Pressure: Atmospheric pressure
5	Lower RH	Humidity: Anode = Cathode = 30% RH All other parameters were same as 4 #
6	Higher Upper Potential	Square wave potential cycle: 10 s at 0.4 V, 10 s at 1.05 V (20 s/cycle) All other parameters were same as 4 #
7	Higher Temperature	Cell Temperature: 90 °C All other parameters were same as 4 #

The beginning of life (BOL) performance curves for cells 1–3 are shown in Figure 2 for H_2/O_2 (a) and for H_2/air (b) using the test conditions described in the cell performance section. The V-I performance curves reported here are corrected for the membrane resistance (*i.e.*, IR-corrected) to allow analysis of the electrode changes. The performance of the 4.9 nm Pt_3Co electrode in O_2 is slightly higher than the 8.1 nm Pt_3Co electrode at current densities <400 mA/cm² (e.g., ~8 mV at 40 mA/cm²), however there is no noticeable voltage benefit at higher current densities under H_2/O_2 condition. Interestingly, the performance of the 8.1 nm electrode provided ~17 mV higher performance than the 4.9 nm Pt_3Co electrode at 1.5 A/cm² under H_2/air conditions. The 14.8 nm Pt_3Co electrode resulted in a

V-I performance that is significantly lower, indicating lower mass activity and/or higher electrode resistance to proton transport. In this study, the catalyst loading was maintained at ~0.2 mg Pt/cm^2 for all electrodes. In an idealized case of spherical particles of uniform particle size, the distance between the particles is expected to scale as $d^{1.5}$, where d is the diameter of the particle. This suggests that the 4.9 nm Pt$_3$Co particles on the carbon support would be ~2 times closer to each other (using the metric of geometric distance) than the 8.1 nm Pt$_3$Co particles and ~4 times closer when compared to the 14.8 nm particles. Fewer particles in electrodes could result in fewer access points for ORR, and lower net oxygen concentration at the catalyst surface, resulting in lower performance with larger particles [39]. The effect of this catalyst particle spacing is expected to be greater on the performance in air than in oxygen, as seen in Figure 2.

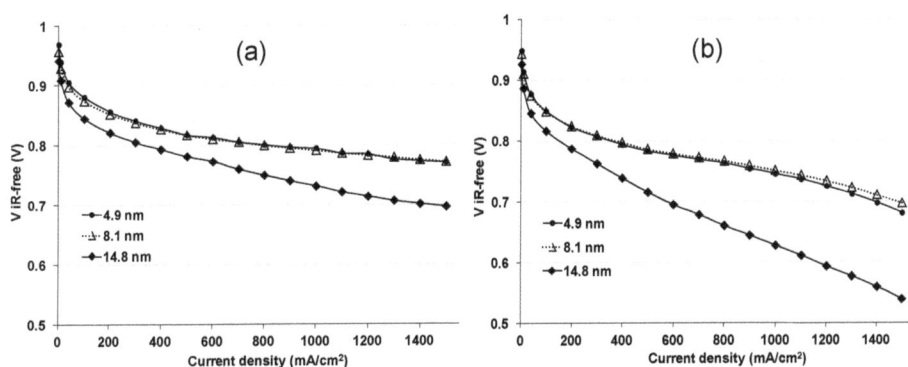

Figure 2. The beginning of life V-I performance of 4.9, 8.1, and 14.8 nm Pt$_3$Co catalyst based MEAs, H$_2$/O$_2$ (**a**) and H$_2$/Air (**b**).

The H$_2$/O$_2$ performance suggests that the electrode ionic resistance is higher for the electrode with the larger catalyst particles (*i.e.*, 14.8 nm Pt$_3$Co). While, comparing to the 8.1 nm Pt$_3$Co catalyst-based electrode, the 4.9 nm Pt$_3$Co catalyst-based electrode also shows slightly higher electrode resistance, which could be due to the increased fraction of cobalt in the ionomer [40]. Decreasing the amount of easily removable Co from the alloy by pre-leaching may decrease the Co content in the ionomer. Hence, a pre-leached 4.9 nm Pt$_3$Co catalyst-based MEA was also tested. The results are shown in Figure 3. The Pt loading in the pre-leached catalyst-based electrode was ~20% lower than the as-made catalyst-based electrode, hence lower performance is seen in the low current regions. The comparison of the H$_2$/O$_2$ performance curves shows that the slope in the middle-high current region is lower for the pre-leached catalysts, suggesting lower ionic resistance in the electrode. The lower Co content in the electrode with the pre-leached catalysts is likely to be the factor for the improved performance observed in the high current region, and it is seen that the pre-leached catalyst resulted in improved performance in the high current region of the H_2/air polarization curve. In summary, both MEAs with as-made large and small Pt_3Co particle sizes showed lower performance at high current region for different reasons; the 4.9 nm Pt_3Co MEAs were likely to have higher Co contamination and the 14.7 nm Pt_3Co MEAs were likely to have higher local oxygen transport resistance issue.

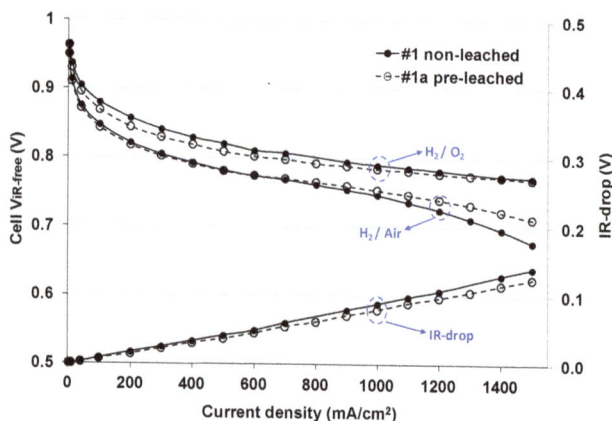

Figure 3. The beginning of life V-I performance of pre-leached and as made 4.9 nm Pt_3Co catalyst based MEAs.

2.1. Impact of Catalyst Particle Size on Cell Performance Degradation

The decay characteristics of the four Pt_3Co electrodes are tracked by monitoring the ECA, mass activity, and H_2/Air performance at 0.8 A/cm^2 and 1.5 A/cm^2. Figure 4a shows the trends of ECA evolution, in which the ECA for 4.9 nm Pt_3Co generally decreases from 1000 to 30,000 cycles while the ECA for the larger particles, 8.1 nm and 14.8 nm, stabilizes after 3000 to 5000 cycles. The smaller particle electrodes exhibited higher initial ECA values as expected, owing to a higher surface area per unit mass. The mass activity decay trends capture the evolution of the catalyst activity for ORR with minimal disturbance from oxygen transport and proton transport resistances. Figure 4b shows that the mass activity decay is highest for 4.9 nm Pt_3Co particles and lowest for 14.8 nm Pt_3Co particles. At the end of 30,000 cycles the MEA with 8.1 nm particle size retained higher mass activity than the MEAs containing the 4.9 nm and 14.8 nm Pt_3Co particles. The relatively stable mass activity and ECA of the catalysts with larger particle sizes suggests minimal changes to the catalyst and to the electrode structure with cycling.

Figure 4c,d show the H_2/Air performance during decay cycling of the four electrodes at 0.8 A/cm^2 and 1.5 A/cm^2, respectively. While the performance of the 8.1 nm Pt_3Co-based electrode improved slightly within the first 5000 cycles, significant improvement is seen in the case of the 14.8 nm Pt$_3$Co-based electrode. This improvement may be attributed to further wet-up conditioning within the cathode from potential cycling and diagnostic tests, such as possibly improving the ionic contact within the electrode. The ORR specific activity calculated by dividing mass activity by the corresponding ECA shows to be higher for the 8.1 nm Pt$_3$Co (0.89 mA/cm^2-Pt) than the 4.9 nm Pt$_3$Co (0.65 mA/cm^2-Pt) at the BOL, and both of which generally decrease with the potential cycles. The ORR specific activity of the 14.8 nm Pt$_3$Co shows to be the lowest (0.43 mA/cm^2-Pt) at the BOL, which gains significant increase from potential cycling and peaks at 0.67 mA/cm^2-Pt after the first 5000 cycles. At the end of 30,000 cycles, the ORR specific activities of 8.1 nm Pt$_3$Co and 14.8 nm Pt$_3$Co retain the same (0.61 mA/cm^2-Pt) and are higher than that of 4.9 nm Pt3Co particles (0.52 mA/cm^2-Pt).

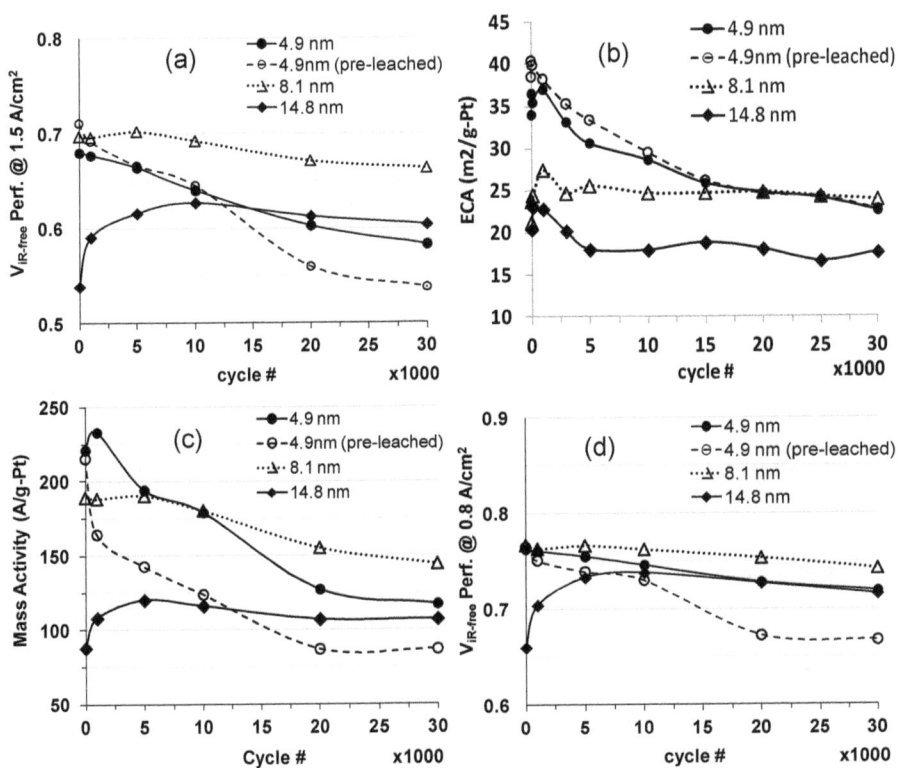

Figure 4. Performance decay with potential cycling for 4.9 nm, 8.1 nm, and 14.8 nm Pt$_3$Co electrodes, (a) ECA, (b) Mass activity (A/g-Pt at 0.90 V), (c) H$_2$/Air performance at 0.8 A/cm^2 and (d) H$_2$/Air performance at 1.5 A/cm^2.

It is clear from the cell performance trends shown in Figure 4 that 8.1 nm Pt_3Co is significantly more durable than the 4.9 nm Pt_3Co and provides significantly better performance than 14.8 nm Pt_3Co throughout the kinetic and mass transport-limited regions. The decay trends of the pre-leached Pt_3Co 4.9 nm based electrode are also shown in Figure 4, with the dashed line. The performance decay rate of the pre-leached catalyst is higher than that observed for the non-leached catalyst of the same particle size. The detailed reason is unclear.

The cross-section post-test EMPA analysis with Pt, S and Co profiling, shown in Figure 5, indicates that dissolution-precipitation occurs near the cathode-membrane interface where $Pt^{2+/4+}$ ions migrate from the cathode into the membrane to be reduced by crossover H_2. A very distinct band of platinum close to the cathode/membrane interface is seen in all samples. The estimated fraction of platinum from the cathode that moved into the membrane after 30,000 cycles is shown in Table 4. The 4.9 nm Pt_3Co-based electrode exhibited ~14% of platinum from the electrode lost to the membrane, while the other two electrodes lost less than 5%. The fraction of ECA lost for the 4.9 nm, 8.1 nm, and 14.8 nm Pt_3Co based electrodes after 30,000 cycles was 39, 13, and 23%, respectively. The loss of platinum from the electrode was only a small contributor to the overall ECA and mass activity losses observed for these electrodes, thus, having insignificant impact on the overall ECA and mass activity decay trends in this study, though all ECAs and mass activities were calculated from the electrodes' initial Pt loading. It is important to note that some entrapment of air occurred during processing of the catalyst ionomer inks of the Pt_3Co/C cathode materials used in this study leading to macroscopic porosity within the electrodes of the final MEAs. The pores appear as dark lenticular features in the EMPA images although it should be noted that similar features can also be introduced during the sample preparation for the EMPA. Therefore, analysis of those features is not attempted in this article. The Co line scans shown in the top plots are re-plotted in the bottom plots with a more sensitive y-axis scale for clarity. The Co plots indicate that some Co^{n+} ions migrate into the membrane, but stop at the anode. Co^{n+} remains in a cationic form after the Pt_3Co dissolves while $Pt^{2+/4+}$ is reduced to metallic Pt (Pt^0) by crossover H_2 within the cathode and membrane ionomer. An interesting observation from these plots is that the Co^{n+} wt.% in the membrane is greatest for the 4.9 nm non-leached case and lowest for the 14.8 nm case. The decreased Co^{n+} content in the membrane and electrode observed via EPMA for the pre-leached 4.9 nm Pt_3Co is consistent with the hypothesis presented earlier.

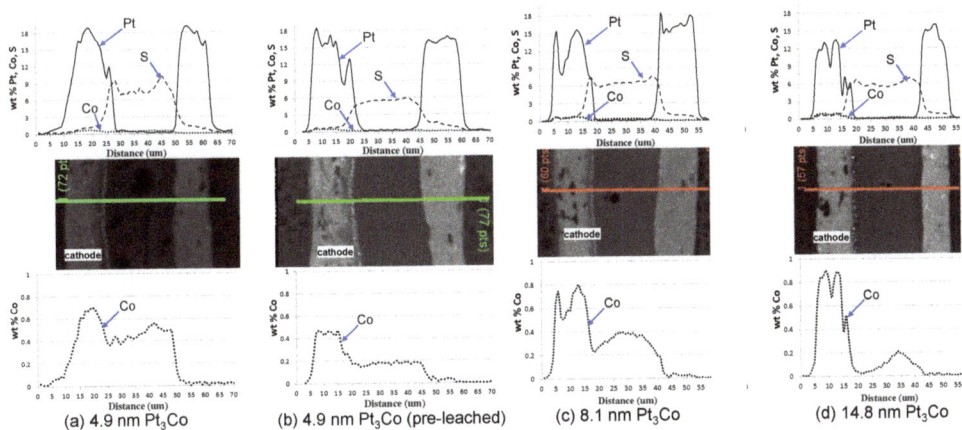

Figure 5. Post-test analysis of Pt₃Co particle size study at subscale cell plan form midpoint shows minor migration of Pt and Co into membrane.

Table 4. The loss of platinum calculated from EMPA, loss of cobalt calculated from the XRF, and post-cycling Pt to Co ratio in electrode from XRF after 30,000 cycles.

Cell #	Mean diameter (nm) (ECA, m²/g-Pt)	% of Pt lost into membrane	% Co most from electrode	Pt/Co ratio
1	4.9 (37)	14.3 ± 2%	63	8.4
2	8.1 (27)	3.7 ± 2%	45	5.5
3	14.8 (23)	3.0 ± 1%	30	4.2

The Pt to Co atomic ratios in the cathodes of the three cycled MEAs, as determined by XRF, are shown in Table 4. This ratio was found to decrease with increasing initial mean Pt₃Co size. This trend reflects decreased extent of Co leaching from the catalysts with increasing particle size, which is in agreement with the trend observed for wt.% Co^{n+} found in the membrane for the three different particle sizes from the EMPA analysis. The observed trends in the particle size dependence of the fraction of Co lost from the catalysts can be understood in terms of the particle size dependence of the fraction of total atoms in a nanoparticle residing on the surface of the particles (*i.e.*, surface/volume $\propto 1/d$, where d is the diameter of the nanoparticle). Assuming that an insignificant amount of Pt lost from the catalyst particles relative to the amount of Co lost, and using the known data, including the initial and post-cycling Pt to Co ratios, the mean particle sizes, and the XRD-determined lattice spacing of the alloy (0.385 nm), the depth of Co loss from the topmost layers of the catalyst particles was calculated to be approximately one atomic layer for the pre-cycled electrodes and three atomic layers for the post-cycled electrodes. The results are in agreement with previous data [24,41,42]. The depth of de-alloying was found to be independent of the initial mean particle size of the catalysts.

431

To gain greater insight into the observed decay trends of MEAs with varied catalyst particle size, the TEM-based catalyst particle size distributions of the pre- and post-cycling electrodes are shown in Figure 6. The TEM results of the 4.9 nm based electrode, Figure 6a, shows significant tailing towards the larger particle sizes at the end of 30,000 decay cycles compared to the pristine catalyst. However, the changes in particle size distribution for the 8.1 nm and 14.8 nm are relatively insignificant with cycling (Figure 6b,c). These results suggest that the loss in mass activity observed for the 4.9 nm catalyst could be a combination of catalyst restructuring as well as particle size changes, while for the 8.1 and 14.8 nm catalysts, the origin of the losses may be primarily due to catalyst restructuring. Based on the literature data and the EMPA and the XRF data presented here, it can be speculated that the restructuring includes dissolution of Co from the Pt_3Co particle surface and sub-layers and subsequent migration into cathode/membrane ionomer as Co^{n+}. This may result in a Pt-rich shell encasing a Pt_3Co core, explaining the enhanced ORR activity observed with Pt-Co alloys [24]. The TEM-derived particle size distributions also show that cycling caused the average particle size of the 14.8 nm catalyst to decrease by approximately 1.8 nm. The observed shrinkage of these particles indicates that there is minimal coarsening of these large particles such that the shrinkage of the particles due to loss of cobalt is observable. The further understanding of the catalysts surface restructuring at atomic level needs more comprehensive characterization which was not attempted in this study.

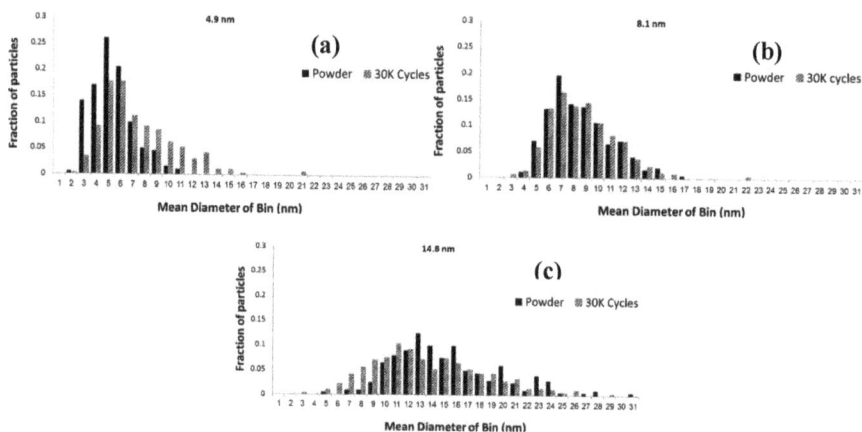

Figure 6. TEM analysis of Pt_3Co particle size distribution of the pristine (**black**) and decayed MEAs (**hashed**), (a) 4.9 nm (b) 8.1 nm and (c) 14.8 nm Pt_3Co supported on carbon.

The initial performance and decay trends of Pt_3Co and Pt over a wide range of particle sizes are compared. The performance and decay characteristics of 1.9, 3.2,

7.1 and 12.3 nm Pt particle based-MEAs is reported in our prior work [35]. Figure 7 shows the mass activity, ECA, specific activity and high current performance of both Pt and Pt$_3$Co-based MEAs along with the decay trends. The initial mass activity values of between 180 and 220 A/g-Pt measured for the 4.9 and 8.3 nm Pt$_3$Co MEAs in this study are slightly lower than those reported previously on similar catalyst materials [1], however the overall trends are comparable, with a peak in mass activity at intermediate particle size and lower mass activity for the very large particles due to the reduced available metal area. The heat treatment conditions during catalyst preparation, the formulations and processes used in catalyst ink preparation, and MEA fabrication all have an important role to play in realizing the initial mass activity benefits reported in literature. Comparable procedures were used for both the Pt and Pt$_3$Co materials in this study, however resulting data show greater improvements of Pt only performance over past data [1] compared to Pt$_3$Co examples, bringing mass activity values closer for the two catalyst types in the current study.

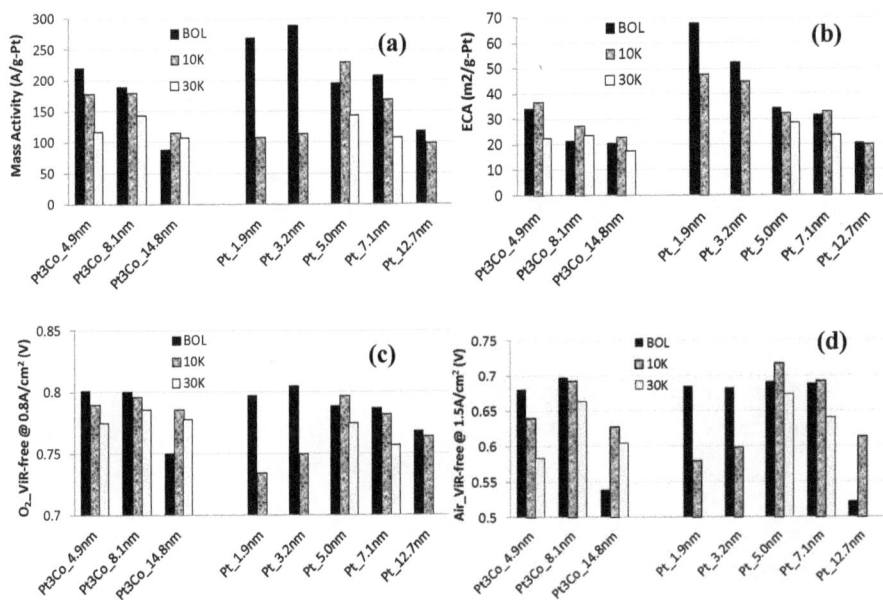

Figure 7. Comparison of Pt *vs.* Pt$_3$Co decay trends as a function of potential cycles. (a) Mass Activity, (b) ECA, (c) H$_2$/O$_2$ performance at 0.8 A/cm^2, (d) H$_2$/Air performance at 1.5 A/cm^2.

The results are shown for initial and after 10,000 and 30,000 (when available) decay cycles. While the experimental errors make the decay rates hard to quantify with great accuracy, a few trends are clearly evident. The decay rate is catalyst size dependent with larger sized catalysts decaying slower. This is due to that larger nanoparticles have larger surface area and therefore better stability (Gibbs-Thompson effect) and larger nanoparticles (>10 nm) are more stable against agglomeration compared to the smaller particles. The mass activity data shown in Figure 7a indicates that the initial mass activity decreases with increase in particle size. However, the 10,000 and 30,000 mass activity data suggests that the best results are obtained with the 8.1 nm Pt_3Co and 5 nm Pt. The ECA decay rate is lower for larger particle sizes as shown in Figure 7b, for both the Pt and Pt_3Co catalysts. Figure 8c,d show the performance at 0.8 A/cm^2 in H_2/O_2 and 1.5 A/cm^2 in H_2/air for all the cathode catalysts and sizes evaluated. A common trend observed is the initial performance of the largest particles is significantly lower than the smaller particle sizes. However, the performance improves during the 10,000 cycles and subsequently decreases. The intermediate size catalysts evaluated in this study provide the best end-of-life performance. In summary, the catalysts with initial mean diameters of ~5 to ~8.1 nm for both Pt and Pt_3Co showed balanced performance and durability, giving the best overall life-averaged performance. For the Pt MEAs, that containing the 5 nm Pt was the best performer and most durable over 30,000 cycles. Analogously, 8.1 nm was the best performer for the Pt_3Co MEAs, within the particle sizes studied.

2.2. Impact of Operating Conditions on Cell Performance Degradation

The parametric study of fuel cell operating conditions, for cells listed in Table 3, was performed with the 4.9 nm Pt_3Co cathodes using a square wave potential cycle, 0.40–0.95V (*vs.* anode), which represents transients between peak and idle power for typical automotive FC operation. The anode was 4% H_2 (balance N_2), which served as a stable reference to the square wave potential cycle imposed on the cathode. Low RH (30%) and high temperature (90 °C) are extreme operating conditions that may impact catalyst degradation. Higher upper potential (1.05 V *vs.* anode) can occur at high fuel utilization conditions and/or during startup/shutdown.

Performance decay to 10,000 cycles for the parametric study at 0.8 A/cm^2 and 1.5 A/cm^2 in H_2/Air is shown in Figure 8c,d respectively, with corresponding ECA and mass activity losses provided in Figure 8a,b. In general, ECA is still declining for all conditions after 5000 cycles. Comparing the ECA and mass activity decay curves, it can be noted that while the 90 °C operation was more detrimental to ECA than the baseline conditions, the mass activity losses were comparable for these two conditions. While ECA loss reflects catalyst changes due to particle growth and Pt loss into the membrane, mass activity loss is also affected by changes in the intrinsic ORR activity of the electrochemically-active surfaces due to catalyst re-structuring and a decrease in the influence of Co on ORR activity (e.g., a loss of Co from the near-surface region of the particles). Comparing Figures 9b and 9c, the mass activity decay trends are similar to performance decay trends in the kinetic region, 0.8 A/cm^2. Higher temperature and baseline (80 °C) showed about the same performance loss in the kinetic region (0.8 A/cm^2) compared to stable performance for 30% RH and greater performance loss for 1.05 V upper potential limit. In this study, higher temperature resulted in greater performance decay in the mass transport region as compared to the baseline conditions. The 30% RH condition shows the least performance decay with minimal ECA loss and moderate mass activity loss. It is speculated that the low ionomer hydration at 30% RH restricts Pt$^{2+/4+}$ or Co^{n+} generation/diffusion and thereby limits Pt Ostwald ripening, loss of Pt into the membrane, and ionomer contamination within the cathode and membrane. Several studies have reported the acceleration of carbon corrosion and Pt dissolution at potentials >0.95V [43].

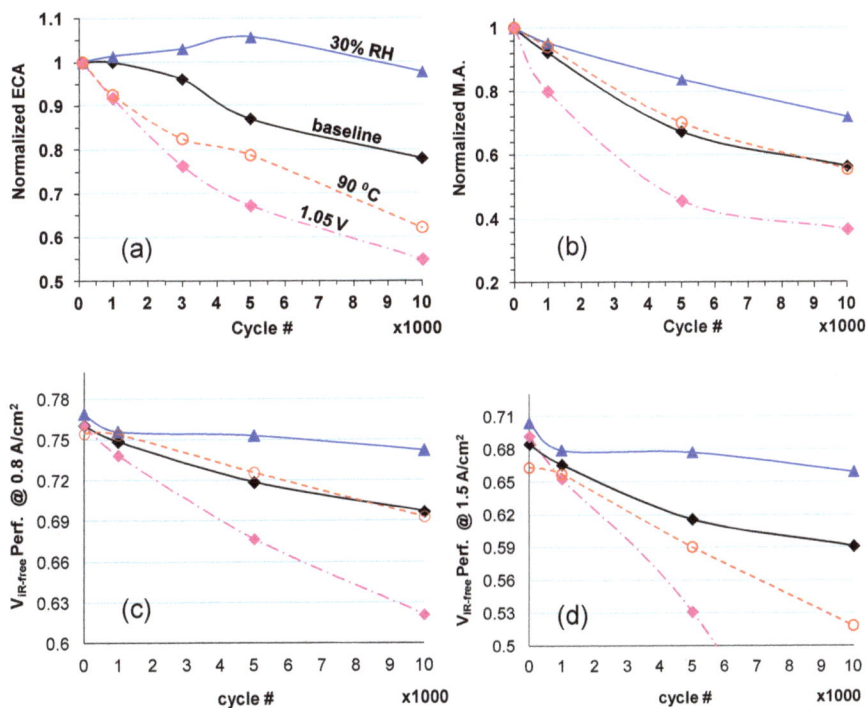

Figure 8. Pt$_3$Co cathode catalysts normalized ECA loss (**a**), normalized mass activity (M.A.) loss (**b**) and H$_2$/Air performance loss at 0.8 A/cm^2 (**c**) and 1.5 A/cm^2 (**d**) under extreme operating conditions after 0.4–0.95 V square wave cycling (20 sec/cycle).

The post-test cross-section microscopy and Pt, Co and S profiling shown in Figure 9 indicate that Pt dissolution-deposition occurred extensively near the cathode side of the membrane for all conditions, except for 30% RH. It should be noted that Pt$^{2+/4+}$ generated during the parametric studies diffuses out of the cathode a farther

435

distance into the membrane before being reduced to metallic Pt by crossover H_2 than during the particle size study due to lower H_2 diffusion rate for 4% H_2 (balance N_2) on the anode compared to 100% H_2 on the anode used for particle size study. Co mapping is also shown in the figure for the four MEAs. The wt.% Co^{n+} in the membranes was in the order 1.05 V > 90 °C > baseline $\approx$ 30% RH.

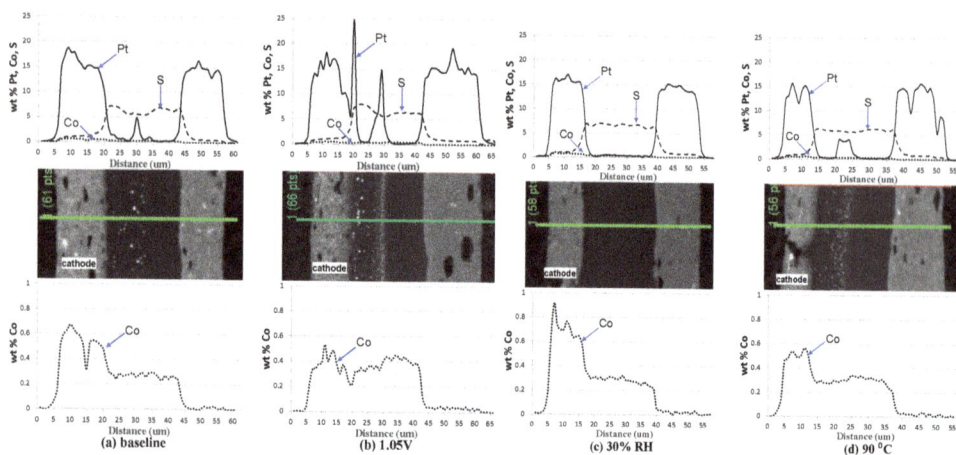

Figure 9. Post-test analysis of Pt_3Co parametric operational study at subscale plan form midpoint consists of cross-section back-scattered SEM images and corresponding Pt, Co and S elemental profiles taken along lines indicated. (a) Baseline, (b) Higher upper potential (1.05 V), (c) Lower RH (30%) and (d) Higher Temperature (90 °C).

For 90 °C condition, both higher diffusion rate of H_2 from the anode and faster deposition kinetics account for the denser focused Pt band in the membrane than the more diffused band for the baseline and high upper potential limit conditions. Ostwald ripening occurred for the 1.05 V condition as indicated by bright Pt spots within the cathode in back-scattered cross-section image, Figure 9b. Bright spots in the membrane seen for both the baseline and 1.05 V cells are indicative of growth of Pt deposits into large Pt particles from $Pt^{2+/4+}$ dissolution-migration-deposition from the cathode. Lastly, accelerated carbon corrosion at 1.05 V may also promote Pt particle migration and coalescence into large aggregate clusters within the cathode electrode. Losses in ECA and mass activity from Pt migration into the membrane for the 90 °C and 1.05 V cells could play significant role in performance decay, especially in the mass transport limited region, 1.5 A/cm^2.

The Pt to Co atomic ratios in the cathodes of the four parametric-study cycled MEAs, as determined by XRF, are shown in Table 5. The trends in Co loss from the cathodes are 30% RH < Baseline $\approx$ 90 °C << 1.05 V. The trends in mass activity decay, Figure 8b, reflect these trends in impact of operating conditions on Co loss

436

from the catalysts indicating that one of the primary sources of performance loss is catalyst restructuring. Using the same assumptions as described in the particle size study section, the depth of de-alloying was calculated to range from two to three monolayers for the 30% RH, baseline, and 90 °C cells and greater than three monolayers for the 1.05 V cell. This illustrates the destructive impact of increasing upper potential limits of cycling on the catalyst structure. Increased rates of Pt dissolution when cycling to upper potentials limits >1.0 V in aqueous electrolyte have been noted for polycrystalline Pt and Pt_3Co nanoparticle catalysts [44–47]. The increased extent of Pt dissolution and/or the multi-layer structure of the oxide formed at these higher potentials [48,49] may expose additional Co in the sub-surface layers to the acidic environment resulting in an increased depth of de-alloying.

To gain greater insights into the observed decay trends of the 4.9 nm Pt_3Co MEAs with varied operating conditions, the TEM-based catalyst particle size distribution analysis of the pre- and post-cycling electrodes is shown in Figure 10. All parametric conditions caused an increase in the mean diameter of the catalyst particles and an increasing in the tailing of the particle size distributions toward larger diameters. The extent of changes in the TEM particle size distributions are 30% RH < BL < 90 °C < 1.05 V. The trends agree with the ECA loss trends shown in Figure 8a. A quantitative comparison of the parametric decay trends between the Pt and Pt_3Co alloys is not attempted, due to the size difference in the catalyst particles (3.2 nm for pure Pt [35] and 4.9 nm for Pt_3Co), however, the qualitative trends are similar with no noticeable difference.

Table 5. The post-cycling Pt to Co ratios in the electrodes of the cells subjected to the parametric study square wave decay protocol.

Cell #	Description	Pt/Co Ratio
4	Baseline (BL)	7.8
5	Lower RH (30% RH)	6.6
6	Higher Upper Potential (1.05 V)	10.2
7	Higher Temperature (90 °C)	7.9

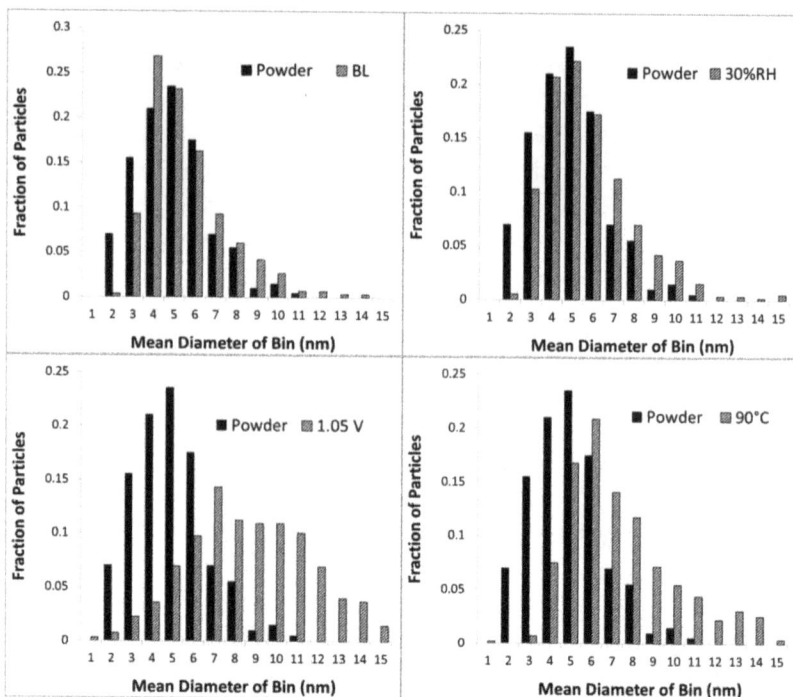

Figure 10. TEM analysis of Pt₃Co particle size distributions of the pristine (**black**) and decayed catalysts (**hashed**) from parametric studies.

3. Experimental Section

3.1. Catalyst Preparation

Catalysts were prepared by deposition of 40 wt.% Pt and Co (3:1 atomic ratio) on Akzo nobel Ketjen EC300J (Amsterdam, The Netherlands) via proprietary methods, then annealed at increasing temperatures (T1, T2 and T3) to achieve alloying and produce catalysts of specific particle sizes. Catalysts were characterized for total metal content by inductively-coupled plasma optical emission spectroscopy (ICP-OES) and X-ray fluorescence and metal surface area by gas-phase CO chemisorption. The results of these characterizations are summarized in Table 1. Powder X-ray diffraction (XRD), using a Bruker AXS D-500 diffractometer (Billerica, MA, USA) with a Cu Kα X-ray source, was used to determine the average Pt₃Co crystallite size (calculated using peak fitting and Rietveld analysis) and degree of alloying (shift in lattice parameter). The results of the XRD analysis are also summarized in Table 1. In some cases, the catalyst was pre-leached to remove the easily leachable Co. Pre-leaching was carried out by chemically treating the Pt₃Co/Ketjen EC 300J catalysts in air-saturated 0.5 M H_2SO_4 at 363 K for 24 h. The metal content of the leached materials was determined via ICP-OES analysis.

3.2. Transmission Electron Microscopy (TEM) Characterization of Pt_3Co/C Catalysts Annealed to Various Particle Sizes

A small portion of Pt_3Co/C catalyst was crushed and dusted onto a holey carbon film on a Cu TEM grid. The samples were analyzed using a JEOL 2010F TEM operated (Peabody, MA, USA) at 200 kV. Energy Dispersive X-ray (EDX) analysis was used to confirm the Pt and Co composition. Particle size distributions were determined from at least 200 particles using a procedure described elsewhere [50]. Figure 11 shows TEM bright-field images of 40 wt.% Pt_3Co/Ketjen EC 300J versions annealed at increasing temperatures (T1, T2 and T3) to generate larger average particle sizes.

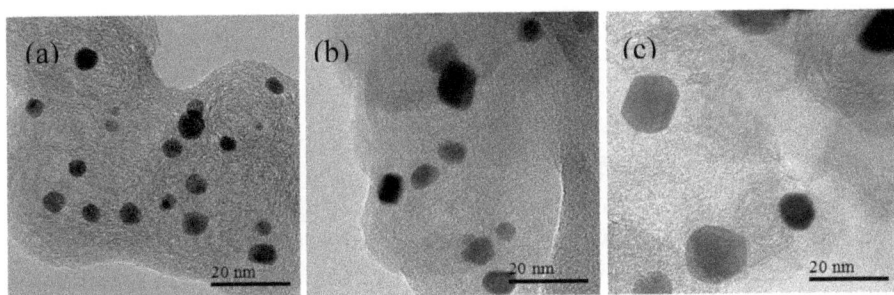

Figure 11. TEM bright-field images of as prepared (**a**) 4.9 nm (**b**) 8.1 nm and (**c**) 14.8 nm Pt_3Co cathode catalysts supported on carbon (Scale bar = 20 nm).

The particle size distributions are summarized in Figure 12, where the bin frequency has been chosen such that all samples can be compared on the same plot. More detailed particle size distribution analysis for individual samples showed mono-modal log normal distributions for all three 40 wt.% Pt_3Co/Ketjen EC 300J materials. Mean particle sizes were calculated from these images and are shown in Table 1.

3.3. Membrane Electrode Assembly (MEA)

In this study, MEAs were fabricated via a developmental process, with identical anode Pt/Ketjen EC 300J catalyst, 35.6 wt.% Pt/C (~2 nm Pt particle size) and the same 25 μm thickness perfluorosulfonic acid (PFSA) membrane. The cathode catalysts were the Pt_3Co/Ketjen EC 300J materials described in Table 1. Cathode electrodes all had the same Pt loading of 0.2 mg ± 0.02 mg Pt/cm^2 while the anode electrodes had a similar loading of 0.22 mg ± 0.05 mg Pt/cm^2.

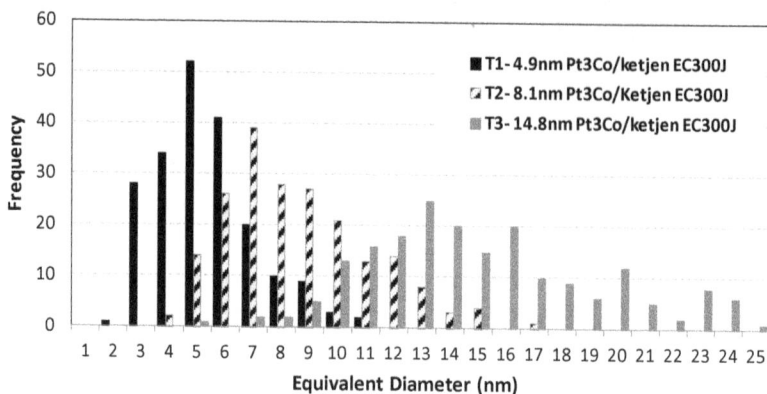

Figure 12. Particle size distributions of annealed Pt₃Co cathode catalysts determined by TEM.

3.4. Fuel Cell Construction

The MEA performance and durability tests utilized a 25 cm² active area fuel cell test fixture from Fuel Cell Technologies Inc. (Albuquerque, NM, USA). Prior to assembly in this fixture, the MEA was sandwiched between anode and cathode gas diffusion layers (GDLs) consisting of porous carbon paper coated on one side with a micro-porous layer (SGL25BC, SGL Group, Wiesbaden, Germany). This cell was then assembled between anode and cathode current collectors that had superimposed serpentine flow channels. The inlet and outlet was configured for co-flow of reactant gases. The assembled single cell was tested for cell resistance, gas crossover, and leakage. Subsequently, the cell was conditioned by scanning the current between 0.1 and 1.5 A/cm², in 0.1 A/cm² increments every 5 min, for 16 h. The cell was maintained at 80 °C and operated at 101 kPa pressure (absolute) and 100% relative humidity (RH). The conditioning scans were performed by alternating between H_2/Air and H_2/O_2 reactant flows, corresponding to 50% utilization (or 0.05 SPLM as minimum flow rates) on both sides.

3.5. Diagnostics and V-I Performance

Diagnostic tests to determine electrochemically-active surface area (ECA) and ORR mass activity (A/g-Pt at 0.9 V) are commonly used to characterize cathode catalyst degradation. Decreases in ECA and mass activity are indicative of Pt nanoparticle growth and $Pt^{2+/4+}$ ion migration and deposition within the cathode or membrane ionomer. Microscopic post-test analysis was used to assess which Pt re-distribution processes dominate for the various standard and accelerated degradation protocols utilized in this study.

440

3.6. Hydrogen Crossover

Molecular H_2 crossover rate from anode to cathode of the MEAs was measured with H_2 on anode and N_2 on cathode at 80 °C, 150 kPa (absolute). A positive linear sweep voltammogram was performed by scanning the potential of the cathode between 0 and 0.45 V (*vs.* anode) with a sweep rate of 1 mV/s. The magnitude of the oxidation current at 0.35 V (*vs.* anode) in the voltammogram was used to determine the H_2 crossover rate.

3.7. Electrochemically-Active Surface Area (ECA)

The ECA of the cathode was measured using a cyclic voltammogram (CV) between 0.03 V and 1.0 V (*vs.* anode). A sweep rate of 10 mV/s was used with 0.5 SLPM of 4% H_2 (balance nitrogen) flowing over the anode (as reference as well as counter electrode) and 0.5 SLPM of N_2 flowing over the cathode (as working electrode) at 80 °C, 101 kPa (absolute). ECA values (m^2/g-Pt) were calculated by integrating the hydrogen adsorption charge in the voltammogram (0.05 to 0.35 V), dividing by cell active area (cm^2), 210 $\mu C/cm^2$ (theoretical hydrogen monolayer adsorption on Pt) and the cathode initial Pt loading (mg-Pt/cm^2).

3.8. Cell Performance

Electrochemical performance of the MEAs, represented by a voltage-current (V-I) curve, was measured with an electronic load box (Model 890, Scribner Associates Inc., Southern Pines, NC, USA), by scanning current from the open circuit voltage (OCV) to 1500 mA/cm^2 and back to the OCV. The cells were purged with H_2 and N_2 on the anode and cathode, respectively, to reduce the OCV to ~0.1 V for ~10 min right before each V-I curve measurement to minimize the effect of platinum oxide, formed on the cathode catalyst at high potentials, on the cell performance. The V-I data collected during the decreasing current scan are reported. To focus the results on the cathode electrode characteristics, the cell voltage corrected for the membrane's ohmic loss (measured by the current interrupt method), the so-called IR-corrected voltage, is reported. The IR-corrected polarization plots are useful for quantifying the kinetic, ohmic and mass transport behaviors of the electrode.

H_2/O_2 performance was measured at 80 °C and 150 kPa (absolute) with H_2 on the anode and O_2 on the cathode with flow rates corresponding to 50% utilization (or 0.05 SPLM as minimum flow rates) on both sides. Mass activity (A/g-Pt) is the measured current (A/cm^2) @ 0.9 V in IR-corrected H_2/O_2 polarization plots, corrected by hydrogen crossover current (A/cm^2) and then normalized to initial Pt loading (g-Pt/cm^2) of cathode electrodes.

H_2/air performance was measured at 80 °C, 150 kPa (absolute) with H_2 on the anode and air on the cathode with flow rates of 1 SLPM and 2 SLPM, respectively.

441

3.9. Decay Protocol for Particle Size Studies

Accelerated cell degradation cycling protocols were imposed on the MEAs using a potentiostat (EG&G 273A, Princeton Applied Research Inc., Oak Ridge, TN, USA). The decay conditions are provided in Table 2. To elucidate the effect of cathode catalyst particle size on performance degradation, MEAs containing the three Pt_3Co cathode catalysts were subjected to a triangle-wave potential cycle between 0.6 V and 1.0 V with 50 mV/s ramp rate (16 s/cycle). The cells were maintained at 80 °C and had fixed gas flows of 0.1 SLPM H_2 on the anode and 0.05 SLPM N_2 on the cathode, both at 100% RH. During the potential cycling, the cathode served as working electrode, while the anode served as both reference and counter electrode. The cyclic decay tests were paused at specific intervals to evaluate the cathode ECA and cell V-I performance.

3.10. Decay Protocol for Parametric Studies

The accelerated cell degradation test conditions to elucidate the effects of cell operating parameters on MEAs containing the 4.9 nm mean diameter Pt_3Co cathode catalyst are shown in Table 3. A square-wave potential cycle with 20 s/cycle was imposed on the cell, using a potentiostat, at defined temperature, RH and voltage window conditions for 10,000 cycles. Fixed gas flows of 0.5 SLPM 4% H_2 (balance N_2) on anode and 0.5 SLPM N_2 on cathode were used. A new MEA with 4.9 nm mean particle size Pt_3Co cathode catalyst was used for each test to evaluate the impact of one parametric condition (high temperature, low RH, or higher upper potential limit) on cell decay behavior. During the potential cycling, the cathode served as working electrode, while the anode served as both reference and counter electrode. The cyclic decay tests were paused at specific intervals to evaluate the cathode ECA and cell V-I performance.

3.11. Electron MicroProbe Analysis (EMPA)

Narrow strips (~5 mm wide) were cut down the midline from the inlet to the outlet from pristine and decayed MEAs (5 cm × 5 cm active area) for cross-section microscopic analysis. The MEA strips were set into epoxy resin, polished, surface-coated with a uniform thin layer of carbon for electron microprobe analysis by a JEOL 8900 Super Probe (Peabody, MA, USA) equipped with a multiple wave length dispersive spectrometer (WDS) for simultaneous profiling of Pt, Co, and S distribution across the MEA cross-sections.

3.12. X-ray Fluorescence Analysis (XRF)

Portions of the cathode catalyst were removed from narrow strips of the fresh and decayed MEAs, identical to those used for EMPA, using adhesive tape.

The atomic ratio of Pt to Co in these catalyst samples were analyzed utilizing an Energy-Dispersive X-ray Fluorescence Spectrometer (Rigaku NEX CG EDXRF Analyzer with Polarization, The Woodlands, TX, USA), the copper and molybdenum targets of the spectrometer, and empirical fitting using the integrated intensity of the Pt L and the Co K.

4. Conclusions

Electrochemical decay protocols were used to accelerate performance loss of Pt_3Co catalyst based electrodes with three distinct mean particle sizes. The decay trends observed for the Pt_3Co alloy particles is compared with that of the similar study reported on platinum particles [36]. Over the broad range of particle sizes evaluated for the first time, a clear trend is emerging. While the initial performance of the smaller particle based catalysts is higher, the durability of the larger particle based catalysts is higher. The intermediate particle sizes of ~5 nm for Pt and ~8 nm for Pt_3Co catalysts seems to provide the best life averaged performance. Furthermore, it is interesting that in the current study the 5 nm size Pt catalyst based electrodes exhibited comparable mass activity as that of 4.9 nm Pt_3Co catalyst based electrodes and the durability is also comparable. This result is rather unexpected, as the literature trends suggest improved mass activity with alloying. The heat treatment conditions during catalyst preparation, catalyst ink preparation and MEA fabrication may have an important role to play in realizing the initial mass activity benefits reported in literature. The fraction of Co leached into the membrane is shown to be a function of the particle size and is lowest for the larger particle size.

The impact of operating conditions on the durability of the 4.9 nm Pt_3Co alloy catalyst based electrodes is also reported. The order of performance decay based on operational conditions, from greatest to least, for the parametric study was 1.05 V upper limit >> 90 °C > baseline > 30% RH. In addition to this qualitative trend, the approaches used are able to quantify the extent of catalyst damage incurred due to the fuel cell operation and the choice of catalysts used.

Acknowledgments: This work was performed under the DOE contract DE-AC02-06CH11357.

Author Contributions: Sarah Ball carried out the preparation of catalysts and MEAs, contributed to the characterization of catalyst powders. Zhiwei Yang and Mallika Gummalla performed the MEAs' in-cell tests and post-test EMPA analysis. Somaye Rasouli, Kang Yu and Paulo Ferreira carried out the TEM and XRD analysis. Deborah Myers performed the XRF measurements. All the authors contributed equally to the data interpretation and discussion. Mallika Gummalla and David Condit drafted the manuscript, and Deborah Myers and Zhiwei Yang revised the final version of paper.

Conflicts of Interest: The authors declare no conflict of interest.

References

1. Ball, S.C.; Hudson, S.L.; Theobald, B.R.C.; Thompsett, D. PtCo, a Durable Catalyst for Automotive PEMFC. *ECS Trans.* **2007**, *11*, 1267–1278.

2. Stamenkovic, V.R.; Mun, B.S.; Arenz, M.; Mayrhofer, K.J.J.; Lucas, C.A.; Wang, G.; Ross, P.N.; Markovic, N.M. Trends in Electrocatalysis on Extended and Nanoscale Pt-Bimetallic Alloy Surfaces. *Nat. Mater.* **2007**, *6*, 241–247.

3. Colón-Mercado, H.R.; Popov, B.N. Stability of Platinum Based Alloy Cathode Catalysts in PEM Fuel Cells. *J. Power Sources* **2006**, *155*, 253–263.

4. Stamenkovic, V.R.; Fowler, B.; Mun, B.S.; Wang, G.; Ross, P.N.; Lucas, C.A.; Markovic, N.M. Improved Oxygen Reduction Activity on Pt3Ni(111) via Increased Surface Site Availability. *Science* **2007**, *315*, 493–497.

5. Van der Vliet, D.F.; Wang, C.; Tripkovic, D.; Strmcnik, D.; Zhang, X.; Debe, M.K.; Atanasoski, R.T.; Markovic, N.M.; Stamenkovic, V.R. Mesostructured Thin Films as Electrocatalysts with Tunable Composition and Surface Morphology. *Nat. Mater.* **2012**, *11*, 1051–1058.

6. Wang, C.; van der Vliet, D.; Chang, K.; You, H.; Strmcnik, D.; Schlueter, J.A.; Markovic, N.M.; Stamenkovic, V.R. Monodisperse Pt3Co Nanoparticles as a Catalyst for the Oxygen Reduction Reaction: Size-Dependent Activity. *J. Phys. Chem.* **2009**, *113*, 19365–19368.

7. Paulusa, U.A.; Wokauna, A.; Scherera, G.G.; Schmidtb, T.J.; Stamenkovicb, V.; Markovicb, N.M.; Rossb, P.N. Oxygen Reduction on High Surface Area Pt-Based Alloy Catalysts in Comparison to Well Defined Smooth Bulk Alloy Electrodes. *Electrochem. Acta* **2002**, *47*, 3787–3798.

8. Huang, Y.; Zhang, J.; Kongkanand, A.; Wagner, F.T.; Li, J.C.M.; Jorné, J. Transient Platinum Oxide Formation and Oxygen Reduction on Carbon-Supported Platinum and Platinum-Cobalt Alloy Electrocatalysts. *J. Electrochem. Soc.* **2014**, *161*, F10–F15.

9. Stamenkovic, V.R.; Mun, B.S.; Mayrhofer, K.J.J.; Ross, P.N.; Markovic, N.M.; Rossmeisl, J.; Greeley, J.; Nørskov, J.K. Changing the Activity of Electrocatalysts for Oxygen Reduction by Tuning the Surface Electronic Structure. *Angew. Chem.* **2006**, *45*, 2897–2901.

10. Stamenkovic, V.R.; Mun, B.S.; Mayrhofer, K.J.J.; Ross, P.N.; Markovic, N.M. Effect of Surface Composition on Electronic Structure, Stability, and Electrocatalytic Properties of Pt-Transition Metal Alloys: Pt-Skin *versus* Pt-Skeleton Surfaces. *J. Am. Chem. Soc.* **2006**, *128*, 8813–8819.

11. Jalan, V.; Taylor, E.J. Importance of Interatomic Spacing in Catalytic Reduction of Oxygen in Phosphoric Acid. *J. Electrochem. Soc.* **1983**, *130*, 2299–2302.

12. Toda, T.; Igarashi, H.; Uchida, H.; Watanabe, M. Enhancement of the Electroreduction of Oxygen on Pt Alloys with Fe, Ni, and Co. *J. Electrochem. Soc.* **1999**, *146*, 3750–3756.

13. Min, M.K.; Cho, J.; Cho, K.; Kim, H. Particle size and Alloying Effects of Pt-Based Alloy Catalysts for Fuel Cell Applications. *Electrochem. Acta* **2000**, *45*, 4211–4217.

14. Mukerjee, S.; Srinivasan, S.; Soriaga, M.; McBreen, J. Role of Structural and Electronic Properties of Pt and Pt Alloys on Electrocatalysis of Oxygen Reduction an *in situ* XANES and EXAFS Investigation. *J. Electrochem. Soc.* **1995**, *142*, 1409–1422.

15. Roques, J.; Anderson, A.B.; Murthi, V.S.; Mukerjee, S. Potential Shift for OH_{ads} Formation on the Pt Skin on Pt_3Co (111) Electrodes in Acid. *J. Electrochem. Soc.* **2005**, *152*, E193–E199.

16. Dubau, L.; Maillard, F.; Chatenet, M.; Guetaz, L.; André, J.; Rossinot, E. Durability of Pt_3Co / C Cathodes in a 16 Cell PEMFC Stack: Macro/Microstructural Changes and Degradation Mechanisms. *J. Electrochem. Soc.* **2010**, *157*, B1887–B1895.

17. Dubaua, L.; Lopez-Haroa, M.; Castanheiraa, L.; Dursta, J.; Chateneta, M.; Bayle-Guillemaudb, P.; Guétazc, L.; Caquéd, N.; Rossinotd, E.; Maillard, F. Probing the Structure, the Composition and the ORR Activity of Pt_3Co/C nanocrystallites during a 3422 h PEMFC Ageing Test. *Appl. Catal. B* **2013**, *142–143*, 801–808.

18. Chen, S.; Gasteiger, H.A.; Hayakawa, K.; Tada, T.; Yang, S. Platinum-Alloy Cathode Catalyst Degradation in Proton Exchange Membrane Fuel Cells: Nanometer-Scale Compositional and Morphological Changes. *J. Electrochem. Soc.* **2010**, *157*, A82–A97.

19. Shao, M.; Peles, A.; Shoemaker, K. Electrocatalysis on Platinum Nanoparticles: Particle Size Effect on Oxygen Reduction Reaction Activity. *Nano Letters* **2011**, *11*, 3714–3719.

20. Kinoshita, K. Particle Size Effects for Oxygen Reduction on Highly Dispersed Platinum in Acid Electrolytes. *J. Electrochem. Soc.* **1990**, *137*, 845–848.

21. Han, B.C.; Miranda, C.R.; Ceder, G. Effect of Particle Size and Surface Structure on Adsorption of O and OH on Platinum Nanoparticles: A First-Principles Study. *Phys. Rev.* **2008**, *B77*, 075410.

22. Tritsaris, G.A.; Greeley, J.; Rossmeisl, J.; Nørskov, J.K. Atomic-Scale Modeling of Particle Size Effects for the Oxygen Reduction Reaction on Pt. *Catal. Lett.* **2011**, *141*, 909–913.

23. Wang, D.; Xin, H.L.; Hovden, R.; Wang, H.; Yu, Y.; Muller, D.A.; DiSalvo1, F.J.; Abruña, H.D. Structurally Ordered Intermetallic Platinum-Cobalt Core-Shell Nanoparticles with Enhanced Activity and Stability as Oxygen Reduction Electrocatalysts. *Nat. Mater.* **2013**, *12*, 81–87.

24. Chen, S.; Sheng, W.; Yabuuchi, N.; Ferreira, P.J.; Allard, L.F.; Yang, S. Origin of Oxygen Reduction Reaction Activity on "Pt_3Co" Nanoparticles: Atomically Resolved Chemical Compositions and Structures. *J. Phys. Chem. C* **2009**, *113*, 1109–1125.

25. Gasteiger, H.A.; Kocha, S.S.; Sompalli, B.; Wagner, F.T. Activity Benchmarks and Requirements for Pt, Pt-Alloy, and Non-Pt Oxygen Reduction Catalysts for PEMFCs. *Appl. Catal. B* **2005**, *56*, 9–35.

26. Nesselberger, M.; Ashton, S.; Meier, J.C.; Katsounaros, I.; Mayrhofer, K.J.J.; Arenz, M. The Particle Size Effect on the Oxygen Reduction Rreaction Activity of Pt Catalysts: Influence of Electrolyte and Relation to Single Crystal Models. *J. Am. Chem. Soc.* **2011**, *133*, 17428–17433.

27. Mayrhofer, K.J.J.; Blizanac, B.B.; Arenz, M.; Stamenkovic, V.R.; Ross, P.N.; Markovic, N.M. The Impact of Geometric and Surface Electronic Properties of Pt-Catalysts on the Particle Size Effect in Electrocatalysis. *J. Phys. Chem. B* **2005**, *109*, 14433–14440.

28. Aindow, T.T.; Bi, W.; Izzo, E.; Motupally, S.; Murthi, V.S.; Perez-Acosta, C. Structure-Activity-Durability relationship of Pt and Pt based Alloy Electrocatalysts. In Proceedings of 215th Meeting of the Electrochemical Society, San Francisco, CA, USA, 24–29 May 2009.

29. Bi, W.; Izzo, E.; Murthi, V.S.; Perez-Acosta, C.; Lisitano, J.; Protsailo, L.V. Durability of Low Temperature Hydrogen PEM Fuel Cells with Pt and Pt Alloy ORR Catalysts. In Proceedings of 218th Meeting of the Electrochemical Society, Las Vegas, NV, USA, 10–15 October 2010.

30. Matsutani, K.; Hayakawa, K.; Tada, T. Effect of Particle Size of Platinum and Platinum-Cobalt Catalysts on Stability against Load Cycling. *Platin. Met. Rev.* **2010**, *54*, 223–232.

31. Strasser, P.; Koh, S.; Anniyev, T.; Greeley, J.; More, K.; Yu, C.; Liu, Z.; Kaya, S.; Nordlund, D.; Ogasawara, H.; Toney, M.F.; Nilsson, A. Lattice-Strain Control of the Activity in Dealloyed Core-Shell Fuel Cell Catalysts. *Nat. Chem.* **2010**, *2*, 454–460.

32. Adzic, R.R.; Zhang, J.; Sasaki, K.; Vukmirovic, M.B.; Shao, M.; Wang, J.X.; Nilekar, A.U.; Mavrikakis, M.; Valerio, J.A.; Uribe, F. Platinum Monolayer Fuel Cell Electrocatalysts. *Top. Catal.* **2007**, *46*, 249–262.

33. Vukmirovic, M.B.; Zhang, J.; Sasaki, K.; Uribe, F.; Mavrikakis, M.; Adzic, R.R. Platinum Monolayer Electrocatalysts for Oxygen Reduction. *Electrochem. Acta* **2007**, *52*, 2257–2263.

34. Sasaki, K.; Wang, J.X.; Naohara, H.; Marinkovic, N.; More, K.; Inada, H.; Adzic, R.R. Recent Advances in Platinum Monolayer Electrocatalysts for Oxygen Reduction Reaction: Scale-up Synthesis, Structure and Activity of Pt Shells on Pd Cores. *Electrochem. Acta* **2010**, *55*, 2645–2652.

35. Yang, Z.; Ball, S.C.; Condit, D.A.; Gummalla, M. Systematic Study on the Impact of Pt Particle Size and Operating Conditions on PEMFC Cathode Catalyst Durability. *J. Electrochem. Soc.* **2011**, *158*, B1439–B1445.

36. Carter, R.N.; Kocha, S.S.; Wagner, F.; Fay, M.; Gasteiger, H.A. Artifacts in Measuring Electrode Catalyst Area of Fuel Cells through Cyclic Voltammetry. *ECS Trans.* **2007**, *11*, 403–410.

37. Shao, M.; Odell, J.H.; Choi, S.; Xia, Y. Electrochemical Surface Area Measurements of Platinum- and Palladium-Based Nanoparticles. *Electrochem. Commun.* **2013**, *31*, 46–48.

38. Van der Vliet, D.F.; Wang, C.; Li, D.; Paulikas, A.P.; Greeley, J.; Rankin, R.B.; Strmcnik, D.; Tripkovic, D.; Markovic, N.M.; Stamenkovic, V.R. Unique Electrochemical Adsorption Properties of Pt-Skin Surfaces. *Angew. Chem.* **2012**, *51*, 3139–3142.

39. Yan, Q.; Wu, J. Modeling of Single Catalyst Particle in Cathode of PEM Fuel Cells. *Energy Convers. Manage.* **2008**, *49*, 2425–2433.

40. Greszler, T.A.; Moylan, T.E.; Gasteiger, H.A. Modeling the Impact of Cation Contamination in a Polymer Electrolyte Membrane Fuel Cell. In *Handbook of Fuel Cells—Fundamentals, Technology and Applications*; John Wiley & Sons Ltd.: Chichester, UK, 2009.

41. Chen, S.; Ferreira, P.J.; Sheng, W.; Yabuuchi, N.; Allard, L.; Yang, S. Enhanced Activity for Oxygen Reduction Reaction on "Pt$_3$Co" Nanoparticles: Direct Evidence of Percolated and Sandwich-Segregation Structures. *J. Am. Chem. Soc.* **2008**, *130*, 13818–13819.

42. Carlton, C.E.; Chen, S.; Ferreira, P.J.; Allard, L.F.; Yang, S. Sub-Nanometer-Resolution Elemental Mapping of "Pt$_3$Co" Nanoparticle Catalyst Degradation in Proton-Exchange Membrane Fuel Cells. *J. Phys. Chem. Lett.* **2012**, *3*, 161–166.

43. Yang, S.; Sheng, W.C.; Chen, S.; Ferreira, P.J.; Holby, E.F.; Morgan, D. Instability of Supported Platinum Nanoparticles in Low-Temperature Fuel Cells. *Top. Catal.* **2007**, *46*, 285–305.

44. Kinoshita, K.; Lundquist, J.T.; Stonehart, P.J. Potential Cycling Effects on Platinum Electrocatalyst Surfaces. *J. Electroanal. Chem. Interfacial Electrochem.* **1973**, *48*, 157–166.

45. Rand, D.A.J.; Woods, R.J. A Study of the Dissolution of Platinum, Palladium, Rhodium and Gold Electrodes in 1 m Sulphuric Acid by Cyclic Voltammetry. *J. Electroanal. Chem. Interfacial Electrochem.* **1972**, *35*, 209–218.

46. Wang, X.; Myers, D.J.; Kariuki, N.; Kumar, R. Dissolution of Platinum and Platinum Alloy PEFC Cathode Electrocatalysts. In Proceedings of 211th Meeting of the Electrochemical Society, Chicago, Illinois, USA, 6–10 May 2007.

47. Wang, X.; Myers, D.J.; Smith, M.C.; Mawdsley, J.; Kumar, R. Dissolution of Platinum-based PEFC Cathode Electrocatalysts. In Proceedings of 214th Meeting of the Electrochemical Society, Honolulu, HI, USA, 12–17 October 2008.

48. Teliska, M.; O'Grady, W.E.; Ramaker, D.E. Determination of O and OH Adsorption Sites and Coverage *in situ* on Pt Electrodes from PtL$_{23}$ X-ray Absorption Spectroscopy. *J. Phys. Chem. B* **2005**, *109*, 8076–8084.

49. Imai, H.; Izumi, K.; Matsumoto, M.; Kubo, Y.; Kato, K.; Imai, Y. *In situ* and Real-Time Monitoring of Oxide Growth in a Few Monolayers at Surfaces of Platinum Nanoparticles in Aqueous Media. *J. Am. Chem. Soc.* **2009**, *131*, 6293–6300.

50. Groom, D.J. The Effect of Nanocatalyst Size on Performance and Degradation in the Cathode of Proton Exchange Membrane Fuel Cells. M.S. Thesis, University of Texas at Austin, December 2011.

Oxygen Reduction Reaction Activity and Durability of Pt Catalysts Supported on Titanium Carbide

Morio Chiwata, Katsuyoshi Kakinuma, Mitsuru Wakisaka, Makoto Uchida, Shigehito Deki, Masahiro Watanabe and Hiroyuki Uchida

Abstract: We have prepared Pt nanoparticles supported on titanium carbide (TiC) (Pt/TiC) as an alternative cathode catalyst with high durability at high potentials for polymer electrolyte fuel cells. The Pt/TiC catalysts with and without heat treatment were characterized by X-ray diffraction (XRD), X-ray photoelectron spectroscopy (XPS), and transmission electron microscopy (TEM). Hemispherical Pt nanocrystals were found to be dispersed uniformly on the TiC support after heat treatment at 600 °C in 1% H_2/N_2 (Pt/TiC-600 °C). The electrochemical properties (cyclic voltammetry, electrochemically active area (ECA), and oxygen reduction reaction (ORR) activity) of Pt/TiC-600 °C and a commercial Pt/carbon black (c-Pt/CB) were evaluated by the rotating disk electrode (RDE) technique in 0.1 M $HClO_4$ solution at 25 °C. It was found that the kinetically controlled mass activity for the ORR on Pt/TiC-600 °C at 0.85 V (507 A g^{-1}) was comparable to that of c-Pt/CB (527 A g^{-1}). Moreover, the durability of Pt/TiC-600 °C examined by a standard potential step protocol (E = 0.9 V↔1.3 V *vs.* RHE, holding 30 s at each E) was much higher than that for c-Pt/CB.

Reprinted from *Catalysts*. Cite as: Chiwata, M.; Kakinuma, K.; Wakisaka, M.; Uchida, M.; Deki, S.; Watanabe, M.; Uchida, H. Oxygen Reduction Reaction Activity and Durability of Pt Catalysts Supported on Titanium Carbide. *Catalysts* **2015**, 5, 966–980.

1. Introduction

Polymer electrolyte fuel cells (PEFCs) have been extensively investigated for potential applications in fuel cell vehicles (FCVs) and residential co-generation systems. The reduction of the amount of Pt used in the cathode catalyst layers (CLs) is indispensable for the large-scale commercialization. To obtain high mass activity (MA) for the oxygen reduction reaction (ORR), it is essential to increase the electrochemically active area (ECA) for the ORR at minimum Pt loading in the CLs. So far, Pt nanoparticles with ECA values as large as 100 m^2 g_{Pt}^{-1} have been dispersed on high-surface-area (HSA) supports such as carbon black (CB, e.g., S_{CB} = 800 m^2 g^{-1}). However, a severe degradation of the CB support of the Pt/CB cathode catalysts has been recognized at high potentials, especially during the start-stop cycles of

448

FCVs [1–6]. It is known that the corrosion rate of carbon itself is low even under PEFC operating conditions, but the rate is accelerated by Pt catalyst loading with increasing temperature and potential [2,7–9]. The corrosion of the carbon support leads to agglomeration (sintering) and/or a detachment of Pt nanoparticles from the surface, together with a reduction of the electronic conductance in the CL [1,10–17]. Thus, the ECA for the ORR decreases significantly. It is, therefore, essential to develop novel cathode catalysts with both high MA for the ORR and high durability at high electrode potentials up to 1.5 V *vs.* reversible hydrogen electrode (RHE) [3–6].

So far, electronic conductive oxides or nitrides have been examined as stable supports for PEFCs, e.g., Pt/SnO_2 [18,19], Pt/TiO_2 [20–23], Pt/Ti_4O_7 [24,25], Pt/TiN [26], among others. The support materials used are typically in the form of nanoparticles with HSA to disperse Pt catalyst particles uniformly, but the use of HSA supports often leads to a high contact resistance between the particles. Recently, Kakinuma *et al.* have developed Sb-, Nb- and Ta-doped $SnO_{2-\delta}$ nanoparticle supports with a fused aggregated structure having both HSA and low contact resistance [27–30]. They reported that Pt-dispersed $Nb–SnO_{2-\delta}$ and $Ta–SnO_{2-\delta}$ exhibited both higher ORR activity and higher durability at high potentials than those for commercial Pt/CB (c-Pt/CB) catalysts. It was also found that the kinetically controlled specific ORR activities on various $Pt/Nb–SnO_{2-\delta}$ catalysts increased with increasing apparent electrical conductivity of the support [29].

Here, we focus on the support material having high electrical conductivity together with chemical stability at high potentials in acidic media. Titanium carbide (TiC) exhibits high electrical conductivity. For example, the conductivity of bulk TiC has been reported to be as high as 1.5×10^4 S cm^{-1} [31], which is approximately one order of magnitude higher than that of bulk $Ta–SnO_{2-\delta}$ [32]. In strong acidic media and high potentials, TiC is chemically and electrochemically stable. Indeed, several reports are available for the application of TiC or TiC-based materials to bipolar plates in phosphoric acid fuel cells (PAFCs) [33], Pt/TiC cathode catalysts for PAFCs [34], and Ir-dispersed TiC as the anode catalyst (O_2 evolution) in a proton exchange membrane water electrolysis system [35].

In this paper, we have examined the ORR activity and durability of Pt supported on TiC nanoparticles (Pt/TiC) by the use of the rotating disk electrode (RDE) technique. PtO nanoparticles were first dispersed on the TiC support by a colloidal method [26,36,37]. After a heat treatment at 600 °C in 1% H_2/N_2, hemispherical Pt nanoparticles with clear lattice fringes were found to be well dispersed on the TiC support (Pt/TiC-600 °C). The Pt/TiC-600 °C thus prepared exhibited high MA for the ORR, comparable to that of a c-Pt/CB, with much higher durability at high potentials.

2. Results and Discussion

2.1. Characterization of Pt/TiC Catalysts

Figure 1 shows X-ray diffraction (XRD) patterns of various Pt/TiC catalysts. The sharp peaks at $2\theta = 36°$ and $42°$ for both samples were assigned to cubic TiC (111) and TiC (200), respectively. The broad peaks at $2\theta = 40°$ and $46°$ for the catalysts heat-treated at 600 °C were assigned to Pt (111) and Pt (200), respectively. The Pt crystallite size d_{XRD}, calculated from Scherrer's equation for the XRD peak at *ca.* 46°, was 3.8 nm for the Pt/TiC-600 °C. However, none of peaks assigned to Pt were observed for the as-prepared Pt/TiC catalyst, suggesting that the supported particles were not metallic platinum.

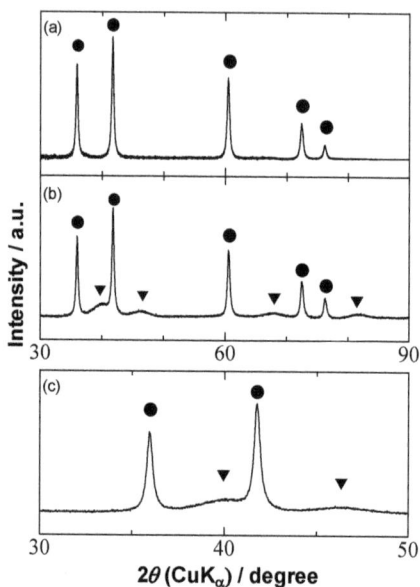

Figure 1. X-ray diffraction patterns for (**a**) Pt/TiC as-prepared and (**b**) (**c**) heat-treated at 600 °C (Pt/TiC-600 °C). The panel (**c**) is the enlarged XRD pattern from 30° to 50° for Pt/TiC-600 °C. The assignment of peaks is shown by (●) cubic TiC and (▾) Pt. The peaks in (**a**) from low diffraction angles to high angles correspond to the lattice distance of TiC (111), (200), (220), (311), and (222). The peaks in (**b**) marked with ▾ correspond to the lattice distance of Pt (111), (200), (220), and (311) from low diffraction angle to high angle, respectively.

The X-ray photoelectron spectra of as-prepared Pt/TiC and Pt/TiC-600 °C are shown in Figure 2. The formation of Pt(II) oxide (PtO) was confirmed for the as-prepared Pt/TiC catalyst from the Pt 4f core-level region in Figure 2a. After the heat treatment at 600 °C in N_2 containing 1% H_2, the peak of metallic Pt (Pt^0)

appeared, with significant diminishing of the PtO peak, which is consistent with the XRD results described above. In Figure 2b, we observed a broad peak assigned to Ti^{4+}, presumably TiO_2, besides the main peak assigned to Ti^{3+} in the TiC phase. The heat treatment at 600 °C in N_2 containing 1% H_2 resulted in a decrease of the Ti^{4+} peak with a low-energy shift. Such a shift has also been ascribed to the reduction of TiO_2 [38].

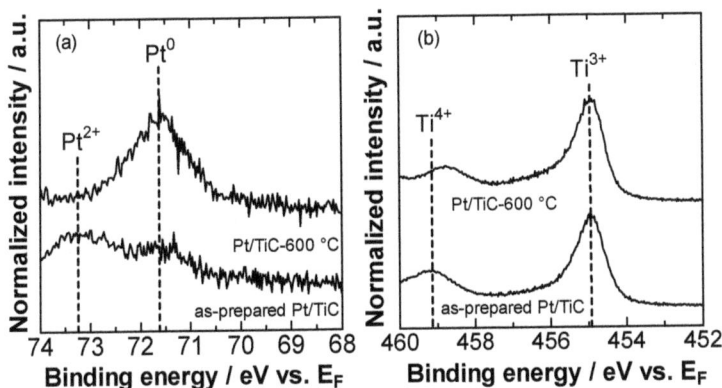

Figure 2. X-ray photoelectron spectra of as-prepared Pt/TiC and Pt/TiC-600 °C in the binding energy regions of (**a**) Pt $4f_{7/2}$ and (**b**) Ti $2p_{3/2}$.

Figure 3 shows TEM images of as-prepared Pt/TiC and Pt/TiC-600 °C, together with the particle size distribution histograms. PtO or Pt particles were well dispersed on the TiC support for both samples. The average Pt particle size d_{TEM} of as-prepared Pt/TiC and Pt/TiC-600 °C were 1.9 ± 0.4 nm and 3.7 ± 1.0 nm, respectively. It was seen in a typical high resolution image (Figure 3b) for the as-prepared Pt/TiC that a dome-shaped particle (presumably PtO) was covered with a thin amorphous layer. After the reduction at 600 °C (Pt/TiC-600 °C, Figure 3d), clear fringes corresponding to the (111) lattice distance of Pt (0.224 nm) were observed, without any thin amorphous layer.

Considering the XPS results shown in Figure 2, the thin amorphous layer observed in Figure 3d can be assigned with certainty to TiO_2, which was reduced at 600 °C in 1% H_2. The d_{TEM} of Pt on Pt/TiC-600 °C accords well with the crystallite sizes d_{XRD}, i.e., each Pt particle observed by TEM was a single crystallite. The Pt loading amount on the Pt/TiC-600 °C was quantified to be 10.3 wt % (see Experimental section). Thus, we clarified that Pt nanocrystals were formed on the TiC support by the reduction of TiO_2-covered PtO particles, followed by agglomeration. It is also noted that most of Pt nanocrystals dispersed on the TiC support were hemispherical as seen in Figure 3d, suggesting a strong interaction between Pt and the support.

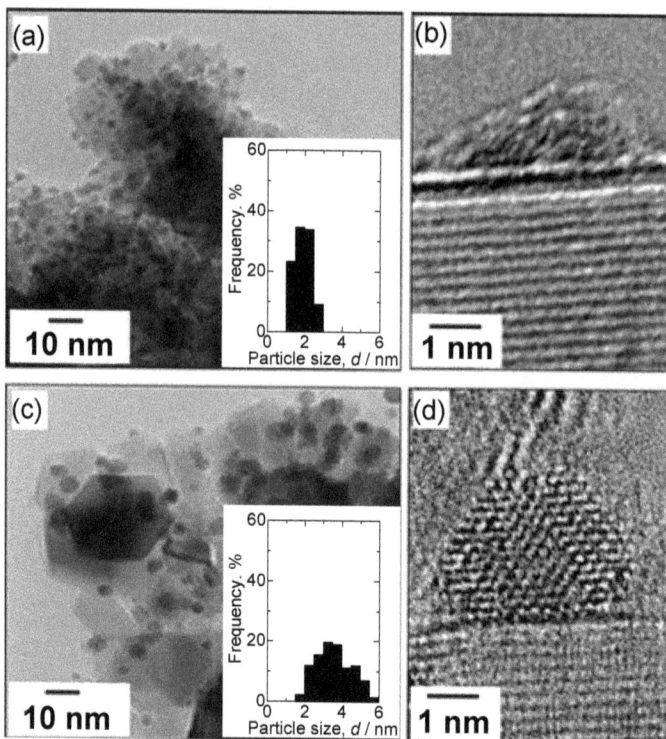

Figure 3. Transmission electron microscopic (TEM) images of as-prepared Pt/TiC (**a**) (**b**) and Pt/TiC-600 °C (**c**) (**d**), together with the Pt particle size distribution histograms.

2.2. Electrochemical Characterization of Pt/TiC Catalysts

Figure 4 shows the cyclic voltammograms (CVs) of the Nafion-coated Pt/TiC-600 °C and c-Pt/CB electrodes in N_2-purged 0.1 M $HClO_4$ solution measured at 25 °C. For both electrodes, the hydrogen adsorption/desorption peaks were clearly observed at potentials below 0.4 V. The oxidation of Pt commenced at approximately 0.8 V in the positive-going scan, while the reduction peak was seen at 0.75 V in the negative-going scan. The ECA values of Pt/TiC-600 °C and c-Pt/CB, which were evaluated from the hydrogen adsorption charge in Figure 4, were 75 m^2 g_{Pt}^{-1} and 80 m^2 g_{Pt}^{-1} [39], respectively. Assuming a spherical shape for the Pt particles with d_{TEM}, the specific surface area was calculated to be 76 m^2 g_{Pt}^{-1} for Pt/TiC-600 °C and 127 m^2 g_{Pt}^{-1} for c-Pt/CB. This suggests that nearly all Pt particles for the Pt/TiC-600 °C catalyst can easily contact the electrolyte solution, whereas an appreciable fraction of the Pt particles in the c-Pt/CB catalyst cannot contact the electrolyte solution. It has been reported that nearly half of the Pt particles for c-Pt/CB were located in the interiors of carbon black particles [40].

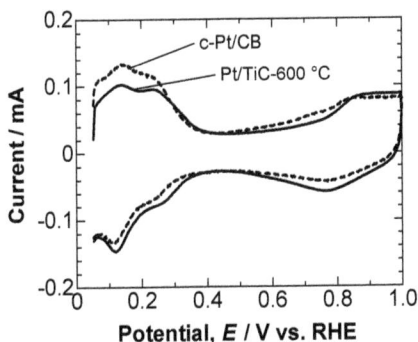

Figure 4. Cyclic voltammograms for Pt/TiC-600 °C and c-Pt/CB in N_2-saturated 0.1 M $HClO_4$ at a sweep rate of 0.1 V s^{-1}.

The ORR was examined by the RDE technique in O_2-saturated 0.1 M $HClO_4$ solution at 25 °C. Hydrodynamic voltammograms for the ORR at Pt/TiC-600 °C and c-Pt/CB electrodes are shown in Figure 5. Both Pt/TiC-600 °C and c-Pt/CB electrodes exhibited nearly identical onset potential (0.98 V) for the ORR. The ORR current reached a diffusion limit at about 0.4 V. Then, the limiting current-corrected current, $I_{LCC} = I \times I_L/(I_L - I)$, was calculated at 1500 rpm. According to the Koutecky-Levich equation, I_{LCC} is equivalent to the kinetically-controlled current I_k.

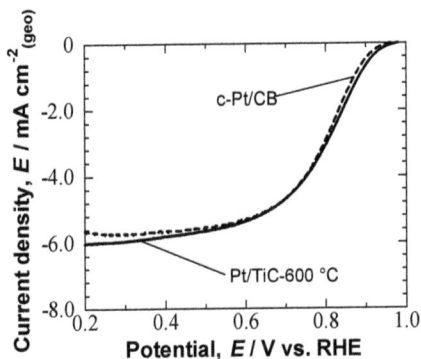

Figure 5. Hydrodynamic voltammograms for the ORR at Nafion-coated Pt/TiC-600 °C and c-Pt/CB in O_2-saturated 0.1 M $HClO_4$ solution at 25 °C. Rotating rate was 1500 rpm, and the potential sweep rate was 5 mV s^{-1}.

Figure 6 shows Tafel plots (E vs. log $|I_{LCC}|$) for the ORR at Pt/TiC-600 °C and c-Pt/CB. The Pt/TiC-600 °C showed two Tafel slope regions, similar to the case of c-Pt/CB: *ca.* -60 mV decade^{-1} in the high potential region $E > 0.9$ V, and *ca.* -120 mV decade^{-1} in the low potential region E 0.85 V, being in agreement with those reported for bulk-Pt or Pt/CB [41]. Therefore, the rate determining step for the ORR at Pt/TiC-600 °C is identical with that at Pt/CB or bulk-Pt.

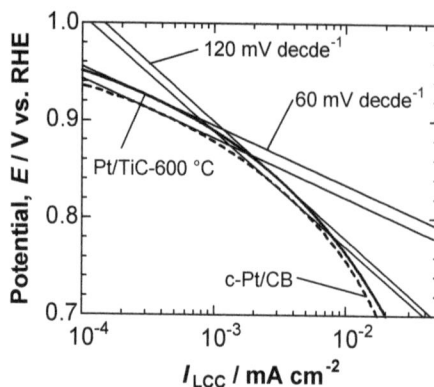

Figure 6. Tafel plots for the ORR at Pt/TiC-600 °C and c-Pt/CB in O_2-saturated 0.1 M HClO$_4$ solution at 25 °C with the rotating rate of 1500 rpm and the potential sweep rate of 5 mV s^{-1}.

The kinetically-controlled currents I_k at given potentials E were determined based on the Koutecky-Levich equation,

$$1/I = 1/I_k + 1/(0.62\, n\, F\, S\, D^{2/3}\, C_O\, \nu^{-1/6}\, \omega^{1/2}) \tag{1}$$

where n is the number of electrons transferred, F is the Faraday constant, S is the effective projected area of the Pt catalyst, D is the diffusion coefficient of O_2, C_O is the oxygen concentration, ν is the viscosity of the electrolyte and ω is the angular velocity. An example of the Koutecky-Levich plot for the ORR on the Nafion-coated Pt/TiC-600 °C is shown in Figure 7. Linear relationships with a constant slope are seen at all of the potentials, 0.85, 0.80 and 0.76 V. By extrapolating $\omega^{-1/2}$ to 0 (infinite mass transport rate), the value of the kinetically controlled current I_k was calculated. The kinetically-controlled specific activity (j_k) and mass activity (MA) were calculated based on the ECA value and the amount of Pt initially loaded on the working electrode, respectively. The value of j_k of Pt/TiC-600 °C at 0.85 V was 0.70 mA cm^{-2}, which was approximately 1.4 times higher than that of c-Pt/CB. Similar enhancement factors of the j_k were also reported for Pt/Nb–SnO$_{2-\delta}$ and Pt/Ta–SnO$_{2-\delta}$ [29,30]. The value of MA of Pt/TiC-600 °C at 0.85 V (507 A g^{-1}) was, however, comparable to that of c-Pt/CB (527 A g^{-1}), since the ECA of Pt/TiC-600 °C was smaller than that of c-Pt/CB.

So far, the MA or j_k at 0.90 V has been evaluated in both RDE cells using 0.1 M HClO$_4$ electrolyte solution and conventional membrane-electrode assemblies (MEAs), e.g., with 0.40 mg$_{Pt}$ cm^{-2} loading operated with air of 150 kPa$_{absolute}$

454

humidified at 100% RH [42]. In contrast, the current density at 0.90 V is not completely kinetically-controlled in recent MEAs with less Pt loading of 0.04 mg_{Pt} cm^{-2} and a thin electrolyte membrane operated under ambient pressure at low humidity (30% RH) [43,44]. We have therefore judged that the MA measured at 0.85 V is more appropriate, considering the actual operating conditions of PEFCs [44]. However, in order to compare the ORR activity of our Pt/TiC-600 °C with values in the literature, we have also evaluated the j_k at 0.90 V and 25 °C with the potential sweep rate of 5 mV s^{-1} to be 0.17 mA cm^{-2}. The value of j_k is consistent with those of Pt/CB catalysts at 0.90 V and 60 °C (with the same sweep rate) [42]. Although the j_k values summarized in the literature were evaluated at higher temperature than the present work, such an accordance is certainly due to a small effect of temperature on j_k for the ORR since an increase in the ORR activity with increasing temperature is almost cancelled by the decrease in O_2 solubility [45]. Recently, Ignaszak et al. prepared a Pt/TiC catalyst with similar Pt size d_{TEM} = 3.1 nm by a microwave-assisted polyol process [46]. They reported a j_k value at 0.90 V of 0.024 mA cm^{-2}, which is only 1/7 of our value. It is also noted that the value of ECA reported was 40 m^2 g_{Pt}^{-1}, which is approximately 1/2 that of our catalyst (76 m^2 g_{Pt}^{-1}). The most important difference, we consider, is that Ignaszak et al. did not carry out any heat treatment after dispersing the Pt on TiC. As described above, the heat treatment in H_2-containing atmosphere was found to be essential to remove the thin amorphous TiO_2 layer from the Pt surface. Because the current density during the ORR is higher than that of the CV (hydrogen adsorption/desorption), it is reasonable that the effect of the oxide layer on Pt and/or the Pt–TiC interface would be more pronounced for the j_k values for the ORR than it would be for the ECA values.

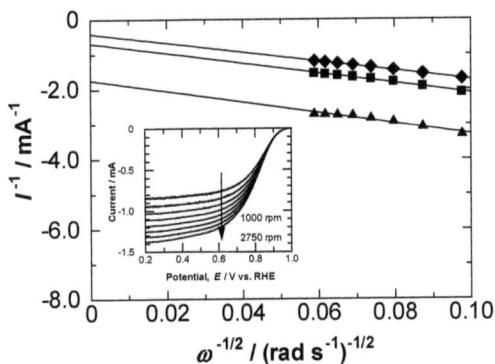

Figure 7. Koutecky-Levich plots obtained from hydrodynamic voltammograms for the ORR (shown in the inset) at (▲) 0.85 V, (●) 0.80 V and (♦) 0.76 V vs. RHE at Nafion-coated Pt/TiC-600 °C electrode in O_2-saturated 0.1 M $HClO_4$ solution at 25 °C.

2.3. Durability of Pt/TiC-600 °C in the Potential Step Cycle Test

Then, we have examined the durability of the Pt/TiC catalyst at high potentials. Figure 8a shows the changes in the ECA values of the Nafion-coated Pt/TiC-600 °C and c-Pt/CB electrodes during the potential step cycle test, simulating the start-stop cycles of the FCV. The ECA values of c-Pt/CB decreased quickly after 100 cycles, whereas the ECA values of Pt/TiC-600 °C decreased slowly. As a measure of the durability, we defined $N_{1/2,ECA}$, i.e., the value of N at which ECA had decreased to 1/2 of the initial value of c-Pt/CB. It is clear that Pt/TiC-600 °C showed a much lower rate of ECA decrease; the $N_{1/2,ECA}$ value for Pt/TiC-600 °C was 12 times larger than that for c-Pt/CB. Figure 8b shows changes in the MA at 0.85 V ($MA_{0.85V}$) for the ORR on the Nafion-coated Pt/TiC-600 °C and c-Pt/CB electrodes as a function of log N. The Pt/TiC-600 °C exhibited a much lower rate of MA decrease than c-Pt/CB. The value of N at which MA had decreased to 1/2 of the initial value, $N_{1/2,MA}$, for Pt/TiC-600 °C was 11 times larger than that for c-Pt/CB. These results suggest that the decrease in the MA of Pt/TiC-600 °C can be ascribed mainly to the decrease in ECA.

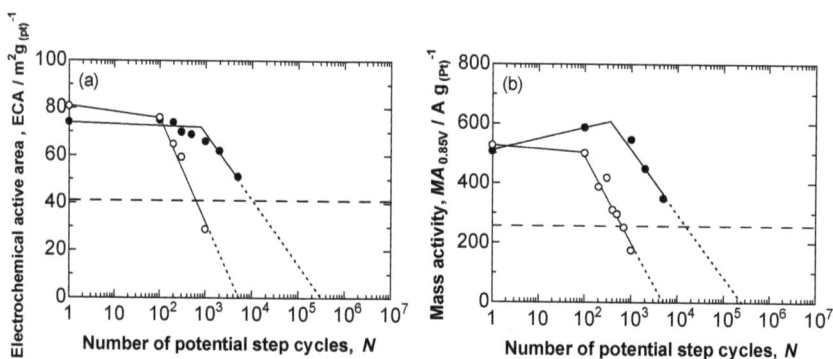

Figure 8. (a) Plots of ECA at Nafion-coated (●) Pt/TiC-600 °C and (○) c-Pt/CB electrodes as a function of log N. (b) Plots of $MA_{0.85V}$ at Nafion-coated (●) Pt/TiC-600 °C and (○) c-Pt/CB electrodes at 0.85 V as a function of log N. Each dashed line indicates 1/2 of the initial value of (a) ECA and (b) $MA_{0.85V}$ for c-Pt/CB.

Figure 9 shows the TEM images of Pt/TiC-600 °C and c-Pt/CB before and after the durability test ($N = 5000$). As is well known, the CB support of c-Pt/CB corrodes severely at high potentials [17,39]. Many Pt particles were found to be detached from the CB support, in addition to the agglomeration of Pt particles. It was found that Pt/TiC-600 °C exhibited high durability at high potentials, and Pt particles were not detached from the TiC support. The slow decrease of the ECA and MA values of Pt/TiC-600 °C can certainly be ascribed to an agglomeration of Pt particles. In contrast, Ignaszak et al. claimed [46] that their Pt/TiC lost 78% of its original

ORR activity after only 500 potential cycles between 0.05 V and 1.2 V at 20 mV s^{-1}. Although the test protocol (upper limit and lower limit potential, potential sweep *vs.* potential step) was different, our Pt/TiC-600 °C catalyst exhibited superior durability, even with a higher potential being used, *i.e.*, 1.3 V. Thus, we have confirmed that the removal of the TiO$_2$ layer, as we performed for Pt/TiC-600 °C, was very important to obtain both high ORR activity and high durability at high potentials The next target will be to examine this catalyst in the MEA.

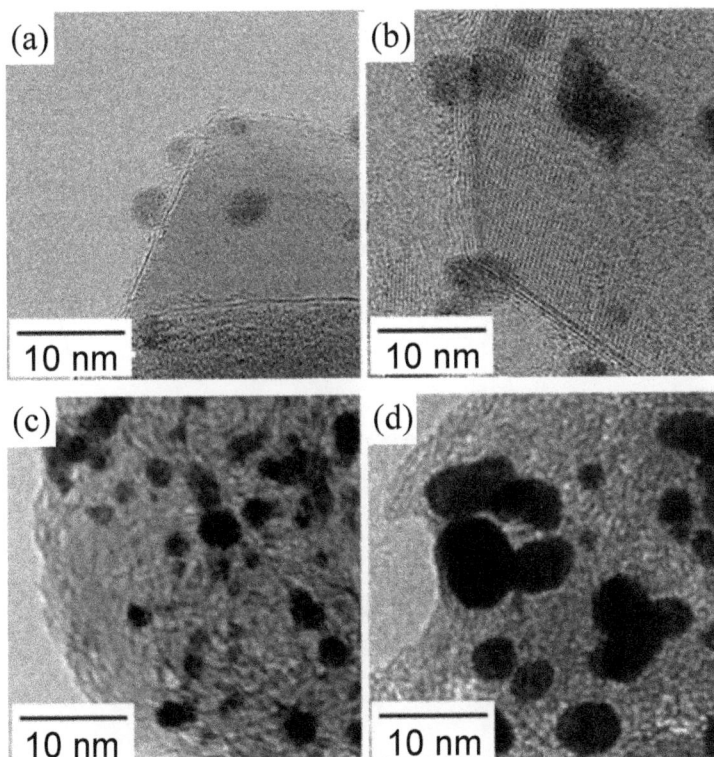

Figure 9. TEM images of Pt/TiC-600 °C (**a**) (**b**) and c-Pt/CB (**c**) (**d**) before (**a**) (**c**) and after (**b**) (**d**) the durability test (N = 5000).

3. Experimental Section

3.1. Preparation of Pt/TiC Catalyst

TiC nanoparticles (average diameter = *ca.* 40 nm, prepared by a radio-frequency plasma method) were supplied by Nisshin Engineering Co. (Tokyo, Japan) The surface area of the TiC nanoparticles was measured to be 77 m^2 g$_{Pt}^{-1}$ by the Brunauer, Emmett and Teller (BET) adsorption method (BELSORP-max, BEL Japan Inc., Osaka, Japan). Platinum nanoparticles were dispersed on the TiC support by the colloidal method [26,36,37]. A calculated amount of hexachloroplatinic acid was dissolved in sodium hydrogen sulfite solution under stirring. In order to prepare a Pt (or PtO_x) colloid, hydrogen peroxide was added to the solution at a rate of 2 mL min^{-1}, and the pH value was held at 5.0 by adding 5 wt. % sodium hydroxide solution. A dispersion of TiC powder, pure water (Milli-Q water, 18.2 MΩ cm, Millipore Japan Co., Ltd., Tokyo, Japan) and catalase (to decompose excess H_2O_2) were added into the Pt colloid solution at room temperature, followed by stirring for 6 h. The powder obtained was filtered and washed thoroughly with pure water. The powder (PtO_x/TiC) was then heat-treated at 600 °C in 1% H_2-containing N_2 atmosphere for 2 h and quenched to room temperature. The amount of Pt loaded on the TiC support was measured by an inductively coupled plasma-mass spectrometric analyzer (ICP-MS, 7500CX, Agilent Technologies Inc., Tokyo, Japan). The Pt loading amount on the Pt/TiC-600 °C was found to be 10.3 wt %. Considering the density of TiC (4.91 g cm^{-3}, based on JCPDS#321383 data) and carbon black (*ca.* 2 g cm^{-3}), we can estimate that the thickness of the catalyst layer with 10.3 wt %-Pt/TiC is comparable to that with *ca.* 25 wt %-Pt/CB under the given Pt amount and the porosity.

3.2. Characterization of Pt/TiC

The crystalline phase of the Pt/TiC catalyst was characterized using X-ray diffraction (XRD, Ultima 4, Rigaku Co., Tokyo, Japan) with monochromated CuKα radiation (0.15406 nm, 40 kV, 40 mA). The morphology of the catalyst was observed by transmission electron microscopy (TEM, H-9500, Hitachi High-Technologies Co., Tokyo, Japan) and scanning transmission electron microscopy (STEM, HD-2700, Hitachi High-Technologies Co.). The Pt (or PtO_x)/TiC catalyst was also analyzed by X-ray photoelectron spectroscopy (XPS, ESCA5800, ULVAC-PHI Inc., Chigasaki, Japan).

3.3. Electrochemical Measurements

The ORR activities of the Pt/TiC and a c-Pt/CB (TEC10E50E, 45.6 wt %-Pt supported on high-surface-area carbon black, Tanaka Kikinzoku Kogyo K.K., Tokyo, Japan) catalysts were examined by the rotating disk electrode (RDE) technique. The working electrode consisted of a thin layer of these catalysts uniformly dispersed on a glassy carbon disk substrate (diameter = 5 mm, geometric area = 0.196 cm^2) at a constant loading of 5.50 μg_{Pt} cm^{-2}, which corresponds to an approximately 2.5-monolayer height of TiC support particles. A thin film of Nafion was coated on the catalyst layer with an average thickness of 0.05 μm [39]. The use of such a thin catalyst layer with a thin Nafion film enables us to evaluate a "real" kinetically-controlled activity of the catalyst for the ORR [47].

A platinum wire and a reversible hydrogen electrode (RHE) were used as the counter and the reference electrodes, respectively. The electrolyte solution of

0.1 M HClO$_4$ was prepared from reagent-grade chemicals (Kanto Chemical Co., Tokyo, Japan) and Milli-Q water. All of the electrode potentials were controlled by a potentiostat (HZ5000, Hokuto Denko Co., Tokyo, Japan). The electrolyte solution was saturated with N$_2$ or O$_2$ gas bubbling for at least 1 h prior to the electrochemical measurements.

The durability testing of the catalysts was performed according to a standard potential step protocol recommended by the Fuel Cell Commercialization Conference of Japan (FCCJ) in 0.1 M HClO$_4$ solution purged with N$_2$ at 25 °C. The potential was stepped between 0.9 V and 1.3 V, with a holding period of 30 s at each potential (1 min per cycle) [48]. After a given number of potential step cycles, changes in the ECA values and ORR activities were examined.

4. Conclusions

We have succeeded in preparing Pt nanoparticles uniformly dispersed on a TiC support (Pt/TiC) by the colloidal method, followed by heat treatment in a hydrogen-containing atmosphere. Such a heat treatment was found to be important in removing a thin amorphous TiO$_2$ layer from the Pt surface, resulting in hemispherical Pt nanocrystals with clear lattice fringes. The heat-treated Pt/TiC at 600 °C (Pt/TiC-600 °C) exhibited high MA for the ORR in O$_2$-saturated 0.1 M HClO$_4$ solution at 25 °C, comparable to that of c-Pt/CB. It was also found that Pt/TiC-600 °C exhibited much higher durability than that of c-Pt/CB in a standard a potential step protocol (E = 0.9 V ↔ 1.3 V). By TEM observation, we have clearly demonstrated that the major reason for such a high durability of Pt/TiC-600 °C was suppression of the detachment of Pt particles from the support, unlike c-Pt/CB. Hence, based on our systematic work using various ceramic supports (TiN [26], doped SnO$_2$ [27–30], and TiC in the present work), the essential factors for the highly active and highly durable cathode catalysts are the use of a chemically and electro chemically stable support with high electrical conductivity, uniform dispersion of Pt nanocrystals on the support, and the removal of an oxide layer, if any, on the Pt surface and/or Pt-ceramic support interface.

Acknowledgments: This work was supported by funds for the Research on Nanotechnology for High Performance Fuel Cells ("HiPer-FC") Project of the New Energy and Industrial Technology Development Organization (NEDO) of Japan.

Author Contributions: This work was coordinated by Hiroyuki Uchida and Masahiro Watanabe. Morio Chiwata carried out the preparation and characterization (XRD, TEM, and ICP-MS) of catalysts, and performed the electrochemical measurements (CVs and RDE). Katsuyoshi Kakinuma contributed to the preparation and characterization (high-resolution TEM and RDE) of catalysts. Mitsuru Wakisaka performed XPS analysis. Makoto Uchida and Shigehito Deki contributed to the durability tests and all of the characterization. All the authors contributed equally to the data interpretation and discussion. Morio Chiwata prepared the manuscript, and Hiroyuki Uchida revised the final version of paper.

Conflicts of Interest: The authors declare no conflict of interest.

References

1. Wilson, M.S.; Garzon, F.H.; Sickafus, K.E.; Gottesfeld, S. Surface Area Loss of Supported Platinum in Polymer Electrolyte Fuel Cells. *J. Electrochem. Soc.* **1993**, *140*, 2872–2877.
2. Willsaw, J.; Heitbaum, J. The Influence of Pt-activation on the Corrosion of Carbon in Gas Diffusion Electrodes—A DEMS Study. *J. Electroanal. Chem.* **1984**, *161*, 93–101.
3. Yu, X.; Ye, S. Recent Advances in Activity and Durability Enhancement of Pt/C Catalytic Cathode in PEMFC: Part II: Degradation Mechanism and Durability Enhancement of Carbon Supported Platinum Catalyst. *J. Power Sources* **2007**, *172*, 145–154.
4. Tang, H.; Qi, Z.G.; Ramani, M.; Elter, J.F. PEM Fuel Cell Cathode Carbon Corrosion Due to the Formation of Air/Fuel Boundary at the Anode. *J. Power Sources* **2006**, *158*, 1306–1312.
5. Meyers, J.P.; Darling, R.M. Model of Carbon Corrosion in PEM Fuel Cells. *J. Electrochem. Soc.* **2006**, *153*, A1432–A1442.
6. Reiser, C.A.; Bregoli, L.; Patterson, T.W.; Yi, J.S.; Yang, D.; Perry, M.L.; Jarvi, T.D. A Reverse-Current Decay Mechanism for Fuel Cells. *Electrochem. Solid-State Lett.* **2005**, *8*, A273–A276.
7. Borup, R.L.; Davey, J.R.; Garzon, H.F.; Wood, D.J.; Inbody, M.A. PEM Fuel Cell Electrocatalyst Durability Measurements. *J. Power Sources* **2006**, *163*, 76–81.
8. Roen, L.M.; Paik, C.H.; Jarvi, T.D. Electrocatalytic Corrosion of Carbon Support in PEMFC Cathodes. *Electrochem. Solid-State Lett.* **2004**, *7*, A19–A22.
9. Passalacqua, E.; Antonucci, P.L.; Vivadi, M.; Patti, A.; Antonucci, V.; Giordano, N.; Kinoshita, K. The Influence of Pt on the Electrooxidation Behaviour of Carbon in Phosphoric Acid. *Electrochim. Acta* **1992**, *37*, 2725–2730.
10. Ferreira, P.J.; Lao, G.J.; Shao-Horn, Y.; Morgan, D.; Makharia, R.; Kocha, S.; Gasteiger, H.A. Instability of Pt/C Electrocatalysts in Proton Exchange Membrane Fuel Cells A Mechanistic Investigation. *J. Electrochem. Soc.* **2005**, *152*, A2256–A2271.
11. Darling, R.M.; Meyers, J.P. Kinetic Model of Platinum Dissolution in PEMFCs. *J. Electrochem. Soc.* **2003**, *150*, A1523–A1527.
12. Xie, J.; Wood, D.L.; Wayne, D.M.; Zawodinski, T.A.; Atanassov, P.; Borup, R.L. Durability of PEFCs at High Humidity Conditions. *J. Electrochem. Soc.* **2005**, *152*, A104–A113.
13. Xie, J.; Wood, D.L.; More, K.L.; Atanassov, P.; Borup, R.L. Microstructural Changes of Membrane Electrode Assemblies during PEFC Durability Testing at High Humidity Conditions. *J. Electrochem. Soc.* **2005**, *152*, A1011–A1020.
14. Stevens, D.A.; Hicks, M.T.; Haugen, G.M.; Dahn, J.R. *Ex situ* and *in situ* Stability Studies of PEMFC Catalysts Effect of Carbon Type and Humidification on Degradation of the Carbon. *J. Electrochem. Soc.* **2005**, *152*, A2309–A2315.
15. Patterson, T.W.; Darling, R.M. Damage to the Cathode Catalyst of a PEM Fuel Cell Caused by Localized Fuel Starvation. *Electrochem. Solid-State Lett.* **2006**, *9*, A183–A185.
16. Yoda, T.; Uchida, H.; Watanabe, M. Effects of Operating Potential and Temperature on Degradation of Electrocatalyst Layer for PEFCs. *Electrochim. Acta* **2007**, *52*, 5997–6006.

17. Hara, M.; Lee, M.; Liu, C.-Y.; Chen, B.-H.; Yamashita, Y.; Uchida, M.; Uchida, H.; Watanabe, M. Electrochemical and Raman Spectroscopic Evaluation of Pt/Graphitized Carbon Black Catalyst Durability for the Start/Stop Operating Condition of Polymer Electrolyte Fuel Cells. *Electrochim. Acta* **2012**, *70*, 171–181.

18. Masao, A.; Noda, S.; Takasaki, F.; Ito, K.; Sasaki, K. Carbon-Free Pt Electrocatalysts Supported on SnO_2 for Polymer Electrolyte Fuel Cells. *Electrochem. Solid-State Lett.* **2009**, *12*, B119–B122.

19. Takasaki, F.; Matsuie, S.; Takabatake, Y.; Noda, Z.; Hayashi, A.; Shiratori, Y.; Ito, K.; Sasaki, K. Carbon-Free Pt Electrocatalysts Supported on SnO_2 for Polymer Electrolyte Fuel Cells: Electrocatalytic Activity and Durability. *J. Electrochem. Soc.* **2011**, *158*, B1270–B1275.

20. Mentus, S.V. Oxygen Reduction on Anodically Formed Titanium Dioxide. *Electrochim. Acta* **2007**, *50*, 27–32.

21. Ioroi, T.; Akita, T.; Yamazaki, S.; Siroma, Z.; Fujiwara, N.; Yasuda, K. Corrosion-Resistant PEMFC Cathode Catalysts Based on a Magnéli-Phase Titanium Oxide Support Synthesized by Pulsed UV Laser Irradiation. *J. Electrochem. Soc.* **2011**, *158*, C329–C334.

22. Huang, S.Y.; Ganesan, P.; Park, S.; Popov, B.N. Development of a Titanium Dioxide-Supported Platinum Catalyst with Ultrahigh Stability for Polymer Electrolyte Membrane Fuel Cell Applications. *J. Am. Chem. Soc.* **2009**, *131*, 13898–13899.

23. Huang, S.Y.; Ganesan, P.; Park, S.; Popov, B.N. Titania Supported Platinum Catalyst with High Electrocatalytic Activity and Stability for Polymer Electrolyte Membrane Fuel Cell. *Appl. Catal. B* **2011**, *102*, 71–77.

24. Ioroi, T.; Senoh, H.; Yamazaki, S.; Siroma, Z.; Fujiwara, N.; Yasuda, K. Stability of Corrosion-Resistant Magnéli-Phase Ti_4O_7-Supported PEMFC Catalysts at High Potentials. *J. Electrochem. Soc.* **2008**, *155*, B321–B326.

25. Ioroi, T.; Siroma, Z.; Fujiwara, N.; Yamazaki, S.; Yasuda, K. Sub-Stoichiometric Titanium Oxide-Supported Platinum Electrocatalyst for Polymer Electrolyte Fuel Cells. *Electrochem. Commun.* **2005**, *7*, 183–188.

26. Kakinuma, K.; Wakasugi, Y.; Uchida, M.; Kamino, T.; Uchida, H.; Deki, S.; Watanabe, M. Preparation of Titanium Nitride-Supported Platinum Catalysts with Well Controlled Morphology and Their Properties Televant to Polymer Electrolyte Fuel Cells. *Electrochim. Acta* **2012**, *77*, 279–284.

27. Kakinuma, K.; Uchida, M.; Kamino, T.; Uchida, H.; Watanabe, M. Synthesis and Electrochemical Characterization of Pt Catalyst Supported on $Sn_{0.96}Sb_{0.04}O_{2-\delta}$ with a Network Structure. *Electrochim. Acta* **2011**, *56*, 2881–2887.

28. Kakinuma, K.; Chino, Y.; Senoo, Y.; Uchida, M.; Kamino, T.; Uchida, H.; Deki, S.; Watanabe, M. Characterization of Pt catalysts on Nb-Doped and Sb-Doped $SnO_{2-\delta}$ Support Materials with Aggregated Structure by Rotating Disk Electrode and Fuel Cell Measurements. *Electrochim. Acta* **2013**, *110*, 316–324.

29. Senoo, Y.; Kakinuma, K.; Uchida, M.; Uchida, H.; Deki, S.; Watanabe, M. Improvements in Electrical and Electrochemical Properties of Nb-Doped SnO_2 Supports for Fuel Cell Cathodes Due to Aggregation and Pt Loading. *RSC Adv.* **2014**, *4*, 32180–32188.

30. Senoo, Y.; Taniguchi, K.; Kakinuma, K.; Uchida, M.; Uchida, H.; Deki, S.; Watanabe, M. Cathodic Performance and High Potential Durability of Ta–SnO$_{2-\delta}$-Supported Pt Catalysts for PEFC Cathodes. *Electrochem. Commun.* **2015**, *51*, 37–40.

31. Oyama, S.T. *The Chemistry of Transition Metal Carbides and Nitrides*; Blackie Academic and Professional: London, UK, 1996; Chapter 1; pp. 9–14.

32. Nakao, S.; Yamada, N.; Hitosugi, T.; Hirose, Y.; Shimada, T.; Hasegawa, T. High Mobility Exceeding 80 cm^2 V^{-1} s^{-1} in Polycrystalline Ta-Doped SnO$_2$ Thin Films on Glass Using Anatase TiO$_2$ Seed Layers. *Appl. Phys. Express* **2010**, *3*, 031102.

33. La Conti, A.B.; Griffith, A.E.; Cropley, C.C.; Kosek, J.A. Titanium Carbide Bipolar Plate for Electrochemical Devices. U.S. Patent 6,083,641, 4 July 2000.

34. Jalan, V.; Frost, D.G. Fuel Cell Electrocatalyst Support Comprising an Ultra-Fine Chainy-Structured Titanium Carbide. U.S. Patent 4,795,684, 3 January 1989.

35. Ma, L.; Sui, S.; Zhai, Y. Preparation and Characterization of Ir/TiC Catalyst for Oxygen Evolution. *J. Power Sources* **2008**, *177*, 470–477.

36. Watanabe, M.; Uchida, M.; Motoo, S. Application of the Gas Diffusion Electrode to a Backward Feed and Exhaust (BFE) Type Methanol Anode. *J. Electroanal. Chem.* **1986**, *199*, 311–322.

37. Watanabe, M.; Uchida, M.; Motoo, S. Preparation of Highly Dispersed Pt + Ru Clusters and the Activity for the Electro-Oxidation of Methanol. *J. Electroanal. Chem.* **1987**, *229*, 395–406.

38. Brambilla, A.; Calloni, A.; Berti, G.; Bussetti, G.; Duò, L.; Ciccacci, F. Growth and Interface Reactivity of Titanium Oxide Thin Films on Fe(001). *J. Phys. Chem. C* **2013**, *117*, 9229–9236.

39. Yano, H.; Akiyama, T.; Bele, P.; Uchida, H.; Watanabe, M. Durability of Pt/Graphitized Carbon Catalysts for the Oxygen Reduction Reaction Prepared by the Nanocapsule Method. *Phys. Chem. Chem. Phys.* **2010**, *12*, 3806–3814.

40. Uchida, M.; Park, Y.-C.; Kakinuma, K.; Yano, H.; Tryk, A.D.; Kamino, T.; Uchida, H.; Watanabe, M. Effect of the State of Distribution of Supported Pt Nanoparticles on Effective Pt Utilization in Polymer Electrolyte Fuel Cells. *Phys. Chem. Chem. Phys.* **2013**, *5*, 11236–11247.

41. Markovic, N.; Adzic, R.; Cahan, B.; Yeager, E. Structural Effects in Electrocatalysis: Oxygen Reduction on Platinum Low Index Single-Crystal Surfaces in Perchloric Acid Solutions. *J. Electroanal. Chem.* **1994**, *377*, 249–259.

42. Gasteiger, H.A.; Kocha, S.S.; Sompalli, B.; Wagner, F.T. Activity Benchmarks and Requirements for Pt, Pt-alloy, and Non-Pt Oxygen Reduction Catalysts for PEMFCs. *Appl. Catal. B Environ.* **2005**, *56*, 9–35.

43. Lee, M.; Uchida, M.; Tryk, D.A.; Uchida, H.; Watanabe, M. The Effectiveness of Platinum/Carbon Electrocatalysts: Dependence on Catalyst Layer Thickness and Pt Alloy Catalytic Effects. *Electrochim. Acta* **2011**, *56*, 4783–4790.

44. Okaya, K.; Yano, H.; Kakinuma, K.; Watanabe, M.; Uchida, H. Temperature Dependence of Oxygen Reduction Reaction Activity at Stabilized Pt Skin-PtCo Alloy/Graphitized Carbon Black Catalysts Prepared by a Modified Nanocapsule Method. *ACS Appl. Mater. Interfaces* **2012**, *4*, 6982–6991.

462

45. Paulus, U.A.; Schmidt, T.J.; Gasteiger, H.A.; Behm, R.J. Oxygen Reduction on a High-surface Area Pt/Vulcan Carbon Catalyst: A Thin-film Rotating Ring-Disk Electrode Study. *J. Electroanal. Chem.* **2001**, *495*, 134–145.

46. Ignaszak, A.; Songa, C.; Zhu, W.; Zhang, J.; Bauer, A.; Baker, R.; Neburchilov, V.; Ye, S.; Campbell, S. Titanium Carbide and Its Core-Shelled Derivative TiC@TiO$_2$ as Catalyst Supports for Proton Exchange Membrane Fuel Cells. *Electrochim. Acta* **2012**, *69*, 397–405.

47. Higuchi, E.; Uchida, H.; Watanabe, M. Effect of Loading Level in Platinum-Dispersed Carbon Black Electrocatalysts on Oxygen Reduction Activity Evaluated by Rotating Disk Electrode. *J. Electroanal. Chem.* **2005**, *583*, 69–76.

48. Iiyama, A.; Shinohara, K.; Iguchi, S.; Daimaru, A. *Handbook of Fuel Cells: Fundamentals, Technology and Applications*; Vielstich, W., Lamm, A., Gasteiger, H.A., Eds.; John Wiley & Sons Ltd.: Hoboken, NJ, USA, 2009; Volume 6.

Novel Mesoporous Carbon Supports for PEMFC Catalysts

Dustin Banham, Fangxia Feng, Tobias Fürstenhaupt, Katie Pei, Siyu Ye and Viola Birss

Abstract: Over the past decade; a significant amount of research has been performed on novel carbon supports for use in proton exchange membrane fuel cells (PEMFCs). Specifically, carbon nanotubes, ordered mesoporous carbon, and colloid imprinted carbons have shown great promise for improving the activity and/or stability of Pt-based nanoparticle catalysts. In this work, a brief overview of these materials is given, followed by an in-depth discussion of our recent work highlighting the importance of carbon wall thickness when designing novel carbon supports for PEMFC applications. Four colloid imprinted carbons (CICs) were synthesized using a silica colloid imprinting method, with the resulting CICs having pores of 15 (CIC-15), 26 (CIC-26), 50 (CIC-50) and 80 (CIC-80) nm. These four CICs were loaded with 10 wt. % Pt and then evaluated as oxygen reduction (ORR) catalysts for use in proton exchange membrane fuel cells. To gain insight into the poorer performance of Pt/CIC-26 *vs.* the other three Pt/CICs, TEM tomography was performed, indicating that CIC-26 had much thinner walls (0–3 nm) than the other CICs and resulting in a higher resistance (leading to distributed potentials) through the catalyst layer during operation. This explanation for the poorer performance of Pt/CIC-26 was supported by theoretical calculations, suggesting that the internal wall thickness of these nanoporous CICs is critical to the future design of porous carbon supports.

Reprinted from *Catalysts*. Cite as: Banham, D.; Feng, F.; Fürstenhaupt, T.; Pei, K.; Ye, S.; Birss, V. Novel Mesoporous Carbon Supports for PEMFC Catalysts. *Catalysts* **2015**, *5*, 1046–1067.

1. Introduction

Proton exchange membrane fuel cells (PEMFCs) are energy conversion devices that are capable of cleanly and efficiently converting the chemical energy of the reactants (typically H_2 and O_2 from air) directly into electrical energy. PEMFCs operate at relatively low temperatures (60 °C–95 °C), making them ideally suited for transportation and portable power applications, and they are also now being investigated for small-scale distributed stationary power generation [1].

For the hydrogen oxidation (HOR) and the oxygen reduction (ORR) reactions to occur at the PEMFC anode and cathode, respectively, a high surface area Pt catalyst is required. However, the intrinsic ORR kinetics are roughly five orders of magnitude slower than the HOR rate [2], and thus nearly 90% of the Pt in a PEMFC is located at the cathode to increase its activity [3,4]. Therefore, it is crucial to understand and optimize the cathode catalyst layer in order to maximize Pt utilization and eventually decrease the Pt loading at the cathode, thus also lowering PEMFC cost.

Presently, the cathode catalyst layer in a PEMFC consists of 2–6 nm Pt or Pt alloy nanoparticles deposited on a high surface area microporous (< 2 nm diameter pores) carbon support (typically Vulcan carbon XC-72R (VC); Cabot, Alpharetta, GA, USA) [5,6]. Unfortunately, the pores of VC are too small to accommodate the Pt nanoparticles and therefore they reside primarily on the outer VC surface. As a result, Pt deposition on VC occurs on only a fraction of the total surface area, making it difficult to achieve small, uniformly distributed Pt nanoparticles on conventional VC at high Pt loadings of $\geqslant$ 20 wt. % [7]. In addition, any Pt that is deposited into a micropore is expected to experience significant mass transport limitations and thus may hardly be used [7,8].

In an effort to improve Pt distribution and utilization and to reduce mass transport losses through the catalyst layer of PEMFCs, many researchers have been investigating mesoporous carbon support materials (2 nm < d < 50 nm), which have both much larger pore diameters and typically larger surface areas than conventional VC. A high surface area is an important property of any potential catalyst support for PEMFCs, as it is known that average Pt particle size decreases with increasing catalyst support area [9], likely due to an increased number of available nucleation sites. Furthermore, it has been demonstrated that mass transport of reactants and products is greatly improved when using carbon supports with pore diameters of > 3 nm [9]. Specifically, three promising families of mesoporous carbons are: (1) carbon nanotubes (CNTs); (2) ordered mesoporous carbons (OMCs); and (3) colloid imprinted carbons (CICs). Other carbon supports that have been investigated include carbon nanofibres [10,11] and carbon nanocoils [12,13]. Although some promising data has been reported for these additional carbon supports, the largest focus has been on the three main families listed above.

While the importance of carbon support pore diameter is now well understood, recent work by our group [14,15] has helped to elucidate the previously under-emphasized importance of carbon support wall thickness on ORR activity. This finding has been observed for both OMC [14] and CIC [15] catalyst supports in our previous studies.

1.1. Carbon Nanotubes (CNTs)

CNTs offer several advantages over conventional microporous carbons as catalyst supports for PEMFCs. Unlike carbon blacks, CNTs are nearly 100% sp2 hybridized, which provides them with a much higher electronic conductivity than carbon black [16]. This is advantageous when electrons must rapidly transport through the support. Multi-walled CNTs typically have a higher electronic conductivity (but lower surface area) than single wall CNTs, and are therefore the most commonly used CNTs in PEMFC research [9]. The high percentage of sp2 hybridization of CNTs also provides these materials with enhanced corrosion resistance *vs.* VC [9,17]. Also, CNTs generally have [17] fewer impurities than carbon blacks, such as VC, which often contain organic sulphur species that may poison the Pt catalyst [18].

While the high degree of sp2 hybridization provides CNTs with many benefits, it unfortunately makes Pt deposition very challenging. This is because, unlike VC, there are very few functional groups available to anchor the Pt nanoparticles to the surface [9]. Therefore, CNTs must first be functionalized, often by refluxing in HNO_3/H_2SO_4 solutions at elevated temperatures (90 °C–140 °C) in order to create hydroxyl and carboxylic acid surface groups [9]. However, this lowers the CNT electronic conductivity, as well as making them more susceptible to oxidation under normal PEMFC conditions [9]. Proper design of CNT supports must therefore take into account the trade-offs between high electronic conductivity and corrosion resistance *versus* Pt dispersion and stability. It has been shown [19–21] that nitrogen-doped CNTs have advantages over CNTs in terms of Pt deposition and catalytic activity. Moreover, first principles calculations indicate that Pt should bond more strongly on nitrogen-doped CNTs than on CNTs [22,23], with a higher Pt stability confirmed experimentally [24].

1.2. Ordered Mesoporous Carbons (OMCs)

Joo *et al.* [25] first reported the use of OMCs for PEMFC applications, using a well-studied ordered mesoporous silica (SBA-15) as a template for OMC synthesis. They filled the pores of SBA-15 with furfuryl alcohol and then heated the sample to 80 °C to induce polymerization of furfuryl alcohol on the acidic sites on the SBA-15 walls. The remaining furfuryl alcohol was then removed under vacuum and the carbon/silica composite was carbonized, followed by removal of the SBA-15 template through refluxing in NaOH. The resulting OMC was composed of hollow tubes, with a pore diameter of 5.9 nm, and 4.2 nm pores between adjacent tubes. It was reported [25] that the outer diameter of the OMC tubes can be controlled by the SBA-15 pore size, while the inner tube diameter can be tuned via the polymerization conditions employed (temperature, time, number of furfuryl alcohol infiltration steps [26]). Importantly, the resultant OMC was found to have a surface area of *ca.* 2000 m^2/g, based on N_2 gas sorption measurements. The authors demonstrated that this large OMC surface area, composed of 5.9 and 4.2 nm pores, allowed for much smaller Pt particles to be deposited *vs.* on a high surface area (1500 m^2/g) carbon black. As a result, the Pt/OMC catalyst was found to greatly outperform the Pt-loaded carbon black catalyst.

While this initial work was very promising, subsequent work [8,27–29] demonstrated that, similar to microporous VC, the ORR occurring at the Pt/OMC catalysts may become transport limited at high current densities, as reactants and products must rapidly transport to and from active Pt sites deep inside the OMC pores. Specifically, it has been suggested that, while Pt may deposit inside the OMC pores, the Nafion proton conductor may not be able to access them, likely then resulting in proton transport limited currents [9].

In previous work by Ambrosio *et al.* [8,27], it was demonstrated that Pt/OMC catalysts outperform Pt/VC at low current densities, but quickly succumb to mass transport losses at higher current densities. Similar to Joo *et al.* [25], they demonstrated that the higher surface area of the OMC support (1080 m^2/g) allowed for smaller, more uniform Pt nanoparticles to be deposited *vs.* similarly loaded VC. While the smaller Pt particle size of the Pt/OMC *vs.* Pt/VC material did provide the Pt/OMC catalyst with a higher electrochemically active surface area, proposed to be responsible for the higher activity of Pt/OMC at low current densities, a poorer activity of the Pt/OMC catalysts was observed at higher current densities. This was likely due to either a low electronic conductivity through the carbon matrix, or proton diffusion limitations within the ~3.5 nm pores. Based on these recent findings, [25] achieving larger pore diameters in the OMCs appears to be an important next step. Unfortunately, the hard templating approach used in OMC synthesis constrains the OMC pore diameters to <7 nm in diameter [30]. This is because it is very difficult to increase the wall thickness (which eventually becomes the OMC pore diameter) of the silica template while still maintaining the desired porosity.

Recently, our group performed a controlled study to further verify the importance of the OMC support dimensions (as opposed to the pore diameter) on ORR activity of Pt/OMC catalysts [14]. Since OMCs represent an inverse replica of the sacrificial porous silica, we have previously described the final OMC structure as consisting of carbon "nano-strings" (where carbon has filled the worm-hole-shaped pores of the silica) separated by pores (previously occupied by the walls of the SiO$_2$ template) [14]. SiO$_2$ templates with pore diameters ranging from 1.5 to 3.1 nm were prepared, all having wall thicknesses of ~2.3 nm.

OMCs were prepared from these SiO$_2$ templates, and loaded with 20 wt. % Pt. Rotating disc electrode (RDE) testing was used to show that the ORR activity correlated well with the diameter of the solid carbon nano-strings. Specifically, Pt/OMC catalysts with narrower nano-string diameters were shown to have higher ohmic losses, resulting in more pronounced problems due to distributed potentials through the catalyst layer on the RDE surface and thus lower ORR activities, *vs.* what was seen for Pt/OMC catalysts with larger nano-string diameters. This work was the first to demonstrate the importance of OMC nano-string diameter on the

ORR activity of Pt/OMC catalysts, which had not previously been considered an important design parameter for Pt/OMC catalysts.

1.3. Colloid Imprinted Carbons (CICs)

It has been previously shown that, while mesoporous carbons can accommodate Pt in their pores, Nafion deposition may prove to be more challenging for pores <40 nm in diameter [9,31,32]. This would result in proton transport limitations to Pt particles that are buried deep within the pores, but are not in direct contact with Nafion. It is thus crucial to develop carbon materials that allow for facile tuning of pore diameter, beyond what is achievable for OMCs, to avoid these problems.

In 2001, Li et al. [30] reported a novel approach to mesoporous carbon design, allowing for control of both pore diameter and length. This synthesis was based on the unique physical properties of a naphthalene-based mesophase pitch precursor, which is a polycyclic aromatic hydrocarbon (PAH) and a common by-product of the petroleum industry. These PAHs interact through long range London dispersion forces and arrange in an ordered, liquid crystalline structure [33]. A synthetic naphthalene-based mesophase pitch (Mitsubishi AR pitch) was chosen by Li et al. [30], as petroleum-based pitches often contain sulphur impurities. The reported softening range of this pitch varies in the literature, with the lower and upper temperatures generally falling between 230 °C and 350 °C [18,30,33].

In their original work, Li et al. [30] demonstrated that SiO_2 colloids coated on the surface of AR pitch can be imprinted into its volume by heating to temperatures within its softening range. Importantly, they were able to show that the depth of imprinting of the SiO_2 colloids can be controlled based on the imprinting temperature that is used. By subsequently carbonizing the silica/pitch composite and removing the SiO_2 colloids by refluxing in NaOH, a porous colloid imprinted carbon (CIC) powder is generated, with pore diameters equal to the size of the SiO_2 colloids that are used and pore lengths controlled by the imprinting temperature. While other colloid-based approaches to mesoporous carbon synthesis have been reported, such as colloid infused resorcinol-formaldehyde resins [34], no other reported synthesis allows for control of both pore depth and diameter [27], making the CICs an ideal material for both enhancing PEMFC performance as well as understanding the impact of pore parameters on ORR performance.

Since this discovery, several groups have investigated CICs as potential catalyst supports for PEMFC applications. In a communication by Fang et al. [35], it was shown that, when loaded with 20 wt. % Pt, a fully imprinted CIC, with a pore diameter of 26 nm (CIC-26), outperformed similarly loaded VC as a PEMFC cathode. This was explained by the higher accessible surface area of CIC-26 vs. VC, leading to a more uniform Pt distribution on CIC-26 vs. VC. Also, the measured electronic conductivity of CIC-26 was indicated to be higher than VC, which would serve to minimize ohmic losses, while the larger pore diameter of CIC-26 *vs.* VC should facilitate mass transport.

The use of CICs as a catalyst support has also been extensively explored by our group [7,15,36]. In previous work, we used 22 nm SiO$_2$ colloids to prepare CICs having a controllable pore diameter of 26 nm, but different pore depths (controlled by varying the imprinting temperature from 250 °C–400 °C) [36]. 3D TEM and TEM tomography were used to demonstrate that Pt had successfully been loaded down the full length of the CIC pores. Importantly, it was shown that the ORR activity of the Pt nanoparticles was independent of their location down the CIC pores. This is likely due to the relatively large (26 nm) pore diameter of these CIC supports, which greatly facilitates mass transport to the Pt nanoparticles inside these pores.

In another study, we demonstrated that a Pt-loaded CIC with 26 nm dia pores (CIC-26) exhibited a lower ORR activity than a similarly loaded CIC with 15 nm pores (CIC-15) [15]. Through the use of TEM tomography, we showed that the poorer ORR activity of Pt/CIC-26 *vs.* Pt/CIC-15 was likely due to its very thin walls (~ 1.5 nm) *vs.* the 5–15 nm walls of Pt/CIC-15. This conclusion is in agreement with our findings on the importance of the diameter of OMC nano-strings on ORR activity [14]. However, only CIC-15 and CIC-26 were evaluated in this initial study [15]. Therefore, the goal of the present work is to verify the importance of carbon wall thickness on the ORR activity of Pt/CIC cathode materials. Control of the CIC wall thickness was achieved by varying the size of the silica colloids used in their synthesis, with larger diameter silica particles (15–80 nm) resulting in larger gaps between adjacent particles, and thus thicker carbon walls. All four CICs (CIC-15, CIC-26, CIC-50, and CIC-80) were loaded with 10 wt. % Pt, and gas sorption, transmission electron microscopy (TEM), and TEM tomography were used to characterize the porous nanostructure of the CICs.

In terms of the observed oxygen reduction (ORR) activity, determined through RDE studies in aerated 0.5 M H$_2$SO$_4$ solutions, it is demonstrated here that the 10 wt. % Pt/CIC-26 catalyst is a much poorer catalyst than all of the other CIC-based materials, due primarily to its notably higher Tafel slope. A theoretical study, based on porous electrode theory, was also carried out, confirming that pore diameter alone cannot be used to predict the performance of these novel Pt-loaded mesoporous carbon catalysts. Overall, this work shows clearly that the internal dimensions, including both pore sizes and wall thicknesses, need to be carefully controlled in the future design of carbon-based catalyst support materials.

2. Results and Discussion

2.1. Determination of CIC Pore Diameter and Pt Nanoparticle Size

The sorption isotherms of the four CICs under study here (expected pore diameters of 15–80 nm) are given in Figure 1a. Each CIC demonstrates a Type IV isotherm with H1 hysteresis, indicative of mesoporous (2 < pore diameter < 50 nm) materials [10,30]. Type H1 hysteresis is associated with uniform diameter and tightly packed spheres, matching well with the synthesis and predicted porous structure of the CICs. Additionally, the steep slopes of the hysteresis branches at P/Po of 0.75-1 reflect a narrow pore size distribution [37]. Thus, the fact that all four CICs demonstrate a sharp uptake of N_2 in their hysteresis curves provides evidence that they each have a narrow pore size distribution.

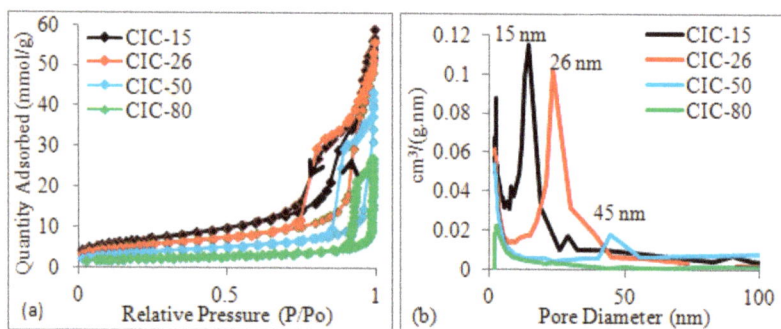

Figure 1. (a) N_2 sorption isotherms and (b) pore size distributions obtained from the adsorption branch of the sorption isotherm calculated using the BJH model [37] for the four CIC supports.

For both CIC-15 and CIC-26, the pore size distributions (Figure 1b) are centered at slightly larger values than expected, based on the anticipated size of the commercially obtained colloids used in their respective synthesis. This was also found by Li *et al.* [30] and can be attributed to a distribution in the size of the silica particles, as well as some possible agglomeration during the synthesis. For CIC-50, the pore size distribution is very broad (Figure 1b), while for CIC-80, no clear average pore size can be observed, likely due to limits of the BJH model and the Kelvin equation upon which it is based [37,38] for materials with such large pore diameters. However, the fact that such a well-defined H1 hysteresis was observed for both CIC-50 and CIC-80 strongly suggests that these materials do possess highly uniform pore sizes.

In order to verify the CIC pore diameters, especially for the larger pore sizes, transmission electron microscopy (TEM) characterization was used. TEM analysis was performed on the CICs after Pt loading, as the presence of Pt gives excellent contrast, thus greatly aiding in discerning the structure of the carbon. As is clearly visible in Figure 2a,b, the pore diameters of CIC-15 and CIC-26 match closely with those obtained from the gas sorption data (Table 1). For CIC-50 and CIC-80, the TEM images (Figure 2c,d) show pores of 50–60 and 80–90 nm in diameter, respectively, which match well with the expected values based on the size of the SiO_2 colloids used in their respective synthesis.

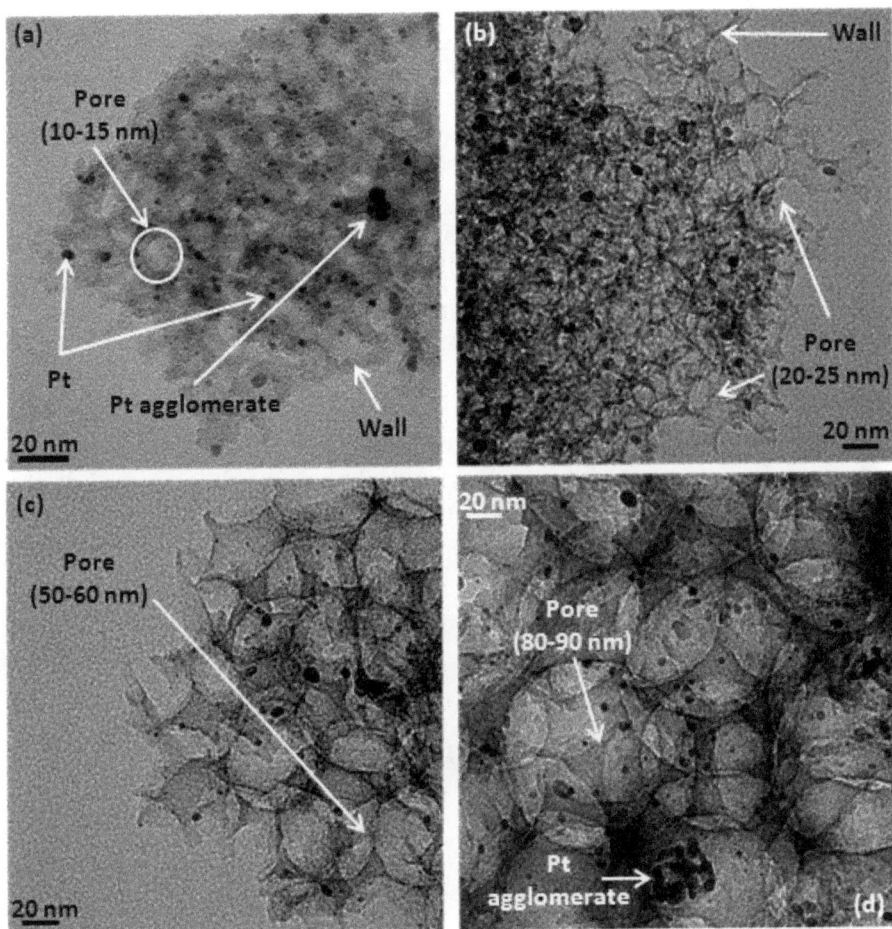

Figure 2. TEM images of 10 wt. % Pt/ (**a**) CIC-15; (**b**) CIC-26; (**c**) CIC-50; and (**d**) CIC-80.

Importantly, Figure 2 also clearly demonstrates that all four Pt-loaded (10 wt. %) CIC catalysts have very similar Pt nanoparticle sizes, centered at ~ 4.5 nm (Table 1). In the absence of any support effects or any transparent limitations, they should therefore all demonstrate very similar oxygen reduction reaction (ORR) activities.

471

A few large (10–20 nm) agglomerates of Pt are seen for each of the four Pt/CICs (examples are shown in Figure 2a,d). These agglomerates are large enough to have blocked some of the 15 nm pores for the Pt/CIC-15 sample, but would likely not have blocked the entrance of the larger pore diameter CICs.

Table 1. Physical properties of the CICs before and after 10 wt. % Pt-loading.

Sample	Pore Diameter [a] (nm)	Pt Particle Size [b] (nm)	Graphite Crystallite Size [c] (nm)	BET Surface Area ($m^2 \cdot g^{-1} \pm 10\%$)	Pore Volume ($mL \cdot g^{-1}) \pm 0.1$ [d]	Wall Thickness (nm) [e]
CIC-15	15 (10–15)	-	1.5	330	1.2	10 ± 5
CIC-26	26 (20–25)	-	1.5	380	1.6	0–3
CIC-50	45 (50–60)	-	1.5	240	1.4	0–20
CIC-80	N/A (80–90)	-	1.5	150	1.4	0–50
Pt/CIC-15	9	4.0 (4.3)	-	280	0.9	-
Pt/CIC-26	24	4.0 (4.7)	-	360	1.6	-
Pt/CIC-50	45	6.0 (4.4)	-	260	1.6	-
Pt/CIC-80	N/A	4.5 (4.5)	-	140	1.3	-

[a] Pore diameter determined from N_2 adsorption isotherm data. The average pore diameter from TEM analysis is given in brackets; [b] Calculated from Pt(111) XRD peak width using the Scherrer equation. The average particle size obtained from TEM analysis is given in brackets; [c] Calculated from the graphite (002) peak; [d] Total pore volume determined at $P/P_o =$ ~0.99; [e] Obtained from TEM tomography; -: This data is not applicable/available.

X-ray diffraction (XRD) was used to determine the size of the Pt particles on the four CIC supports by applying the Scherrer equation to the Pt(111) peak (Figure 3). While the relative intensities of the Pt(111) peaks are not all the same, all of the catalysts show Pt peaks with a similar full width at half maximum, consistent with Pt crystallite sizes of ~4.5 nm, in good agreement with the TEM data (Table 1). By tracking the graphite 002 peak (Table 1), all four CICs were found to have graphite crystallite sizes of ~1.5 nm, similar to what is observed for many conventional carbon blacks [16]. Therefore, the intrinsic resistivity of the four CICs should be similar to each other and to carbon black.

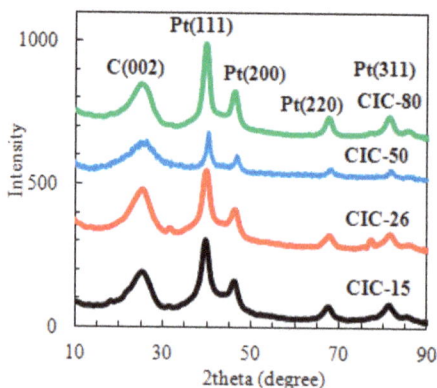

Figure 3. XRD patterns of the four 10 wt. % Pt-loaded CICs.2.1.1. This Is Subsection Heading.

2.2. Surface Area of Pt-Free and Pt-Loaded CICs

The BET surface areas and pore volumes of the four CICs are also reported in Table 1. It is seen that, with the exception of CIC-15, the specific surface area (m^2/g) decreases with increasing pore radius, while the specific pore volume (cm^3/g) remains relatively constant. The trends in these parameters (Table 1) are not surprising, since these should both correlate with the size of the colloids used during the imprinting step. While the specific surface area of a spherical colloidal particle will depend inversely on its radius, the specific volume will be independent of radius, as shown in Equation (1) (r = particle radius, ρ = particle density).

$$\text{Specified surface area: } \frac{m2}{g} = \frac{4\pi r^2}{\frac{4}{3}\pi r^3 \rho} = \frac{3}{\gamma\rho}$$

$$\text{Specified volume: } \frac{m3}{g} = \frac{\frac{4}{3}\pi r^3}{\frac{4}{3}\pi r^3 \rho} = \frac{1}{\rho}$$

(1)

While the gas sorption isotherms (Figure 1a), pore size distributions (Figure 1b), and TEM images (Figure 2) all indicate that the synthesis of the four CICs was successful, a comparison of the measured (BET areas reported in Table 1) *vs.* predicted CIC surface areas allows for further verification of the assumed CIC porous structure. Hexagonal close packing is the most spatially efficient configuration for the packing of spheres, with a packing density of 0.74 [39]. During the solvent evaporation stage of the CIC synthesis, the silica nanoparticles self-assemble into a hexagonally close packed (hcp) configuration in order to maximize their contact with nearest neighbours and reduce their surface energy. Since CICs are inverse replicas of the hcp silica colloids, the maximum theoretical specific surface area (m^2/g) of the CICs can be calculated from the surface area of the silica colloids, based on their size, packing arrangement, and the density of the carbon walls (1.6 g/mL). It should be noted that there is a range of carbon density values reported in the literature for carbon materials. For our purposes, we have used the same value reported by Li *et al.* in their original work with CICs [40]. To increase the accuracy of this calculation, the TEM-measured CIC pore diameters were used (Figure 2 and Table 1), rather than the expected pore diameters, based on the assumed size of the imprinting silica colloids.

Figure 4 clearly shows that, aside from CIC-15, the measured BET surface areas of the CICs match very closely with the expected values based on Equation (1), confirming the assumed hcp configuration of their pores. Importantly, the fact that CIC-15 has a much lower surface area than expected suggests that close packing of the 15 nm silica particles did not occur for CIC-15 and that CIC-15 must have a more disordered porous structure than the other three CICs, consistent with the TEM results (Figure 2).

Looking first at Figure 2a, pores of ~15 nm diameter are present, but they are not arranged in a hcp configuration, as predicted above. This may be due to a disordered arrangement of the 12 nm silica colloids around the mesophase pitch carbon during synthesis or due to the partial collapse of the carbon structure during carbonization, which has been previously suggested to occur for imprinted carbons with pores <50 nm in diameter [41,42]. The fact that the pores of CIC-15 are not tightly packed would produce a lower surface area than expected, thus confirming the results shown in Figure 4. It should also be noted that CIC-15 was synthesized multiple times, with the measured surface area always being lower than expected.

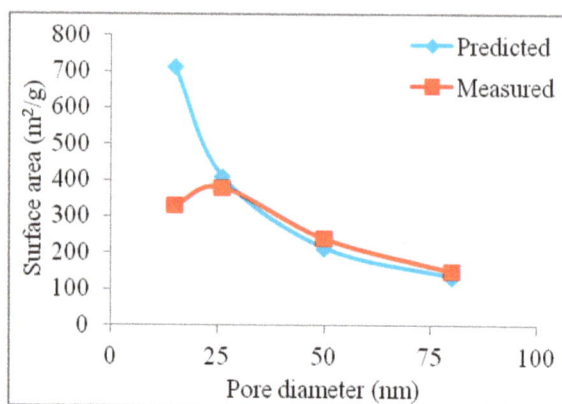

Figure 4. BET-determined and predicted (Equation (1)) surface area of the four CICs *vs.* their TEM measured pore diameter.

A decrease in both the surface area and pore volume of CIC-15 was observed (Table 1) after Pt loading, possibly due to the larger Pt agglomerates (Figure 2) blocking the entrance to some of the pores. However, for the three CIC catalysts with the larger pores, within error, very little change in surface area and pore volume was observed after Pt loading (Table 1), suggesting that the pore mouths remained open.

2.3. Wall Thickness of CIC Supports

As one of the main goals of the present work was to investigate how the wall thickness of the Pt-loaded CIC catalyst supports influences their oxygen reduction reaction (ORR) activity, it was critical to accurately determine this variable. Therefore, TEM tomography was used to characterize each of the four Pt/CIC catalysts, with the results shown in Figure 5.

It is seen for the CIC-(26-80) materials that the close packing of their pores (Figure 5b–d) resulted in wall thicknesses ranging from non-existent (reported as 0 nm) to a maximum value, as shown schematically in Figure 6a and also given in Table 1. For an ordered packing of the spherical SiO_2 particles around the pitch particles in the first step of CIC synthesis, thicker carbon walls are expected if larger diameter silica colloids are used and then removed (Figure 6b), explaining why CIC-26 (prepared from 22 nm diameter SiO_2 colloids) has thinner walls than both CIC-50 and CIC-80, which were prepared using the 50 and 80 nm colloids, respectively. However, the disordered packing of the 12 nm diameter SiO_2 particles in CIC-15 (Figure 5a) has led to much thicker walls (~10 nm) than expected, had the 12 nm colloids been tightly packed together. This is because HCP is the most spatially efficient packing of spheres and thus any other arrangement will result in more available space between them. This will result in thicker carbon walls after filling with the mesophase pitch precursor and its subsequent carbonization to form carbon, as is shown schematically in Figure 6c.

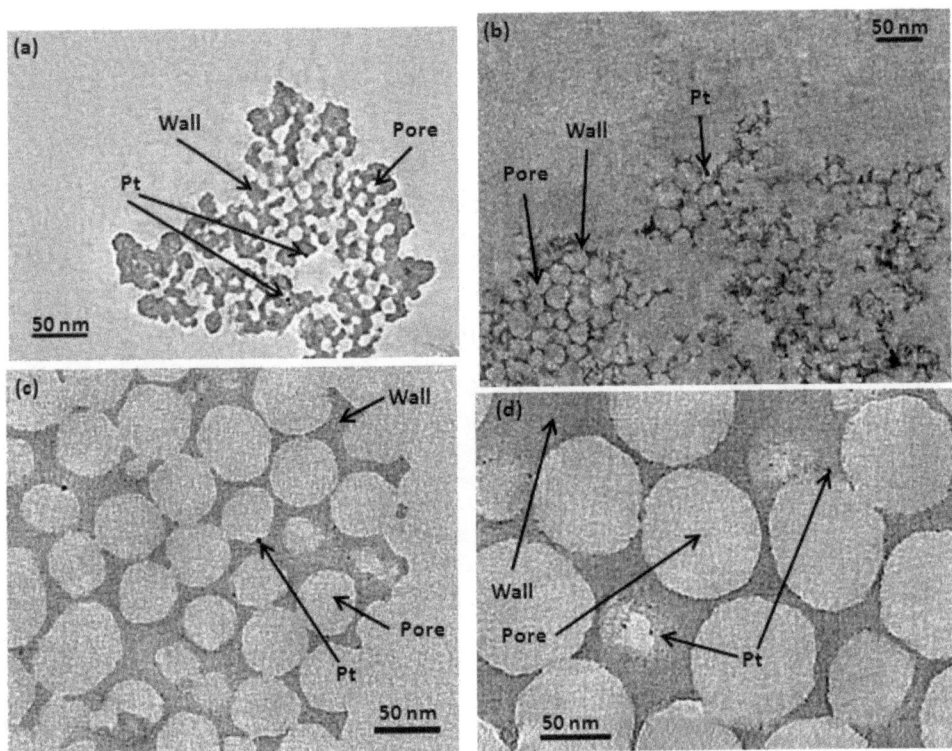

Figure 5. TEM tomography image slices of 10 wt. % Pt supported on (a) CIC-15; (b) CIC-26; (c) CIC-50; and (d) CIC-80, with the dark grey areas representing the CIC walls and the light grey regions being the open pores. All four images were obtained roughly halfway through the CIC particles.

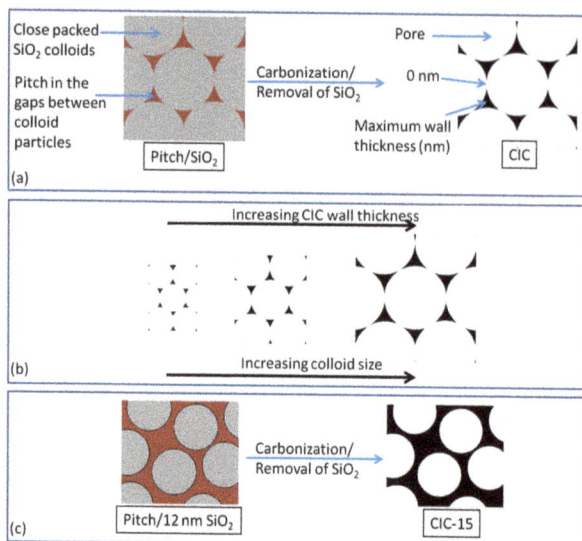

Figure 6. 2-D cartoon showing (**a**) the origin of the range of wall thicknesses for CIC-26, CIC-50 and CIC-80 and revealing; (**b**) how the CIC wall thickness will increase as the silica colloid diameter increases (after removal of SiO_2); (**c**) shows that the disordered packing of the 12 nm colloids results in thicker walls than expected for CIC-15.

2.4. Effect of CIC Nanostructure on ORR at Pt/CIC Catalysts

Prior to performing the oxygen reduction reaction (ORR) studies, cyclic voltammograms (CVs) of the four 10 wt. % Pt/CIC catalysts were collected under a N_2 environment (Figure 7). The electrochemically active Pt surface area (ECSA) was obtained by calculating the total charge passed in the hydrogen underpotential (HUPD) region (0–0.37 V) after subtracting the charging current and assuming a value of 210 $\mu C/cm^2$ for one monolayer of H_{ads} on Pt [16,43,44]. Importantly, Table 2 shows that all four Pt/CIC catalysts have similar Pt ECSAs and thus should have a very similar activity towards the ORR, assuming that the CIC support or its nanostructure has no additional influence on activity.

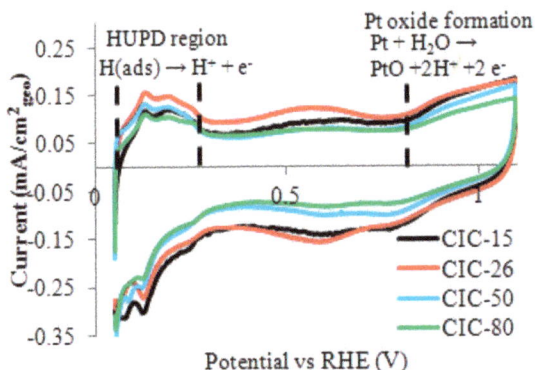

Figure 7. CV response (10 mV/s) of the four 10 wt. % Pt/CIC catalysts, cast on the glassy carbon RDE surface, in room temperature, deaerated 0.5 M H_2SO_4.

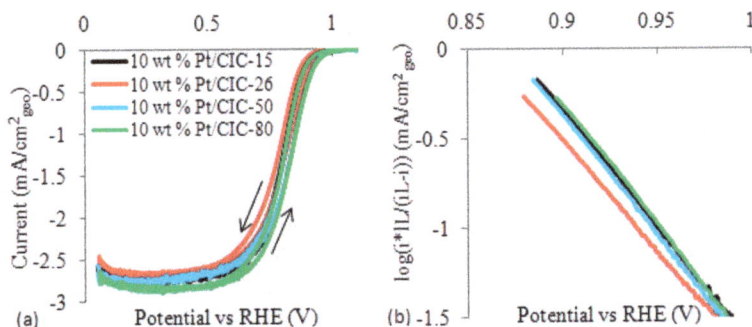

Figure 8. (a) ORR response at 10 mV/s of the four 10 wt. % Pt/CICs in room temperature 0.5 M H_2SO_4, all at 1000 rpm; and (b) the corresponding ORR Tafel plots.

The ORR responses of the four 10 wt. % Pt-loaded CICs are shown in Figure 8a. The current is seen to be fully kinetically controlled (Butler-Volmer kinetics) at low overpotentials (>0.9 V), as expected, followed by mixed kinetic/diffusion control (0.6–0.9 V) and finally diffusion control at <0.6 V (when the limiting current is reached). The Tafel plots in Figure 8b clearly show that the ORR activity (per geometric area of the electrode) of the 10 wt. % Pt/CIC-26 catalyst is lower than that of the other three 10 wt. % Pt/CICs, as is also shown in Table 2. In fact, while Pt/CIC-15, -50 and -80 all demonstrate the expected ORR Tafel slope (in the anodic sweep) of ~70 mV/dec [45], Pt/CIC-26 exhibits an anomolously high Tafel slope of 85 mV/dec (Table 2). As a difference in the ORR mechanism, or slow step, is not expected at these very similar Pt/C materials, the higher Tafel slope of the Pt/CIC-26 catalyst likely reflects the onset of transport limitations in the CIC-26 material with increasing overpotential [46,47]. However, since CIC-26 should have wider pores than CIC-15, it is highly unlikely that the observed difference in Tafel slope is due to transport limitations of protons or dissolved oxygen through the pores of the CICs.

Table 2. Electrochemical results obtained using the 10 wt. % Pt/CIC catalysts.

Sample	ECSA[a] (m²/g) ± 10%	ORR Current at 0.9 V [b] (mA/cm²)	Tafel Slope (mV/dec)
10 wt. % Pt/CIC-15	45	0.40	70
10 wt. % Pt/CIC-26	55	0.32	85
10 wt. % Pt/CIC-50	55	0.40	70
10 wt. % Pt/CIC-80	45	0.40	70

[a] Electrochemical surface area (ECSA) calculated from the CV hydrogen underpotential deposition (HUPD) charge, assuming a value of 210 $\mu C/cm^2$ for one monolayer of adsorbed hydrogen on the Pt surface [16,43,44]. [b] Obtained in 0.5 M H_2SO_4 using 10 mV/s, 1000 rpm, and room temperature conditions.

As further evidence of this, Figure 9a shows a plot of ORR activity (current at 0.9 V *vs.* RHE) *vs.* pore diameter. If diffusion limitations in the pores were dominating the response, the ORR activity (current) would be expected to increase with increasing CIC pore diameter, which is clearly not the case.

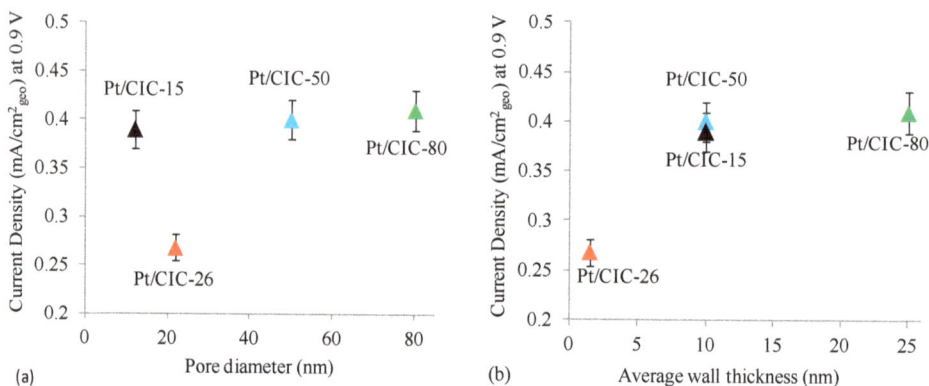

(a)

(b)

Figure 9. ORR current densities (per geometric area of the catalyst layer deposited on the RDE) at 0.9 V for the four Pt/CIC catalysts *vs.* (a) CIC pore diameter and (b) average CIC wall thickness.

The poor ORR activity of Pt/CIC-26 is therefore proposed to be due to higher ohmic losses through its walls *vs.* the other three Pt/CIC-X catalysts. As the XRD data (Table 1) provide evidence that the walls of the four CICs are all equally graphitic, their inherent wall conductivity is expected to be the same. However, the very thin walls of CIC-26 (Figure 5) *vs.* the other three CIC supports, as is clearly visible in Figure 9b (Table 2), is believed to result in high ohmic losses. At a wall thickness of ³10 nm (CIC-15, CIC-50, and CIC-80), there is little difference in the ORR activity of the four CIC catalysts. For CIC-26, with an average wall thickness of 1.5 nm, a significant performance drop is clearly observed. This shows that the ORR activity is significantly compromised when the walls of the carbon support become thinner than 10 nm. This would explain the higher Tafel slope observed for the Pt/CIC-26 catalyst in Figure 8b [48].

The results in Figure 9b clearly show a precipitous drop in ORR activity when the CIC wall thickness is <10 nm, but control of the CIC wall thickness lacks the precision to further study the effect of wall thicknesses between 1.5 and 10 nm. However, in parallel work focused on ordered mesoporous carbons (OMCs, see Section 1.2), we have been able to precisely tune carbon support wall thicknesses (described as "nano-strings") in this desired range [14]. Consistent with the results in Figure 8b, our earlier results have confirmed that the ORR activity is dramatically lower for carbon nano-string thicknesses of <3 nm [14].

2.5. Verification of Effect of Carbon Wall Thickness on ORR Activity by Theoretical Modeling

As further evidence supporting the role of carbon support wall thickness on the observed ORR activity trends (Figure 9), porous electrode theory [46,48–51] was used to calculate theoretical Tafel slopes for the ORR at the four Pt/CIC catalysts and then compare them with the experimentally obtained results (Figure 8). Only the limiting case of ohmic (migration control) overpotentials was considered here, neglecting the contributions from diffusion (i.e., concentration overpotentials) [46]. It is known from earlier porous electrode studies and modelling that electrode porosity will have a similar influence on electrochemistry when either migration or diffusion control is present [46,47].

Here, it was assumed that the measured overpotential ($\eta_{meas.}$) contains contributions only from activation (η_{act}) and ohmic (η_{ohm}) overpotentials, with the concentration overpotential assumed to be non-existent. The ohmic overpotential will depend on the pore length (L) and pore radius (r) (or the wall length and wall thickness), the electronic or ionic conductivity (κ) of the carbon matrix or electrolyte, respectively, and the exchange current density (i_o), as shown in Equation (2). The parameter A is given by Equation (3) (where R is the gas constant, F is Faraday's constant, and T is the temperature) [49,51].

$$\eta_{measured} = \eta_{act} + \underbrace{\frac{4RT}{F} + \ln\sec(A)}_{\eta_{ohm}} \tag{2}$$

$$A = \left(\frac{i_0 L^2 F}{2\kappa RTr}\right)^{\frac{1}{2}} \exp\left(\frac{\eta_{act}F}{4RT}\right) \tag{3}$$

The current generated down the length of the pore depends on parameter A, as shown in Equation (4) [49], is then:

$$I = \left(\frac{4RT\kappa\pi r^2}{LF} \right) A\tan A \qquad (4)$$

It is important to note that, when using Equations (2) and (4) to predict trends in Tafel data, only relative changes in the magnitude of r, κ, L, and i_o are important, and thus it was not necessary to know their exact value. In fact, only r (the CIC pore radius) is known with certainty in this work and thus the current and potential scales in Figure 10 are arbitrary. To aid in the comparison of the experimental and theoretical Tafel plots, the calculated Tafel plots were first normalized to the experimental Pt/CIC-80 Tafel plot (Figure 8). The pore radius was then varied to determine the effect on the calculated Tafel slope.

In the approach used here, the average pore length was assumed to be equal to the average thickness of the dried catalyst film. It should be noted that the pores in the catalyst layer are expected to be quite tortuous, resulting in an average length much greater than the average film thickness. However, only relative differences in average pore length will influence the theoretical Tafel plots and the tortuosity factor (assumed to be the same for each film) was assumed to be constant for the four catalyst layers and thus ignored. For the same reason, any increase in the average pore length due to swelling of the catalyst layers after immersion in solution (due at least partly to Nafion hydration) was assumed to be experienced equally by all four catalyst layers, and was therefore also neglected. The average film thickness (3.5 μm) was obtained for three different catalyst layers (on the RDE surface) at 15 different locations using an optical microscope. The conductivity (κ) of the solution in the pores was assumed to be 10 S/m (based on previous measurements of 0.5 M H_2SO_4) [48]. The same i_o for the ORR was assumed, as all four CICs have the same 10 wt. % Pt loading and very similar Pt nanoparticle sizes (Table 1).

To understand what the predicted effect of solution resistance in the CIC pores (impact of pore radius) would be on the Tafel data, theoretical Tafel plots, allowing only r in Equations (3) and (4) to vary, were produced (Figure 10). At low overpotentials, all four Pt/CICs are predicted to show the same activity (Tafel slope of ca. 70 mV) since, at these correspondingly low currents, mass transfer losses would not yet have become significant and thus the pores of all four CICs would be fully utilized for the ORR. As the overpotential is increased, however, the catalyst layers with smaller diameter pores begin to show an increasing Tafel slope, while the Pt/CICs having larger diameter pores are shown to retain the original theoretical 70 mV slope (Figure 10). Thus, if the pore radius were the only factor influencing ORR activity, the experimentally observed activity of the four 10 wt. % Pt/CIC catalysts should increase in the following sequence: CIC-15 < CIC-26 < CIC-50 < CIC-80, which is clearly not observed in the real experimental data shown in Figure 8. These results, as well as the theoretical predictions made in Figure 10, are summarized in Table 3, clearly showing that the experimental data cannot be simulated if only pore diameter effects are considered.

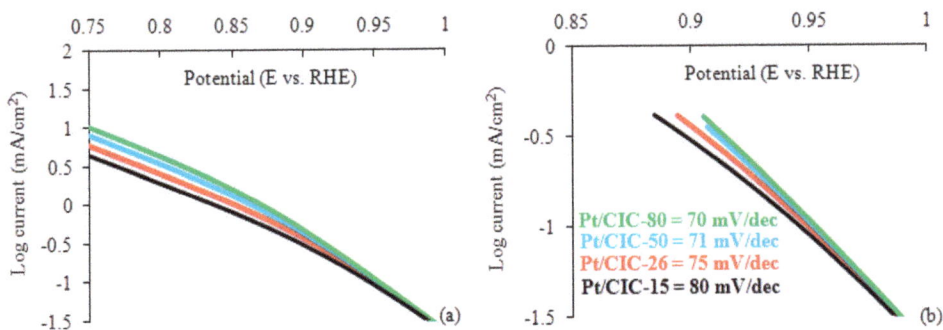

Figure 10. (a) Theoretical Tafel plots for the ORR at the four 10 wt. % Pt/CICs, with (b) showing the simulated data shown in (a), but over a narrower potential range, also giving the theoretical (from Equations (2) and (4)) Tafel slopes over this range of potential. The parameters used to calculate the theoretical Tafel slope are $i_0 = 1 \times 10^{-5}$ A/m^2, $\kappa = 10$ S/m, $L = 3.5$ μm (the assumed thickness of the catalyst layer), and r (radius of CIC pore) = 15, 26, 50, or 80 nm.

Table 3. Comparison of measured *vs.* calculated ORR Tafel slope for the four Pt/CIC catalysts.

Catalysts	Experimental ORR Tafel Slope (Figure 8) [a]	Modelled Tafel Slope [b] (Figure 10)
10 wt. % Pt/CIC-15	70	80
10 wt. % Pt/CIC-26	85	75
10 wt. % Pt/CIC-50	70	71
10 wt. % Pt/CIC-80	70	70

[a] Obtained in 0.5 M H$_2$SO$_4$ at 10 mV/s, 1000 rpm, room temperature conditions.
[b] Obtained from Equations (3) and (4), considering only the effects of pore diameter and assuming migrational mass transport limitations.

3. Experimental Section

3.1. Synthesis of Silica Colloid Imprinted Carbons (CICs)

The synthesis of the silica colloid-imprinted carbon (CIC) was based on a previously reported procedure [30]. Briefly, 1 g of a mesophase carbon pitch carbon (Mitsubishi, Tokyo, Japan) was dispersed in 20 mL of EtOH:H$_2$O (60:40 in volume), followed by the drop-wise addition of a colloidal SiO$_2$ suspension (colloid sizes of 12 nm (Ludox-HS-40), 22 nm (Ludux-AS-40; Sigma-Aldrich, St. Louis, MO, USA), 50 nm, and 80 nm (Chemical Products Corporation, Cartersville, GA, USA) with vigorous stirring at room temperature. A mass ratio of 10.4:1 SiO$_2$:carbon was used for the synthesis of the CICs, with the temperature of this mixture then raised to 50 °C to promote solvent evaporation. The SiO$_2$/C composite was then heated to 360 °C for 2 h (the temperature used at this stage determines the final depth of colloid penetration and thus the surface area of the CIC material) [14], followed by carbonization at 900 °C for 2 h, all under N$_2$. The solid product was then refluxed in 3 M NaOH for 24 h at 100 °C to remove the SiO$_2$ from the pores.

3.2. Pt Loading of CIC-X (Where X Defines Pore Diameter) Supports

The CIC samples were loaded with 10 wt. % Pt using wet-impregnation [25], with H$_2$PtCl$_6$ 6H$_2$O used as the Pt precursor and H$_2$ as the reducing agent. Briefly, 0.67 g of H$_2$PtCl$_6$ 6H$_2$O were dissolved in 10 mL of acetone and added drop-wise to 1 g of carbon (CIC) with vigorous stirring. The Pt/CIC samples were then dried at 60 °C overnight. Reduction of H$_2$PtCl$_6$ 6 H$_2$O was achieved by heating the sample under H$_2$ from room temperature (RT) to 300 °C over a period of 2 h. The samples were then kept at 300 °C for 2 h under N$_2$ in order to remove any adsorbed hydrogen and then allowed to cool to RT.

3.3. Electrochemical Evaluation of Oxygen Reduction Activity at Pt/CIC Catalysts

2 mg of the Pt/CIC powder was dispersed into 2 mL of a 80:20 (in weight) H$_2$O:isopropanol mixture. 20 μL of a 5 wt. % Nafion/isopropanol solution was then added to this suspension, followed by sonication for 45 minutes. An Eppendorf pipette was used to deposit 20 μL of the resulting ink onto a 7 mm diameter glassy carbon (GC) rotating disc (RDE) working electrode (WE), followed by drying at room temperature. A second 20 μL aliquot was then deposited onto the GC electrode, followed by air drying, giving a total catalyst (Pt/CIC) loading of ca. 105 μg/cm^2.

Electrochemical evaluation of the catalysts was performed in a three-electrode cell containing a platinised Pt mesh counter electrode, a reversible hydrogen reference electrode (RHE), and the 7 mm diameter glassy carbon RDE as the WE, cast with the catalyst film. Cyclic voltammetry (CV) was carried out using an EG&G 173 potentiostat in conjunction with an EG&G PARC 175 function generator. The cell solution was 0.5 M H$_2$SO$_4$, purged with vigorous bubbling of either N$_2$ (Praxair 99%) or O$_2$ (Praxair medical grade).

Prior to evaluation, the catalyst layers were first electrochemically cleaned by scanning between -0.05 V and 1.3 V at 100 mV/s for 14 cycles, followed by CV analysis (0.05 to 1.1 V vs. RHE) in a N$_2$-saturated aqueous solution with no electrode rotation. The ORR electrochemistry was then examined at 10 mV/s in an O$_2$-saturated cell and at a WE rotation rate of 1000 rpm, using a Pine analytical rotor (Model ASR-2; Cisco, San Jose, CA, USA). The baseline CVs in N$_2$-saturated conditions were subtracted from the CVs collected under aerated conditions to remove the non-Faradaic component of the current. Chart 5 by PowerLab (ADInstruments, Colorado Springs, CO, USA) was used for data acquisition.

3.4. Catalyst Characterization

X-ray powder diffraction (XRD) patterns were obtained using CuK_α radiation (λ = 0.15406 nm) at 40 kV (20 mA) using a Rigaku Multiflex X-ray diffractometer (Department of Geosciences, University of Calgary, Calgary, Canada), with the data processed using Jade software (Jade 6.5; Softonic, Barcelona, Spain). N_2 adsorption-desorption isotherms were obtained at $-196\,^{\circ}C$ (Tristar 3000 Analyzer; Micromeritics, Norcross, GA, USA). Prior to analysis, samples were out-gassed in N_2 at 250 °C for 4 h. The specific surface area of the CICs was obtained using the Brunauer-Emmett-Teller (BET) plot ($0.05 < P/P_o < 0.30$), where P and P_o are the partial pressure and vapour pressure of the adsorbate gas, respectively. The total pore volume was calculated at P/P_o = 0.99, while the pore size distribution curves were determined from the adsorption branch of the isotherm using the Barrett-Joyner-Halenda (BJH) mode.

All Transmission Electron Microscopy (TEM) work was carried out using a Tecnai TF20 G2 FEG-TEM (FEI, Hillsboro, OR, USA) in the Microscopy and Imaging Facility (Health Sciences Centre) at the University of Calgary, with a Fischione 2040 Dual-Axis Tomography Holder (Fischione Instruments, Export, PA, USA). The catalysts were suspended in ethanol and sonicated for 5 minutes. A droplet of this suspension was placed on one side of a TEM Slot Grid that was covered with a 40 nm continuous Formvar film (EMS, Hatfield, PA, USA) and then left to dry for several minutes. In some cases, a Lacey Carbon Grid (EMS) was used. Colloidal Au particles (10 nm diameter, Cell Microscopy Center, University Medical Center Utrecht, Utrecht, The Netherlands) were placed on the opposite side of the grid to serve as fiducial markers. Finally, a thin carbon coating was applied to both sides of the grid for mechanical stabilization and to reduce electrical charging in the microscope. All TEM images were captured on a 1024 × 1024 pixel Gatan GIF 794 CCD (Gatan, Pleasanton, CA, USA). Dual axis tilt images were taken with the SerialEM software [52] at a tilt range from $-63°$ to $+63°$ in 1° increments. Tomographic reconstruction was achieved by weighted back-projection with the IMOD software package [53,54]. The same software was used for visualization and analysis.

4. Conclusions

Significant research over the past decade has focused on the development of novel mesoporous carbons materials for use as Pt supports in proton exchange membrane fuel cell (PEMFC) electrodes. Specifically, carbon nanotubes (CNTs), ordered mesoporous carbons (OMCs), and colloid imprinted carbons (CICs) have all shown promise as possible alternatives to conventional microporous carbons.

While the majority of the research efforts in this field have focused on the impact of pore diameter on the oxygen reduction reaction (ORR) activity of Pt/C catalysts, our group has recently highlighted the importance of carbon wall ("nano-string") diameter in OMCs and carbon wall thickness in CICs on ORR activity.

In the present study, four colloid imprinted carbons (CICs) with pore diameters of 15, 26, 50 or 80 nm in diameter were synthesized and then loaded with 10 wt. % Pt, all showing particle sizes of ~4.5 nm. Through rotating disc electrode studies in fully aerated 0.5 M H_2SO_4 solutions, it was demonstrated that the CIC wall thickness, as opposed to pore diameter, is the most critical factor in determining the ORR activity of these Pt/CIC catalysts. This was further confirmed through modeling efforts that demonstrated that the experimentally observed Tafel slopes cannot be predicted based on changes in pore diameter alone.

While the majority of the research efforts in designing porous carbon supports have focused on controlling and optimizing pore diameter, our work has again shown that careful considerations of wall thickness must also be made. It is likely that the relative importance of wall thickness *vs.* pore diameter depends on many factors, including the actual dimensions of the pores and walls and the catalyst layer components (*i.e.*, ionomer content, Pt loading, *etc.*). However, our experimental and modeling work have clearly demonstrated the significant impact that carbon wall thickness can have on the ORR activity of Pt/carbon catalysts, and the importance of fully characterizing wall thickness when attempting to interpret ORR data obtained at Pt-loaded mesoporous carbon materials.

Acknowledgments: The authors gratefully acknowledge the Natural Sciences and Engineering Research Council of Canada (NSERC) Strategic Research Project Program and Ballard Power Systems for their financial support of this work. We would also like to thank Alberta Ingenuity, now part of Alberta Innovates-Technology Futures, and NSERC for their scholarship support of D.B. Acknowledgements are also made to J. Hill and P. Pereira (University of Calgary) for access to N_2 sorption instrumentation. Finally, the authors thank Scott Paulson for many helpful discussions.

Author Contributions: Dustin Banham, Fangxia Feng and Viola Birss conceived and designed the experiments. Tobias Fürstenhaupt performed the TEM imaging and helped in interpreting the data. Katie Pei performed electrochemical and materials science characterization and helped in planning the experiments. Siyu Ye provided guidance on the project and helped in interpreting the data. Dustin Banham and Viola Birss wrote the paper.

Conflicts of Interest: The authors declare no conflict of interest.

References

1.   Barbir, F.; Yazici, S. Status and development of PEM fuel cell technology. *Int. J. Energy Res.* **2008**, *32*, 369–378.
2.   Barbir, F. *PEM Fuel Cells: Theory and Practice*; Elsevier Academic Press: San Diego, CA, USA, 2005.

3.   Gasteiger, H.A.; Panels, J.E.; Yan, S.G. Dependence of PEM fuel cell performance on catalyst loading. *J. Power Source* **2004**, *127*, 162–171.

4.   Herranz, J.; Jaouen, F.; Lefèvre, M.; Kramm, U.I.; Proietti, E.; Dodelet, J.-P.; Bogdanoff, P.; Fiechter, S.; Abs-Wurmbach, I.; Bertrand, P.; *et al.* Unveiling *N*-protonation and anion-binding effects on Fe/N/C catalysts for O_2 reduction in proton-exchange-membrane fuel cells. *J. Phys. Chem. C* **2011**, *115*, 16087–16097.

5.   Joo, S.H.; Kwon, K.; You, D.J.; Pak, C.; Chang, H.; Kim, J.M. Preparation of high loading Pt nanoparticles on ordered mesoporous carbon with a controlled Pt size and its effects on oxygen reduction and methanol oxidation reactions. *Electrochim. Acta* **2009**, *54*, 5746–5753.

6.   Wikander, K.; Ekström, H.; Palmqvist, A.E.C.; Lindbergh, G. On the influence of Pt particle size on the PEMFC cathode performance. *Electrochim. Acta* **2007**, *52*, 6848–6855.

7.   Banham, D.; Feng, F.; Fürstenhaupt, T.; Pei, K.; Ye, S.; Birss, V. Effect of Pt-loaded carbon support nanostructure on oxygen reduction catalysis. *J. Power Source* **2011**, *196*, 5438–5445.

8.   Ambrosio, E.P.; Francia, C.; Manzoli, M.; Penazzi, N.; Spinelli, P. Platinum catalyst supported on mesoporous carbon for PEMFC. *Int. J. Hydrogen Energy* **2008**, *33*, 3142–3145.

9.   Antolini, E. Carbon supports for low-temperature fuel cell catalysts. *Appl. Catal. B* **2009**, *88*, 1–24.

10. Sebastián, D.; Ruíz, A.G.; Suelves, I.; Moliner, R.; Lázaro, M.J.; Baglio, V.; Stassi, A.; Aricò, A.S. Enhanced oxygen reduction activity and durability of Pt catalysts supported on carbon nanofibers. *Appl. Catal. B* **2012**, *115–116*, 269–275.

11. Sebastián, D.; Lázaro, M.J.; Suelves, I.; Moliner, R.; Baglio, V.; Stassi, A.; Aricò, A.S. The influence of carbon nanofiber support properties on the oxygen reduction behavior in proton conducting electrolyte-based direct methanol fuel cells. *Int. J. Hydrogen Energy* **2012**, *37*, 6253–6260.

12. Zhang, J.; Tang, S.; Liao, L.; Yu, W.; Li, J.; Seland, F.; Haarberg, G.M. Improved catalytic activity of mixed platinum catalysts supported on various carbon nanomaterials. *J. Power Source* **2014**, *267*, 706–713.

13. Celorrio, V.; Flórez-Montaño, J.; Moliner, R.; Pastor, E.; Lázaro, M.J. Fuel cell performance of Pt electrocatalysts supported on carbon nanocoils. *Int. J. Hydrog. Energy* **2014**, *39*, 5371–5377.

14. Banham, D.; Feng, F.; Pei, K.; Ye, S.; Birss, V. Effect of carbon support nanostructure on the oxygen reduction activity of Pt/C catalysts. *J. Mater. Chem. A* **2013**, *1*, 2812–2820.

15. Pei, K.; Banham, D.; Feng, F.; Fürstenhaupt, T.; Ye, S.; Birss, V. Oxygen reduction activity dependence on the mesoporous structure of imprinted carbon supports. *Electrochem. Commun.* **2010**, *12*, 1666–1669.

16. Serp, P.; Figueiredo, J.L. *Carbon Materials for Catalysis*; John Wiley & Sons, Inc.: Hoboken, NJ, USA, 2008.

17. Shao, Y.; Liu, J.; Wang, Y.; Lin, Y. Novel catalyst support materials for PEM fuel cells: Current status and future prospects. *J. Mater. Chem.* **2009**, *19*, 46–59.

18. Shao, Y.; Yin, G.; Wang, J.; Gao, Y.; Shi, P. Multi-walled carbon nanotubes based Pt electrodes prepared with *in situ* ion exchange method for oxygen reduction. *J. Power Source* **2006**, *161*, 47–53.

19. Saha, M.S.; Li, R.; Sun, X.; Ye, S. 3-D composite electrodes for high performance PEM fuel cells composed of Pt supported on nitrogen-doped carbon nanotubes grown on carbon paper. *Electrochem. Commun.* **2009**, *11*, 438–441.

20. Vijayaraghavan, G.; Stevenson, K.J. Synergistic assembly of dendrimer-templated platinum catalysts on nitrogen-doped carbon nanotube electrodes for oxygen reduction. *Langmuir* **2007**, *23*, 5279–5282.

21. Sun, C.-L.; Chen, L.-C.; Su, M.-C.; Hong, L.-S.; Chyan, O.; Hsu, C.-Y.; Chen, K.-H.; Chang, T.-F.; Chang, L. Ultrafine platinum nanoparticles uniformly dispersed on arrayed CNx nanotubes with high electrochemical activity. *Chem. Mater.* **2005**, *17*, 3749–3753.

22. Li, Y.-H.; Hung, T.-H.; Chen, C.-W. A first-principles study of nitrogen- and boron-assisted platinum adsorption on carbon nanotubes. *Carbon* **2009**, *47*, 850–855.

23. An, W.; Turner, C.H. Chemisorption of transition-metal atoms on boron- and nitrogen-doped carbon nanotubes: Energetics and geometric and electronic structures. *J. Phys. Chem. C* **2009**, *113*, 7069–7078.

24. Chen, Y.; Wang, J.; Liu, H.; Li, R.; Sun, X.; Ye, S.; Knights, S. Enhanced stability of Pt electrocatalysts by nitrogen doping in CNTs for PEM fuel cells. *Electrochem. Commun.* **2009**, *11*, 2071–2076.

25. Joo, S.H.; Choi, S.J.; Oh, I.; Kwak, J.; Liu, Z.; Terasaki, O.; Ryoo, R. Ordered nanoporous arrays of carbon supporting high dispersions of platinum nanoparticles. *Nature* **2001**, *412*, 169.

26. Marie, J.; Berthon-Fabry, S.; Achard, P.; Chatenet, M.; Pradourat, A.; Chainet, E. Highly dispersed platinum on carbon aerogels as supported catalysts for PEM fuel cell-electrodes: Comparison of two different synthesis paths. *J. Non-Cryst. Solids* **2004**, *350*, 88–96.

27. Ambrosio, E.; Francia, C.; Gerbaldi, C.; Penazzi, N.; Spinelli, P.; Manzoli, M.; Ghiotti, G. Mesoporous carbons as low temperature fuel cell platinum catalyst supports. *J. Appl. Electrochem.* **2008**, *38*, 1019–1027.

28. Lebedeva, N.P.; Booij, A.S.; Janssen, G.J. Cathodes for proton-exchange-membrane fuel cells based on ordered mesoporous carbon supports. *ECS Trans.* **2008**, *16*, 2083–2092.

29. Su, F.; Poh, C.K.; Tian, Z.; Xu, G.; Koh, G.; Wang, Z.; Liu, Z.; Lin, J. Electrochemical behavior of Pt nanoparticles supported on meso- and microporous carbons for fuel cells. *Energy Fuels* **2010**, *24*, 3727–3732.

30. Li, Z.; Jaroniec, M. Colloidal imprinting: A novel approach to the synthesis of mesoporous carbons. *J. Am. Chem. Soc.* **2001**, *123*, 9208–9209.

31. Uchida, M.; Fukuoka, Y.; Sugawara, Y.; Ohara, H.; Ohta, A. Improved preparation process of very-low-platinum-loading electrodes for polymer electrolyte fuel cells. *J. Electrochem. Soc.* **1998**, *145*, 3708–3713.

32. Uchida, M.; Fukuoka, Y.; Sugawara, Y.; Eda, N.; Ohta, A. Effects of microstructure of carbon support in the catalyst layer on the performance of polymer-electrolyte fuel cells. *J. Electrochem. Soc.* **1996**, *143*, 2245–2252.

33. Hurt, R.; Krammer, G.; Crawford, G.; Jian, K.; Rulison, C. Polyaromatic assembly mechanisms and structure selection in carbon materials. *Chem. Mater.* **2002**, *14*, 4558–4565.

34. Joo, J.B.; Kim, P.; Kim, W.; Yi, J. Preparation of Pt supported on mesoporous carbons for the reduction of oxygen in polymer electrolyte membrane fuel cell (PEMFC). *J. Electroceram.* **2006**, *17*, 713–718.

35. Fang, B.; Kim, J.H.; Yu, J.-S. Colloid-imprinted carbon with superb nanostructure as an efficient cathode electrocatalyst support in proton exchange membrane fuel cell. *Electrochem. Commun.* **2008**, *10*, 659–662.

36. Banham, D.; Feng, F.; Furstenhaupt, T.; Ye, S.; Birss, V. First time investigation of Pt nanocatalysts deposited inside carbon mesopores of controlled length and diameter. *J. Mater. Chem.* **2012**, *22*, 7164–7171.

37. Lowell, S.; Shields, J.E.; Thomas, M.A.; Thommes, M. *Characterization of Porous Solids and Powders: Surface Area, Pore Size and Density*, 1st ed.; Springer: Dordrecht, The Netherlands, 2006; p. 347.

38. Haynes, J.M. Pore size analysis according to the kelvin equation. *Mater. Construct.* **1973**, *6*, 209–213.

39. Mooney, M. The viscosity of a concentrated suspension of spherical particles. *J. Colloid Sci.* **1951**, *6*, 162–170.

40. Li, Z.; Jaroniec, M. Synthesis and adsorption properties of colloid-imprinted carbons with surface and volume mesoporosity. *Chem. Mater.* **2003**, *15*, 1327–1333.

41. Gierszal, K.P.; Yoon, S.B.; Yu, J.-S.; Jaroniec, M. Adsorption and structural properties of mesoporous carbons obtained from mesophase pitch and phenol-formaldehyde carbon precursors using porous templates prepared from colloidal silica. *J. Mater. Chem.* **2006**, *16*, 2819–2823.

42. Chai, G.S.; Yoon, S.B.; Yu, J.-S.; Choi, J.-H.; Sung, Y.-E. Ordered porous carbons with tunable pore sizes as catalyst supports in direct methanol fuel cell. *J. Phys. Chem. B* **2004**, *108*, 7074–7079.

43. Chen, M.-H.; Jiang, Y.-X.; Chen, S.-R.; Huang, R.; Lin, J.-L.; Chen, S.-P.; Sun, S.-G. Synthesis and durability of highly dispersed platinum nanoparticles supported on ordered mesoporous carbon and their electrocatalytic properties for ethanol oxidation. *J. Phys. Chem. C* **2010**, *114*, 19055–19061.

44. Kimijima, K.; Hayashi, A.; Umemura, S.; Miyamoto, J.; Sekizawa, K.; Yoshida, T.; Yagi, I. Oxygen reduction reactivity of precisely controlled nanostructured model catalysts. *J. Phys. Chem. C* **2010**, *114*, 14675–14683.

45. Shan, J.; Pickup, P.G. Characterization of polymer supported catalysts by cyclic voltammetry and rotating disk voltammetry. *Electrochim. Acta* **2000**, *46*, 119–125.

46. De Levie, R. *Electrochemical Response of Porous and Rough Electrodes*; John Wiley & Sons: New York, NY, USA, 1967; Volume 6.

47. Perry, M.L.; Newman, J.; Cairns, E.J. Mass transport in gas-diffusion electrodes: A diagnostic tool for fuel-cell cathodes. *J. Electrochem. Soc.* **1998**, *145*, 5–15.

48. Banham, D.W.; Soderberg, J.N.; Birss, V.I. Pt/carbon catalyst layer microstructural effects on measured and predicted tafel slopes for the oxygen reduction reaction. *J. Phys. Chem. C* **2009**, *113*, 10103–10111.

49. Bockris, J.O.M.; Srinivasan, S. *Fuel Cells: Their Electrochemistry*; McGraw-Hill: New York, NY, USA, 1969.

50. Soderberg, J.N.; Co, A.C.; Sirk, A.H.C.; Birss, V.I. Impact of porous electrode properties on the electrochemical transfer coefficient. *J. Phys. Chem. B* **2006**, *110*, 10401–10410.

51. Srinivasan, S.; Hurwitz, H.D.; Bockris, J.O.M. Fundamental equations of electrochemical kinetics at porous gas-diffusion electrodes. *J. Chem. Phys.* **1967**, *46*, 3108–3122.

52. Mastronarde, D.N. Automated electron microscope tomography using robust prediction of specimen movements. *J. Struct. Biol.* **2005**, *152*, 36–51.

53. Kremer, J.R.; Mastronarde, D.N.; McIntosh, J.R. Computer visualization of three-dimensional image data using imod. *J. Struct. Biol.* **1996**, *116*, 71–76.

54. Mastronarde, D.N. Dual-axis tomography: An approach with alignment methods that preserve resolution. *J. Struct. Biol.* **1997**, *120*, 343–352.

Electrocatalytic Activity and Durability of Pt-Decorated Non-Covalently Functionalized Graphitic Structures

Emanuela Negro, Alessandro Stassi, Vincenzo Baglio, Antonino S. Aricò and Ger J.M. Koper

Abstract: Carbon graphitic structures that differ in morphology, graphiticity and specific surface area were used as support for platinum for Oxygen Reduction Reaction (ORR) in low temperature fuel cells. Graphitic supports were first non-covalently functionalized with pyrene carboxylic acid (PCA) and, subsequently, platinum nanoparticles were nucleated on the surface following procedures found in previous studies. Non-covalent functionalization has been proven to be advantageous because it allows for a better control of particle size and monodispersity, it prevents particle agglomeration since particles are bonded to the surface, and it does not affect the chemical and physical resistance of the support. Synthesized electrocatalysts were characterized by electrochemical half-cell studies, in order to evaluate the Electrochemically Active Surface Area (ECSA), ORR activity, and durability to potential cycling and corrosion resistance.

Reprinted from *Catalysts*. Cite as: Negro, E.; Stassi, A.; Baglio, V.; Aricò, A.S.; Koper, G.J.M. Electrocatalytic Activity and Durability of Pt-Decorated Non-Covalently Functionalized Graphitic Structures. *Catalysts* **2015**, *5*, 1622–1635.

1. Introduction

Fuel cells operating at low temperature and employing polymer electrolyte membranes are very promising as sustainable power sources for portable, automotive and stationary applications because of their high efficiency and low CO_2 emission [1–3]. However, for large scale distribution of these devices, it is necessary to reduce the cost and, at the same time, increase the durability of the catalyst. In fact, the latter is commonly based on platinum nanoparticles (NPs) supported on high surface area carbon [4–7]. The low durability of the conventional Pt-based electrode materials is due to several phenomena, such as sintering, corrosion and dissolution of catalyst metal particles, that take place especially in non-ideal conditions such as potential and temperature cycling and fuel starvation [8]. The major degradation mechanism has been identified in carbon corrosion at high potentials (>0.8 V *vs.* RHE). Carbon corrosion can cause, among others, loss of hydrophobicity leading to electrode flooding, catalyst detachment leading to loss of the Electrochemically Active Surface Area (ECSA), loss of porosity, with consequent mass transport problems [4,9].

489

Moreover, as clearly reported by Siroma *et al.* [10], the presence of platinum onto the carbon surface significantly increases the degradation of the latter. Recent studies have proven that graphitic materials such as carbon nanotubes (CNTs), nanofibers, *etc.* are significantly more resistant to carbon corrosion than the widely used carbon black due to the higher stability of the sp^2-hybridized carbon [3,11–18].

Within the TU Delft group, we developed a novel carbon material that consists of networked carbon nanostructures (CNNs), currently produced by the TU Delft spin-off company CarbonX (formerly Minus9) [11,19]. CNNs are 3D hyper-branched carbon graphitic structures organized in a nano-scale pattern. They can be easily produced by Chemical Vapour Deposition (CVD) of ethene over transition metal catalyst, used as nucleation elements, and synthetized in bicontinuous microemulsions (BME) [20–22]. The carbonization of the surfactant, being the primary carbon source, leads to the formation of networked, sponge-like, carbon graphitic structures (CNNs), which show promising properties for application as fuel cell supports, such as high electrical conductivity, great oxidation resistance, high specific surface area, micro- and meso-porosity, surface defects increasing the material ability to disperse in solution [11–14,19]. Previous studies showed that CNNs are more durable supports for platinum catalysts in the ORR compared to commercial carbon supports, while the simplicity and versatility of the synthesis route allows a cheaper production than CNTs [12–14].

The relatively inert surface of graphitic carbon supports has been addressed as a drawback for the use of these materials because of lower interaction with the catalyst NPs and thus the weaker bonding [23]. Defects on the surface have a controversial effect: they are considered to be beneficial since they act as anchoring points for the platinum NPs, reducing migration and coalescence and thus loss of ECSA, whereas they can decrease the carbon resistance to corrosion by making it more prone to oxidation [24–26]. Introduction of defects on the surface can be done chemically or electrochemically, e.g., with acid treatment or potential cycling [27], leading to covalently functionalized surfaces, or physically, by absorption of small molecules containing functional groups [28], leading to non-covalently functionalized surfaces. In covalent functionalization, the graphitic surface is functionalized using an oxidative process, such as an acid or plasma treatment. Surface groups such as hydroxyl (-OH), carboxyl (-COOH) and carbonyl (-C=O) groups are created. These groups will act as anchor points for Pt NPs; however, they also represent defects in the graphitic structure and thus they might decrease CNNs chemical resistance. Non-covalent functionalization of graphitic surfaces is based on π-π stacking between the surface and a linking molecule. A linker molecule consists of a benzyl or pyrene group that attaches to the surface of the CNTs/CNNs and a thiol, amine or carboxylic acid group that anchors the Pt ion [28]. Recent work by Oh *et al.* [28] used a 1-pyrene carboxylic acid (1-PCA) as a linker molecule and the standard polyol method for Pt deposition. PCA turned out not to be poisonous for the surface of the Pt catalyst, and not only was no carbon corrosion measured (indirectly via produced CO_2) [28], but also PCA seemed to protect the graphitic surface from degradation, inhibiting corrosion by indirect contact of the Pt catalyst with the carbon surface, particle migration and coalescence.

In the present study, we aim to compare electrocatalysts synthetized with the PCA functionalization combined with the polyol method over graphitic materials with different physical-chemical properties, using two batches of CNN supports produced in-house and a batch of commercial CNTs.

2. Results and Discussion

2.1. Carbon Support Synthesis, Functionalization and Characterization

A summary of the properties of different carbon supports is reported in Table 1. The batches of CNN used for this work exhibit a lower porosity (63 and 39 $m^2 \cdot g^{-1}$ for CNN50 and CNN80 respectively) than commercial CNTs (110 $m^2 \cdot g^{-1}$), Table 1. This can be easily explained by the ticker average diameter, 50 nm and 80 nm, respectively, for CNN50 and CNN80 compared to 20 nm for CNT20, Table 1. This limits the possibility to achieve high Pt loading without affecting the durability of the catalyst. However, recent studies demonstrated that lowering Pt loading improves catalyst utilization and performance [29]. Previous studies reveal that CNN material is mainly meso-porous (2–50 nm), which is beneficial for catalysis as it assures an optimal mass transport [14]. For CNTs, the porosity largely depends on the packing parameter that can strongly vary if the carbon surface is functionalized [14].

Table 1. Carbon supports' physical-chemical properties.

Sample	Average Diameter/nm	Specific Surface Area/$m^2 g^{-1}$	$T_{ox}/°C$	I_D/I_G	PCA/%
CNT20	20	110	700	0.7	7.5
CNN50	50	63	630	0.9	8.3
CNN80	80	39	610	1	9

Figure 1a reports the weight loss of the samples as a function of temperature from TGA analysis in air. Table 1 reports the temperature of oxidation, 700 °C, 630 °C and 610 °C for CNT20, CNN50 and CNN80, respectively. All the samples show higher oxidation resistance compared to for example carbon black or Vulcan, the most widely used catalyst supports [4,9,14]. The difference in oxidation resistance can be attributed to the presence of defects. Defects can be detected and quantified with Raman Spectroscopy measurements, Figure 1b. Table 1 also reports the I_D/I_G values (ratio between the disordered and the graphitic carbon structure) estimated from the ratio between the intensity of the D band at circa 1350 cm^{-1} and G band at circa 1580 cm^{-1} [30]. Raman spectroscopy analysis showed a higher I_D/I_G for CNN80 and CNN50 than for CNT20, implying a higher content of defects in the graphitic structure for the CNN, Figure 1b. This is due to the oxygen and sulphur that has been integrated into the structure as a consequence of the precursor used and was confirmed by EDX in previous work [14]. The amount of defects increases with the average diameter of the carbon nanostructures, from 0.7 to 1. A similar trend can be observed from the broadness of the peak corresponding to graphitic crystalline structure in XRD measurements, Figure 1c. The broader the peak, the lower the level of order indicating smaller crystalline domains [15,16]. Part of these defects consists of quinone groups on the surface, which in previous studies were found to improve ORR or even to be the active catalyst for it [11,14].

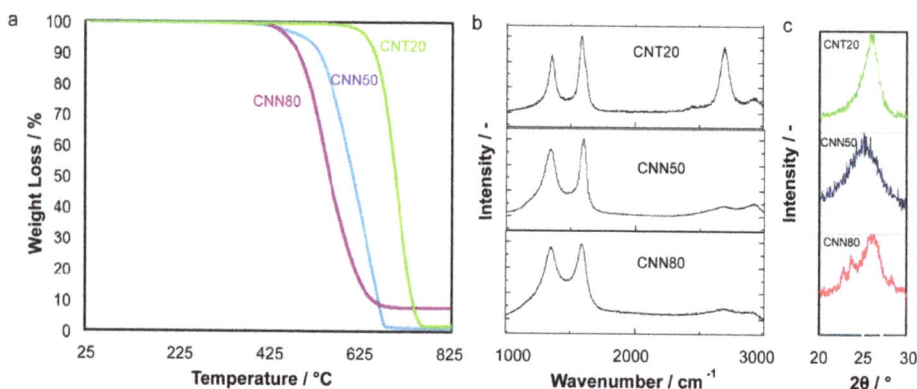

Figure 1. (a) TGA traces of decomposition in air; (b) Raman spectrum and (c) XRD measurements for the graphitic carbon supports.

Carbon nanostructures were functionalized with PCA. PCA is a fluorescent molecule. The confocal microscopy image in Figure 2a shows that PCA is uniformly distributed on the surface of the graphitic carbon nanostructures. PCA concentration in ethanol was measured by UV-Vis spectroscopy.

After functionalization and filtration of carbon nanostructures, the collected PCA in ethanol was diluted 100 times and measured by UV-Vis. A calibration curve was calculated before using Lambert-Beer law and monitoring the absorbance peak at 345 nm. Figure 2b shows that concentration was higher after functionalization of respectively CNT20, CNN50 and CNN80, indicating lower adsorption efficiency on the surface of the graphitic carbon supports. This might be due to the more defected carbon (see Table 1 and Figure 1b) into ethanol, that was also shown being beneficial for dispersion [11]. Additionally, PCA might more easily adhere on less-curved surfaces, better approximating a planar configuration. PCA interaction with a more curved surface as CNT20 might be less strong since optimal adhesion would require a strain in the molecule.

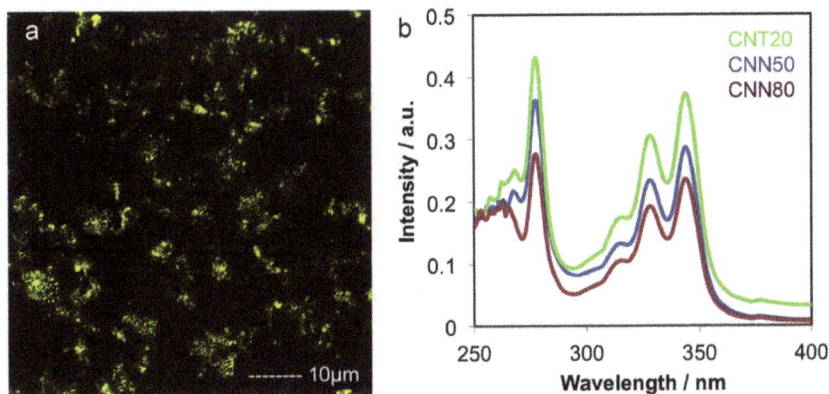

Figure 2. (a) Confocal Microscopy image of CNN80 after functionalization; (b) UV-Vis spectra of PCA/ethanol collected after functionalization and filtration and diluted 100 times.

2.2. Platinum Deposition

The platinum loading was calculated from TGA and confirmed by EDX. For all the samples, the Pt loading result was lower than the target 15%, *i.e.*, 7%, 9% and 11% for Pt-CNT20, Pt-CNN50 and Pt-CNN80 respectively (see Table 2). Interestingly, the Pt loading was proportional to the PCA loading, Table 1, 7.5%, 8.3% and 9% for CNT20, CNN50 and CNN80 respectively. The more functionalized the graphitic carbon support, the better the Pt NPs adsorption. In fact, carboxylic groups and defects in general can both act as nucleation points for Pt reduction and provide anchoring points for the formed NPs [12,14,28].

Table 2. Electrocatalyst physical chemical properties.

Sample	Pt Loading/%	Particle Size TEM/nm	Particle Size XRD/nm
CNT20	7	2.3 ± 0.2	3.9
CNN50	9	2.3 ± 0.2	3.1
CNN80	11	2.3 ± 0.2	2.4

Figure 3 reports the TEM images for the synthetized electrocatalysts. In all the samples, the Pt NPs size calculated from the measurement of at least 200 particles was 2.3 ± 0.2 nm. However, NP size calculated from XRD patterns using the Scherrer equation for the peak corresponding to Pt (220), which is not affected by the interference of other peaks, differed from TEM measurements. Results are reported in Table 2. The NP size measured was higher for CNT20 and CNN50, respectively

3.9 nm and 3.1 nm, indicating the presence of larger NPs probably resulting from coalescence due to poor NP interactions with the support because of a lower amount of functional groups. These results are indeed consistent with different PCA loadings achieved on the supports. These larger NPs could not be visualized by TEM analysis.

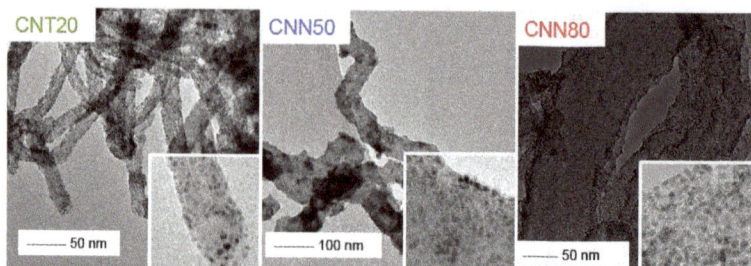

Figure 3. TEM images of the synthetized electrocatalysts.

2.3. Electrochemical Characterization

2.3.1. ECSA and ORR Activity before and after ADT

An accelerated test procedure consisting of 1000 cycles at potentials between 0.6 and 1.2 V and a scan rate of $20 \ mV \cdot s^{-1}$ in 0.5 M H_2SO_4 saturated with N_2 was carried out to evaluate catalyst stability under potential cycling conditions. The cyclic voltammetry profiles for the different electrodes before and after the potential cycling procedure are reported in Figure 4a–c. It can be observed how the PtO_x reduction peak shifts to higher potentials and decreases in intensity after the cycling process, which can be attributed to an increase of particle size with a consequent reduction of the electrochemically active surface area (ECSA) and increase of intrinsic catalytic activity for oxygen reduction. It is clear also from the hydrogen adsorption region that the ECSA decreases after the accelerated test, due to sintering or dissolution of Pt particles [31]. The resulting ECSA values before and after ADTs are reported in Table 3. The ECSA is consistent with the average particle size. Due to the smallest crystallite size and the highest functionalization degree, the Pt/CNN80 catalyst showed the highest ECSA among the considered samples both before and after the ADT (Table 3 and Figure 5). However, the loss of ECSA is quite high for all samples (more than 40%) but still comparable to or higher than the values reported for catalyst synthetized for similar studies as well as for many commercial catalysts [6,16,17,32]. In addition, the current density in the activation region (Figure 4d) decreases after the ADTs for all catalysts. The activity at the beginning of life is higher for CNN50 (due to the optimal particle size of 3 nm, which is a compromise between surface area and specific surface activity); whereas, after the ADT, the Pt/CNN80 shows the best behaviour probably due to the largest ECSA. Figure 5 shows the XRD patterns of the catalysts before and after the ADTs; also, the patterns after the carbon corrosion tests are reported for comparison. An increase of crystallite size was observed for all samples after both tests. This is also evident in the TEM images of Figure 6d (only images for Pt-CNT20 sample are reported as an example), in which a certain degree of sintering in comparison to the fresh catalyst is observed. This catalyst is also affected by the Ostwald ripening process, which leads to the dissolution and re-precipitation of Pt particles onto other Pt particles leading to big agglomerates. The results regarding the crystallite size determination before and after the AD and carbon corrosion tests are summarized in Figure 6e.

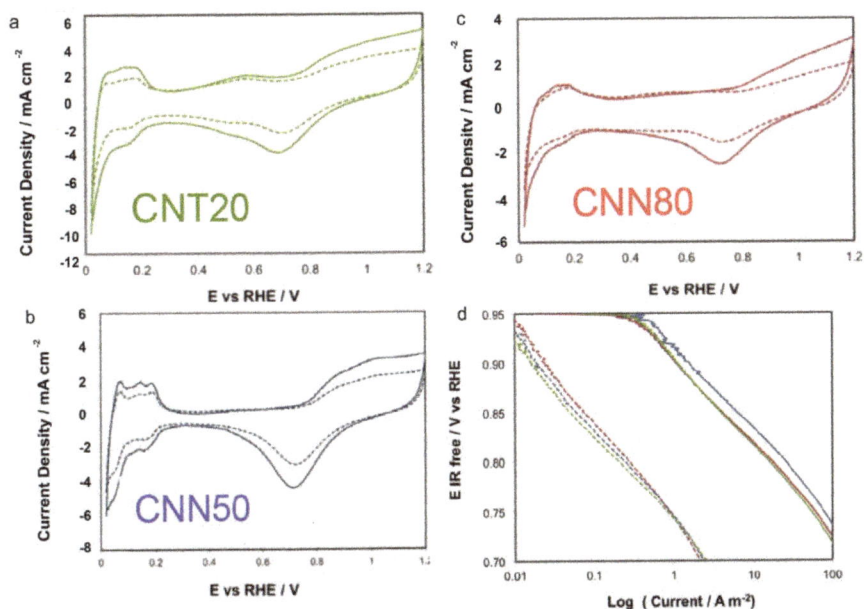

Figure 4. Cyclic Voltametry performed in 0.5 M H_2SO_4, N_2 saturated, scan rate 20 $mV \cdot s^{-1}$, for (**a**) Pt-CNT20; (**b**) CNN50 and (**c**) CNN80. Before (line) and after ADT tests (dotted); (**d**) Tafel plots IR corrected. Dotted lines represent the measurements after ADT tests.

Table 3. ECSA before and after ADT, particle size measured from XRD after ADT and corrosion test, DLC, DLC after over before corrosion, carbon corrosion parameters.

Sample	ECSA/m² g⁻¹	ECSA after ADT/m² g⁻¹	ECSA Loss with ADT/%	Particle Size XRD after ADT/nm	DLC₀/F g⁻¹	DLCf/DLC₀	k	n	Particle Size XRD after Corrosion/nm
CNT20	44.2	22.4	−49	5.1	99	1.11	30.97	0.88	6.7
CNN50	52.1	30.2	−42	5.9	66	1.26	6.46	0.98	5
CNN80	58.3	32.5	−44	6.1	77	1.39	30.7	0.92	5.8

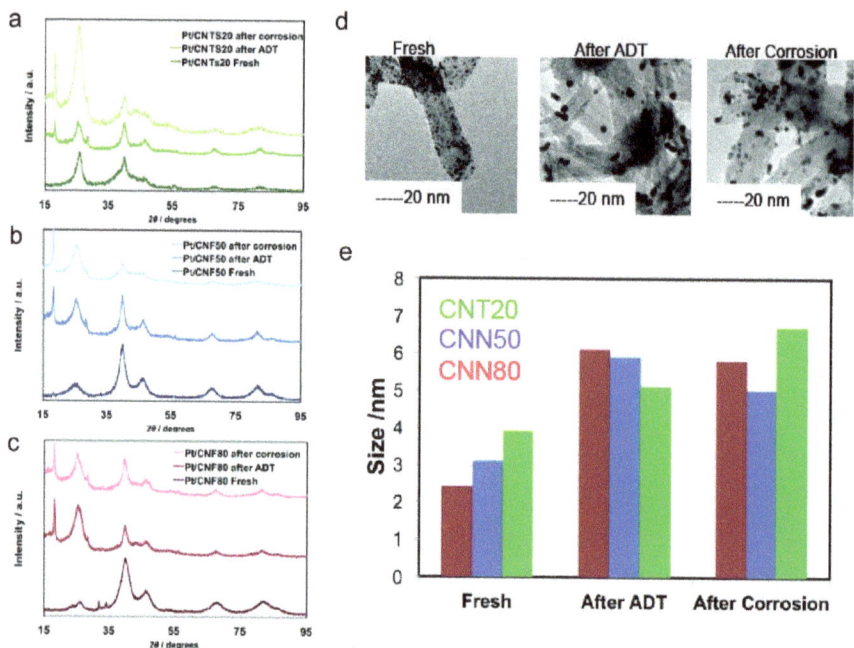

Figure 5. XRD measurements of electrocatalysts fresh, after ADT and after corrosion: (**a**) Pt/CNT20; (**b**) Pt/CNN50; (**c**) Pt/CNN80; (**d**) TEM images for Pt-CNT20 fresh, after ADT and after corrosion; (**e**) Pt particle size measured from XRD measurements for fresh, after ADT and after corrosion electrocatalysts.

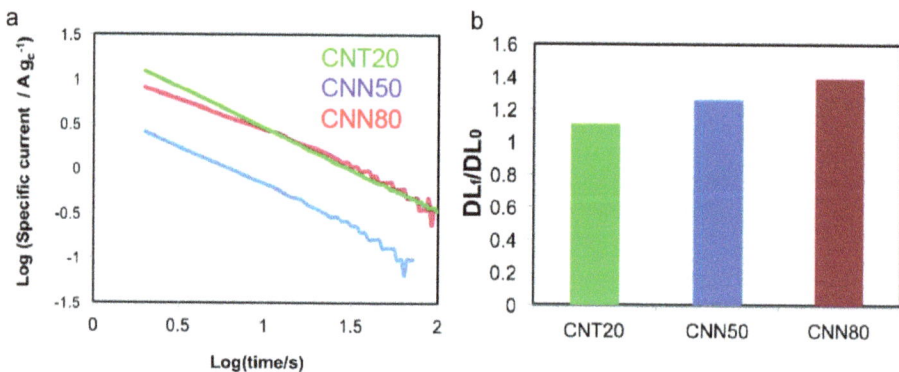

Figure 6. (**a**) Current density–time curves of carbon samples corrosion in double logarithmic scales, performed in half-cell at 1.4 V *vs.* RHE, fed with nitrogen and at room temperature; (**b**) Ratio of the double layer capacitance before and after and corrosion for the electrocatalysts.

2.3.2. Corrosion Tests

Corrosion resistance was evaluated by electrochemical methods. Carbon corrosion experiments were conducted in a three-electrode cell by means of potential holding (1.4 V *vs.* RHE) for 60 min and measuring exchange current density as a function of time. Carbon corrosion takes place at potentials larger than 0.207 V *vs.* RHE according to the following reactions:

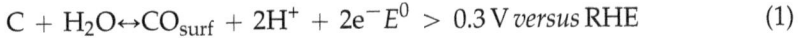

$$C + H_2O \leftrightarrow CO_{surf} + 2H^+ + 2e^- \quad E^0 > 0.3\ \text{V } versus\ \text{RHE} \tag{1}$$

$$CO_{surf} + H_2O \leftrightarrow CO_2 + 2H^+ + 2e^- \quad E^0 > 0.8\ \text{V } versus\ \text{RHE} \tag{2}$$

Generally, the overall reaction is expressed as:

$$C + 2H_2O \rightarrow CO_2 + 4H^+ + 4e^- \quad E_0 = 0.207\ \text{V vs RHE} \tag{3}$$

whose current-time, $j - t$, behavior is generally described according to the following equation:

$$j = kt^{-n} \tag{4}$$

where k and n are the corrosion rate coefficient and the order of corrosion reaction respectively [16]. Figure 6a shows $\log j$ *vs.* $\log t$ curves from which the parameters, k and n, were extracted. The charge exchanged over 60 min and the extracted parameters k and n are summarized in Table 3.

From Figure 6a, one may observe that the corrosion rate for Pt/CNT20 is higher than Pt/CNN80. The lowest corrosion rate is achieved for Pt/CNN50. Generally, corrosion depends on the amount of defects present in the CNN samples and on the Pt-carbon contact surface area, since Pt is a strong catalyst for carbon corrosion. Previous studies show that corrosion in PEM electrodes, measured as CO_2 evolution, is proportional to the Pt-carbon contact area [4]. Since the Pt loading does not vary significantly for the different samples, one would expect that the most defected carbon (CNN80) would be the less resistant to corrosion. The different behavior of our catalysts is probably related to the presence of PCA, which acts as a protection from carbon corrosion, preventing direct contact with Pt. Additionally, the platinum-support interaction plays an important role in improving the long-term stability, as much as 20% as reported in previous studies [33]. The amount of defects on the carbon surface might then have a controversial role: they result in a support more prone to corrosion but they favor PCA adhesion that acts as a protection. CNN50 might then have an optimum amount of defects.

After corrosion, the Double Layer Capacitance (DLC) increased due to an increased amount of defects on the surface and an increased roughening of the surface. The initial DLC, calculated at the net of GDL contributions and neglecting the contribution of Pt, and the ratio of DLC after and before corrosion are reported in Table 3 and Figure 6b, respectively, for the various samples. For CNT the corrosion is expected to take place according to Equations (1) and (2), leading to an increase of DLC due to the creation of more surface defects, but at the same time, the ratio DLC_f/DLC_0 does not increase so much due to a loss of carbon. For CNN, carbon loss plays a less significant role, since DLC increase is much more significant.

3. Experimental Section

3.1. Chemicals

Sodium bis(2-ethyhexyl) sulphosuccinate as surfactant, also known as Na-AOT ($C_{20}H_{37}NaO_7S$, 99%), n-heptane (99.9%) as solvent, ethanol (99.5%), methanol (MeOH, 99.8%), sulphuric acid (97%–98%), chloroplatinic acid hexahydrate ($\geqslant$99.9%), perchloric acid (70%),) sodium hydroxide (>97%), 1-pyrenecarboxylic acid (97%) were purchased from Sigma-Aldrich, Milan, Italy. Iron(II) acetate (FeAc, 97%) metal source was purchased from Strem Chemicals. Ethene, nitrogen, hydrogen and oxygen gases were supplied in cylinders by Siad with 99.999% purity. Multi-walled CNTs (NTX3) were purchased from Nanothinx S.A, Rio Patras, Greece, ethylene glycol (99.5%) and acetone (>99.8%) from Fluka, Milan, Italy, Nafion® solution in aliphatic alcohols (5%) was purchased from Sigma-Aldrich, Milan, Italy, Carbon Cloth coated with a gas diffusion layer was purchased from E-TEK (Boston, MA, USA). All aqueous solutions were prepared using ultrapure water obtained from a Millipore Milli-Q system with resistivity >18 M$\Omega\cdot$cm^{-1}. All chemicals were used as received from suppliers.

3.2. Carbon Support Synthesis and Functionalization

3.2.1. Preparation of CNNs

Two batches of CNNs, labelled CNN50 and CNN80, were synthesized by catalytic CVD, as previously described [11,19].

3.2.2. PCA Functionalization of Graphitic Carbon Supports

Graphitic Carbon supports were non-covalently functionalized by 1-pyrenecarboxylic acid (PCA). The PCA-functionalized carbons were prepared by adding the raw carbons to concentrated ethanol containing PCA (1 mM). The mixture was ultrasonicated for 20 min and then refluxed for 3 h at 25 °C under vigorous stirring. PCA-functionalized Carbons were recovered by centrifugation and washed three times with ethanol. PCA concentration in ethanol at the end of functionalization was measured by UV-Vis in order to calculate the PCA deposited, having previously built a calibration curve with known concentrations.

3.2.3. Catalyst Deposition

Platinum NPs were prepared by a colloidal route described in previous works [34]. Briefly, H_2PtCl_6 and carbon support were mixed in ethylene glycol with a concentration of 0.5 mg· Pt mL^{-1} and 3.5 mg· mL^{-1} respectively. The pH was adjusted to 8–10 by addition of 1M NaOH solution. Subsequently, the temperature was increased to 125 °C and kept for 2 h, under N_2 reflux and rigorous stirring. The pH was adjusted to 2 by addition of sulphuric acid, in order to enhance Pt/carbon support interactions as described in previous work [28,34], and was mixed for 12 h. Afterwards, the mixture was filtrated, washed, centrifuged and dried at 100 °C.

3.3. *Supports and Electrocatalyst Characterization*

UV-Vis spectra were recorded as a function of time with a UV-1800 spectrophotometer from Shimadzu Corporation, Kyoto, Japan. Water was used as a reference because of its stable absorption. Confocal Microscopy images were acquired using a liquid cell.

Transmission Electron Microscopy (TEM) was accomplished using a CM300UT-FEG electron microscope, Philips, Amsterdam, The Netherlands, with a point resolution of 0.17 nm, information limit of 0.1 nm, which was operated at 200 kV, in which images were acquired with a TVIPS CCD camera. For TEM-measurements, samples were prepared by immersing a Quantifoil R copper microgrid in a dispersion of powder consisting of electrocatalyst prepared in ethanol. Size distributions were obtained by measuring at least 200 NPs per sample. Elemental analysis by means of energy-dispersive X-ray spectroscopy (EDX) was carried out on all the samples. For textural characterization, an Autosorb-1c setup (Quantachrome Instruments, Boynton Beach, FL, USA) was used. All samples were outgassed at 350 °C for 17 h in vacuum. Nitrogen (N_2) adsorption isotherms were obtained at 77 K. Specific surface area was obtained from the N_2 isotherm using the Brunauer-Emmett-Teller (BET) method. Micropore volumes were calculated using the Dubinin Radushkevich equation. Raman spectroscopy was performed with a Raman imaging microscope, System 2000 from Renishaw Public Limited Company, Wotton-under-Edge, UK, operated with a 20 mW Argon ion laser of wavelength 514 nm. Samples were cast on a silicon wafer and measured over 60 s. TGA measurements were carried on a TGA7, Perkin Elmer, Waltham, MA, USA. The weight loss in air was measured when increasing the temperature from 25 °C to 900 °C, with a rate of 10 °C min^{-1}. Raman spectroscopy was performed with a Raman imaging microscope, System 2000 from Renishaw Public Limited Company, operated with a 20 mW Argon ion laser of wavelength 514 nm. Samples were cast on a silicon wafer and measured over 60 s.

The catalysts were characterized by XRD using an X-pert 3710 X-ray diffractometer (Philips, Amsterdam, The Netherlands) with Cu Kα radiation operating at 40 kV and 30 mA. The peak profile of the (220) reflection in the face centered cubic structure of Pt catalysts was analyzed by using the Marquardt algorithm, and it was used to calculate the crystallite size by the Debye-Scherrer equation. Instrumental broadening was determined by using a standard Pt sample.

3.4. Electrochemical Characterization

Half-cell tests were carried out in a three electrode cell consisting of a gas diffusion electrode as a working electrode, a mercury-mercurous sulphate (Hg/Hg_2SO_4, sat.) as a reference electrode and a platinum grid as counter electrode. The gas diffusion electrode was prepared according to a procedure described in previous work [16]. Briefly, hydrophobic carbon cloth covered with a diffusion layer (LT 1200 W Elat, ETEK, Boston, MA, USA) was used as backing layer on which the catalytic layer was distributed with a blade. The catalytic layer was composed of 33 wt% Nafion and 67 wt. % catalyst. Electrocatalyst loading was 1.3 ± 0.3 mg·cm^{-2}. The electrode area was 1.5 cm^2 and 0.5 M H_2SO_4 solution was used as electrolyte. Gases (oxygen or nitrogen) were fed from the backside of the electrode in order to perform electrochemical test as described. An Autolab (Metrohm, Utrecht, The Netherlands) potentiostat/galvanostat was used to perform the measurements. ORR activities of the prepared catalysts were evaluated at room temperature by means of the linear sweep voltammetry at 10 mVs^{-1} between the Open Circuit Voltage (OCV) and 0.1 V *versus* a Reversible Hydrogen Electrode (RHE). Kinetic current, mass specific activity i_m and specific activity i_s were evaluated at 0.9 V. Ohmic resistance correction calculated with Impedance Spectroscopy (IS) was applied. ORR activities after degradation were evaluated in the same manner. Accelerated Durability Tests (ADTs) were carried out by scanning the potential between 0.6 and 1.2 V at a scan rate of 20 mV·s^{-1} in 0.5 M H_2SO_4 saturated with N_2 by performing 1000 cycles [16]. Uncompensated resistance was obtained from high frequency impedance. Cyclic Voltammetry (CV) was carried out before and after ADT at room temperature, 25 °C, with a scan rate of 50 mV·s^{-1} in a potential window of 0.02–1.2 V *vs.* RHE. ORR activities before and after ADT were evaluated by means of Linear Sweep Voltammetry (LSV) at 10 mV·s^{-1} between the OCV and 0.1 V *vs.* RHE. Electrochemical carbon corrosion experiments were conducted in the half-cell system by means of potential holding (1.4 V *vs.* RHE).

4. Conclusions

We have successfully synthetized a Pt-catalyst supported on different graphitic supports, two synthetized in-house (CNN) and one commercial (CNT), previously non-covalently functionalized with PCA. Supports differ in geometry and thus specific surface area, content of defects and thus oxidation resistance. We have showed that Pt deposition over non-covalently functionalized graphitic structure is efficient and reliable. However, PCA functionalization efficiency depends on the number of defects that enhance the dispersion properties of the otherwise inert graphitic surfaces. The presence of defects then facilitates PCA adsorption and subsequently the Pt deposition. From accelerated durability tests and corrosion tests, catalyst durability seems to depend on a synergy between carbon support properties and deposition methods.

Acknowledgments: A.S., A.S.A. and V.B. acknowledge the financial support of the PRIN 2010–2011 project "Advanced nanocomposite membranes and innovative electrocatalysts for durable polymer electrolyte membrane fuel cells (NAMED-PEM)". We thank Louw Florusse for TEM micrographs and Roman Latsuzbaia for discussion. We acknowledge financial support from the Advanced Dutch Energy Materials (ADEM) Program, the Ministry of Economic Affairs in the Netherlands in the framework of IOP-Self Healing Materials (SHM) Program, and the COST CM1101 Action.

Author Contributions: E.N., V.B. and G.J.M.K. conceived and designed the experiments; E.N. prepared and characterized the catalysts; A.S. performed the XRD analyses; E.N., V.B., A.S.A. and G.J.M.K. analyzed the data; E.N. wrote the paper; V.B., A.S.A. and G.J.M.K. revised the paper; all authors provided feedback during preparation of the manuscript.

Conflicts of Interest: The authors declare no conflict of interest.

References

1. Dillon, R.; Srinivasan, S.; Aricò, A.S.; Antonucci, V. International activities in DMFC R&D: Status of technologies and potential applications. *J. Power Sources* **2004**, *127*, 112–126.

2. Specchia, S.; Francia, C.; Spinelli, P. Polymer electrolyte membrane fuel cells. In *Electrochemical Technologies for Energy Storage and Conversion*; Wiley-VCH Verlag GmbH & Co. KGaA: Weinheim, Germany, 2011; pp. 601–670.

3. Aricò, A.S.; Baglio, V.; Antonucci, V. *Direct Methanol Fuel Cells*; Nova Science Publishers: Hauppauge, NY, USA, 2010.

4. De Bruijn, F.A.; Dam, V.A.T.; Janssen, G.J.M. Review: Durability and degradation issues of PEM fuel cell components. *Fuel Cells* **2008**, *8*, 3–22.

5. Zeng, J.; Francia, C.; Dumitrescu, M.A.; Videla, A.H.A.M.; Ijeri, V.S.; Specchia, S.; Spinelli, P. Electrochemical performance of Pt-based catalysts supported on different ordered mesoporous carbons (Pt/OMCs) for oxygen reduction reaction. *Ind. Eng. Chem. Res.* **2011**, *51*, 7500–7509.

6. Gasteiger, H.A.; Kocha, S.S.; Sompalli, B.; Wagner, F.T. Activity benchmarks and requirements for Pt, Pt-alloy, and non-Pt oxygen reduction catalysts for PEMFCs. *Appl. Catal. B* **2005**, *56*, 9–35.

7. Galvez, M.E.; Calvillo, L.; Alegre, C.; Sebastian, D.; Suelves, I.; Perez-Rodriguez, S.; Celorrio, V.; Pastor, E.; Pardo, J.I.; Moliner, R.; *et al.* Nanostructured carbon materials as supports in the preparation of direct methanol fuel cell electrocatalysts. *Catalysts* **2013**, *3*, 671–682.

8. Ferrandon, M.; Wang, X.; Kropf, A.J.; Myers, D.J.; Wu, G.; Johnston, C.M.; Zelenay, P. Stability of iron species in heat-treated polyaniline-iron-carbon polymer electrolyte fuel cell cathode catalysts. *Electrochim. Acta* **2013**, *110*, 282–291.

9. Borup, R.; Meyers, J.; Pivovar, B.; Kim, Y.S.; Mukundan, R.; Garland, N.; Myers, D.; Wilson, M.; Garzon, F.; Wood, D.; *et al.* Scientific aspects of polymer electrolyte fuel cell durability and degradation. *Chem. Rev.* **2007**, *107*, 3904–3951.

10. Siroma, Z.; Ishii, K.; Yasuda, K.; Miyazaki, Y.; Inaba, M.; Tasaka, A. Imaging of highly oriented pyrolytic graphite corrosion accelerated by Pt particles. *Electrochem. Commun.* **2005**, *7*, 1153–1156.

11. Negro, E.; Dieci, M.; Sordi, D.; Kowlgi, K.; Makkee, M.; Koper, G.J.M. High yield, controlled synthesis of graphitic networks from dense micro emulsions. *Chem. Commun.* **2014**, *50*, 11848–11851.

12. Negro, E.; Latsuzbaia, R.; Dieci, M.; Boshuizen, I.; Koper, G.J.M. Pt electrodeposited over carbon nano-networks grown on carbon paper as durable catalyst for PEM fuel cells. *Appl. Catal. B* **2015**, *166–167*, 155–165.

13. Negro, E.; Videla, A.H.A.M.; Baglio, V.; Aricò, A.S.; Specchia, S.; Koper, G.J.M. Fe-N supported on graphitic carbon nano-networks grown from cobalt as oxygen reduction catalysts for low-temperature fuel cells. *Appl. Catal. B* **2015**, *166–167*, 75–83.

14. Negro, E.; Vries, M.A.D.; Latsuzbaia, R.; Koper, G.J.M. Networked graphitic structures as durable catalyst support for PEM electrodes. *Fuel Cells* **2014**, *14*, 350–356.

15. Sebastián, D.; Lázaro, M.J.; Suelves, I.; Moliner, R.; Baglio, V.; Stassi, A.; Aricò, A.S. The influence of carbon nanofiber support properties on the oxygen reduction behavior in proton conducting electrolyte-based direct methanol fuel cells. *Int. J. Hydrogen Energy* **2012**, *37*, 6253–6260.

16. Sebastián, D.; Ruíz, A.G.; Suelves, I.; Moliner, R.; Lázaro, M.J.; Baglio, V.; Stassi, A.; Aricò, A.S. Enhanced oxygen reduction activity and durability of Pt catalysts supported on carbon nanofibers. *Appl. Catal. B* **2012**, *115–116*, 269–275.

17. Sebastián, D.; Suelves, I.; Moliner, R.; Lázaro, M.J.; Stassi, A.; Baglio, V.; Aricò, A.S. Optimizing the synthesis of carbon nanofiber based electrocatalysts for fuel cells. *Appl. Catal. B* **2013**, *132–133*, 22–27.

18. Carrera-Cerritos, R.; Baglio, V.; Aricò, A.S.; Ledesma-García, J.; Sgroi, M.F.; Pullini, D.; Pruna, A.J.; Mataix, D.B.; Fuentes-Ramírez, R.; Arriaga, L.G. Improved Pd electro-catalysis for oxygen reduction reaction in direct methanol fuel cell by reduced graphene oxide. *Appl. Catal. B* **2014**, *144*, 554–560.

19. Kowlgi, K.N.K.; Koper, G.J.M.; van Raalten, R.A.D. Carbon nanostructures and networks produced by chemical vapor deposition. US Patent 20130244023 A1, 19 September 2013.

20. Kowlgi, K.; Lafont, U.; Rappolt, M.; Koper, G. Uniform metal nanoparticles produced at high yield in dense microemulsions. *J. Colloid Interf. Sci.* **2012**, *372*, 16–23.

21. Latsuzbaia, R.; Negro, E.; Koper, G. Bicontinuous microemulsions for high yield, wet synthesis of ultrafine nanoparticles: A general approach. *Faraday Discuss.* **2015**, *181*, 37–48.

22. Negro, E.; Latsuzbaia, R.; Koper, G.J.M. Bicontinuous microemulsions for high yield wet synthesis of ultrafine platinum nanoparticles: Effect of precursors and kinetics. *Langmuir* **2014**, *30*, 8300–8307.

23. Saha, M.S.; Kundu, A. Functionalizing carbon nanotubes for proton exchange membrane fuel cells electrode. *J. Power Sources* **2010**, *195*, 6255–6261.

24. Wu, H.; Wexler, D.; Wang, G.; Liu, H. Pt/C catalysts using different carbon supports for the cathode of PEM fuel cells. *Adv. Sci. Lett.* **2011**, *4*, 115–120.

25. Xu, C.; Chen, J.; Cui, Y.; Han, Q.; Choo, H.; Liaw, P.K.; Wu, D. Influence of the surface treatment on the deposition of platinum nanoparticles on the carbon nanotubes. *Adv. Eng. Mater.* **2006**, *8*, 73–77.

26. Xing, Y. Synthesis and electrochemical characterization of uniformly-dispersed high loading Pt nanoparticles on sonochemically-treated carbon nanotubes. *J. Phys. Chem. B* **2004**, *108*, 19255–19259.

27. Latsuzbaia, R.; Negro, E.; Koper, G.J.M. Environmentally friendly carbon-preserving recovery of noble metals from supported fuel cell catalysts. *ChemSusChem* **2015**, *8*, 1926–1934.

28. Oh, H.-S.; Kim, H. Efficient synthesis of Pt nanoparticles supported on hydrophobic graphitized carbon nanofibers for electrocatalysts using noncovalent functionalization. *Adv. Funct. Mater.* **2011**, *21*, 3954–3960.

29. Esmaeilifar, A.; Rowshanzamir, S.; Eikani, M.H.; Ghazanfari, E. Synthesis methods of low-Pt-loading electrocatalysts for proton exchange membrane fuel cell systems. *Energy* **2010**, *35*, 3941–3957.

30. Zou, J.; Zeng, X.; Xiong, X.; Tang, H.; Li, L.; Liu, Q.; Li, Z. Preparation of vapor grown carbon fibers by microwave pyrolysis chemical vapor deposition. *Carbon* **2007**, *45*, 828–832.

31. Stassi, A.; Modica, E.; Antonucci, V.; Aricò, A.S. A half cell study of performance and degradation of oxygen reduction catalysts for application in low temperature fuel cells. *Fuel Cells* **2009**, *9*, 201–208.

32. Curnick, O.J.; Mendes, P.M.; Pollet, B.G. Enhanced durability of a Pt/C electrocatalyst derived from nafion-stabilised colloidal platinum nanoparticles. *Electrochem. Commun.* **2010**, *12*, 1017–1020.

33. Stamatin, S.N.; Borghei, M.; Dhiman, R.; Andersen, S.M.; Ruiz, V.; Kauppinen, E.; Skou, E.M. Activity and stability studies of platinized multi-walled carbon nanotubes as fuel cell electrocatalysts. *Appl. Catal. B* **2015**, *162*, 289–299.

34. Oh, H.-S.; Oh, J.-G.; Kim, H. Modification of polyol process for synthesis of highly platinum loaded platinum-carbon catalysts for fuel cells. *J. Power Sources* **2008**, *183*, 600–603.

Application of a Coated Film Catalyst Layer Model to a High Temperature Polymer Electrolyte Membrane Fuel Cell with Low Catalyst Loading Produced by Reactive Spray Deposition Technology

Timothy D. Myles, Siwon Kim, Radenka Maric and William E. Mustain

Abstract: In this study, a semi-empirical model is presented that correlates to previously obtained experimental overpotential data for a high temperature polymer electrolyte membrane fuel cell (HT-PEMFC). The goal is to reinforce the understanding of the performance of the cell from a modeling perspective. The HT-PEMFC membrane electrode assemblies (MEAs) were constructed utilizing an 85 wt. % phosphoric acid doped Advent TPS® membranes for the electrolyte and gas diffusion electrodes (GDEs) manufactured by Reactive Spray Deposition Technology (RSDT). MEAs with varying ratios of PTFE binder to carbon support material (I/C ratio) were manufactured and their performance at various operating temperatures was recorded. The semi-empirical model derivation was based on the coated film catalyst layer approach and was calibrated to the experimental data by a least squares method. The behavior of important physical parameters as a function of I/C ratio and operating temperature were explored.

Reprinted from *Catalysts*. Cite as: Myles, T.D.; Kim, S.; Maric, R.; Mustain, W.E. Application of a Coated Film Catalyst Layer Model to a High Temperature Polymer Electrolyte Membrane Fuel Cell with Low Catalyst Loading Produced by Reactive Spray Deposition Technology. *Catalysts* **2015**, *5*, 1673–1691.

1. Introduction

Three of the most significant challenges facing the wide commercialization of proton exchange membrane fuel cells (PEMFCs) are: (1) improving catalyst tolerance to impurities [1–6]; (2) simplifying water and thermal management schemes [1–4,6]; and (3) enhancing the kinetics of the oxygen reduction reaction (ORR) at the cathode [1,2,6–8]. Recent research has focused on mitigating the above challenges by increasing the operating temperature of PEMFCs from ~80 °C to $\geqslant$ 120 °C, resulting in so-called high temperature PEMFCs (HT-PEMFCs). At high operating temperatures, carbon monoxide adsorption onto the catalyst surface, which greatly reduces cell performance, is not favored. It was found for temperatures above 160 °C that upwards of 3% carbon monoxide in the feed stream could be tolerated [9,10]. The elevated temperature of the HT-PEMFC also means that there is a greater thermal gradient between the cell and the environment. This simplifies the balance of plant related to thermal management due to faster heat rejection [2]. Operating at elevated temperature does introduce new difficulties in maintaining adequate hydration of proton exchange membranes such as Nafion®, though this issue has been addressed with the use of phosphoric acid doped membranes which do not require external humidification [1,11,12]. Operating without external humidification not only simplifies balance of plant but also implies single phase, gaseous, transport in the gas diffusion layers (GDL) and flow channels, which makes reactant diffusion processes more facile [3,4]. To date, phosphoric acid-doped polybenzimidazole (PBI) membranes have been among the most successful acid doped membranes for HT-PEMFC applications [11–20]. PBI outperforms conventional membranes (*i.e.*, Nafion®) by self-solvating protons to allow charge migration, hence minimizing the reliance on water for proton transport.

Despite their advantages, widespread implementation of HT-PEMFCs still has significant roadblocks. Material performance and durability can be compromised at elevated temperature [6]. The use of phosphoric acid doped membranes to combat dehydration issues results in a highly acidic environment that can increase component degradation [2]. Several authors have noted a rapid degradation of cell performance for PBI-based HT-PEMFCs [21–25]. Two key processes related to this rapid degradation have been identified: catalyst agglomeration in the cathode due the presence of phosphoric acid and oxygen coupled with the high potential, and the direct degradation of the polymer. In addition, as with their low temperature counterpart, cost of production is a major concern for HT-PEMFCs [26]. As such, Reactive Spray Deposition Technology (RSDT) has been examined in recent publications as a low cost method for producing HT-PEMFC MEAs [27,28]. RSDT is a flame-based method suitable for nanoscale particle production and deposition in a single step process. RSDT relies on combustion of a fuel and solvent as a thermal energy source that drives particle nucleation. Annealing occurs either by reaction of precursor gases (gas-to-particle conversion) or by evaporation and/or reaction of suspended precursor particles or droplets (particle-to-particle conversion) in gas streams. The RSDT system avoids the wet chemistry byproducts and the associated nanoparticle separation/purification steps necessary for separate catalyst formation and deposition. It also combines catalyst production and electrode fabrication in one step.

Previous work with RSDT in HT-PEMFCs explored the proof of concept in manufacturing catalyst-coated membranes (CCMs) using PBI based membranes [27], as well as investigating the manufacturing of gas diffusion electrodes (GDEs) using the more recently developed TPS® membrane produced by Advent Technologies [28].

The TPS® membrane was chosen in the more recent study due to its similar behavior to pure phosphoric acid at elevated temperatures [29].

It is important to correlate the resulting performance of the HT-PEMFC to physical properties of the RSDT electrode such as the binder to carbon support ratio (I/C), catalyst loading, pore size distribution, catalyst roughness, and critical pore radius. For this reason a semi-empirical model was developed in this work. There are several commonly used approaches to modeling the behavior of the catalyst layer in a PEMFC. The simplest approach is the interface approximation, which treats the catalyst layer as an ultra-thin reactive boundary between the membrane electrolyte and the GDL [30,31]. The issue with this approach is that it does not glean any relevant structural information about the catalyst layer and, additionally, it generally over predicts the performance since mass transport limitations are not taken into account [32]. A second approach involves a few different scenarios that can be collectively referred to as thin film modeling [33–41]. These models are more complex than the interface approach, but still relatively simple and easy to implement. In this approach, the catalyst layer is treated as a thin film where the porous structure contains either water (in the case of a polymer electrolyte) or phosphoric acid (in the case a phosphoric acid doped membrane is used). The nature of how the liquid occupies the catalyst layer is where the distinction is made between approaches. Several works have assumed the liquid floods the catalyst layer and reactant transport takes place through the flooded porous media [33–35]. Others have assumed there are large, gas phase pores that exist due to hydrophobic binders such as PTFE coupled with smaller hydrophilic pores [36–39]. Many of these works focus on gas phase diffusion through the catalyst layer as the primary mass transport resistance. A further subtlety may be applied when the larger, gas filled pores are characterized as having a thin coating of liquid on the walls [40,41]. Reactants must then dissolve into and diffuse through this coating to reach the catalyst embedded in the pore walls. This type of model will be referred to as a coated film model in this work. In the case of phosphoric acid fuel cells, it was argued by Scott *et al.*, that the gas phase diffusion resistance was negligible compared to the diffusion through the coated film [41,42]. The final approach is the catalyst agglomerate model [32,40,41,43–48]. This approach is the most complex and generally considered to be the most complete for most catalyst layer structures. It considers the catalyst layer to be composed of spherical agglomerates of catalyst material where the space between the agglomerates is filled with a mixture of the electrolyte and reactant gases. Gas dissolves and diffuses to the center of the agglomerates while continuously reacting.

The coated film model was selected in this work for modeling the transport in the catalyst layer of the HT-PEMFC produced by RSDT for several reasons: (1) the balance of simplicity and detail makes it a good candidate to capture the behavior of the performance data; (2) it has been stated that it is appropriate for catalyst

506

layer thicknesses $\leqslant 5$ µm [45]. The typical catalyst layer thickness produced by RSDT ranges from 500 nm up to 5 µm; and (3) RSDT provides excellent dispersion and particle size distributions when I/C ratios are optimized, so the agglomerate model may not be geometrically appropriate for catalyst layers produced via RSDT. Experimental data for the Advent TPS® membrane MEAs reported in [28] exposed to pure oxygen have been correlated with the developed model. The resulting calibrated parameters were examined as a function of varying ionomer(PTFE)/carbon(Vulcan XC-72R) (I/C) ratio and as a function of operating temperature of the cell. The effect of varying I/C ratio for HT-PEMFCs has been shown to be important in previous studies by Lobato *et al.* for (PBI)-H_3PO_4/Vulcan XC-72R in the catalyst layer and PTFE/Vulcan XC-72R in the microporous layers [49,50]. It was observed that the optimum performance of the RSDT produced HT-PEMFC was obtained for I/C = 0.9 and T = 190 °C (determined by the current density achieved at a cell voltage of 0.6 V) [28]. This work seeks to build upon the previous experimental study by applying a physical model to reinforce the understanding of the cell performance as a function of temperature and I/C ratio. Interpretations of the trends in the model parameters calculated in this study have been provided to explain the observed behavior of the experimental polarization results with specific focus on how the distribution of phosphoric acid within the catalyst layer relates to cell performance.

2. Results and Discussion

2.1. Correlation of Model and Experimental Data

Figure 1 shows the comparison between the model and the experimental data collected in our group's previous work [28]. The model compares to the data very well over the full range of current densities regardless of the I/C ratio (average normalized root mean square deviation (NRMSD) was 0.020 with a standard deviation of 0.004). Detailed discussions of these trends are available in [28] from an experimental point of view and will be briefly summarized here for convenience. It appears that increasing the I/C ratio, up to 0.9, results in an improvement in the overall performance relating to activation overpotential, ohmic losses, and mass transport limiting behavior. The best performance was obtained for I/C = 0.9 [28]. This was due, in part, to the excellent dispersion of the platinum nanoparticles achieved at that ratio. The TEM images in Figure 2 show increased platinum agglomeration and poor coverage of some of the binder for the less optimal I/C ratios, particularly at the low I/C (I/C = 0.1). Increasing I/C to 1.0 resulted in a decrease in the resulting performance. This trend persisted for all operating temperatures studied. The effect of temperature is somewhat more obscure with what appears to be an initial increase in performance across all I/C ratios followed by a decrease at the

507

higher temperatures of 190 and 200 °C depending on the I/C ratio considered. These trends will be further elucidated in the discussion of the modeling results below.

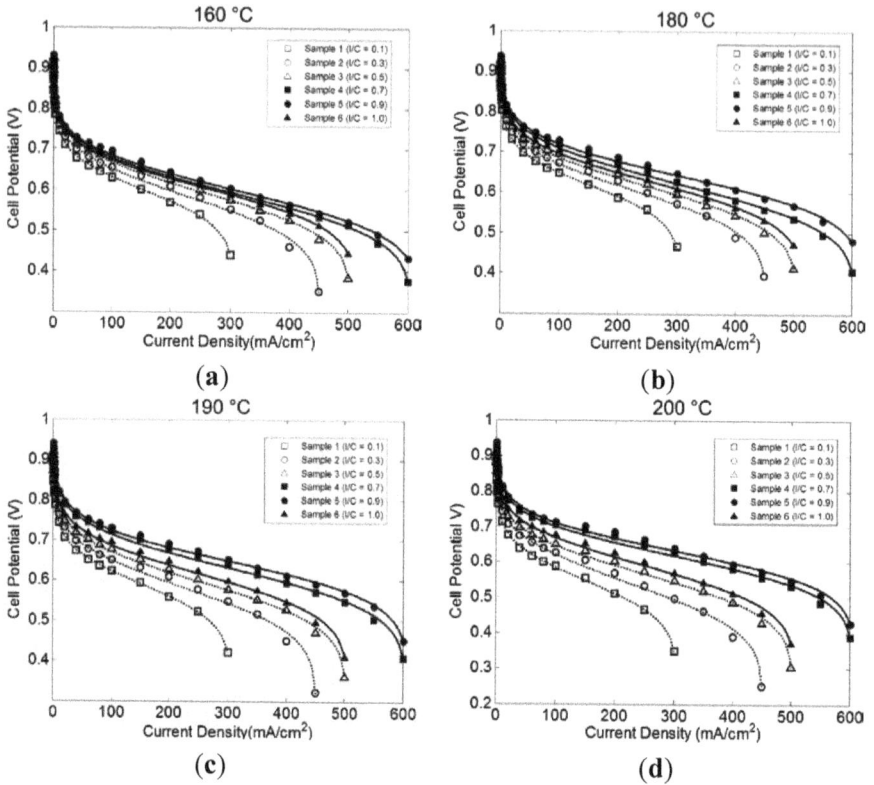

Figure 1. Correlation between experimental data and the coated film model developed in this work (curves are modeling results and points are experimental data from [28], reproduced with permission, with I/C ratios indicated in the figure legends). Cell operating temperatures are (**a**) 160 °C, (**b**) 180 °C, (**c**) 180 °C, and (**d**) 200 °C.

2.2. Effect of I/C Ratio

Three parameters were used to describe the experimental data: $j'_{0,c}$, κ, and Γ. More details on the modeling approach are provided in Section 3.2. Each of the parameters is representative of one of the characteristic regions of a typical fuel cell polarization curve: the activation loss region, the ohmic loss region, and the mass transport limiting region, respectively. Figure 3 shows the results for each parameter as a function of I/C at various temperatures.

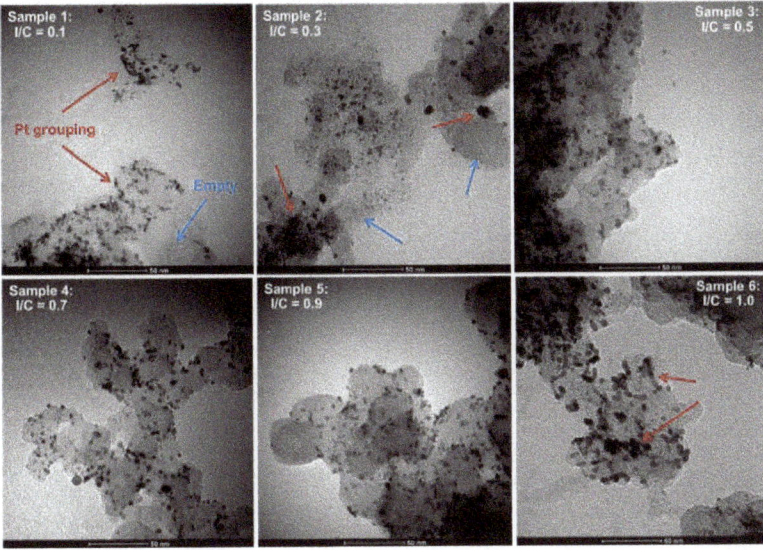

Figure 2. TEM images showing evolution of platinum distribution as a function of varying ratios of PTFE binder to carbon support material (I/C). Red arrows indicate areas where platinum has the grouped whereas blue arrows indicate areas where platinum is absent from the support.

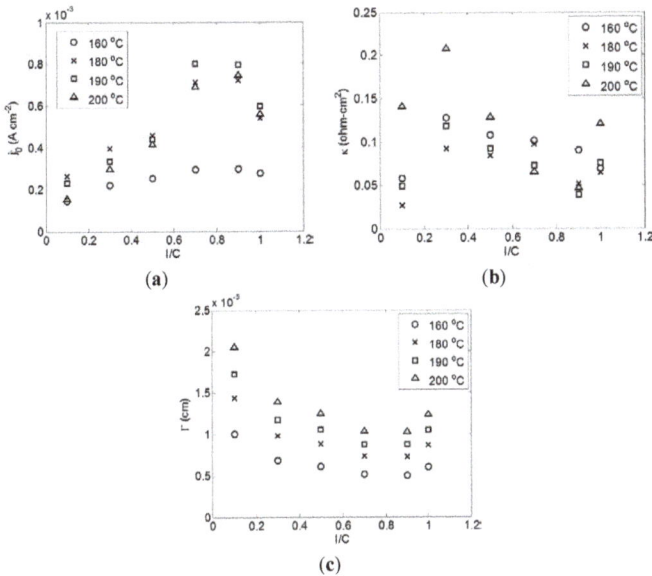

Figure 3. Correlation parameters as a function of I/C for different temperatures. Inset (**a**) represents the exchange current density parameter, $j'_{0,c}$, (**b**) represents the area specific resistance parameter, κ, and (**c**) represents the geometric parameter, Γ.

When examining the behavior of the parameters in Figure 3, it is observed that $j'_{0,c}$ and Γ have similar trends towards increasing performance with increasing I/C up to 0.9 at which point the trend starts to invert. However, κ has a slightly different behavior where there is an initial increase between I/C = 0.1 and 0.3. At elevated temperatures, there is a steady decrease between I/C = 0.3 and 0.9 followed by an increase again at I/C = 1.0. At lower temperatures, however, this trend is obscured. The observed behavior can be understood by considering the pore size distribution (obtained via mercury intrusion porosimetry using a Micromeritics AutoPore IV 9500) reported in our previous work [28]. As the I/C ratio was increased, there was a shift in the average pore size towards smaller pores in the 0–1000 nm range (the relevant range for the catalyst layer). Figure 4 shows this shift in the pore size distribution.

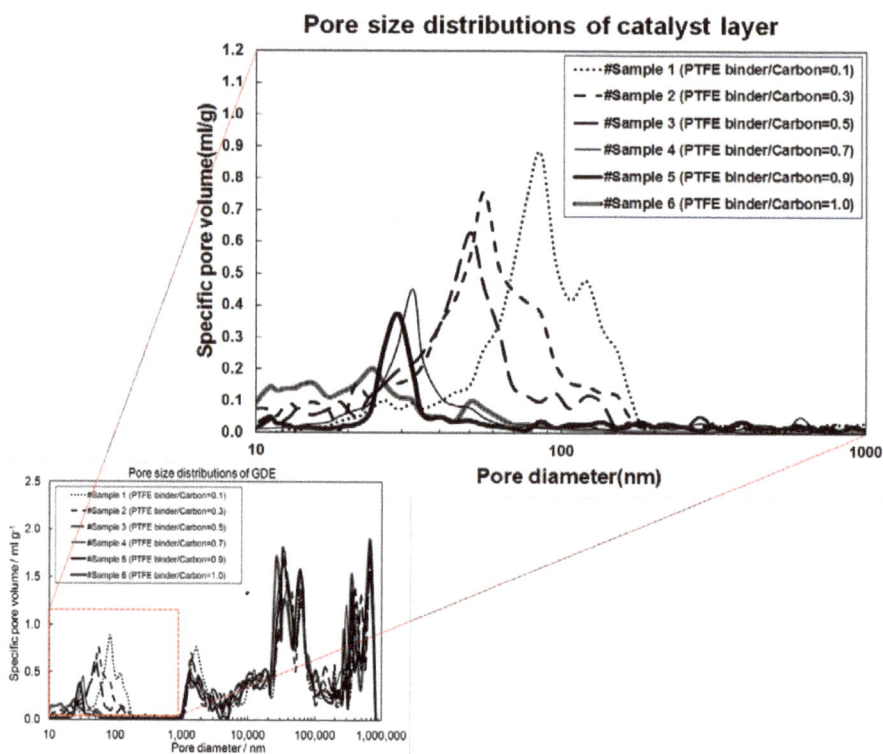

Figure 4. Pore size distribution of the gas diffusion electrode (GDE) (adapted from [28] with permission). Blow up shows the range from 0–1000 nm which is attributed to the catalyst layer.

Li *et al.* showed that a critical radius exists within a porous catalyst layer below which the pores are completely flooded with acid and above which they are gas filled with a coating of acid on the pore walls [40]. Due to the shift in the pore size

510

distribution towards smaller pore radii, the number of acid flooded pores can be expected to increase. This behavior can be used to explain the observed trends in the three correlation parameters shown in Figure 3. Furthermore, as discussed in Section 3.2, it is assumed that the gas phase mass transport resistance has a negligible effect on cell performance for this particular HT-PEMFC configuration. The fact that an increase in cell performance occurs despite a decrease in overall pore size and porosity with increasing I/C ratio is supportive of this assumption since a reduction in porosity would tend to negatively impact gas phase transport.

Beginning with the exchange current density parameter, $j'_{0,c}$, there is a general upward trend in its value until a maximum is obtained between I/C = 0.7 and 0.9 at which point the trend starts to decrease. This behavior can be rationalized by considering the catalyst roughness, a_c, (Equation (3) in Section 3.2). As the I/C ratio increases and smaller pores are favored, causing increased flooding of those pores in the catalyst layer with phosphoric acid, there will be less catalyst surface area exposed directly to the reactants since it is assumed that reactions occur primarily in the acid films coating the walls of the larger, gas filled pores. As pointed out by Li et al. [40], the coated film approach assumes that only the catalyst material within these larger pores are electrochemically active while the catalyst material in the depths of the completely flooded pores is unutilized. Simultaneously, the ionic interconnectivity between the larger gas filled pores will likely be improved, which will activate otherwise isolated catalyst material. The net effect is an initial increase in the effective surface area of the catalyst until the pores become too small and the catalyst layer begins to flood completely causing a drop in the surface roughness, and consequently, the exchange current density at I/C ratios above 0.9. It is also possible that excessive PTFE binder begins to cover the catalyst sites at high I/C ratios creating the same effect.

The behavior of the area specific resistance parameter, κ, is more complicated than the others. As mentioned above, there is an initial increase in κ between I/C = 0.1 and 0.3, followed by a general decrease with increasing I/C up to I/C = 0.9. At elevated temperatures, 190 °C and 200 °C, there is a second increase between I/C = 0.9 and 1.0. As explained in Section 3.2, it is believed that the ohmic resistance parameter is a combined effect of the catalyst layer ionic resistance, the contact resistance between the membrane and the GDE, and the electrolyte resistance. When the I/C ratio is changed at constant temperature it is not expected that the electrolyte resistance should change and thus the behavior of κ should be related to changes in contact resistance and catalyst layer resistance. The initial increase in the cell resistance can likely be explained by the very low I/C ratio. For such a low ratio, there is an abundance of hydrophilic carbon present in the catalyst layer. This, coupled with the greater porosity as evident in Figure 4, will lead to a much greater level of phosphoric acid in the electrodes. There will be large scale flooding of the cathode, which will have a negative impact on the other cell parameters, but will lead to excellent ionic conductivity of the catalyst layer. When the I/C ratio is increased, this high level of flooding will be reduced causing the increased cell resistance observed in Figure 3. The resistance is then observed to decrease with increasing I/C between 0.3 and 0.9 most likely for similar reasons discussed in relation to the exchange current density. As the shift towards smaller pore sizes continues, more pores will be sized below the critical pore radius and there will be an improvement in the phosphoric acid percolation network. As the I/C ratio increases beyond 0.9, Figure 4 indicates a breakdown in the uniformity of the pores, which may cause increased contact resistance. It is also possible that excessive levels of hydrophobic binder can begin to impede capillary action. These two scenarios would then seem to be exacerbated by increased temperature explaining the strong increase in resistance for higher temperatures at I/C = 1.0.

The behavior of Γ can be explained similarly to the exchange current density in terms of the effect on the active catalyst surface area. However, it is conceivable that increased flooding begins to cause an increase in the thickness of the phosphoric acid film in the larger gas pores. This would tend to hinder the transport of the reactants. The trend observed in Figure 3 would then be the result of a balancing of the two opposing effects.

It is worth mentioning that the improvement to the catalyst roughness can also be attributed to improved distribution of the platinum nanoparticles as discussed in Section 2.1 and [28]. The optimum I/C ratio gives the most uniform platinum distribution in the catalyst layer, which helps maximize utilization. However, this improved distribution would not directly affect the behavior of κ indicating there is an effect of changing phosphoric acid distribution in the catalyst layer.

2.3. Effect of Temperature

Aside from the effect of the I/C ratio, there is an interesting behavior of the three parameters with operating temperature, which are shown in Figure 5.

The exchange current density parameter, $j'_{0,c}$, does not display the Arrhenius relationship one might initially expect. This departure from linear behavior is caused by the catalyst roughness, which decreases with increasing temperature. Similarly, the parameter Γ increases with temperature since the surface roughness influences it as well. As discussed in Section 2.2, an increase in the phosphoric acid content of the catalyst layer, to a point, increased the catalyst electrochemically active surface area. The trend observed in Figure 5 seems to indicate a reduction of phosphoric acid saturation within the catalyst layer with increasing temperature. The reason for this phenomenon requires further exploration, but it may be possible that there is a shift to a lower critical radius with increasing temperature or the TPS® membrane has a higher phosphoric acid retention at elevated temperatures leading to decreased migration into the catalyst layer. It is interesting to note that the negative influence on $j'_{0,c}$ is enhanced at the less optimum I/C ratios (*i.e.*, there is a maximum influence for I/C = 0.1 and a minimum for I/C = 0.9). The reason for this behavior can be attributed to better interconnectivity of the pores at the optimum I/C ratio. Even as the phosphoric acid content decreases, the remaining phosphoric acid has better percolation, which leads to a higher active catalyst surface area. The effect of the reduced surface roughness appears to be less pronounced for the Γ parameter, which can be explained by a simultaneous reduction in film thickness as the overall phosphoric acid content in the electrodes decreases.

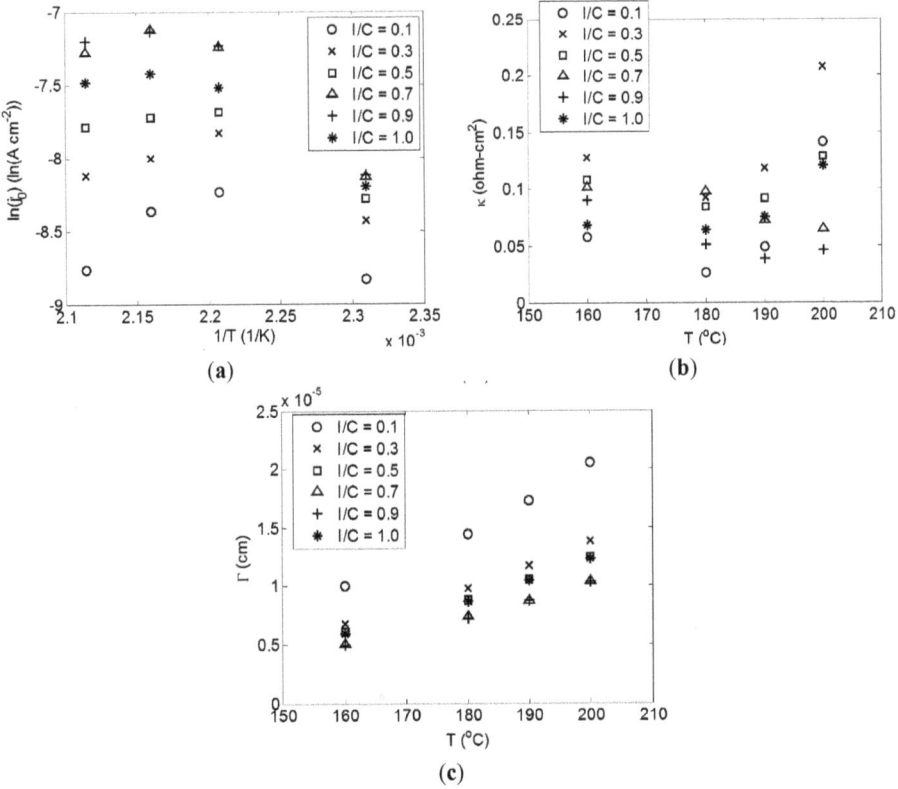

(a)

(b)

(c)

Figure 5. Correlation parameters as a function of temperature for I/C ratios. Inset (a) represents the exchange current density parameter, $j'_{0,c}$, (b) represents the area specific resistance parameter, κ, and (c) represents the geometric parameter, Γ.

Regarding κ, there are unique behaviors with temperature depending on the I/C ratio. For I/C = 0.1, 0.3, and 0.5, there is an observed initial decrease in the resistance with increasing temperature followed by an increase at the higher temperatures. For I/C = 0.7 and 0.9, the resistance tends to more or less decrease with

513

increasing temperature. Finally, for I/C = 1.0, the resistance increases as temperature increases. It is important to keep in mind here that there are several competing effects at play. The membrane conductivity, which was unchanged with respect to I/C, should be expected to have increased ionic conductivity as temperature increases. This will be partly responsible for any observed decreases in resistance with increasing temperature. The hypothesized reduction of phosphoric acid migration into the catalyst layer will reduce the effective conductivity of the catalyst layer with increasing temperature. Lastly, the contact resistance may be expected to change depending on the relative thermal expansion rates of the fluorinated ethylene propylene (FEP) gasket and the TPS® membrane as illustrated in Figure 6. The different trends observed for different I/C ratios are then a product of the porous structures unique to each I/C ratio and how those influence phosphoric acid distribution. For the more optimum structures, I/C = 0.7 and 0.9, the negative influences are reduced and increasing phosphoric acid conductivity outweighs reduced migration or potential increased contact resistance. For less optimum structures, the decreased migration of phosphoric acid into catalyst layer overtakes increasing acid conductivity to result in a net increase in cell resistance.

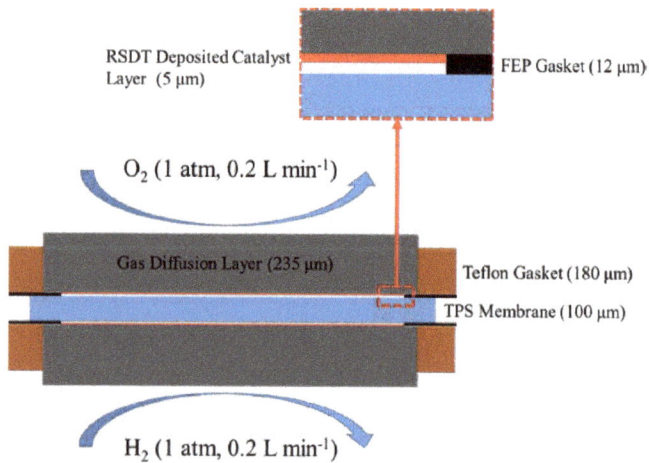

Figure 6. Schematic of the single cell cross section. Measurements in parenthesis indicate thicknesses on the specified component.

3. Methods

3.1. Experimental Section

The experimental details related to the HT-PEMFC assembly and testing are available in detail in [28]. Additional RSDT publications provide further details on the device and effects of process parameters [51,52]. A brief summary of conditions follows. GDEs were fabricated by RSDT. RSDT is an open atmosphere, flame based, deposition process that utilizes the enthalpy of combustion of highly flammable solvents to decompose metal-organic precursors. In this work, platinum(II) acetylacetonate (Pt(acac)$_2$, Colonial Metals, Elkton, MD, USA) was the Pt precursor, which was dissolved directly into the solvent. The solvent was a mixture of xylene, acetone (Sigma Aldrich, St. Louis, MO, USA), and thiol-free propane (Airgas East Inc., Cheshire, CT, USA). The precursor solution was then pumped through an atomizing nozzle and the resulting droplets were continuously ignited with a pilot flame, facilitating the decomposition of the precursor metal in the high temperature reaction zone of the flame (1000–2000 °C). The support material was introduced post-combustion by a set of two secondary nozzles that sprayed a slurry consisting of carbon (Vulcan XC-72R, Cabot Corp., Boston, MA, USA) and various concentrations of polytetrafluoroethylene (PTFE) binder dissolved in dimethylformamide (DMF, Sigma Aldrich, St. Louis, MO, USA). These secondary nozzles were placed on either side of the primary nozzle and angled such that the resulting spray intersected within the post luminous zone of the flame. The resulting spray (consisting of the platinum, carbon, and PTFE) was directed at the GDL substrate (SIGRACET® GDL25BC) to manufacture the GDE in a single step process. The nominal platinum loading for each electrode was kept constant at 0.05 mg· cm^{-2} giving a combined loading of 0.1 mg· cm^{-2}, which achieves the DoE 2017 target of 0.125 mg· cm^{-2}. The I/C ratios were 0.1, 0.3, 0.5, 0.7, 0.9, and 1.0 for Samples 1–6, respectively.

As mentioned above, Advent TPS® membranes were utilized as the polymer electrolyte material for the MEA. The membranes were doped with 85 wt. % phosphoric acid (the remaining 15 wt. % being water). The doping process involved immersing the membrane for 16 h (open to air) in acid heated to 120 °C. The membranes were weighed before and after doping and an average mass increase of 218% was observed (standard dev., $\sigma = 6.24\%$). Assembly of the MEA was done with a Carver hot press where the procedure consisted of: 60 °C, 10 min, 2500 lbs. loading; 75 °C, 10 min, 2500 lbs. loading; 90 °C, 20 min, 2500 lbs. loading; 110 °C, 15 min, 2500 lbs. loading; 150 °C, 10 min, 2500 lbs. loading; 150 °C, 15 min, 5000 lbs. loading; Cool down rapidly at 45 °C.

The assembled MEAs (active area 5 × 5 cm^2) were tested in a single cell configuration. FEP gaskets were used to seal the edges of the cell to prevent leakage of the phosphoric acid (Figure 6). While the use of these gaskets is necessary, it is believed this may create some elevated contact resistance between the GDE and the membrane. The contact resistance was explicitly accounted for in the model.

Prior to performing fuel cell polarization tests, the cell was held at 0.6 V for 3 h for performance break-in. Typically, around 24 h is necessary for a break-in period for PEMFC technology but it was found that stable voltages were achieved after 3 h in this case. This may be due to the lack of external humidification, which is required to hydrate a low temperature PEMFC that requires long periods of time to reach steady state. The cell operating temperature was varied between 160 and 200 °C. The anode was exposed to 1 atm of hydrogen gas at a flow rate of $0.2 \, \text{L} \cdot \text{min}^{-1}$ while the cathode was exposed to 1 atm of oxygen at a flow rate of $0.2 \, \text{L} \cdot \text{min}^{-2}$. Again, no external humidification was provided.

3.2. Performance Model

In order to predict the performance of the single cell, a semi-empirical model was developed. The derivation begins with a modified expression for the overall cell voltage.

$$\Delta E = E_{OCV} - E_{cell} = |\eta_{a,act}| + |\eta_{c,act}| + j\kappa \tag{1}$$

Due to the difficulty of predicting the OCV of the cell and to avoid a miscalculation of the exchange current density, the cell overpotential defined by ΔE in Equation (1) was used to calibrate the model to the experimental data. The OCV can typically be theoretically calculated using the Nernst equation and a crossover current term to account for fuel permeation through the electrolyte, but it was found that this approach did not yield satisfactory comparisons. This may be due to additional factors such as carbon corrosion and surface structure, as well as effects of locally varying acidity making it difficult to predict the true activity of the constituents. The terms on the right hand side of Equation (1) represent the activation overpotential of the anode and cathode, and the ohmic resistance losses of the cell, respectively. For the HT-PEMFC, the activation overpotential of the anode is assumed to be negligible due to the relatively facile reaction kinetics for hydrogen oxidation [53]. The cathode overpotential can be determined using Tafel approximation.

$$\eta_{c,act} = -\frac{RT}{\alpha F} \ln \left(\frac{j}{j_{0,c}} \right) \tag{2}$$

The cathodic exchange current density, $j_{0,c}$, can be expanded where the parameter $j'_{0,c}$ groups the reference exchange current density, Arrhenius effect, and surface roughness together.

$$j_{0,c} = j_{0,c}^{ref} a_c \left(\frac{C_{O_2}}{C_{O_2,sat}} \right)^{\gamma} \exp\left[-\frac{E_{act}}{RT} \left(1 - \frac{T}{T_{ref}} \right) \right] = \left(\frac{C_{O_2}}{C_{O_2,sat}} \right)^{\gamma} j'_{0,c} \tag{3}$$

For the ohmic overpotential, it is assumed that the major contributors to the overall cell resistance are the membrane electrolyte, the ionic resistance of the catalyst layers, and the contact resistance between the GDE and the polymer electrolyte. This implies all electronic resistance contributions from the cell are negligible.

$$\kappa = \frac{l_{\text{m}}}{\sigma_{\text{m}}} + \frac{2l_{\text{cl}}}{\sigma_{\text{cl}}} + R_C \qquad (4)$$

In this study, the overall area specific resistance, κ, is treated as an adjustable parameter. Combining the above expressions gives the relatively simple result:

$$\Delta E = \left| \frac{-RT}{\alpha_c F} \ln \left[\frac{j}{(C_{O_2}/C_{O_2,\text{sat}})^\gamma j'_{0,c}} \right] \right| + j\kappa \qquad (5)$$

At this point, it is necessary to introduce the physical description of the catalyst layer in order to calculate the oxygen concentrations in Equation (5). For this work, the coated film model has been adopted as schematically presented in Figure 7.

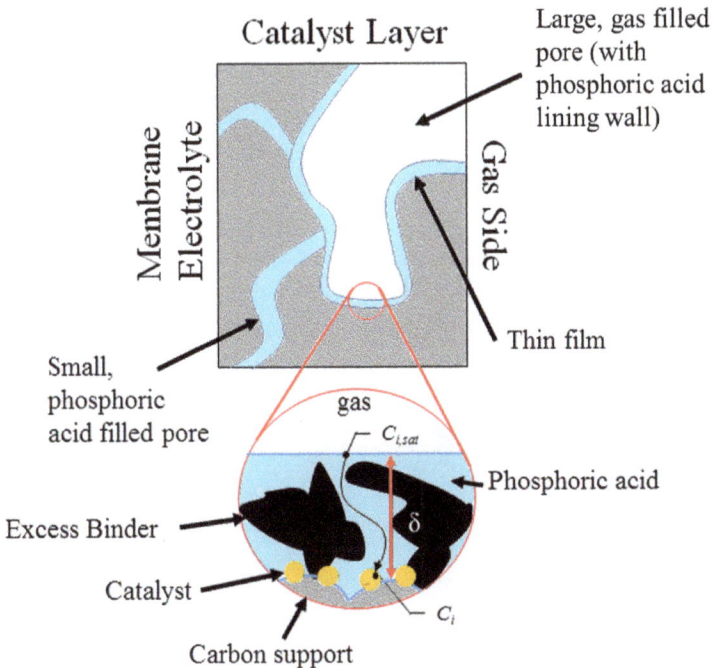

Figure 7. Depiction of the coated film model for the porous electrode (not drawn to scale).

This model describes a porous catalyst layer consisting of a mix of small phosphoric acid filled pores and larger gas filled pores. The smaller pores create a capillary action that draws phosphoric acid to the larger pores while also serving as a pathway for ion conduction to and from the membrane electrolyte. Due to the capillary effect feeding the larger pores, the pores become coated with a film of phosphoric acid. The walls of the pores are assumed to be lined with the catalyst material and the reactant gases must first diffuse through this coated film before reacting. It is assumed in this case that gas phase transport is negligible compared to the transport through the coated film. This assumption was validated previously by Scott *et al.* for HT-PEMFCs [41,42]. Additionally, the current experimental setup considered in this work utilizes pure oxygen in the reactant feed stream meaning there is no interdiffusion with other gases such as nitrogen, and no external humidification is applied, which can lead to complex multiphase transport in the GDL. Taking this into consideration concentrations in Equation (7) are calculated assuming 1-D Fickian diffusion through the coated film where the saturation concentrations are assumed to correlate with the Henry's law saturation behavior.

$$C_{O_2} = C_{O_2,\text{sat}} - \frac{j\delta}{nFa_c m D_{O_2,H_3PO_4}} = C_{O_2,\text{sat}} - \frac{\Gamma j}{nF D_{O_2,H_3PO_4}} \tag{6}$$

The parameter m in Equation (6) corrects for the effective diffusivity of the coated film and accounts for any obstructions through the coated film. Additionally, the experimental current density, j, must be corrected in Equation (6) to account for the active surface area of the catalyst rather than the MEA active area (5×5 cm^2). This is the reason for the appearance of the roughness factor, a_c. For increasing surface roughness the local molar flux of reactants through the coated film would be expected to decrease for the same experimental current density, which is measured relative to the planform area of the cell. Without the addition of the roughness factor, the calculation of the local concentration of reactants near the catalyst surface through the coated film would be lower than expected. This same correction has also been employed by Mamlouk *et al.* [54]. The geometric parameters in Equation (6) are grouped together as the parameter Γ.

Using the coated film model, Equation (5) can be solved for comparison to the experimental data. Equations (5) and (6) contain three adjustable parameters: $j'_{0,c}$, κ, and Γ. The necessary physical parameters used to solve the governing equations are contained in Table 1. The mathematical model was correlated to the experimental data using Matlab's least squares routine (lsqcurvefit).

Table 1. Parameters used in Equations (5) and (6).

Parameter (Units)	Value
α_c [55]	0.94
γ [15]	1
$C_{O_2,\text{sat}}$ (mol·cm^{-3}) [56]	$1.7410 \times 10^{-8}\exp\left[\frac{-9.4866 \times 10^3}{RT}\right]$
D_{O_2,H_2PO_4} (cm^2·s^{-1}) [56]	$2.0402\exp\left[\frac{-3.9729 \times 10^4}{RT}\right]$

4. Conclusions

The focus of this study was the development of a simple, yet accurate, semi-empirical model to calibrate against performance data from a HT-PEMFC with GDEs manufactured by RSDT. The model used the coated film approach to approximate transport in the catalyst layer. The effects of I/C ratio of the GDEs as well as operating temperature were explored. The evidence suggested that increasing the I/C ratio increased performance to a point due to the decreasing average pore size and the more uniform pores. This caused increased phosphoric acid migration into the electrodes, which improved the effective catalyst utilization and improved effective ionic conductivity. If the I/C ratio was too high or extremely low, this caused electrode flooding and decreased performance. Based on the trends of the correlation parameters, the increased temperature led to a decrease in phosphoric acid content of the electrodes, which led to greater ohmic resistance and reduced active catalyst surface area. These effects were partly counteracted by Arrhenius behavior of the exchange current density and enhanced ionic conductivity of phosphoric acid in the electrodes. As expected, the general conclusions reached in the previous experimental study have not changed; however, the modeling approach has provided some new insight into the nature of the performance variations as a function of I/C and temperature. The model provides new theories regarding cell performance behavior, which will be the subject of continuing work. Future work will seek to improve the fidelity of the model and begin to explore the influence of the higher temperature on the reduced acid migration. Further theory will be developed in an attempt to predict the adjustable parameters utilized in this study. Additionally, studies of the durability of the manufactured cells will be investigated.

Acknowledgments: The authors would like to acknowledge the financial support of National Science Foundation, NSF-GOALI (contract No. CCM1-1265893). The authors also express their gratitude to Advent Technologies Inc. for the supply of the membranes used in this study.

Author Contributions: T.D.M and W.E.M. developed the model presented and carried out the correlation to experimental data. S.K. performed the RSDT experiments and single cell tests. T.D.M, S.K., W.E.M., and R.M. participated in the design of the study and assisted in drafting the manuscript. All authors read and approved the manuscript prior to submission.

Conflicts of Interest: The authors declare no conflict of interest.

Nomenclature

a_c	Catalyst surface roughness factor (catalyst surface area/electrode geometric area)
C_i	Molar concentration of species i at the catalyst surface $(\text{mol} \cdot \text{cm}^{-3})$
$C_{i,\text{sat}}$	Saturation concentration of species i $(\text{mol} \cdot \text{cm}^{-3})$
$D_{i,\text{H}_3\text{PO}_4}$	Diffusivity of species i in phosphoric acid $(\text{cm}^2 \cdot \text{s}^{-1})$
E_{act}	Activation energy $(\text{kJ} \cdot \text{mol}^{-1})$
E_{cell}	Cell potential (V)
E_{OCV}	Open circuit potential (V)
F	Faraday's constant $(\text{C} \cdot \text{mol}^{-1})$
j	Current density $(\text{A} \cdot \text{cm}^{-2})$
$j_{0,c}$	Exchange current density $(\text{A} \cdot \text{cm}^{-2})$
$j_{0,c}^{\text{ref}}$	Reference exchange current density $(\text{A} \cdot \text{cm}^{-2})$
$j'_{0,c}$	Parameter defined by Equation (3) $(\text{A} \cdot \text{cm}^{-2})$
l_{cl}	Thickness of catalyst layer (cm)
l_{m}	Thickness of membrane electrolyte (cm)
m	Diffusivity correction factor
n	Number of transfer electrons
R	Ideal gas constant $(\text{kJ} \cdot \text{mol}^{-1} \cdot \text{K}^{-1})$
R_C	Contact resistance $(\Omega \cdot \text{cm}^2)$
T	Operating temperature (K)
T_{ref}	Reference temperature (K)
Greek	
α_c	Transfer coefficient
γ	Pressure coefficient
Γ	Parameter defined by Equation (6) (cm)
δ	Film thickness (cm)
ΔE	Cell potential drop (V)
ΔE_{Nernst}	Change in Nernst potential (V)
$\eta_{a,\text{act}}$	Anode activation overpotential (V)
$\eta_{c,\text{act}}$	Cathode activation overpotential (V)
κ	Parameter defined by Equation (4) $(\Omega \cdot \text{cm}^2)$
σ_{cl}	Conductivity of the catalyst layer $(\text{S} \cdot \text{cm}^{-1})$
σ_{m}	Conductivity of the membrane electrolyte $(\text{S} \cdot \text{cm}^{-1})$

References

1. Asensio, J.A.; Sanchez, E.M.; Gomez-Romero, P. Proton-conducting Membranes Based on Benzimidazole Polymers for High-temperature PEM Fuel Cells. A Chemical Quest. *Chem. Soc. Rev.* **2010**, *39*, 3210–3239.

2. Chandan, A.; Hattenberger, M.; El-kharouf, A.; Du, S.; Dhir, A.; Self, V.; Pollet, B.G.; Ingram, A.; Bujalski, W. High Temperature (HT) Polymer Electrolyte Membrane Fuel Cells (PEMFC)—A Review. *J. Power Sources* **2013**, *231*, 264–278.

3. Costamagna, P.; Yang, C.; Bocarsly, A.B.; Srinivasan, S. Nafion® 115/Zirconium Phosphate Composite Membranes for Operation of PEMFCs above 100 °C. *Electrochim. Acta* **2002**, *47*, 1023–1033.

4. Reichman, S.; Ulus, A.; Peled, E. PTFE-Based Solid Polymer Electrolyte Membrane for High-Temperature Fuel Cell Applications. *J. Electrochem. Soc.* **2007**, *154*, B327–B333.

5. Rikukawa, M.; Sanui, K. Proton-Conducting Polymer Electrolyte Membranes Based on Hydrocarbon Polymers. *Prog. Polym. Sci.* **2000**, *25*, 1463–1502.

6. Zhang, J.; Xie, Z.; Zhang, J.; Tang, Y.; Song, C.; Navessin, T.; Shi, Z.; Song, D.; Wang, H.; Wilkinson, D.P.; *et al.* High Temperature PEM Fuel Cells. *J. Power Sources* **2006**, *160*, 872–891.

7. Ahluwalia, R.K.; Doss, E.D.; Kumar, R. Performance of High Temperature Polymer Electrolyte Fuel Cell Systems. *J. Power Sources* **2003**, *117*, 45–60.

8. Parthasarathy, A.; Srinivasan, A.; Appleby, A.J.; Martin, C.R. Temperature Dependence of the Electrode Kinetics of Oxygen Reduction at the Platinum/Nafion® Interface—A Microelectrode Investigation. *J. Electrochem. Soc.* **1992**, *139*, 2530–2537.

9. Pan, C.; He, R.; Li, Q.; Jensen, J.O.; Bjerrum, N.J.; Hjulmand, H.A.; Jensen, A.B. Integration of High Temperature PEM Fuel Cells with a Methanol Reformer. *J. Power Sources* **2005**, *145*, 392–398.

10. Li, Q.; He, R.; Gao, J.-A.; Jensen, J.O.; Bjerrum, N.J. The CO Poisoning Effect in PEMFCs Operational at Temperatures up to 200 °C. *J. Electrochem. Soc.* **2003**, *150*, A1599–A1605.

11. Li, Q.; He, R.; Jensen, J.O.; Bjerrum, N.J. PBI-Based Polymer Membranes for High Temperature Fuel Cells—Preparation, Characterization and Fuel Cell Demonstration. *Fuel Cells* **2004**, *4*, 147–159.

12. Li, Q.; Hjuler, H.A.; Bjerrum, N.J. Phosphoric Acid Doped Polybenzimidazole Membranes: Physiochemical Characterization and Fuel Cell Applications. *J. Appl. Electrochem.* **2001**, *31*, 773–779.

13. Hasiotis, C.; Li, Q.; Deimede, V.; Kallistis, J.K.; Kontoyannis, C.G.; Bjerrum, N.J. Development and Characterization of Acid-Doped Polybenzimidazole/Sulfonated Polysulfone Blend Polymer Electrolytes for Fuel Cells. *J. Electrochem. Soc.* **2001**, *148*, A513–A519.

14. Kim, H.-J.; An, S.J.; Kim, J.-Y.; Moon, J.K.; Cho, S.Y.; Eun, Y.C.; Yoon, H.-K.; Park, Y.; Kweon, H.-J.; Shin, E.-M. Polybenzimidazoles for High Temperature Fuel Cell Applications. *Macromol. Rapid Commun.* **2004**, *25*, 1410–1413.

15. Liu, Z.Y.; Wainright, J.S.; Litt, M.H.; Savinell, R.F. Study of the Oxygen Reduction Reaction (ORR) at Pt Interfaced with Phosphoric Acid Doped Polybenzimidazole at Elevated Temperature and Low Relative Humidity. *Electrochim. Acta* **2006**, *51*, 3914–3923.

16. Lobato, J.; Cañizares, P.; Rodrigo, M.A.; Linares, J.J. Study of Different Bimetallic Anodic Catalysts Supported on Carbon for a High Temperature Polybenzimidazole-Based Direct Ethanol Fuel Cell. *Appl. Catal. B* **2009**, *91*, 269–274.

521

17. Savadogo, O.; Varela, F.J.R. Low-Temperature Direct Propane Polymer Electrolyte Membranes Fuel Cell (DPFC). *J. New Mater. Electrochem. Syst.* **2001**, *4*, 93–97.
18. Wainright, J.S.; Wang, J.-T.; Weng, D.; Savinell, R.F.; Litt, M. Acid Doped Polybenzimidazoles, A New Polymer Electrolyte. *J. Electrochem. Soc.* **1995**, *142*, L121–L123.
19. Wang, J.T.; Lin, W.F.; Weber, M.; Wasmus, S.; Savinell, R.F. Trimethoxymethane as an Alternative Fuel for a Direct Oxidation PBI Polymer Electrolyte Fuel Cell. *Electrochim. Acta* **1998**, *43*, 3821–3828.
20. Wang, J.T.; Savinell, R.F.; Wainright, J.S.; Litt, M.; Yu, H. A H_2/O_2 Fuel Cell using Acid Doped Polybenzimidazoles as a Polymer Electrolyte. *Electrochim. Acta* **1996**, *41*, 193–197.
21. Hu, J.; Zhang, H.; Zhai, Y.; Liu, G.; Hu, J.; Yi, B. Performance Degradation studies on PBI/H_3PO_4 High Temperature PEMFC and One-dimensional numerical analysis. *Electrochim. Acta* **2006**, *52*, 394–401.
22. Liao, J.H.; Li, Q.F.; Rudbeck, H.C.; Jensen, J.O.; Chromik, A.; Bjerrum, N.J.; Kerres, J.; Xing, W. Oxidative Degradation of Polybenzimidazole Membranes as Electrolytes for High Temperature Proton Exchange Membrane Fuel Cells. *Fuel Cells* **2011**, *11*, 745–755.
23. Liu, G.; Zhang, H.; Hu, J.; Zhai, Y.; Xu, D.; Shao, Z.G. Studies of Performance Degradation of a High Temperature PEMFC Based on H_3PO_4-doped PBI. *J. Power Sources* **2006**, *162*, 547–552.
24. Modestov, A.D.; Tarasevich, M.R.; Filimonov, V.Y.; Zagudaeva, N.M. Degradation of High Temperature MEA with $PBI-H_3PO_4$ Membrane in a Life Test. *Electrochim. Acta* **2009**, *54*, 7121–7127.
25. Ubeda, D.; Canizares, P.; Rodrigo, M.A.; Pinar, F.J.; Lobato, J. Durability Study of HTPEMFC Through Current Distribution Measurements and the Application of a Model. *Int. J. Hydrogen Energy* **2014**, *39*, 21678–21687.
26. Mench, M.M. Other Fuel Cells. In *Fuel Cell Engines*; John Wiley & Sons Inc.: Hoboken, NJ, USA, 2008; p. 410.
27. Yu, H.; Roller, J.; Kim, S.; Wang, Y.; Kwak, D.; Maric, R. One-Step Deposition of Catalyst Layers for High Temperature Proton Exchange Membrane Fuel Cells (PEMFC). *J. Electrochem. Soc.* **2014**, *161*, F622–F627.
28. Kim, S.; Myles, T.D.; Kunz, H.R.; Kwak, D.; Wang, Y.; Maric, R. The Effect of Binder Content on the Performance of a High Temperature Polymer Electrolyte Membrane Fuel Cell Produced with Reactive Spray Deposition Technology. *Electrochim. Acta* **2015**, *177*, 190–200.
29. Dale, M.K.; Geomezi, M.; Vogli, E.; Voyiatzis, G.A.; Neophytides, S.G. The Interaction of H_3PO_4 and Steam with PBI and TPS Polymeric Membranes. A TGA and Raman Study. *J. Mater. Chem. A* **2014**, *2*, 1117–1127.
30. Berning, T.; Lu, D.M.; Djilali, N. Three-Dimensional Computational Analysis of Transport Phenomena in a PEM Fuel Cell. *J. Power Sources* **2002**, *106*, 284–294.
31. Harvey, D.; Pharoah, J.G.; Karen, K. A Comparison of Different Approaches to Modelling the PEMFC Catalyst Layer. *J. Power Sources* **2008**, *179*, 209–219.

32. Sousa, T.; Mamlouk, M.; Scott, K. An Isothermal Model of a Laboratory Intermediate Temperature Fuel Cell Using PBI Doped Phosphoric Acid Membranes. *Chem. Eng. Sci.* **2010**, *65*, 2513–2530.

33. Bernardi, D.M.; Verbrugge, M.W. Mathematical Model of a Gas Diffusion Electrode Bonded to a Polymer Electrolyte. *AIChE J.* **1991**, *37*, 1151–1163.

34. Marr, C.; Li, X. Composition and Performance Modelling of Catalyst Layer in a Proton Exchange Membrane Fuel Cell. *J. Power Sources* **1999**, *77*, 17–27.

35. Song, D.; Wang, Q.; Liu, Z.; Navessin, T.; Eikerling, M.; Holdcraft, S. Numerical Optimization Study of the Catalyst Layer of PEM Fuel Cell Cathode. *J. Power Sources* **2004**, *126*, 104–111.

36. Eikerling, M.; Kornyshev, A.A. Modelling the Performance of the Cathode Catalyst Layer of Polymer Electrolyte Fuel Cells. *J. Electroanal. Chem.* **1998**, *453*, 89–106.

37. Bevers, D.; Wohr, M.; Yasuda, K.; Oguro, K. Simulation of Polymer Electrolyte Fuel Cell Electrode. *J. Appl. Electrochem.* **1997**, *27*, 1254–1264.

38. Kulikovsky, A.A.; Divisek, J.; Kornyshev, A.A. Modeling Cathode Compartment of Polymer Electrolyte Fuel Cells: Dead and Active Reaction Zones. *J. Electrochem. Soc.* **1999**, *146*, 3981–3991.

39. You, L.; Liu, H. A Parametric Study of the Cathode Catalyst Layer of PEM Fuel Cells using a Pseudo-homogeneous Model. *Int. J. Hydrogen Energy* **2001**, *26*, 991–999.

40. Li, Q.; Xiao, G.; Hjuler, H.A.; Berg, R.W.; Bjerrum, N.J. Limiting Current of Oxygen Reduction on Gas-Diffusion Electrodes for Phosphoric Acid Fuel Cells. *J. Electrochem. Soc.* **1994**, *141*, 3114–3119.

41. Scott, K.; Mamlouk, M. A Cell Voltage Equation for an Intermediate Temperature Proton Exchange Membrane Fuel Cell. *Int. J. Hydrogen Energy* **2009**, *34*, 9195–9202.

42. Scott, K.; Pilditch, S.; Mamlouk, M. Modelling and Experimental Validation of a High Temperature Polymer Electrolyte Fuel Cell. *J. Appl. Electrochem.* **2007**, *37*, 1245–1259.

43. Broka, K.; Ekdunge, P. Modelling the PEM Fuel Cell Cathode. *J. Appl. Electrochem.* **1997**, *27*, 281–289.

44. Cetinbas, F.C.; Advani, S.G.; Prasad, A.K. A Modified Agglomerate Model with Discrete Catalyst Particles for the PEM Fuel Cell Catalyst Layer. *J. Electrochem. Soc.* **2013**, *160*, F750–F756.

45. Sun, W.; Peppley, B.A.; Karan, K. An Improved Two-Dimensional Agglomerate Cathode Model to Study the Influence of Catalyst Layer Structural Parameters. *Electrochim. Acta* **2005**, *50*, 3359–3374.

46. Moein-Jahromi, M.; Kermani, M.J. Performance Prediction of PEM Fuel Cell Cathode Catalyst Layer Using Agglomerate Model. *Int. J. Hydrogen Energy* **2012**, *37*, 17954–17966.

47. Roa, R.M.; Rengaswamy, R. Optimization Study of an Agglomerate Model for Platinum Reduction and Performance in PEM Fuel Cell Cathode. *Chem. Eng. Res. Des.* **2006**, *84*, 952–964.

48. Wang, Q.; Eikerling, M.; Song, D.; Liu, Z. Structure and Performance of Different Types of Agglomerates in Cathode Catalyst Layers of PEM Fuel Cells. *J. Electroanal. Chem.* **2004**, *573*, 61–69.

49. Lobato, J.; Cañizares, P.; Rodrigo, M.A.; Linares, J.J.; Pinar, F.J. Study of the Influence of the Amount of PBI-H$_3$PO$_4$ in the Catalytic Layer of a High Temperature PEMFC. *Int. J. Hydrogen Energy* **2010**, *35*, 1347–1355.
50. Lobato, J.; Cañizares, P.; Rodrigo, M.A.; Úbeda, D.; Pinar, F.J.; Linares, J.J. Optimisation of the Microporous Layer for a Polybenzimidazole-Based High Temperature PEMFC—Effect of Carbon Content. *Fuel Cells* **2010**, *10*, 770–777.
51. Roller, J.M.; Arellano-Jiménez, M.J.; Jain, R.; Yu, H.; Carter, C.B.; Maric, R. Oxygen Evolution duringWater Electrolysis from Thin Films Using Bimetallic Oxides of Ir–Pt and Ir–Ru. *J. Electrochem. Soc.* **2013**, *160*, F716–F730.
52. Roller, J.M.; Renner, J.; Yu, H.; Capuano, C.; Kwak, T.; Wang, Y.; Carter, C.B.; Ayers, K.; Mustain, W.E.; Maric, R. Flame-Based Processing as a Practical Approach for Manufacturing Hydrogen Evolution Electrodes. *J. Power Sources* **2014**, *271*, 366–376.
53. Korsgaard, A.R.; Refshauge, R.; Nielsen, M.P.; Bang, M.; Kaer, S.K. Experimental Characterization and Modeling of Commercial Polybenzimidazole-Based MEA Performance. *J. Power Sources* **2006**, *162*, 239–245.
54. Mamlouk, M.; Sousa, T.; Scott, K. A High Temperature Polymer Electrolyte Membrane Fuel Cell Model for Reformate Gas. *Int. J. Electrochem.* **2011**, *2011*, 1–18.
55. Kunz, H.R.; Gruver, G.A. The Catalytic Activity of Platinum Supported on Carbon for Electrochemical Oxygen Reduction in Phosphoric Acid. *J. Electrochem. Soc.* **1975**, *122*, 1279–1287.
56. Klinedinst, K.; Bett, J.A.S.; Macdonald, J.; Stonehart, P. Oxygen Solubility and Diffusivity in Hot Concentrated H$_3$PO$_4$. *J. Electroanal. Chem. Interfacial Electrochem.* **1974**, *57*, 281–289.

The Use of C-MnO$_2$ as Hybrid Precursor Support for a Pt/C-Mn$_x$O$_{1+x}$ Catalyst with Enhanced Activity for the Methanol Oxidation Reaction (MOR)

Alessandro H.A. Monteverde Videla, Luigi Osmieri,
Reza Alipour Moghadam Esfahani, Juqin Zeng, Carlotta Francia and
Stefania Specchia

Abstract: Platinum (Pt) nanoparticles are deposited on a hybrid support (C-MnO$_2$) according to a polyol method. The home-made catalyst, resulted as Pt/C-Mn$_x$O$_{1+x}$, is compared with two different commercial platinum based materials (Pt/C and PtRu/C). The synthesized catalyst is characterized by means of FESEM, XRD, ICP-MS, XPS and μRS analyses. MnO$_2$ is synthesized and deposited over a commercial grade of carbon (Vulcan XC72) by facile reduction of potassium permanganate in acidic solution. Pt nanoparticles are synthesized on the hybrid support by a polyol thermal assisted method (microwave irradiation), followed by an annealing at 600 °C. The obtained catalyst displays a support constituted by a mixture of manganese oxides (Mn$_2$O$_3$ and Mn$_3$O$_4$) with a Pt loading of 19 wt. %. The electro-catalytic activity towards MOR is assessed by RDE in acid conditions (0.5 M H$_2$SO$_4$), evaluating the ability to oxidize methanol in 1 M concentration. The synthesized Pt/C-Mn$_x$O$_{1+x}$ catalyst shows good activity as well as good stability compared to the commercial Pt/C based catalyst.

Reprinted from *Catalysts*. Cite as: Videla, A.H.A.M.; Osmieri, L.; Esfahani, R.A.M.; Zeng, J.; Francia, C.; Specchia, S. The Use of C-MnO$_2$ as Hybrid Precursor Support for a Pt/C-Mn$_x$O$_{1+x}$ Catalyst with Enhanced Activity for the Methanol Oxidation Reaction (MOR). *Catalysts* **2015**, *5*, 1399–1416.

1. Introduction

Fuel cells are electrochemical devices that produce electricity from the energy of a fuel through a highly efficient conversion process, resulting in low emissions and low environmental impact [1]. Between the different types of fuel cell, Direct Alcohol Fuel Cells (DAFC) and more specifically, Direct Methanol Fuel Cells (DMFC), represent a valid alternative for small portable electronic devices and auxiliary power units, due to the high energy density of alcohols, their lightweight and compact nature and their ability for fast recharging [2,3].

Platinum is the most widely used catalyst for both the anodic methanol oxidation reaction (MOR) and the cathodic oxygen reduction reaction (ORR) [4,5]. Pt is considered the most suitable electro-catalyst for MOR due to its high activity and stability, especially in acidic media [6]. However, one of the main barriers to the commercialization of DMFC technologies is still the high cost of Pt. To reduce the cost, an improvement of the performance of conventional Pt-based catalysts is necessary. This would lead to a reduction of the total Pt loading on the electrode. For this purpose, reaction rates need to be enhanced (*i.e.*, the overvoltage needs to be decreased) by modifying the catalyst composition or structure, to produce a more active electro-catalytic material [7].

A common approach to enhance the activity of Pt involves well dispersed nanoparticle structures, avoiding agglomeration and increasing utilization [7]. Further optimization of Pt-based electro-catalysts has been achieved through the formation of bi-metallic alloys such as PtCo and PtNi (for cathodic ORR) and PtRu (for anodic MOR) [8,9]. It has also been demonstrated that efficiency can be further improved by promotion of methanol electro-oxidation by means of various metal oxides-, carbides- and nitrides-promoted electro-catalysts. Non-noble metal oxides $(M-O_x)$ such as WO_3, CeO_2, V_2O_5, Nb_2O_5, MoO_x, ZrO_2, TiO_2, MgO and MnO_2, exhibit suitable surface properties which can efficiently promote the methanol and ethanol electro-oxidation reactions combined with Pt/C [10]. Therefore, a good strategy to improve the catalytic activity of Pt-based catalysts for MOR is to use metal oxides in the catalyst supports, as a hybrid structure $(C + M-O_x)$ [11]. Pure Pt, in fact, is readily poisoned by strongly-adsorbed intermediates, of which CO is consistently considered as one of the main poisoning species at low operating temperature [12].

The use of metal oxide-containing Pt/C electro-catalysts has been reported to effectively enhance the electro-oxidation of methanol by the spillover of CO on Pt sites to the adjacent metal oxides. These oxides are supposedly capable of adsorbing large quantities of –OH species, which are then donated to the neighboring Pt sites where stepwise methanol dehydrogenation occurs. Metal oxides also provide suitable functional groups which strongly interact with small Pt crystallites, impeding their random growth and agglomeration during device operation for longer duration. In particular, high surface area metal oxides used as supports or matrices, are capable of physically separating metal particles (to diminish their tendency to undergo degradation by agglomeration) and of interacting mutually with them, thus affecting their chemisorptive and catalytic properties. Oxides are often thought as insulating or semi-conducting materials but certain non-stoichiometric oxides existing in various valance states exhibit conductivity not much lower than that of metals and possess appreciable catalytic activity [7]. In electro-catalysis, the reactions occur at the interface, so surface reactivity is very important. Redox reactions of metal oxides involve both ion and electron transfer processes. The electron transfer reactions are influenced by the distribution of electronic states in the electrolyte and within the oxide. When oxides are in contact with aqueous solutions, their surfaces are covered with –OH groups; their actual population depends on the nature of the oxide and its specific crystal face. Some metal oxides are more hydrous than the others. The hydrous behavior, which varies from oxide to oxide, favors proton mobility and affects overall reactivity [7].

MnO_2 has been used in a wide range of applications such as catalyst, molecular-sieves, ion-sieves, batteries and magnetic materials due to its excellent physicochemical properties [13]. Manganese oxides were widely used as catalyst support for fuel cells due to their promoting effects in the oxidation of small organic molecules, such as the excellent proton conductivity, the increase of catalyst utilization and the synergistic effect between catalysts and manganese oxides [6]. In particular, MnO_2 possesses good proton-electron intercalation properties and is known to show good electro-chemical properties under various operating conditions [10]. Mn possesses a wide range of oxidation states and such oxo-manganese species are generally strong chemical oxidants. Due to the possibility of the Mn^{4+}/Mn^{3+} redox couple and the presence of labile oxygen, MnO_2 shows high promoting and anti-poisoning activities for alcohol electro-oxidation [7]. The main reasons for effectiveness of MnO_2 as a promoter are attributed to its surface area, tunnel structure and crystal phase. It is known that oxides with one-dimensional structures such as nanorods, nanowires and nanotubes possess distinctive crystalline phase states, as compared to their bulk counterparts [10]. It is also known that the interaction between metal crystallites and an oxide surface is influenced by the nature of interfacial contact and the crystalline characteristic of the oxide. MnO_2 with smaller and uniform crystalline orientation as well as suitable surface morphology should offer apposite active sites for facile interaction with Pt crystallites, which can provide optimized synergistic effect for alcohol electro-oxidation [14]. However, the effect of microstructure/morphology of manganese oxides on the nature of Pt dispersion on Mn_xO_{1+x}/carbon-based electrocatalysts has not been extensively investigated so far [10].

In this work, α-MnO_2 was synthesized and deposited on a commercial carbon black (Vulcan XC-72). Then, Pt nanoparticles were deposited on the formed hybrid support (C-MnO_2), called Pt/C-Mn_xO_{1+x}, by a microwave-assisted polyol method followed by a thermal treatment in inert atmosphere at 600 °C. The synthesized catalyst was compared with two commercial Pt-based catalysts characterized and tested for MOR in acidic medium.

2. Results and Discussion

2.1. Physical-Chemical Characterization

The XRD pattern of the prepared C-MnO_2 is given in Figure 1A. The broad peak at about 23.5° is attributed to the graphitic carbon support. All other peaks are clearly indexed to the pure tetragonal phase of α-MnO$_2$ (JCPDS card #44-0141), with lattice constants of a = 9.73 Å and c = 2.84 Å. No peaks were observed for other types of crystals or amorphous MnO$_2$ which confirmed the purity of the prepared sample. The intensive diffraction peaks appeared at 12.46°, 18.08°, 28.83°, 37.00°, 37.66°, 41.95°, 50.13°, 60.36°, 66.30°, 72.87°, respectively, which are characteristic peaks of α-MnO$_2$ with the major peaks intensity at 18.08° [15,16]. α-MnO$_2$ is constructed from the double chains of edge-sharing MnO$_6$ octahedra, which are linked at the corners to form tunnel structures [17].

Figure 1. C-MnO$_2$ support: (**A**) XRD patterns; (**B,C**) FESEM images at different magnification.

The morphology of prepared C-MnO$_2$ was investigated by FESEM and the corresponding micrographs are shown in Figure 1B,C. MnO$_2$ nanocrystals grow up on the carbon black, forming sphere-like micro-particles with an average diameter of about 1 μm (Figure 1B). At higher magnification (Figure 1C), the outside part of the particles appeared to be urchins, homogeneously composed of densely aligned nanorods with uniform diameter of about 35 nm. The percentage of MnO$_2$ in the hybrid support evaluated by ICP-MS technique is 64.8 wt. %. This value is very close to the theoretical wt. % percentage of MnO$_2$ expected for the adopted synthesis, which is equal to 68.4 wt. %. Thus, the synthesis method adopted allows a very precise control of the mass loading level of nanostructured MnO$_2$ onto the carbon material by controlling the ratio between KMnO$_4$ and Vulcan XC-72 carbon.

XRD patterns were acquired for all catalysts supported on carbon (Figure 2). X-ray spectra exhibited the characteristic peaks corresponding to the Pt face-centered cubic (*fcc*) polycrystalline structure (111, 200, 220 and 311 reflection planes), consistent with the XRD pattern of JCPDS card #00-4-0802 (2θ = 39.76°, 46.24°, 67.45°, 81.28°). A signal near to 25° (2θ) was obtained, which corresponds to the (002) graphite basal planes for Pt/C and PtRu/C [18]. This peak is not appreciable on Pt/C-Mn$_x$O$_{1+x}$ because of the high manganese oxides content. No metallic Ru diffraction peaks were detected in the commercial PtRu/C which is an indication of alloyed PtRu as reported in the literature [19]. As can be seen in Figure 2, Pt diffraction peaks of the commercial PtRu/C shifted to a positive 2θ value compared with that of Pt/C

528

which reveals alloy formation. Formation of solid solution between Pt and Ru by replacing of Pt with smaller Ru atoms in the lattice points of Pt *fcc* structure results in the reduction of lattice parameter and positive shift of *fcc* diffraction signals [20,21]. It is well known that the alloying of Pt with Ru leads to a decrease in the interatomic bond length because of the smaller Ru atomic radius [22].

Particle diameters were calculated by the Debye-Scherrer equation for all catalyst used. 1.7, 4.5 and 4 nm were obtained for $Pt/C-Mn_xO_{1+x}$, Pt/C and PtRu/C, respectively, with lattice parameter of 3.78, 3.93 and 3.82 Å, respectively. According to ICP-MS analysis, the Pt wt. % loading for the three $Pt/C-Mn_xO_{1+x}$, Pt/C and PtRu/C was equal to 19, 20 and 38 wt. %, respectively.

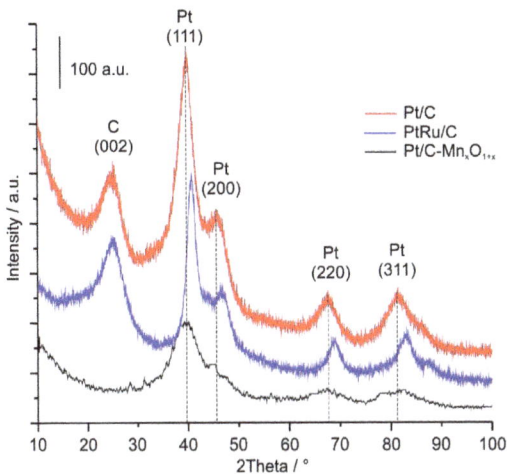

Figure 2. XRD patterns of Pt/C, PtRu/C and Pt/C-Mn$_x$O$_{1+x}$, with the characteristic Miller indexes of Pt *fcc* (JCPSD card #00-4-0802).

The Mn 2p XPS spectrum in Figure 3A exhibited Mn $2p_{1/2}$ at 653.52 eV and Mn $2p_{3/2}$ at 641.71 eV, in which the spin-energy separation of 11.79 eV which indicates that the element manganese in the sample exists in the chemical state of Mn^{2+} and Mn^{4+} and therefore the formation of MnO_2, Mn_2O_3 and Mn_3O_4 [23–25]. As shown in Figure 3B, the Pt 4f spectrum of the $Pt/C-Mn_xO_{1+x}$ was deconvoluted into two doublet peaks, corresponding to a spin-orbit splitting of $4f_{7/2}$ and $4f_{5/2}$ states of ca. 3.33 eV. The most intense doublet at 71.45 ($4f_{7/2}$) and 74.78 eV ($4f_{5/2}$) was due to the metallic Pt, corresponding to metallic platinum particles (Pt^0) [4,6,10,13,17].

To assess the chemical structure of the $Pt/C-Mn_xO_{1+x}$ catalyst, an extra sample of MnO_2 annealed at 600 °C (same temperature used to anchor Pt nanoparticles on the $C-MnO_2$ support) was examined by μRS. Raman spectra in different points of the annealed oxide (Figure 4) show the presence of two types of oxides structure, the Mn_3O_4 hausmannite spinel-like and the Mn_2O_3 bixbyite. Specifically, five characteristic Raman peaks for the spinel structure [$v1 = 310$ cm^{-1}, $v2 = 357$ cm^{-1}, $v3 = 485$ cm^{-1}, $v4 = 579$ cm^{-1}, $v5 = 653$ cm^{-1}] and for the bixbyite [$v1 = 263$ cm^{-1}, $v2 = 308$ cm^{-1}, $v3 = 512$ cm^{-1}, $v4 = 631$ cm^{-1}, $v5 = 670$ cm^{-1}] were detected, suggesting the presence of a mix of oxides [26]. In fact, according to the literature [27–29], in inert atmosphere at around 500 °C MnO_2 is reduced to Mn_2O_3 and further reduced to Mn_3O_4 at 900 °C. These results are in line with the presence of the mixture of hausmannite and bixbyite manganese oxides on the final Pt/C-Mn_xO_{1+x} catalyst, annealed at 600 °C in nitrogen atmosphere, as pointed out by XPS analysis.

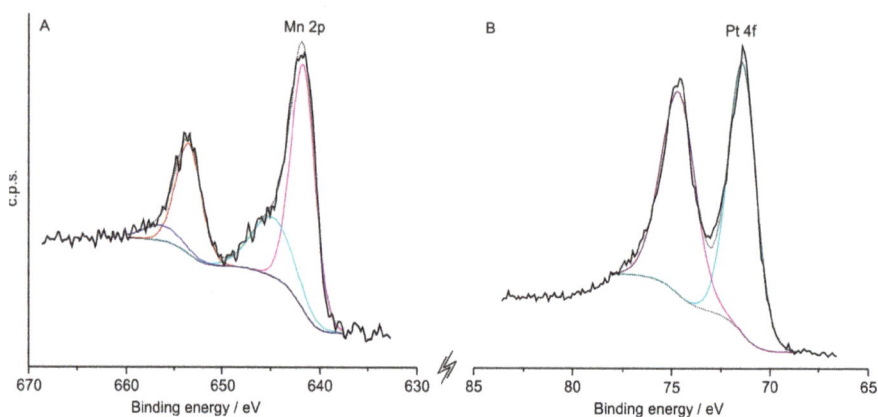

Figure 3. XPS deconvolution of Pt/C-Mn_xO_{1+x}: (**A**) high resolution Mn 2p spectrum; (**B**) high resolution Pt 4f spectrum.

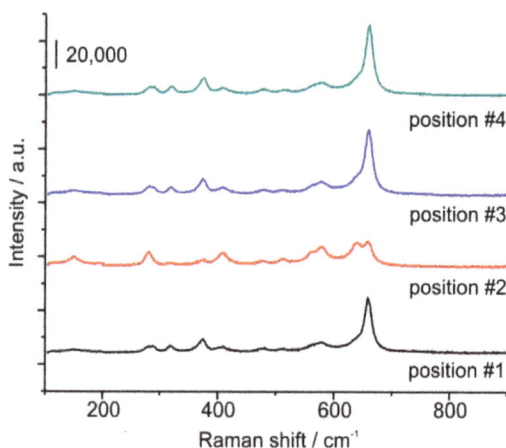

Figure 4. Raman spectra ($\lambda = 785$ nm) of MnO_2 annealed at 600 °C in four different areas of the examined sample (denoted as position #1–#4).

2.2. Electro-Chemical Characterization

The CV profiles recorded in N_2-saturated 0.5 M H_2SO_4 (Figure 5) exhibit defined regions of hydrogen underpotential, adsorption/desorption and platinum oxide formation/reduction for all the samples. In particular, the hydrogen adsorption/desorption peaks shapes are similar for the Pt/C-Mn_xO_{1+x} and the commercial Pt/C, with the latter exhibiting a higher hydrogen desorption area. The electrochemical active surface area (ECSA) resulted 44.1 $m^2 \cdot g^{-1}$ for Pt/C-Mn_xO_{1+x} and 45.4 $m^2 \cdot g^{-1}$ for Pt/C, respectively. For the PtRu/C catalyst the peak shape is different from the previous ones, with a sharper hydrogen desorption peak typical for the PtRu based catalysts [30,31] and ECSA of 69.8 $m^2 \cdot g^{-1}$. Regarding the Pt oxides reduction peak, for the Pt/C-Mn_xO_{1+x} it is shifted towards more positive potentials of about 100 mV in comparison with the commercial Pt/C catalyst. This could indicate a lower oxygen reduction overpotential if this catalyst should be used as a cathode catalyst for a PEMFC [32]. Otherwise, for the PtRu/C catalyst, the same peak is about 200 mV shifted to more negative potentials. This is also typical for PtRu based catalysts [21]. The similar ECSA values for the Pt/C-Mn_xO_{1+x} and Pt/C catalysts could be due to the similar Pt content of these catalysts, equal to 19 wt. % and 20 wt. %, respectively, according to ICP-MS analysis, whereas the measured Pt content of PtRu/C was almost double compared to the other two catalysts. Moreover, the presence of a high amount of manganese oxides on the surface of Pt/C-Mn_xO_{1+x} (see previous discussion on XPS and µRS) could result in a low electrical conductivity of the electrode [13,33,34].

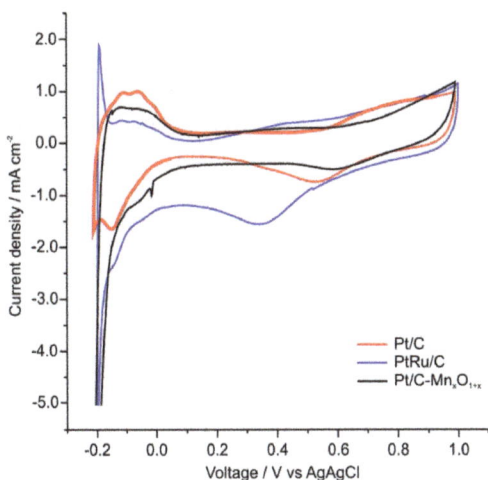

Figure 5. CV of Pt/C, PtRu/C and Pt/C-Mn_xO_{1+x} recorded at 10 $mV \cdot s^{-1}$ (N_2-saturated in 0.5 M H_2SO_4, 0.1 ITC mass ratio, 20 $mg_{Pt} \cdot cm^{-2}$ catalyst loading).

CO stripping voltammetries are shown in Figure 6. The first cycle of the CO stripping voltammetry profile of each one of the three samples shows the CO electro-oxidation peak. Then, in the second cycle, only typical hydrogen underpotential deposition and Pt oxides formation/reduction phenomena are evident. This essentially shows the complete oxidation of adsorbed CO during the first voltammetry scan leaving the active Pt surface clean [10,13,35,36].

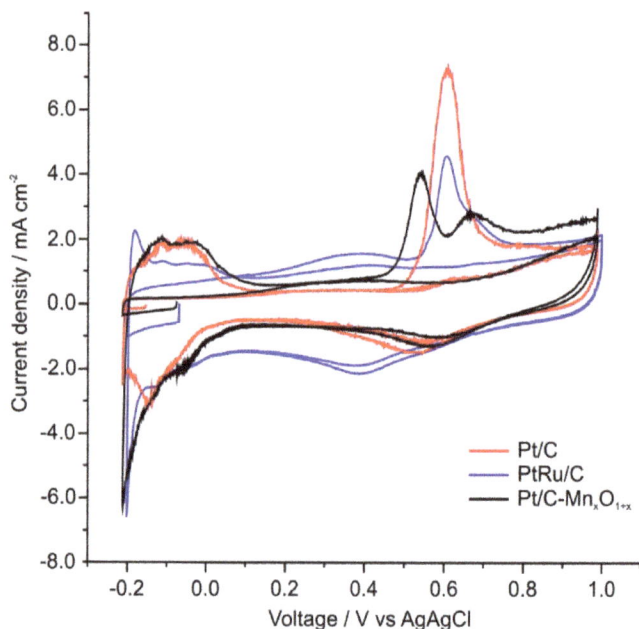

Figure 6. CO stripping voltammetries of Pt/C, PtRu/C and Pt/C-Mn$_x$O$_{1+x}$ recorded at 20 mV·s^{-1} (0.5 M H$_2$SO$_4$, 0.1 ITC mass ratio, 20 mg$_{Pt}$·cm^{-2} catalyst loading).

The onset potential of CO electro-oxidation, taken as the potential at which 5% of the maximum current was reached, was of 0.61, 0.59 and 0.54 V for the commercial Pt/C, the PtRu/C and the Pt/C-Mn$_x$O$_{1+x}$ catalysts, respectively (see Table 1). A negative shift of this onset potential indicates an enhanced catalytic activity for CO oxidation. This has been extensively reported for PtRu catalysts [18,20,21]. In the case of Pt-transition metal oxide based catalysts this effect can also be observed [14]. In particular, for our Pt/C-Mn$_x$O$_{1+x}$ catalyst, the onset potential is 70 mV lower than for the commercial Pt/C catalyst.

Table 1. CO stripping potential, MOR peak potential of the forward scan and I_f/I_b ratio in the CV curves of Pt/C-Mn$_x$O$_{1+x}$, Pt/C and Pt-Ru/C in 0.5 M H$_2$SO$_4$ and 1 M MeOH.

Catalysts	CO Stripping Peak Position V *vs.* Ag/AgCl	MOR forward Peak Position V *vs.* Ag/AgCl	MOR I_f/I_b
Pt/C-Mn$_x$O$_{1+x}$	0.54	0.66	1.54
Pt/C	0.61	0.80	0.92
PtRu/C	0.59	0.66	1.92

Analyzing more in detail the CO stripping peak's shape, it can be observed that the commercial Pt/C peak has a symmetric shape, while the Pt/C-Mn$_x$O$_{1+x}$ catalyst exhibits a second lower and broader peak at higher potentials than the main peak. The presence of the double peak could be due to the presence of both Pt predominant crystal face or Pt agglomerates [37,38]. In fact, the oxidation of a monolayer of CO is strongly influenced by the particle size of Pt. Pt agglomerates show a notable activity towards CO oxidation compared to isolated Pt particles. Moreover, the presence of the double peak for Pt/MnO$_2$/C hybrid catalysts has been noticed by other authors as well [10,39]. In fact, the CO stripping from the Pt surface in the presence of MnO$_2$ occurs via a kind of synergic effect between the OH$_{ads}$ on MnO$_2$ and the CO$_{ads}$ on Pt [10]. The double peak due to CO stripping from the Pt/C-Mn$_x$O$_{1+x}$ could be linked with the possible formation of labile OH species on the triple-phase interface between the Pt, the oxide and the electrolyte, which provides electronic suitability for the oxidation of CO species on the Pt surface [39]. It has been demonstrated that MnO$_2$ nanorods promoted Pt/C catalysts showed larger negative shift in the CO electro-oxidation peak potential due to OH$_{ads}$ species on the MnO$_2$ that tend to electronically weaken the Pt-CO bond and promote the oxidation of CO to CO$_2$ [10]. The presence of MnO$_2$ in a MnO$_2$-Pt/C composite electrode primarily plays a catalytic role in the ORR. It enhances the catalytic behavior of Pt for the ORR by substituting for oxygen as an electron-acceptor in the case of oxygen starvation [40]. Furthermore, based on studies on Pt/Mn$_3$O$_4$-MWCNT [33], the Mn$_3$O$_4$ leads to uniform and small Pt nanoparticle deposition, with enhanced CO-tolerance and excellent stability in methanol oxidation. In fact, Mn$_3$O$_4$ nanoparticles promote the dissociation of coordinated water and further oxidize CO$_{ads}$ to release more Pt active sites [33]. The hydrous Mn$_3$O$_4$ would use its inherent Mn$_3$O$_4$–OH bonds to directly donate the hydroxide species to the Pt sites and oxidize the adsorbed CO species [41]. Mn$_3$O$_4$ is also known for favoring the growth of Pt nanoparticles with high index facets [33,42]. Consequently, in our Pt/C-Mn$_x$O$_{1+x}$ catalyst, where a mix of manganese oxides is present, both mechanisms could be involved. The promotional CO stripping from the Pt surface in the presence of manganese oxides occurs via a type of synergic effect as reported in the literature [10,39]. Overall, the following reaction mechanism can be assumed:

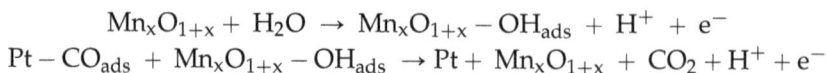

$$Mn_xO_{1+x} + H_2O \rightarrow Mn_xO_{1+x} - OH_{ads} + H^+ + e^-$$
$$Pt - CO_{ads} + Mn_xO_{1+x} - OH_{ads} \rightarrow Pt + Mn_xO_{1+x} + CO_2 + H^+ + e^-$$

The PtRu/C catalyst peak appears to be more asymmetric, with a broadening towards higher overpotentials. Thus, the presence of MnO_2 on the surface of the carbon support causes a co-catalytic action which promotes the CO oxidation catalytic effect of Pt.

The electro-catalytic activity toward MOR of the three catalysts was investigated by cyclic voltammetry using a 0.5 M H_2SO_4 and 1.0 M MeOH solution as electrolyte (Figure 7). For all of the catalysts, the voltammograms exhibit two characteristics oxidation peaks. The first peak is observed during the anodic potential sweep and it is characteristic of the oxidation of methanol adsorbed on the Pt surface. This oxidation occurs in multiple steps, with the production of carboxyl intermediates and strongly adsorbed CO species, as discussed in the literature [43]. Mainly formic acid and formaldehyde have been found during methanol oxidation on Pt surface, as intermediate products [44]. With the potential increase, after the forward peak, the oxidation current decreases, due to the poisoning effect of the CO species strongly adsorbed on Pt and to the Pt oxides formation, which passivates the Pt surface [45]. The second peak appears during the cathodic potential sweep (reverse scan) and it is attributed to the oxidation of adsorbed CO species and/or to the oxidation of further methanol on the Pt oxide surface formed during the anodic scan [10]. During electro-oxidation of MeOH, strongly adsorbed carbonaceous species inhibit further adsorption of MeOH on the catalyst surface, which causes a positive shift in the onset potential and a decrease in the current at a specific potential. Therefore, the more the forward scan peak is shifted to negative potentials, the greater the promotion effect of the electro-catalyst [18].

The forward peak potentials for the Pt/C-Mn_xO_{1+x}, Pt/C and PtRu/C catalysts obtained from the CV are shown in Table 1. The value of the ratio between the forward peak maximum current density (I_f) and the backward peak maximum current density (I_b) is also shown in Table 1. The higher this I_f/I_b ratio is, the greater the electro-catalyst's resistance to poisoning is [46]. Hence, the value of I_f/I_b can be viewed as an index of the tolerance of a catalyst to poisoning species, i.e., adsorbed CO molecules. From these results, it can be concluded that the Pt/C-Mn_xO_{1+x} catalyst is a stronger promoter than the commercial Pt/C catalyst for the MOR and this is in agreement with the CO stripping results. In much of the literature, adsorbed CO is considered as a poisoning intermediate for MOR on pure Pt surface. To remove CO from the Pt surface, adsorbed OH species generated from water activation are indispensable. However, a high potential is needed to activate water on the Pt surface. In PtRu catalysts, water activation can occur at lower potential on Ru-sites. Therefore, MOR activity on PtRu catalysts can be enhanced through the bi-functional mechanism [44].

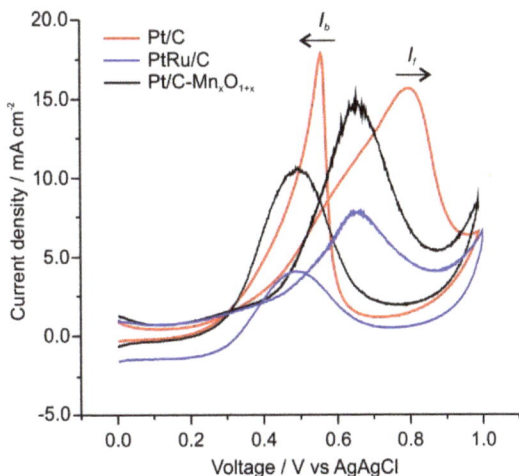

Figure 7. CV of Pt/C, PtRu/C and Pt/C-Mn$_x$O$_{1+x}$ towards MOR 1 M MeOH recorded at 20 mV·s^{-1} (0.5 M H$_2$SO$_4$, 1 M MeOH, 0.1 ITC mass ratio, 20 mg$_{Pt}$·cm^{-2} catalyst loading).

The durability of the electrodes was measured via an accelerated durability test (ADT) at room temperature up to 5000 consecutive cycles for the Pt/C-Mn$_x$O$_{1+x}$ catalyst and up to 4000 cycles for the Pt/C one. A noticeable corrosion of carbon nanostructures as Pt supports, confirmed from the reduction of the catalyst thickness from oxidation of carbon, agglomeration and detaching of Pt, had already been reported by other researcher groups, as well [23]. Figure 8 compares voltammograms before and after stability tests. The results show that Pt/C-Mn$_x$O$_{1+x}$ is more stable compared to Pt/C, showing an increase of ECSA (+29% after 5000 cycles, Figure 8A). In fact, its CV increased progressively up to 1000 cycles, remaining then stable up to the end of the ADT at 5000 cycles. On the contrary, the Pt/C lost continuously stability cycle after cycle, with an overall decrease of ECSA equal to −9% after 4000 cycles (Figure 8B). According to the literature [47], the ECSA increase of Pt/C-Mn$_x$O$_{1+x}$ could be due to a re-arrangement of Pt over carbon. This re-arrangement is not a stable condition, but a reversible process. In fact, as observed for Pt/C-Mn$_x$O$_{1+x}$ during CO stripping analysis (Figure 6), Pt nanoparticles agglomeration can evolve to more disperse Pt nanoparticles or different Pt nano-shape islands depending on the stress cycling adopted for accelerated degradation procedure [47,48].

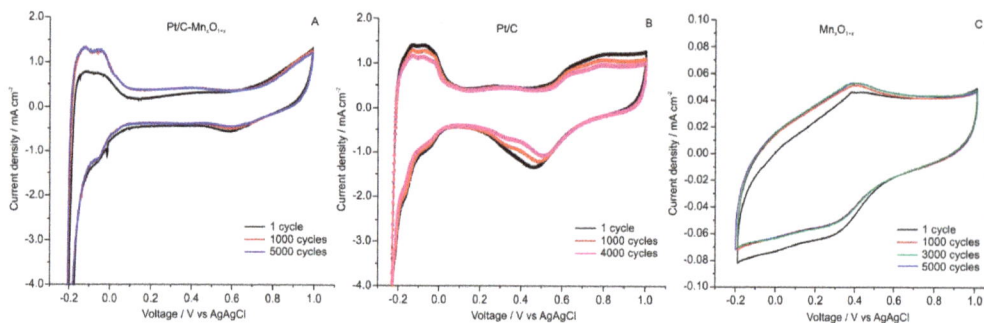

Figure 8. CV after consecutive potential cycling recorded at 50 mV·s^{-1}: (A) Pt/C-Mn$_x$O$_{1+x}$; (B) Pt/C; (C) Mn$_x$O$_{1+x}$ (0.5 M H$_2$SO$_4$, 0.1 ITC mass ratio, 20 mg$_{Pt}$·cm^{-2} catalyst loading).

To better check stability of the Pt/C-Mn$_x$O$_{1+x}$ and in particular of the Mn$_x$O$_{1+x}$ support, the extra Mn$_x$O$_{1+x}$ sample used for µRS analysis (Figure 4) was used to assess its stability in acid conditions. Specifically an RDE prepared with pure Mn$_x$O$_{1+x}$, was subjected to ADT by cycling it 5000 times in the same conditions used for the Pt/C-Mn$_x$O$_{1+x}$ (50 mV·s^{-1} between 0.4 and 0.8 V *vs.* Ag/AgCl, in a 0.5 M H$_2$SO$_4$ solution). Results from CV degradation (Figure 8C) show very little degradation, sign that Mn$_x$O$_{1+x}$ is a stable support in acid environment, as reported in the literature as well [29,40,48,49]. Moreover, this RDE was analyzed directly by SEM coupled with EDX detector before and after ADT. Images of this RDE are shown in Figure S1 of the supporting info. From a visual point of view, the Mn$_x$O$_{1+x}$ on RDE after cycling showed a rearrangement compared to the fresh configuration: it appears more agglomerated near the edges of the disk (Figure S1C,D), whether in the fresh configuration the electrode appears more homogeneous (Figure S1A,B). EDX elementary analyses on the overall Mn atomic quantity available on the RDE before and after ADT enlightened that after cycling the Mn overall content diminished by 18%. Thus, Mn$_x$O$_{1+x}$, can be considered a stable support in acidic environment. The presence of manganese oxides in a Pt/C composite electrode plays a catalytic role in the ORR by enhancing the catalytic behavior of Pt for the ORR [40]. In fact, manganese oxides in the composite electrode can be considered as substitute for oxygen as an electron-acceptor in the case of oxygen starvation.

3. Experimental Section

3.1. Chemicals

Vulcan XC-72 was purchased from Cabot. Chloroplatinic acid hexahydrate (H$_2$PtCl$_6$·6H$_2$O) ⩾ 37.50% Pt basis, potassium permanganate (KMnO$_4$), potassium hydroxide 85 wt. % (KOH), ethylene glycol 98 wt. % (EG, HOCH$_2$CH$_2$OH),

isopropyl alcohol 99.7 wt. % $((CH_3)_2CHOH)$, sulfuric acid (H_2SO_4) 98 wt. %, Nafion®
perfuorinated resin 5 wt. % hydro-alcoholic solution and methanol (MeOH, CH_3OH)
99.8 wt. % were purchased from Sigma Aldrich Italia (Milano, Italy). Commercial
20 wt. % Pt/Vulcan XC-72 electrocatalyst (QuinTech QuinTech e.K., Göppingen,
Germany) and commercial PtRu 1:1 at % (Hispec 6000, Alfa Aesar GmbH & Co KG,
Karlsruhe, Germany) were used for comparison tests. Nitrogen (99.999% purity) and
diluted carbon monoxide (10 vol % CO in Ar) gases were supplied in cylinders by
SIAD S.p.A. (Bergamo, Italy). All aqueous solutions were prepared using ultrapure
water obtained from a Millipore Milli-Q system (Merck KGaA, Darmstadt, Germany)
with resistivity > 18 $m\Omega\cdot cm^{-1}$.

3.2. Synthesis of the Hybrid Support $C-MnO_2$

To prepare the $C-MnO_2$, 3.9 g of $KMnO_4$ and 12.6 g of H_2SO_4 were added into
130 g of deionized water under magnetic stirring to form the precursor solution.
Then, 1.0 g of Vulcan XC-72 was added into this precursor solution. Subsequently, the
formed suspension was heated up to 80 °C and kept at 80 °C for 6 h under magnetic
stirring. The precipitates were filtered and washed with distilled water. Finally, the
obtained powder was dried at 120 °C for 6 h under vacuum. Assuming that all the
$KMnO_4$ used in the synthesis can be reduced to MnO_2, the theoretical wt. % of MnO_2
in the $C-MnO_2$ support is equal to 68.4 wt. %.

3.3. Synthesis of the $Pt/C-Mn_xO_{1+x}$ Catalyst by Thermal Method

For the synthesis of $Pt/C-Mn_xO_{1+x}$ catalyst, 200 mg of the previously prepared
$C-MnO_2$ was added to 50 mL EG and the mixture was stirred for 30 min. Then,
$H_2PtCl_6\cdot6H_2O$ was dissolved into the EG solution under stirring. The pH was
adjusted to 12, by the addition of 1 M KOH in EG solution. Microwave irradiation
was applied to the solution at 700 W for 2 min, in order to reduce the Pt^{4+} ions to
metallic Pt^0. The solution was left to cool naturally to room temperature. After
cooling, some drops of acetone were added to the solution and the $Pt/C-Mn_xO_{1+x}$
catalyst was washed thoroughly with abundant water. Finally, the catalyst was
annealed under nitrogen atmosphere for 2 h at 600 °C.

3.4. Synthesis of the PtRu/C Catalyst

PtRu/C catalyst was prepared by adding 60 wt. % [50] commercial PtRu 1:1
at % on functionalized Vulcan XC72 into a water-isopropyl alcohol solution under
stirring for 24 h. Then the PtRu/C catalyst was centrifuged and dried.

3.5. Chemical-Physical Characterization

Field-emission scanning electron microscopy (FESEM JEOL-JSM-6700F instrument, FEI Europe, Eindhoven, The Netherlands) coupled with an Energy Dispersive X-ray Spectrometry Detector (EDX OXFORD INCA, EDAX Inc., Mahwah, NJ, U.S.A.) and scanning electron microscopy (SEM-EXD FEI-QuantaTM Inspect 200, FEI Europe, Eindhoven, The Netherlands, with EDAX PV 9900 instrument, working at 15 kV, EDAX Inc., Mahwah, NJ, USA) were performed to analyze the morphology and check the amount of Pt and Mn.

The MnO_2 and platinum-to-carbon weight percentage in the catalysts was determined by inductively coupled plasma atomic mass spectroscopy (ICP-MS ICAP-Q instrument, ThermoFisher Scientific Inc., Waltham, MA, USA). Prior to analysis, the samples were digested in hot concentrated HCl/HNO_3 3:1 mixture with some droplets of H_2SO_4.

The XRD reflections were recorded on a PANalytical X'Pert PRO diffractometer with a PIXcel detector (PANalytical B.V., Almelo, The Netherlands), using Cu Kα radiation, under the conditions of $2\theta = 10°-100°$ and 2θ step size = 0.03, in order to examine the different polymorphs.

X-ray photoelectron spectroscopy (XPS) was performed to determine the elemental surface composition of the catalysts. The analysis was carried out using a Physical Electronics PHI 5000 Versa Probe electron spectrometer system (Physical Electronics Inc., Chanhassen, MN, USA) with monochromated Al Kα X-ray source (1486.60 eV) run at 15 kV and 1 mA anode current. The survey spectra were collected from 0 to 1200 eV. The narrow Mn 2p spectra were collected from 635 to 665 eV, the narrow Pt 4f spectra from 66 to 86 eV and the narrow C 1s spectra from 280 to 293 eV. All of the spectra were calibrated against a value of the C 1s binding energy of 284.5 eV. Multipak 9.0 software (Physical Electronics Inc., Chanhassen, MN, USA) was used for obtaining semi-quantitative atomic percentage compositions, using Gauss-Lorentz equations with Shirley-type background. A Gaussian/Lorentzian 70%/30% line shape was used to evaluate peak positions and areas of the high resolution Pt 4f and Mn 2p spectra, with a standard deviation in locating the peaks equal to 0.3 eV.

The chemical structure of the support was analyzed by a μ-Raman Spectroscopy (μRS Renishaw InVia spectrometer equipped with a Leica DMLM confocal microscope and a CCD detector with an excitation wavelength of 785 nm, Renishaw plc, Gloucestershire, United Kingdom). The Raman scattered light was collected in the spectral range 100–1000 cm^{-1}. At least ten scans were accumulated in four different positions of the catalyst to ensure a sufficient signal to noise ratio.

3.6. Electro-Chemical Characterization

The prepared electro-catalysts were tested in a conventional three-compartment electrochemical cell using a multi-potentiostat (Bio-Logic SP150, Bio-Logic Science Instruments SAS, Claix, France) and a rotating ring-disk electrode instrument (RRDE-3A ALS Model 2323, ALS Co. Ltd, Tokyo, Japan). The electrolyte was 0.5 M H_2SO_4 aqueous solution saturated with either N_2 or CO 10% v/v in Ar by direct bubbling the gas into the solution. For RDE measurements, the cell was equipped with a glassy carbon (GC) disk working electrode (0.1256 cm^2 geometric area), a Pt helical wire counter electrode and a silver chloride electrode (Ag/AgCl) as reference electrode. Glassy carbon (GC) electrodes were polished with alumina powder, ultrasonic washing and blow drying, before dropping the catalyst ink. Different GC disk electrodes were arranged by preparing the ink using an ionomer-to-catalyst (ITC) mass ratio (mg of Nafion® over mg of catalyst) equal to 0.1 and catalyst loading of 20 $\mu g_{Pt} \cdot cm^{-2}$ [51]. The working electrode was surface-polished with 1 and 0.06 μm alumina powders to a mirror-like finish its surface and sonicated to remove alumina particles before each experiment. Cyclic voltammograms (CV) with either N_2 or CO 10 vol % in Ar were recorded at 10 $mV \cdot s^{-1}$ and 20 $mV \cdot s^{-1}$, respectively.

CO stripping voltammetry was performed in 0.5 M H_2SO_4 at a scan rate of 20 $mV \cdot s^{-1}$. Prior to analysis a flow rate of 10 vol % CO in Ar was pre-adsorbed for 30 min while maintaining the working electrode at the constant potential of -0.19 V (vs. Ag/AgCl) and rotating disk speed of 900 rpm. Afterwards, a flow rate of pure N_2 was used for 15 min to remove the CO reversibly adsorbed onto the surface and the excess CO dissolved in the solution.

Cyclic voltammetries for the methanol oxidation reaction in acid conditions were carried out in a 0.5 M H_2SO_4 solution with 1 M MeOH. The scan rate was 20 $mV \cdot s^{-1}$ and the potential window was 0.0–1.0 V vs. Ag/AgCl. The highest initial activity was usually obtained within ~20 cycles and then the experiment was stopped [52].

The electrocatalyst stability was performed by ADT cycling up catalysts to 5000 times between 0.4 and 0.8 V vs. Ag/AgCl forwards and backwards at a scan rate of 50 $mV \cdot s^{-1}$ in N_2-saturated 0.5 M H_2SO_4 solution. Such a potential range for accelerated degradation tests should enlighten any problem related to the corrosion of carbon supports as well as the sintering of Pt nanoparticles based on the protocol suggested by DoE [53].

4. Conclusions

In this work a C-MnO_2 hybrid support was coated with platinum nanoparticles followed by a annealing at 600 °C, in order to promote the methanol oxidation reaction. The enhancement of the electrochemical performance of the Pt/C-Mn_xO_{1+x} was mainly due to the optimized dispersion and smaller particle size of Pt nanoparticles favored by the presence of a mixture of Mn_2O_3 and Mn_3O_4, as well as synergistic integration of nanomaterials. $Pt/C-Mn_xO_{1+x}$ shows better activity than the commercial Pt/C catalyst. However, its performance still falls short of the most commonly used commercial $PtRu/C$ catalyst, due to the presence of some Pt agglomerates. The aspiration that this hybrid support can be optimized and then go on to replace the current PtRu based catalysts can be realized by understanding the real function of this kind of hybrid support and by reducing the presence of Pt agglomerates. All results suggested that the $Pt/C-Mn_xO_{1+x}$ can act as promising catalysts for fuel cells.

Acknowledgments: The authors gratefully acknowledge the Italian project PRIN NAMEDPEM ("Advanced nanocomposite membranes and innovative electrocatalysts for durable polymer electrolyte membrane fuel cells", protocol n. 2010CYTWAW) funded by the Italian Ministry of Education, University and Research. S. Guastella and M. Raimondo from the Politecnico di Torino (Italy), P. Stelmachowski from the Jagellonian University in Krakow (Poland) and R. Doherty from the University of Strathclyde (United Kingdom) are gratefully acknowledged for XPS/FESEM-EDX/μRS analyses and enriching discussion.

Author Contributions: A.H.A.M.V. and L.O. conceived and designed the experiments; L.O. and R.A.M.E. performed the experiments; J.Z. and C.F. prepared and characterized the catalysts; A.H.A.M.V., L.O., and S.S. analyzed data, wrote and revised the paper.

Conflicts of Interest: The authors declare no conflict of interest.

References

1. Specchia, S.; Francia, C.; Spinelli, P. Polymer Eelectrolyte Membrane Fuel Cells. In *Electrochemical Technologies for Energy Storage and Conversion*, 1st ed.; Liu, R.S., Zhang, L., Sun, X., Liu, H., Zhang, J., Eds.; Wiley: Weinheim, Germany, 2010; Volume 1, pp. 601–670.
2. Sebastián, D.; Lázaro, M.J.; Moliner, R.; Suelves, I.; Aricò, A.S.; Baglio, V. Oxidized carbon nanofibers supporting PtRu nanoparticles for direct methanol fuel cells. *Int. J. Hydrogen Energy* **2014**, *39*, 5414–5423.
3. Santasalo-Aarnio, A.; Borghei, M.; Anoshkin, I.V.; Nasibulin, A.G.; Kauppinen, E.I.; Ruiz, V.; Kallio, T. Durability of different carbon nanomaterial supports with PtRu catalyst in a direct methanol fuel cell. *Int. J. Hydrogen Energy* **2012**, *37*, 3415–3424.
4. Monteverde Videla, A.H.A.; Alipour Moghadam Esfahani, R.; Peter, I.; Specchia, S. Influence of the preparation method on Pt_3Cu/C electrocatalysts for the oxygen reduction reaction. *Electrochim. Acta* **2015**.
5. Gasteiger, H.A.; Kocha, S.S.; Sompalli, B.; Wagner, F.T. Activity benchmarks and requirements for Pt Pt-alloy and non-Pt oxygen reduction catalysts for PEMFCs. *Appl. Catal. B* **2005**, *56*, 9–35.
6. Cai, J.; Huang, Y.; Huang, B.; Zheng, S.; Guo, Y. Enhanced activity of Pt nanoparticle catalysts supported on manganese oxide-carbon nanotubes for ethanol oxidation. *Int. J. Hydrogen Energy* **2014**, *39*, 798–807.

7. Kulesza, P.J.; Pieta, I.S.; Rutkowska, I.A.; Wadas, A.; Marks, D.; Klak, K.; Stobinski, L.; Cox, J.A. Electrocatalytic oxidation of small organic molecules in acid medium: Enhancement of activity of noble metal nanoparticles and their alloys by supporting or modifying them with metal oxides. *Electrochim. Acta* **2013**, *110*, 474–483.

8. Mani, P.; Srivastava, R.; Strasser, P. Dealloyed binary PtM$_3$ (M = Cu, Co, Ni) and ternary PtNi$_3$M (M = Cu, Co, Fe, Cr) electrocatalysts for the oxygen reduction reaction: Performance in polymer electrolyte membrane fuel cells. *J. Power Sources* **2011**, *196*, 666–673.

9. Sundarrajan, S.; Allakhverdiev, S.I.; Ramakrishna, S. Progress and perspectives in micro direct methanol fuel cell. *Int. J. Hydrogen Energy* **2012**, *37*, 8765–8786.

10. Meher, S.K.; Rao, G.R. Morphology-controlled promoting activity of nanostructured MnO$_2$ for methanol and ethanol electrooxidation on Pt/C. *J. Phys. Chem. C* **2013**, *117*, 4888–4900.

11. Wu, M.; Han, M.; Li, M.; Li, Y.; Zeng, J.; Liao, S. Preparation and characterizations of platinum electrocatalysts supported on thermally treated CeO$_2$–C composite support for polymer electrolyte membrane fuel cells. *Electrochim. Acta* **2014**, *139*, 308–314.

12. Zhou, W.; Zhou, Z.; Song, S.; Li, W.; Sun, G.; Tsiakaras, P.; Xin, Q. Pt based anode catalysts for direct ethanol fuel cells. *Appl. Catal. B* **2003**, *46*, 273–285.

13. Zhou, C.; Wang, H.; Peng, F.; Liang, J.; Yu, H.; Yang, J. MnO$_2$/CNT supported Pt and PtRu nanocatalysts for direct methanol fuel cells. *Langmuir* **2009**, *25*, 7711–7717.

14. Boucher, M.B.; Goergen, S.; Yi, N.; Flytzani-Stephanopoulos, M. "Shape effects" in metal oxide supported nanoscale gold catalysts. *Phys. Chem. Chem. Phys.* **2011**, *13*, 2517–2527.

15. Xu, M.W.; Bao, S.J. Nanostructured MnO$_2$ for electrochemical capacitor, energy storage in the emerging era of smart grids. In *Energy Storage in the Emerging Era of Smart Grids*; Carbone, R., Ed.; InTech: Rijeka, Croatia, 2011; Available online: http://cdn.intechopen.com/pdfs-wm/20372.pdf (accessed on 9 March 2015).

16. Huang, X.; Lv, D.; Yue, H.; Attia, A.; Yang, Y. Controllable synthesis of α- and β-MnO$_2$: Cationic effect on hydrothermal crystallization. *Nanotechnology* **2008**, *19*, 225606.

17. Xiao, W.; Wang, D.; Lou, X.W. Shape-controlled Synthesis of MnO$_2$ nanostructures with enhanced electrocatalytic activity for oxygen reduction. *J. Phys. Chem. C* **2010**, *114*, 1699–1700.

18. Calderón, J.C.; Mahata, N.; Pereira, M.F.R.; Figueiredo, J.L.; Fernandes, V.R.; Rangel, C.M.; Calvillo, L.; Lázaro, M.J.; Pastor, E. Pt-Ru catalysts supported on carbon xerogels for PEM fuel Cells. *Int. J. Hydrogen Energy* **2012**, *37*, 7200–7211.

19. Fu, X.Z.; Liang, Y.; Chen, S.P.; Lin, J.D.; Liao, D.W. Pt-rich shell coated Ni nanoparticles as catalysts for methanol electro-oxidation in alkaline media. *Catal. Commun.* **2009**, *10*, 1893–1897.

20. Rahsepar, M.; Pakshir, M.; Piao, Y.; Kim, H. Synthesis and electrocatalytic performance of high loading active PtRu multiwalled carbon nanotube catalyst for methanol oxidation. *Electrochim. Acta* **2012**, *71*, 246–251.

21. Woo, S.; Lee, J.; Park, S.K.; Kim, H.; Chung, T.D.; Piao, Y. Enhanced electrocatalysis of PtRu onto graphene separated by Vulcan carbon spacer. *J. Power Sources* **2013**, *222*, 261–266.

22. Wang, Z.B.; Yin, G.P.; Shi, P.F. The influence of acidic and alkaline precursors on Pt-Ru/C catalyst performance for a direct methanol fuel cell. *J. Power Sources* **2007**, *163*, 688–694.

23. Ban, S.; Malek, K.; Huang, C. A molecular simulation study of Pt stability on oxidized carbon nanoparticles. *J. Power Sources* **2013**, *221*, 21–27.

24. Han, B.; Zhang, F.; Feng, Z.; Liu, S.; Deng, S.; Wang, Y.; Wang, Y. A designed Mn_2O_3/MCM-41 nanoporous composite for methylene blue and rhodamine B removal with high efficiency. *Ceram. Int.* **2014**, *40*, 8093–8101.

25. Moses Ezhil Raj, A.; Grace Victoria, S.; Bena Jothy, V.; Ravidhas, C.; Wollschläger, J.; Suendorf, M.; Neumann, M.; Jayachandran, M.; Sanjeeviraj, C. XRD and XPS characterization of mixed valence Mn_3O_4 hausmannite thin films prepared by chemical spray pyrolysis technique. *Appl. Surf. Sci.* **2010**, *256*, 2920–2926.

26. Julien, C.M.; Massot, M.; Poinsignon, C. Lattice vibrations of manganese oxides Part I. Periodic structures. *Spectrochim. Acta A* **2004**, *60*, 689–700.

27. Stobbe, E.R.; de Boer, B.A.; Geus, J.W. The reduction and oxidation behaviour of manganese oxides. *Catal. Today* **1999**, *47*, 161–167.

28. Baturina, O.A.; Aubuchon, S.R.; Wynne, K.J. Thermal stability in air of Pt/C catalysts and PEM fuel cell catalyst layers. *Chem. Mater.* **2006**, *18*, 1498–1504.

29. Zamanzad Ghavidel, M.R.; Bradley Easton, E. Thermally induced changes in the structure and ethanol oxidation activity of $Pt_{0.25}Mn_{0.75}$/C. *Appl. Catal. B* **2015**, *176–177*, 150–159.

30. Liu, H.X.; Tian, N.; Brandon, M.P.; Zhou, Z.Y.; Lin, J.L.; Hardacre, C.; Lin, W.F.; Sun, S.G. Tetrahexahedral Pt nanocrystal catalysts decorated with Ru adatoms and their enhanced activity in methanol electrooxidation. *ACS Catal.* **2012**, *2*, 708–715.

31. Li, L.; Xing, Y. Pt-Ru nanoparticles supported on carbon nanotubes as methanol fuel cell catalysts. *J. Phys. Chem. C* **2007**, *111*, 2803–2808.

32. Kang, Y.; Murray, C.B. Synthesis and electrocatalytic properties of cubic Mn-Pt nanocrystals (Nanocubes). *J. Am. Chem. Soc.* **2010**, *132*, 7568–7569.

33. Yang, X.; Wang, X.; Zhang, G.; Zheng, J.; Wang, T.; Liu, X.; Shu, C.; Jiang, L.; Wang, C. Enhanced electrocatalytic performance for methanol oxidation of Pt nanoparticles on Mn_3O_4-modified multi-walled carbon nanotubes. *Int. J. Hydrogen Energy* **2012**, *37*, 11167–11175.

34. Xu, C.; Shen, P.K. Electrochemical oxidation of ethanol on Pt-CeO_2/C catalysts. *J. Power Sources* **2005**, *142*, 27–29.

35. Zeng, J.; Francia, C.; Gerbaldi, C.; Baglio, V.; Specchia, S.; Aricò, A.S.; Spinelli, P. Hybrid ordered mesoporous carbons doped with tungsten trioxide as supports for Pt electrocatalysts for methanol oxidation reaction. *Electrochim. Acta* **2013**, *94*, 80–91.

36. Alegre, C.; Gálvez, M.E.; Baquedano, E.; Pastor, E.; Moliner, R.; Lázaro, M.J. Influence of support's oxygen functionalization on the activity of Pt/carbon xerogels catalysts for methanol electro-oxidation. *Int. J. Hydrogen Energy* **2012**, *37*, 7180–7191.

37. Maillard, F.; Schreier, S.; Savinova, E.R.; Weinkauf, S.; Stimming, U. Influence of particle agglomeration on the catalytic activity of carbon-supported Pt nanoparticles in CO monolayer oxidation. *Phys. Chem. Chem. Phys.* **2005**, *7*, 385–393.

38. Wang, H.; Abruña, H.D. Origin of multiple peaks in the potentiodynamic oxidation of CO adlayers on Pt and Ru-modified Pt electrodes. *J. Phys. Chem. Lett.* **2015**, *6*, 1899–1906.

39. Meher, S.K.; Rao, G.R. Polymer-assisted hydrothermal synthesis of highly reducible shuttle-shaped CeO_2: Microstructural effect on promoting Pt/C for methanol electrooxidation. *ACS Catal.* **2012**, *2*, 2795–2809.

40. Wei, Z.D.; Ji, M.B.; Hong, Y.; Sun, C.X.; Chan, S.H.; Shen, P.K. MnO_2–Pt/C composite electrodes for preventing voltage reversal effects with polymer electrolyte membrane fuel cells. *J. Power Sources* **2006**, *160*, 246–251.

41. Lee, S.W.; Chen, S.; Sheng, W.; Yabuuchi, N.; Kim, Y.-T.; Mitani, T. Roles of surface steps on Pt nanoparticles in electro-oxidation of carbon monoxide and methanol. *J. Am. Chem. Soc.* **2009**, *131*, 15669–15677.

42. Gong, X.; Yang, Y.; Huang, S. Mn_3O_4 catalyzed growth of polycrystalline Pt nanoparticles and single crystalline Pt nanorods with high index facets. *Chem. Commun.* **2011**, *47*, 1009–1011.

43. Aricò, A.S.; Srinivasan, S.; Antonucci, V. DMFCs from fundamental aspects to technology development. *Fuel Cells* **2001**, *1*, 133–161.

44. Velázquez-Palenzuela, A.; Centellas, F.; Garrido, J.A.; Arias, C.; Rodríguez, R.M.; Brillas, E.; Cabot, P.L. Kinetic analysis of carbon monoxide and methanol oxidation on high performance carbon-supported Pt–Ru electrocatalyst for direct methanol fuel cells. *J. Power Sources* **2011**, *196*, 3503–3512.

45. Raoof, J.B.; Ojani, R.; Hosseini, S.R. Electrochemical fabrication of novel Pt/poly (m-toluidine)/Triton X-100 composite catalyst at the surface of carbon nano-tube paste electrode and its application for methanol oxidation. *Int. J. Hydrogen Energy* **2011**, *36*, 52–63.

46. Ye, K.-H.; Zhou, S.-A.; Zhu, X.-C.; Xu, C.-W.; Shen, P.K. Stability analysis of oxide (CeO_2, NiO, Co_3O_4 and Mn_3O_4) effect on Pd/C for methanol oxidation in alkaline medium. *Electrochim. Acta* **2013**, *90*, 108–111.

47. Mao, L.; Zhang, K.; Chan, H.S.O.; Wu, J.S. Nanostructured MnO_2/graphene composites for supercapacitor electrodes: The effect of morphology, crystallinity and composition. *J. Mater. Chem.* **2012**, *22*, 1845–1851.

48. Trogadas, P.; Ramani, V. Pt/C/MnO_2 hybrid electrocatalysts for degradation mitigation in polymer electrolyte fuel cells. *J. Power Sources* **2007**, *174*, 159–163.

49. Huang, H.; Chen, Q.; He, M.; Sun, X.; Wang, X. A ternary Pt/MnO_2/graphene nanohybrid with an ultrahigh electrocatalytic activity toward methanol oxidation. *J. Power Sources* **2013**, *239*, 189–195.

50. Aricò, A.S.; Baglio, V.; di Blasi, A.; Modica, E.; Antonucci, P.L.; Antonucci, V. Analysis of the high-temperature methanol oxidation behaviour at carbon-supported Pt–Ru catalysts. *J. Electroanal. Chem.* **2003**, *557*, 167–176.

51. Garsany, Y.; Ge, J.; St-Pierre, J.; Rocheleau, R.; Swider-Lyons, K.E. Analytical procedure for accurate comparison of rotating disk electrode results for the oxygen reduction activity of Pt/C. *J. Electrochem. Soc.* **2014**, *161*, F628–F640.
52. Sneed, B.T.; Young, A.P.; Jalalpoor, D.; Golden, M.C.; Mao, S.; Jiang, Y.; Wang, Y.; Tsung, C.K. Shaped PdNiPt core-sandwich-shell nanoparticles: Influence of Ni sandwich layers on catalytic electrooxidations. *ACS Nano* **2014**, *7*, 7239–7250.
53. US DoE 2014 Annual Progress Report V. Fuel Cells. Available online: http://www. hydrogen.energy.gov/annual_progress14_fuelcells.html#a (accessed on 12 January 2015).

Facile Electrodeposition of Flower-Like PMo$_{12}$-Pt/rGO Composite with Enhanced Electrocatalytic Activity towards Methanol Oxidation

Xiaoying Wang, Xiaofeng Zhang, Xiaolei He, Ai Ma, Lijuan Le and Shen Lin

Abstract: A facile, rapid and green method based on potentiostatic electrodeposition is developed to synthesize a novel H$_3$PMo$_{12}$O$_{40}$-Pt/reduced graphene oxide (denoted as PMo$_{12}$-Pt/rGO) composite. The as-prepared PMo$_{12}$-Pt/rGO is characterized by X-ray diffraction (XRD), scanning electron microscopy (SEM) and X-ray photoelectron spectroscopy (XPS). The results reveal that graphene oxide (GO) is reduced to the rGO by electrochemical method and POMs clusters are successfully located on the rGO as the modifier. Furthermore, the PMo$_{12}$-Pt/rGO composite shows higher electrocatalytic activity, better tolerance towards CO and better stability than the conventional pure Pt catalyst.

Reprinted from *Catalysts*. Cite as: Wang, X.; Zhang, X.; He, X.; Ma, A.; Le, L.; Lin, S. Facile Electrodeposition of Flower-Like PMo$_{12}$-Pt/rGO Composite with Enhanced Electrocatalytic Activity towards Methanol Oxidation. *Catalysts* **2015**, *5*, 1275–1288.

1. Introduction

Direct methanol fuel cells (DMFCs) have drawn increasing attention due to their simple operation, high energy density, low pollutant emission, low operating temperature (60–100 °C) and ease of handling liquid fuel [1–3]. It is widely agreed that, as a single component catalyst, platinum shows significant electrocatalytic activity for methanol oxidation at lower temperatures. However, there are two key problems inhibiting its utilization in DMFCs: (1) high cost of precious platinum and (2) pure Pt electrocatalysts are prone to deactivation/ poisoning by the reaction intermediates (mainly CO), which generate from incomplete oxidation of methanol and chemically adsorb onto the Pt surface and block the active sites [4].

In order to decrease the usage of pure Pt electrocatalysts, various nanostructured carbon materials have been used to effectively disperse metal nanoparticles. In particular, reduced graphene oxide (rGO) has been found as a promising candidate for catalyst support in DMFCs [5]. Reduced graphene oxide (rGO) is gradually attracting more scientific and technological research interests due to its unique mechanical and electronic properties and wide applications [6,7]. On the other hand, polyoxometalates (POMs) are early-row transition metal oxygen anionic clusters with a remarkable redox and photo-electrochemical properties [8]. It was

545

demonstrated [9] that Keggin-type $PMo_{12}O_{40}{}^{3-}$ anions in an aqueous solution could effectively convert carbon monoxide to carbon dioxide over catalysts, as represented by following equation.

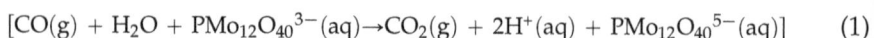

$$[CO(g) + H_2O + PMo_{12}O_{40}{}^{3-}(aq) \rightarrow CO_2(g) + 2H^+(aq) + PMo_{12}O_{40}{}^{5-}(aq)] \qquad (1)$$

In order to enhance CO tolerance in methanol oxidation and improve the durability of the Pt electrocatalysts, our group have studied the effects of silicotungstic acid ($H_4SiW_{12}O_{40}$) on the electrocatalytic activity of Pt catalysts towards methanol oxidation, and found that silicotungstic acid can promote the further oxidation of intermediates such as CO and supplies enough active sites for methanol oxidation [10]. Moreover, $H_3PMo_{12}O_{40}$ can also enhance electrocatalysis of Pd toward formic acid electrooxidation. The addition of $H_3PMo_{12}O_{40}$ contributes to converting CO into CO_2, which reduces the poisoning effects of CO over Pd catalyst [11]. All of these positive studies provide evidence that POMs could enhance antipoisoning ability of Pt in the methanol electrooxidation process on fuel cell anodes.

Up to now, many electrochemical methods have been used to reduce graphene oxide (GO) into reduced graphene oxide (rGO), such as cyclic voltammograms [12], potentiostatic electro deposition methods [13] and differential pulse voltammetry (DPV) [14]. The experiment results reveal that the electrochemical approach is a relatively economic, fast and environmental friendly method to prepare graphene avoiding toxic and hazardous chemicals such as hydrazine or dimethylhydrazine in the reduction process [15].

In this study, we firstly report a facile, fast, scalable, economic and environmentally benign pathway to prepare PMo_{12}-Pt/rGO composites. The electrochemical prepared approach can be undertaken via two steps (Figure 1): the first step involves direct electrochemical reduction of GO in suspension onto the substrate. Then, PMo_{12}-Pt clusters on the substrate surface were also deposited by electrodeposition method *in situ*. As expected, the as-prepared PMo_{12}-Pt/rGO composite exhibits superior catalytic activity on the electrochemical catalysis of methanol and CO oxidation.

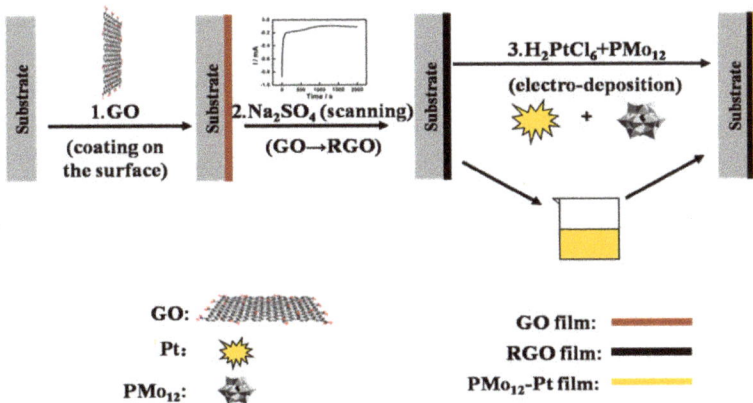

Figure 1. Schematic preparation of PMo_{12}-Pt/rGO composites.

2. Results and Discussion

Figure 2 displays the X-ray diffraction (XRD) patterns of rGO/indium tin oxide (rGO/ITO), Pt/rGO/ITO and PMo_{12}-Pt/rGO/ITO, respectively. As shown in Figure 2a, the diffraction peaks located at 30.23°, 35.16°, 50.47° and 60.02° can be considered as (222), (400), (440) and (622) crystal planes of ITO [16]. According to the ICDD PDF 04-0802, the diffraction peaks at 40.00°, 46.54°, 67.91° and 81.48° can be indexed to the (111), (200), (220) and (311) planes for Pt. These diffraction peaks are found in Figure 2b,c, which suggest that the successful formation of Pt on the rGO film by electrodeposition *in situ*. However, the diffraction peaks of Pt crystal planes (curve c) slightly shift comparing with curve (b). It may be as a result of the interaction among rGO, PMo_{12} and Pt. There are no distinct diffraction peaks of PMo_{12}, which may be due to the characteristic diffraction pattern of crystalline PMo_{12} being absent, which further implies that PMo_{12} clusters do not exist in the crystalline state but in the dispersed state [17].

The presence of Pt, P, Mo, C, N, and O elements on the surface of the composite is confirmed in the full-spectra of X-ray photoelectron spectroscopy (XPS) (Figure 3a). As shown in Figure 3b, the C1s XPS spectrum of the prepared composite show that there are four kinds of carbon atoms in different functional groups: C–C/C=C bonds (284.6 eV), C–O bands (286.7 eV), C=O bands (287.5 eV) and O–C=O bands (288.5 eV) [18,19]. The C1s spectrum of GO shows the presence of two typical carbon bonds: C–C/C=C (284.6 eV) and C–O (286.7 eV) (Figure 3e). After electrochemical reduction, only the C–C/C=C bands remain dominant, which implies that the functional groups such as carboxyl groups, hydroxyl groups, and epoxy groups are reduced and detached from graphene surface. In Pt (4f) XPS of the composite (Figure 3c), the principle peaks are attributed to Pt° at 71.2 eV ($4f_{7/2}$) and 74.6 eV ($4f_{5/2}$) [20], while peaks at 72.1, 75.9 and 74.4, 77.5 eV are assigned to Pt in +2 and +4 states [21,22], respectively. The results of different Pt species are calculated based on above data and listed in Table S1. After electrolytic deposition by cyclic voltammetry, the relative intensity of Pt°, Pt^{2+} and Pt^{4+} are calculated to be 62.92%, 26.57% and 10.51% for prepared composite, respectively. However, in contrast, the proportion of Pt° on the surface is only 36.3% via chemical synchronous reduction [23]. Thus, the preparation method we used can effectively improve the content of Pt°. Moreover, the Mo 3d core level spectrum displays two peaks at binding energies of 232.8 eV and 236.0 eV, corresponding to the Mo $3d_{3/2}$ and Mo $3d_{5/2}$ spin-orbit states of PMo_{12}, respectively (Figure 3d) [24], which indicated the presence of PMo_{12} in the composite.

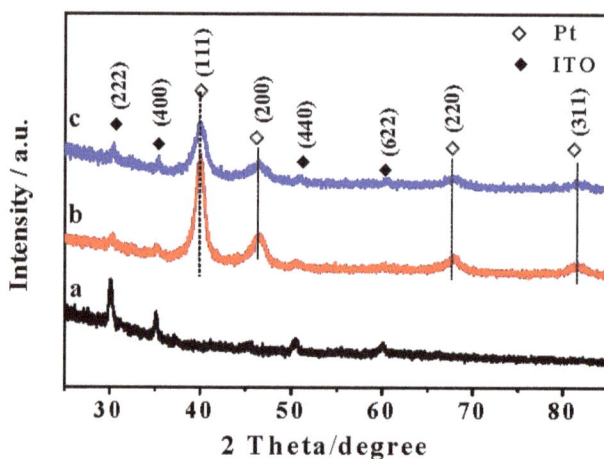

Figure 2. XRD patterns of the (**a**) rGO/ITO, (**b**) Pt/rGO/ITO and (**c**) PMo_{12}-Pt/rGO/ITO.

Raman spectroscopy is a powerful nondestructive technique that is widely used to distinguish order and disorder in the crystal structure of carbon [25]. Figure 4 presents the Raman spectra of GO and PMo_{12}-Pt/rGO, respectively. Two groups of typical characteristic peaks of D bands and G bands can be observed at about ~1320 and ~1590 cm^{-1}, respectively. The D band originates from the disordered structural defects or edge areas, and the G band is associated with the in-plane vibration of sp^2 bonded carbon atoms [26]. Meanwhile, the intensity ratio of D and G bands (I_D/I_G) can be used to evaluate the extent of defects in carbonaceous materials. The I_D/I_G value of PMo_{12}-Pt/rGO is estimated about 1.80, which is higher than that of GO (1.33). The increase suggests the realization of deoxygenation during the reduction of GO [27].

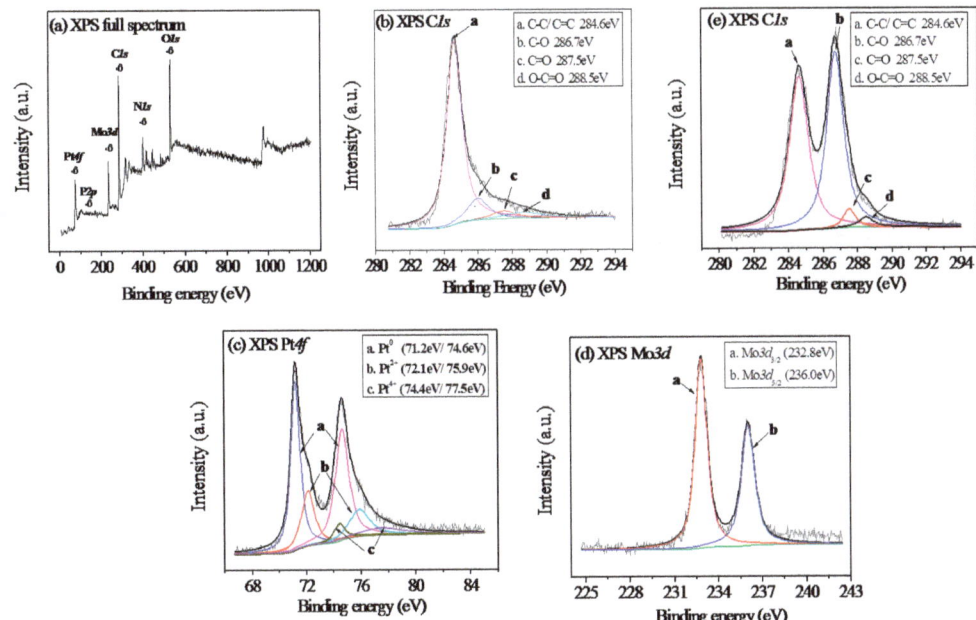

Figure 3. XPS spectra: (**a**) Full scan of PMo$_{12}$-Pt/rGO composites; (**b**) (**e**) C1s spectrum of PMo$_{12}$-Pt/rGO composites and GO; (**c**) (**d**) Pt 4f and Mo 3d spectrum of PMo$_{12}$-Pt/rGO composites.

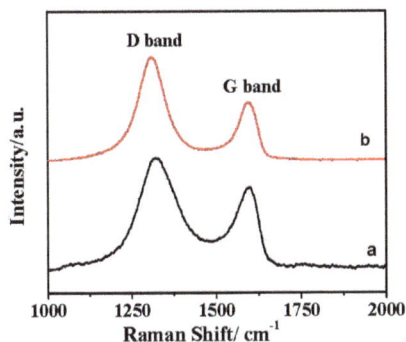

Figure 4. Raman spectra of different samples: (**a**) GO; (**b**) PMo$_{12}$-Pt/rGO.

Figure 5 displays the SEM images of different composites. Figure 5a–c are the surface morphology of PMo$_{12}$-Pt/ rGO/ITO composite, which are deposited at different electrode potentials. As shown in Figure 5a, when the deposition potential is −0.2 V, the coral-like clusters shape up on the rGO surface with the 100–800 nm diameters and less aggregation. When the deposition potential is −0.3 V, they are composed of flower-like clusters (Figure 5b) and the diameter is in range from 450 to

900 nm. Each flower-like cluster (Figure 5b inset) is three-dimensional, spear-shaped and multi-faceted. The mean diameter is of ~100 nm. These special structures may provide a larger specific surface area compared with other morphologies. When the deposition potential is decreased to −0.4 V, the PMo_{12}-Pt/rGO/ITO composite is in an irregular shape (Figure 5c).

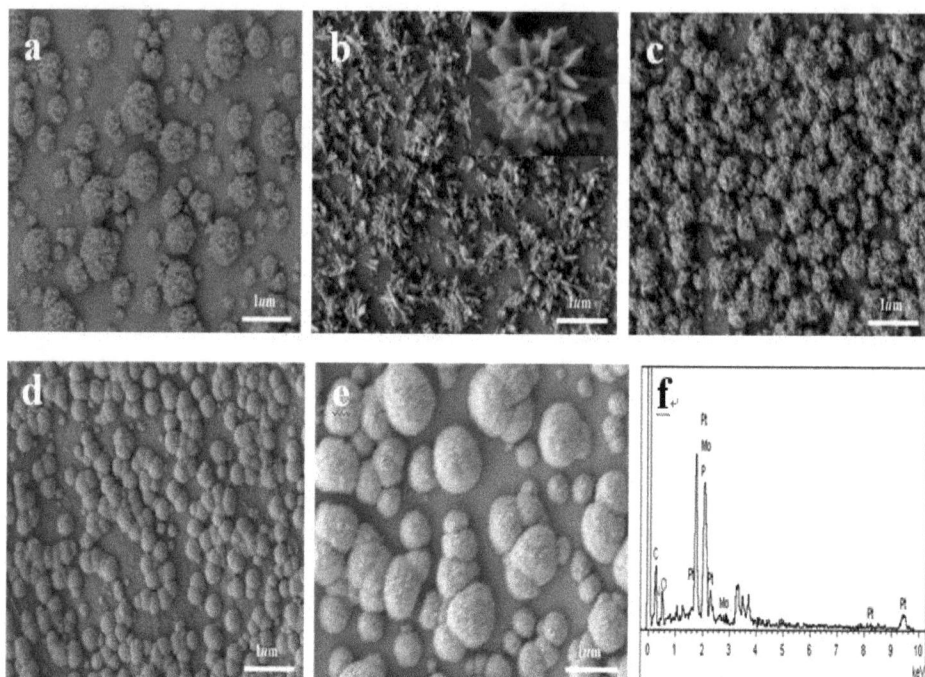

Figure 5. SEM images of different modified electrodes: (a) PMo_{12}-Pt/rGO/ITO composites at −0.2V; (b) PMo_{12}-Pt/rGO/ITO composites at −0.3V; (c) PMo_{12}-Pt/rGO/ITO composites at −0.4V; (d) Pt/rGO/ITO composites at −0.3V; (e) PMo_{12}-Pt/ITO composites at −0.3V. Scan time: 600 s; (f) EDS of PMo_{12}-Pt/rGO/ITO modified electrode at the deposition potential of −0.3V for 600 s.

Figure 5b,d and e presents the SEM images of PMo_{12}-Pt/rGO/ITO (b), Pt/rGO/ITO (d) and PMo_{12}-Pt/ITO (e) obtained from deposition potential of −0.3 V, respectively. As shown in Figure 5d, Pt clusters are spherical with diameters in the range of 100–450 nm and less aggregation. By contrast, PMo_{12}-Pt clusters are the same morphology as Pt/rGO with big size distribution (150–900 nm) in Figure 5e. It suggests that the introduction of the PMo_{12} and rGO may have an impact on the formation of the structure of clusters. Energy dispersive spectroscopy (EDS) analysis in Figure 5f identifies the presence of Pt, P, Mo, C, N and O on the PMo_{12}-Pt/rGO/ITO electrode and further confirms that PMo_{12}, Pt, and rGO are present in the composite.

The catalytic activity of the different modified electrodes was studied in a conventional three electrode system in 0.5 M H_2SO_4 + 1 M CH_3OH electrolyte solutions at a scan rate of 100 mV·s^{-1}. Figure 6 presents the steady state cyclic voltammograms of Pt, Pt/rGO, PMo_{12}-Pt/rGO deposited on glassy carbon electrodes, referring to the Ag/AgCl electrode. The forward scan current density (I_f) of PMo_{12}-Pt/rGO/glass carbon electrode (PMo_{12}-Pt/rGO/GCE) was 269.1 mA·cm^{-2}·mg^{-1} Pt, but it was 176.6 mA·cm^{-2} mg^{-1} Pt for Pt/rGO/GCE catalysts and only 147.9 mA·cm^{-2} mg^{-1} Pt for Pt/GCE catalysts. It is evident that the forward peak current value of PMo_{12}-Pt/rGO/GCE (about 269.1 mA·cm^{-2}·mg^{-1} Pt) is 1.52 times higher than that of Pt/rGO/GCE (about 176.6 mA·cm^{-2} mg^{-1} Pt), which indicates that PMo_{12}-Pt clusters have better catalytic activity for methanol electrooxidation. The result may be explained as following: the poisonous intermediates such as CO that are absorbed on the active sites of Pt nanoparticles significantly can be catalytically oxidized by POMs, which results in the increased electrocatalytic activity of PMo_{12}-Pt clusters [28]. Therefore, PMo_{12}-Pt/rGO/GCE modified electrode for the oxidation of methanol in acidic medium shows better catalytic activity than Pt/rGO/GCE and Pt/GCE.

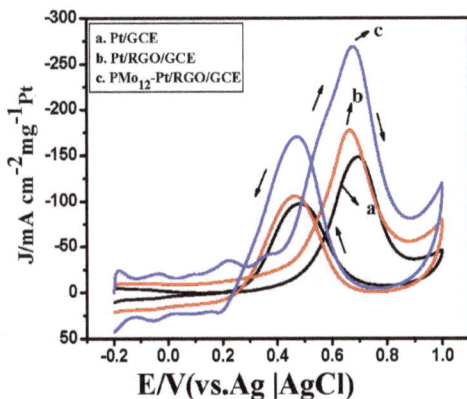

Figure 6. Cyclic voltammograms of different modified electrodes: (a) Pt/GCE; (b) Pt/rGO/GCE; (c) PMo_{12}-Pt/rGO/GCE in 0.5 M H_2SO_4 + 1 M CH_3OH solution. Scan rate: 100 mV·s^{-1}.

The short-term stability of the catalysts was investigated by accelerated aging tests, which were performed by running the 1st time and 100 times between −0.20 V and 1.0 V with a scan rate of 0.10 V·s^{-1} in 0.5 M H_2SO_4 and 1 M CH_3OH aqueous solution for the catalysts are presented in Figure 7a–c. The peak for the Pt/GCE have the same change characteristics as that for the other electrodes with the increasing cycling number, but the electricity density for the Pt/GCE declines faster than the other electrodes (Figure 7a). However, for the Pt/rGO/GCE (Figure 7b) and PMo$_{12}$-Pt/rGO/GCE composite (Figure 7c), the current densities decline 11.33% and 6.99% (Table S2), respectively. Therefore, the PMo$_{12}$-Pt/rGO/GCE has better short-term stability than the other two.

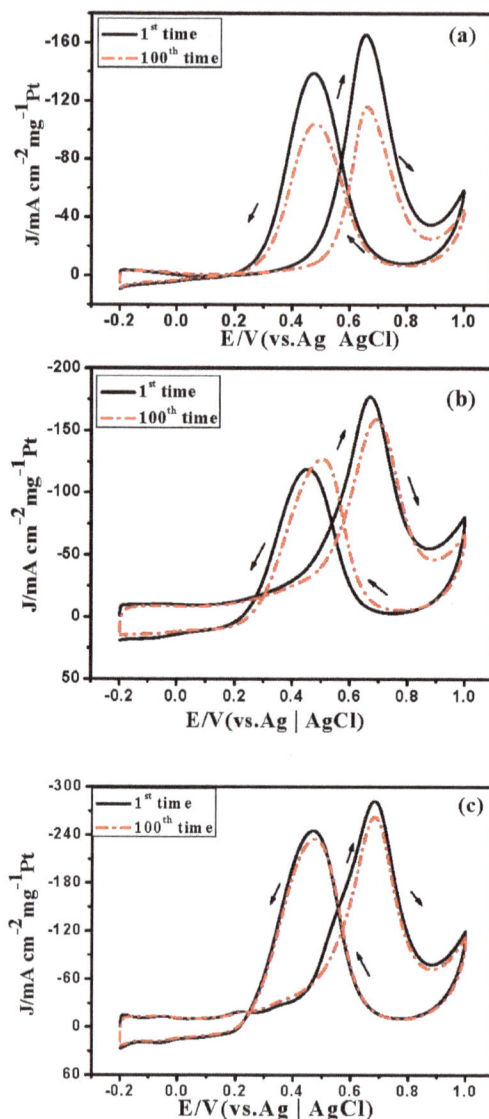

Figure 7. Comparative cyclic voltammograms of the different modified GCE electrodes: (a) Pt/GCE; (b) Pt/rGO/GCE; (c) PMo$_{12}$-Pt/rGO/GCE at 1st time and 100th times in 0.5 M H$_2$SO$_4$ + 1 M CH$_3$OH solution. Scan rate: 100 mV·s^{-1}.

To evaluate the long-term performance of the three electrodes for methanol oxidation (in 0.5 M H_2SO_4 + 1 M CH_3OH solution), they were polarized at 0.68 V for 7200 s. As shown in Figure 8, a rapid initial current density decay is observed, due to the formation of some intermediate species (mainly CO_{ads}) during the methanol oxidation reaction [29]. Then the currents slowly decrease and reach a quasi-stationary state within 7200 s. As observed from Figure 9, the current densities at 7200 s are 6.18, 4.60, 2.81 mA·$cm^{-2}mg^{-1}$·Pt towards methanol oxidation, respectively. The maximum steady-state oxidation current density for PMo_{12}-Pt/rGO/GCE is the largest compared to those of other electrodes. Thus, it confirms that the combination of PMo_{12}-Pt and rGO enhance electrocatalytic performance of the Pt catalyst.

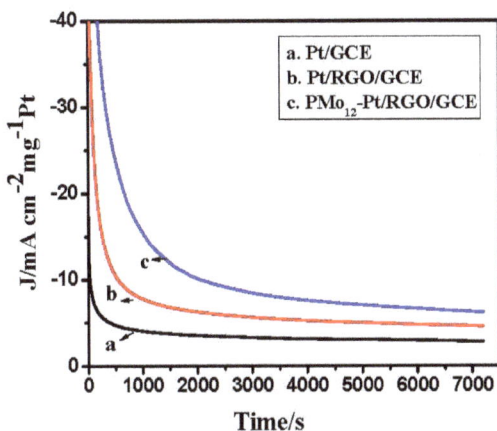

Figure 8. Chronoamperometric curves of different modified electrodes: **(a)** Pt/GCE; **(b)** Pt/rGO/GCE; **(c)** PMo_{12}-Pt/rGO/GCE in 0.5 M H_2SO_4 + 1 M CH_3OH solution at a fixed potential of 0.68 V for 2 h.

Electrochemical impedance spectroscopy (EIS) was used to further investigate the intrinsic behavior of the anodic process. The Nyquist plots of EIS for Pt/GCE (curve a), Pt/rGO/GCE (curve b), PMo_{12}-Pt/rGO/GCE (curve c) in 1 M CH_3OH + 0.5 M H_2SO_4 are shown in Figure 9. The diameter of the primary semicircle can be used to analyze the charge transfer resistance of the catalyst, and describe the rate of charge transfer during the methanol oxidation reaction [30]. The semicircle radius on the Nyquist plots of EIS for PMo_{12}-Pt/rGO/GCE is much smaller than that of Pt/GCE and Pt/rGO/GCE, clearly authenticating that the incorporation of PMo_{12} and rGO results in the improved conductivity of PMo_{12}-Pt/rGO/GCE.

Efficient elimination of the poisoning species such as CO from the catalyst is very important for assessing the catalyst performance in DMFCs. Figure 10 shows the voltammograms of CO oxidation on the different modified electrodes. As seen in Figure 10c, the distinct CO oxidation peak appears during the first forward scan, whereas it disappears in the second forward scan, indicating that the adsorbed CO on the surface of PMo_{12}-Pt/rGO nanoparticles has been oxidized during the first forward scan [31]. The PMo_{12}-Pt/rGO/GCE exhibited the more negative peak potential (0.647 V), while the peak potential of Pt/rGO/GCE and Pt/GCE only 0.687 V and 0.729 V, respectively. The negatively shifted peak potentials indicate that CO species on the PMo_{12}-Pt/rGO interfaces are more easily transformed to CO_2 due to the oxidation ability of PMo_{12} [32]. Therefore, the active sites on Pt are released for further electrochemical reaction.

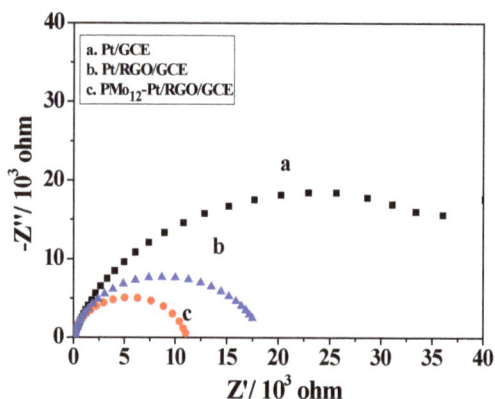

Figure 9. Nyquist plots of EIS with different films modified electrodes: (**a**) Pt/GCE; (**b**) Pt/rGO/GCE; (**c**) PMo_{12}-Pt/rGO/GCE in 0.5 M H_2SO_4 + 1 M CH_3OH solution.

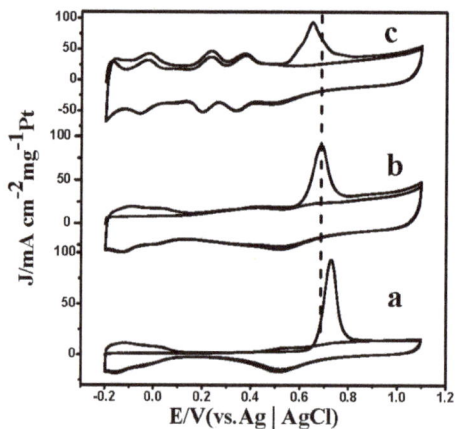

Figure 10. CO-stripping curves of different modified electrodes: (**a**) Pt/GCE; (**b**) Pt/rGO/GCE; (**c**) PMo_{12}-Pt/rGO/GCE in 0.5 M H_2SO_4 solution. Scan rate: 100 mV·s^{-1}.

3. Experimental Section

3.1. Materials

Graphite powder (−325 mesh, 99.9995%) was purchased from Alfa Aesar (Shanghai, China). $Cu(Ac)_2$, K_2PtCl_4 (99%), $KMnO_4$, H_2O_2 (30%), $K_2S_2O_8$, P_2O_5, H_2SO_4, methanol, and ethanol were all purchased from Sinopharm Chemical Reagent Co., Ltd. (Shanghai, China) and used without further purification. Deionized water was used throughout the experiments.

3.2. Preparation of Reduced Graphene Oxide (rGO) Modified Electrode

Graphite oxide was prepared from graphite powder by a modified Hummers' method [33,34]. To obtain a homogeneous suspension, Graphite oxide was mixed with deionized water and ultrasonicated for 2 h. The as-prepared GO suspension was coated onto an indium tin oxide (ITO) glass or glass carbon electrode (GCE) to form GO film. The rGO film was prepared upon electrochemically reduction of GO film in 0.1 M Na_2SO_4 at a constant potential of −1.5 V for 600 s. (Figure 1).

3.3. Synthesis of PMo_{12}-Pt/rGO

PMo_{12}-Pt clusters were deposited in situ on the surface of the rGO film-modified GCE (3 mm in diameter) or ITO electrode (4 mm in width) by potentiostatic electrodeposition in a 0.5 M H_2SO_4 solution containing 2 mM H_2PtCl_6 and 0.2 mM PMo_{12} at −0.3 V for 600 s. After deposition, the working electrode was rinsed with distilled water and dried under an infrared lamp. For comparison, the electrodeposition of platinum on the surface of bare GCE or ITO, and rGO was also performed under the same conditions.

3.4. Characterization

XPS was performed at room temperature with monochromatic Al Kα radiation (1486.6 eV) using a Quantum 2000 system (PHI, Chanhassen, MN, USA). XRD patterns were measured on an X'pert Pro diffractometer (Philips, Almelo, The Netherlands), using Cu Kα radiation. Field emission scanning electron microscopy (FE-SEM) images were observed on a JSM-7500F field emission scanning electron microanalyzer (JEOL, Tokyo, Japan). EDS was used to confirm the existence of Pt particles. Ramam spectra was measured using a Renishaw-in-Via Raman (Renishaw, London, UK) micro-spectrometer equipped with 514 nm diode laser excitation on a 300 lines·mm^{-1} grating. The actual amount of Pt loadings of the catalysts was determined by inductively coupled plasma-mass spectroscopy (ICP-MS, X Series 2, Thermo Scientific, Waltham, MA, USA).

3.5. Electrochemical Measurements

Electrochemical measurements were conducted on a CHI660 Electrochemical Workstation (Chenhua, Shanghai, China) using a conventional three-electrode electrochemical system. The working electrode was a glassy carbon electrode (geometric area, 0.07 cm^2) modified with the catalysts; Ag/AgCl electrode and Pt wire were used as the counter and reference electrode, respectively. Cyclic voltammetric, EIS and chronoamperometric experiments were carried out in 0.5 M H$_2$SO$_4$ in the absence and presence of 1 M methanol. The electrolyte solution was deaerated with ultrahigh-purity N$_2$ before scanning. The CO stripping voltammograms were measured by oxidation of preadsorbed CO (CO$_{ad}$) in the 0.5 M H$_2$SO$_4$ solution at a scan rate of 100 mV·s^{-1}. CO was bubbled for 30 min to allow the complete adsorption of CO onto the composites when the potential was kept at 0.1 V. Excess CO in the electrolyte was then purged out with N$_2$ for 15 min.

CO-stripping curves of Pt/GCE, Pt/rGO/GCE, and PMo$_{12}$-Pt/rGO/GCE in 0.5 M H$_2$SO$_4$ solution is collected to evaluate the electrochemical surface areas (ECSA). All the composites show characteristic CO oxidation peak in the first forward scan, suggesting the presence of electrochemically active Pt. The ECSA were calculated by the integrated charge (Q) in the CO oxidation region. According to the equation ECSA = $Q/(420 \ \mu C \cdot cm^{-2} \times$ Pt loading), we have added the Pt load for each sample in Table S3.

4. Conclusions

In this work, a unique flower-like PMo$_{12}$-Pt/rGO composite has been successfully synthesized by the electrochemical reduction method and used as an electrocatalyst for methanol oxidation. Cyclic voltammetry, chronoamperometry and CO stripping voltammetry were used to study electrocatalytic properties of PMo$_{12}$-Pt/rGO composite in acidic medium for methanol oxidation. The PMo$_{12}$-Pt/rGO composite modified electrode shows higher catalytic activity, better electrochemical stability and resistance to CO poisoning, which may be attributed to the synergistic effect of the special morphology of the composite, excellent conductivity of rGO and superior redox properties of PMo$_{12}$. These findings suggest that the PMo$_{12}$-Pt/rGO composite can be considered as a good electrocatalyst material for DMFCs.

Acknowledgments: This project was financially supported by the National Natural Science Foundation of China (No. 21171037), the Natural Science Foundation of Fujian Province (No. 2014J01033) and a key item of Education Department of Fujian Province (No. JA13085 and JB13009).

Author Contributions: Xiaoying Wang, Xiaolei He and Lijuan Le did the experiments, Xiaoying Wang wrote the first draft, Ai Ma and Xiaofeng Zhang took part in the discussion of experiments results and revision of paper, Shen Lin was responsible for the research work and paper revision.

Conflicts of Interest: The authors declare no conflict of interest.

References

1. Kumar, P.; Dutta, K.; Das, S.; Kundu, P.P. An overview of unsolved deficiencies of direct methanol fuel cell technology: Factors and parameters affecting its widespread use. *Int. J. Energy Res.* **2014**, *38*, 1367–1390.

2. Li, Z.S.; Ji, S.; Pollet, B.G.; Shen, P.K. Supported 3-D Pt nanostructures: The straightforward synthesis and enhanced electrochemical performance for methanol oxidation in an acidic medium. *J. Nanopart. Res.* **2013**, *15*, 1959.

3. Singh, R.N.; Awasthi, R.; Sharma, C.S. An Overview of Recent Development of Platinum-Based Cathode Materials for Direct Methanol Fuel Cells. *Int. J. Electrochem. Sci.* **2014**, *9*, 5607–5639.

4. Yuan, T.; Yang, J.; Wang, Y.; Ding, H.; Li, X.; Liu, L.; Yang, H. Anodic diffusion layer with graphene-carbon nanotubes composite material for passive direct methanol fuel cell. *Electrochim. Acta* **2014**, *147*, 265–270.

5. Huang, Y.Q.; Huang, H.L.; Gao, Q.Z.; Gan, C.F.; Liu, Y.G.; Fang, Y.P. Electroless synthesis of two-dimensional sandwich-like Pt/Mn$_3$O$_4$/reduced-graphene-oxide nanocomposites with enhanced electrochemical performance for methanol oxidation. *Electrochim. Acta* **2014**, *149*, 34–41.

6. Liu, J.Q.; Liu, Z.; Barrow, C.J.; Yang, W.R. Molecularly engineered graphene surfaces for sensing applications: A review. *Anal. Chim. Acta* **2015**, *859*, 1–19.

7. Wang, X.; Liu, B.; Lu, Q.P.; Qu, Q.S. Graphene-based materials: Fabrication and application for adsorption in analytical chemistry. *J. Chromatogr. A* **2014**, *1362*, 1–15.

8. Khadempir, S.; Ahmadpour, A.; Ashraf, N.; Bamoharram, F.F.; Mitchell, S.G.; de la Fuente, J.M. A polyoxometalate-assisted approach for synthesis of Pd nanoparticles on graphene nanosheets: Synergistic behaviour for enhanced electrocatalytic activity. *RSC Adv.* **2015**, *5*, 24319–24326.

9. Seo, M.H.; Choi, S.M.; Kim, H.J.; Kim, J.H.; Cho, B.K.; Kim, W.B. A polyoxometalate-deposited Pt/CNT electrocatalyst via chemical synthesis for methanol electrooxidation. *J. Power Sources* **2008**, *179*, 81–86.

10. Zhang, X.F.; Huang, Q.F.; Li, Z.S.; Ma, A.; He, X.L.; Lin, S. Effects of silicotungstic acid on the physical stability and electrocatalytic activity of platinum nanoparticles assembled on graphene. *Mater. Res. Bull.* **2014**, *60*, 57–63.

11. Ma, A.; Zhang, X.F.; Li, Z.S.; Wang, X.Y.; Ye, L.T.; Lin, S. Graphene and Polyoxometalate Synergistically Enhance Electro-Catalysis of Pd toward Formic Acid Electro-Oxidation. *J. Electrochem. Soc.* **2014**, *161*, F1224–F1230.

12. Shao, Y.Y.; Wang, J.; Engelhard, M.; Wang, C.M.; Lin, Y.H. Facile and controllable electrochemical reduction of graphene oxide and its applications. *J. Mater. Chem.* **2010**, *20*, 743–748.

13. Ping, J.F.; Wang, Y.X.; Fan, K.; Wu, J.; Ying, Y.B. Direct electrochemical reduction of graphene oxide on ionic liquid doped screen-printed electrode and its electrochemical biosensing application. *Biosens. Bioelectron.* **2011**, *28*, 204–209.

14. Guo, H.L.; Wang, X.F.; Qian, Q.Y.; Wang, F.B.; Xia, X.H. A Green Approach to the Synthesis of Graphene Nanosheets. *ACS Nano* **2009**, *3*, 2653–2659.

15. Wang, Z.J.; Wu, S.X.; Zhang, J.; Chen, P.; Yang, G.C.; Zhou, X.Z.; Zhang, Q.C.; Yan, Q.Y.; Zhang, H. Comparative studies on single-layer reduced graphene oxide films obtained by electrochemical reduction and hydrazine vapor reduction. *Nanoscale Res. Lett.* **2012**, *7*, 1–7.

16. Sun, Z.Y.; He, J.B.; Kumbhar, A.; Fang, J.Y. Nonaqueous Synthesis and Photoluminescence of ITO Nanoparticles. *Langmuir* **2010**, *26*, 4246–4250.

17. Wang, S.; Li, H.L.; Li, S.; Liu, F.; Wu, L.X. Electrochemical-Reduction-Assisted Assembly of a Polyoxometalate/Graphene Nanocomposite and Its Enhanced Lithium-Storage Performance. *Chem. Eur. J.* **2013**, *19*, 10895–10902.

18. Li, H.L.; Pang, S.P.; Wu, S.; Feng, X.L.; Müllen, K.; Bubeck, C. Layer-by-Layer Assembly and UV Photoreduction of Graphene Polyoxometalate Composite Films for Electronics. *J. Am. Chem. Soc.* **2011**, *133*, 9423–9429.

19. Liu, R.J.; Li, S.W.; Yu, X.L.; Zhang, G.J.; Zhang, S.J.; Yao, J.N.; Zhi, L.J. A general green strategy for fabricating metal nanoparticles/polyoxometalate/graphene tri-component nanohybrids: Enhanced electrocatalytic properties. *J. Mater. Chem.* **2012**, *22*, 3319–3322.

20. Han, D.M.; Guo, Z.P.; Zeng, R.; Kim, C.J.; Meng, Y.Z.; Liu, H.K. Multiwalled carbon nanotube-supported Pt/Sn and Pt/Sn/PMo$_{12}$ electrocatalysts for methanol electro-oxidation. *Int. J. Hydrogen Energy* **2009**, *34*, 2426–2434.

21. Roth, C.; Goetz, M.; Fuess, H. Synthesis and characterization of carbon-supported Pt–Ru–WO$_x$ catalysts by spectroscopic and diffraction methods. *J. Appl. Electrochem.* **2001**, *31*, 793–798.

22. Liu, Z.L.; Guo, B.; Hong, L.; Lim, T.H. Microwave heated polyol synthesis of carbon-supported PtSn nanoparticles for methanol electrooxidation. *Electrochem. Commun.* **2005**, *8*, 83–90.

23. Xin, Y.; Liu, J.G.; Zhou, Y.; Liu, W.; Gao, J.; Xie, Y.; Zou, Z. Preparation and characterization of Pt supported on graphene with enhanced electrocatalytic activity in fuel cell. *J. Power Sources* **2011**, *196*, 1012–1018.

24. Bhattacharyya, K.; Majeed, J.; Dey, K.K.; Ayyub, P.; Tyagi, A.K.; Bharadwaj, S.R. Effect of Mo-Incorporation in the TiO$_2$ Lattice: A Mechanistic Basis for Photocatalytic Dye Degradation. *J. Phys. Chem. C* **2014**, *118*, 15946–15962.

25. Ji, Z.Y.; Shen, X.P.; Zhu, G.X.; Chen, K.M.; Fu, G.H.; Tong, L. Enhanced electrocatalytic performance of Pt-based nanoparticles on reduced graphene oxide for methanol oxidation. *J. Electroanal. Chem.* **2012**, *682*, 95–100.

26. Ding, Y.H.; Zhang, P.; Zhuo, Q.; Ren, H.M.; Yang, Z.M.; Jiang, Y. A green approach to the synthesis of reduced graphene oxide nanosheets under UV irradiation. *Nanotechnology* **2011**, *22*, 215601–215605.

27. Wang, C.Q.; Jiang, F.X.; Yue, R.R.; Wang, H.W.; Du, Y.K. Enhanced photo-electrocatalytic performance of Pt/RGO/TiO$_2$ on carbon fiber towards methanol oxidation in alkaline media. *J. Solid State Electrochem.* **2014**, *18*, 515–522.

28. Kang, Z.H.; Wang, Y.B.; Wang, E.B.; Lian, S.Y.; Gao, L.; You, W.S.; Hu, C.W.; Xu, L. Polyoxometalates nanoparticles: Synthesis, characterization and carbon nanotube modification. *Solid State Commun.* **2004**, *29*, 559–564.

29. Dios, M.; Salgueirino, V.; Perez-Lorenzo, M.; Correa-Duarte, M.A. Synthesis of Carbon Nanotube-Inorganic Hybrid Nanocomposites: An Instructional Experiment in Nanomaterials Chemistry. *J. Chem. Educ.* **2012**, *89*, 280–283.

30. Ye, L.T.; Li, Z.S.; Zhang, L.; Lei, F.L.; Lin, S. A green one-pot synthesis of Pt/TiO$_2$/Graphene composites and its electro-photo-synergistic catalytic properties for methanol oxidation. *J. Colloid Interface Sci.* **2014**, *433*, 156–162.

31. Kim, M.S.; Fang, B.Z.; Chaudhari, N.K.; Song, M.Y.; Bae, T.S.; Yu, J.S. A highly efficient synthesis approach of supported Pt-Ru catalyst for direct methanol fuel cell. *Electrochim. Acta* **2010**, *55*, 4543–4550.

32. Zhao, X.; Zhu, J.B.; Liang, L.; Liu, C.P.; Liao, J.H. Enhanced electroactivity of Pd nanocrystals supported on H$_3$PMo$_{12}$O$_{40}$/carbon for formic acid electrooxidation. *J. Power Sources* **2012**, *210*, 392–396.

33. Zeng, Q.; Cheng, J.S.; Tang, L.H.; Liu, X.F.; Liu, Y.Z.; Li, J.H.; Jiang, J.H. Self-Assembled Graphene-Enzyme Hierarchical Nanostructures for Electrochemical Biosensing. *Adv. Funct. Mater.* **2010**, *20*, 3366–3372.

34. Hummers, W.S., Jr.; Offeman, R.E. Preparation of graphitic oxide. *J. Am. Chem. Soc.* **1958**, *80*, 1339.

Copolymers Based on Indole-6-Carboxylic Acid and 3,4-Ethylenedioxythiophene as Platinum Catalyst Support for Methanol Oxidation

Tzi-Yi Wu, Chung-Wen Kuo, Yu-Lun Chen and Jeng-Kuei Chang

Abstract: Indole-6-carboxylic acid (ICA) and 3,4-ethylenedioxythiophene (EDOT) are copolymerized electrochemically on a stainless steel (SS) electrode to obtain poly(indole-6-carboxylic acid-co-3,4-ethylenedioxythiophene)s (P(ICA-co-EDOT))s. The morphology of P(ICA-co-EDOT)s is checked using scanning electron microscopy (SEM), and the SEM images reveal that these films are composed of highly porous fibers when the feed molar ratio of ICA/EDOT is greater than 3/2. Platinum particles can be electrochemically deposited into the P(ICA-co-EDOT)s and PICA films to obtain P(ICA-co-EDOT)s-Pt and PICA-Pt composite electrodes, respectively. These composite electrodes are further characterized using X-ray photoelectron spectroscopy (XPS), SEM, X-ray diffraction analysis (XRD), and cyclic voltammetry (CV). The SEM result indicates that Pt particles disperse more uniformly into the highly porous P(ICA3-co-EDOT2) fibers (feed molar ratio of ICA/EDOT = 3/2). The P(ICA3-co-EDOT2)-Pt nanocomposite electrode exhibited excellent catalytic activity for the electrooxidation of methanol in these electrodes, which reveals that P(ICA3-co-EDOT2)-Pt nanocomposite electrodes are more promising for application in an electrocatalyst as a support material.

Reprinted from *Catalysts*. Cite as: Wu, T.-Y.; Kuo, C.-W.; Chen, Y.-L.; Chang, J.-K. Copolymers Based on Indole-6-Carboxylic Acid and 3,4-Ethylenedioxythiophene as Platinum Catalyst Support for Methanol Oxidation. *Catalysts* **2015**, *5*, 1657–1672.

1. Introduction

Electrochemical oxidation of methanol has been widely studied in the last decades due to their application for electrochemical energy conversion in direct methanol fuel cells (DMFC) [1–3]. The DMFC is considered a highly promising power source as an alternative to conventional energy converting devices due to its high power density, high energy-conversion efficiency, low emission of pollutants, and good fuel availability [4,5]. Despite the many efforts devoted to DMFC development, the usefulness of DMFCs is limited by the requirement of expensive platinum electrocatalyst for the methanol oxidation; platinum reserves on earth are limited, and the electrocatalytic efficiency is restricted by poisonous CO species on the Pt surface. These reasons limit the development and commercialization of DMFCs.

One of the practical solutions has been to decrease the use of pure platinum and enhance the catalytic efficiency of the entire catalyst (Pt and support) for methanol oxidation, such as with Pt-based bimetallic catalysts (Pt-Ru, Pt-Cu, Pt-Pd, Pt-WO$_3$, Pt-Co, Pt-Sn, Pt-Pb, Pt-Rh, and Pt-Au) [6–10]. Bimetallic catalysts are composed of two distinct metal elements, which significantly reduce the over-potential of methanol oxidation and offer considerable improvement in the catalytic properties relative to the individual metals. Moreover, the Pt-based bimetallic catalysts decrease the high Pt cost.

In recent years, conducting polymers (CPs) have been demonstrated to be suitable host materials for dispersing metallic particles [11–14]. CP matrices with porous structures, high accessible surface area, low chemical resistance and high stability are attractive and favorable supports for incorporation of the catalyst particles [15–17]. Moreover, this support structure avoids the agglomeration and reduces the Pt loading under the condition of keeping high catalytic activity.

Nowadays, the most common conducting polymers, such as polyaniline (PANI) [18], polypyrrole (PPy) [19], polythiophene (PTh) [20], polycarbazole [21], polyindole (PIn) [22], and their derivatives, have been successfully used as catalyst supports for methanol oxidation. Among many promising CPs, polypyrrole has the advantages of good electrical conductivity and ease of anodic electrodeposition of freestanding polypyrrole films. Indole has both a benzene ring and a pyrrole ring, the incorporation of a benzene unit link to the indole unit increases the chemical stability of polyindole. Poly(3,4-ethylenedioxythiophene) (PEDOT) is an important polythiophene derivative with two electron-donating oxygen atoms on 3,4-positions of thiophene PEDOT has good chemical and electrochemical properties in comparison with other kinds of polythiophene derivatives.

Copolymerization is an easy, facile method for preparing a specific polymer with different properties than those of their corresponding homopolymers. Synthesis of conjugated copolymer involves chemical and electrochemical polymerization [23,24]. Electrochemical copolymerization can be carried out at room temperature and homogeneous copolymer films can be formed directly at the electrode surface. During the past few years, copolymers have received increasing attention because they allow the preservation of homopolymers' properties and display specific electrochemical and physicochemical properties [25]. In the present work, poly(indole-6-carboxylic acid)-based homopolymers and copolymers are used as catalyst supports, and carboxylic acid groups are incorporated into the polymer backbone to help the uptake of Pt^{4+} ions and prevent the aggregation of Pt particles. P(ICA-co-EDOT)s are prepared on the stainless steel (SS) electrode using the electrochemical copolymerization of indole-6-carboxylic acid (ICA) and EDOT, and the feed molar percentage of ICA/(ICA+EDOT) is 100, 80, 60, and 40%. The characteristics of deposited homopolymer and copolymer films are characterized by

Fourier transform infrared spectroscopy (FT-IR) and scanning electron microscopy (SEM). Platinum particles were deposited onto the homopolymer and copolymer films using H_2PtCl_6 as the precursor to prepare PICA-Pt, P(ICA4-*co*-EDOT1)-Pt, P(ICA3-*co*-EDOT2)-Pt, and P(ICA2-*co*-EDOT3)-Pt composite catalyst supports, and their morphology and platinum particle size were investigated by SEM and X-ray diffraction (XRD), respectively. The electrochemical surface areas, electrocatalytic properties, and long-time stability toward methanol oxidation of the as-prepared composite catalyst were obtained by implementing cyclic voltammetry and chronoamperometry measurements in 0.5 M methanol +0.5 M H_2SO_4 solution. The present work focuses on the preparation of homopolymer and copolymer films using various feed molar percentages of ICA/EDOT and investigates the electrocatalytic activities of their Pt catalyst supports towards methanol oxidation.

2. Results and Discussion

2.1. Electrochemical Polymerization and Characterizations

The anodic polarization curves of 0.02 M EDOT and 0.02 M ICA in an acetonitrile (ACN) solution containing 0.1 M $LiClO_4$ as the supporting electrolytes are shown in Figure 1. The onset oxidation potential (E_{onset}) of EDOT and ICA in the solution is about +0.77 and +0.83 V (*vs.* Ag/AgCl), respectively.

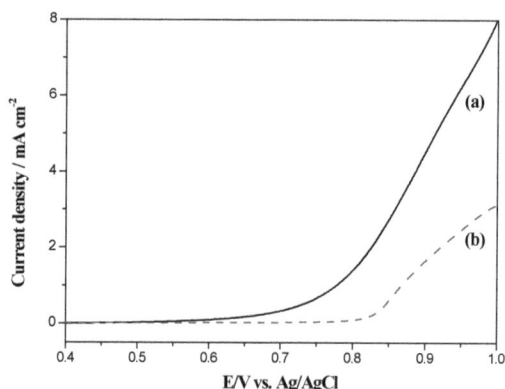

Figure 1. Anodic polarization curves of (**a**) 0.02 M EDOT and (**b**) 0.02 M ICA in ACN + 0.1 M $LiClO_4$. Scanning rates: 50 mV s^{-1}.

It is well known that successful electrochemical copolymerization of different monomers is due to the fact that the E_{onset} values of the monomers are close to each other [26]. The difference of the onset oxidation potential between EDOT and ICA monomers is 0.06 V, implying that the electrochemical copolymerization may happen at the proper potential [27]. The schematic illustration for the formation of P(ICA-*co*-EDOT) is shown in Figure 2.

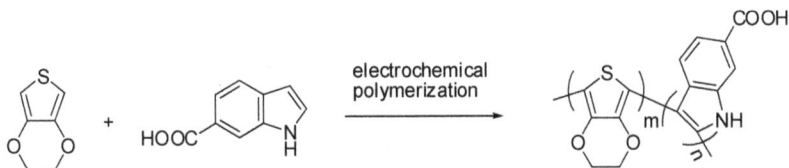

Figure 2. Schematic illustration for the formation of P(ICA-*co*-EDOT).

Figure 3a–d shows the FT-IR spectra of PICA, P(ICA4-*co*-EDOT1), P(ICA3-*co*-EDOT2), and P(ICA2-*co*-EDOT3) films. The PICA (curve a) exhibits some peaks, the –C–N stretching vibration of PICA is observed clearly at 1295 cm^{-1}, and the absorption peak at 1615 cm^{-1} is attributed to C=C stretching. The characteristic –C=O groups of PICA are observed at 1695 cm^{-1} [24]. These characteristic peaks of PICA can also be seen in copolymers (curve b–d). Compare with PICA homopolymer, the absorption peaks of the C-S bond in P(ICA4-*co*-EDOT1), P(ICA3-*co*-EDOT2), and P(ICA2-*co*-EDOT3) films can be observed at 860 and 698 cm^{-1} [28]. This implies that the 3,4-ethylenedioxythiophene units are incorporated into the copolymer chain. The absorption peaks of PICA are located at 1695 cm^{-1}, whereas the absorption peaks of these copolymers shifted to short wavenumber position, indicating the formation of P(ICA4-co-EDOT1), P(ICA3-*co*-EDOT2), and P(ICA2-*co*-EDOT3) copolymers by the electrochemical polymerization. For instance, the characteristic peak shifts to 1680 cm^{-1} (curve d) when the feed molar ratio of ICA/EDOT is 2/3.

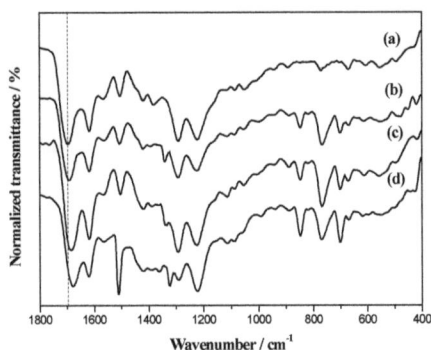

Figure 3. FT-IR spectra of (**a**) PICA; (**b**) P(ICA4-*co*-EDOT1); (**c**) P(ICA3-*co*-EDOT2); and (**d**) P(ICA2-*co*-EDOT3).

XPS is a quantitative spectroscopic method for the element analysis. Figure 4a shows the survey scan of PICA-Pt, P(ICA4-*co*-EDOT1)-Pt, P(ICA3-*co*-EDOT2)-Pt, and P(ICA2-*co*-EDOT3)-Pt. Signals of C_{1s}, N_{1s}, O_{1s}, Pt_{4f}, and S_{2p} can be seen from Figure 4a. The Pt core-level spectra of E-Pt (Pt particles were deposited onto the stainless steel electrode), PICA-Pt, P(ICA4-*co*-EDOT1)-Pt, P(ICA3-*co*-EDOT2)-Pt, and P(ICA2-*co*-EDOT3)-Pt are shown in Figure 4b. The intensive Pt_{4f} binding energy peaks appeared at 71.3 and 74.7 eV are metallic Pt [29]. The depositions of Pt particle on PICA, P(ICA4-*co*-EDOT1), P(ICA3-*co*-EDOT2), and P(ICA2-*co*-EDOT3) films influence the N_{1s} orbital of polymers and Pt_{4f} orbital of Pt, this could change the Pt_{4f} energy level for the PICA-Pt, P(ICA4-*co*-EDOT1)-Pt, P(ICA3-*co*-EDOT2)-Pt, and P(ICA2-*co*-EDOT3)-Pt electrodes. For instance, the Pt_{4f} peaks of E-Pt (71.3 and 74.7 eV) shifted to 70.5 and 73.9 eV (P(ICA3-*co*-EDOT2)-Pt (curve IV in Figure 4b), respectively.

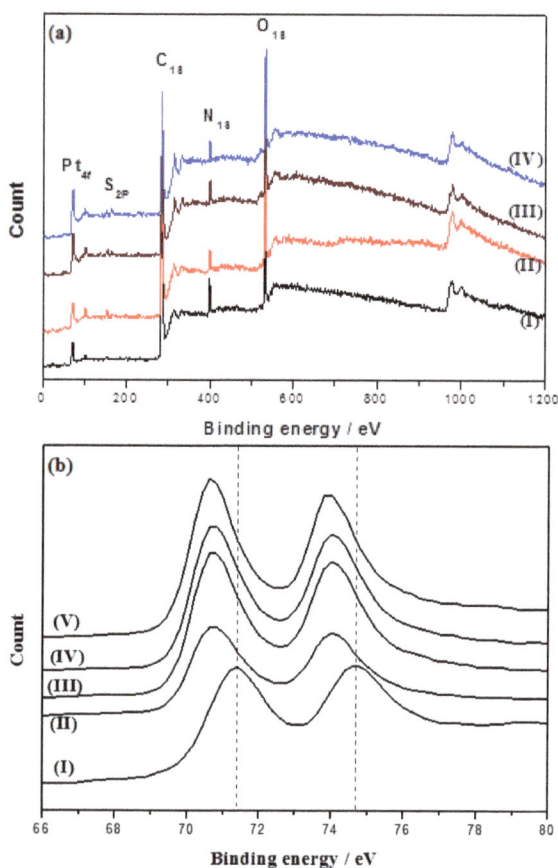

Figure 4. (a) XPS spectra of the survey scan (I) PICA-Pt, (II) P(ICA4-*co*-EDOT1)-Pt, (III) P(ICA3-*co*-EDOT2)-Pt, and (IV) P(ICA2-*co*-EDOT3)-Pt; (b) Pt_{4f} XPS core-level spectra of (I) E-Pt, (II) PICA-Pt, (III) P(ICA4-*co*-EDOT1)-Pt, (IV) P(ICA3-*co*-EDOT2)-Pt, and (V) P(ICA2-*co*-EDOT3)-Pt.

2.2. Surface Morphology

Figure 5a–h shows the scanning electron microscopy (SEM) analysis of surface morphology of PICA, PICA-Pt, P(ICA4-*co*-EDOT1), P(ICA4-*co*-EDOT1)-Pt, P(ICA3-*co*-EDOT2), P(ICA3-*co*-EDOT2)-Pt, P(ICA2-*co*-EDOT3), and P(ICA2-*co*-EDOT3)-Pt composite electrodes. The SEM images reveal that these films without Pt are composed of highly porous fibers when the feed molar ratio of ICA/EDOT is greater than 3/2, and the fibers have an average diameter of 50–200 nm.

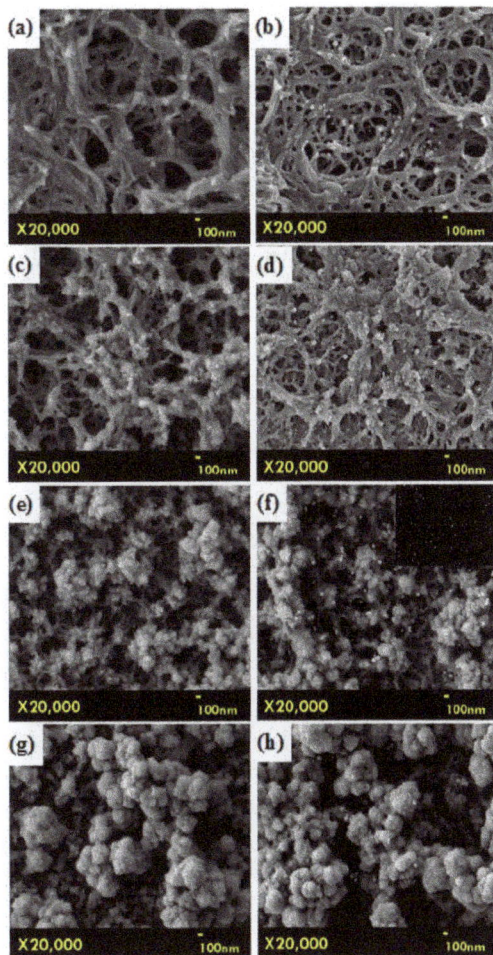

Figure 5. SEM images of (**a**) PICA; (**b**) PICA-Pt; (**c**) P(ICA4-*co*-EDOT1); (**d**) P(ICA4-*co*-EDOT1)-Pt; (**e**) P(ICA3-*co*-EDOT2); (**f**) P(ICA3-*co*-EDOT2)-Pt; (**g**) P(ICA2-*co*-EDOT3); and (**h**) P(ICA2-*co*-EDOT3)-Pt composite films; inset in (**f**) is X-ray map (bright spots indicate Pt).

The fibers with cloud shape morphology of P(ICA3-*co*-EDOT2) film can be clearly seen in Figure 5e. However, P(ICA2-*co*-EDOT3) films show aggregative cloud shape morphology in Figure 5g, implying excess EDOT feed ratio gives rise to the formation of cloud shape structures. The fiber morphology of these films provides a large surface area for the subsequent deposition of Pt particles. Pt particles were incorporated into these films by electrochemical deposition at a constant potential of −0.2 V (*vs.* Ag/AgCl).

As shown in Figure 5b,d,f,h, the incorporation of Pt into PICA, P(ICA4-*co*-EDOT1), P(ICA3-*co*-EDOT2), P(ICA2-*co*-EDOT3) did not alter their morphologies. Pt particles can be clearly seen on these composite electrodes from SEM images. Pt particles with a size of about 30–100 nm can be seen on the PICA-Pt electrode, whereas Pt particles are about 30–60 nm on the P(ICA4-*co*-EDOT1)-Pt and P(ICA2-*co*-EDOT3)-Pt electrodes. The particle size of Pt (20–50 nm) for the P(ICA3-*co*-EDOT2)-Pt electrode is smaller than those for other three electrodes, which indicates that the P(ICA3-*co*-EDOT2)-Pt electrode presents a higher active surface area. The uniform distribution of Pt in the P(ICA3-*co*-EDOT2) spatial network structure may increase the utilization of Pt for methanol oxidation. The inset in Figure 5f shows the result of Pt in a P(ICA3-*co*-EDOT2)-Pt composite electrode. The bright spots indicate the existence of platinum in the electrode.

2.3. XRD Patterns

The crystalline structure of Pt particles incorporated into PICA, P(ICA4-co-EDOT1), P(ICA3-*co*-EDOT2), and P(ICA2-*co*-EDOT3) composite electrodes are examined using XRD analysis, and the XRD patterns are displayed in Figure 6. These electrodes show intensive peaks at $2\theta = 43°$ and $51°$, indicating the diffraction peaks of the SS electrode. In addition, the characteristic diffraction peaks of face-centered cubic (fcc) platinum for the four electrodes are observed at $40°$, $46°$, and $68°$, corresponding to Pt(111), Pt(200), and Pt(220) planes, respectively [8]. Because the Pt(111) peak is isolated with the diffraction peaks of PICA, P(ICA4-*co*-EDOT1), P(ICA3-*co*-EDOT2), and P(ICA2-*co*-EDOT3) composite electrodes, the average size of Pt particles can be calculated from this peak according to Scherrer's formula, [30].

$$d = \frac{0.9\lambda}{\beta\cos\theta} \tag{1}$$

where d is the average size of the Pt particles, λ is the X-ray wavelength (Cu Kα $\lambda = 1.54178$ Å), θ_{max} is the diffraction angle at the peak position, and β is the half-peak width for Pt(111) in radians. The average sizes of Pt particles for PICA-Pt, P(ICA4-*co*-EDOT1)-Pt, P(ICA3-*co*-EDOT2)-Pt, and P(ICA2-*co*-EDOT3)-Pt calculated using the Scherrer's equation were 10, 8, 8, and 9 nm, respectively. The calculated size of Pt particles from Scherrer equation is smaller than the average diameter measured by SEM, which can be attributed to the fact that the particle size estimated from XRD and the Scherrer equation is the primary particle size. However, the particle size observed from SEM results is the secondary particle size (or aggregated particle size).

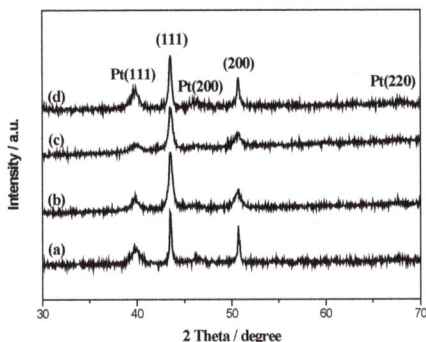

Figure 6. XRD patterns of (**a**) PICA-Pt; (**b**) P(ICA4-*co*-EDOT1)-Pt; (**c**) P(ICA3-*co*-EDOT2)-Pt; and (**d**) P(ICA2-*co*-EDOT3)-Pt.

2.4. Electrocatalytic Activity of Electrodes for Methanol Oxidation

Figure 7a–d show the incorporation of Pt particles into in P(Id3-*co*-Ed2) films via electrochemical deposition at a constant potential of −0.2 V from 0.5 M CH_3OH + 0.5 M H_2SO_4 solution with various deposition charges of 0.15, 0.20, 0.25, and 0.30 C. The maximum anodic peak current density (I_{pa}) for the oxidation of methanol observed at the P(ICA3-*co*-EDOT2) electrode (deposition charge of 0.20 C) is 58 mA·cm^{-2}·mg^{-1}, which is higher than those of other electrodes (deposition charges of 0.15, 0.25, and 0.30 C).

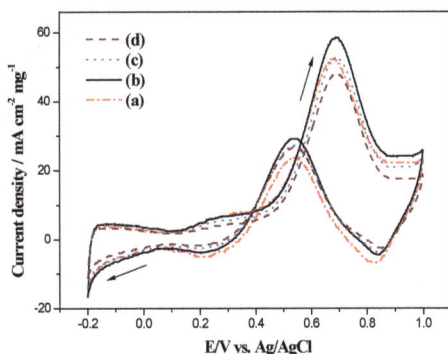

Figure 7. Cyclic voltammograms of P(ICA3-*co*-EDOT2) with various Pt deposition charges of (**a**) 0.15; (**b**) 0.20; (**c**) 0.25; and (**d**) 0.30 C in 0.5 M CH_3OH + 0.5 M H_2SO_4 solution, scan rate = 50 mV·s^{-1}.

567

Cyclic voltammograms (CVs) of PICA-Pt, P(ICA4-*co*-EDOT1)-Pt, P(ICA3-*co*-EDOT2)-Pt, and P(ICA2-*co*-EDOT3)-Pt electrodes under the same Pt deposition charge of 0.20 C recorded with a scan rate of 50 mV·s^{-1} in 0.5 M H$_2$SO$_4$ are presented in Figure 8. These composite electrodes show obvious hydrogen adsorption/desorption differences in the range of 0.1 to −0.2 V. It is known that the integrated area represents the number of Pt sites available for hydrogen adsorption and desorption. Among these composite electrodes, P(ICA3-*co*-EDOT2)-Pt film shows the largest charge for hydrogen adsorption and desorption, and this may be ascribed to the uniform dispersion of Pt particles in P(ICA3-*co*-EDOT2) spatial network structure. For example, the charge for hydrogen absorption and desorption on the P(ICA3-*co*-EDOT2)-Pt film is 71.4 mC·cm^{-2}·mg^{-1}, which is 1.5 times larger than that on the PICA-Pt surface (48.9 mC·cm^{-2}·mg^{-1}). The electrochemical surface area (ESA) of Pt can be calculated from the area of hydrogen adsorption-desorption peaks using the following equation [31],

$$ESA = \frac{Q_H}{0.21\,[Pt]} \tag{2}$$

where Q_H (mC·cm^{-2}) represents the mean value between the amounts of charge exchanged during the electro-adsorption (Q_1) and desorption (Q_2) of H$_2$ on Pt sites, [Pt] is the Pt loading (mg·cm^{-2}) on the electrode, and 0.21 (mC·cm^{-2}) represents the charge required to oxidize a monolayer of H$_2$ on clean Pt. The contribution from the double layer capacitance is deduced while calculating the ESA. As shown in Table 1, the ESA values of Pt supported on the PICA-Pt, P(ICA4-*co*-EDOT1)-Pt, P(ICA3-*co*-EDOT2)-Pt, and P(ICA2-*co*-EDOT3)-Pt electrodes are calculated to be 230, 286, 340, and 283 cm^2·mg^{-1}, respectively.

Table 1. The ESA values of Pt supported on the electrodes.

Electrodes	Q_H (mC cm^{-2}·mg^{-1})	ESA (cm^2·mg^{-1})
PICA-Pt	48.9	230
P(ICA4-*co*-EDOT1)-Pt	60.1	286
P(ICA3-*co*-EDOT2)-Pt	71.4	340
P(ICA2-*co*-EDOT3)-Pt	60.0	283

The Pt particles are loaded onto PICA, P(ICA4-*co*-EDOT1), P(ICA3-*co*-EDOT2), P(ICA2-*co*-EDOT3)-Pt, and PEDOT-Pt composite electrodes under the same Pt deposition charge of 0.20 C and tested for their electrocatalytic activity of methanol oxidation by cyclic voltammetry. As shown in Figure 9, the current for the methanol oxidation increases slowly below 0.5 V in the forward sweep for these electrodes, which can be ascribed to the formation of reaction intermediates. The current increases quickly and reaches a peak at around 0.7 V, and this can be attributed to the partial oxidation of Pt surface, which helps the transformation of intermediates to carbon dioxide.

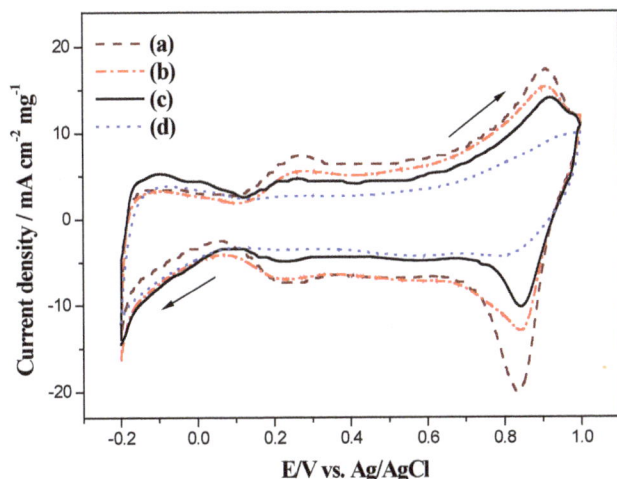

Figure 8. Cyclic voltammograms of (**a**) PICA-Pt; (**b**) P(ICA4-*co*-EDOT1)-Pt; (**c**) P(ICA3-*co*-EDOT2)-Pt; and (**d**) P(ICA2-*co*-EDOT3)-Pt in 0.5 M H_2SO_4 solution, the deposition charge of Pt is 0.20 C, scan rate = 50 mV·s^{-1}.

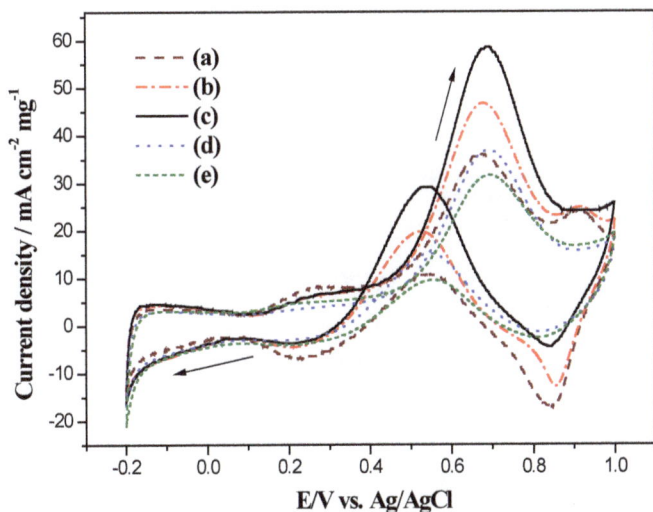

Figure 9. Cyclic voltammograms of (**a**) PICA-Pt; (**b**) P(ICA4-*co*-EDOT1)-Pt; (**c**) P(ICA3-*co*-EDOT2)-Pt; (**d**) P(ICA2-*co*-EDOT3)-Pt; and (**e**) PEDOT-Pt in 0.5 M CH_3OH + 0.5 M H_2SO_4 solution; the deposition charge of Pt is 0.20 C, scan rate = 50 mV·s^{-1}.

The anodic peak current density for the methanol oxidation of PICA-Pt and P(ICA2-*co*-EDOT3)-Pt electrodes (curve a and d) is about 36 mA·cm^{-2}·mg^{-1}, which is lower than P(ICA4-*co*-EDOT1)-Pt and P(ICA3-*co*-EDOT2)-Pt electrodes. Comparing the CV results of these electrodes, the highest oxidation current (I_{pa}) toward methanol oxidation is observed for the P(ICA3-*co*-EDOT2)-Pt electrode, confirming the crucial effect of P(ICA3-*co*-EDOT2)-Pt on the enhancement of platinum particle efficiency towards the catalytic oxidation of methanol. The highest current density for the P(ICA3-*co*-EDOT2)-Pt electrode may be due to suitable amounts of CO_2^- groups helping with the uptake of Pt^{4+} ions. A uniform distribution of Pt particles into P(ICA3-*co*-EDOT2) can twist to form a spatial 3D matrix. As shown in Table 2, the methanol oxidation on P(ICA3-*co*-EDOT2)-Pt electrode shows higher current density (I_{pa}) than those reported for PANI-300PSS-Pt [32], Pt/PANI/MGCE [33], Pt/Nano-PDAN/MGCE [34], and PANI-PSS-Pt [11]. However, P(ICA3-*co*-EDOT2)-Pt electrode shows lower I_{pa} than that reported for nanotube-Pt [35].

Table 2. Comparisons of the methanol oxidation data in H_2SO_4 solution at the P(ICA3-*co*-EDOT2)-Pt composite electrode with some modified electrodes.

Electrodes	$C_{H2SO4}/C_{Methanol}$ (M/M)	ν (mV·s^{-1})	E_{onset} (V) [c]	I_{pa} (mA·mg^{-1})	Ref.
PANI-300PSS-Pt [b]	0.5/0.1	10	0.4	19	[32]
Pt/PANI/MGCE [a]	0.5/0.5	5	0.3	32	[33]
Pt/Nano-PDAN/MGCE [a]	0.5/2.4	50	0.2	28	[34]
PANI-PSS-Pt [b]	0.5/1.0	10	0.4	31	[11]
nanotube-Pt [b]	0.5/1.0	50	-	141	[35]
P(ICA3-*co*-EDOT2)-Pt [b]	0.5/0.5	50	0.4	58	This work

[a] The potentials were referred to SCE; [b] The potentials were referred to Ag/AgCl; [c] The methanol oxidation onset potential.

The effect of methanol concentration on the electrocatalytic activity of these composite electrodes is examined (Figure 10). It can be clearly observed that the anodic current increases with increasing methanol concentration and levels off at concentrations higher than 1.5 M. This effect can be attributed to the saturation of active sites at the surface of the electrode, and the optimum concentration of methanol for a higher current density may be considered to be about 1.5 M.

2.5. Electrocatalytic Long-Term Stability of Electrodes for Methanol Oxidation

The performance of four electrodes towards the methanol oxidation reaction after long term operation was tested using chronoamperometry. Figure 11 shows the current–time responses of PICA-Pt, P(ICA4-*co*-EDOT1)-Pt, P(ICA3-*co*-EDOT2)-Pt, and P(ICA2-*co*-EDOT3)-Pt electrodes recorded at 0.6 V in

0.5 M CH$_3$OH + 0.5 M H$_2$SO$_4$ solution for 5 h. After long term operation, a steady state current density is achieved. It can also be observed that the electrocatalyst of P(ICA3-co-EDOT2)-Pt maintained the highest current density in these electrodes. The methanol oxidation currents at 3 h follow this order: P(ICA3-*co*-EDOT2)-Pt (6.8 mA·cm^{-2}·mg^{-1}) > P(ICA4-*co*-EDOT1)-Pt (4.8 mA·cm^{-2}·mg^{-1}) > P(ICA2-*co*-EDOT3)-Pt (3.2 mA·cm^{-2}·mg^{-1}) > PICA-Pt (2.6 mA·cm^{-2}·mg^{-1}).

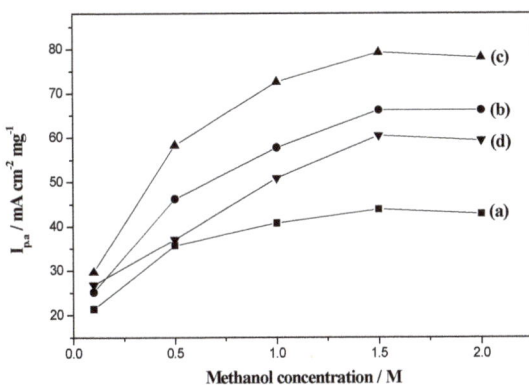

Figure 10. Plot of anodic peak current as a methanol concentration for (**a**) PICA-Pt; (**b**) P(ICA4-*co*-EDOT1)-Pt; (**c**) P(ICA3-*co*-EDOT2)-Pt; and (**d**) P(ICA2-*co*-EDOT3)-Pt. Supporting electrolyte: 0.5 M H$_2$SO$_4$. Scan rate: 50 mV·s^{-1}.

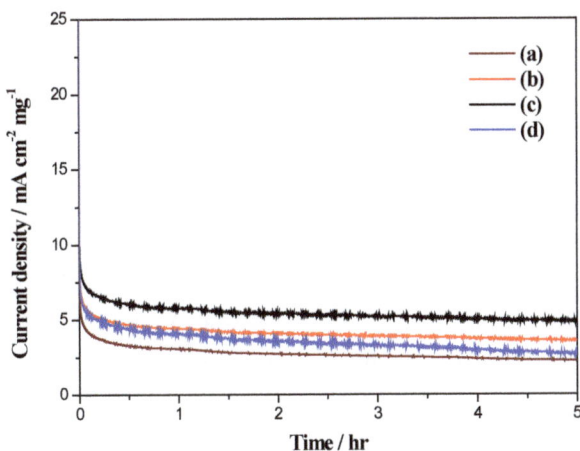

Figure 11. Chronoamperometric responses of (**a**) PICA-Pt; (**b**) P(ICA4-*co*-EDOT1)-Pt; (**c**) P(ICA3-*co*-EDOT2)-Pt; and (**d**) P(ICA2-*co*-EDOT3)-Pt at 0.6 V (*vs.* Ag/AgCl) in 0.5 M CH$_3$OH + 0.5 M H$_2$SO$_4$ solution.

3. Experimental Section

3.1. Preparation of PICA, P(ICA4-co-EDOT1), P(ICA3-co-EDOT2), and P(ICA2-co-EDOT3) Films

Before electrochemical studies, stainless steel (SS) substrates were cleaned in an ultrasonic bath using detergent, isopropanol, and deionized water, and all solutions were degassed with N_2. The PICA, P(ICA4-co-EDOT1), P(ICA3-co-EDOT2), and P(ICA2-co-EDOT3) films were deposited from a 20 mM ICA, 16 mM ICA + 4 mM EDOT, 12 mM ICA + 8 mM EDOT, and 8 mM ICA + 12 mM EDOT in a 0.1 M LiClO$_4$/ACN solution, respectively. Electrochemical deposition was carried out potentiostatically at 1.4 V (vs. Ag/AgCl electrode) for 0.08 C· cm^{-2}.

3.2. Deposition of Pt into PICA, P(ICA4-co-EDOT1), P(ICA3-co-EDOT2), and P(ICA2-co-EDOT3) Matrices

Pt particles were incorporated into PICA, P(ICA4-co-EDOT1), P(ICA3-co-EDOT2), and P(ICA2-co-EDOT3) polymer films via electrochemical deposition from 5 mM H_2PtCl_6 containing 0.5 M H_2SO_4 solution with a constant deposition charge of 0.20 C at -0.2 V (vs. Ag/AgCl). After the incorporation of Pt particles, the composite electrodes were washed with deionized water for 5 min and then dried at 120 °C for 3 min. The weights of Pt loaded into PICA, P(ICA4-co-EDOT1), P(ICA3-co-EDOT2), P(ICA2-co-EDOT3), and PEDOT films are calculated using the TGA, and they are 125, 115, 128, 118, and 121 µg· cm^{-2}, respectively. The Pt wt. % in PICA-Pt, P(ICA4-co-EDOT1)-Pt, P(ICA3-co-EDOT2)-Pt, P(ICA2-co-EDOT3)-Pt, and PEDOT-Pt catalysts are 32.8, 33.6, 34.8, 34.1, and 30.6%, respectively.

3.3. Physical and Electrochemical Characterizations

The FT-IR spectra of PICA, P(ICA4-co-EDOT1), P(ICA3-co-EDOT2), and P(ICA2-co-EDOT3) in KBr pellets were measured using a Perkin Elmer infrared spectrophotometer (Perkin Elmer, Waltham, MA, USA) with 16 scans at a resolution of 4 cm^{-1} and in the range of 400–4000 cm^{-1}. An XPS study was carried out using an ESCA 210 spectrometer (VG Scientific, Waltham, MA, USA) with Mg Kα ($h\nu = 1253.6$ eV) irradiation as the light source. The primary tension and the pressure were 12 kV and ca. 10^{-10} mbar, respectively. The surface morphologies of as-prepared electrodes were studied using a scanning electron microscope (SEM) (JEOL, Boston, MA, USA) equipped with an energy-dispersive X-ray spectroscopy (EDS) detector (JEOL, Boston, MA, USA). X-ray diffraction spectra (XRD) for the as-prepared electrodes were obtained by exposing the samples to a Bruker D8 Discover (Bruker, Billerica, MA, USA) SSS X-ray source with Cu Kα ($\lambda = 0.154$ nm) as a target, at diffraction angles (2θ) ranging from 5° to 90° with a scan rate of 4° min^{-1}.

Electrochemical characterizations of PICA-Pt, P(ICA4-*co*-EDOT1)-Pt, P(ICA3-*co*-EDOT2)-Pt, and P(ICA2-*co*-EDOT3)-Pt composite electrodes were implemented using a CHI627D electrochemical analyzer (Antec Leyden BV, Zoeterwoude, Netherlands). All experiments were carried out in a three-constituent cell. A Pt wire, SS (area = 1 cm^2), and Ag/AgCl electrode (in 3 M KCl) were used as the counter, working and reference electrodes, respectively.

3.4. Methanol Electro-Oxidation and Stability of Composite Electrodes

The catalytic activities of composite electrodes were examined by CV at 50 mV·s^{-1} in the range of −0.2 to 1.0 V. Chronoamperometric response curves were obtained at 0.6 V in 0.5 M CH$_3$OH + 0.5 M H$_2$SO$_4$ solution. All the electrochemical experiments were carried out at room temperature.

4. Conclusions

Copolymers based on ICA and EDOT were successfully synthesized by electrochemical oxidation with various feed molar ratios of ICA/EDOT in ACN solution containing 0.1 M LiClO$_4$. The existence of -CO$_2{}^-$ groups in P(ICA3-*co*-EDOT2) spatial structure assists in holding Pt^{4+} ions in the polymer matrix, resulting in the homogenous distribution of Pt in P(ICA3-*co*-EDOT2). Among the as-prepared catalysts, the P(ICA3-*co*-EDOT2)-Pt electrode exhibits the highest current density and the best stability toward methanol oxidation, demonstrating that the P(ICA3-*co*-EDOT2)-Pt composite electrode is a promising material as a catalyst for methanol oxidation.

Acknowledgments: The authors would like to thank the Ministry of Science and Technology of Republic of China for financially supporting this project under grants MOST 103-2221-E-151-051 and MOST 103-2221-E-224-058-MY3.

Author Contributions: C.-W.K. and T.-Y.W. conceived and designed the experiments; Y.-L.C. performed the experiments; C.-W.K. and J.-K.C. analyzed the data; T.-Y.W. wrote the paper.

Conflicts of Interest: The authors declare no conflict of interest.

References

1. Alegre, C.; Gálvez, M.E.; Moliner, R.; Lázaro, M.J. Influence of the synthesis method for Pt catalysts supported on highly mesoporous carbon xerogel and vulcan carbon black on the electro-oxidation of methanol. *Catalysts* **2015**, *5*, 392–405.
2. Chen, X.; Wang, H.; Wang, Y.; Bai, Q.; Gao, Y.; Zhang, Z. Synthesis and electrocatalytic performance of multi-component nanoporous PtRuCuW alloy for direct methanol fuel cells. *Catalysts* **2015**, *5*, 1003–1015.
3. Wu, T.Y.; Chen, B.K.; Chang, J.K.; Chen, P.R.; Kuo, C.W. Nanostructured poly(aniline-*co*-metanilic acid) as platinum catalyst support for electro-oxidation of methanol. *Int. J. Hydrogen Energy* **2015**, *40*, 2631–2640.

4. Zhao, Y.L.; Wang, Y.H.; Zang, J.B.; Lu, J.; Xu, X.P. A novel support of nano titania modified graphitized nanodiamond for Pt electrocatalyst in direct methanol fuel cell. *Int. J. Hydrogen Energy* **2015**, *40*, 4540–4547.

5. Kamarudina, S.K.; Achmada, F.; Daud, W.R.W. Overview on the application of direct methanol fuel cell (DMFC) for portable electronic devices. *Int. J. Hydrogen Energy* **2010**, *34*, 6902–6916.

6. Chou, H.Y.; Yeh, T.K.; Tsai, C.H. Electrodeposited Pt and PtRu Nanoparticles without hydrogen evolution reaction on mesoporous carbon for methanol oxidation. *Int. J. Electrochem. Sci.* **2014**, *9*, 5763–5775.

7. Shen, P.K.; Tseung, A.C.C. Anodic oxidation of methanol on Pt/WO$_3$ in acidic media. *J. Electrochem. Soc.* **1994**, *141*, 3082–3090.

8. Wu, T.Y.; Kuo, Z.Y.; Jow, J.J.; Kuo, C.W.; Tsai, C.J.; Chen, P.R.; Chen, H.R. Co-electrodeposition of platinum and rhodium in poly(3,4-ethylenedioxythiophene)-poly(styrene sulfonic acid) as electrocatalyst for methanol oxidation. *Int. J. Electrochem. Sci.* **2012**, *7*, 8076–8090.

9. Caballero-Manrique, G.; Brillas, E.; Centellas, F.; Garrido, J.A.; Rodríguez, R.M.; Cabot, P.-L. Electrochemical oxidation of the carbon support to synthesize Pt(Cu) and Pt-Ru(Cu) core-shell electrocatalysts for low-temperature fuel cells. *Catalysts* **2015**, *5*, 815–837.

10. Qin, H.; Qian, X.; Meng, T.; Lin, Y.; Ma, Z. Pt/MO$_x$/SiO$_2$, Pt/MO$_x$/TiO$_2$, and Pt/MO$_x$/Al$_2$O$_3$ catalysts for CO oxidation. *Catalysts* **2015**, *5*, 606–633.

11. Kuo, C.W.; Chen, B.K.; Tseng, Y.H.; Hsieh, T.H.; Ho, K.S.; Wu, T.Y.; Chen, H.R. A comparative study of poly(acrylic acid) and poly(styrenesulfonic acid) doped into polyaniline as platinum catalyst support for methanol electro-oxidation. *J. Taiwan Inst. Chem. Eng.* **2012**, *43*, 798–805.

12. Habibi, B.; Pournaghi-Azar, M.H.; Abdolmohammad-Zadeh, H.; Razmi, H. Electrocatalytic oxidation of methanol on mono and bimetallic composite films: Pt and Pt-M (M = Ru, Ir and Sn) nano-particles in poly(o-aminophenol). *Int. J. Hydrogen Energy* **2009**, *34*, 2880–2892.

13. Kuo, C.W.; Tsai, C.J.; Chen, W.P.; Chen, P.R.; Wu, T.Y.; Tseng, C.G. Nano-composite based on platinum particles and modified polyaniline for methanol, formic acid, and ethanol oxidation. *J. Chin. Chem. Soc.* **2014**, *61*, 819–826.

14. Sun, C.L.; Su, J.S.; Tang, J.H.; Lin, M.C.; Wu, J.J.; Pu, N.W.; Shi, G.N.; Ger, M.D. Investigation of the adsorption of size-selected Pt colloidal nanoparticles on high-surface-area graphene powders for methanol oxidation reaction. *J. Taiwan Inst. Chem. Eng.* **2014**, *45*, 1025–1030.

15. Maiyalagan, T.; Mahendiran, C.; Chaitanya, K.; Tyagi, R.; Nawaz Khan, F. Electro-catalytic performance of Pt-supported poly (o-phenylenediamine) microrods for methanol oxidation reaction. *Res. Chem. Intermed.* **2012**, *38*, 383–391.

16. Maiyalagan, T.; Dong, X.; Chen, P.; Wang, X. Electrodeposited Pt on three-dimensional interconnected graphene as a free-standing electrode for fuel cell application. *J. Mater. Chem.* **2012**, *22*, 5286–5290.

17. Maiyalagan, T.; Viswanathan, B. Synthesis, characterization and electrocatalytic activity of Pt supported on poly(3,4-ethylenedioxythiophene)-V$_2$O$_5$ nanocomposites electrodes for methanol oxidation. *Mater. Chem. Phys.* **2010**, *121*, 165–171.

18. Yang, C.C.; Wu, T.Y.; Chen, H.R.; Hsieh, T.H.; Ho, K.S.; Kuo, C.W. Platinum particles embedded into nanowires of polyaniline doped with poly(acrylic acid-*co*-maleic acid) as electrocatalyst for methanol oxidation. *Int. J. Electrochem. Sci.* **2011**, *6*, 1642–1654.

19. Selvaraj, V.; Alagar, M.; Hamerton, I. Electrocatalytic properties of monometallic and bimetallic nanoparticles-incorporated polypyrrole films for electro-oxidation of methanol. *J. Power Sources* **2006**, *160*, 940–948.

20. Fernandez-Blanco, C.; Ibanez, D.; Colina, A.; Ruiz, V.; Heras, A. Spectroelectrochemical study of the electrosynthesis of Pt nanoparticles/poly(3,4-(ethylenedioxythiophene) composite. *Electrochim. Acta* **2014**, *145*, 139–147.

21. Wu, T.Y.; Tsai, C.J.; Tseng, L.Y.; Chen, S.J.; Hsieh, T.H.; Kuo, C.W. Nanocomposite of platinum particles embedded into nanosheets of polycarbazole for methanol oxidation. *J. Chin. Chem. Soc.* **2014**, *61*, 860–866.

22. Nagashree, K.L.; Raviraj, N.H.; Ahmed, M.F. Carbon paste electrodes modified by Pt and Pt-Ni microparticles dispersed in polyindole film for electrocatalytic oxidation of methanol. *Electrochim. Acta* **2010**, *55*, 2629–2635.

23. Wu, T.Y.; Chen, Y. Synthesis and optical and electrochemical properties of novel copolymers containing alternate 2,3-quinoxaline and hole-transporting units. *J. Polym. Sci. A* **2002**, *40*, 4570–4580.

24. Kuo, C.W.; Hsieh, T.H.; Hsieh, C.K.; Liao, J.W.; Wu, T.Y. Electrosynthesis and characterization of four electrochromic polymers based on carbazole and indole-6-carboxylic acid and their applications in high-contrast electrochromic devices. *J. Electrochem. Soc.* **2014**, *161*, D782–D790.

25. Mikkelsen, K.; Cassidy, B.; Hofstetter, N.; Bergquist, L.; Taylor, A.; Rider, D.A. Block copolymer template synthesis of core-shell PtAu bimetallic nanocatalysts for the methanol oxidation reaction. *Chem. Mater.* **2014**, *26*, 6928–6940.

26. Kham, K.; Sadki, S.; Chevrot, C. Oxidative electropolymerizations of carbazole derivatives in the presence of bithiophene. *Synth. Met.* **2004**, *145*, 135–140.

27. Gaupp, C.L.; Reynolds, J.R. Multichromic copolymers based on 3,6-bis(2-(3,4-ethylenedioxythiophene))-*N*-alkylcarbazole derivatives. *Macromolecules* **2003**, *36*, 6305–6315.

28. Nie, T.; Leng, J.; Bai, L.; Lu, L.; Xu, J.; Zhang, K. Synthesis and characterization of benzene sulfonate derivatives doped poly(3,4-ethylenedioxythiophene) films and their application in electrocatalysis. *Synth. Met.* **2014**, *189*, 161–172.

29. Kuo, C.W.; Sivakumar, C.; Wen, T.C. Nanoparticles of Pt/H$_x$MoO$_3$ electrodeposited in poly(3,4-ethylenedioxythiophene)-poly(styrene sulfonic acid) as the electrocatalyst for methanol oxidation. *J. Power Sources* **2008**, *185*, 807–814.

30. Radmilovic, V.; Gasteiger, H.A.; Ross, P.N. Structure and chemical composition of a supported Pt-Ru electrocatalyst for methanol oxidation. *J. Catal.* **1995**, *154*, 98–106.

31. Qiu, L.H.; Liu, B.Q.; Peng, Y.J.; Yan, F. Fabrication of ionic liquid-functionalized polypyrrole nanotubes decorated with platinum nanoparticles and their electrocatalytic oxidation of methanol. *Chem. Commun.* **2011**, *47*, 2934–2936.

32. Kuo, C.W.; Chen, S.J.; Chen, P.R.; Wu, T.Y.; Tsai, W.T.; Tseng, C.G. Doping process effect of polyaniline doped with poly(styrenesulfonic acid) supported platinum for methanol oxidation. *J. Taiwan Inst. Chem. Eng.* **2013**, *44*, 497–504.

33. Niu, L.; Li, Q.; Wei, F.; Wu, S.; Liu, P.; Cao, X. Electrocatalytic behavior of Pt-modified polyaniline electrode for methanol oxidation: Effect of Pt deposition modes. *J. Electroanal. Chem.* **2005**, *578*, 331–337.

34. Raoof, J.B.; Ojani, R.; Hosseini, S.R. Electrocatalytic oxidation of methanol onto platinum particles decorated nanostructured poly(1,5-diaminonaphthalene) film. *J. Solid State Electrochem.* **2012**, *16*, 2699–2708.

35. Maiyalagan, T. Electrochemical synthesis, characterization and electro-oxidation of methanol on platinum nanoparticles supported poly(o-phenylenediamine) nanotubes. *J. Power Sources* **2008**, *179*, 443–450.

Synthesis and Electrocatalytic Performance of Multi-Component Nanoporous PtRuCuW Alloy for Direct Methanol Fuel Cells

Xiaoting Chen, Hao Wang, Ying Wang, Qingguo Bai, Yulai Gao and Zhonghua Zhang

Abstract: We have prepared a multi-component nanoporous PtRuCuW (np-PtRuCuW) electrocatalyst via a combined chemical dealloying and mechanical alloying process. The X-ray diffraction (XRD), transmission electron microscopy (TEM) and electrochemical measurements have been applied to characterize the microstructure and electrocatalytic activities of the np-PtRuCuW. The np-PtRuCuW catalyst has a unique three-dimensional bi-continuous ligament structure and the length scale is 2.0 ± 0.3 nm. The np-PtRuCuW catalyst shows a relatively high level of activity normalized to mass (467.1 mA mg_{Pt}^{-1}) and electrochemically active surface area (1.8 mA cm^{-2}) compared to the state-of-the-art commercial PtC and PtRu catalyst at anode. Although the CO stripping peak of np-PtRuCuW 0.47 V (*vs.* saturated calomel electrode, SCE) is more positive than PtRu, there is a 200 mV negative shift compared to PtC (0.67 V *vs.* SCE). In addition, the half-wave potential and specific activity towards oxygen reduction of np-PtRuCuW are 0.877 V (*vs.* reversible hydrogen electrode, RHE) and 0.26 mA cm^{-2}, indicating a great enhancement towards oxygen reduction than the commercial PtC.

Reprinted from *Catalysts*. Cite as: Chen, X.; Wang, H.; Wang, Y.; Bai, Q.; Gao, Y.; Zhang, Z. Synthesis and Electrocatalytic Performance of Multi-Component Nanoporous PtRuCuW Alloy for Direct Methanol Fuel Cells. *Catalysts* **2015**, 5, 1003–1015.

1. Introduction

In recent decades, there are increasing requirements for high-efficiency and eco-friendly energy to cope with environment problems, such as the depletion of energy and pollution. Direct methanol fuel cells (DMFCs) meet the above requirement and are important for the automotive industry [1,2]. As an important component, the state-of-the-art electrocatalysts must ensure the methanol oxidation reaction (MOR) at anode and oxygen reduction reaction (ORR) at cathode to generate electricity with water and carbon dioxide as the byproducts [1,3]. The use of current catalysts has several disadvantages including the low catalytic efficiency as well as CO tolerance, and sluggish kinetics towards ORR. Much effort has been dedicated to alloying Pt with other metals (e.g., Fe [4], Co [5–7], Ni [5,8,9], Cu [10],

577

Sn [11], *etc.*) to improve electrochemical activities. For high-activity applications of DMFCs, PtRu alloy remains the most active electrocatalyst due to its unique reaction mechanism [12,13]. Based upon the bi-functional and electronic effects [14–16], the electrocatalytic performance of catalysts could be significantly enhanced through compositional design of catalysts. For instance, W/Mo-modified PtRu/C showed improved electro-oxidation activity in comparison to the un-modified PtRu/C [17]. The PtRuOsIr showed better activity compared to PtRuOs and PtRuIr [3].

Traditional preparation methods for Pt-based alloys have been focusing on microemulsions [18,19], microwave irradiation [20], electrodeposition [21] and chemical reduction [3,22]. These processes are limited to the tunability of alloy composition, cockamamie, and not suitable for batch production, which hampers the commercialization of DMFCs. Thus, although there are obvious advantages in multicomponent catalysts, most of the prior studies were restricted to the synthesis of binary and ternary alloys. In contrast, facile dealloying has shown its advantage in preparing nanoporous metals/alloys. Thus, materials with unique nanoporous structure possess intriguing physical and chemical properties to generate promising potentials for various important applications such as sensors [23], mechanical actuators [24] and catalysis [25,26].

From the viewpoint of activity and accessibility, Cu are W are superior candidates as indispensible component in catalysts. In previous work, PtRuCu [27,28] and PtRuW [29,30] ternary alloys have received unremitting interest. As part of the continuing effort in new catalyst exploration, the nanoporous PtRuCuW (np-PtRuCuW) alloy was synthesized through mechanical alloying and subsequent mild chemical dealloying process in the present paper. This new catalyst was characterized with electrochemical measurements at the anode as well as cathode for DMFCs. Our results show that the np-PtRuCuW catalyst performs better than the commercial PtC and PtRu catalysts. Furthermore, the catalyst can be synthesized in gram-scale, which makes repeatable experiments in laboratory possible and batch preparation in factory reliable. Given the advantages, including easy access to component design, simplicity in the fabrication process and the enhanced activity, we hope that the combination of mechanical alloying with dealloying provides an efficient way to design multicomponent catalyst materials.

2. Results and Discussion

2.1. Microstructural Characterization of np-PtRuCuW

Figure 1 shows the X-ray diffraction (XRD) patterns of the $Al_{66}Cu_{30}(Pt_{53}Ru_{32}W_{15})_4$ precursor alloy and the as-dealloyed samples. In Figure 1a, a number of diffraction peaks appear on the pattern of the as-milled $Al_{66}Cu_{30}(Pt_{53}Ru_{32}W_{15})_4$ precursor, which can be ascribed to a Al_4Cu_9-type (PDF

No. 65-3347) intermetallic phase and Ru (PDF No. 06-0663). In addition, there is a sharp peak around scattering angles (2θ) of *ca.* 40° which can be ascribed to PtRu. A little amount of Ru dissolved during the dealloying process, leading to the reduction of peak intensity and increase of peak width around 40° [31]. The XRD spectrum presents obviously broadened Bragg peaks at scattering angles (2θ) of *ca.* 40.8°, 47.1° and 69.4° (Figure 1b) compared to the precursor alloy. These signals consist with the face-centered cubic (f.c.c.) Pt (PDF No. 04-0802) in spite of the shifting of Bragg peaks. In addition, the Ru diffraction peaks still exist after dealloying (Figure 1b). Due to the minor addition of Ru (only 1.28 at %) into the precursor, the strong diffraction peaks of Ru also suggest the alloying of other elements (Pt, W and Al) with Ru to form a solid solution. Also, it is understandable to assume that the alloying of Pt with Ru as well as Cu and W results in the shift towards higher Bragg angles [32]. The chemical component of the as-dealloyed samples was characterized by EDX and one typical spectrum is presented in Figure 1c. The corresponding results reveal that the sample is composed of Pt (56.2 at %), Ru (18.7 at %), Cu (14.2 at %) and W (10.9 at %), with a minor residual Al (only few atom percent) could be detected in the as-dealloyed samples.

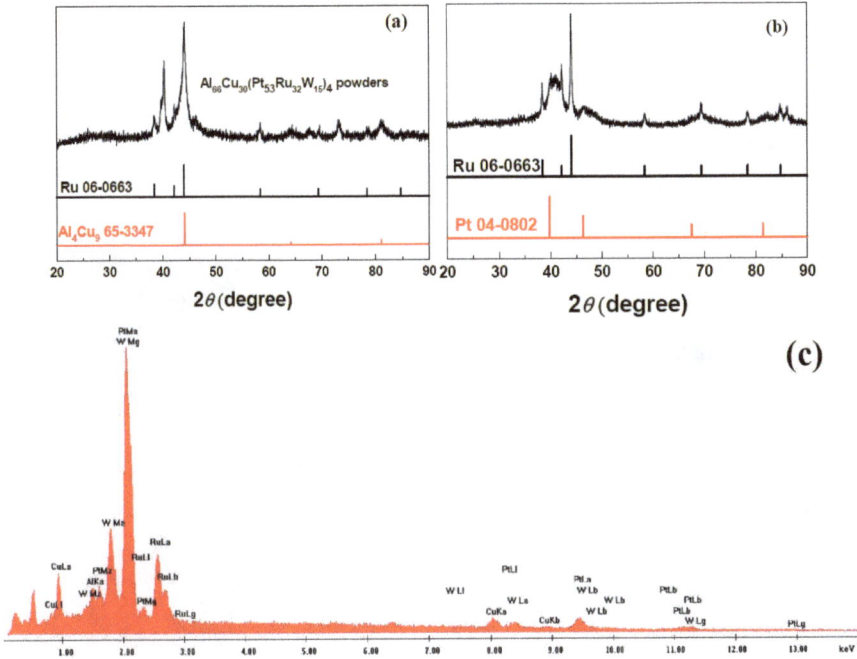

Figure 1. XRD spectrum of (a) the mechanically alloyed Al66Cu30(Pt$_{53}$Ru$_{32}$W$_{15}$)$_4$ powders; (b) the precursor alloy after dealloying in the 1M HNO$_3$ solution; and (c) corresponding EDX result of the as-dealloyed samples.

As shown in Figure 2a,b, the final samples display a nanoporous structure composed of interconnected nanoscaled ligaments (2.0 ± 0.3 nm in size) and bi-continuous channels. Besides the ultrafine ligament/channel structure, nanoparticles embedded in the nanoporous matrix are observed, and one particle is marked by a red arrow in Figure 2b. The unique nanostructure results from the dealloying process and similar phenomenon has been studied before [33,34]. The diffraction rings come from the nanoporous matrix and correspond to (111), (200), (220) and (311) reflections of f.c.c. Pt and the diffraction spots originate from the embedded nanoparticles corresponding to the Ru phase (inset of Figure 2a). Overall, the transmission electron microscopy (TEM) results are consistent with the XRD results from Figure 1b.

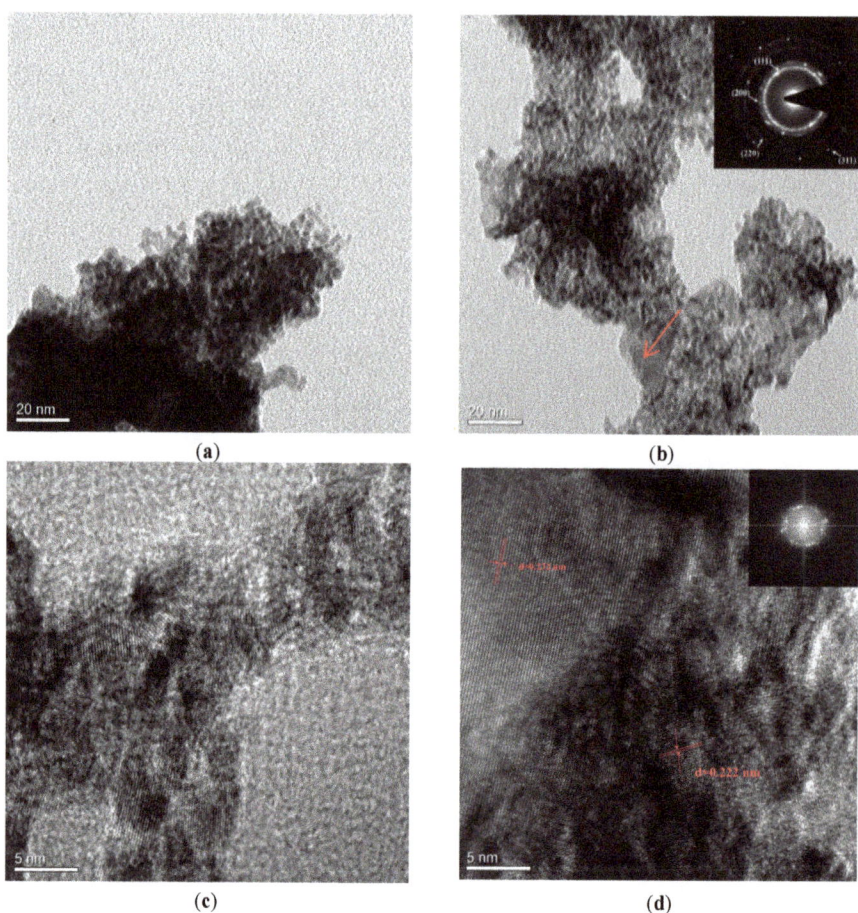

Figure 2. (a,b) TEM and (c,d) HRTEM images of the np-PtRuCuW alloy with nanoporous microstructure; Insets in (b,d): typical SAED and FFT spectrums, respectively.

The high-resolution TEM (HRTEM) images show typical bi-continuous ligament-channel structure of the as-dealloyed samples (Figure 2c,d). The ligaments are composed of small nanocrystals of several nanometers and the spacing of some lattice fringes is 0.222 nm (as indicated in Figure 2d), which is similar to the (111) crystal plane of f.c.c. Pt (~0.226 nm). The corresponding fast Fourier transform (FFT) pattern further verifies the nanocrystalline character of the selected area (presented in the inset of Figure 2d). In addition, Figure 2d also shows the HRTEM image of another area. The lattice spacing of these regular lattice fringes running across the area is 0.232 nm, which is close to the value of Ru (100) crystal plane (0.234 nm). Both the XRD and TEM results confirm that partial Ru failed to be alloyed with Pt and the final as-dealloyed samples include Ru and Pt solid solution. The np-PtRuCuW can be used as abbreviation for the obtained samples.

2.2. Catalytic Activity of np-PtRuCuW at Anode

The activation step results in the catalyst surface cleaning and typical electrochemical features including the typical hydrogen ad/desorption, double layer and metallic redox region [35]. In addition, less-noble metals dissolution would happen when applying potential cycling to multi-metallic alloys [32,36,37]. For np-PtRuCuW, the CV features are different from the PtC and PtRu catalysts (Figure 3a). Firstly, there are broader double electric layer and a more featureless shoulder region than PtC. Moreover, the reduction peak of Pt oxides shifts to a more positive direction compared to the PtC and PtRu catalysts (as highlighted by dotted line in Figure 3a). The characteristics are typical to multi-component alloys [7,27,32,36,37]. Normally, the electrochemically active surface area (ECSA) can be obtained from the equation $ECSA_{Pt}$ (m^2/g) = $Q_H/(2.1 \times m_{Pt})$ by integrating the hydrogen ad/desorption charge and using the value of 2.1 C m^{-2} for the oxidation of a monolayer of hydrogen on a polycrystalline Pt electrode [35,38]. The ECSAs of the np-PtRuCuW, PtC and PtRu catalysts were determined to be 26, 47 and 40 m^2 g^{-1}.

Figure 3b,c shows the ECSA- and Pt mass-normalized results for the np-PtRuCuW, PtC and PtRu catalysts in the H$_2$SO$_4$ solution contain 0.5 M CH$_3$OH. As shown in Figure 3b, the specific activity of the np-PtRuCuW is 1.8 mA cm^{-2}, which is 3.6 and 2.9 times that of the PtC and PtRu catalysts (0.5 and 0.63 mA cm^{-2}), respectively. Further comparison according to Pt mass (Figure 3c) indicates that the np-PtRuCuW catalyst shows higher mass activity of 467.1 mA mg$_{Pt}$$^{-1}$, which is about 2.0 and 1.6 times of that for PtC and PtRu (229.5 and 287.0 mA mg$_{Pt}$$^{-1}$). It can be seen that the peak potentials are comparable for all catalysts while the ratio of the forward anodic peak current density (I_f) to the reverse anodic peak current density (I_b) is different. It has been studied that the electro-oxidation of methanol molecules results in the formation of current peak in the forward scan and the removal of the incompletely oxidized carbonaceous species contributes to the current peak in the reverse scan [39]. Hence, the ratio I_f/I_b can be used to measure the tolerance to carbonaceous species [4,36,40]. The values for the np-PtRuCuW, PtC and PtRu are 1.29, 0.91 and 1.75, respectively, which indicates the higher CO tolerance of np-PtRuCuW than PtC while shows disadvantage compared to PtRu.

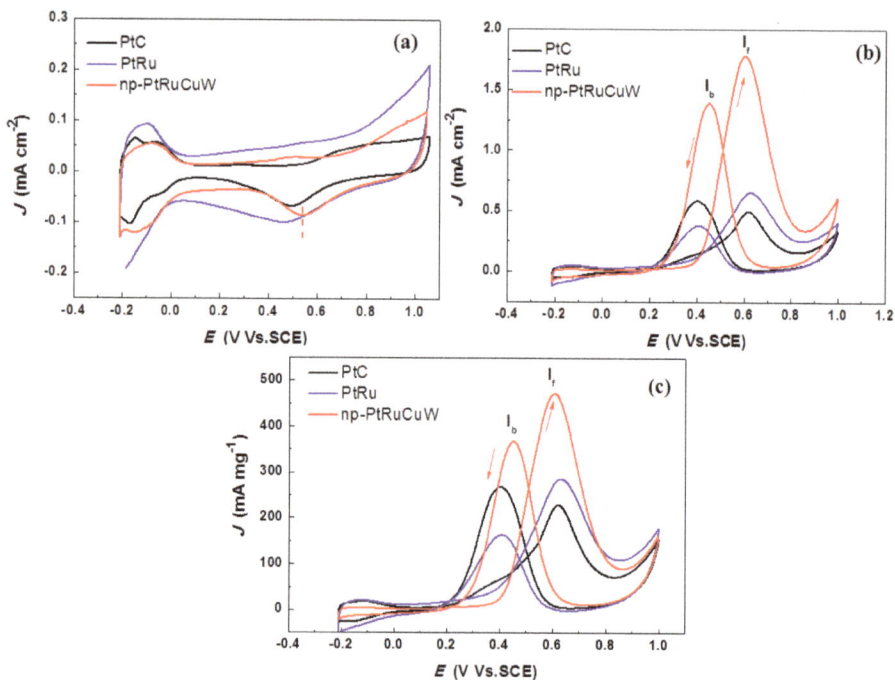

Figure 3. (a) The stable CVs of the np-PtRuCuW, PtC and PtRu catalysts in the N2 purged 0.5 M H_2SO_4 solution (Scan rate: 50 mV s^{-1}); and (b) ECSA- and (c) mass-normalized activities of the np-PtRuCuW, PtC and PtRu catalysts for methanol oxidation. (Scan rate: 50 mV s^{-1}).

The subsequent CO tolerance experiment was carried out to characterize our np-PtRuCuW catalyst directly. As shown in Figure 4, the CO stripping peak of np-PtRuCuW (0.47 V *vs*. SCE) is more positive than PtRu catalyst (0.33 V *vs*. SCE). However, there is a remarkable negative shift (about 200 mV) compared to the PtC (0.67 V *vs*. SCE). It is interesting to observe that the stripping curve of np-PtRuCuW presents two CO oxidation peaks, a first peak centered at about 0.47 V (*vs*. SCE) and the second one close to 0.55 V (*vs*. SCE). Moreover, the onset potential for CO stripping is located at around 0.35 V (*vs*. SCE), which is comparable to the peak potential of PtRu. This also suggests the good CO tolerance of our np-PtRuCuW catalyst. Maillard *et al.* [41] observed that catalysts comprising Pt nanoparticles with 2 to 6 nm size exhibited better CO tolerance. Their observations are consistent with our results, which indicate that Pt-based alloys with ligament size of 2.0 ± 0.3 nm have been successfully fabricated.

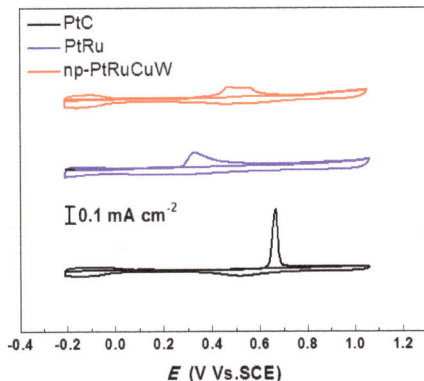

Figure 4. Electrochemical CO-stripping curves of the np-PtRuCuW, PtC and PtRu catalysts in the 0.5 M H_2SO_4 solution (Scan rate: 20 mV s^{-1}).

During the oxidation of methanol on np-PtRuCuW, Pt accomplishes the dissociative chemisorption of methanol and the alloyed metals (Ru, Cu and W) form oxyhydroxide, which extracts an active oxygen to oxidize the carbonaceous residues to CO_2 [42,43]. The three-dimensional bi-continuous structure with interconnected channels and nano-sized ligaments in np-PtRuCuW can facilitate the transportation of molecules and electrons, which greatly improves the reaction kinetics. On the basis of these effects, the present np-PtRuCuW alloy shows superior activity to the commercial PtC and PtRu catalysts. The result also indicates enhanced electrocatalytic performance compared to trimetallic (PtRuCu [27,28] and PtRuW [29,30]).

2.3. Catalytic Activity of np-PtRuCuW towards ORR

Figure 5a displays the CVs of the np-PtRuCuW and PtC catalysts in the N_2-saturated 0.1 M $HClO_4$ solution. Similar to our experiments at anode, the reduction peak of Pt oxides shows a slight positive shift for the np-PtRuCuW catalyst than that of PtC, indicating earlier onset of Pt-O(H) reduction [44,45]. The polarization results for the ORR on the np-PtRuCuW and PtC electrodes are shown in Figure 5b. It is obvious that the polarization curve of the np-PtRuCuW catalyst shifts to a more positive position than PtC. At cathode, the diffusion-limited region (below 0.8 V $vs.$ RHE) suggests a four-electron transfer reaction and indicates low hydrogen peroxide formation [46]. The half-wave potential of the np-PtRuCuW alloy is located at 0.877 V $vs.$ RHE, 13 mV positive shift compared to PtC catalyst (0.864 V $vs.$ RHE). The present results indicate a greatly enhanced ORR activity of np-PtRuCuW.

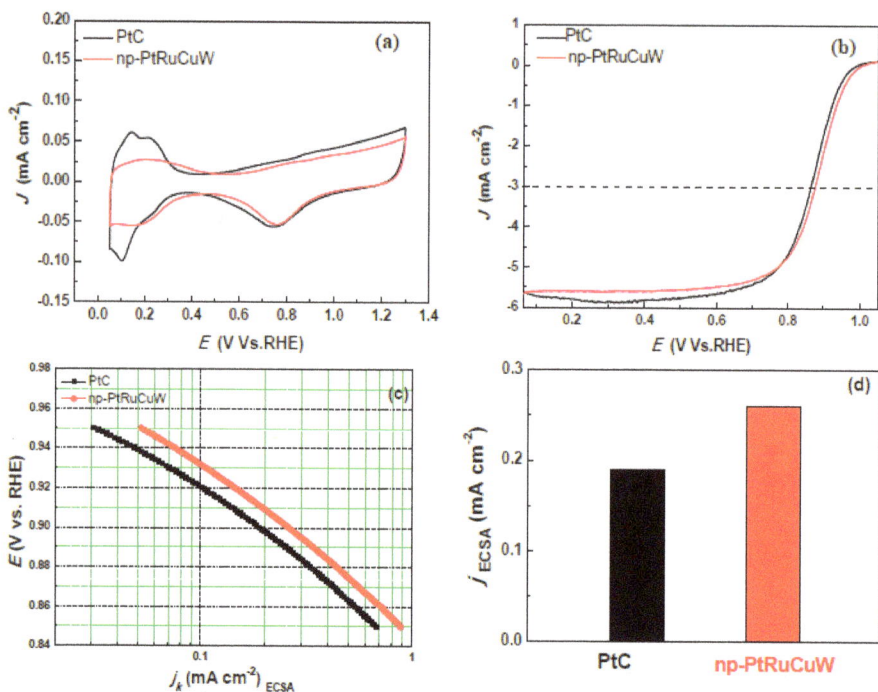

Figure 5. (a) CV curves of the np-PtRuCuW and PtC catalysts in the N2-purged 0.1 M HClO4 solution (Scan rate: 50 mV s^{-1}); (b) ORR results in the O2-saturated 0.1 M HClO$_4$ solution at 1600 rpm (Scan rate: 10 mV s^{-1}); (c) ECSA-normalized specific kinetic current densities (jk); and (d) the ECSA-normalized specific kinetic current densities for the np-PtRuCuW and PtC catalysts at 0.90 V (*vs.* RHE).

According to the Tafel plots (Figure 5c), the np-PtRuCuW catalyst shows better specific activity than the commercial PtC at the whole selected potential region (0.85~0.95 V *vs.* RHE). Additionally, the Tafel curve of the np-PtRuCuW catalyst is parallel to that of the commercial PtC, which can be observed directly from the plots. This potential region corresponds to ORR when the hydroxyl species (OH$_{ads}$) at Pt active sites and determines the electrode activity. Therefore, the plots demonstrate the similar ORR processes on the np-PtRuCuW and commercial PtC catalysts in acidic media [47,48]. The kinetic current densities at 0.90 V *vs.* RHE represent the activities towards ORR and were calculated by the Koutecky–Levich equation from the ORR polarization curves. The kinetic current densities were normalized by the ECSA to obtain the specific activity in Figure 5d, respectively. The np-PtRuCuW exhibits a higher specific activity of 0.26 mA cm^{-2}, which is 1.37 times of the commercial PtC catalyst (0.19 mA cm^{-2}).

The ORR consists of four proton and electron transfer ($O_2 + 4H^+ + 4e^- \rightarrow 2H_2O$), O–O breaking and OH$_{ads}$ intermediates formation [46,49]. After alloying with Ru,

Cu and W, the lattice parameter of Pt changes which would bring about ligand and strain effects. This will accelerate the scission of the O–O bond and the formation of the OH_{ads}, thus increasing the O_2 reaction rate [50,51]. On the other hand, the hydrogenation rates of OH_{ads} intermediates enhanced and reduced the coverage of Pt surface because of alloy effect [52]. These could rationalize that the present np-PtRuCuW alloy shows improved catalytic performance towards ORR relative to the commercial PtC catalyst.

3. Experimental Section

3.1. Synthesis and Characterizations of np-PtRuCuW

A multi-component $Al_{66}Cu_{30}(Pt_{53}Ru_{32}W_{15})_4$ alloy (nominal composition, at %) was chosen as the precursor and prepared by milling a mixture of pure elemental powders (i.e., Al, Cu, Pt, Ru and W with 99.9 wt. % purity) as before [34,53]. Then, dealloying of the $Al_{66}Cu_{30}(Pt_{53}Ru_{32}W_{15})_4$ alloy powders was carried out in a 1 M HNO_3 solution at room temperature until no obvious bubbles emerged. After the facile dealloying, the samples were rinsed using distilled water and dehydrated alcohol to gain the final np-PtRuCuW catalyst.

X-ray diffractometer (XRD, Rigaku D/max-rB, Osaka, Japan) with Cu Kα radiation was used to analysis the X-ray diffractograms of the $Al_{66}Cu_{30}(Pt_{53}Ru_{32}W_{15})_4$ precursor and as-dealloyed samples. The chemical compositions of the as-dealloyed samples were obtained by an energy-dispersive X-ray (EDX) analyzer in an area-analysis mode (a typical area of 50 μm × 50 μm). Transmission electron microscope (TEM, FEI Tecnai G2, Pleasanton, CA, USA) and high-resolution TEM (HRTEM, FEI Tecnai G2, CA, USA) were also applied to characterize the microstructure. In addition, selected-area electron diffraction (SAED) and fast Fourier transform (FFT) patterns were obtained from the corresponding images.

3.2. Electrochemical Characterization

The ORR and MOR measurements were performed using a standard three-electrode cell with a CHI 760E Potentiostat (CH Instruments, Shanghai, China). The reference electrode was saturated calomel electrode (SCE) and the counter electrode was a bright Pt plate. The catalyst ink preparation and measurement conditions were consistent with our previous report [34,53]. For comparison, we benchmarked the electrochemical properties of the np-PtRuCuW against the commercial (Johnson's Matthey, Pennsylvania, PA, USA) PtC and PtRu catalysts under identical experimental conditions.

4. Conclusions

In conclusion, a novel np-PtRuCuW catalyst has been explored through the combination of a mechanical alloying with the subsequent simple chemical dealloying step. The facile and green technique shows great advantages in the design of multiple component nanostructured alloy electrocatalysts. The np-PtRuCuW catalyst with a unique ligament/channel structure shows enhanced catalytic activity for methanol oxidation at anode compared to PtC and PtRu catalysts. Furthermore, the catalyst also indicates enhanced catalytic activity towards oxygen reduction reaction at cathode. Our results provide a novel strategy for the design of precursor alloys and fabrication of multi-component nanoporous alloy catalysts for DMFCs.

Acknowledgments: The authors gratefully acknowledge financial support by National Natural Science Foundation of China (51371106), National Basic Research Program of China (973, 2012CB932800), Cross Disciplinary Training Project of Shandong University (2014JC004), Specialized Research Found for the Doctoral Program of Higher Education of China (20120131110017), and Young Tip-Top Talent Support Project (the Organization Department of the Central Committee of the CPC). We also acknowledge the experimental assistance from Ruhr University Bochum, Germany. Y.L. Gao acknowledges the support by Program for Professor of Special Appointment (Eastern Scholar) at Shanghai Institutions of Higher Learning (No. TP2014042).

Author Contributions: We thank our collaborators for their contributions to the manuscript. Thank Ying Wang and Qingguo Bai for the fabrication of precursor alloy. Great valuable discussions for the results were supported by Yulai Gao. Xiaoting Chen and Hao Wang carried out most of the experiment and completed documentation and the manuscript. Zhonghua Zhang is in charge of the whole work.

Conflicts of Interest: The authors declare no conflict of interest.

References

1. Kamarudin, S.K.; Achmad, F.; Daud, W.R.W. Overview on the application of direct methanol fuel cell (DMFC) for portable electronic devices. *Int. J. Hyd. Energy* **2009**, *34*, 6902–6916.

2. Zhang, G.; Xia, B.Y.; Wang, X. Strongly Coupled NiCo$_2$O$_4$-Rgo Hybrid Nanosheets as a Methanol-Tolerant Electrocatalyst for the Oxygen Reduction Reaction. *Adv. Mater.* **2014**, *26*, 2408–2412.

3. Reddington, E.; Sapienza, A.; Gurau, B.; Viswanathan, R.; Sarangapani, S.; Smotkin, E.S.; Mallouk, T.E. Combinatorial electrochemistry: A highly parallel, optical screening method for discovery of better electrocatalysts. *Science* **1998**, *280*, 1735–1737.

4. Ma, X.; Luo, L.; Zhu, L.; Yu, L.; Sheng, L.; An, K.; Ando, Y.; Zhao, X. Pt–Fe catalyst nanoparticles supported on single-wall carbon nanotubes: Direct synthesis and electrochemical performance for methanol oxidation. *J. Power Sour.* **2013**, *241*, 274–280.

5. Wang, C.; Markovic, N.M.; Stamenkovic, V.R. Advanced platinum alloy electrocatalysts for the oxygen reduction reaction. *ACS Catal.* **2012**, *2*, 891–898.

6. Ahmadi, R.; Amini, M.; Bennett, J. Pt–Co alloy nanoparticles synthesized on sulfur-modified carbon nanotubes as electrocatalysts for methanol electrooxidation reaction. *J. Catal.* **2012**, *292*, 81–89.

7. Xu, C.; Hou, J.; Pang, X.; Li, X.; Zhu, M.; Tang, B. Nanoporous PtCo and PtNi alloy ribbons for methanol electrooxidation. *Int. J. Hyd. Energy* **2012**, *37*, 10489–10498.

8. Choi, S.-I.; Xie, S.; Shao, M.; Odell, J.H.; Lu, N.; Peng, H.-C.; Protsailo, L.; Guerrero, S.; Park, J.; Xia, X. Synthesis and characterization of 9 nm Pt–Ni octahedra with a record high activity of 3.3 A/mgPt for the oxygen reduction reaction. *Nano Lett.* **2013**, *13*, 3420–3425.

9. Cui, C.; Gan, L.; Heggen, M.; Rudi, S.; Strasser, P. Compositional segregation in shaped Pt alloy nanoparticles and their structural behaviour during electrocatalysis. *Nat. Mater.* **2013**, *12*, 765–771.

10. Zhang, X.; Li, D.; Dong, D.; Wang, H.; Webley, P.A. One-step fabrication of ordered Pt–Cu alloy nanotube arrays for ethanol electrooxidation. *Mater. Lett.* **2010**, *64*, 1169–1172.

11. Sun, S.; Zhang, G.; Geng, D.; Chen, Y.; Banis, M.N.; Li, R.; Cai, M.; Sun, X. Direct growth of single-crystal Pt nanowires on Sn@CNT nanocable: 3D electrodes for highly active electrocatalysts. *Chem. A Eur. J.* **2010**, *16*, 829–835.

12. Zou, L.; Guo, J.; Liu, J.; Zou, Z.; Akins, D.L.; Yang, H. Highly alloyed PtRu black electrocatalysts for methanol oxidation prepared using magnesia nanoparticles as sacrificial templates. *J. Power Sources.* **2014**, *248*, 356–362.

13. Guo, J.; Zhao, T.; Prabhuram, J.; Chen, R.; Wong, C. Preparation and characterization of a PtRu/C nanocatalyst for direct methanol fuel cells. *Electrochim. Acta* **2005**, *51*, 754–763.

14. Lipkowski, J.; Ross, P.N. *Electrocatalysis*; John Wiley & Sons: New York, NY, USA, 1998; Volume 3, pp. 70–200.

15. Kua, J.; Goddard, W.A. Oxidation of methanol on 2nd and 3rd row group viii transition metals (Pt, Ir, Os, Pd, Rh, and Ru): Application to direct methanol fuel cells. *J. Am. Chem. Soc.* **1999**, *121*, 10928–10941.

16. Tong, Y.; Rice, C.; Wieckowski, A.; Oldfield, E. A detailed NMR-based model for CO on Pt catalysts in an electrochemical environment: Shifts, relaxation, back-bonding, and the Fermi-Level local density of states. *J. Am. Chem. Soc.* **2000**, *122*, 1123–1129.

17. Zhou, W.; Zhou, Z.; Song, S.; Li, W.; Sun, G.; Tsiakaras, P.; Xin, Q. Pt based anode catalysts for direct ethanol fuel cells. *Appl. Catal. B* **2003**, *46*, 273–285.

18. Liu, H.; Song, C.; Zhang, L.; Zhang, J.; Wang, H.; Wilkinson, D.P. A review of anode catalysis in the direct methanol fuel cell. *J. Power Sour.* **2006**, *155*, 95–110.

19. Liu, Z.; Lee, J.Y.; Han, M.; Chen, W.; Gan, L.M. Synthesis and characterization of PtRu/C catalysts from microemulsions and emulsions. *J. Mater. Chem.* **2002**, *12*, 2453–2458.

20. Almeida, T.; Palma, L.; Leonello, P.; Morais, C.; Kokoh, K.; de Andrade, A. An optimization study of PtSn/C catalysts applied to direct ethanol fuel cell: Effect of the preparation method on the electrocatalytic activity of the catalysts. *J. Power Sour.* **2012**, *215*, 53–62.

21. Liu, L.; Huang, Z.; Wang, D.; Scholz, R.; Pippel, E. The fabrication of nanoporous Pt-based multimetallic alloy nanowires and their improved electrochemical durability. *Nanotechnology* **2011**, *22*, 105604.

22. Chen, Z.; Waje, M.; Li, W.; Yan, Y. Supportless Pt and PtPd nanotubes as electrocatalysts for oxygen-reduction reactions. *Angew. Chem. Int. Ed.* **2007**, *46*, 4060–4063.

23. You, T.; Niwa, O.; Tomita, M.; Hirono, S. Characterization of platinum nanoparticle-embedded carbon film electrode and its detection of hydrogen peroxide. *Anal. Chem.* **2003**, *75*, 2080–2085.

24. Weissmüller, J.; Viswanath, R.; Kramer, D.; Zimmer, P.; Würschum, R.; Gleiter, H. Charge-induced reversible strain in a metal. *Science* **2003**, *300*, 312–315.

25. Qi, Z.; Geng, H.; Wang, X.; Zhao, C.; Ji, H.; Zhang, C.; Xu, J.; Zhang, Z. Novel nanocrystalline PdNi alloy catalyst for methanol and ethanol electro-oxidation in alkaline media. *J. Power Sour.* **2011**, *196*, 5823–5828.

26. Xu, C.; Liu, Y.; Hao, Q.; Duan, H. Nanoporous PdNi alloys as highly active and methanol-tolerant electrocatalysts towards oxygen reduction reaction. *J. Mater. Chem. A* **2013**, *1*, 13542–13548.

27. Jeon, M.K.; Cooper, J.S.; McGinn, P.J. Methanol electro-oxidation by a ternary Pt–Ru–Cu catalyst identified by a combinatorial approach. *J. Power Sour.* **2008**, *185*, 913–916.

28. Naidoo, Q.-L.; Naidoo, S.; Petrik, L.; Nechaev, A.; Ndungu, P.; Vaivars, G. Synthesis highly active platinum tri-metallic electrocatalysts using "one-step" organometallic chemical vapour deposition technique for methanol oxidation process. *IOP Conf. Ser. Mater. Sci. Eng.* **2012**.

29. Kang, D.K.; Noh, C.S.; Park, S.T.; Sohn, J.M.; Kim, S.K.; Park, Y.-K. The effect of PtRuW ternary electrocatalysts on methanol oxidation reaction in direct methanol fuel cells. *Korean J. Chem. Eng.* **2010**, *27*, 802–806.

30. Jeon, M.K.; Lee, K.R.; Woo, S.I. Ternary $Pt_{45}Ru_{45}M_{10}$/C (M = Mn, Mo and W) catalysts for methanol and ethanol electro-oxidation. *Korean J. Chem. Eng.* **2009**, *26*, 1028–1033.

31. Aricò, A.S.; Antonucci, P.L.; Modica, E.; Baglio, V.; Kim, H.; Antonucci, V. Effect of PtRu alloy composition on high-temperature methanol electro-oxidation. *Electrochim. Acta* **2002**, *47*, 3723–3732.

32. Ammam, M.; Easton, E.B. Quaternary PtMnCuX/C (X = Fe, Co, Ni, and Sn) and PtMnMoX/C (X = Fe, Co, Ni, Cu and Sn) alloys catalysts: Synthesis, characterization and activity towards ethanol electrooxidation. *J. Power Sour.* **2012**, *215*, 188–198.

33. Qian, L.; Chen, M. Ultrafine nanoporous gold by low-temperature dealloying and kinetics of nanopore formation. *Appl. Phys. Lett.* **2007**, *91*, 083105.

34. Chen, X.; Jiang, Y.; Sun, J.; Jin, C.; Zhang, Z. Highly active nanoporous Pt-based alloy as anode and cathode catalyst for direct methanol fuel cells. *J. Power Sour.* **2014**, *267*, 212–218.

35. Pozio, A.; de Francesco, M.; Cemmi, A.; Cardellini, F.; Giorgi, L. Comparison of high surface Pt/C catalysts by cyclic voltammetry. *J. Power Sour.* **2002**, *105*, 13–19.

36. Lee, Y.-W.; Ko, A.-R.; Han, S.-B.; Kim, H.-S.; Park, K.-W. Synthesis of octahedral Pt–Pd alloy nanoparticles for improved catalytic activity and stability in methanol electrooxidation. *Phys. Chem. Chem. Phys.* **2011**, *13*, 5569–5572.

37. Li, H.H.; Cui, C.H.; Zhao, S.; Yao, H.B.; Gao, M.R.; Fan, F.J.; Yu, S.H. Mixed-PtPd-shell PtPdCu nanoparticle nanotubes templated from copper nanowires as efficient and highly durable electrocatalysts. *Adv. Energy Mater.* **2012**, *2*, 1182–1187.

38. He, C.; Liang, Y.; Fu, R.; Wu, D.; Song, S.; Cai, R. Nanopores array of ordered mesoporous carbons determine Pt's activity towards alcohol electrooxidation. *J. Mater. Chem.* **2011**, *21*, 16357–16364.

39. Mancharan, R.; Goodenough, J.B. Methanol oxidation in acid on ordered NiTi. *J. Mater. Chem.* **1992**, *2*, 875–887.

40. Xu, C.; Wang, L.; Wang, R.; Wang, K.; Zhang, Y.; Tian, F.; Ding, Y. Nanotubular mesoporous bimetallic nanostructures with enhanced electrocatalytic performance. *Adv. Mater.* **2009**, *21*, 2165–2169.

41. Maillard, F.; Schreier, S.; Hanzlik, M.; Savinova, E.R.; Weinkauf, S.; Stimming, U. Influence of particle agglomeration on the catalytic activity of carbon-supported Pt nanoparticles in Co monolayer oxidation. *Phys. Chem. Chem. Phys.* **2005**, *7*, 385–393.

42. Ley, K.L.; Liu, R.; Pu, C.; Fan, Q.; Leyarovska, N.; Segre, C.; Smotkin, E. Methanol oxidation on single-phase Pt-Ru-Os ternary alloys. *J. Electrochem. Soc.* **1997**, *144*, 1543–1548.

43. Lei, H.-W.; Suh, S.; Gurau, B.; Workie, B.; Liu, R.; Smotkin, E.S. Deuterium isotope analysis of methanol oxidation on mixed metal anode catalysts. *Electrochim. Acta* **2002**, *47*, 2913–2919.

44. Zhang, X.; Choi, I.; Qu, D.; Wang, L.; Lee, C.-W.J. Coverage-dependent electro-catalytic activity of Pt sub-monolayer/Au bi-metallic catalyst toward methanol oxidation. *Int. J. Hyd. Energy* **2013**, *38*, 5665–5670.

45. Hodnik, N.; Jeyabharathi, C.; Meier, J.C.; Kostka, A.; Phani, K.L.; Rečnik, A.; Bele, M.; Hočevar, S.; Gaberšček, M.; Mayrhofer, K.J. Effect of ordering of PtCu$_3$ nanoparticle structure on the activity and stability for the oxygen reduction reaction. *Phys. Chem. Chem. Phys.* **2014**, *16*, 13610–13615.

46. Stephens, I.E.L.; Bondarenko, A.S.; Grønbjerg, U.; Rossmeisl, J.; Chorkendorff, I. Understanding the electrocatalysis of oxygen reduction on platinum and its alloys. *Energy Environ. Sci.* **2012**, *5*, 6744–6762.

47. Hu, Y.; Jensen, J.O.; Zhang, W.; Cleemann, L.N.; Xing, W.; Bjerrum, N.J.; Li, Q. Hollow spheres of iron carbide nanoparticles encased in graphitic layers as oxygen reduction catalysts. *Angew. Chem. Int. Ed.* **2014**, *53*, 3675–3679.

48. Kongkanand, A.; Kuwabata, S.; Girishkumar, G.; Kamat, P. Single-wall carbon nanotubes supported platinum nanoparticles with improved electrocatalytic activity for oxygen reduction reaction. *Langmuir* **2006**, *22*, 2392–2396.

49. Guo, S.; Zhang, S.; Sun, S. Tuning nanoparticle catalysis for the oxygen reduction reaction. *Angew. Chem. Int. Ed.* **2013**, *52*, 8526–8544.

50. Toda, T.; Igarashi, H.; Uchida, H.; Watanabe, M. Enhancement of the electroreduction of oxygen on Pt alloys with Fe, Ni, and Co. *J. Electrochem. Soc.* **1999**, *146*, 3750–3756.

51. Toda, T.; Igarashi, H.; Watanabe, M. Role of electronic property of Pt and Pt alloys on electrocatalytic reduction of oxygen. *J. Electrochem. Soc.* **1998**, *145*, 4185–4188.

52. Zhang, J.; Vukmirovic, M.B.; Xu, Y.; Mavrikakis, M.; Adzic, R.R. Controlling the catalytic activity of platinum-monolayer electrocatalysts for oxygen reduction with different substrates. *Angew. Chem. Int. Ed.* **2005**, *44*, 2132–2135.

53. Chen, X.; Si, C.; Gao, Y.; Frenzel, J.; Sun, J.; Eggeler, G.; Zhang, Z. Multi-component nanoporous platinum-ruthenium-copper-osmium-iridium alloy with enhanced electrocatalytic activity towards methanol oxidation and oxygen reduction. *J. Power Sour.* **2015**, *273*, 324–332.

Preparation and Electrocatalytic Characteristics of PdW/C Catalyst for Ethanol Oxidation

Qi Liu, Mingshuang Liu, Qiaoxia Li and Qunjie Xu

Abstract: A series of PdW alloy supported on Vulcan XC-72 Carbon (PdW/C) with total 20 wt. % as electrocatalyst are prepared for ethanol oxidation by an ethylene glycol assisted method. Transmission electron microscopy (TEM) characterization shows that PdW nanoparticles with an average size of 3.6 nm are well dispersed on the surface of Vulcan XC-72 Carbon. It is found that the catalytic activity and stability of the PdW/C catalysts are strongly dependent on Pd/W ratios, an optimal Pd/W composition at 1/1 ratio revealed the highest catalytic activity toward ethanol oxidation, which is much better than commercial Pd/C catalysts.

Reprinted from *Catalysts*. Cite as: Liu, Q.; Liu, M.; Li, Q.; Xu, Q. Preparation and Electrocatalytic Characteristics of PdW/C Catalyst for Ethanol Oxidation. *Catalysts* **2015**, *5*, 1068–1078.

1. Introduction

Development of novel catalysts with high electrocatalytic activity for ethanol oxidation has received much attention because the electroactivity of anodic materials is one of the main factors influencing the practical application of direct ethanol fuel cells (DEFCs) [1,2]. Palladium (Pd), as one of platinum(Pt) group elements, could hold high electro-oxidation catalytic activity and has larger abundance and lower price compared to Pt [3,4]. Pd catalyst does not exhibit electroactivity for ethanol electro-oxidation in acid solutions, while it displays high electroactivity for ethanol electro-oxidation in alkaline solutions, such as NaOH and KOH [5,6].

Recently, Pd nanoparticle has attracted much attention due to their distinguished advantages, such as significantly large surface areas and high stability [7]. The interest in Pd metals is not only for lowering the cost of catalysts, but also for improving the catalytic activities [8]. One method to promote the catalytic activity of Pd is alloyed with other metals, including Ag [9,10], Fe [11] and Sn [12,13]. Many binary or ternary composite catalysts involved in Pd have been developed to enhance the electroactivity of the Pd catalyst for ethanol oxidation [14,15], such as Pd–Ru [16], Pd–Ni–P [17], Pd–Co [18], Pd–Pt [19,20], Pd–Au [21], and so on.

So, the addition of a second metal with Pd, to enhance its activity for ethanol electro-oxidation, is effective approach, but the durability with time of such electrode needs further improvement [22]. It has been claimed that tungsten (W) oxide was a suitable promoter for noble metal catalyst, leading to a significant decrease in poisoning species (CO) [23,24]. The presence of W species is expected to assist in the electro-oxidation of poisonous reaction intermediates adsorbed on the active Pd sites [25,26].

In this work, PdW/C catalysts with different Pd/W ratios were successfully prepared by an ethylene glycol assisted method. The catalytic activity and stability of PdW/C catalysts towards ethanol oxidation reaction (EOR) in alkaline solution were examined. The electrochemical properties of PdW electrocatalysts were also probed to explore their potential applications in DEFCs.

2. Results and Discussion

2.1. TEM

Figure 1 shows a typical TEM image of the prepared PdW/C and Pd/C catalysts. The nanoparticle sizes of PdW/C catalyst were primarily distributed within the range of 2–6 nm. The average PdW nanoparticles size of PdW/C was approximately 3.6 nm, whereas Pd nanoparticles of Pd/C were 5.2 nm. It should be pointed out that ethylene glycol as the reducing agent and dispersing agent, could effectively disperse the Pd nanoparticles. At the same time, the PdW nanoparticles size of PdW/C is smaller than Pd nanoparticles of Pd/C, indicating that demonstrating PdW/C is more beneficial for ethanol electro-oxidation in alkaline medium [27,28]. As shown in Figure 1, PdW/C catalyst was spherical and homogeneously dispersed on Vulcan XC-72 Carbon with no remarkable observation of agglomerations compared with Pd/C. High Resolution Transmission Electron Microscopy (HRTEM) image clearly shows the lattice fringe image of (1 1 1) planes with the interplanar distance of 0.2 nm. In Figure 1d, EDS of PdW/C shows the existence of Pd, W and C elements, illustrating the formation of W metal in as-obtained materials.

2.2. XRD

The XRD patterns of W/C, Pd/C and PdW/C catalysts were shown in Figure 2. The typical diffraction peaks of WC around $2\theta = 26°$ and $43°$, is attributed to the C (002) and C (004), which are not W typical diffraction, as shown in the XRD patterns of Pd/C and PdW/C, and meanwhile C diffraction peaks also appeared. Sharp and well-defined peaks of Pd/C was observed at 2θ values of $40.14°$, $46.69°$, $68.17°$, $82.17°$, and $86.69°$, corresponding to the planes of (1 1 1), (2 0 0), (2 2 0), (3 1 1), and (2 2 2), respectively, according to JCPDS No.65-6174. The strong diffraction peak of PdW/C catalyst was also found at $40.14°$, corresponding to the plane of (1 1 1). No significant peak shift is observed for the PdW/C (1:1) [29]. The average particle size of the prepared PdW/C nanoparticles (d) was estimated by using the Scherrer

Equation [30] after background subtraction from Pd (1 1 1) peak at 2θ of 40°, agreeing with TEM results, which is as shown in Table 1.

$$d = \frac{k\lambda}{\beta\cos\theta} \tag{1}$$

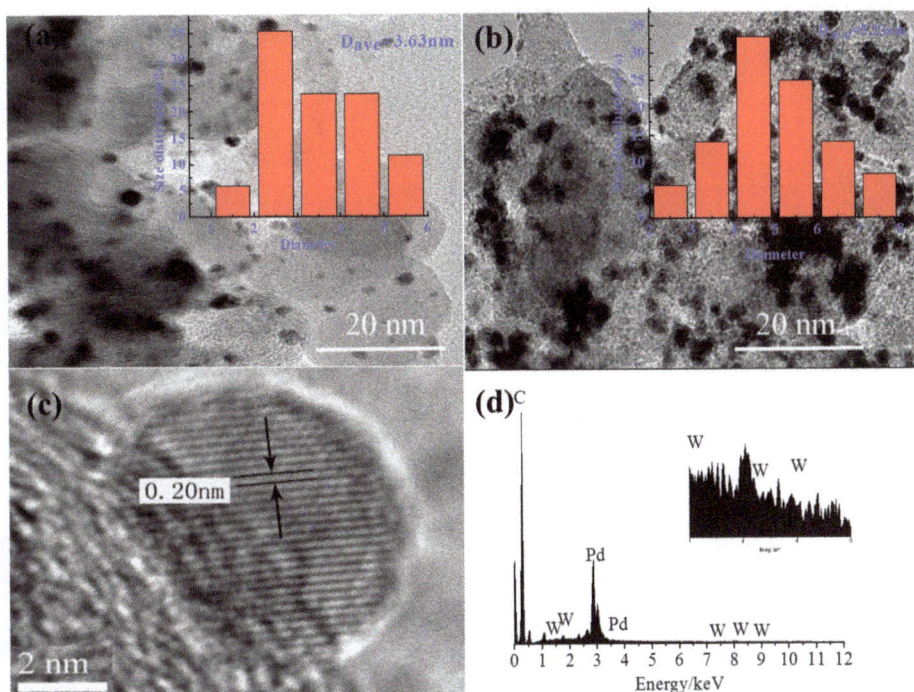

Figure 1. TEM images and their corresponding particle size distribution histograms (**a**) PdW/C; (**b**) Pd/C catalyst, respectively. (**c**) HRTEM images of PdW/C catalyst. (**d**) EDS of PdW/C catalyst.

Table 1. Summary of physical properties of PdW/C and Pd/C catalyst.

Catalysts	Pd Metal Loading Detected by ICP-AES	Diameter Calculated Form XRD/nm	Diameter Measured by TEM/nm	EASA/m²·g⁻¹
PdW/C	10.6%	3.9	3.6	144.1
Pd/C	19.3%	5.74	5.32	71.2

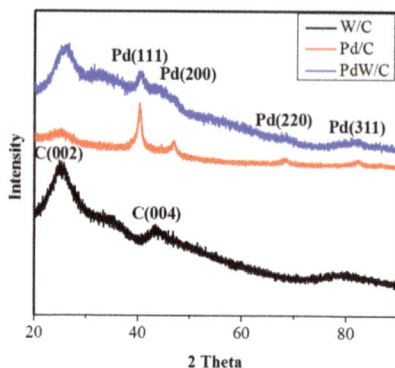

Figure 2. XRD patterns of PdW/C (1:1) and Pd/C.

2.3. Electrochemical Measurements

The cyclic voltammetry of PdW/C (1:1) and Pd/C in the absence of ethanol is shown in Figure 3. It is noted that they all exhibit significantly high anodic and cathodic current densities. The oxidation peak at lower anodic potential during the forward scan is ascribed to the formation of the adsorbed hydroxyl OH_{ads} while the peaks at high positive potential are related to the formation of Pd oxides [31,32]. The potential region from -1.1 to -0.6 V *versus* SCE on the CV curve of the catalyst is associated with the hydrogen adsorption/desorption. The potential region from -0.3 V to 0.3 V can be attributed to the formation of the palladium oxide layer on the surface of the PdW/C catalyst, and OH^- ions are first chemisorbed in the initial stage of the oxide formation at higher potentials, which are transformed into higher valence oxides. The electrochemical active surface areas (EASA) of PdW/C (1:1) and Pd/C was calculated to be 144.1 and 71.2 $m^2 \cdot g^{-1}$, respectively (Table 2), indicating that PdW/C (1:1) has a higher electrochemical activity [33].

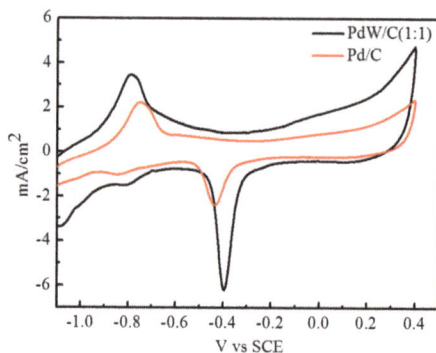

Figure 3. Voltammetric curves of PdW/C (1:1) and Pd/C in 1 M KOH solution at 50 mV·s^{-1}.

Table 2. Comparison of electrochemical performances on the prepared Pd-based catalysts.

Catalysts	E_{onset}/V	E_p/V	i_p/mA·cm^{-2}	i (after 3600 s)/mA·cm^{-2}
PdW/C (1:1)	−0.71	−0.27	62.29	6.94
PdW/C (1:2)	−0.63	−0.32	14.02	0.42
PdW/C (2:1)	−0.68	−0.29	37.39	0.19
PdW/C (4:1)	−0.70	−0.40	15.37	0.03
Pd/C	−0.64	−0.25	39.58	1.58
Pd/C (JM)	−0.65	−0.17	48.70	2.29

Cyclic voltammetry was used to quantify the electrocatalytic activities of the Pd-based catalysts prepared at room temperature. Figure 4 shows the CV results detected in 1 M KOH + 1 M C_2H_5OH solution. The scan rate was selected at 50 mV·s^{-1} in the potential range from −0.8 to 0.4 V. The oxidation peak in the forward scan corresponds to the oxidation of freshly chemisorbed species from ethanol adsorption. At a higher potential, the formation of PdO will block further adsorption of reactive species and lead to a remarkable decrease in current. During the negative-going sweep, the previously formed PdO will be reduced to catalytic active Pd, leading to the recovery of EOR current. Corresponding reactions are shown in Equations (2) and (3) [34]:

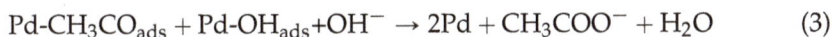

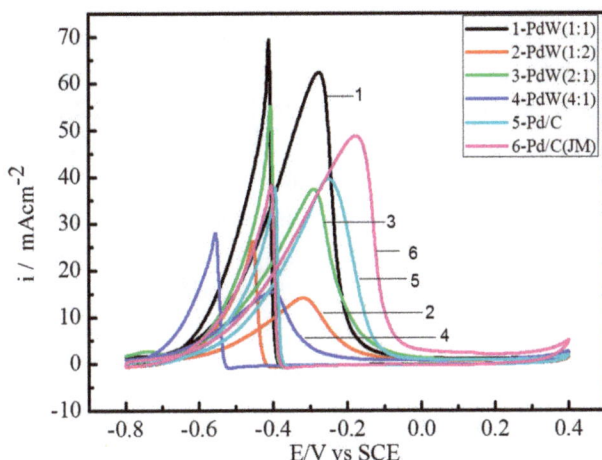

$$Pd+C_2H_5OH+3OH^- \leftrightarrow Pd\text{-}CH_3CO_{ads} + 3H_2O + 3e \qquad (2)$$

$$Pd\text{-}CH_3CO_{ads} + Pd\text{-}OH_{ads}+OH^- \rightarrow 2Pd + CH_3COO^- + H_2O \qquad (3)$$

Figure 4. Voltammetric curves of Pd-based catalysts in 1 M KOH +1 M C_2H_5OH solution at 50 mV·s^{-1}.

It is observed that the peak current density on PdW/C (1:1) is higher than those on other Pd-based catalysts (Figure 4), which could indicates that PdW/C (1:1) catalyst has the highest catalytic activity toward ethanol. In the forward scan, the onset potential (E_{onset}) of PdW/C (1:1) is −0.71 V, which has a negative shift of ~60 mV compared to that of Pd/C (JM) (−0.65 V). The peak current densities are 62.29 and 48.70 mA·cm^{-2} (the area is the surface area of the electrode) for PdW/C (1:1) and Pd/C (JM), respectively, while their peak potentials are −0.27 and −0.17 V. The parameters, including the onset potential, the forward peak potential (E_p) and the forward peak current intensity (i_p) are shown in Table 2.

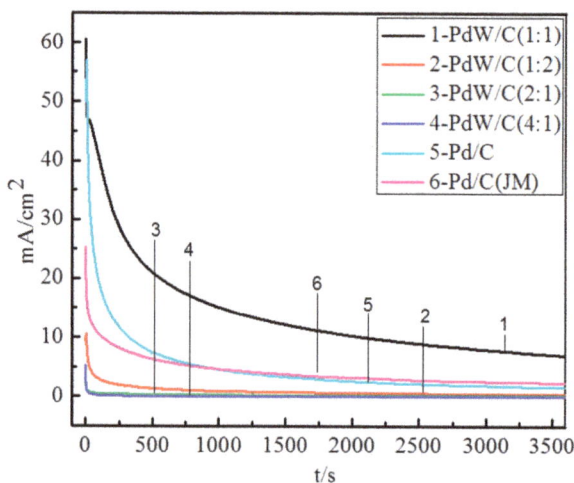

Figure 5. Chronoamperometry curves of Pd-based catalysts in 1 M KOH +1 M C$_2$H$_5$OH solution at a fixed potential of −0.3 V at 50 mV s^{-1}.

Chronoamperometry was employed to evaluate the stability of the Pd-based catalysts. As shown in Figure 5, the current densities represent less decay at the applied constant potentials for 3600 s. The current density of ethanol electro-oxidation on the PdW/C (1:1) catalyst is 6.94 mA·cm^{-2}, which is the highest among all the Pd-based catalysts, indicating that the PdW/C (1:1) exhibits a more stable electrocatalytic activity towards ethanol oxidation in the alkaline media than other catalysts. These results are in good accordance with the CV results.

3. Experimental Section

3.1. Materials

PdCl$_2$ was purchased from Shanghai Institute of Fine Chemical Materials (Shanghai, China); Vulcan XC-72 Carbon was supplied by Cabot Co. Ltd. (Boston, MA, USA); Tungsten hexachloride (99.5%, WCl$_6$), Ethylene glycol (AR, C$_2$H$_6$O$_2$),

Sodium hydroxide (AR, Nelectro-oxidation H) and ethanol were obtained from Sinopharm Chemical Reagent Co. Ltd. (Shanghai, China) 5% Nafion® solution was provided by DuPont Co. Ltd (Wilmington, DE, USA); All reagents were of analytical reagent grade and used without further purification. The water utilized in the studies was double-distilled and deionized.

3.2. Catalyst Preparation

Vulcan XC-72 Carbon was treated in 5 M HNO_3 solution with vigorous stirring. A certain amounts with different ratios of WCl_6 and $PdCl_2$ were dissolved in 50 mL ethylene glycol. Subsequently, the pH of the solution was adjusted to 9 using 1 M NaOH solution .The mixtures were stirred for 1 h at 80 °C. The prepared carbon ethylene glycol solution was added in the mixture. After stirred for 3 h, the mixtures were filtered and washed several times with deionized water. The remaining solids were dried in a vacuum oven for 24 h at 80 °C. The weight percentages of metal were 20% in all catalysts. 20% Pd/C was prepared by the similar method. 20% commercial Pd/C (JM) catalyst was purchased from Johnson Matthet Company (Shanghai, China).

3.3. Characterization

XRD patterns were collected using a Bruker D8-Advance Powder X-ray diffractometer (Cu KR radiation, wavelength 1.5418 Å, Bruker, Germany). Transmission electron microscopy (TEM) images were characterized with a JEM-2100F HR-TEM model (JEM, Tokyo, Japan) using an accelerating voltage of 80 and 200 KV. 10.0 mg catalyst was dissolved in aqua regia (a strong acid mixture with HCl: HNO_3 volume ratio of 3:1) to form a Pd aqueous solution, and ICP-AES (Varian, Anaheim, CA, USA) was performed to detect the catalyst metal loading. All electrochemical measurements were performed in a standard three-electrode cell using a CHI 660C Electrochemical Analyzer (Chenhua Company, Shanghai, China).

3.4. Electrochemical Test

Cyclic voltammetry (CV) and chronoamperometry measurements were collected in 1 M KOH + 1 M C_2H_5OH solution at a scan rate of 50 mV·s^{-1}. The working electrodes ware prepared, dropping 4 μL of the catalyst onto glassy carbon electrode (GCE, 0.07 cm^2). The ink was prepared by ultrasonically mixing 5 mg of electrocatalyst sample in a mixture of 1 mL of ethanol and 120 μL of 5% Nafion® solution. The counter electrode was Pt foils and the reference electrode was saturated calomel electrode (SCE). The CV tests were carried out in the potential range of −0.8 to 0.4 V. Before experiments, pure nitrogen gas (99.99%) was bubbled through the solution at least 30 min to remove the dissolved oxygen in the solution.

4. Conclusions

In summary, Vulcan XC-72 Carbon supported 20 wt. % PdW/C catalysts with different Pd/W ratios were prepared by an ethylene glycol method. Among them, PdW/C (1:1) catalysts have a small average diameter (3.6 nm) and large electrochemical surface areas (144.1 $m^2 \cdot g^{-1}$). It could also exhibit a higher reactivity toward EOR in alkaline electrolyte, compared to other PdW/C electrocatalysts. The peak current densities of PdW/C (1:1) (62.29 $mA \cdot cm^{-2}$) is higher than that of Pd/C (JM) (48.70 $mA \cdot cm^{-2}$). PdW/C (1:1) also exhibits more stable electrocatalytic activity than Pd/C (JM) towards ethanol oxidation in the alkaline media.

Acknowledgments: The project was supported by the National Science Foundation of China (21473039) and Shanghai Science and Technology Committee (No. 14DZ2261000).

Author Contributions: Qi Liu and Mingshuang Liu commonly carried out the catalyst preparation, electrochemical test and draft the manuscript. Qiaoxia Li and Qunjie Xu participated in the design of the study and helped to modify the manuscript. All authors read and approved the final manuscript.

Conflicts of Interest: The authors declare no conflict of interest.

References

1. Cai, Z.-X.; Liu, C.-C.; Wu, G.-H.; Chen, X.-M.; Chen, X. Green synthesis of Pt-on-Pd bimetallic nanodendrites on graphene via *in situ* reduction, and their enhanced electrocatalytic activity for methanol oxidation. *Electrochim. Acta* **2014**, *127*, 377–383.

2. Liu, Q.; Xu, Q.J.; Fan, J.C.; Zhou, Y.; Wang, L.L. A Review of Graphene Supported Electrocatalysts for Direct Methanol Fuel Cells. *Adv. Mater. Res.* **2015**, *1070–1072*, 492–496.

3. Li, Q.X.; Liu, M.S.; Xu, Q.J.; Mao, H.M. Preparation and Electrocatalytic Characteristics Research of Pd/C Catalyst for Direct Ethanol Fuel Cell. *J. Chem.* **2013**, *2013*, 1–6.

4. Ahmed, M.S.; Jeon, S. Highly Active Graphene-Supported Ni_xPd_{100-x} Binary Alloyed Catalysts for Electro-Oxidation of Ethanol in an Alkaline Media. *ACS Catal.* **2014**, *4*, 1830–1837.

5. Wang, Y.; Shi, F.-F.; Yang, Y.-Y.; Cai, W.-B. Carbon supported Pd–Ni–P nanoalloy as an efficient catalyst for ethanol electro-oxidation in alkaline media. *J. Power Sources* **2013**, *243*, 369–373.

6. Liang, Z.X.; Zhao, T.S.; Xu, J.B.; Zhu, L.D. Mechanism study of the ethanol oxidation reaction on palladium in alkaline media. *Electrochim. Acta* **2009**, *54*, 2203–2208.

7. Feng, L.; Zhang, J.; Cai, W.; Liang, L.; Xing, W.; Liu, C. Single passive direct methanol fuel cell supplied with pure methanol. *J. Power Sources* **2011**, *196*, 2750–2753.

8. Wang, Y.; He, Q.; Guo, J.; Wei, H.; Ding, K.; Lin, H.; Bhana, S.; Huang, X.; Luo, Z.; Shen, T.D.; *et al.* Carboxyl Multiwalled Carbon-Nanotube-Stabilized Palladium Nanocatalysts toward Improved Methanol Oxidation Reaction. *ChemElectroChem* **2015**, *2*, 559–570.

9. Lim, E.J.; Choi, S.M.; Seo, M.H.; Kim, Y.; Lee, S.; Kim, W.B. Highly dispersed Ag nanoparticles on nanosheets of reduced graphene oxide for oxygen reduction reaction in alkaline media. *Electrochem. Commun.* **2013**, *28*, 100–103.

10. Oliveira, M.C.; Rego, R.; Fernandes, L.S.; Tavares, P.B. Evaluation of the catalytic activity of Pd–Ag alloys on ethanol oxidation and oxygen reduction reactions in alkaline medium. *J. Power Sources* **2011**, *196*, 6092–6098.

11. Guo, S.; Sun, S. FePt nanoparticles assembled on graphene as enhanced catalyst for oxygen reduction reaction. *J. Am. Chem. Soc.* **2012**, *134*, 2492–2495.

12. Kim, J.; Momma, T.; Osaka, T. Cell performance of Pd–Sn catalyst in passive direct methanol alkaline fuel cell using anion exchange membrane. *J. Power Sources* **2009**, *189*, 999–1002.

13. Mao, H.; Wang, L.; Zhu, P.; Xu, Q.; Li, Q. Carbon-supported PdSn–SnO2 catalyst for ethanol electro-oxidation in alkaline media. *Int. J. Hydrogen Energy* **2014**, *39*, 17583–17588.

14. Ding, K.; Wang, Y.; Yang, H.; Zheng, C.; Cao, Y.; Wei, H.; Wang, Y.; Guo, Z. Electrocatalytic activity of multi-walled carbon nanotubes-supported Pt_xPd_y catalysts prepared by a pyrolysis process toward ethanol oxidation reaction. *Electrochim. Acta* **2013**, *100*, 147–156.

15. Jiang, K.; Cai, W.-B. Carbon supported Pd–Pt–Cu nanocatalysts for formic acid electrooxidation: Synthetic screening and componental functions. *App. Catal.* **2014**, *147*, 185–192.

16. Sieben, J.M.; Comignani, V.; Alvarez, A.E.; Duarte, M.M.E. Synthesis and characterization of Cu core Pt–Ru shell nanoparticles for the electro-oxidation of alcohols. *Int. J. Hydrogen Energy* **2014**, *39*, 8667–8674.

17. Ding, L.X.; Wang, A.L.; Li, G.R.; Liu, Z.Q.; Zhao, W.X.; Su, C.Y.; Tong, Y.X. Porous Pt–Ni–P composite nanotube arrays: Highly electroactive and durable catalysts for methanol electrooxidation. *J. Am. Chem. Soc.* **2012**, *134*, 5730–5733.

18. Wang, Y.; Zhao, Y.; Yin, J.; Liu, M.; Dong, Q.; Su, Y. Synthesis and electrocatalytic alcohol oxidation performance of Pd–Co bimetallic nanoparticles supported on graphene. *Int. J. Hydrogen Energy* **2014**, *39*, 1325–1335.

19. Kim, Y.; Noh, Y.; Lim, E.J.; Lee, S.; Choi, S.M.; Kim, W.B. Star-shaped Pd@Pt core-shell catalysts supported on reduced graphene oxide with superior electrocatalytic performance. *J. Mater. Chem. A* **2014**, *2*, 6976.

20. Zhang, Y.; Chang, G.; Shu, H.; Oyama, M.; Liu, X.; He, Y. Synthesis of Pt–Pd bimetallic nanoparticles anchored on graphene for highly active methanol electro-oxidation. *J. Power Sources* **2014**, *262*, 279–285.

21. Datta, J.; Dutta, A.; Mukherjee, S. The Beneficial Role of the Cometals Pd and Au in the Carbon-Supported PtPdAu Catalyst Toward Promoting Ethanol Oxidation Kinetics in Alkaline Fuel Cells: Temperature Effect and Reaction Mechanism. *J. Phys. Chem. C* **2011**, *115*, 15324–15334.

22. Feng, L.; Yan, L.; Cui, Z.; Liu, C.; Xing, W. High activity of Pd–WO3/C catalyst as anodic catalyst for direct formic acid fuel cell. *J. Power Sources* **2011**, *196*, 2469–2474.

23. Wang, Z.-B.; Zuo, P.-J.; Yin, G.-P. Effect of W on activity of Pt–Ru/C catalyst for methanol electrooxidation in acidic medium. *J. Alloy. Compd.* **2009**, *479*, 395–400.

24. Mellinger, Z.J.; Kelly, T.G.; Chen, J.G. Pd-Modified Tungsten Carbide for Methanol Electro-oxidation: From Surface Science Studies to Electrochemical Evaluation. *ACS Catal.* **2012**, *2*, 751–758.

25. Moon, J.-S.; Lee, Y.-W.; Han, S.-B.; Park, K.-W. Pd nanoparticles on mesoporous tungsten carbide as a non-Pt electrocatalyst for methanol electrooxidation reaction in alkaline solution. *Int. J. Hydrogen Energy* **2014**, *39*, 7798–7804.

26. Lu, Y.; Jiang, Y.; Gao, X.; Wang, X.; Chen, W. Strongly coupled Pd nanotetrahedron/tungsten oxide nanosheet hybrids with enhanced catalytic activity and stability as oxygen reduction electrocatalysts. *J. Am. Chem. Soc.* **2014**, *136*, 11687–11697.

27. Fan, Y.; Zhao, Y.; Chen, D.; Wang, X.; Peng, X.; Tian, J. Synthesis of Pd nanoparticles supported on PDDA functionalized graphene for ethanol electro-oxidation. *Int. J. Hydrogen Energy* **2015**, *40*, 322–329.

28. Hong, W.; Fang, Y.; Wang, J.; Wang, E. One-step and rapid synthesis of porous Pd nanoparticles with superior catalytic activity toward ethanol/formic acid electrooxidation. *J. Power Sources* **2014**, *248*, 553–559.

29. Wang, Y.; He, Q.; Ding, K.; Wei, H.; Guo, J.; Wang, Q.; O'Connor, R.; Huang, X.; Luo, Z.; Shen, T.D.; *et al.* Multiwalled Carbon Nanotubes Composited with Palladium Nanocatalysts for Highly Efficient Ethanol Oxidation. *J. Electrochem. Soc.* **2015**, *162*, F755–F763.

30. Monshi, A.; Foroughi, M.R.; Monshi, M.R. Modified Scherrer Equation to Estimate More Accurately Nano-Crystallite Size Using XRD. *World J. Nano Sci. Eng.* **2012**, *2*, 154–160.

31. Zhang, Z.; Xin, L.; Sun, K.; Li, W. Pd–Ni electrocatalysts for efficient ethanol oxidation reaction in alkaline electrolyte. *Int. J. Hydrogen Energy* **2011**, *36*, 12686–12697.

32. Maghsodi, A.; Milani Hoseini, M.R.; Dehghani Mobarakeh, M.; Kheirmand, M.; Samiee, L.; Shoghi, F.; Kameli, M. Exploration of bimetallic Pt-Pd/C nanoparticles as an electrocatalyst for oxygen reduction reaction. *Appl. Surf. Sci.* **2011**, *257*, 6353–6357.

33. Shi, J.; Ci, P.; Wang, F.; Peng, H.; Yang, P.; Wang, L.; Wang, Q.; Chu, P.K. Pd/Ni/Si-microchannel-plate-based amperometric sensor for ethanol detection. *Electrochim. Acta* **2011**, *56*, 4197–4202.

34. Yi, Q.; Niu, F.; Sun, L. Fabrication of novel porous Pd particles and their electroactivity towards ethanol oxidation in alkaline media. *Fuel* **2011**, *90*, 2617–2623.

A Facile Synthesis of Hollow Palladium/Copper Alloy Nanocubes Supported on N-Doped Graphene for Ethanol Electrooxidation Catalyst

Zhengyu Bai, Rumeng Huang, Lu Niu, Qing Zhang, Lin Yang and Jiujun Zhang

Abstract: In this paper, a catalyst of hollow PdCu alloy nanocubes supported on nitrogen-doped graphene support (H-PdCu/ppy-NG) is successfully synthesized using a simple one-pot template-free method. Two other catalyst materials such as solid PdCu alloy particles supported on this same nitrogen-doped graphene support (PdCu/ppy-NG) and hollow PdCu alloy nanocubes supported on the reduced graphene oxide support (H-PdCu/RGO) are also prepared using the similar synthesis conditions for comparison. It is found that, among these three catalyst materials, H-PdCu/ppy-NG gives the highest electrochemical active area and both the most uniformity and dispersibility of H-PdCu particles. Electrochemical tests show that the H-PdCu/ppy-NG catalyst can give the best electrocatalytic activity and stability towards the ethanol electrooxidation when compared to other two catalysts. Therefore, H-PdCu/ppy-NG should be a promising catalyst candidate for anodic ethanol oxidation in direct ethanol fuel cells.

Reprinted from *Catalysts*. Cite as: Bai, Z.; Huang, R.; Niu, L.; Zhang, Q.; Yang, L.; Zhang, J. A Facile Synthesis of Hollow Palladium/Copper Alloy Nanocubes Supported on N-Doped Graphene for Ethanol Electrooxidation Catalyst. *Catalysts* **2015**, *5*, 747–758.

1. Introduction

As a kind of sustainable clean energy technology, fuel cells have been demonstrated and recognized as the feasible option for energy conversion for power generation due to their high efficiency and zero/low emissions [1,2]. In several types of fuel cells, direct ethanol fuel cells (DEFCs) are considered to be one of the important options for automotive and portable electronic applications, owing to their high energy density, low operating temperature, and liquid fuel feeding operation [3,4]. Compared with the direct methanol fuel cells (DMFCs), DEFCs also have some advantages including lower fuel cost, lower toxicity, lower fuel crossover effect, and higher theoretical mass energy density (8 kWh· kg^{-1} *vs.* 6.1 kWh· kg^{-1}) [5]. Furthermore, ethanol can be easily produced in large scale from agricultural products or biomass [6].

Unfortunately, the low reaction activity and difficult C-C bond breaking of ethanol electrooxidation are the major drawbacks hindering DEFCs' practical applications [7,8]. Currently, the most effective catalysts used for ethanol electrooxidation are Pt-based materials, which are high-cost and also insufficient in overcoming both the low catalyst activity and difficulty of breaking C-C bond of ethanol. To overcome these challenges, tremendous efforts have been made to explore alternative catalysts which hopefully could give high activity/selectivity/stability, and be low cost.

With respect to this, some less expensive and more abundant non-platinum catalysts with acceptable performance have been widely explored. For instance, Pd-based catalysts have been found to have good a performance in ethanol electrooxidation, and therefore are considered to be good candidates for DEFCs [9,10]. Pd alloying with non-noble metals (Fe, Co, Ni, *etc.*) to produce multiple-component catalysts has also tested to be one of the effective approaches in enhancing the catalytic activity, decreasing the loading of noble metals, and then reducing the cost of Pd catalysts [11,12]. Furthermore, the nanostructure types of Pd-alloy catalysts have also been identified to play a considerable role in improving the catalyst's performance towards the ethanol electrooxidation [13]. Among the different nanostructures of Pd-alloy materials, hollow nanostructure represents a new type of catalyst because of their high surface area, low density, easy recovery, self-supporting capacity, and high surface permeability [14,15]. In this regard, various hollow nanostructures including hollow nanospheres [16] and hollow nanotubes [17] have been reported in literature. Meanwhile, the ideal catalyst supports with large surface areas, good conductivity and strong adsorption of metals have been demonstrated to have the ability to improve the dispersion of metal nanoparticles, and thereby enhance the utilization and efficiency of the noble metal electrocatalysts [18,19]. Among different catalyst supports, graphene-based materials have been considered to be one of the ideal catalyst supports because they possess a large surface area, good thermal and chemical stability as well as great electrical conductivity [20,21]. However, when metal or metal alloy particles are supported on the graphene surface, they tend to aggregate together due to the inefficient binding sites on the pristine graphene surface for anchoring metal nanoparticles. To improve the binding interaction between the catalyst particles and the graphene surface, some doping strategies to create more binding sites has been developed. For example, when graphene is doped with nitrogen to form N-doped graphene (NG), the binding interaction can be significantly improved [22]. It was observed that with the introduction of nitrogen into graphene support material, the metal nanoparticles could be homogeneously anchored onto the support, leading to the generation of MeN_x (Me Co, Fe) active sites, and thereby enhancing the electrocatalytic activity and utilization efficiency of the catalysts [23]. The studies showed that the N species on the graphene surface could play an important role in controlling and regulating the shape and size of metal nanoparticles [24]. Therefore, design and synthesis of hollow Pd-alloy nanosphere catalysts supported on N-doped graphene represent a new way to improve the performance and utilization of catalysts with the reduced cost.

In this paper, hollow PdCu-alloy nanocube catalysts supported on N-doped graphene (H-PdCu/ppy-NG) are successfully synthesized by a facile and low-cost method. The results indicate that the successfully synthesized N-doped graphene can tightly support hollow PdCu-alloy nanocubes with a uniform dispersion on the support and a relatively narrow distribution of catalyst particle size. Electrochemical characterizations reveal that the H-PdCu/ppy-NG catalyst has both excellent catalytic activity and stability toward ethanol electrooxidation in alkaline electrolyte, demonstrating that this catalyst would be a promising anode catalyst for DEFCs.

2. Results and Discussion

Figure 1 shows the TEM images and the size frequency curve of the resulting sample from the typical experiment. From Figure 1a, it can be seen that the hollow PdCu nanocubes are uniformly dispersed on the N-doped graphene surface with a uniform dispersion and a narrow particle size distribution. The catalyst particle diameters from the amplificatory TEM image vary from 35 to 53 nm, and the mean size calculated by the lognormal distribution is about 46.2 nm (Figure 1d). As observed in Figure 1c, the metal shell is clearly visible due to its higher contrast compared to the central cavity region. The contrast difference can prove the existence of the hollow structure. To better investigate the formation mechanism of hollow nanocubes in our system, two sets of control experiments were carried out. Figure 2 shows the TEM images from the control experiments. Control A was carried out under the same conditions described as the typical experiment, apart from 140 °C as the reaction tempeture. When a lower temperature (*i.e.*, 140 °C) was adopted, only solid PdCu nanoparticles supported on ppy-NG could be observed (Figure 2a), which might be due to that the low temperature was not conducive to oriented attachment process in the dynamics. As shown in Figure 2a, the PdCu alloy nanoparticles have near-spherical shapes and showed a slight agglomeration. It demonstrates that reaction tempeture is important in controlling and regulating the shape and size of the hollow nanospheres. To further study the effect of N doping, control B was done in the same conditions described as the typical experiment, apart from the RGO as the support. Figure 2b shows the TEM image of the H-PdCu/RGO, in which a small quantity of hollow PdCu nanocubes was immobilized on the RGO compared with the typical experiment. This indicates that the N doping is a key factor to absorb PdCu nanoparticles onto the surface of the support. From the above results, it can be believed that a much more uniform size and distribution of H-PdCu nanoparticles relies on the cooperation of the appropriate reaction temperature and N doping.

Figure 3 shows the XRD patterns of H-PdCu/ppy-NG, PdCu/ppy-NG and H-PdCu/RGO catalysts, respectively. As displayed in Figure 3, four peaks at 39.8°, 46.1°, 68.6° and 81.9° are characteristics of face-centered-cubic (fcc) crystalline Pd, which are corresponding to the facets (111), (200), (220), and (311), respectively. Obviously, the peak positions for H-PdCu/RGO slightly shift to higher angles when compared to Pd/RGO, which is ascribed to the formation of PdCu alloy. Additionally, the peak at 21.5° in each case can be attributed to the (002) planes of RGO, which is different from that sharp peak centered at 10.2° for GO, indicating the decreased interlayer distance from 0.71 to 0.34 nm [25]. This is due to the removal of oxygen-containing functional groups from the RGO. These observations demonstrate that the GO was efficiently transformed to RGO. Moreover, the XRD pattern of H-PdCu/ppy-NG is consistent with that of H-PdCu/RGO. It can be concluded that the crystal structure of RGO is not changed after N doping.

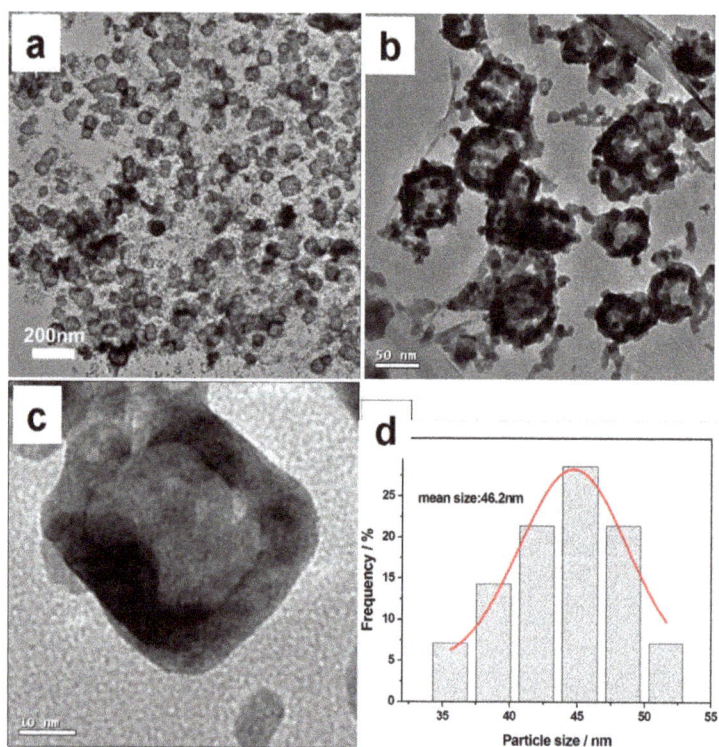

Figure 1. TEM images of TEM images (**a–c**) and the size frequency curve (**d**) of H-PdCu/ppy-NG.

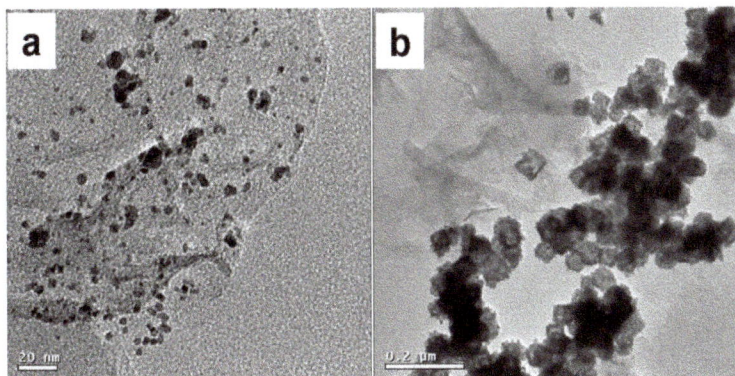

Figure 2. TEM images from PdCu/ppy-NG (**a**) and H-PdCu/RGO (**b**) catalysts.

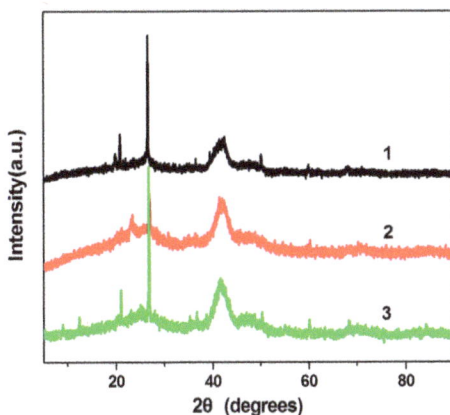

Figure 3. XRD patterns of H-PdCu/ppy-NG (*Curve* 1), PdCu/ppy-NG (*Curve* 2) and H-PdCu/RGO (*Curve* 3) catalysts.

The surface chemical states and elemental compositions of ppy-NG support were analyzed by XPS. Figure 4 shows the XPS spectra of the ppy-NG and the corresponding high-resolution N1s spectrum. The survey-scan spectrum of ppy-NG support is mainly dominated by the signals of C 1s, N 1s and O 1s elements. The presence of N1s peak at about 400 eV demonstrates the successful incorporation of nitrogen in NG supports. From Figure 4b, the high-resolution N1s spectrum can be deconvoluted into three peaks, which correspond to three individual N-containing species. The peak at about 400.1 eV can be assigned to pyrrolic N species from the pentagonal ring of ppy, 398.1 eV to pyridinic N, and 401.1 eV to graphitic N, respectively. From the sizes of the peaks, it can be calculated that the total N content in ppy-NG is 1.62%, which contains 15% of pyrrolic N, 43% of pyridinic N and 42% of graphitic N. More pyridinic N and graphitic N species on ppy-NG surfaces

605

should be helpful to load more H-PdCu for enhancing the catalytic activity of the H-PdCu/ppy-NG.

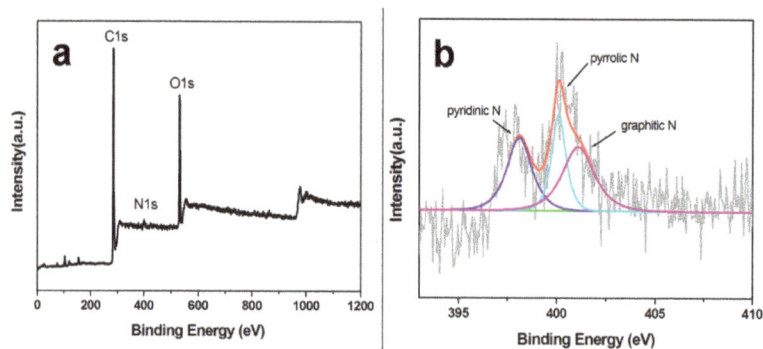

Figure 4. XPS spectra of ppy-NG support (**a**) and its corresponding high-resolution N1s spectrum (**b**).

In order to investigate the distribution of the different elements in the catalysts, elemental mapping measurements were also performed. Figure 5 shows the SEM images and the corresponding elemental mapping of the as-prepared catalysts from H-PdCu/ppy-NG (a) and PdCu/ppy-NG (b) catalysts. From the mapping images of the samples, a homogeneous distribution of N, Pd and Cu elements can be clearly observed, except from the C element. It can be seen that Pd and Cu are uniformly distributed in the mappings, which is in agreement with the TEM results. The results reveal that the graphene has been successfully doped by N and the N atoms are all homogeneous distributed in ppy-NG, which is in good accordance with the results of XPS spectra. In the process of N dope, the N atoms can provide highly effective functional groups on the surface of graphene, which contribute to the subsequent deposition of PdCu nanoparticles with a much more uniform size and distribution.

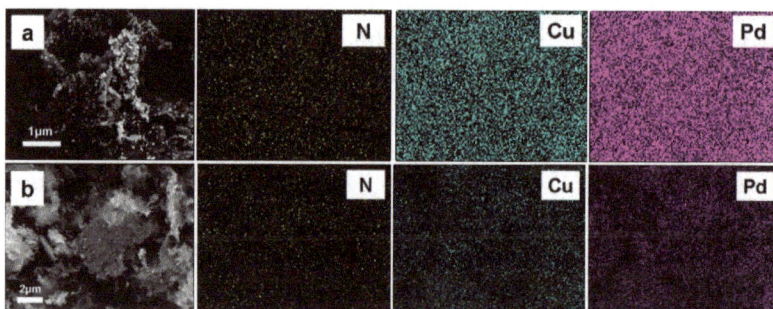

Figure 5. FESEM images and the corresponding chemical mapping of H-PdCu/ppy-NG (**a**) and PdCu/ppy-NG (**b**) catalyst.

606

For evaluating the electrochemically-active surface areas (ECSA) of the catalysts, CO-stripping experiments were carried out in N_2-saturated 0.5 M H_2SO_4 electrolyte at a scan rate of 50 mV·s^{-1}. Figure 6 shows the CO-stripping cyclic voltammograms (CVs) for three different catalysts (H-PdCu/ppy-NG (a), PdCu/ppy-NG (b) and H-PdCu/RGO (c), respectively). The corresponding ECSA values were calculated using Equation (1) [26]:

$$ECSA = \frac{Q}{G \times 420} \tag{1}$$

where Q is the charge of the CO desorption-electrooxidation in microcoulomb (μC), G represents the total amount of Pd (μg) on the electrode, and 420 is the charge required to oxidize a monolayer of CO on the catalyst in μC·cm^{-2}. The calculated ECSA values are 268, 202, and 138 m^2·g^{-1} for H-PdCu/ppy-NG, PdCu/ppy-NG and H-PdCu/RGO catalysts, respectively. Clearly, the ECSA value for the H-PdCu/ppy-NG catalyst is much larger than those of the other two, probably suggesting that H-PdCu/ppy-NG may be more active than both PdCu/ppy-NG and H-PdCu/RGO. Obviously, this ECSA value further demonstrates that N-doped graphene can effectively increase the active sites, and thereby may be able to enhance the catalytic activity and stability of the electrocatalysts.

The electrocatalytic activities for ethanol oxidation using the synthesized electrocatalysts were also analyzed by CV measurement in N_2-saturated 1.0 M KOH containing 1.0 M CH_3CH_2OH aqueous solution under the half-cell conditions at a scan rate of 50 mV·s^{-1}. Figure 7 compares the CV curves of three catalysts of H-PdCu/ppy-NG, PdCu/ppy-NG and H-PdCu/RGO, respectively. In general, the ethanol electrooxidation can be characterized by two well-defined current peaks at the forward and reverse scans. In the forward scan, the oxidation peak in Figure 7 is corresponding to the oxidation of freshly chemisorbed species which come from ethanol adsorption. The reverse scan peak is primarily associated with removal of carbonaceous species which are not completely oxidized in the forward scan. The value of the peak current in the forward scan represents the electrocatalytic activities of the electrocatalysts. From Figure 7, two main peaks for ethanol oxidation in both forward and reverse scan directions can be observed at all three electrodes coated by three catalysts separately. The corresponding anodic peak current density of H-PdCu/ppy-NG is about 650 mA·mg^{-1}, much higher than those of the PdCu/ppy-NG (*ca.* 320 mA·mg^{-1}) and H-PdCu/RGO (*ca.* 150 mA·mg^{-1}). This demonstrates that H-PdCu/ppy-NG modified electrode can give an extraordinarily higher electrocatalytic activity than the other two for ethanol electrooxidation.

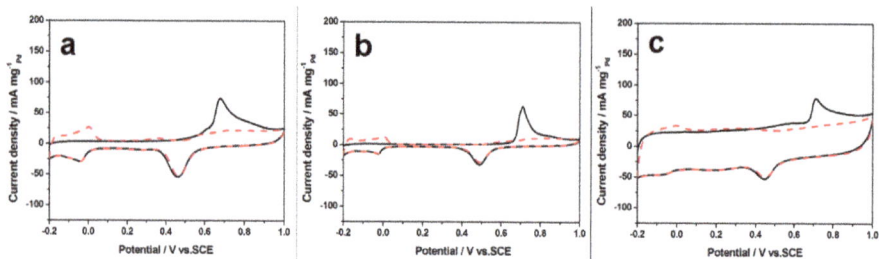

Figure 6. Cyclic voltammograms (CVs) of the electrooxidation of pre-adsorbed CO on H-PdCu/ppy-NG (**a**); PdCu/ppy-NG (**b**) and H-PdCu/RGO (**c**) catalysts coated glassy carbon electrodes in N_2-saturated 0.5 M H_2SO_4 aqueous solution with a scan rate of 50 mV·s^{-1} at 25 °C. Dashed curves were CVs for these catalyzed electrodes without CO adsorption.

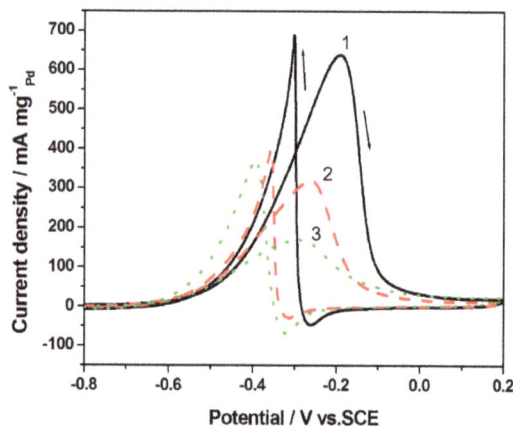

Figure 7. Cyclic voltammograms of H-PdCu/ppy-NG (*Curve* 1), PdCu/ppy-NG (*Curve* 2) and H-PdCu/RGO (*Curve* 3) coated glassy carbon electrodes. Electrolyte: N_2-saturated 1.0 M KOH containing 1.0 M CH_3CH_2OH aqueous solution at 25 °C, potential scan rate: 50 mV·s^{-1}.

In order to compare the electrochemical stability of the catalysts for alcohol oxidation, chronoamperometric tests were carried out at −0.3 V for 6000 s in N_2-saturated 1 M NaOH solution containing 1 M ethanol (Figure 8). Evidently, the H-PdCu/ppy-NG catalyst shows a much higher anodic current and a much slower degradation than other two catalysts. The result further demonstrates that the NG using ppy as a N source can significantly enhance both the activity and stability of the catalyst toward ethanol electrooxidation.

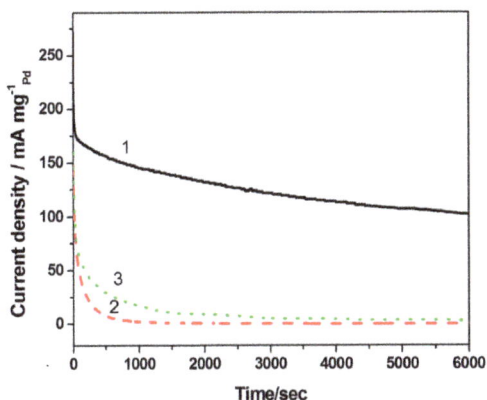

Figure 8. Chronoamperometric curves of H-PdCu/ppy-NG (*Curve* 1), PdCu/ppy-NG (*Curve* 2) and H-PdCu/RGO (*Curve* 3) coated glassy carbon electrodes. Electrolyte: N_2-saturated 1.0 M KOH containing 1.0 M CH_3CH_2OH aqueous solution at 25 °C. Electrode potential held at −0.3 V.

3. Experimental Section

3.1. Preparation of Different N-Doped Graphenes

Graphene oxide (GO) was prepared based on the modified Hummers' method [27]. The N-doped graphene using polypyrrole as nitrogen source (abbreviated as ppy-NG) was synthesized by *in-situ* chemical oxidative polymerization of pyrrole monomer and carbonization of the polypyrrole. In a typical synthesis, GO (0.4 g) was dispersed in 50 mL of ethanol aqueous solution (volume ratio 1:1) by ultrasonic treatment for 30 min to form a suspension. 80 mg of pyrrole monomer was then added to this suspension and stirred for 10 min to form a mixture solution. Then 50 mL of $Na_2S_2O_8$ (2.289 g) aqueous solution was added slowly to this mixture solution with a constant stirring for 12 h in ice-water bath. After the reaction, the obtained ppy-GO powder was dried and heated at 800 °C for 2 h under the protection of N_2. The formed product is labeled as ppy-NG and used as the catalyst support in this paper.

3.2. Synthesis of Hollow PdCu-Alloy Nanocube Catalysts

In a typical synthesis, an aqueous solution of $PdCl_2$ (3.3 mg·mL^{-1}, 3.8 mL), 20 mg of $CuSO_4 \cdot 5H_2O$, and 50 mg of glutamate were mixed together in 40 mL of ethylene glycol (EG). The solution pH was adjusted to 11 by dropwise addition of 8 wt.% KOH/EG solution with vigorous stirring. Then, 30 mg of the as-prepared ppy-NG above was added into the solution with ultrasonicately stirred for 2 h to obtain a homogeneous suspension. Upon completion, the suspension was transferred

609

into a 50 mL Teflon-lined stainless-steel autoclave. The autoclave was sealed, heated at 160 °C for 6 h, and then air-cooled to room temperature. Finally, the product was collected by filtration and washed several times with double distilled water. The catalyst was dried at 40 °C under vacuum for 8 h. The catalyst thus obtained is denoted as H-PdCu/ppy-NG in this paper.

3.3. Comparison Experiments

For comparison, solid PdCu nanoparticles supported on ppy-NG (PdCu/ppy-NG) was also prepared at 140 °C under nearly identical conditions as theose for H-PdCu/ppy-NG. For a further comparison, the graphene made from the reduction of GO without N-doping (expressed as RGO) was also used as a support to prepare a RGO-supported hollow PdCu nanotube catalyst, abbreviated as H-PdCu/RGO in this paper.

3.4. Material Characterization

The morphology of the catalyst samples was characterized by transmission electron microscopy (TEM) (JEOL-100CX) at 200 kV. The crystal structure of the products was analyzed by X-ray diffraction (XRD) recorded on a D/max-2200/PC X-ray diffractometer with Cu Kα radiation source. Field emission scanning electron microscope (FESEM) images and energy dispersive X-ray spectroscopy (EDX) results were obtained with ZEISS SUPRA 40 and X-MAX 20. The X-ray photoelectron spectroscopy (XPS) measurements were made using ESCALABMKLL electron spectroscope from VG Scientific (West Sussex, UK).

The electrochemical measurements in this study were conducted with a conventional three-electrode electrochemical cell using a CHI 660E electrochemical workstation. A glassy carbon electrode (3 mm o.d.) coated with catalyst was used as the working electrode, a saturated calomel electrode (SCE) as the reference electrode, and a Pt foil (1 cm^2) as the counter electrode. The cyclic voltammonograms (CVs) and chronoamperometric curves for ethanol electrooxidation experiments were recorded in N$_2$-saturated 1 M KOH containing 1 M ethanol. Electrochemical CO-stripping voltammograms were conducted by bubbling CO into 0.5 M H$_2$SO$_4$ for 30 min at a potential of 0.1 V (*vs.* SCE electrode). All electrochemical experiments were performed at 25 $\pm$ 1 °C.

4. Conclusions

In this work, the hollow PdCu alloy nanocubes supported on nitrogen-doped graphene support (H-PdCu/ppy-NG) were successfully synthesized using a simple one-pot template-free method. For comparison, two other catalyst materials such as solid PdCu alloy particles supported on nitrogen-doped graphene support (PdCu/ppy-NG) and hollow PdCu alloy nanocubes supported on the reduced graphene oxide support (H-PdCu/RGO) were also prepared under the similar synthetic conditions. Among these three catalyst materials (H-PdCu/ppy-NG, PdCu/ppy-NG, and H-PdCu/RGO), H-PdCu/ppy-NG showed the highest electrochemical active sites and both the most uniformity and dispersibility of H-PdCu particles, and the electrochemical tests showed that H-PdCu/ppy-NG catalyst could give the highest electrocatalytic activity and stability towards the ethanol electrooxidation. Therefore, H-PdCu/ppy-NG should be a promising catalyst candidate for anodic ethanol oxidation in direct ethanol fuel cells.

Acknowledgments: This work was financially supported by the National Natural Science Foundation of China (grant nos. 21301051), and Basic and Frontier Research Program of Henan Province (grant No. 132300410016).

Author Contributions: L.Y. and J.Z. conceived and designed the experiments; Z.B., R.H. and L.N. performed the experiments; Z.B. and Q.Z. analyzed the data; Z.B. and J.Z. wrote the paper.

Conflicts of Interest: The authors declare no conflict of interest.

References

1. Wang, L.; Nemoto, Y.; Yamauchi, Y. Direct Synthesis of Spatially-Controlled Pt-on-Pd Bimetallic Nanodendrites with Superior Electrocatalytic Activity. *J. Am. Chem. Soc.* **2011**, *133*, 9674–9677.
2. Rajesh, B.; Piotr, Z. A Class of Non-Precious Metal Composite Catalysts for Fuel Cells. *Nature* **2006**, *443*, 63–66.
3. Ma, L.; Chu, D.; Chen, R.R. Comparison of Ethanol Electro-Oxidation on Pt/C and Pd/C Catalysts in Alkaline Media. *Int. J. Hydrogen Energy* **2012**, *37*, 11185–11194.
4. Liu, J.P.; Ye, J.Q.; Xu, C.W.; Jiang, S.P.; Tong, Y.X. Kinetics of Ethanol Electrooxidation at Pd Electrodeposited on Ti. *Electrochem. Commun.* **2007**, *9*, 2334–2339.
5. Hsin, Y.L.; Hwang, K.C.; Yeh, C.T. Poly (vinylpyrrolidone)-Modified Graphite Carbon Nanofibers as Promising Supports for PtRu Catalysts in Direct Methanol Fuel Cells. *J. Am. Chem. Soc.* **2007**, *129*, 9999–10010.
6. Chen, X.T.; Jiang, Y.Y.; Sun, J.Z.; Jin, C.H.; Zhang, Z.H. Highly Active Nanoporous Pt-Based Alloy as Anode and Cathode Catalyst for Direct Methanol Fuel Cells. *J. Power Sources* **2014**, *267*, 212–218.
7. Bian, C.N.; Shen, P.K. Palladium-based Electrocatalysts for Alcohol Oxidation in Half Cells and in Direct Alcohol Fuel Cells. *Chem. Rev.* **2009**, *109*, 4183–4206.
8. Dong, Q.; Zhao, Y.; Han, X.; Wang, Y.; Liu, M.C.; Li, Y. Pd/Cu Bimetallic Nanoparticles Supported on Graphene Nanosheets: Facile Synthesis and Application as Novel Electrocatalyst for Ethanol Oxidation in Alkaline Media. *Int. J. Hydrogen Energy* **2014**, *39*, 14669–14679.
9. Xu, C.W.; Shen, P.K.; Liu, Y.L. Ethanol Electrooxidation on Pt/C and Pd/C Catalysts Promoted with Oxide. *J. Power Sources* **2007**, *164*, 527–531.

10. Li, L.Z.; Chen, M.X.; Huang, G.B.; Yang, N.; Zhang, L.; Wang, H.; Liu, Y.; Wang, W.; Gao, J.P. A Green Method to Prepare Pd–Ag Nanoparticles Supported on Reduced Graphene Oxide and Their Electrochemical Catalysis of Methanol and Ethanol Oxidation. *J. Power Sources* **2014**, *263*, 13–21.

11. Yang, Z.S.; Wu, J.J. Pd/Co Bimetallic Nanoparticles: Coelectrodeposition under Protection of PVP and Enhanced Electrocatalytic Activity for Ethanol Electrooxidation. *Fuel Cells* **2012**, *12*, 420–425.

12. Feng, Y.Y.; Liu, Z.H.; Xu, Y.; Wang, P.; Wang, W.H.; Kong, D.S. Highly Active PdAu Alloy Catalysts for Ethanol Electro-Oxidation. *J. Power Sources* **2013**, *232*, 99–105.

13. Qi, Z.; Geng, H.R.; Wang, X.G.; Zhao, C.C.; Ji, H.; Zhang, C.; Xu, J.L.; Zhang, Z.H. Novel nanocrystalline PdNi alloy catalyst for methanol and ethanol electro-oxidation in alkaline media. *J. Power Sources* **2011**, *196*, 5823–5828.

14. Lang, H.F.; Maldonado, S.; Stevenson, K.J.; Chandler, B.D. Synthesis and Characterization of Dendrimer Templated Supported Bimetallic Pt–Au Nanoparticles. *J. Am. Chem. Soc.* **2004**, *126*, 12949–12956.

15. Zhang, H.; Hao, Q.; Geng, H.R.; Xu, C.X. Nanoporous PdCu Alloys as Highly Active and Methanol-Tolerant Oxygen Reduction Electrocatalysts. *Int. J. Hydrogen Energy* **2013**, *38*, 10029–10038.

16. Stamenkovic, V.R.; Mun, B.S.; Arenz, M.K.; Mayrhofer, J.J.; Lucas, C.A.; Wang, G.; Ross, P.N.; Markovic, N.M. Trends in Electrocatalysis on Extended and Nanoscale Pt-Bimetallic Alloy Surfaces. *Nat. Mater.* **2007**, *6*, 241–247.

17. Lv, J.J.; Zheng, J.N.; Wang, Y.Y.; Wang, A.J.; Chen, L.L.; Feng, J.J. A Simple One-Pot Strategy to Platinum–Palladium@Palladium Core–Shell Nanostructures with High Electrocatalytic Activity. *J. Power Sources* **2014**, *265*, 231–238.

18. Liang, H.P.; Guo, Y.G.; Zhang, H.M.; Hu, J.S.; Wan, L.J.; Bai, C.L. Controllable AuPt Bimetallic Hollow Nanostructures. *Chem. Comm.* **2004**, *13*, 1496–1497.

19. Liu, Z.L.; Zhao, B.; Guo, C.L.; Sun, Y.J.; Xu, F.G.; Yang, H.B.; Li, Z. Novel Hybrid Electrocatalyst with Enhanced Performance in Alkaline Media: Hollow Au/Pd Core/Shell Nanostructures with a Raspberry Surface. *J. Phys. Chem. C* **2009**, *113*, 16766–16771.

20. Tan, C.; Huang, X.; Zhang, H. Synthesis and Applications of Graphene-Based Noblemetal Nanostructures. *Mater. Today* **2013**, *16*, 29–36.

21. Chen, X.M.; Wu, G.H.; Chen, J.M.; Chen, X.; Xie, Z.X.; Wang, X.R. Synthesis of "Clean" and Well-Dispersive Pd Nanoparticles with Excellent Electrocatalytic Property on Graphene Oxide. *J. Am. Chem. Soc.* **2011**, *133*, 3693–3695.

22. Zhang, L.S.; Liang, X.Q.; Song, W.G.; Wu, Z.Y. Identification of the Nitrogen Species on N-Doped Graphene Layers and Pt/NG Composite Catalyst for Direct Methanol Fuel Cell. *Phys. Chem. Chem. Phys.* **2010**, *12*, 12055–12059.

23. Shao, Y.Y.; Zhang, S.; Engelhard, M.H.; Li, G.S.; Shao, G.C.; Wang, Y.; Liu, J.; Aksay, I.A.; Lin, Y.H. Nitrogen-Doped Graphene and Its Electrochemical Applications. *J. Mater. Chem.* **2010**, *20*, 7491–7496.

612

24. Favaro, M.; Agnoli, S.; Perini, L.; Durante, C.; Gennaro, A.; Granozzi, G. Palladium Nanoparticles Supported on Nitrogen-Doped HOPG: A Surface Science and Electrochemical Study. *Phys. Chem. Chem. Phys.* **2013**, *15*, 2923–2931.

25. Li, S.S.; Hu, Y.Y.; Feng, J.J.; Lv, Z.Y.; Chen, J.R.; Wang, A.J. Rapid Room-Temperature Synthesis of Pd Nanodendrites on Reduced Graphene Oxide For catalytic Oxidation of Ethylene Glycol and Glycerol. *Int. J. Hydrogen Energy* **2014**, *39*, 3730–3738.

26. Bai, Z.Y.; Guo, Y.M.; Yang, L.; Li, L.; Li, W.J.; Xu, P.L.; Hu, C.G.; Wang, K. Highly Dispersed Pd Nanoparticles Supported on 1,10-Phenanthroline-Functionalized Multi-walled Carbon Nanotubes for Electrooxidation of Formic Acid. *J. Power Sources* **2011**, *196*, 6232–6237.

27. Geng, D.S.; Chen, Y.; Chen, Y.G.; Li, Y.L.; Li, R.Y.; Sun, X.L.; Ye, S.Y.; Knights, S. High Oxygen-Reduction Activity and Durability of Nitrogen-Doped Graphene. *Energy Environ. Sci.* **2011**, *3*, 760–764.

Improving the Ethanol Oxidation Activity of Pt-Mn Alloys through the Use of Additives during Deposition

Mohammadreza Zamanzad Ghavidel and E. Bradley Easton

Abstract: In this work, sodium citrate (SC) was used as an additive to control the particle size and dispersion of Pt-Mn alloy nanoparticles deposited on a carbon support. SC was chosen, since it was the only additive tested that did not prevent Mn from co-depositing with Pt. The influence of solution pH during deposition and post-deposition heat treatment on the physical and electrochemical properties of the Pt-Mn alloy was examined. It was determined that careful control over pH is required, since above a pH of four, metal deposition was suppressed. Below pH 4, the presence of sodium citrate reduced the particle size and improved the particle dispersion. This also resulted in larger electrochemically-active surface areas and greater activity towards the ethanol oxidation reaction (EOR). Heat treatment of catalysts prepared using the SC additive led to a significant enhancement in EOR activity, eclipsing the highest activity of our best Pt-Mn/C prepared in the absence of SC. XRD studies verified the formation of the Pt-Mn intermetallic phase upon heat treatment. Furthermore, transmission electron microscopy studies revealed that catalysts prepared using the SC additive were more resistant to particle size growth during heat treatment.

Reprinted from *Catalysts*. Cite as: Ghavidel, M.Z.; Easton, E.B. Improving the Ethanol Oxidation Activity of Pt-Mn Alloys through the Use of Additives during Deposition. *Catalysts* **2015**, *5*, 1016–1033.

1. Introduction

Problems associated with the costs and efficiency of fuel cells are a great barrier for industrial and consumer applications [1–4]. Direct alcohol fuel cells (DAFCs) are one of the most promising candidates for portable power applications in electronic devices and vehicles [4]. The cost and performance of the DAFCs are mainly controlled by the catalysts used at each electrode. Pt is the most commonly utilized electrocatalyst, which is quite expensive [3]. Furthermore, strongly adsorbing species, such as CO, which are formed during the alcohol oxidation process on pure Pt particles, result in severe activity and efficiency losses [4,5]. The development of Pt alloy catalysts offers the potential of greater tolerance to poisoning and significant cost reduction. Likewise, decreasing Pt alloy particle sizes and improving particle dispersion can further increase performance [3], although there is some debate about the effects of particle size on catalytic activity [6,7].

The Pt-Mn alloy system has recently been identified by our group as having enhanced activity towards the ethanol oxidation reaction (EOR) [8,9]. Alloy formation was confirmed with X-ray powder diffraction (XRD) analysis, and the most active alloys contained less than 25 at% Pt, which is beneficial from a cost standpoint. The results showed that the presence of Mn affects both particle size and the intrinsic activity of the catalysts. Further study by the authors also showed that post-heat treatment had a great impact on the activity of the Pt-Mn alloys, and the main reason for enhancing the EOR activity was the formation of Pt-Mn intermetallic phase [10]. However, particle size growth during heat treatment was an unwanted consequence of heat treatment. Sintering can happen by the migration and coalescence of the catalyst particles or by evaporation and condensation of the atoms from small crystallites [11]. While the benefits of thermally treating alloy nanoparticles out-weighed any activity losses that may occur due to particle size growth, it would be desirable to find a way to produce smaller particles with better dispersion that are resistant to particle growth during heat treatment.

Common strategies used to deposit small and well-dispersed metal nanoparticles on carbon include functionalization of the support [12], using the polyol [13] or microemulsion [14,15] deposition methods [11,16] employing surfactants [17,18], which are significant strategies to improve particle dispersion and to reduce particle sizes. It has been shown that oxygenated surface groups on the carbon support can enhance the dispersion and the stability of Pt/C catalysts [11]. However, oxygen containing groups can be reduced during the reduction step, which can result in the redistribution of platinum particles and less favorable Pt dispersion [11]. Studies have shown that nitrogen functionalization on carbon can improve cathode performance [11,19]. While Dinotto and Negro [20,21] have produced some carbon nitride-based electrocatalysts at lower temperatures, nitrogen groups are more commonly introduced via high temperature processes that can also alter the porosity and microstructure of the support [11]. Unfortunately, the efforts in our group to produce Pt-Mn alloys from polyol and microemulsion methods were not successful because of very negative reduction potential for Mn ions in these solutions, which prevented Mn co-deposition [22].

Several studies [17,18,23,24] have shown that adding surfactants reduces the particle sizes of Pt and Pt alloys nanoparticles and also improves their dispersion on the support [25]. Sodium citrate (SC) is a common surfactant used in both aqueous and organic solutions by numerous researchers to produce Pt [18], Pt-Au [26], PtRuIr [27] and Pt-Co [28] nano-particles that were small (2–6 nm) with a narrow size distribution. However, to the best of our knowledge, SC has not been used to prepare Pt-Mn alloys [18,26–28].

In this paper, the effect of sodium citrate on particle size, dispersion, structure and EOR activity of Pt-Mn was investigated. In addition, the influence of solution pH and heat treatment on the crystalline structure, the uniformity of alloyed phases and the activity of the catalysts has been examined.

2. Results and Discussion

2.1. Material Characterization

Citric acid is a polyprotic acid, with pKa values of 3.14, 4.76 and 6.40 for each acid site. As such, the charge on SC will be influenced by solution pH, which can influence both Mn and Pt deposition, as well as the resulting particle sizes. Our preliminary studies showed that pH had an impact on Mn and Pt deposition. When the solution pH was above three, the metal loading and, as a consequence, the electrochemical activity of the sample dropped, and different trends were seen from sample to sample at higher pHs (Figure S1). We believe that at higher pHs, there is a stronger interaction between the citrate ions and Mn^{2+} ions in the precursor solution, which prevents them from deposition. Therefore, in this paper, all of the samples have been produced at pH 3.

Table 1 contains the post-chemical reduction composition of the Pt-Mn catalyst samples and the residual solutions, which were determined by inductively-coupled plasma optical emission spectroscopy (ICP-OES). Catalysts were prepared using SC to metal weight ratios of 1:1, 2:1 and 3:1, which are hereafter referred to as 1X, 2X and 3X, respectively. These results showed that the Pt-Mn alloys were produced with a molar ratio close to the calculated values. By adding SC, a small increase in the amount of metal ions in the residual solution was observed, especially the Pt content.

Table 1. Composition of the samples and concentration of Pt and Mn in filtrated solution, which was measured by ICP, along with grain size measured by TEM.

Samples	Alloy molar ratios measured by ICP		Ions concentration in filtered solution by ICP		Grain size measured by TEM (nm)
	Pt (%)	Mn (%)	Pt (ppm)	Mn (ppm)	
$Pt_{0.25}Mn_{0.75}$	22.18	77.82	nil	nil	4.5
$Pt_{0.25}Mn_{0.75}$-1X	20.30	79.70	0.49	nil	2.6
$Pt_{0.25}Mn_{0.75}$-2X	20.56	79.44	8.23	nil	2.8
$Pt_{0.25}Mn_{0.75}$-3X	21.94	78.06	11.70	0.20	2.9
$Pt_{0.25}Mn_{0.75}$-2X-500-1 h	-	-	-	-	5.7
$Pt_{0.25}Mn_{0.75}$-2X-700-1 h	-	-	-	-	5.8
$Pt_{0.25}Mn_{0.75}$-2X-875-1 h	-	-	-	-	6.0
$Pt_{0.25}Mn_{0.75}$-2X-950-1 h	-	-	-	-	6.6

The thermogravimetry (TG) and derivative thermogravimetry (DTG) of Pt-Mn alloys, which were synthesized on a Vulcan carbon support in the presence and absence of SC, are shown in Figure 1a,b. Five distinctive mass loss regions are observed in Figure 1. The mass loss between 100 °C and 250 °C was attributed to the thermal decomposition of residual, weak carbon functional groups and water evaporation in the powders [29,30]. A second major mass loss began at 300 °C, which was related to the oxidation of carbon black by the oxygen or the air trapped within the powder particles [30]. The mass loss at 700–800 °C is attributed to the loss of various functionalized groups on the carbon surface and graphitization [29]. The mass loss at 577 and 928 °C, for the sample prepared without additive, and at 460 and 919 °C, for the sample prepared in the presence of SC, was attributed to Mn oxide phase modifications and a reduction in the amount of oxygen [31]. It has been shown that pure MnO_2 is reduced to Mn_2O_3 at 500 °C and further reduced to Mn_3O_4 at 900 °C [30]. The source of mass loss observed at ~1076 °C was not identified. However, from the TG and DTG diagrams of the samples prepared with and without SC, it could be concluded that the temperature required for most of the phase transitions was moved to lower temperatures and facilitated by adding SC.

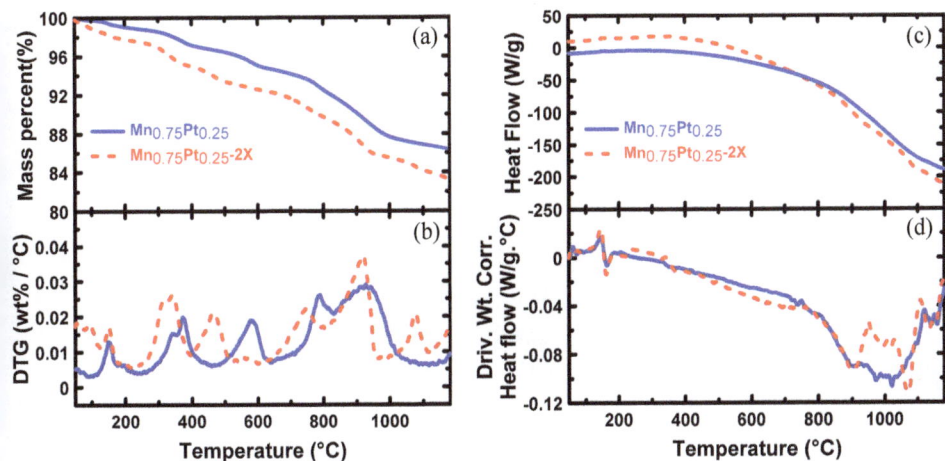

Figure 1. (**a**) TG, (**b**) DTG, (**c**) DSC and (**d**) derivative weight-corrected DSC for Pt-Mn alloys, which were synthesized on Vulcan carbon support in the presence and absence of sodium citrate.

Figure 1b illustrates the DSC curves obtained for Pt-Mn alloys prepared in the presence and absence of SC. Most of the reactions were endothermic, except those related to carbon oxidation at 300–400 °C. The derivative heat flow diagrams from 700–1200 °C showed that the heat flow in the presence of the additive was divided into two separate peaks. The first peak was related to an expected phase transformation from the Pt-Mn phase diagram [32] or Mn oxide phase modifications. The second peak might be attributed to the alloy melting or unknown phase transformation. As catalysts were not prepared at heat treatments above 950 °C, this was not examined in detail. The measured heat between 700–1000 °C (3.97 mW/g) for the Pt-Mn alloy prepared without SC is higher than that measured for the sample synthesized by SC (2.89 mW/g). Therefore, adding SC decreased the heat required for the phase transformation and facilitated the alloying process. Based on the DSC results, heat treatment temperatures of 500, 700, 875 and 950 °C were selected to compare the electrochemical and structural changes of Pt-Mn samples prepared with and without SC.

Figure 2 shows the TGA diagrams and mass loss in air. The mass lost at around 400 °C was due to carbon combustion. The Vulcan carbon ignition was around 600 °C, which was facilitated in the presence of metal. The data calculated from Figure 2 are presented in Table 2. The results showed that when SC is used, the combustion temperature is reduced by *ca.* 80 °C. This implies that the Pt-alloy particle size was reduced and the particle dispersion was improved. It was previously observed that by increasing Pt loading and available Pt surface area, the combustion temperature of carbon black was decreased because of a higher oxygen and carbon reaction rate [30,33]. Finally, the residual mass above 600 °C indicated that the metal loading in all samples was close to the expected 20 wt%.

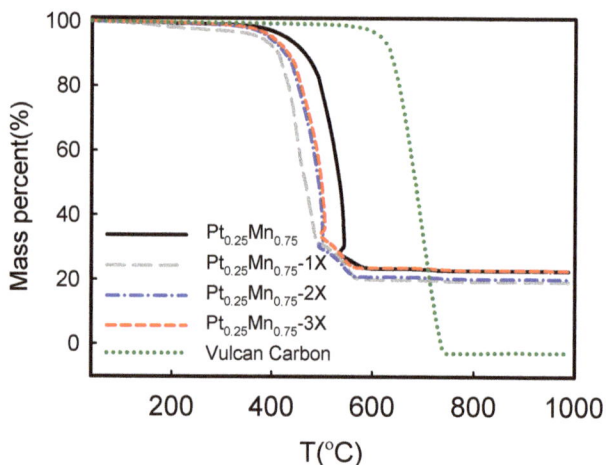

Figure 2. The effect of sodium citrate concentration on the weight loss of Pt-Mn alloys, which were synthesized on Vulcan carbon support.

Table 2. The metal loading and the Vulcan carbon combustion temperature measured from Figure 2.

Samples	Carbon black combustion Temperature (°C)	Metal loading (wt. %)
$Pt_{0.25}Mn_{0.75}$	429.2	22.6
$Pt_{0.25}Mn_{0.75}$-1X	348.4	19
$Pt_{0.25}Mn_{0.75}$-2X	350.7	20.1
$Pt_{0.25}Mn_{0.75}$-3X	359.5	22.8
Vulcan carbon	633.4	-

XRD patterns obtained for Pt-Mn alloy catalysts prepared with varying amounts of SC are shown in Figure 3. We have previously reported that as-deposited catalysts contained a mixed structure of Pt-Mn alloys and non-alloyed phases [10]. Broad peaks indicate that alloy particles with small grain sizes were produced. The broadening of the peaks can also be due to the presence of oxide phases and/or non-uniform alloys. The diffractogram displayed the characteristics of the face-centered cubic (fcc) structure of Pt, and the peaks were shifted to higher angles, indicating the incorporation of Mn in the fcc structure. Additionally, there is a peak at 36.2°, which is likely related to Mn-rich phases. Moreover, it can be concluded that, by adding SC, the peaks became broader, which was the result of the particle size reduction. Increasing the concentration of sodium citrate up to 2X reduced the amount of oxide phases. However, the oxide phases reappeared after the amount of sodium citrate was increased to 3X. It seems that the optimum amount for the SC concentration is 2X. Therefore, catalysts prepared with a 2X ratio were selected for a more detailed heat-treatment investigation.

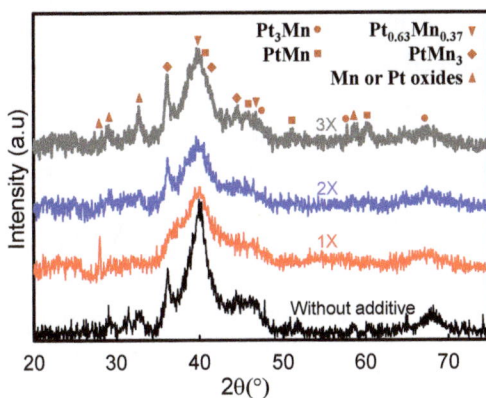

Figure 3. XRD analysis of the samples, which were prepared in the presence and absence of sodium citrate.

The XRD patterns obtained for Pt-Mn catalysts prepared with 2X SC that were heat treated at different temperatures are presented in Figure 4. As has been shown previously [10], heat treatment has a great impact on the activity and the crystallite structure of the Pt-Mn samples produced without the additive. Here, we have observed similar results for the samples prepared in the presence of SC. The peak at 36.2° associated with the Mn-rich phases in the XRD pattern disappeared after heat treatment at 500 °C for 1 h and the intensity of the remaining peaks increased with crystallization. However, the predominant structure was still the Pt face-centered cubic (fcc) structure.

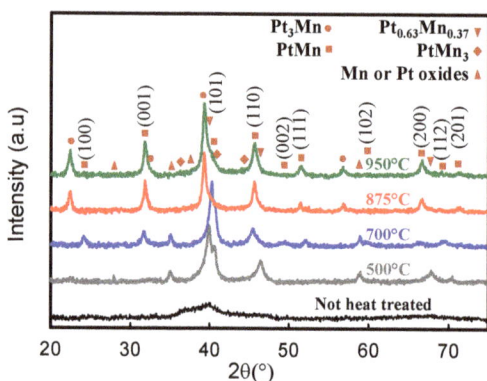

Figure 4. The XRD patterns of the sample prepared in the presence 2X SC and after heat treatment at different temperatures.

Upon further increasing the heat treatment temperature to 700 °C, substantial changes were observed. New peaks at lower diffraction angles (22°–40°) indicate that the ordered Pt-Mn intermetallic phase was formed at 700 °C. The Pt-Mn intermetallic phase [34] has a tetragonal structure; therefore, new peaks, (001) and (100), were demonstrated at lower diffraction angles of ~24.1° and ~37.1°, respectively. As a result of Pt and Mn further alloying and intermetallic phase formation, the peak shifts from 39.8° up to 40.2° and 46.5° down to 45.5° were observed when heat treatment temperature increased from 500 up to 700 °C. The shift and the intensity decline of the peaks at 39.8° and 46.5° can be assigned to the completion of phase modifications at 700 °C [10]. In Figure 5, the XRD spectra of samples prepared in the presence of SC and heat treated at 700 °C for different periods are compared. The spectra show that the phase transformation was completed after 1 h of heat treatment, and further increasing of the heat treatment time has no effect on the structure of the samples. The optimum heat treatment period for the samples prepared without SC was 4 h [10]. Presumably, smaller particle sizes undergo a faster phase transformation, which has also been found by the TGA and DSC analysis. This fast phase transformation is very beneficial, since a shorter treatment time should minimize particle size growth, yielding a higher active surface area and potentially improved electrochemical activity.

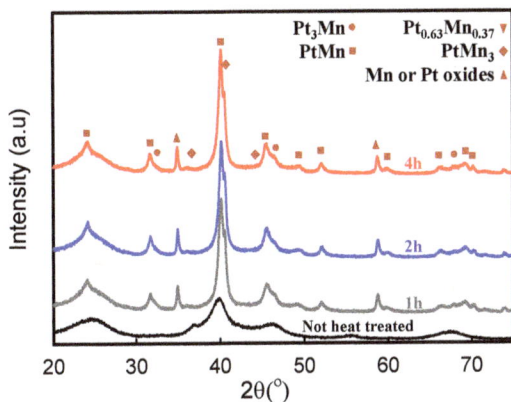

Figure 5. The XRD patterns of the sample prepared in the presence of 2X sodium citrate (SC) and after heat treatment at 700 °C for different periods.

In Figure 4, when the heat treatment temperature increased to 875 °C and 950 °C, additional peaks at 22.4° and 39.4° were observed. This variation in the crystalline structure of the Pt-Mn samples is due to a phase separation and the formation of phases with higher Pt content, such as Pt_3Mn. This phenomenon is also observed for the Pt-Mn samples prepared without SC [10].

TEM images of as-synthesized samples with and without the additive are presented in Figure 6. The mean particle sizes measured by using the TEM images are given in Table 1. These TEM images showed that the catalysts with nanosized metal particles were synthesized, and adding sodium citrate dramatically decreased the particle sizes and reduced the agglomeration of alloy particles. Correspondingly, the particle dispersion was improved in the presence of sodium citrate. However, increasing the sodium citrate ratio to 3X amplified the agglomeration and deteriorated the particle dispersion. It seems that adding further sodium citrate blocked the particle nucleation sites on the surface of carbon black particles, which directed the metal particle deposition toward the grain boundaries of carbon black particles. Therefore, the metal particle agglomeration was observed in between carbon particles. Furthermore, TEM images once more proved that smaller particle sizes and better particle dispersion in the presence of SC were responsible for facilitating the phase transformation during heat treatment and changing the start temperature of thermally-activated processes, which were observed in the TGA and DSC analyses.

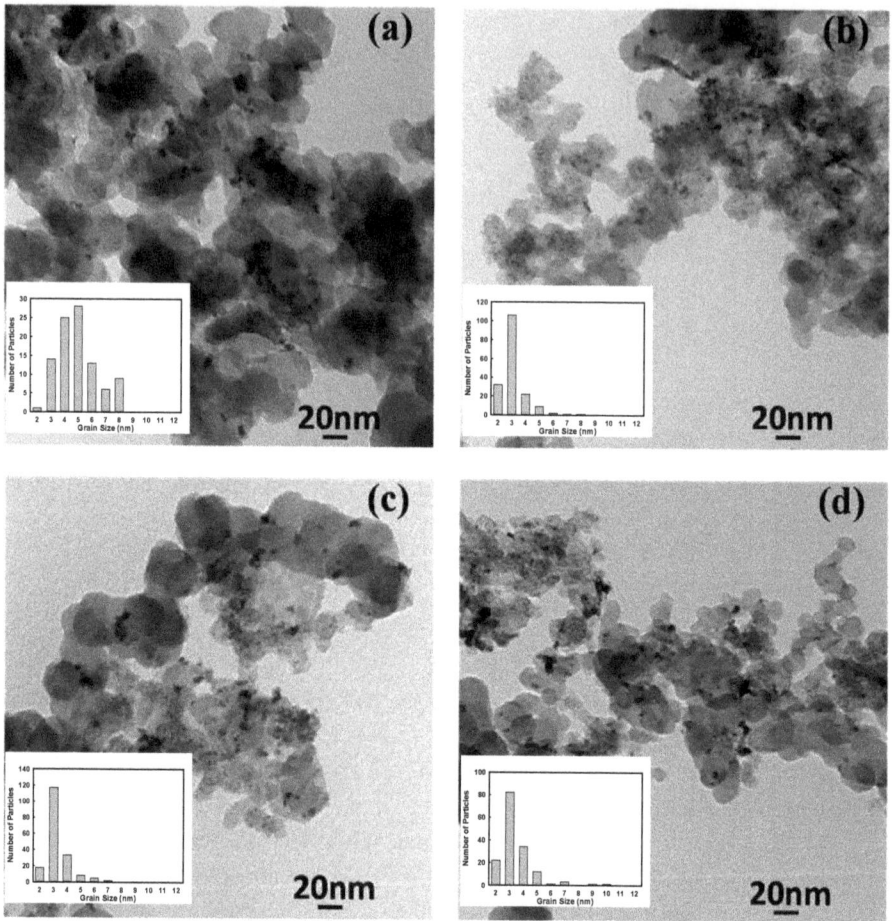

Figure 6. TEM image of $Pt_{0.75}Mn_{0.25}$ samples prepared in the presence of different contents of sodium citrate: (**a**) no additive; (**b**) 1X; (**c**) 2X; and (**d**) 3X.

Figure 7 displays TEM images of samples synthesized in the presence of SC that were heat treated at different temperatures. The particle sizes and details calculated from the TEM images are summarized in Table 1. The heat treatment increased the particle sizes, including some very large particles with a radius of more than 10 nm. However, the average particle size was enlarged by only 3 nm by increasing the heat treatment temperature to 950 °C. This means that Pt-Mn alloys were resistant to particle growth, which has also been observed for other alloys [35] and Pt-Mn samples produced without SC [10]. In Figure 8, the TEM images of two samples, which were prepared with and without SC and heat treated for 1 h at 700 °C, are compared. It can be concluded that the presence of SC improved the alloy particle dispersion and prevented the particle enlargement to a great extent during heat treatment. Therefore, it is expected that samples prepared with SC should show enhanced activity compared to samples prepared without additives or heat treated for a longer time.

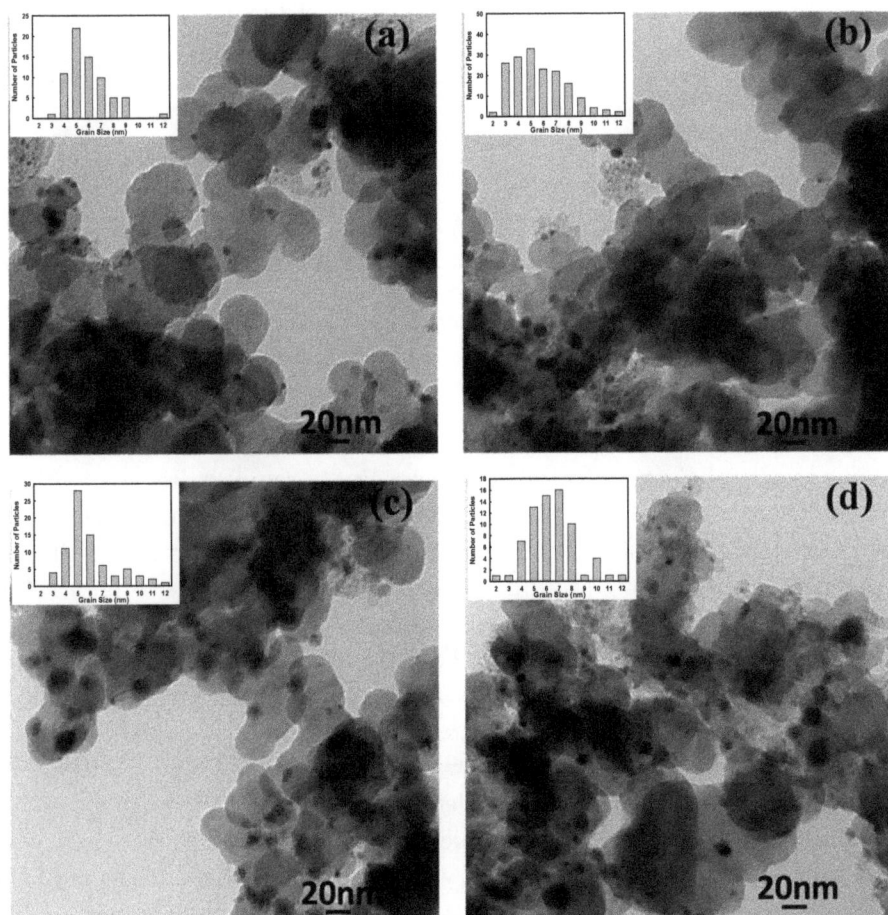

Figure 7. TEM image of $Pt_{0.75}Mn_{0.25}$ samples prepared in the presence 2X of sodium citrate and after heat treatment at different temperatures: (**a**) 500 °C; (**b**) 700 °C; (**c**) 875 °C; and (**d**) 950 °C.

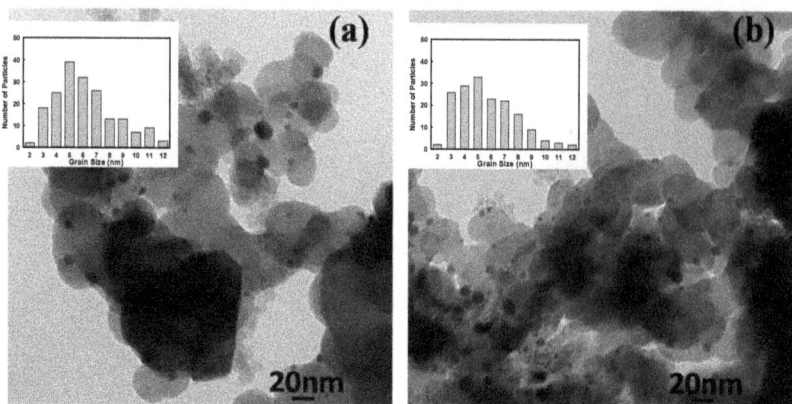

Figure 8. TEM images of $Pt_{0.75}Mn_{0.25}$ samples prepared (**a**) without additive and (**b**) in the presence 2X of sodium citrate and heat treated at 700 °C for 1 h.

2.2. Electrochemical Characterization

The results of electrochemical studies for Pt-Mn samples prepared with and without SC are presented in Figure 9a. The experiments were conducted in the deaerated 0.5 M H_2SO_4 solution. The common Pt cyclic voltammetry (CV) shape with hydrogen adsorption/desorption peaks at lower potentials was observed for all samples. The electrochemical active surface area (ECSA) was calculated by integrating the charge under the hydrogen adsorption peaks and is compiled in Table 3 and Figure 10. The results show that the ECSA values were dramatically increased by adding SC from 10.5–20.6 m^2/g_{Pt} (Figure 10), which is in good agreement with the TEM and XRD results. The ECSA improvement is attributed to the superior particle dispersion and smaller grain sizes in the samples prepared with SC. The maximum ECSA value was achieved with the sample prepared using the 2X SC content. When the amount of SC was increased to 3X, the measured ECSA was lower than that measured for 2X. Based on the TEM analysis (Figure 5d), this reduced ECSA is the result of uneven particle dispersion on the carbon support.

CV obtained for samples prepared with 2X SC and heat treated at different temperatures is illustrated in Figure 9b. Heat treatment at 500 °C reduced the ECSA values for Pt-Mn samples, but the ECSA for the Pt-Mn samples improved after increasing the heat treatment temperature to above 700 °C. The ECSA values never reached the same value as the untreated sample. Particle size growth is the main cause of the drop in ECSA. However, the optimal ECSA was achieved after heat treatment at 700 °C for 1 h and led to an ECSA that was only 7% lower than that of the untreated sample. These results show that Pt-Mn samples prepared with SC are more resistant to particle growth compared to samples prepared without SC. Previously, it has been shown that the formation of the ordered structure and the generation of a higher alloying degree of Pt and Mn by changing the surface composition and structure enhanced the ECSA [10]. This resistance to ECSA loss is most likely due to a roughening of the alloy particle surface upon dissolution of the surface oxide layer [36] and also because of better particle dispersion and smaller grain sizes achieved in the presence of SC [37–39]. Additionally, the CV obtained with samples that were heat treated for different periods is shown in Figure 9c. Increasing the time of heat treatment to 4 h at 700 °C reduced the ECSA values (Table 3) (not shown in Figure 9c). A similar trend was also observed by increasing the heat treatment time at 500 °C.

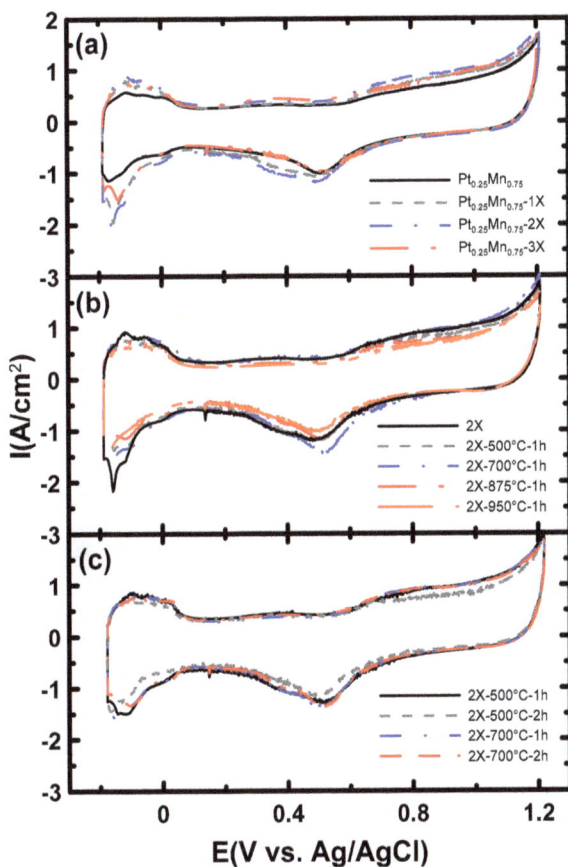

Figure 9. Cyclic voltammetry (CV) in the 0.5 M H_2SO_4 solution at a scan rate of 20 mV/s for samples prepared (**a**) with different sodium citrate contents and then heat treated (**b**) at different temperatures for 1 h and (**c**) at 500 and 700 °C for different periods.

Table 3. Summary of ethanol oxidation reaction (EOR) activity parameters of the Pt-Mn samples. Also listed are the measured electrochemical active surface area (ECSA) values for each sample.

Samples	Onset potential (mV)	Current (mA.cm^{-2}) at 350 mV	Peak potential (mV)	Peak current (mA.cm^{-2})	ECSA (m^2/g$_{Pt}$)
Pt$_{0.25}$Mn$_{0.75}$	248	0.461	665	2.49	10.5
Pt$_{0.25}$Mn$_{0.75}$-1x	247	0.472	678	2.72	16.8
Pt$_{0.25}$Mn$_{0.75}$-2x	249	0.489	690	2.81	20.6
Pt$_{0.25}$Mn$_{0.75}$-3x	266	0.355	695	2.14	15.7
Pt$_{0.25}$Mn$_{0.75}$-2X-500°C-1 h	248	0.880	703	6.49	16.2
Pt$_{0.25}$Mn$_{0.75}$-2X-700°C-1 h	240	0.628	710	8.40	19.1
Pt$_{0.25}$Mn$_{0.75}$-2X-875°C-1 h	237	0.602	729	4.50	17.1
Pt$_{0.25}$Mn$_{0.75}$-2X-950°C-1 h	233	0.638	703	4.76	14.5
Pt$_{0.25}$Mn$_{0.75}$-2X-500°C-2 h	245	0.655	662	4.58	13.8
Pt$_{0.25}$Mn$_{0.75}$-2X-500°C-4 h	241	0.709	671	3.87	15.4
Pt$_{0.25}$Mn$_{0.75}$-2X-700°C-2 h	240	0.588	682	6.30	17.1
Pt$_{0.25}$Mn$_{0.75}$-2X-700°C-4 h	241	0.556	698	6.35	15.7

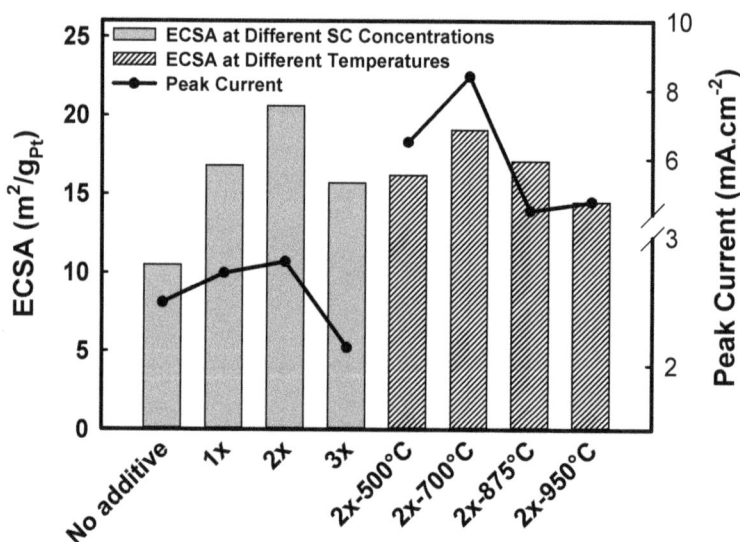

Figure 10. ECSA changes at different SC concentrations and after heat treatment at different temperatures, measured from Figure 9, and the ethanol oxidation peak current, measured from Figure 11.

The EOR activity of the samples prepared with and without sodium citrate is presented in Figure 11a. In addition, Pt/C, which was synthesized without SC, was also tested in ethanol solution and is referred to as Pt$_{100}$. Comparing the electroactivity of the samples (Table 3 and Figure 10) showed that, by increasing the amount of sodium citrate to the 2X concentration, the activity of the samples was improved. The improvement in activity was a result of the higher ECSA and better particle dispersion, which was concluded from the TEM images and CV analysis in 0.5 M H_2SO_4 solution. However, by increasing the amount of sodium citrate to 3X, the EOR activity was reduced. Based on XRD and TEM results, the lower electrochemical activity of samples prepared with 3X SC was related to the increase in the quantity of the oxide phase and, importantly, to uneven particle dispersion, which resulted in lower ECSA. Furthermore, the linear sweep voltammetry (LSV) in the ethanol solution showed that by adding SC, the onset potential of the samples was almost constant (247–249 mV) up to the optimum ratio (2X), but increased to 266 mV when higher ratios of SC were used.

The EOR activity of the samples after heat treatment is illustrated in Figure 11b. After performing the heat treatment at different temperatures, the activity was improved in all samples compared to the as-produced sample. The activity of heat-treated samples is reported in Table 3 and compared in Figure 10. The most active samples were produced at 700 °C, at which temperature the Pt-Mn ordered phase was formed. Upon increasing the heat treatment temperature to 850 and 950 °C, the EOR activity of the samples was decreased because of particle growth and forming new phases. However, the EOR activity of the heat-treated samples in all temperatures was greater than the untreated sample.

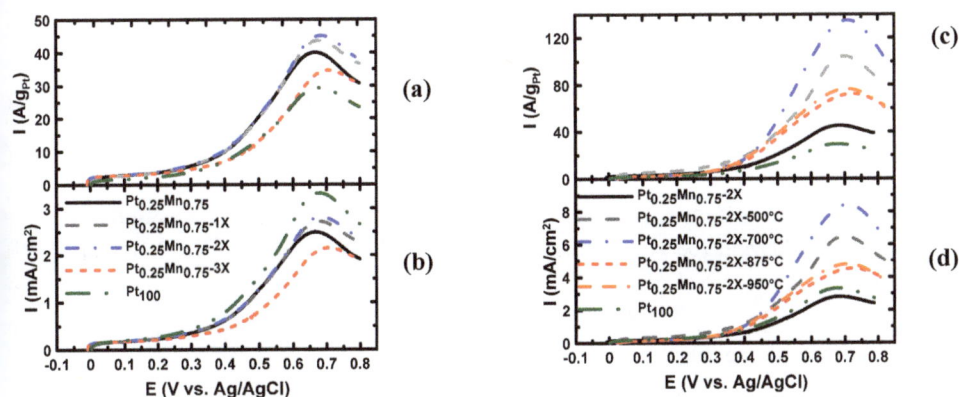

Figure 11. Linear sweep voltammetry (LSV) in the 0.5 M H_2SO_4 + 0.1 M ethanol solution at a scan rate of 20 mV/s for: (**a**) the samples prepared with different sodium citrate contents; (**c**) the sample produced with 2X sodium citrate content and heat treated at different temperatures. (**b**) The same as (**d**), normalized for platinum content of (**a**) and (**d**), respectively.

The impact of heat treatment time at 500 °C and 700 °C on the EOR activity is shown in Figure 12. At 500 °C and 700 °C, increasing the heat treatment time beyond 1 h had a negligible impact on the EOR activity, which is in good agreement with the XRD and TGA results. The XRD patterns indicated that phase transformations occurred at 700 °C, and this does not change by increasing the time of heat treatment from 1 h to 4 h (Figure 5). Therefore, increasing the time of treatment can only affect the particle sizes, and as shown at 700 °C, the EOR activity slightly reduced by extending the time of the heat treatment.

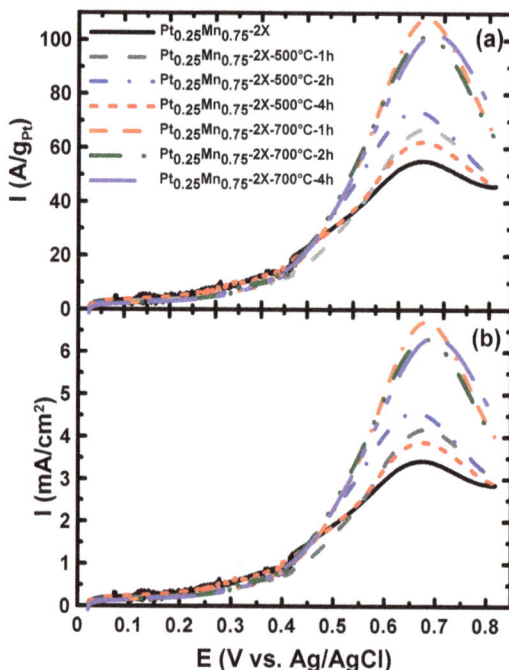

Figure 12. LSV of the samples prepared in the presence of sodium citrate and heat treated at 500 and 700 °C for different periods, in the 0.5 M H_2SO_4 + 0.1 M ethanol solution at a scan rate of 20 mV/s.

3. Experimental Section

3.1. Catalyst Synthesis

The impregnation method was used to prepare Pt-Mn/C catalysts with a composition of $Pt_{0.25}Mn_{0.75}$, similar to that previously published [4]. The metal precursors were $H_2PtCl_6 \cdot H_2O$ (Aldrich, Oakville, ON, Canada) and $MnCl_2 \cdot 4H_2O$ (Aldrich, Oakville, ON, Canada). Trisodium citrate (SC) (Aldrich) and Vulcan XC72R carbon black (Cabot Corp., Billerica, MA, USA) were used as an additive and metal support, respectively. The total metal loading was kept constant at ~20 wt. % for all samples. The SC was added in a weight ratio of 1:1, 2:1 and 3:1 to the metal ratios, which are represented in this article by 1X, 2X and 3X, respectively. The pH of the solution was adjusted using a HCl solution (15 v/v %) and/or a 1 M NaOH solution.

628

$NaBH_4$ powder was used as the reduction agent. The weight ratio of $NaBH_4$ to the metal content was 3:1. In addition, a control sample of 20 wt. % Pt/C was synthesized with the same method without SC, which was referred to as Pt_{100}. The powders were collected by suction filtration, washed with isopropanol alcohol (IPA), acetone and deionized water and, finally, dried in an oven at 80 °C overnight.

Heat treatment was performed in a nitrogen atmosphere at either 500, 700, 875 or 950 °C in a Barnstead Thermolyne tube furnace with a quartz tube. The period of heat treatment was kept constant (1 h) for all samples and temperatures, unless otherwise specified. After the heat treatment, the samples were cooled down under a constant flow of nitrogen gas and were preserved inside the furnace until room temperature was reached.

3.2. Materials Characterization

The chemical composition of the Pt-Mn samples was examined by inductively-coupled plasma optical emission spectroscopy (ICP-OES, Varian Vista-MPX, Mississauga, ON, Canada). Aqua Regia solution was used to dissolve metal powders, and diluted solutions were consumed for ICP-OES analyses. ICP-OES instrument was calibrated by four standard solutions of Pt and Mn with concentrations of 1, 5, 10 and 20 ppm.

The carbon/metal weight ratio and the temperature of phase changes were determined by thermogravimetric analysis (TGA) and differential scanning calorimetry (DSC). Thermal analysis was performed using a TA Instruments Q600 SDT system (TA Instruments, New Castle, DE, USA). Measurements were made in both argon and air atmospheres, using a heating ramp of 5 °C/min and 20 °C/min, respectively.

Powder X-ray diffraction (XRD) patterns were obtained for each catalyst. Measurements were made using either a Bruker D8 Advance powder X-ray diffractometer (Bruker, East Milton, ON, Canada) equipped with a germanium monochromator (provided by Bruker) or a Rigaku Ultima IV X-ray diffractometer (Rigaku, Toronto, ON, Canada) equipped with a graphite monochromator (provided by Rigaku). Both instruments employ a Cu $K_{\alpha 1}$ X-ray source.

A Philips CM 10 instrument equipped with an AMT digital camera system was used for transmission electron microscopy (TEM, Philips, Andover, MA, USA) analysis. Samples for TEM analysis were dispersed in a mixture of water and isopropanol and applied to nickel 400 mesh reinforced grids coated by carbon and allowed to dry under air before being introduced into the chamber. The mean particle sizes were determined by measuring the diameter of 100–200 metal particles.

3.3. Electrochemical Characterization

The electrochemical activity of the samples was studied after applying a thin layer of catalyst on glassy carbon (GC) electrodes. The ink of samples was produced by mixing 10 mg of catalyst with 100 μL Nafion solution (5% in alcohols, Dupont) and a 400 μL 50:50 mixture of isopropyl alcohol and water. A uniform suspension was achieved after sonicating for 45 min. A 2-μL droplet of the well-dispersed catalyst ink was deposited onto a clean and polished GC electrode (diameter = 3 mm, CH instruments) and dried in air at room temperature prior to electrochemical tests. The total metal loading of the catalyst layer was 0.11 mg/cm^2. Cyclic voltammetry (CV) and linear sweep voltammetry (LSV) were performed in a N_2-purged solution. A 0.5 M H_2SO_4 solution was used to determine the electrochemical active surface area (ESCA). A 0.5 M H_2SO_4 + 0.1 M ethanol solution was employed to study the EOR activity of the catalysts. Measurements were made in a 3-electrode cell with a Pt wire counter electrode and a Ag/AgCl reference electrode. The LSV and CV for all samples were collected at a scan rate of 20 mV/s. Besides, the sample surfaces were cleaned prior to recording the final electrochemical test, by scanning at a scan rate of 100 mV/s and then at the scan rate of 20 mV/s, until we got a clean and reproducible CV.

4. Conclusions

This work has shown how the addition of sodium citrate (SC) influences the particle dispersion and grain size of Pt-Mn particles and facilitates the crystalline phase transformation. The results indicated that adding SC to the impregnation solution improved particle dispersion, decreased particle sizes, reduced the heat treatment time from 4 h to 1 h and increased the ECSA. Therefore, the EOR activity of the Pt-Mn alloy catalysts was enhanced. However, the weight ratio of SC to metal loading should be kept lower than 2X, because a higher weight ratio hindered the metal particle dispersion. Furthermore, this investigation proved that the SC had a positive impact on the EOR activity of Pt-Mn alloys when the pH of the impregnation solution was lower than four. Moreover, the heat-treated samples showed superior activity toward ethanol oxidation in comparison with the as-synthesized samples. The EOR activity was the highest for the sample heat treated at 700 °C for 1 h. The XRD analysis illustrated that Pt-Mn intermetallic was formed at the same temperature, and this was the main reason for the superior activity.

Acknowledgments: This work was supported by the Natural Sciences and Engineering Research Council (NSERC) of Canada and University of Ontario Institute of Technology (UOIT). The authors also acknowledge equipment support from the Canada Foundation for Innovation. We thank Wen He Gong (McMaster University) for the XRD data and Richard B. Gardiner (University of Western Ontario) for the TEM images.

Author Contributions: M.R.Z.G. and E.B.E. conceived of and designed the experiment. M.R.Z.G. performed the experiment and primary data analysis. E.B.E. contributed reagents/materials/analysis tools. M.R.Z.G. and E.B.E. wrote the paper.

Conflicts of Interest: The authors declare no conflict of interest. The founding sponsors had no role in the design of the study; in the collection, analyses or interpretation of data; in the writing of the manuscript; nor in the decision to publish the results.

References

1.   Antolini, E. Formation of carbon-supported PtM alloys for low temperature fuel cells: A review. *Mater. Chem. Phys.* **2003**, *78*, 563–573.

2.   Watanabe, A.; Uchida, H. Catalysts for the Electro-Oxidation of Small Molecules. In *Handbook of Fuel Cells—Fundamentals, Technology and Applications*; Vielstich, W., Lamm, A., Gasteiger, H., Eds.; John Wiley & Sons, Ltd.: Chichester, UK, 2010.

3.   Liu, H.; Song, C.; Zhang, L.; Zhang, J.; Wang, H.; Wilkinson, D.P. A review of anode catalysis in the direct methanol fuel cell. *J. Power Sources* **2006**, *155*, 95–110.

4.   Ammam, M.; Prest, L.E.; Pauric, A.D.; Easton, E.B. Synthesis, Characterization and Catalytic Activity of Binary PtMn/C Alloy Catalysts towards Ethanol Oxidation. *J. Electrochem. Soc.* **2012**, *159*, B195–B200.

5.   Castro Luna, A.M.; Camara, G.A.; Paganin, V.A.; Ticianelli, E.A.; Gonzalez, E.R. Effect of thermal treatment on the performance of CO-tolerant anodes for polymer electrolyte fuel cells. *Electrochem. Communs.* **2000**, *2*, 222–225.

6.   Cherstiouk, O.V.; Gavrilov, A.N.; Plyasova, L.M.; Molina, I.Y.; Tsirlina, G.A.; Savinova, E.R. Influence of structural defects on the electrocatalytic activity of platinum. *J. Solid State Electrochem.* **2008**, *12*, 497–509.

7.   Vigier, F.; Rousseau, S.; Coutanceau, C.; Leger, J.; Lamy, C. Electrocatalysis for the direct alcohol fuel cell. *Topics Catal.* **2006**, *40*, 111–121.

8.   Ammam, M.; Easton, E.B. Ternary PtMnX/C (X = Fe, Co, Ni, Cu, Mo and, Sn) Alloy Catalysts for Ethanol Electrooxidation. *J. Electrochem. Soc.* **2012**, *159*, B635–B640.

9.   Ammam, M.; Easton, E.B. Quaternary PtMnCuX/C (X = Fe, Co, Ni, and Sn) and PtMnMoX/C (X = Fe, Co, Ni, Cu and Sn) alloys catalysts: Synthesis, characterization and activity towards ethanol electrooxidation. *J. Power Sources* **2012**, *215*, 188–198.

10. Zamanzad Ghavidel, M.R.; Easton, E.B. Thermally induced changes in the structure and ethanol oxidation activity of $Pt_{0.25}Mn_{0.75}$/C. *Appl. Catal. B* **2015**, *176-177*, 150–159.

11. Bezerra, C.W.B.; Zhang, L.; Liu, H.; Lee, K.; Marques, A.A.L.B.; Marques, E.P.; Wang, H.; Zhang, J. A review of heat-treatment effects on activity and stability of PEM fuel cell catalysts for oxygen reduction reaction. *J. Power Sources* **2007**, *173*, 891–908.

12. Lee, K.; Zhang, J.; Wang, H.; Wilkinson, D.P. Progress in the synthesis of carbon nanotube- and nanofiber-supported Pt electrocatalysts for PEM fuel cell catalysis. *J. Appl. Electrochem.* **2006**, *36*, 507–522.

13. Liu, Z.; Lee, J.Y.; Chen, W.; Han, M.; Gan, L.M. Physical and Electrochemical Characterizations of Microwave-Assisted Polyol Preparation of Carbon-Supported PtRu Nanoparticles. *Langmuir* **2003**, *20*, 181–187.

14. Bonnemann, H.; Richards, R.-M. Nanoscopic Metal Particles- Synthetic Methods and Potential Applications. *Eur. J. Inorg. Chem.* **2001**, *2001*, 2455–2480.
15. Xiong, L.; Manthiram, A. Nanostructured Pt-M/C (M = Fe and Co) catalysts prepared by a microemulsion method for oxygen reduction in proton exchange membrane fuel cells. *Electrochim. Acta* **2005**, *50*, 2323–2329.
16. Antonucci, P.L.; Alderucci, V.; Giordano, N.; Cocke, D.L.; Kim, H. On the role of surface functional groups in Pt carbon interaction. *J. Appl. Electrochem.* **1994**, *24*, 58–65.
17. Hui, C.L.; Li, X.G.; Hsing, I.M. Well-dispersed surfactant-stabilized Pt/C nanocatalysts for fuel cell application: Dispersion control and surfactant removal. *Electrochim. Acta* **2005**, *51*, 711–719.
18. Moghaddam, R.B.; Pickup, P.G. Support effects on the oxidation of ethanol at Pt nanoparticles. *Electrochim. Acta* **2012**, *65*, 210–215.
19. Tian, J.H.; Wang, F.B.; Shan, Z.H.Q.; Wang, R.J.; Zhang, J.Y. Effect of Preparation Conditions of Pt/C Catalysts on Oxygen Electrode Performance in Proton Exchange Membrane Fuel Cells. *J. Appl. Electrochem.* **2004**, *34*, 461–467.
20. Di Noto, V.; Negro, E. Pt-Fe and Pt-Ni Carbon Nitride-Based "Core-Shell" ORR Electrocatalysts for Polymer Electrolyte Membrane Fuel Cells. *Fuel Cells* **2010**, *10*, 234–244
21. Di Noto, V.; Negro, E. Development of nano-electrocatalysts based on carbon nitride supports for the ORR processes in PEM fuel cells. *Electrochim. Acta* **2010**, *55*, 7564–7574.
22. Easton, E.B.; Zamanzad Ghavidel, M.R.; Reid, O.R.; Ammam, M.; Prest, L.E. Limiting the Amount of Oxides in Pt-Mn Alloy Catalysts for Ethanol Oxidation. In Proceedings of the 223rd ECS Meeting, Toronto, ON, Canada, 12–16 May 2013; MA2013–01(39), p. 1366.
23. Wang, X.; Hsing, I.M. Surfactant stabilized Pt and Pt alloy electrocatalyst for polymer electrolyte fuel cells. *Electrochim. Acta* **2002**, *47*, 2981–2987.
24. Yang, J.; Lee, J.Y.; Too, H.P. Size effect in thiol and amine binding to small Pt nanoparticles. *Analytica. Chimica. Acta* **2006**, *571*, 206–210.
25. Gasteiger, H.A.; Kocha, S.S.; Sompalli, B.; Wagner, F.T. Activity benchmarks and requirements for Pt, Pt-alloy, and non-Pt oxygen reduction catalysts for PEMFCs. *Appl. Catal. B* **2005**, *56*, 9–35.
26. Zeng, J.; Yang, J.; Lee, J.Y.; Zhou, W. Preparation of Carbon-Supported Core-Shell Au-Pt Nanoparticles for Methanol Oxidation Reaction: The Promotional Effect of the Au Core. *J. Phys. Chem. B* **2006**, *110*, 24606–24611.
27. Liao, S.; Holmes, K.A.; Tsaprailis, H.; Birss, V.I. High Performance PtRuIr Catalysts Supported on Carbon Nanotubes for the Anodic Oxidation of Methanol. *J. Am. Chem. Soc.* **2006**, *128*, 3504–3505.
28. Zeng, J.; Lee, J.Y. Effects of preparation conditions on performance of carbon-supported nanosize Pt-Co catalysts for methanol electro-oxidation under acidic conditions. *J. Power Sources* **2005**, *140*, 268–273.
29. Kangasniemi, K.H.; Condit, D.A.; Jarvi, T.D. Characterization of Vulcan Electrochemically Oxidized under Simulated PEM Fuel Cell Conditions. *J. Electrochem. Soc.* **2004**, *151*, E125–E132.

30. Baturina, O.A.; Aubuchon, S.R.; Wynne, K.J. Thermal Stability in Air of Pt/C Catalysts and PEM Fuel Cell Catalyst Layers. *Chem. Mater.* **2006**, *18*, 1498–1504.

31. Stobbe, E.R.; de Boer, B.A.; Geus, J.W. The reduction and oxidation behaviour of manganese oxides. *Catal. Today* **1999**, *47*, 161–167.

32. Ji, C.X.; Ladwig, P.; Ott, R.; Yang, Y.; Yang, J.; Chang, Y.A.; Linville, E.; Gao, J.; Pant, B. An investigation of phase transformation behavior in sputter-deposited PtMn thin films. *JOM* **2006**, *58*, 50–54.

33. Easton, E.B.; Yang, R.; Bonakdarpour, A.; Dahn, J.R. Thermal Evolution of the Structure and Activity of Magnetron-Sputtered TM-C-N (TM = Fe, Co) Oxygen Reduction Catalysts. *Electrochem. Solid-State Lett.* **2007**, *10*, B6–B10.

34. Ghosh, T.; Leonard, B.M.; Zhou, Q.; DiSalvo, F.J. Pt Alloy and Intermetallic Phases with V, Cr, Mn, Ni, and Cu: Synthesis As Nanomaterials and Possible Applications As Fuel Cell Catalysts. *Chem. Mater.* **2010**, *22*, 2190–2202.

35. Antolini, E. Formation, microstructural characteristics and stability of carbon supported platinum catalysts for low temperature fuel cells. *J. Mater. Sci.* **2003**, *38*, 2995–3005.

36. Watanabe, M.; Tsurumi, K.; Mizukami, T.; Nakamura, T.; Stonehart, P. Activity and Stability of Ordered and Disordered Co-Pt Alloys for Phosphoric Acid Fuel Cells. *J. Electrochem. Soc.* **1994**, *141*, 2659–2668.

37. Harlow, J.E.; Stevens, D.A.; Sanderson, R.J.; Liu, G.C.K.; Lohstreter, L.B.; Vernstrom, G.D.; Atanasoski, R.T.; Debe, M.K.; Dahn, J.R. Structural Changes Induced by Mn Mobility in a Pt1-xMnx Binary Composition-Spread Catalyst. *J. Electrochem. Soc.* **2012**, *159*, B670–B676.

38. Stevens, D.A.; Mehrotra, R.; Sanderson, R.J.; Vernstrom, G.D.; Atanasoski, R.T.; Debe, M.K.; Dahn, J.R. Dissolution of Ni from High Ni Content $Pt_{1-x}Ni_x$ Alloys. *J. Electrochem. Soc.* **2011**, *158*, B905–B909.

39. Chen, C.; Kang, Y.; Huo, Z.; Zhu, Z.; Huang, W.; Xin, H. L.; Snyder, J.D.; Li, D.; Herron, J.A.; Mavrikakis, M.; Chi, M.; More, K.L.; Li, Y.; Markovic, N.M.; Somorjai, G.A.; Yang, P.; Stamenkovic, V.R. Highly Crystalline Multimetallic Nanoframes with Three-Dimensional Electrocatalytic Surfaces. *Science* **2014**, *343*, 1339–1343.

Sb Surface Modification of Pd by Mimetic Underpotential Deposition for Formic Acid Oxidation

Long-Long Wang, Xiao-Lu Cao, Ya-Jun Wang and Qiao-Xia Li

Abstract: The newly proposed mimetic underpotential deposition (MUPD) technique was extended to modify Pd surfaces with Sb through immersing a Pd film electrode or dispersing Pd/C powder in a Sb(III)-containing solution blended with ascorbic acid (AA). The introduction of AA shifts down the open circuit potential of Pd substrate available to achieve suitable Sb modification. The electrocatalytic activity and long-term stability towards HCOOH electrooxidation of the Sb modified Pd surfaces (film electrode or powder catalyst) by MUPD is superior than that of unmodified Pd and Sb modified Pd surfaces by conventional UPD method. The enhancement of electrocatalytic performance is due to the third body effect and electronic effect, as well as bi-functional mechanism induced by Sb modification which result in increased resistance against CO poisoning.

Reprinted from *Catalysts*. Cite as: Wang, L.-L.; Cao, X.-L.; Wang, Y.-J.; Li, Q.-X. Sb Surface Modification of Pd by Mimetic Underpotential Deposition for Formic Acid Oxidation. *Catalysts* **2015**, *5*, 1388–1398.

1. Introduction

Direct formic acid fuel cells (DFAFCs) are considered promising power sources of clean and environment-friendly energy for miniature and portable electronic devices because of excellent performance, such as high power density [1,2]. The direct formic acid fuel cell has a theoretical open circuit potential of 1.48 V, higher than that of direct methanol fuel cell (1.18 V) [3]. The improvement of performance of DFAFCs depends on fabrication of high-efficient electrocatalysts. The commonly-used anodic catalyst for DFAFCs is platinum black, on which the formic acid electrooxidation occurs via a dual pathway [4,5], which consists of the direct pathway without CO poison and the indirect pathway with the formation of CO as poisonous intermediates. The resulting CO intermediates are strongly adsorbed on the Pt surface and block the active sites, then decrease the activity. In this regards, platinum is not so favorable for practical formic acid fuel cell application because of the CO intermediates build-up, poisoning the catalysts, and degenerating the fuel cell performance gradually [3,6].

Many studies have confirmed that palladium is a more efficient catalyst with higher catalytic activity for the electrooxidation of formic acid [6–11]. The excellent property derives from the extraordinary formic acid oxidation mechanism on Pd which is different from the dual pathway mechanism on Pt. Briefly, on the Pd surface, the electrooxidation of formic acid occurs via a dominantly direct pathway with a minimized buildup of CO on the surface (The formation of CO_{ad} on a Pd electrode in formic acid solutions at the OCP and practical working potentials has been confirmed by Wang et al., using in situ high-sensitivity attenuated-total-reflection surface-enhanced infrared absorption spectroscopy (ATR-SEIRAS), proposing that the reduction of the FA dehydrogenation product CO_2 should be mainly responsible for the above CO_{ad} formation [12]). Unfortunately, the activity of Pd is unstable and deactivation exists during formic acid oxidation due to gas build-up on the anodic side of a fuel cell [13,14], catalyst leaching or impurities in the formic acid or intermediate species [15]. However, a majority of literature proved that it is mainly CO-like intermediates accumulated on Pd surface that degenerate the activity of Pd, and it reached a consensus among most research workers [16,17]. On this issue, much effort has been made to improve the catalytic activity and stability through alloying or surface modification with metallic adatoms, such as Sb. Yu et al. [18] fabricated carbon supported PdSb alloy catalysts which show much better resistance to poisoning (deactivation) and decrease the accumulation of CO on the catalyst surface during formic acid oxidation. Masel et al. [19] have studied the effects of Sb adatoms on the performance of a DFAFC. They showed that electrochemical surface modification of Pd by Sb adatoms enhances the oxidation of formic acid by more than two-fold in an electrochemical cell. For Sb modification, previous approaches, such as irreversible adsorption (IRA) and traditional UPD method required external potential controlled desorption of partial Sb, which were not suitable for scaled synthesis or upgrading of practical powder catalysts [20–22]. Among the known Sb surface modification method, mimetic underpotential deposition (MUPD) technique was a newly proposed electroless approach to achieve sufficient surface modification [23]. Compared with underpotential deposition followed by potential controlled desorption of partial Sb adatoms usually applied in Sb modification on Pt, MUPD requires no external potential control and is a versatile electroless approach extended for surface nanoengineering of electrocatalysts [24].

In this work, we extended the MUPD strategy to the modification of Sb on Pd surface of film electrode and Pd/C powder by introducing ascorbic acid (AA) as a mild reductant to Sb(III)-containing modification solution. Besides, for comparison, Pd film substrate was also modified by Sb UPD. We studied the influence of Sb modification on Pd surface for the electro-oxidation of formic acid by cyclic oltammetry and chronoamperometry together with anodic stripping voltammetry of pre-adsorbed CO.

2. Results and Discussion

2.1. Sb UPD on Pd Film Electrodes

Different from Sb UPD on Pt surfaces [20,25], coverage of Sb (θ_{Sb}, a coefficient based on the ratio of Pd sites filled and not filled) on Pd surface was tuned by controlling the UPD time. In this paper, Sb UPD on fresh Pd films was performed for 10 s, 20 s, 30 s, respectively. The modified Pd film electrodes were marked as Sb/Pd(UPD). Figure 1a depicts cyclic voltammograms of unmodified Pd film and Sb/Pd(UPD) electrodes in 0.5 M H_2SO_4. It can be seen that the hydrogen adsorption/desorption region was restrained for Sb modified Pd film electrodes. A conspicuous peak near 0.66 V in positive scan is due to the oxidative dissolution of Sb modifiers on Pd surface. The peak of negative scan in high potential is ascribed to reduction of oxygenous species formed on positive scan. Based on hydrogen adsorption-desorption charge with or without Sb modifiers, θ_{Sb} on Pd surface can be evaluated through the following equation [22]:

$$\theta_{Sb} = (Q_{O-H} - Q_{Sb-H})/Q_{O-H} \tag{1}$$

where Q_{O-H} and Q_{Sb-H} is the charge for oxidation of adsorbed hydrogen on unmodified and Sb modified Pd, respectively. By calculating, it is found that the Pd electrodes through UPD for 10 s, 20 s, 30 s enable θ_{Sb} to reach a value of 0.52, 0.62 and 0.66, respectively. This revealed that coverage of Sb on Pd surface increased with UPD time. Formic acid oxidation was chose as a probe reaction to compare the catalytic activity of Sb/Pd(UPD) electrodes with various θ_{Sb}. Figure 1b showed that peak current density of HCOOH electrooxidation on Sb/Pd(UPD) increased with θ_{Sb} from 0.52 to 0.62 and then dropped down with θ_{Sb} from 0.62 to 0.66, which might follow a volcano-like relationship between θ_{Sb} and electrocatalytic activity of Sb/Pd(UPD) (seen from inset in Figure 1b).

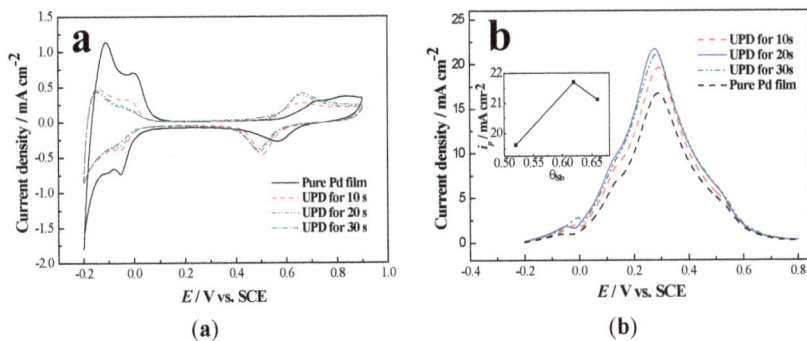

Figure 1. (a) CV curves of Sb/Pd(UPD) electrodes with various θ_{Sb} in 0.5 mol L^{-1} H$_2$SO$_4$ at 50 mV s^{-1}; (b) The comparison of catalytic activity of Sb/Pd(UPD) electrodes with various θ_{Sb} towards HCOOH electrooxidation in 0.5 mol L^{-1} H$_2$SO$_4$ + 0.5 mol L^{-1} HCOOH at 50 mV s^{-1}. The inset presents a plot showing the direct relation between catalytic activity and θ_{Sb}.

2.2. Sb MUPD on Pd Film Electrodes

Figure 2 compared the open circuit potential recorded on the Pd film electrode in 0.1 mM APT solution with or without 20 mM AA. A high open circuit potential of *ca.* 0.29 V at 30 s was seen in single 0.1 mM APT (curve a) due to the oxygen-containing species that spontaneously formed on Pd surface, thus limited effective modification of Sb on Pd. With addition of 20 mM AA to 0.1 mM APT aqueous solution (curve b), the OCP negatively shifted to 0.12 V at 30 s because the ascorbic acid served as mild reductant removed the oxygen-containing species to ensure freshly reduced Pd surfaces for better Sb modification [23]. Upon this, Sb MUPD was carried out through immersing Pd film electrode into modification solution for 30 s which was the optimal MUPD time reported by Cai *et al.* [23] when modifying bulk Pt electrode and powder catalyst by MUPD.

Hydrogen region properties of Sb/Pd(MUPD) electrode were studied by cyclic voltammetry in 0.5 mol L^{-1} H$_2$SO$_4$. Observed in Figure 3a, after pure Pd film electrode was immersed in Sb containing solution for just 30 s, the area of hydrogen region was severely shrinked due to Sb coverage on Pd surface, thus leading to restriction of hydrogen adsorption/desorption, and, therefore, θ_{Sb} herein reached 0.67.

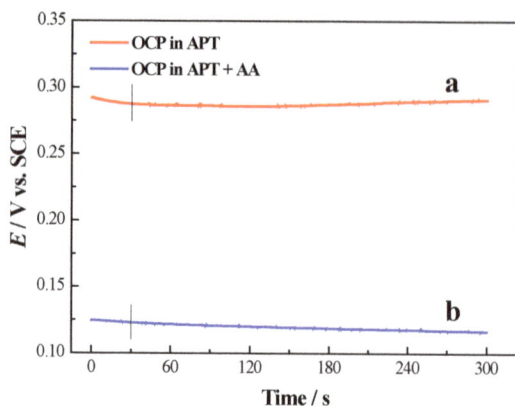

Figure 2. Open circuit potential (OCP) curves recorded on a Pd film electrode in 0.1 mM APT aqueous solution without (**a**) or with (**b**) 20 mM AA.

To compare the electrocatalytic activity of modified and unmodified Pd film electrodes, linear sweep voltammograms for formic acid oxidation are recorded on pure Pd film, optimal Sb/Pd(UPD) with UPD time for 20 s and Sb/Pd(MUPD) in 0.5 M H_2SO_4 containing 0.5 M HCOOH. As can be seen in Figure 3b, the formic acid oxidation current density in low potential was weak on unmodified pure Pd film. A small anodic peak was observed below 0 V which might be assigned to oxidative desorption of hydrogen produced in decomposition of formic acid over Pd surface at open circuit [26] and the main larger peak centered at 0.3 V was attributed to the direct oxidation of formic acid to CO_2 (black curve). For Sb/Pd(UPD), the shape of LSV was all the same except that the current density of formic acid oxidation was higher than that of unmodified Pd film (red curve). In the case of Sb/Pd(MUPD) not only the peak current density was further increased, but the main peak potential and onset potential of formic acid oxidation shifted negatively by 100 mV and 80 mV, respectively (blue curve). It indicated that the electrocatalysis of formic acid oxidation was significantly enhanced at low potentials on Sb/Pd(MUPD) compared with unmodified Pd and Sb/Pd(UPD).

The long-term electrocatalytic activities of the modified or unmodified Pd film electrodes are explored by polarizing pure Pd film, Sb/Pd(UPD) and Sb/Pd(MUPD) electrodes at 0.2 V in 0.5 M H_2SO_4 + 0.5 M HCOOH for 3600 s. Figure 3c showed the corresponding curves. The current density on pure Pd film was intensively decayed in the initial stage (black curve) because of Pd surface poisoning by CO intermediates produced in self-decomposition of formic acid. For Sb/Pd(UPD) and Sb/Pd(MUPD), the current density for HCOOH electrooxidation was enhanced maximally and the decay became weak. During the whole testing (3600 s), the current density followed the order of Sb/Pd(MUPD) > Sb/Pd(UPD) > unmodified

638

Pd which was consistent with the results in Figure 3b. Namely, the electrocatalytic performances were improved on Pd film electrodes by Sb modification due to the so-called third body effect, which accelerated formic acid oxidation through direct pathway [19,23]. Specifically, Sb modification can break adjacent Pd active sites which is favorable to dehydration of formic acid molecules to produce water and CO, thus CO poisoning on Pd surfaces were inhibited to some extend.

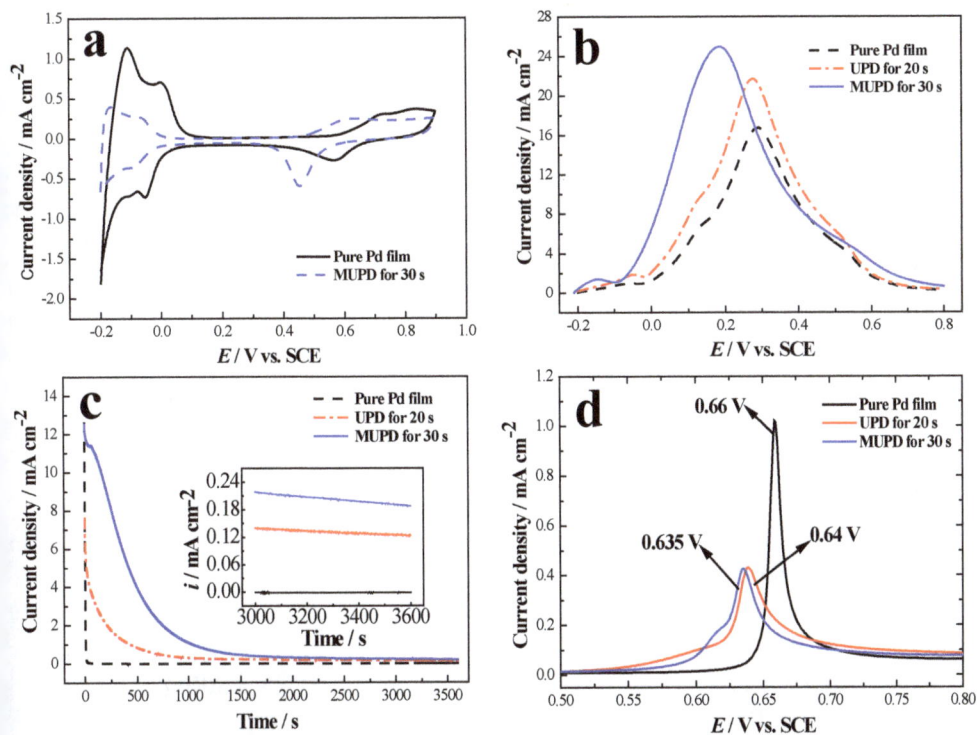

Figure 3. (a) CVs of pure Pd film and Sb/Pd(MUPD) electrodes in 0.5 mol L^{-1} H_2SO_4 at 50 mV s^{-1}; (b) Linear sweep voltammograms and (c) chronoamperometric i vs. t curves of HCOOH oxidation at the constant potential of 0.2 V on pure Pd film, Sb/Pd(UPD) and Sb/Pd(MUPD) electrodes in 0.5 mol L^{-1} H_2SO_4 + 0.5 mol L^{-1} HCOOH; (d) Pre-adsorbed CO stripping voltammograms on pure Pd film, Sb/Pd(UPD) and Sb/Pd(MUPD) in 0.5 mol L^{-1} H_2SO_4 at 10 mV s^{-1}. The inset in (c) is partial enlarged image.

To further explore the poisoning resistant effect after Sb modification, pre-adsorbed CO stripping voltammetry was performed. Figure 3d showed CO stripping voltammograms for pure Pd film, Sb/Pd(UPD) and Sb/Pd(MUPD) in 0.5 mol L^{-1} H_2SO_4. It can be observed that on pure Pd film, the oxidative stripping peak was normally sharp and located at 0.66 V. Otherwise, both the onset and the peak potential of CO oxidation were significantly shifted to lower potentials on Sb/Pd(UPD) and Sb/Pd(MUPD) compared to unmodified pure Pd film. Therefore, the presence of Sb promoted the oxidation of CO adsorbed on Pd [18]. Seen from Figure 3d, the promotion can be explained by electronic effect or bi-functional mechanism induced by Sb. The electronic effect leads to a weakening of the CO-Pd interaction [27] and makes the direct pathway of formic acid oxidation being predominant [23]. In bi-functional mechanism, Sb adatoms provide active sites for –OH formation at lower potentials than on pure Pd, and –OH promotes the oxidative removal of adsorbed poisoning intermediates during formic acid oxidation [18,20,28]

2.3. Sb MUPD on Pd/C Powder Catalyst

The MUPD approach was further applied to modify 40 wt. % Pd/C(BASF) powder catalyst. Figure 4a shows XRD patterns of Pd/C before or after Sb modification by MUPD. The main diffraction peaks at $40.06°$, $46.68°$, $68.08°$, $82.08°$, $86.60°$ are characteristic peaks of Pd(111), (200), (220), (311), (222) plane which suggest face-centered cubic structure of metallic Pd. The peak located at $33.92°$ is attributed to palladium oxide on Pd/C (BASF), but the peak disappeared after immersed in the MUPD modification solution, which may be attributed to reduction of PdO by ascorbic acid. Any peak was observed for Sb or PdSb alloy among the XRD peaks because the Sb content is extremely low, as well as Sb is highly dispersed in active structure on Pd surface or Sb exists as amorphous structure.

Figure 4b shows cyclic voltammograms within hydrogen adsorption/desorption region on unmodified Pd/C and Sb modified Pd/C by MUPD. It was found that the hydrogen region was partially restrained, as well on Pd/C(MUPD) with a relatively small θ_{Sb} of 0.26, which was far below optimal θ_{Sb} value around 0.6. This may be due to partial adsorption of Sb on active carbon supports leading to limited Sb modification on Pd and less higher electrocatalytic activity for Pd/C(MUPD) (seen from Figure 4c). Despite of this, Figure 4d showed that the catalytic stability was enhanced further. CO anodic stripping voltammograms (Figure 4e) revealed negative shift of both peak potential and onset potential of CO oxidation by 70 mV and 50 mV, respectively. It is suggested that the resistance against CO poisoning was enhanced on Pd/C after Sb MUPD modification.

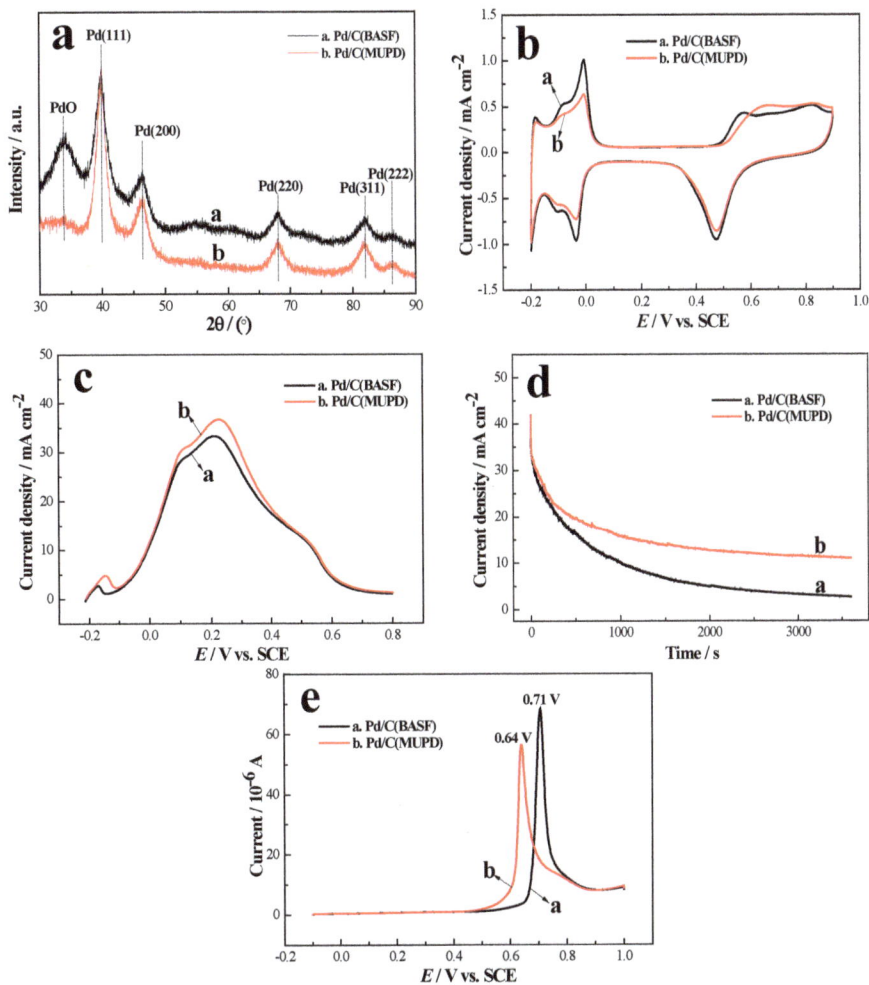

Figure 4. XRD patterns (**a**), cyclic voltammograms (**b**) in 0.5 mol L^{-1} H$_2$SO$_4$ scanned at 50 mV s^{-1}, anodic linear sweep voltammograms (**c**) and chronoamperometric *i-t* curves (**d**) in 0.5 mol L^{-1} H$_2$SO$_4$ + 0.5 mol· L^{-1} HCOOH scanned at 50 mV s^{-1}, anodic stripping voltammograms (**e**) of pre-adsorbed CO monolayer in 0.5 mol L^{-1} H$_2$SO$_4$ scanned at 10 mV s^{-1} on Pd/C before (**a**) and after (**b**) Sb modification by MUPD.

3. Experimental Section

3.1. Modification of Pd Surfaces

Palladium chloride (PdCl$_2$, analytical reagent), formic acid (analytical reagent) were obtained from Sinopharm Chemical Reagent Co. Ltd (SCRC, Shanghai, China).

Perchloric acid (HClO$_4$, analytical reagent), sulfuric acid (H$_2$SO$_4$, analytical reagent), antimony potassium tartrate (analytical reagent) and ascorbic acid (analytical reagent) were obtained from Aladdin. Pd/C (40 wt. %) and 5 wt. % Nafion solution were purchased from BASF (Ludwigshafen, Germany) and Cabot Co. (Boston, MA, USA), respectively.

The Sb modification on Pd surface was carried out by the recently proposed MUPD method. For the film substrate, Pd thin film was first electrodeposited on a electrochemically cleaned glass carbon (GC, Φ = 3 mm) electrode by cyclic voltammetric scanning in the potential range between -0.15 V and 0.42 V $vs.$ SCE in the electrolyte of 0.1 M HClO$_4$ and 5 mM PdCl$_2$. Then, to achieve Sb MUPD the Pd film electrode was immersed in the aqueous solution containing 0.1 mM antimony potassium tartrate (APT) and 20 mM ascorbic acid (AA) for 30 s, rinsed with ultrapure Milli-Q water. To make a comparison, the fresh Pd film was modified via Sb UPD process in which the electrode was modified in 0.5 M H$_2$SO$_4$ containing 0.1 mM APT at the UPD potential of 0.25 V $vs.$ SCE for certain time. For the powder catalyst, catalyst ink was prepared by dispersing 2 mg of Pd/C (40 wt. %, BASF) in 1 mL of ethanol with 120 μL of Nafion (5 wt. %) under sonication. An aliquot of the catalyst ink was transferred onto GC electrode with a Pd loading of 28 μg cm^{-2}. After the ink was dried in air, the catalyst coating was modified with the same procedure as Pd film substrate for MUPD.

3.2. X-ray Diffraction

X-ray diffraction (XRD) for Sb modified Pd/C was performed using a Bruker D8-Advance X-ray diffractometer (Karlsruhe, Germany) equipped with Cu kα radiation (λ = 0.15406 nm), employing a scanning rate of 0.02° s^{-1} in the 2θ range from 20° to 90°.

3.3. Electrochemical Measurements

Electrochemical measurements were performed in a conventional three-electrode cell with a CHI 660E workstation (CH Instruments, Shanghai Chenhua, Shanghai, China) in 0.5 M H$_2$SO$_4$ without or with 0.5 M HCOOH solution deaerated by bubbling pure nitrogen (99.999%). The unmodified or modified Pd film electrode or powder catalyst covered GC electrode served as the working electrodes. A platinum guaze was used as the counter electrode, a saturated calomel electrode (SCE) as the reference electrode. For CO anodic stripping voltammetry, CO was pre-adsorbed on the Pd surface at the potential of -0.1 V in CO saturated 0.5 M H$_2$SO$_4$ and then oxidized (stripped) with an anodic potential scan. The values of current density in this paper are normalized by electrode geometric surface area (0.07065 cm^2). All electrochemical measurements were performed at room temperature.

4. Conclusions

The facile electroless MUPD method has been extended to the modification of Sb on Pd surfaces by immersing Pd film substrate and dispersing Pd/C powder into modification aqueous solution containing Sb(III) and ascorbic acid. As a mild reducing agent, ascorbic acid removed oxygenous species to shift down the open circuit potential of Pd substrate for achieving a sub-monolayer of Sb. The Sb/Pd(MUPD) exhibited enhanced electrocatalytic activity towards formic acid oxidation compared to unmodified pure Pd film and Sb modified Pd film by conventional UPD method. As for Pd/C powder catalyst, the electrocatalytic activity was improved by Sb MUPD. These improvements or enhancements derive from the third body effect, electronic effect and bi-functional mechanism resulting in stronger resistance against poisoning by CO poisons.

Acknowledgments: This work was financially supported by the Natural Science Foundation of China (21103107), Key Project of Shanghai Committee of Science and Technology, China (10160502300), Science and Technology Commission of Shanghai Municipality (No: 14DZ2261000).

Author Contributions: Long-Long Wang and Qiao-Xia Li conceived and designed the experiments; Long-Long Wang performed the experiments; all authors analyzed the data; Long-Long Wang and Qiao-Xia Li wrote the paper. All authors read and approved the final manuscript.

Conflicts of Interest: The authors declare no conflict of interest.

References

1. Rice, C.; Ha, S.; Masel, R.I.; Waszczuk, P.; Wieckowski, A.; Barnard, T. Direct formic acid fuel cells. *J. Power Sources* **2002**, *111*, 83–89.
2. Yu, X.W.; Pickup, P.G. Recent advances in Direct Formic Acid Fuel Cells (DFAFC). *J. Power Sources* **2008**, *182*, 124–132.
3. Rice, C.; Ha, S.; Masel, R.I.; Wieckowski, A. Catalysts for direct formic acid fuel cells. *J. Power Sources* **2003**, *115*, 229–235.
4. Markovic, N.M.; Gasteiger, H.; Ross, P.N.; Jian, X.; Villegas, I.; Weaver, M. Electrooxidation mechanism of methanol and formic acid on Pt–Ru alloy surfaces. *Electrochim. Acta* **1995**, *40*, 91–98.
5. Markovic, N.M.; Ross, P.N. Surface science studies of model fuel cell electrocatalysts. *Surf. Sci. Rep.* **2002**, *45*, 117–229.
6. Waszczuk, P.; Barnard, T.M.; Rice, C.; Masel, R.I.; Wieckowski, A. A nanoparticle catalyst with superior activity for electrooxidation of formic acid. *Electrochem. Commun.* **2002**, *4*, 599–603.
7. Zhou, W.P.; Lewera, A.; Larsen, R.; Masel, R.I.; Bagus, P.S.; Wieckowski, A. Size Effects in Electronic and Catalytic Properties of Unsupported Palladium Nanoparticles in Electrooxidation of Formic Acid. *J. Phys. Chem. B* **2006**, *110*, 13393–13398.

8. Zhou, W.J.; Lee, J.Y. Particle Size Effects in Pd-Catalyzed Electrooxidation of Formic Acid *J. Phys. Chem. C* **2008**, *112*, 3789–3793.

9. Zhu, Y.; Kang, Y.Y.; Zou, Z.Q.; Zhou, Q.; Zheng, J.W.; Xia, B.J.; Yang, H. A facile preparation of carbon-supported Pd nanoparticles for electrocatalytic oxidation of formic acid. *Electrochem. Commun.* **2008**, *10*, 802–805.

10. Ge, J.J.; Xing, W.; Xue, X.Z.; Liu, C.P.; Lu, T.H.; Liao, J.H. Controllable Synthesis of Pd Nanocatalysts for Direct Formic Acid Fuel Cell (DFAFC) Application: From Pd Hollow Nanospheres to Pd Nanoparticles. *J. Phys. Chem. C* **2007**, *111*, 17305–17310.

11. Zhang, J.T.; Qiu, C.C.; Ma, H.Y.; Liu, X.Y. Facile Fabrication and Unexpected Electrocatalytic Activity of Palladium Thin Films with Hierarchical Architectures. *J. Phys Chem. C* **2008**, *112*, 13970–13975.

12. Wang, J.Y.; Zhang, H.X.; Jiang, K.; Cai, W.B. From HCOOH to CO at Pd Electrodes: A Surface-Enhanced Infrared Spectroscopy Study. *J. Am. Chem. Soc.* **2011**, *133*, 14876–14879

13. Mikolajczuk, A.; Borodzinski, A.; Kedzierzawski, P.; Stobinski, L.; Mierzwa, B.; Dziura, R Deactivation of Carbon Supported Palladium Catalyst in Direct Formic Acid Fuel Cell *Appl. Surf. Sci.* **2011**, *257*, 8211–8214.

14. Nitze, F.; Sandstrom, R.; Barzegar, H.R.; Hu, G.; Mazurkiewicz, M.; Malolepszy, A. Stobinski, L.; Wagberg, T. Direct Support Mixture Painting, Using Pd(0) Organo-Metallic Compounds—An Easy and Environmentally Sound Approach to Combine Decoration and Electrode Preparation for Fuel Cells. *J. Mater. Chem. A* **2014**, *2*, 20973–20979.

15. Zhou, Y.; Liu, J.G.; Ye, J.L.; Zou, Z.G.; Ye, J.H.; Gu, J.; Yu, T.; Yang, A.D. Poisoning and Regeneration of Pd Catalyst in Direct Formic Acid Fuel Cell. *Electrochim. Acta* **2010**, *55*, 5024–5027.

16. Yu, X.W.; Pickup, P.G. Mechanistic Study of the Deactivation of Carbon Supported Pd during Formic Acid Oxidation. *Electrochem. Commun.* **2009**, *11*, 2012–2014.

17. Miyake, H.; Okada, T.; Osawa, G.S.M. Formic acid electrooxidation on Pd in acidic solutions studied by surface enhanced infrared absorption spectroscopy. *Phys. Chem. Chem. Phys.* **2008**, *10*, 3662–3669.

18. Yu, X.W.; Pickup, P.G. Deactivation resistant PdSb/C catalysts for direct formic acid fuel cells. *Electrochem. Commun.* **2010**, *12*, 800–803.

19. Haan, J.L.; Stafford, K.M.; Morgan, R.D.; Masel, R.I. Performance of the direct formic acid fuel cell with electrochemically modified palladium-antimony anode catalyst *Electrochim. Acta* **2010**, *55*, 2477–2481.

20. Watanabe, M.; Horiuchi, M.; Motoo, S. Electrocatalysis by ad-atoms: Part XXIII. Design of platinum ad-electrodes for formic acid fuel cells with ad-atoms of the IVth and the Vth groups. *J. Electroanal. Chem. Interfacial Electrochem.* **1988**, *250*, 117–125.

21. Fernandez-Vega, A.; Feliu, J.M.; Aldaz, A.; Claviller, J. Heterogeneous Electrocatalysis on Well Defined Platinum Surfaces Modified by Controlled Amounts of Irreversibly Adsorbed Adatoms: Part II. Formic Acid Oxidation on the Pt(100)–Sb System. *J. Electroanal. Chem.* **1989**, *258*, 101–113.

22. Yang, Y.Y.; Sun, S.G.; Gu, Y.J.; Zhou, Z.Y.; Zhen, C.H. Surface modification and electrocatalytic properties of Pt(100), Pt(110), Pt(320) and Pt(331) electrodes with Sb towards HCOOH oxidation. *Electrochim. Acta* **2001**, *46*, 4339–4348.

23. Peng, B.; Wang, J.Y.; Zhang, H.X.; Lin, Y.H.; Cai, W.B. A versatile electroless approach to controlled modification of Sb on Pt surfaces towards efficient electrocatalysis of formic acid. *Electrochem. Commun.* **2009**, *11*, 831–833.

24. Wang, S.H.; Zhang, H.X.; Cai, W.B. Mimetic underpotential deposition technique extended for surface nanoengineering of electrocatalysts. *J. Power Sources* **2012**, *212*, 100–104.

25. Lee, J.K.; Jeon, H.; Uhm, S.Y.; Lee, J.Y. Influence of underpotentially deposited Sb onto Pt anode surface on the performance of direct formic acid fuel cells. *Electrochim. Acta* **2008**, *53*, 6089–6092.

26. Brandt, K.; Steinhausen, M.; Wandelt, K. Catalytic and electrocatalytic oxidation of formic acid on the pure and Cu-modified Pd(111)-surface. *J. Electroanal. Chem.* **2008**, *616*, 27–37.

27. Peng, B.; Wang, H.F.; Liu, Z.P.; Cai, W.B. Combined Surface-Enhanced Infrared Spectroscopy and First-Principles Study on Electro-Oxidation of Formic Acid at Sb-Modified Pt Electrodes. *J. Phys. Chem. C* **2010**, *114*, 3102–3107.

28. Sadiki, A.; Vo, P.; Hu, S.Z.; Copenhaver, T.S.; Scudiero, L.; Ha, S.; Haan, J.L. Increased Electrochemical Oxidation Rate of Alcohols in Alkaline Media on Palladium Surfaces Electrochemically Modified by Antimony, Lead, and Tin. *Electrochim. Acta* **2014**, *139*, 302–307.

Sacrificial Template-Based Synthesis of Unified Hollow Porous Palladium Nanospheres for Formic Acid Electro-Oxidation

Xiaoyu Qiu, Hanyue Zhang, Yuxuan Dai, Fengqi Zhang, Peishan Wu, Pin Wu and Yawen Tang

Abstract: Large scale syntheses of uniform metal nanoparticles with hollow porous structure have attracted much attention owning to their high surface area, abundant active sites and relatively efficient catalytic activity. Herein, we report a general method to synthesize hollow porous Pd nanospheres (Pd HPNSs) by templating sacrificial SiO_2 nanoparticles with the assistance of polyallylamine hydrochloride (PAH) through layer-by-layer self-assembly. The chemically inert PAH is acting as an efficient stabilizer and complex agent to control the synthesis of Pd HPNSs probably accounting for its long aliphatic alkyl chains, excellent coordination capability and good hydrophilic property. The physicochemical properties of Pd HPNSs are thoroughly characterized by various techniques, such as transmission electron microscopy, X-ray diffraction, X-ray photoelectron spectroscopy. The growth mechanism of Pd HPNSs is studied based on the analysis of diverse experimental observations. The as-prepared Pd HPNSs exhibit clearly enhanced electrocatalytic activity and durability for the formic oxidation reaction (FAOR) in acid medium compared with commercial Pd black.

Reprinted from *Catalysts*. Cite as: Qiu, X.; Zhang, H.; Dai, Y.; Zhang, F.; Wu, P.S. Wu, P.; Tang, Y. Sacrificial Template-Based Synthesis of Unified Hollow Porous Palladium Nanospheres for Formic Acid Electro-Oxidation. *Catalysts* **2015**, *5*, 992–1002.

1. Introduction

Palladium (Pd) plays an important role in a variety of chemical reactions [1–4] such as serving as an efficient electrocatalyst towards formic acid oxidation reaction (FAOR) in direct formic acid fuel cells (DFAFCs) [5–7]. Compared to Pt catalysts, Pd has lower cost with relatively higher abundance, emerging as a promise substitute of Pt especially in the development of DFAFCs [8,9]. Obviously, the morphology of Pd nanocrystals powerfully influences their electrocatalytic activity and stability because of structural effects. For example, G. Fu and his group have successfully synthesized Pd nanoparticles with nanocubes and icosahedra structure [10,11]. For hollow Pd nanosphere with porous structure, the advantage is prominent, as follows [12]:

(1) The hollow porous structure gives birth to a combination of high specific surface area, low mass-density and enhanced reaction kinetics due to extra confined internal reaction space [13,14]. (2) Such hollow porous nanostructures not only supply abundant active sites owning to ample edges and corners, but also improve the mass transfer promoting the electrocatalytic activity obviously [15]. (3) Homogeneous hollow porous nanocrystals are less vulnerable to Ostwald ripening, dissolution and aggregation, thus probably restraining the attenuation of the catalyst activity [16].

In this work, positively charged colloidal modified silica serve as a sacrificial template to synthesize the hollow porous Pd nanospheres (Pd HPNSs). The layer-by-layer (LBL) approach is used to functionalize the SiO_2 template with evenly-spread positive charges around the outside surface (PAH/PSS/PAH-SiO_2) [17]. The positively charged silica could powerfully adsorb the $PdCl_4^{2-}$ and BH_4^-, which are negatively charged metal ions, resulting in the reduction reaction mostly taking place on the surface of the SiO_2. The outermost PAH not only functions as polyelectrolyte to strongly adsorb negatively charged precursors and reductants, but also serves as a stabilizer and complex agent to effectively avoid the aggregation of the Pd nanoparticle owning to its long aliphatic alkyl chain and excellent hydrophilic property. $NaBH_4$ is chosen as a strong reductant to rapidly form the uniform Pd nanospheres. After removing the SiO_2 sacrificial template, the morphology of the outermost Pd nanospheres is maintained to obtain the hollow porous nanostructure. The as-prepared Pd HPNSs exhibit observably enhanced electrocatalytic activity and durability for FAOR compared with commercial Pd black [18–21].

2. Results and Discussion

2.1. Physicochemical Characterization of Pd HPNSs

Figure 1 shows the schematic representation for the synthesis of Pd HPNSs. SiO_2 nanosphere is chosen as a sacrificial template owning to its hydrophilia, mechanical stability, and controllability of size and morphology [22–25]. After the layer-by-layer self-assembly of charged polyelectrolyte on SiO_2 template, $PdCl_4^{2-}$ can be uniformly adsorbed owning to electrostatic attraction, and then be *in-situ* reduced by $NaBH_4$, mostly taking place on the surface of the SiO_2. After removing the SiO_2 sacrificial template, the hollow porous nanospheres structure of Pd HPNSs can be obtained and remains unchanged even after a long time storage. Photographs are taken to prove the powerful adsorption of PAH/PSS/PAH-SiO_2 in solution (Figure 2). After two-phase centrifuging, the precipitate of modified solid SiO_2 nanoparticles in white, and in transparent supernatant liquor, is observed (Figure 2A). In contrast, when $PdCl_4^{2-}$ was dropped into the mixture, forming a well-distributed mixture solution, the precipitate turns t yellow and the supernatant liquor stays transparent, indicating the completely adsorption of PdCl$_4{}^{2-}$. This will make sure the follow-up reduction reaction takes place on the surface of SiO$_2$ templates (Figure 2B).

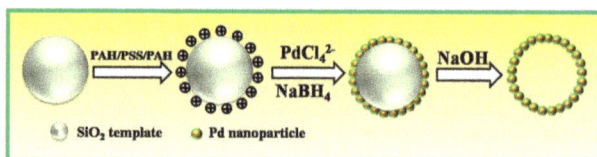

Figure 1. Schematic illustration of the fabrication procedure to produce hollow porous Pd nanospheres (Pd HPNSs).

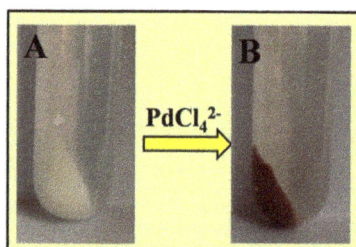

Figure 2. Photographs of (**A**) PAH/PSS/PAH-modified SiO$_2$ solution after centrifuging and (**B**) Mixture of PdCl$_4{}^{2-}$ and PAH/PSS/PAH-modified SiO$_2$ solution after centrifuging.

The morphology of the as-prepared nanoparticles at different stages is investigated by SEM and TEM. Figure 3 represents the SEM/TEM images of SiO$_2$, Pd-SiO$_2$ and Pd HPNSs. As shown in Figure 3A,B, the SiO$_2$ particles possess a similar size of ca. 200 nm. The images of Pd-SiO$_2$ as shown in Figure 3C,D confirm that when Pd was sequentially deposited on the surface of polyelectrolyte-modified SiO$_2$ nanoparticles, the surface becomes less smooth and rougher without obvious morphology change. No individual Pd nanoparticles can be found except on the surface of SiO$_2$ templates (shown in the yellow area in Figure 3D), demonstrating the *in-situ* formation of Pd nanoparticles. After removing the SiO$_2$ template, hollow porous nanospheres structure of Pd HPNSs can be obtained (Figure 3E, F). The yellow arrow highlights one of the broken hollow nanospheres, from which it is clear that the inner space in the as-prepared Pd HPNSs was vacant, indicating that the template of SiO$_2$ was removed, leading to a hollow structure. The TEM image of Pd HPNSs in Figure 3F also demonstrates that the structural integrity of most produced Pd HPNSs with narrow distribution is well maintained even after sonication for a long period of time, indicating superior mechanical properties of the as-prepared Pd HPNSs.

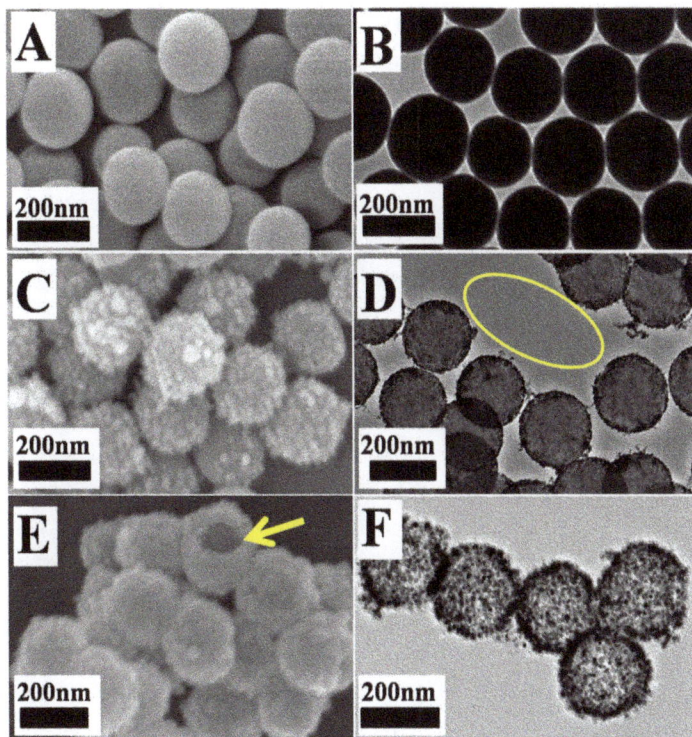

Figure 3. SEM images (**A, C** and **E**) and large-area TEM images (**B, D** and **F**) of SiO$_2$ (**A** and **B**), Pd-SiO$_2$ (**C** and **D**) and Pd HPNSs (**E** and **F**).

Representative large-area TEM image (Figure 4A) and middle-resolution TEM image (Figure 4B) of an individual Pd HPNS clearly show the inner and outer surfaces of the hollow porous spheres, displaying the interconnection through shared outside surface and good dispersibility of the Pd nanoparticles (~4–5nm). Further magnified HRTEM image (Figure 4C) from yellow region of Figure 4B shows an interplanar spacing with 0.225 nm, which is close to the {111} lattice spacing of face-centered cubic (fcc) Pd. The selected-area electron diffraction (SAED) image of an individual Pd HPNS shows diffraction rings corresponding to various facets of face-centered cubic (fcc) (inset in Figure 4C), demonstrating that Pd HPNSs have polycrystalline structure. To observe more clearly about the distribution of Pd nanoparticles, the EDS mapping (Figure 4D) are performed. The resulting patterns show the uniform distribution of Pd throughout the whole Pd HPNS, a strong evidence for the formation of Pd HPNSs as well. Then, the product of different stages is further investigated by XRD (Figure 4E). XRD pattern demonstrates that SiO$_2$-Pd nanospheres have both diffraction peaks of SiO$_2$ and Pd, while the diffraction peak of SiO$_2$ disappears from the XRD pattern of Pd HPNSs, demonstrating the completely remove of SiO_2 template. What's more, from XRD pattern of the Pd HPNSs, fcc structure can be identified as the diffraction peaks of the products are located at the same position as those of pure fcc Pd (PDF#46-1043) crystal phases. The average particle size of the small Pd nanoparticles is 4.3 nm calculated from the peak width of the Pd (111) diffraction according to the Scherrer's equation. It is well known that the near-surface composition plays a critical role on the electrocatalytic behavior of noble-metal catalysts. Thus, the near-surface compositional feature of the Pd HPNSs is examined by XPS (Figure 4F). As observed, no obvious signals of Si and O can be observed and the signals of Pd are strong. Further, the Pd 3d signals are deconvoluted into two components: Pd $3d_{3/2}$ (340.6 eV), Pd $3d_{5/2}$ (335.2 eV) and Pd $3d_{3/2}$ (341.7 eV), Pd $3d_{5/2}$ (336.3 eV), which are assigned to Pd^0 and Pd^{II} species respectively. By measuring the relative peak areas, the percentage of Pd^0 species in the Pd HPNSs is calculated to be 92.1% ($Pd^0/(Pd^0+Pd^{II})$), much higher than the reported value of Pd nanoparticles.

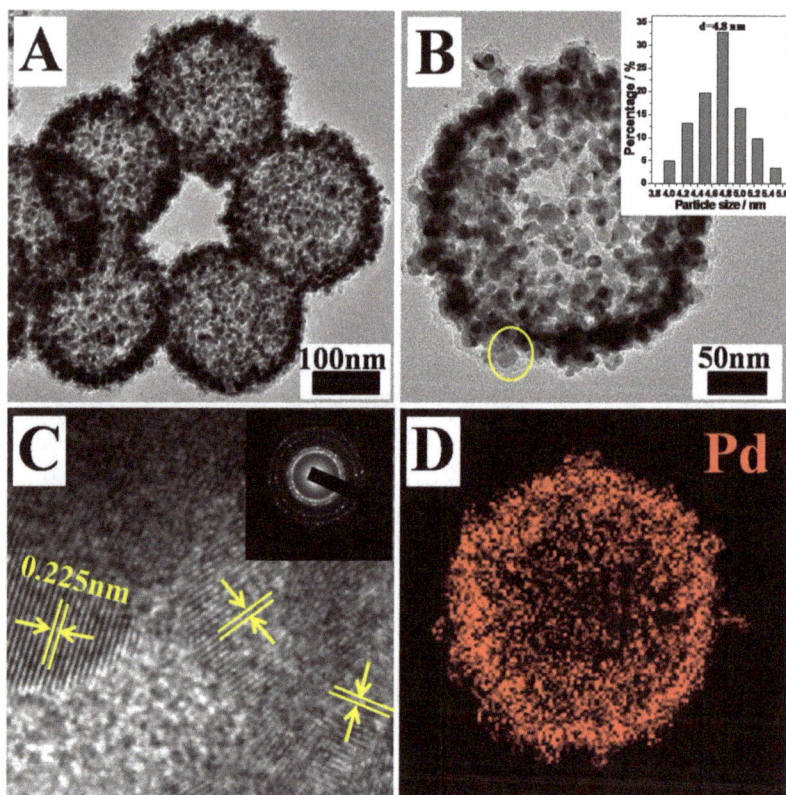

Figure 4. *Cont.*

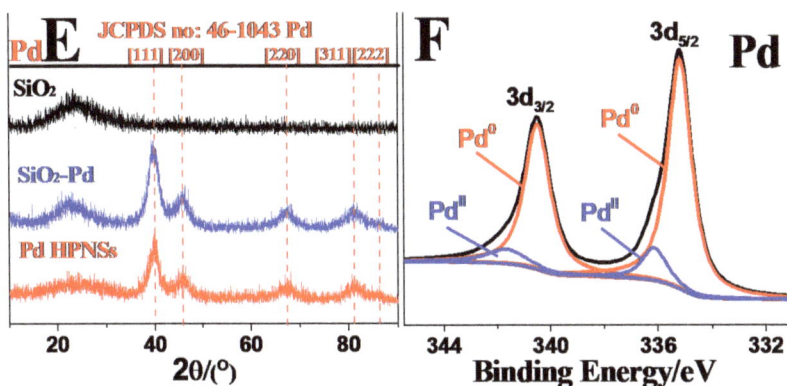

Figure 4. (**A**) Representative large-area transmission electron microscopy (TEM) image of Pd HPNSs. (**B**) Middle-resolution TEM image of an individual Pd HPNSs. Insert: histograms of the particle size distribution (**C**) Magnified HRTEM images recorded from Figure 4B. Inset in Fig. C: selected-area electron diffraction (SAED) pattern of an individual Pd HPNSs. (**D**) EDX elemental mapping patterns of Pd HPNSs. (**E**) X-ray diffraction (XRD) patterns of the SiO$_2$ templates, SiO$_2$-Pd nanospheres, and Pd HPNSs. (**F**) XPS spectra of Pd HPNSs in the Pd 3d regions.

2.2. Electrocatalytic Tests

The electrocatalytic properties of Pd HPNSs for the FAOR were examined and compared with commercial Pd black. The electrochemically active surface areas (ECSA) of Pd HPNSs and commercial Pd black were measured by CO-stripping measurements (Figure 5A). It is observed that the ECSA of Pd HPNSs on glassy carbon electrode is 1.4 cm^2, 1.20 times higher than that of the commercial Pd black (1.2 cm^2), which can be ascribed to the small particle size and hollow porous structure of Pd HPNSs. The mass-normalized cyclic voltammogram shows the FAOR peak potential on Pd HPNSs negatively shifts 110 mV compared to that of commercial Pd black in the forward scan (Figure 5B). Moreover, FAOR peak current on Pd HPNSs reaches a value of 203.2 mA mg^{-1}, which is about 1.4 times higher than that of commercial Pd black (140.6 mA mg^{-1}). It is well-known that the specific kinetic activity (normalized to the ECSA) of a catalyst can effectively evaluate the actual value of the intrinsic activity. Further ECSA-normalized cyclic voltammograms show FAOR peak current on Pd HPNSs is 1.2 times higher than that on commercial Pd black (Figure 5C). The lower FAOR onset oxidation potential and peak potential, the bigger mass-activity and specific activity demonstrate that Pd HPNSs have good electrocatalytic performance for the FAOR, holding promise as potentially practical electrocatalysts for the FAOR. The improved electrocatalytic performance may mainly originate from the unique hollow porous structure. The electrochemical stability of Pd HPNSs for the FAOR is investigated by chronoamperometry at 0.1 V potential

651

(Figure 5D). FAOR current on Pd HPNSs is higher than commercial Pd black during the whole reaction process. At 3000 s, formic oxidation currents on the Pd HPNSs and Pd black decrease to 30.76% and 5.15% of their initial values (taken at 20 s to avoid the contribution of the double-layer discharge and hydrogen adsorption), indicating Pd HPNSs have superior durability for the FAOR.

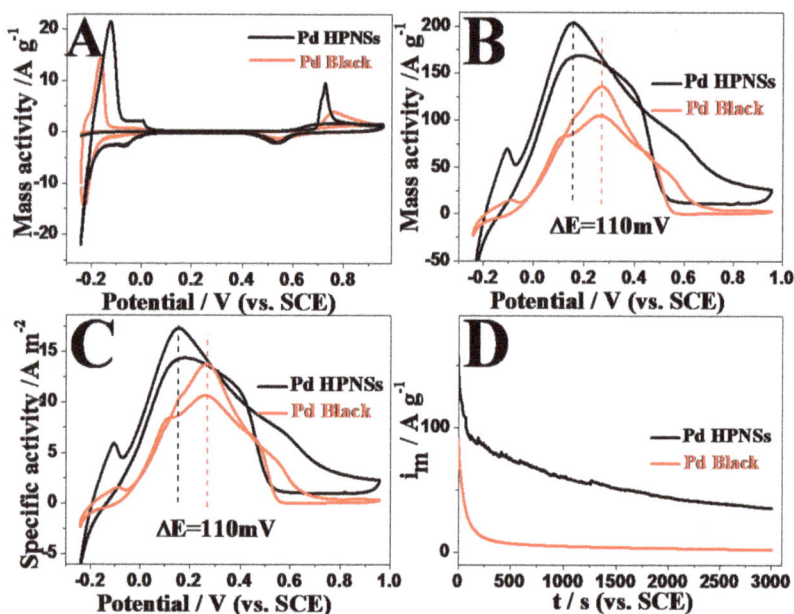

Figure 5. (**A**) CV curves of the Pd HPNSs and Pd black in N_2-saturated 0.5 M H_2SO_4 solution at 50 mV s^{-1}. (**B**) The metal mass-normalized cyclic voltammograms for the Pd HPNSs and Pd black in solution of 0.5 M HCOOH + 0.5 M H_2SO_4 at a scan rate of 50 mV s^{-1}. (**C**) ESCA-normalized cyclic voltammograms of Pd HPNSs and Pd black in solution of 0.5 M HCOOH + 0.5 M H_2SO_4 at a scan rate of 50 mV s^{-1}. (**D**) Chronoamperometry curves for the Pd HPNSs and Pd black in solution of 0.5 M HCOOH + 0.5 M H_2SO_4 for 3000 s at 0.1 V potential.

3. Experimental Section

3.1. Reagents and Chemicals

Poly (allylamine hydrochloride) (PAH, weight-average molecular weight 150 000) was supplied from Nitto Boseki Co., L t d. (Tokyo, Japan). Poly (sodium 4-styrenesulfonate) (PSS, M_w < 700 000 Da), was purchased from Alfa. Aesar Co. Ltd. (Tokyo, Japan). Potassium tetrachloropalladite(II) (K_2PdCl_4) and sodium borohydride ($NaBH_4$) were purchased from Sinopharm Chemical Reagent Co., Ltd (Shanghai, China). Commercial Pd black were purchased from Johnson Matthey

Corporation (London, UK). All the reagents were of analytical reagent grade and used without further purification.

3.2. Synthesis of Hollow porous Pd Nanospheres (Pd HPNSs)

SiO_2 sphere templates with a diameter of ca. 200 nm were synthesized by tetraethyl orthosilicate (TEOS) hydrolyzation in alkaline condition. Typically, 100 mL ethanol, 6 mL ammonium hydroxide, 6 mL H_2O, 3 mL TEOS were mixed and mechanically stirred for 5 h. After the reaction, the obtained SiO_2 templates were separated by centrifugation at 8500 rpm for 5 min, washed several times with water, and then dried at 60 °C for 5 h in a vacuum dryer. Then, the positively charged modified SiO_2 was prepared through a layer-by-layer self-assembly method via electrostatic attraction between charged species. SiO_2 templates were treated with Poly (allylamine hydrochloride) (PAH) and poly (sodium 4-styrenesulfonate) (PSS) in sequence, yielding positively charged PAH/PSS/PAH-modified SiO_2 templates [22]. To obtain the Pd HPNSs, 30 mg PAH/PSS/PAH-modified SiO_2 and 30 mg K_2PdCl_4 were added into 40 mL water and then sonicated for 30 min. After sonication, 10 mg $NaBH_4$ was added into the mixture and mechanically stirred for 1 hour at room temperature to obtain SiO_2-Pd nanospheres. Then, 15 mL 2 M NaOH solution was used to remove the SiO_2 sacrificial template, followed by centrifuging, washing with distilled water and ethanol, and then dried in a vacuum oven at 50 °C to obtain the Pd HPNSs.

3.3. Electrochemical Instrument

All electrochemical experiments were measured with a CHI 660 C electrochemical analyzer (CH Instruments, Shanghai, Chenghua Co.). All electrochemical measurements were carried out at 30 ±1 °C. A standard three-electrode system (consisted of a saturated calomel reference electrode (SCE), a catalyst modified glassy carbon electrode as the working electrode, and a platinum wire as the auxiliary electrode) was used to test all electrochemical experiments. An evenly distributed suspension of catalyst was prepared by ultrasonic the mixture of 10 mg catalyst and 5 mL H_2O for 30 min, and 6 μL of the resulting suspension was loaded on the surface of the glassy carbon electrode (3 mm diameter, 0.07 cm^2). Thus, the working electrode was obtained, and the total mass loading of catalyst on the electrode was about 12 μg. We used the same amount of total metal of Pd HPNSs and commercial Pd Black to make a comparison. Throughout the cyclic voltammetry experiment, cyclic voltammetry tests were performed in N_2-saturated 0.5 M H_2SO_4 solution with or without 0.5 M HCOOH at a scan rate of 50 mV s^{-1}. Chronoamperometry curves were obtained in N_2-saturated 0.5 M HCOOH + 0.5 M H_2SO_4 mixture solution for 3000 s at 0.1 V applied potential.

3.4. Instruments

Transmission electron microscopy (TEM) images were surveyed from a JEOL JEM-2100F transmission electron microscopy operated at an accelerating voltage of 200 kV. X-ray diffraction (XRD) patterns were obtained from a Model D/max-rC X-ray diffractometer by using Cu Kα radiation source (λ = 1.5406 Å), operating at 40 kV and 100 mA. X-ray photoelectron spectroscopy (XPS) measurements were carried out on a Thermo VG Scientific ESCALAB 250 spectrometer with an Al Kα radiator. The vacuum in the analysis chamber was maintained at about 10^{-9} mbar and the binding energy was calibrated by means of the C 1s peak energy of 284.6 eV.

4. Conclusions

In summary, the hollow porous Pd nanospheres with high surface area, low mass-density and abundant active sites are synthesized by a sacrificial template method over PAH. The layer-by-layer (LBL) approach is used to modify the SiO$_2$ template in order to make it positively charged. PAH not only functions as polyelectrolyte to strongly adsorb negatively charged precursors, ensuring the reduction reaction completely take place on the surface of the SiO$_2$, but also serves as stabilizer and complex agent to effectively avoid the aggregation and collapse of the Pd HPNSs owning to its coordination capability, good hydrophilic property, and high chemical stability. Undoubtedly, this method is more promising from an environmental standpoint, adding advantage over the use of high-temperature reaction and toxic organic solvents. Electrochemical measurements demonstrate that Pd HPNSs exhibit superior electrocatalytic activity and long-term durability compared to commercial Pd black. Thus, the superior electrocatalytic performance of Pd HPNSs provides a promising support for good electrocatalyst in DFAFC applications.

Acknowledgments: The authors are grateful for the financial support of NSFC (21376122 and 21273116), United Fund of NSFC and Yunnan Province (U1137602), the National Basic Research Program of China (973 Program, 2012CB215500), Postgraduate Research and Innovation Project in Jiangsu Province (CXLX13-369), and a project funded by the Priority Academic Program Development of Jiangsu Higher Education Institutions, and National and Local Joint Engineering Research Centre of Biomedical Functional Materials.

Author Contributions: Xiaoyu Qiu prepared the samples, performed the experiment and wrote manuscript. Hanyue Zhang and Yuxuan Dai supported the experiments. Fengqi Zhang and Peishan Wu carried out the XRD analysis. Ping Wu and Yawen Tang revised the final version of paper.

Conflicts of Interest: The authors declare no conflict of interest.

References

1. Nelson, N.; Manzano, J.; Sadow, A.; Overbury, S.; Slowing, I. Selective Hydrogenation of Phenol Catalyzed by Palladium on High-Surface-Area Ceria at Room Temperature and Ambient Pressure. *ACS Catal.* **2015**, *5*, 2051–2061.

2. Cervantes, C.; Alamo, M.; Garcia, J. Hydrogenation of Biomass-Derived Levulinic Acid into γ-Valerolactone Catalyzed by Palladium Complexes. *ACS Catal.* **2015**, *5*, 1424–1431.

3. Zhu, F.; Wang, Z. Palladium-Catalyzed Coupling of Azoles or Thiazoles with Aryl Thioethers via C−H/C−S Activation. *Org. Lett.* **2015**, *2*.

4. Zhao, G.; Chen, C.; Yue, Y.; Yu, Y.; Peng, J. Palladium(II)-Catalyzed Sequential C−H Arylation/Aerobic Oxidative C−H Amination: One-Pot Synthesis of Benzimidazole-FusedPhenanthridines from 2-Arylbenzimidazoles and Aryl Halides JOC. *J. Org. Chem.* **2014**, *11*.

5. Yang, S.; Shen, C.; Lu, X.; Tong, H.; Zhu, J.; Zhang, X.; Gao, H. Preparation and electrochemistry of graphene nanosheets–multiwalled carbon nanotubes hybrid nanomaterials as Pd electrocatalyst support for formic acid oxidation. *Electrochim. Acta* **2012**, *62*, 242–249.

6. Yang, S.; Dong, J.; Yao, Z.; Shen, C.; Shi, X.; Tian, Y.; Lin, S.; Zhang, X. One-Pot Synthesis of Graphene-Supported Monodisperse Pd Nanoparticles as Catalyst for Formic Acid Electro-oxidation. *Sci. Rep.* **2014**, *3*.

7. Wang, Y.; Liu, H.; Wang, L.; Wang, H.; Du, X.; Wang, F.; Qi, T.; Lee, J.; Wang, X. Pd catalyst supported on a chitosan-functionalized large-area 3D reduced graphene oxide for formic acid electrooxidation reaction. *J. Mater. Chem. A* **2013**, *4*.

8. Chang, J.; Sun, X.; Feng, L.; Xing, W.; Qin, X.; Shao, G. Effect of nitrogen-doped acetylene carbon black supported Pd nanocatalyst on formic acid electrooxidation. *J. Power Sources* **2013**, *239*, 94–102.

9. Bai, Z.; Yan, H.; Wang, F.; Yang, L.; Jiang, K. Electrooxidation of formic acid catalyzed by Pd Nanoparticles supported on multi-walled carbonnanotubes with sodium oxalate. *Ionics* **2013**, *19*, 543–548.

10. Fu, G.; Jiang, X.; Tao, L.; Chen, Y.; Lin, J.; Zhou, Y.; Tang, Y.; Lu, T. Polyallylamine Functionalized Palladium Icosahedra: One-Pot Water-Based Synthesis and Their Superior Electrocatalytic Activity and Ethanol Tolerant Ability in Alkaline Media. *Langmuir* **2013**, *29*, 4413–4420.

11. Fu, G.; Wu, K.; Jiang, X.; Tao, L.; Chen, Y.; Lin, J.; Zhou, Y.; Wei, S.; Tang, Y.; Lu, T.; Xia, X. Polyallylamine-directed green synthesis of platinum nanocubes. Shape and electronic effect codependent enhanced electrocatalytic activity. *Phys. Chem. Chem. Phys.* **2013**, *15*, 3793–3802.

12. Yang, Y.; Wang, F.; Yang, Q.; Hu, Y.; Yan, H.; Chen, Y.-Z.; Liu, H.; Zhang, G.; Lu, J.; Jiang, H.-L.; Xu, H. Hollow Metal–Organic Framework Nanospheres via Emulsion-Based Interfacial Synthesis and Their Application in Size-Selective Catalysis. *ACS Appl. Mater. Interfaces* **2014**, *6*, 18163–18171.

13. Yang, Z.; Li, Z.; Yang, Y.; Xu, Z.J. Optimization of $Zn_xFe_{3-x}O_4$ Hollow Spheres for Enhanced Microwave Attenuation. *ACS Appl. Mater. Interfaces* **2014**, *6*, 21911–21915.

14. Wang, M.; Zhang, W.; Wang, J.; Wexler, D.; Poynton, S.D.; Slade, R.C. T.; Liu, H. Winther-Jensen, B.; Kerr, R.; Shi, D.; Chen, J. PdNi Hollow Nanoparticles for Improved Electrocatalytic Oxygen Reduction in Alkaline Environments. *ACS Appl. Mater. Interfaces* **2013**, *5*, 12708–12715.

15. Wang, L.; Imura, M.; Yamauchi, Y. Tailored Design of Architecturally Controlled Pt Nanoparticles with Huge Surface Areas toward Superior Unsupported Pt Electrocatalysts *ACS Appl. Mater. Interfaces* **2012**, *4*, 2865–2869.

16. Chen, D.; Cui, P.; He, H.; Liu, H.; Yang, J. Highly Catalytic Hollow Palladium Nanoparticles Derived From Silver@silver–palladium Core–shell Nanostructures for the Oxidation of Formic Acid. *J. Power Sources* **2014**, *272*, 152–159.

17. Wu, P.; Wang, H.; Tang, Y.; Zhou, Y.; Lu, T. Three-Dimensional Interconnected Network of Graphene-Wrapped Porous Silicon Spheres: In Situ Magnesiothermic-Reduction Synthesis and Enhanced Lithium-Storage Capabilities. *ACS Appl. Mater. Interfaces* **2014**, *6*, 3546–3552.

18. Malolepszy, A.; Mazurkiewicz, M.; Mikolajczuk, A.; Stobinski, L.; Borodzinski, A; Mierzwa, B.; Lesiak, B.; Zemek, J.; Jiricek, P. Influence of Pd-Au/MWCNTs surface treatment on catalytic activity in the formic acid electrooxidation. *Phys. Status Solidi C* **2011**, *8*, 3195–3199.

19. Qu, K.; Wu, L.; Ren, J.; Qu, X. Natural DNA-Modified Graphene/Pd Nanoparticles as Highly Active Catalyst for Formic Acid Electro-Oxidation and for the Suzuki Reaction. *ACS Appl. Mater. Interfaces* **2012**, *4*, 5001–5009.

20. Zhang, L.; Wan, L.; Ma, Y.; Chen, Y.; Zhou, Y.; Tang, Y.; Lu, T. Crystalline Palladium–cobalt Alloy Nanoassemblies with Enhanced Activity and Stability for The Formic Acid Oxidation Reaction. *Appl. Catal. B* **2013**, 229–235.

21. Zhang, L.; Sui, Q.; Tang, T.; Chen, Y.; Zhou, Y.; Tang, Y.; Lu, T. Surfactant-free palladium nanodendrite assemblies with enhanced electrocatalytic performance for formic acid oxidation. *Electrochem. Commun.* **2013**, *32*, 43–46.

22. Ott, A.; Yu, X.; Hartmann, R.; Rejman, J.; Schütz, A.; Ochs, M.; J. Parak, W.; Carregal-Romero, S. Light-Addressable and Degradable Silica Capsules for Delivery of Molecular Cargo to the Cytosol of Cells. *Chem. Mater.* **2015**.

23. Dahlberg, K.A.; Schwank, J.W. Synthesis of Ni@SiO$_2$ Nanotube Particles in a Water-in-Oil Microemulsion Template. *Chem. Mater.* **2012**, *24*, 2635–2644.

24. Kim, S.M.; Jeon, M.; Kim, K.W.; Park, J.; Lee, I.S. Postsynthetic Functionalization of a Hollow Silica Nanoreactor with Manganese Oxide-Immobilized Metal Nanocrystals Inside the Cavity. *J. Am. Chem. Soc.* **2013**, *135*, 15714–15717.

25. Deng, T.-S.; Marlow, F. Synthesis of Monodisperse Polystyrene@Vinyl-SiO$_2$ Core–Shell Particles and Hollow SiO$_2$ Spheres. *Chem. Mater.* **2011**, *24*, 536–542.

MDPI AG

Klybeckstrasse 64

4057 Basel, Switzerland

Tel. +41 61 683 77 34

Fax +41 61 302 89 18

http://www.mdpi.com/

Catalysts Editorial Office

E-mail: catalysts@mdpi.com

http://www.mdpi.com/journal/catalysts

www.ingramcontent.com/pod-product-compliance
Lightning Source LLC
Chambersburg PA
CBHW050346230326
41458CB00102B/6431